Obesity

Obesity
A Ticking Time Bomb for Reproductive Health

Edited by

Tahir Mahmood
Victoria Hospital
Kirkcaldy, UK

Sabaratnam Arulkumaran
St George's University of London
London, UK

ELSEVIER

AMSTERDAM • BOSTON • HEIDELBERG • LONDON • NEW YORK • OXFORD
PARIS • SAN DIEGO • SAN FRANCISCO • SINGAPORE • SYDNEY • TOKYO

Elsevier
32 Jamestown Road, London NW1 7BY
225 Wyman Street, Waltham, MA 02451, USA

First edition 2013

Notices

Knowledge and best practice in this field are constantly changing. As new research and experience broaden our understanding, changes in research methods, professional practices, or medical treatment may become necessary.

Practitioners and researchers must always rely on their own experience and knowledge in evaluating and using any information, methods, compounds, or experiments described herein. In using such information or methods they should be mindful of their own safety and the safety of others, including parties for whom they have a professional responsibility.

To the fullest extent of the law, neither the Publisher nor the authors, contributors, or editors, assume any liability for any injury and/or damage to persons or property as a matter of products liability, negligence or otherwise, or from any use or operation of any methods, products, instructions, or ideas contained in the material herein.

British Library Cataloguing-in-Publication Data

A catalogue record for this book is available from the British Library

Library of Congress Cataloging-in-Publication Data

A catalog record for this book is available from the Library of Congress

ISBN: 978-0-12-416045-3

This book has been manufactured using Print On Demand technology. Each copy is produced to order and is limited to black ink. The online version of this book will show color figures where appropriate.

Contents

Foreword

Obesity affects our mental, physical and emotional health. It increases the prevalence of diseases like hypertension, diabetes, sexual dysfunction, infertility and cardiovascular disease leading to impaired health and a lower quality of life. It affects our locomotor system and exacerbates other illnesses such as asthma, arthritis and degenerative disease. It reduces the ability for one to contribute to society and draws on the society to manage the condition of those affected. It is a silent killer and is increasing in prevalence and is becoming a pandemic of great public concern. It affects all age groups and is a problem that affects the individual, the family, the society and the world at large. The prevalence of obesity is increasing across the globe affecting not only well-resourced countries but is also seen in the less resourced due to the dietary habits – the well to do over eating and being sedentary and the less well to do consuming large amounts of carbohydrates and fats due to inadequate finances to have a balanced diet. The adverse effects of female obesity on pregnancy outcome has been repeatedly highlighted by the confidential enquires on maternal deaths in the United Kingdom.

This book looks at the origins of obesity, which starts in utero and its impact on reproductive health. The authors have also analysed the complex interplay of genetic, environmental and psychological factors which underpin the eating habits of the human race. The interaction between female obesity and insulin and its resultant effect on fertility are well known and are manifested by polycystic ovarian disease. Severe obesity in the male is associated with erectile dysfunction and alteration in sperm parameters, thus affecting fertility status. This epidemic of increasing obesity also affects pregnancy rates at assisted conception treatment cycles.

Once pregnant there is a significantly increased risk of early pregnancy loss, recurrent miscarriage and foetal developmental abnormalities. Obesity increases the risk of developing gestational diabetes and pre-eclampsia during pregnancy and its antecedent complications for mothers and foetus. The incidence of serious complication like deep venous thrombosis is increased. There is an increased risk of dysfunctional labour, operative deliveries and risk of stillbirth. Obesity per se is an independent risk factor for stillbirth and maternal death during pregnancy.

The problem of obesity is not purely due to poor eating habits – it seems to have its origin in utero and has an influence by the genetic composition. Although the causation may be complex, there are chances to prevent or identify and control the progression before it is too late when difficulties are encountered with disease process or controlling obesity. The society should tackle obesity in children and adolescents by developing a co-ordinated public health and medical intervention strategy. This book reviews the potential threat to the human race caused by increasing prevalence of obesity by bringing together experts from different fields to share their knowledge and experience. The material presented in the book is evidence based and deals with the epidemiology, explores the biological and genetic causes and reviews the core issues that relates to both sexes. We have explored in-depth issues related to service delivery, health care planning and identified areas for future research which could inform the basis for health care service provision and service organisation.

This book is a valuable resource for postgraduate students studying for higher qualification in the specialities of obstetrics and gynaecology, reproductive medicine, general medicine with a special interest in diabetes, general surgery, public health and health care management. Specialists in genito-urinary medicine, infertility experts and those involved in the provision of safe contraceptive practices will find this book useful. It will also be of interest to dieticians, midwives and women themselves.

Tahir Mahmood
Sabaratnam Arulkumaran

About the Editors

Dr Tahir Mahmood, CBE, MD, FRCOG, FRCPI, MBA, FACOG (Hon), Past Vice-President Standards, Royal College of Obstetricians and Gynaecologists (2007–2010); President Elect, European Board and College of Obstetrics and Gynaecology (EBCOG, 2011–2014); Consultant Obstetrician and Gynaecologist, Victoria Hospital, NHS Fife, Kirkcaldy, Scotland, UK; Honorary Senior Lecturer, School of Biological and Medical Sciences, University of St. Andrews, and Honorary Senior Lecture, Division of Maternal and Childhealth Sciences, University of Dundee, UK

Professor **Sir Sabaratnam Arulkumaran**, MD, PhD, FRCOG, FRCS, FACOG, Past President, Royal College of Obstetricians and Gynaecologists (2007–2010); President Elect FIGO (2009–2012) and President FIGO (2012–2015); Head, Department of Obstetrics and Gynaecology, St. George's University Hospital, London, UK

Contributors

Annie S. Anderson Centre for Public Health Nutrition Research, Population Health Sciences, Medical Research Institute, Ninewells Hospital and Medical School, The University of Dundee, Dundee, UK

Richard A. Anderson MRC Centre for Reproductive Health, The Queen's Medical Research Institute, University of Edinburgh, Edinburgh, UK

Sabaratnam Arulkumaran St George's University of London, London, UK

J. Balani Department of Diabetes and Endocrinology, Epsom and St Helier University Hospitals, NHS Trust, Surrey, UK

Alexander Baldacchino School of Medicine, Dundee University, Dundee, UK

Sarah Beake Research Associate, Florence Nightingale School of Nursing and Midwifery, King's College London, London, UK

Harish Malappa Bhandari Department of Obstetrics and Gynaecology, Centre for Reproductive Medicine, University Hospitals of Coventry and Warwickshire, NHS Trust, Coventry, UK

Siladitya Bhattacharya Division of Applied Health Sciences, University of Aberdeen, Aberdeen, UK

Debra Bick Florence Nightingale School of Nursing and Midwifery, King's College, London, UK

Mairead Black Division of Applied Health Sciences, University of Aberdeen, Aberdeen, UK

Alexander Bolyakov Department of Urology, Weill Cornell Medical College, New York, NY, USA, Consulting Research Services, Inc, Red Bank, NJ, USA

Savita Brito-Mutunayagam MRC Centre for Reproductive Health, University/ BHF Centre for Cardiovascular Science, The University of Edinburgh, The Queen's Medical Research Institute, Edinburgh, UK

Fiona Broughton-Pipkin Emeritus Professor of Perinatal Physiology, Faculty of Medicine and Health Sciences, Queen's Medical Centre, University of Nottingham, Nottingham, UK

Christy Burden Obstetrics and Gynaecology, Gloucester Royal Hospital, Gloucestershire, UK, Research into Safety and Quality (RiSQ), Southmead Hospital, Bristol, UK

Gail Busby St Mary's Hospital, Manchester, UK

Sharon Cameron Consultant Obstetrician & Gynaecologist, Centre for reproductive Health, Edinburgh Royal Infirmary, Edinburgh, UK

Carolyn Chiswick MRC/University of Edinburgh Centre for Reproductive Health, The Queen's Medical Research Institute, Edinburgh, UK

David Churchill New Cross Hospital, Wolverhampton, UK

W. Colin Duncan MRC Centre for Reproductive Health, University/BHF Centre for Cardiovascular Science, The University of Edinburgh, The Queen's Medical Research Institute, Edinburgh, UK

Angela M. Craigie Centre for Public Health Nutrition Research, Population Health Sciences, Medical Research Institute, Ninewells Hospital and Medical School, The University of Dundee, Dundee, UK

Hilary O.D. Critchley MRC Centre for Reproductive Health, University/BHF Centre for Cardiovascular Science, The University of Edinburgh, The Queen's Medical Research Institute, Edinburgh, UK

Konstantinos Dafopoulos Department of Obstetrics and Gynaecology, Medical School, University of Thessalia, Larissa, Greece

Sujeetha Damodaran Iswarya Women's Hospital and Fertility Centre, Madurai, Tamil Nadu, India

Dilip Dan Departments of Clinical Medical and Surgical Sciences, University of the West Indies, St Augustine, Trinidad and Tobago, West Indies

Debbie M. Smith School of Psychological Sciences, The University of Manchester, Manchester, UK

Zsolt Demetrovics Institute of Psychology, Eötvös Loránd University, Budapest, Hungary

Fiona C. Denison MRC/University of Edinburgh Centre for Reproductive Health, The Queen's Medical Research Institute, Edinburgh, UK

Kavita Deonarine Departments of Clinical Medical and Surgical Sciences, University of the West Indies, St Augustine, Trinidad and Tobago, West Indies

Anjum Doshan University Hospitals of Leicester NHS trust, Leicester, UK

Tim Draycott Consultant Obstetrician & Gynaecologist, Southmead Hospital, North Bristol NHS Trust, Bristol, and the University of Bristol, Bristol, UK

Leroy C. Edozien Manchester Academic Health Sciences Centre, St Mary's Hospital, Manchester, UK

Lindsay Edwards Department of Obstetrics and Gynaecology, Royal Hobart Hospital, Hobart, Australia

Joanne Ellison Consultant Obstetrician & Gynaecologist, Royal Alexandra Hospital, Paisley, UK

Margaret J. Evans Department of Pathology, Royal Infirmary of Edinburgh, Little France Crescent, Edinburgh, UK

Gemma Forbes Specialist Registrar, West Midland Deanery, New Cross Hospital, Wolverhampton, UK

J.T. George MRC Centre for Reproductive Health, The Queen's Medical Research Institute, University of Edinburgh, Edinburgh, UK

Peter D. Gluckman Liggins Institute, University of Auckland, Auckland, New Zealand and Singapore Institute of Clinical Sciences, Singapore

Cindy M. Gray Research Fellow, Institute of Healthand Wellbeing, University of Glasgow, UK

Mark Hamilton Aberdeen Fertility Centre, Department of Obstetrics & Gynaecology, Aberdeen Maternity Hospital, Aberdeen, Scotland, UK

Mark A. Hanson Institute of Developmental Sciences, University of Southampton, Southampton, UK

Adnan Hasan Department of Obstetrics & Gynaecology, Victoria Hospital, Kirkcaldy, Fife, Scotland, UK

Rohana N. Haththotuwa Ninewells Care Mother and Baby Hospital, Colombo, Sri Lanka

Pak Chung Ho Department of Obstetrics and Gynaecology, The University of Hong Kong, Queen Mary Hospital, Hong Kong, People's Republic of China

Peter Hornnes Gynækologisk/Obstetrisk, Hvidovre Hospital, Copenhagen, Denmark and Treasurer European Board & College of obstetrics and Gynaecology (EBCOG)

Shahzya S. Huda Department of Obstetrics & Gynaecology, Forth Valley Hospital, Larbert, UK

Stephen Hyer Department of Diabetes and Endocrinology, Epsom and St Helier University Hospitals, NHS Trust, Surrey, UK

Ioannidis Ioannis Diabetes and Obesity Unit, 2nd Department of Internal Medicine, Konstantopoulio Hospital, Athens, Greece

Amanda Jefferys Clinical Research Fellow, Department of obstetrics, Southmead Hospital Bristol, North Bristol NHS Trust, Bristol, UK

Vanessa J. Kay Assisted Conception Unit and Developmental Biology Group, Division of Maternal and Child Health Sciences, Ninewells Hospital, University of Dundee, Dundee, UK

Khalid S. Khan Women's Health Research Unit, Centre for Primary Care and Public Health, Barts and the London School of Medicine and Dentistry, Queen Mary University of London, London, UK

Gyöngyi Kökönyei Institute of Psychology, Eötvös Loránd University, Budapest, Hungary

Justin C. Konje Reproductive Sciences Section, Robert Kilpatrick Clinical Sciences Building, Leicester Royal Infirmary, Leicester, UK

Karen Siu Ling Lam Department of Medicine, The University of Hong Kong, Queen Mary Hospital, Hong Kong, People's Republic of China

Jeannet Lauenborg Gynækologisk/Obstetrisk Afdeling, Hvidovre Hospital, Copenhagen, Denmark

Tina Lavender School of Nursing, Midwifery and Social Work, The University of Manchester, Manchester, UK

Alistair Lee Department of Anaesthesia, Royal Infirmary, Edinburgh, UK

Chin Peng Lee Department of Obstetrics and Gynaecology, The University of Hong Kong, Queen Mary Hospital, Hong Kong, People's Republic of China

Hang Wun Raymond Li Department of Obstetrics and Gynaecology, The University of Hong Kong, Queen Mary Hospital, Hong Kong, People's Republic of China

Boon H. Lim Department of Obstetrics and Gynaecology, Royal Hobart Hospital, Hobart, Australia

Chu Lim Department of Obstetrics & Gynaecology, Victoria Hospital, Kirkcaldy, Fife, UK

R.C.W. Ma Department of Medicine and Therapeutics, Chinese University of Hong Kong and Hong Kong Institute of Diabetes and Obesity, Hong Kong, People's Republic of China

Kate Maclaran West London Menopause and PMS Centre, Queen Charlotte's & Chelsea and Chelsea & Westminster Hospitals, London, UK

Abha Maheshwari Aberdeen Fertility Centre, Department of Obstetrics & Gynaecology, Aberdeen Maternity Hospital, Aberdeen, UK

Tahir Mahmood Victoria Hospital, NHS Fife, Kirkcaldy, UK. Office of Research and Clinical Audit, Lindsay Stewart R&D Centre, Royal College of Obstetricians and Gynaecologists, London, UK

Mani Malarselvi University Hospitals Coventry and Warwickshire, Coventry, UK

Sarah Martins da Silva Assisted Conception Unit and Developmental Biology Group, Division of Maternal and Child Health Sciences, Ninewells Hospital, University of Dundee, Dundee, UK

Fionnuala M. McAuliffe School of Medicine and Medical Science, University College Dublin, National Maternity Hospital, Dublin, Republic of Ireland

Rhona J. McInnes School of Nursing, Midwifery & Health, University of Stirling, Stirling, UK

Sarah McRobbie Aberdeen Fertility Centre, Department of Obstetrics & Gynaecology, Aberdeen Maternity Hospital, Aberdeen, UK

Mohamed K. Mehasseb Department of Gynaecological Oncology, Addenbrooke's Hospital, Cambridge Biomedical Campus, Cambridge, UK

Christina I. Messini Department of Obstetrics and Gynaecology, Medical School, University of Thessalia, Larissa, Greece

Ioannis E. Messinis Department of Obstetrics and Gynaecology, Medical School, University of Thessalia, Larissa, Greece

Alistair Milne Department of Anaesthesia, Royal Infirmary, Edinburgh, UK

Scott M. Nelson School of Medicine, University of Glasgow, Glasgow, UK

Darius A. Paduch Department of Urology, Weill Cornell Medical College, New York, NY, USA, Consulting Research Services, Inc, Red Bank, NJ, USA

Nick Panay Imperial College, London, UK

Grigoropoulou Pinelopi Diabetes and Obesity Unit, 2nd Department of Internal Medicine, Konstantopoulio Hospital, Athens, Greece

Siobhan Quenby Clinical Sciences Research Institute, Warwick Medical School, University of Warwick, Coventry, UK, Division of Reproductive Health, Clinical Science Laboratories, Warwick Medical School, Coventry, UK

Rebecca M. Reynolds Endocrinology Unit, University/BHF Centre for Cardiovascular Science, The University of Edinburgh, The Queen's Medical Research Institute, Edinburgh, UK

Yana Richens Public Health, University College London Hospital, London, UK

Mourad W. Seif Academic Unit of Obstetrics and Gynaecology, St. Mary's Hospital, Manchester, UK

Upul Senarath Department of Community Medicine, Faculty of Medicine, University of Colombo, Colombo, Sri Lanka

Mahmood I. Shafi Department of Gynaecological Oncology, Addenbrooke's Hospital, Cambridge Biomedical Campus, Cambridge, UK

Hassan Shehata Department of Maternal Medicine, Epsom and St Helier University Hospitals, NHS Trust, Surrey, UK

Krishnan Swaminathan Apollo Speciality Hospitals, Madurai, Tamil Nadu, India

Surujpal Teelucksingh Departments of Clinical Medical and Surgical Sciences, University of the West Indies, St Augustine, Trinidad and Tobago, West Indies

Shakila Thangaratinam Women's Health Research Unit, Centre for Primary Care and Public Health, Barts and the London School of Medicine and Dentistry, Queen Mary University of London, London, UK

Omar Thanoon Department of Obstetrics & Gynaecology, Victoria Hospital, Kirkcaldy, Fife, UK

Andrew Thomson Royal Alexandra Hospital, Paisley, UK

Douglas G. Tincello Reproductive Science Section, CSMM University of Leicester, Leicester, UK

Róbert Urbán Institute of Psychology, Eötvös Loránd University, Budapest, Hungary

Laurent Vaucher Department of Urology, Lausanne, Switzerland

Gerard H.A. Visser Department of Obstetrics, University Medical Center, Utrecht, the Netherlands

Sanjay Vyas Department of Obstetrics & Gynaecology, Southmead Hospital, North Bristol Trust, Bristol, UK

M. Wagner Department of Maternal Medicine, Epsom and St Helier University Hospitals, NHS Trust, Surrey, UK

Jennifer M. Walsh School of Medicine and Medical Science, University College Dublin, National Maternity Hospital, Dublin, Republic of Ireland

Chandrika N. Wijeyaratne Department of Obstetrics and Gynaecology, Faculty of Medicine, University of Colombo, Colombo, Sri Lanka

Yariv Yogev Rabin Medical Center, Tel Aviv, Israel

Section 1

Epidemiology

1 Worldwide Epidemic of Obesity

Rohana N. Haththotuwa[1], *Chandrika N. Wijeyaratne*[2] *and Upul Senarath*[3]

[1]Ninewells Care Mother and Baby Hospital, Colombo, Sri Lanka, [2]Department of Obstetrics and Gynaecology, Faculty of Medicine, University of Colombo, Colombo, Sri Lanka, [3]Department of Community Medicine, Faculty of Medicine, University of Colombo, Colombo, Sri Lanka

Introduction

Obesity has reached epidemic proportions globally and has more than doubled worldwide since 1980. According to the estimates by the World Health Organization (WHO) in 2008, more than 1.5 billion adults were overweight, and of those, over 200 million men and nearly 300 million women were obese [1]. WHO in 1997 formally recognised obesity as a global epidemic by designating obesity as a major public health problem [2]. As obesity is becoming an epidemic, concern over its significant health and economic consequences has also grown. The problem of obesity is a major contributor to the global burden of chronic disease and disability. Often coexisting with undernutrition in developing countries, obesity is a complex condition, with serious social and psychological implications that affects virtually all ages and socio-economic groups [3].

Although overweight and obesity were once perceived as a sign of affluence, it has been increasingly recognised as an important health risk that affects men and women, particularly in low- and lower-middle income countries. This has been largely due to the effects of globalisation and rapid and unplanned urbanisation leading to major behavioural changes in the world's populations. Obesity is now recognised as a health risk that is not limited to affluence but affecting the lower socio-economic strata, and among the poorer groups particularly women. The troublesome aspect of this problem is the parallel emergence of childhood and adolescent obesity in many parts of the world. The latter brings into focus the role of obesity in women of reproductive years, which is also linked to obesity in their offspring. Although the main impact of obesity is its cardiometabolic risks that lead to greater occurrence of cardiovascular morbidity and mortality that has reached a crisis point on the socio-economic status of the affected populations, it has also been identified as a major contributory factor for infertility among both men and women.

Obesity. DOI: http://dx.doi.org/10.1016/B978-0-12-416045-3.00001-7

This chapter collates the available epidemiological data on overweight and obesity affecting women the world over. The epidemiological features of obesity, including global prevalence, secular trends and associated factors, and burden of illness related to obesity are addressed.

Definitions of Overweight and Obesity

Obesity is defined as an abnormal or extensive fat accumulation that may impair the health of an individual. Because fat is stored throughout the body, it cannot be measured directly. Body weight itself cannot provide an indication of fat stores. Instead, other measurements including body mass index (BMI), waist circumference, waist/hip ratio, skinfold thickness and bio-impedance are used to assess obesity and overweight.

The prevalence of overweight and obesity is commonly assessed by using BMI, defined as the weight in kilograms divided by the square of the height in metres (kg/m^2). BMI is classified according to the criteria laid by the National Institute of Health in the United States and is recommended by WHO. In general, BMI less than 18.5 kg/m^2 is considered underweight, 18.5–24.9 kg/m^2 is normal, 25 kg/m^2 or greater is overweight and 30 kg/m^2 or greater is obese [4]. Table 1.1 summarises the details of the international classification of adult underweight, overweight and

Table 1.1 The International Classification of Adult Underweight, Overweight and Obesity According to BMI

Classification	BMI (kg/m^2)	
	Principal Cut-Off Points	**Additional Cut-Off Points**
Underweight	**< 18.50**	**< 18.50**
Severe thinness	<16.00	<16.00
Moderate thinness	16.00–16.99	16.00–16.99
Mild thinness	17.00–18.49	17.00–18.49
Normal	**18.5–24.9**	**18.50–22.99**
		23.00–24.99
Overweight	**≥ 25.00**	**≥ 25.00**
Pre-obese	25.00–29.99	25.00–27.49
		27.50–29.99
Obese	**≥ 30.00**	**≥ 30.00**
Obese class I	30.00–34.99	30.00–32.49
		32.50–34.99
Obese class II	35.00–39.99	35.00–37.49
		37.50–39.99
Obese class III	≥40.00	≥40.00

*Values in bold represent underweight, normal, overweight and obese according to BMI
Source: Ref. [4].

obesity according to BMI. The WHO definition of obesity includes central obesity as waist circumference greater than 102 cm in men and 88 cm in women [2].

There is less clarity whether these criteria are applicable for all racial groups due to ethnic variation in body composition and for children and the older groups who have a decline in muscle mass. Nevertheless, the international convention based on the two major WHO technical consultations in 1995 and 2000 endorsed the use of a common BMI scheme for adults irrespective of sex or age [2].

Manifestation of the consequences of insulin resistance, particularly in Asian populations, due to abdominal visceral fat with lower values of BMI warrants due attention to be paid to central obesity. Abdominal fat is closely linked with a greater metabolic risk of diabetes, hypertension, ischaemic heart disease, strokes and gall bladder disease; these diseases have an ethnic variation possibly due to socio-cultural influences [5–8]. Therefore, measurement of waist circumference must be included in the overall assessment of an individual's health-risk categorisation of overweight and obesity. The recommended measure of waist circumference is at the level of the mid-point between the 12th rib and upper margin of the iliac crest when standing (determined from the posterior aspect) in mid inspiration using a non-expansible tape measure.

It was proposed to lower cut-off points for BMI and waist circumference for use in Asian communities [9]. Furthermore, evidence has emerged that Afro-Caribbean, Hispanic Americans, Asian Indians, Chinese and Latin Americans have a greater risk of accumulation of excess adipose tissue in the abdominal viscera than white Caucasians [10]. The basis for this recommendation of using differing cut-offs for ethnic groups was based on the observation that they consistently manifested metabolic problems at a lower BMI [11–14]. A new analysis has also proposed different values for the Chinese [15]. Therefore, the current clinical consensuses are to identify obesity at a lower BMI in ethnic groups originating from Asia in particular and also to pay due attention to waist circumference.

The Extent of the Problem

The estimates made by the WHO indicate that the problem of obesity extends globally. In 2008, 1.5 billion adults, 20 and older, were overweight. Of those, over 200 million men and nearly 300 million women were obese [1]. According to the global database on BMI compiled by the WHO, the prevalence of obesity varied widely across countries. The obesity levels ($BMI \geq 30\, kg/m^2$) range from below 5% collectively in China, Japan, India, Indonesia and certain African nations to over 75% in Samoa and Nauru [4]. But even in relatively low-prevalence countries like China and Japan, rates were almost 20% in some cities [3,16]. The most recent available data reveal that the prevalence of obesity in adults in the United States is 33.9% compared to 23.1% in Canada, 22.7% in the United Kingdom, 16.9% in France and 16.4% in Australia [4]. However, it is important to note that the data presented might not be directly comparable since they vary in terms of sampling procedures, age ranges and the year of data collection. The incidence of obesity reported across different countries should be interpreted with great caution as there are different age structures in the populations, measurement techniques used are not comparable and surveys may not be population-based.

Childhood obesity is already an epidemic in some countries and is on the rise in others. Approximately, 22 million children under 5 years of age are estimated to be overweight worldwide. In the United States, the number of overweight children has doubled and the number of overweight adolescents has trebled since 1980. The prevalence of obese children aged 6–11 years has more than doubled since the 1960s. Obesity prevalence in adolescents aged 12–17 years has increased dramatically from 5% to 13% in boys and from 5% to 9% in girls between 1966–1970 and 1988–1991 in the United States. The problem is global and increasingly extends into the developing world; for example, the prevalence of obesity in 5- to 12-year-old children in Thailand rose from 12.2% to 15.6% in just 2 years [3]. Sedentary lifestyle and rapidly changing dietary practices have led to increasing prevalence of obesity in children aged 5–19 years in developing countries [17,18]. A recent review found huge variation in the prevalence of childhood obesity: 41.8% in Mexico, 22.1% in Brazil, 22.0% in India and 19.3% in Argentina. Moreover, secular trends indicate increasing prevalence rates in these countries: 4.1–13.9% in Brazil during 1974–1997, 12.2–15.6% in Thailand during 1991–1993 and 9.8–11.7% in India during 2006–2009 [17].

In Mexico, the prevalence of overweight and obesity is 16.7% in preschool children, 26.2% in school children and 30.9% in adolescents. For adults, the prevalence of overweight and obesity is 39.7% and 29.9% respectively [19].

Temporal Trends

Worldwide, the prevalence of obesity has increased dramatically during the last three decades. In the United States, the prevalence of excess weight is increasing rapidly across the country, and almost 65% of the adult population was overweight or obese by the year 2000 [20]. Compared with the period 1976–1980, the prevalence of overweight (BMI $\geq$ 25 kg/m^2) had increased by 40% (from 46.0% to 64.5%) and the prevalence of obesity (BMI $\geq$ 30 kg/m^2) had risen by 110% (from 14.5% to 30.5%) by 1999–2000 [20,21]. An alarming increase in weight among the youths in the United States has also been observed. More than 10% aged 2–5 years and 15% aged 6–19 year are overweight (BMI $\geq$ 95th percentile for age and gender) [22]. This represents a near-doubling of overweight children and a near-tripling of overweight adolescents over the last two decades. Although some segments of the population are more likely to be overweight or obese than others, people of all ages, socio-economic levels, ethnicities, races and geographic areas are experiencing a substantial increase in weight [23].

International data indicate that the rising trend is not only confined to the United States and other developed countries but it is in fact a global health problem as the prevalence of obesity is also rising in less-affluent countries as well. The prevalence of obesity increased dramatically during the last decades [2,4,24–26] in the developed countries as well as in the developing countries. Furthermore, predictions have been made that if the trend continues, a majority of the world's adult population will be either overweight or obese by 2030 [27].

Despite the rapid rise of obesity in the past decades, a systematic review of literature and web-based sources of 52 studies from 25 different countries found that there is an overall leveling off of the epidemic in children and adolescents from Australia, Europe, Japan and the United States [28]. In adults, stability was found in the United States, while increases were still observed in some European and Asian countries. Some evidence for heterogeneity in the obesity trends across socio-economic status groups were found with leveling off being less evident in the lower socio-economic group, but no obvious differences in the leveling off between genders were identified.

Factors Associated with Overweight and Obesity

Interaction of many factors including genetic, metabolic, behavioural and environmental influences has resulted in the rise of overweight and obesity. The rapidity with which obesity is increasing suggests that behavioural and environmental influences rather than the biological changes have directly contributed to the epidemic [23].

Epidemiological data indicate that prevalence of obesity in adults is higher in women than in men. For example, a study carried out in Spain revealed that the prevalence of obesity (i.e. $BMI \geq 30\ kg/m^2$) and abdominal obesity (i.e. a waist circumference >102 cm in men or >88 cm in women) was higher in women, at 23.2% (95% CI, 20.9–25.5%), than in men, at 20.4% (95% CI, 18.0–22.7%). The prevalence of abdominal obesity was also higher in women, at 50.1% (95% CI, 47–53.1%), than in men, at 22.8% (95% CI, 20.3–25.2%) [29].

Many surveys had explored the prevalence of obesity among different socio-demographic categories in different parts of the world. The National Health and Nutrition Examination Survey (NHANES) is a nationally representative health examination survey conducted by the National Center for Health Statistics division of the US Centers for Disease Control and Prevention, Atlanta, GA. According to NHANES, for the 2007–2008 periods, 68% of adults in the United States aged 20–74 years were overweight or obese. For men, the age-adjusted prevalence of obesity was 32%, while for women it was 36% [30]. Racial or ethnic differences have been reported in some studies [31]. People living in urban areas have higher rate of overweight and obesity than those in rural areas, possibly due to more sedentary lifestyles and consumption of energy-dense high-fat diets [32].

The energy spent for daily living has reduced over the years promoting obesity, with the development of machinery and modes of transport reducing the dependency on walking and cycling. In the household as well as in the workplaces, the energy requirements have dropped due to labour saving devices and mechanised labour aids.

Until relatively recently, obesity was considered a condition associated with high socio-economic status. Indeed, early in the twentieth century, most populations in which obesity became a public health problem were in the developed world, primarily in the United States and Europe [33,34]. In more recent decades, available data show that the most dramatic increases in obesity are in developing countries such as Mexico, China and Thailand [34].

A recent review using data from 39 countries about women aged 18–49 years with young children indicated that in a substantial number of countries the rise in the prevalence of overweight individuals over time has been greater in the groups with lowest wealth and education compared with those with the highest: 14% when wealth is used as the indicator of socio-economic status and 28% for education [35]. Gross domestic product per capita was associated with a higher overweight prevalence growth rate for the group with lowest wealth compared with those with the highest [35]. Furthermore, it has been concluded that groups with higher (vs. lower) wealth and education had higher overweight prevalence across most developing countries. However, some countries show a faster growth rate in overweight in the groups with lowest (vs. highest) wealth and education, which is indicative of an increasing burden of overweight among the groups with lower wealth and education in the lower income countries.

Disease Burden and Impact on Health and Economy

Overweight and obesity lead to adverse metabolic effects on blood pressure, cholesterol, triglycerides and insulin resistance. The likelihood of developing type-2 diabetes and hypertension rises steeply with increasing body fatness. Approximately 85% of people with diabetes are type 2, and of those, 90% are obese or overweight [3]. It has been estimated that 20–30% of deaths due to cardiovascular disease may be attributable to overweight and obesity [36]. Hypertension is closely related to excess body weight, and a strong linear relationship is seen between the BMI and the blood pressure [37]. Obese individuals have almost 10 times the risk of developing diabetes, when compared with their non-obese counterparts [38].

It has been estimated that overweight and obesity may account for 14% of all cancer deaths in men and 20% in women [39]. According to the International Agency for Research on Cancer, overweight and obesity cause 9% of postmenopausal breast cancer, 11% of colon cancer, 25% of renal cancer, 37% of oesophageal cancer and 39% of endometrial cancer [2]. In addition, overweight and obesity increase the risk of many other conditions including cerebrovascular disease, gallstones, osteoarthritis, sleep apnoea, physical functioning, psychological illness and social discrimination [23].

Furthermore, obesity has become a major challenge to male and female infertility and it affects fertility throughout a woman's life, with obese girls frequently experiencing early menarche and later irregular menstrual cycles and anovulation resulting in infertility.

Obesity accounts for 2–6% of total health care costs in several developed countries; some estimates put the figure as high as 7%. The true costs are undoubtedly much greater as not all obesity-related conditions are included in the calculations [3]. In the United States, it is estimated that obesity costs US$117 billion each year [40]. However, the actual cost of the current epidemic of overweight and obesity will be much higher than this estimated cost [23].

International Response to the Epidemic

The International Obesity Task Force (IOTF) was officially launched in May 1996 to create a coherent action on tackling the global epidemic of obesity. In June 1997, a Consultation on Obesity was convened with the aim of reviewing global prevalence and trends of obesity among children and adults. It was also aimed to draw up recommendations for developing public health policies and programmes for prevention and control of obesity worldwide [32]. The Consultation on Obesity emphasised that overweight and obesity represent a rapidly growing threat to the health of populations, and obesity was recognised as a disease in its own right [2,32].

Conclusions

Worldwide, the prevalence of obesity increased dramatically during the last three decades. Whereas some segments of the population are more likely to be overweight or obese than others, people of all ages, races, ethnicities, socio-economic levels and geographic areas are experiencing a substantial increase in weight. The prevention and treatment of excess weight is critical for the health of both individuals and our society. Health care providers can play an important role in monitoring patients' weights, assist with dietary advice and provide counselling as regards physical activity. Multi-level interventions are needed to address the epidemic of obesity and overweight and prevent its growing negative consequences on the overall health and welfare of the human race.

References

1. World Health Organization. Fact sheet: obesity and overweight. <http://www.who.int/mediacentre/factsheets/fs311/en/2012/>; 2012 (Accessed 28 January 2012).
2. World Health Organization. *Obesity: Preventing and Managing the Global Epidemic. Report of a WHO Consultation*. Geneva: World Health Organization; 2000.
3. World Health Organization. Global strategy on diet, physical activity and health. <http://www.who.int/dietphysicalactivity/publications/facts/obesity/en/>; 2003 (Accessed 16 January 2012).

4. World Health Organization. Global database on body mass index. <http://apps.who.int/bmi/index.jsp?introPage=intro_3.html/>; 2012 (Accessed 28 January 2012).
5. Okosun IS. Ethnic differences in the risk of type 2 diabetes attributable to differences in abdominal adiposity in American women. *J Cardiovasc Risk*. 2000;7:425–430.
6. Okosun IS, Liao Y, Rotimi CN, Choi S, Cooper RS. Predictive values of waist circumference for dyslipidemia, type 2 diabetes and hypertension in overweight White, Black, and Hispanic American adults. *J Clin Epidemiol*. 2000;53:401–408.
7. Okosun IS, Rotimi CN, Forrester TE, et al. Predictive value of abdominal obesity cut-off points for hypertension in blacks from west African and Caribbean island nations. *Int J Obes Relat Metab Disord*. 2000;24:180–186.
8. Okosun IS, Cooper RS, Rotimi CN, Osotimehin B, Forrester T. Association of waist circumference with risk of hypertension and type 2 diabetes in Nigerians, Jamaicans, and African–Americans. *Diabetes Care*. 1998;21:1836–1842.
9. WHO Expert Consultation. Appropriate body-mass index for Asian populations and its implications for policy and intervention strategies. *Lancet*. 2004;363:157–163.
10. Sanchez-Castillo CP, Velazquez-Monroy O, Berber A, et al. Anthropometric cutoff points for predicting chronic diseases in the Mexican National Health Survey 2000. *Obes Res*. 2003;11:442–451.
11. Jafar TH, Chaturvedi N, Pappas G. Prevalence of overweight and obesity and their association with hypertension and diabetes mellitus in an Indo-Asian population. *CMAJ*. 2006;175:1071–1077.
12. Lear SA, Toma M, Birmingham CL, Frohlich JJ. Modification of the relationship between simple anthropometric indices and risk factors by ethnic background. *Metabolism*. 2003;52:1295–1301.
13. Razak F, Anand SS, Shannon H, et al. Defining obesity cut points in a multiethnic population. *Circulation*. 2007;115:2111–2118.
14. WHO/IASO/IOTF. Health Communications Australia, Melbourne, Australia; 2000.
15. Expert panel on the Identification, Evaluation and Treatment of overweight in Adults. Clinical guidelines on the identification, evaluation, and treatment of overweight and obesity in adults. *WMJ*. 1998;**97**:20–21, 24, 25, 27–37.
16. Yoshiike N, Kaneda F, Takimoto H. Epidemiology of obesity and public health strategies for its control in Japan. *Asia Pac J Clin Nutr*. 2002;11(suppl 8):S727–S731.
17. Gupta N, Goel K, Shah P, Misra A. Childhood obesity in developing countries: epidemiology, determinants, and prevention. *Endocrinol Rev*. 2012;33:48–70.
18. Hasanbegovic S, Mesihovic-Dinarevic S, Cuplov M, et al. Epidemiology and etiology of obesity in children and youth of Sarajevo Canton. *Bosn J Basic Med Sci*. 2010;10:140–146.
19. Barquera Cervera S, Campos-Nonato I, Rojas R, Rivera J. Obesity in Mexico: epidemiology and health policies for its control and prevention. *Gac Med Mex*. 2010;146:397–407.
20. Flegal KM, Carroll MD, Ogden CL, Johnson CL. Prevalence and trends in obesity among US adults, 1999–2000. *JAMA*. 2002;288:1723–1727.
21. Flegal KM, Carroll MD, Kuczmarski RJ, Johnson CL. Overweight and obesity in the United States: prevalence and trends, 1960–1994. *Int J Obes Relat Metab Disord*. 1998;22:39–47.
22. Flegal KM, Ogden CL, Wei R, Kuczmarski RL, Johnson CL. Prevalence of overweight in US children: comparison of US growth charts from the Centers for Disease Control and Prevention with other reference values for body mass index. *Am J Clin Nutr*. 2001;73:1086–1093.

23. Stein CJ, Colditz GA. The epidemic of obesity. *J Clin Endocrinol Metab*. 2004;89:2522–2525.
24. Lobstein T, Baur L, Uauy R, TaskForce IIO. Obesity in children and young people: a crisis in public health. *Obes Rev*. 2004;5(suppl 1):4–104.
25. Norton K, Dollman J, Martin M, Harten N. Descriptive epidemiology of childhood overweight and obesity in Australia: 1901–2003. *Int J Pediatr Obes*. 2006;1:232–238.
26. Seidell JC, Flegal KM. Assessing obesity: classification and epidemiology. *Br Med Bull*. 1997;53:238–252.
27. Kelly T, Yang W, Chen CS, Reynolds K, He J. Global burden of obesity in 2005 and projections to 2030. *Int J Obes*. 2008;32:1431–1437.
28. Rokholm B, Baker JL, Sørensen TIA. The levelling off of the obesity epidemic since the year 1999 – a review of evidence and perspectives. *Obes Rev*. 2010;11:835–846.
29. Escribano Garcia S, Vega Alonso AT, Lozano Alonso J, et al. Obesity in Castile and Leon, Spain: epidemiology and association with other cardiovascular risk factors. *Rev Esp Cardiol*. 2011;64:63–66.
30. Selassie M, Sinha AC. The epidemiology and aetiology of obesity: a global challenge. *Best Pract Res Clin Anaesthesiol*. 2011;25:1–9.
31. Gavin AR, Rue T, Takeuchi D. Racial/ethnic differences in the association between obesity and major depressive disorder: findings from the Comprehensive Psychiatric Epidemiology Surveys. *Public Health Rep*. 2010;125:698–708.
32. Gill TP, Antipatis VJ, James WPT. The global epidemic of obesity. *Asia Pac J Clin Nutr*. 1999;8:75–81.
33. Caballero B. The global epidemic of obesity: an overview. *Epidemiol Rev*. 2007;29:1–5.
34. Popkin BM, Gordon-Larsen P. The nutrition transition: worldwide obesity dynamics and their determinants. *Int J Obes Relat Metab Disord*. 2004;28:S2–S9.
35. Jones-Smith JC, Gordon-Larsen P, Siddiqi A, Popkin BM. Is the burden of overweight shifting to the poor across the globe? Time trends among women in 39 low- and middle-income countries (1991–2008). *Int J Obes*. 2011. doi:10.1038/ijo.2011.179.
36. Seidell JC, Verschuren WM, van Leer EM, Kromhout D. Overweight, underweight, and mortality. A prospective study of 48,287 men and women. *Arch Intern Med*. 1996;156:958–963.
37. Witteman JC, Willett WC, Stampfer MJ, et al. A prospective study of nutritional factors and hypertension among US women. *Circulation*. 1989;80:1320–1327.
38. Colditz GA, Willett WC, Rotnitzky A, Manson JE. Weight gain as a risk factor for clinical diabetes mellitus in women. *Ann Intern Med*. 1995;122:481–486.
39. Calle EE, Rodriguez C, Walker-Thurmond K, Thun MJ. Overweight, obesity, and mortality from cancer in a prospectively studied cohort of U.S. adults. *N Engl J Med*. 2003;348:1625–1638.
40. U.S. Department of Health and Human Services. Public Health Services, Office of the Surgeon General, Rockville, MD; 2001.

2 Social and Ethnic Determinants of Obesity

Chu Lim and Omar Thanoon

Department of Obstetrics and Gynaecology, Victoria Hospital, Kirkcaldy, Fife, UK

Introduction

Obesity is a recognised medical problem worldwide with rates on the rise in both developed and developing countries.

The World Health Organization considers obesity as a disease and defines it as a condition of excess body fat to a degree where it causes impairment to the health of an individual.

Overweight and obesity are known risk factors for a number of chronic conditions including non-insulin-dependent diabetes mellitus, coronary heart disease, hypertension, hyperlipidaemia, osteoarthritis and certain cancers [1].

The World Health Organization uses the body mass index (BMI) (weight (kg)/height (m)^2) to define overweight and obesity. Overweight is defined as a BMI of 25 kg/m^2 or more, while obesity is defined as a BMI of 30 kg/m^2 or more.

Obesity usually results from the interaction between genetic and environmental variables.

Ethnic Factors

The relationship between obesity and ethnicity in United Kingdom is highly complex. There is very little nationally representative data on obesity prevalence for adults from ethnic minority groups in United Kingdom. The data are even scarcer or do not even exist for many smaller ethnic groups

The concept of 'ethnicity' or 'ethnic group' is difficult to define. It is multi-dimensional and is not fixed in time. In the world of migration and mixing, it is difficult to create a single, mutually exclusive category of ethnicity. Definitions of ethnic groups often differ between studies, and ethnic classification of dataset can change over time.

Data from the most recent UK census (2001) showed 53% increase in ethnic minority since 1991 [2]. The ethnic minority groups comprise almost 8% of the total population, at about 4.6 million. The largest ethnic minority group was Indian (8%), followed by Pakistani (1.3%), mixed ethnicity (1.2%) and Black Caribbean (1.0%).

Obesity. DOI: http://dx.doi.org/10.1016/B978-0-12-416045-3.00002-9

Health Survey for England (HSE) 2004 contained the most recent robust data on adult obesity prevalence by ethnic group. Even though the subject of measurement for obesity had been debated, the survey provides a comprehensive picture of obesity using BMI as a measure. Obesity prevalence is lower among men from the communities of Black African, Indian, Pakistani, and most markedly Bangladeshi and Chinese. Among women, obesity prevalence appears to be higher for those from Black African, Black Caribbean and Pakistani groups than for women in the general population and lower for women from the Chinese ethnic group.

Disparity in health between ethnic minority and majority had been well documented. Such disparity has been demonstrated in respiratory medicine [3], hypertension [4], heart disease [4–6] and diabetes mellitus [3,4,6]. Studies have also demonstrated ethnic inequalities in obesity risk [7–9], but this literature remains disparate and disorganised.

Ethnic minorities in the United Kingdom tend to be younger and be of lower social economical status [10,11]. They are also more likely to be unemployed [11]. There is also increasing hostility towards immigrants and multi-culturalism among the British public [12], and recent report highlighted increased crimes against ethnic minority groups [13]. Research on morbidity had suggested the inequalities in socio-economic wellbeing [14] as well as other factors such as the area one lives in [15], and rising tension between ethnic minority and majority groups makes a major contribution to ethnic differences on health.

Physical activity levels are major determinants for obesity [16,17]. There is a marked variation in the level of physical activity among ethnic minority populations in the United Kingdom [18]. Similar findings were reported on the HSE 2004 report. A combination of personal, socio-economic, cultural and environmental barriers may discourage the people of ethnic minority groups from engaging in physical activity. Barriers include dress codes, modesty and lack of single-sex facilities [19]. Younger generations of South Asian women were actively discouraged by their parents to involve in physical activity which were deemed incompatible with femininity [20].

There is a perception that ethnic minority groups have healthier eating patterns than the white population. However, there is a wide variation across and within ethnic groups. Factors including food availability, level of income, food beliefs, dietary laws, religion, cultural patterns and customs may have impact on eating patterns. A recent systemic review on nutritional study by Rees et al. [21] showed a higher energy intake among South Asian children and a lower fat intake among Black African and Black Caribbean Children. The MRC DASH study [22] found poor dietary behaviour among the adolescents of ethnic minority who were born in the United Kingdom. There is also a trend for the ethnic minorities to assimilate the fast food aspect of the Western culture into their eating habit [23]. Age and generations were two major factors determining the extent to which diets changed, with processed foods more likely to be consumed by the younger generations than the older generations [24].

Different ethnic groups have different physiological responses to fat storage. South Asian adults may accumulate more weight around abdomen and have greater adiposity in general. The studies that measured ethnicity and obesity had consistently showed larger waist/hip ratios compared to their Caucasian counterparts [4,25,26]. In contrary to South Asian population, the Black Caribbean adults accumulate less weight around the abdomen than Caucasian [27,28]. The weight-for-height based metrics of obesity such as BMI, may systematically overestimate the prevalence of obesity in Black Caribbean population.

The conventional threshold for classification of obesity derived primarily for European populations to correspond to risk thresholds for wide range of chronic disease and mortality. It is now generally accepted that South Asian populations are at greater risk of ill health at lower BMI levels than the European population. This finding led the World Health Organization to revise its recommendations about the cut-off levels of BMI for appropriate overweight and obesity among South Asian populations [29]. A lower threshold of 23 kg/m^2 has been recommended by the South Asian Health foundation in the United Kingdom. In the absence of universal agreement, The National Institute for Health and Clinical excellence (NICE) continues to advise the same threshold for the general population. The Chinese populations have also shown to have elevated blood pressure at significantly lower BMI [30], but no specific BMI threshold had been agreed.

Social Factors

It is very important to understand the relation between obesity and different social factors in order to develop effective obesity-preventing strategies.

The relationship between social factors and obesity is complex because social factors may influence obesity, obesity may influence socio-economic status, or common factor(s) may influence both obesity and socio-economic status.

Evidence suggests that obesity is socially linked, with certain social groups at increased risk. Strong evidence proves that there is an inverse relation between social class and body weight and risk of obesity among women, and less consistently among men in industrialised countries [31–34]. In the following sections, we will look at different social factors such as social class, physical activity, dietary factors, early life factors and smoking and their role in obesity.

Socio-economic Status

Socio-economic status is defined as the position of an individual on a social-economic scale that measures factors such as education, income, type of occupation, place of residence, and, in some populations, heritage and religion [35].

Historically, in the developing world, the prevalence of obesity has been greater in the more advantaged socio-economic groups, whereas in the developed countries obesity is more common in individuals with a low socio-economic status.

A review of the literature looking at the association between socio-economic status and obesity in men, women and children in both the developed and developing worlds was published by Sobal and Stunkard; this included 144 published study between 1960 and 1980 [31]. The findings of this extensive literature review showed an inverse association for women in developed countries, with an increased chance of obesity among women in lower socio-economic societies. The relation for men and children in developed societies was inconsistent. In developing countries, a strong direct relation was seen for women, men and children, with a higher likelihood of obesity among individuals in higher socio-economic class.

In an attempt to update and build on the work of Sobal and Stunkard a more recent literature review was published by McLaren in 2007. This review looked at the data that was published between 1988 and 2004 addressing the relation between socio-economic status and obesity [36]. The primary findings of this review showed that there is a gradual reversal of the social gradient in weight. As one moved from less developed to developing countries, the proportion of positive associations increased and the proportion of negative associations decreased, for both men and women. With regard to sex, this updated review revealed a predominance of negative associations for women in countries with a high development status [36].

A bulletin published by the World Health Organization under the title 'Socio-economic status and obesity in adult populations of developing countries', which was based on a number of studies published between 1998 and 2003 looking at socio-economic status and obesity in developing countries, stated that obesity in the developing world can no longer be considered solely a disease within the groups of higher socio-economic status, and the burden of obesity in each developing country tends to shift towards the groups with lower socio-economic status as the country's gross national product increases [37].

One of the important known indicators of socio-economic status is the household income. According to the national obesity observatory brief on obesity and socio-economic status, the relation between obesity and household income for women is an almost linear relationship, with obesity rising steadily as household income falls. However, the trend levels off as income falls. For men, the pattern is less clear. There is little difference between obesity prevalence in the highest and lowest income group.

Another important parameter of socio-economic status is the educational achievement. The WHO MONICA Project (Monitoring Trends and Determinants in Cardiovascular Disease) which examined educational level, relative body weight and changes in their association over 10 years in 39 participating centres divided between 26 countries showed that there was a statistically significant inverse association between educational level and BMI for women in almost all populations. Women with higher education were leaner than those with lower education. Among men, about one-fourth of the study populations in the initial survey and about half in the final survey also showed such a statistically significant inverse association. Only two populations and one population in the initial and final surveys respectively had a statistically significant positive association [38]. In addition, in about two-thirds of the populations, the difference in mean BMI between the highest and lowest educational levels increased during the 10-year study period. Similar results were published by the National Obesity Observatory in England which showed that obesity prevalence varies with the levels of educational attainment showing a general trend of rising obesity prevalence with decreasing level of education. Both men and women with degree-level qualifications have significantly lower rates of obesity than all others. For men, those with no qualifications have the highest rates of obesity, whilst for women, those with NVQ1 equivalent or no qualifications have the highest rates.

The relation between obesity and marital state is not clearly understood. A recent study found no relation between marital status and obesity level. In earlier reports, a positive relation has been found between obesity levels in husband and wife, but it is possible that obesity in itself has no specific effect on marital status and that psychological factors are important in explaining the relationship between obesity and marital status [39].

Pregnancy is often associated with changes in women's general health [40]. It is not unusual for women to experience an increase in weight with each pregnancy. Weight gained in pregnancy is often not entirely lost afterwards. A study of concordant twins showed a positive relation between pregnancies and obesity [41]. This finding was not noted in randomly selected women. However, the present data support the concept that pregnancy might be a trigger for the development of obesity.

Dietary Factors

Obesity develops when energy intake continuously exceeds energy expenditure, causing a fundamental chronic energy imbalance [42]. Social and behavioural changes over the last decades are held responsible for the considerable increase in sedentary lifestyles [43].

A positive association between dietary fat and obesity has been observed in some studies but not in others [44].

The association of dietary intake and obesity has been difficult to establish in a number of studies because of the bias in reporting dietary intake [42]. This is usually because of under-recording and under-eating during the time of study; a much less frequent bias was due to over-recording [45].

A cross-sectional study published by Duvigneaud et al. [42] in 2007 looking at the association between dietary factors and obesity showed that overweight and obese men reported a higher consumption of fats and proteins, whereas their energy percentages from carbohydrates and fibres were lower compared to their normal weight counterparts. In overweight and obese women, a higher intake of all macronutrients was observed compared to leaner women. The sex differences for dietary intake between obese men and women might reflect the generally higher health consciousness of women.

Alcohol is a rich source of energy production, as 1 gram of alcohol intake produces 7 kcal/gm. However, the relationships between reported alcohol intake and BMI show a mixed pattern [46]. Metabolic studies show that isocaloric substitution of alcohol for food, causes weight loss, whereas the addition of alcohol will not cause weight gain [47]. A review of the epidemiological evidence revealed that 25 studies showed a positive relation between alcohol consumption and obesity, 18 showed a negative relation and 11 showed no relationship [48]. A review of the epidemiological studies linking a high alcohol intake with abdominal fat distribution performed by Emery et al. [49] concluded that the evidence for a relationship was moderate for men and suggestive for women.

Breastfeeding has been shown in some studies to be protective against obesity in childhood and this is important because preventing childhood obesity can result in preventing adult obesity. A review by Butte examined 18 studies (6 retrospective, 10 prospective, 1 cohort, 1 case–control) published up to 1999 with a total of nearly 20,000 subjects. There was a wide time span (1945–1999) and the dentitions of breastfeeding and obesity and the length of follow up were all highly variable. Two of the studies found a positive association between breastfeeding and later obesity and four found a negative relationship (i.e. an apparent protective effect of breastfeeding). The remainder found no differences. The largest study (*n* 9357 children aged 5 to 6 years) found a prevalence of obesity among 2.8% of the breastfed children compared to 4.5% of the children who were never breastfed and there was an apparent dose–response in relation to the duration of breastfeeding. A similar study carried out among 3731 6-year-old British children, however, found no such relationships [50–52].

Conclusion

In conclusion, while associations between social factors and obesity have been studied extensively, the underlying mechanism behind this relation is not fully understood. The present evidences suggest that this is probably gender specific and highly complex. Further research is required to investigate causal mechanisms, as some evidence suggests that obesity influences social status rather than *vice versa* [53,54].

References

1. World Health Organization. *Obesity: Preventing and Managing the Global Epidemic*. Geneva: World Health Organization; 1997.
2. Census 2001. *National Report for England and Wales*. London: Office for National Statistics; 2004.
3. Duran-Tauleria E, Rona RJ, Chinn S, Burney P. Influence of ethnic group on asthma treatment in children in 1990–1991: national cross sectional study. *BMJ*. 1996;313:148–152.
4. McKeigue PM, Shah B, Marmot MG. Relation of central obesity and insulin resistance with high diabetes prevalence and cardiovascular risk in South Asians. *Lancet*. 1991;337:382–386.
5. McKeigue PM, Ferrie JE, Pierpoint T, Marmot MG. Association of early-onset coronary heart disease in South Asian men with glucose intolerance and hyperinsulinemia. *Circulation*. 1993;87:152–161.

6. Chaturvedi N, McKeigue PM, Marmot MG. Resting and ambulatory blood pressure differences in Afro-Caribbeans and Europeans. *Hypertension.* 1993;22:90–96.
7. Rennie KL, Jebb SA. Prevalence of obesity in Great Britain. *Obes Rev.* 2005;6:11–12.
8. Wardle J, Brodersen NH, Cole TJ, Jarvis MJ, Boniface DR. Development of adiposity in adolescence: five year longitudinal study of an ethnically and socioeconomically diverse sample of young people in Britain. *BMJ.* 2006;332:1130–1135.
9. Saxena S, Ambler G, Cole TJ, Majeed A. Ethnic group differences in overweight and obese children and young people in England: cross sectional survey. *Arch Dis Child.* 2004;89:30–36.
10. Fitzpatrick J, Jacobson B, Aspinall P. *Ethnicity and Health: Executive Summary.* York: Association of Public Health Observatories; 2007.
11. Bhattacharyya G, Ison L, Blair M. *Minority Ethnic Attainment and Participation in Education and Training: The Evidence.* Nottingham: Department for Education and Skills; 2003:RTP01-03.
12. Ipsos MORI. *Trend Briefing #1: Doubting Multiculturalism.* London: Ipsos MORI; 2009: <http://centres.exeter.ac.uk/emrc/publications/Islamophobia_and_Anti-Muslim_Hate_Crime.pdf/>
13. Githens-Mazer J, Lambert R. *Islamophobia and Anti-Muslim Hate Crime: A London Case Study.* Exeter: European Muslim Research Center, University of Exeter; 2010: <http://muslimsafetyforum.org/docs/Islamophobia_and_Anti-Muslim_Hate_Crime.pdf/>
14. Nazroo JY. *The Health of Britain's Ethnic Minorities: Findings from a National Survey.* London: Policy Studies Institute; 1997.
15. Bécares L, Nazroo J, Stafford M. The buffering effects of ethnic density on experienced racism and health. *Health Place.* 2009;15(3):700–708.
16. Butland B, Jebb S, Kopelman P, et al. *Foresight Report: Tackling Obesities: Future Choices – Project Report.* London: Government Office for Science; 2007.
17. Department of Health, 2004. *At Least Five a Week: Evidence on the Impact of Physical Activity and Its relationship to Health Gateway Ref: 2389.* London: Department of Health, Physical Activity, Health Improvement and Prevention; 2004.
18. Fischbacher CM, Hunt S, Alexander L. How physically active are South Asians in the United Kingdom? A literature review. *J Public Health.* 2004;26(3):250–258.
19. Long J, Hylton K, Spracklen K, Ratna A, Bailey S. *Sporting Equals. A Systematic Review of the Literature on Black and Minority Ethnic Communities on Sport and Physical Recreation.* Leeds: Carnegie Institute; February 2009.
20. Rojas A, Storch EA, Meriaux BG, Berg M, Hellstrom A-L. Psychological complications of obesity. Everyday experiences of life, body and well-being in children with overweight. *Pediatr Ann.* 2010;39(3):174–180.
21. Rees R, Oliver K, Woodman J, Thomas J. *Children's Views about Obesity, Body Size, Shape and Weight: A Systematic Review.* London: EPPI Centre; 2009.
22. Harding S, Teyhan A, Maynard MJ, Cruickshank JK. Ethnic differences in overweight and obesity in early adolescence in the MRC DASH study: the role of adolescent and parental lifestyle. *Int J Epidemiol.* 2008;37(1):162–172.
23. Lawrence JM, Devlin E, Macaskill S, et al. Factors that affect the food choices made by girls and young women, from minority ethnic groups, living in the UK. *J Hum Nutr Diet.* 2007;20(4):311–319.
24. Gilbert PA, Khokhar S. Changing dietary habits of ethnic groups in Europe and implications for health. *Nutr Rev.* 2008;66(4):203–215.
25. Bose K. Generalised obesity and regional adiposity in adult white and migrant Muslim males from Pakistan in Peterborough. *J R Soc Health.* 1996;116:161–167.

26. Bose K. A comparative study of generalised obesity and anatomical distribution of subcutaneous fat in adult white and Pakistani migrant males in Peterborough. *J R Soc Health*. 1995;115:90–95.
27. Vyas A, Greenhalgh A, Cade J, et al. Nutrient intakes of an adult Pakistani, European and African–Caribbean community in inner city Britain. *J Hum Nutr Diet*. 2003;16:327–337.
28. Miller MA, Cappuccio FP. Cellular adhesion molecules and their relationship with measures of obesity and metabolic syndrome in a multiethnic population. *Int J Obes*. 2006;30:1176–1182.
29. WHO Expert consultation. Appropriate body-mass index for Asian populations and its implications for policy and intervention strategies. *Lancet*. 2004;363:157–163.
30. Razak F, Anand SS, Shannon H, et al. Defining obesity cut points in a multiethnic population. *Circulation*. 2007;115(16):2111–2118.
31. Sobal J, Stunkard AJ. Socioeconomic status and obesity: a review of the literature. *Psychol Bull*. 1989;105:260–275.
32. Power C, Moynihan C. Social class and changes in weight-for-height between childhood and early adulthood. *Int J Obes Relat Metab Disord*. 1988;12:445–453.
33. Martinez JA, Kearney JM, Kafatos A, Paquet S, Martinez-Gonzalez MA. Variables independently associated with self-reported obesity in the European Union. *Public Health Nutr*. 1999;2:125–133.
34. Kuzsmarski RJ. Prevalence of overweight and weight gain in the United States. *Am J Clin Nutr*. 1992;55:S495–S502.
35. *Mosby's Medical Dictionary*. 8th ed. Elsevier; 2009.
36. Mclaren L. Socioeconomic status and obesity. *Epidemiol Rev*. 2007;29:29–48.
37. World Health Organization, Bulletin of WHO. Geneva. 2004;82(12):891–970.
38. Molarius A, Seidell JC, Sans S, Tuomilehto J, Kuulasmaa K. Educational level, relative body weight, and changes in their association over 10 years: an international perspective from the WHO MONICA project. *J Clin Epidemiol*. 1988;41(2):105–14.
39. Noppa H, Bengtsson C. Obesity in relation to socioeconomic status. A population study of women in Goteborg, Sweden. *J Epidemiol Community Health*. 1980;34:139–142.
40. Baric L, MacArthur C. Health norms in pregnancy. *Br J Prev Soc Med*. 1977;31:30–38.
41. Cederlof R, Kaij L. The effect of childbearing on body weight. A twin control study. *Acta Psychiatr Scand*. 1970;219(suppl):47–49.
42. Duvigneaud N, Wijndaele K, Matton L, et al. Dietary factors associated with obesity indicators and level of sports participation in Flemish adults: a cross-sectional study. *Nutr J*. 2007;6:26. doi:10.1186/1475-2891-6-26.
43. Manson JE, Skerrett PJ, Greenland P, VanItallie TB. The escalating pandemics of obesity and sedentary lifestyle. A call to action for clinicians. *Arch Intern Med*. 2004;164:249–258.
44. Lissner L, Heitmann BL. Dietary fat and obesity: evidence from epidemiology. *Eur J Clin Nutr*. 1995;49:79–90.
45. McCrory MA, Hajduk CL, Roberts SB. Procedures for screening out inaccurate reports of dietary energy intake. *Public Health Nutr*. 2002;5:873–882.
46. Swinburn BA, Caterson Ia, Seidell JCa, James WPT. Nutrition and the prevention of excess weight gain and obesity. *Public Health Nutr*. 2004;7(1A):123–146.
47. Lands WE. Alcohol and energy intake. *Am J Clin Nutr*. 1995;62(suppl 5):1101S–1106S.

48. Suter PM, Hasler E, Vetter W. Effects of alcohol on energy metabolism and body weight regulation: is alcohol a risk factor for obesity? *Nutr Rev.* 1997;55(5):157–171.
49. Emery EM, Schmid TL, Kahn HS, Filozof PP. A review of the association between abdominal fat distribution, health outcome measures, and modifiable risk factors. *Am J Health Promot.* 1993;7(5):342–353.
50. Dietz WH. Breastfeeding may help prevent childhood overweight. *JAMA.* 2001;285(19): 2506–2507.
51. Butte NF. The role of breastfeeding in obesity. *Pediatr Clin North Am.* 2001;48(1):189–198.
52. Wadsworth M, Marshall S, Hardy R, Paul A. Breast feeding and obesity. Relation may be accounted for by social factors. *BMJ.* 1999;319(7224):1576.
53. Sargent JD, Blanchflower DG. Obesity and stature in adolescence and earnings in young adulthood. Analysis of a British birth cohort. *Arch Pediatr Adolesc Med.* 1994; 148:681–687.
54. Gortmaker SL, Must A, Perrin JM, Sobol AM, Dietz WH. Social economic consequences of overweight in adolescence and young adulthood. *N Engl J Med.* 1993;329:1008–1012.
55. Harland JO, Unwin N, Bhopal RS, et al. Low levels of cardiovascular risk factors and coronary heart disease in a UK Chinese population. *J Epidemiol Community Health.* 1997;51:636–642.

3 Genetic and Molecular Basis of Obesity

Adnan Hasan and Tahir Mahmood

Department of Obstetrics & Gynaecology, Victoria Hospital, NHSS Fife, Kirkcaldy, UK

Introduction

Obesity is recognised as a global epidemic and is rising in children and in both men and women up to the age of 60. In 2008, the WHO reported that among adults, 20 and older, more than 200 million men and nearly 300 million women were obese. The rate of obesity in the United Kingdom has increased by fourfold over the past 30 years, an incidence of 22% in men and 24% in women in 2008/2009 [1]. Similarly, in the United States, comparable figures were 35.5% and 35.8% in 2009/2010 [2].

Obesity (BMI $>$ 30 kg/m^2) is a heterogeneous group of disorders that results from the imbalance in energy homeostasis, whereby food energy intake exceeds energy expenditure in genetically susceptible individuals. Such excess energy is stored in the most efficient form as triglycerides in white adipose tissues.

Obesity poses serious health problems and reduces longevity from associated co-morbidities such as insulin resistance, type-2 diabetes, hyperlipidaemia, hypertension and various cancers [3,4]. Obesity thus may become the most challenging health problem with huge economic impact on health services. From women's perspective, obesity further predisposes them to hyperandrogenicity with oligomenorrhoea/amenorrhoea and reduction in fertility in addition to health risks related to pregnancy and childbirth.

For most people, the human body maintains its normal weight so that at times of excess nutrients, intake is dissipated through increased physical activity, enhanced basal metabolic rate (BMR) and adaptive thermogenesis in response to daily fluctuation in food intake. The hypothalamus regulates energy homeostasis through integrating neural, hormonal and metabolic signals from peripheral and central nervous system (CNS), endocrine glands, including adipose tissue, and blood metabolites such as glucose and fatty acids. Disruption of the control mechanism whether congenital or acquired results in accumulation of excess energy as lipids.

Obesity. DOI: http://dx.doi.org/10.1016/B978-0-12-416045-3.00003-0

Hereditary Factors Influencing BMI

Studies from twins and adoption that have compared phenotypic correlations for body fat between groups of individuals, varying in genetic relatedness, have yielded interesting observations.

It appears that fat mass is genetically influenced among individuals who are more genetically similar such as in the case of monozygotic twins than dizygotic twins [5]. Even for twins reared apart, the estimated heritability was at 65–75% for BMI [6].

Furthermore, results of longitudinal behaviour genetic studies suggest that there are age-specific genetic effects on BMI, such that different obesity-promoting genes may become active at different ages across the lifespan [7]. The 'Human Obesity Gene Map' has been annually updated since 1994 [8]. This comprehensive compendium summarises evidences from the four classes of human studies:

a. Obesity due to a single gene in digenic mutation.
b. Obesity associated with Mendelian disorders such as Prader–Willi syndrome or Bardot–Biedle syndrome.
c. Associated studies that test whether candidate genes are associated with obesity phenotypes among samples of unrelated participants.
d. Linkage studies that test for causal association between genomic regions and obesity phenotypes in cohort of families.

This compendium has shown that the number of genes associated with obesity-related traits has increased dramatically over the past decade, thereby suggesting that the platform of specific genes that might contribute to obesity is large and involves loci throughout the genome.

Genetic–Molecular Interaction of Obesity

Severe obesity in humans can rarely result from the monogenic mutations of genes involved in hypothalamic appetite control pathways. The genes which encode leptin receptor [9], proopiomelanocortin (POMC) [10], prohormone convertase1 [11], leptin [12], melanocortin 3 receptors (MC3R) [13] and melanocortin 4 receptors (MC4R) [14] are involved. As noted above, human obesity id polygenic in origin [8] and involves a network of genes with several hundred loci located on nearly all the chromosomes including Y chromosome thus affecting obesity-related phenotypes. Monogenic mutations are not involved in the development of common forms of obesity where subtle interaction between numerous related genetic variants and environmental factors, such as diet and exercise, results in the overexpression of molecules that affect energy homeostasis leading to a positive or negative impact. Chronic exposure to environmental factors leads to single nucleotide polymorphism [15] or structural changes in DNA through methylation that promotes gene expression in susceptible individuals [16]. In future, this opens up the possibility of genotyping the individuals in order to identify susceptible individuals to obesogenic stimuli, that may allow for preventive measures to be implemented at individual level.

Regulation of Energy Balance – Molecular Basis

The main biochemical processes involved in energy regulation are those related to the control of energy intake, energy output (expenditure) and adipogenesis. These inter-related processes are governed at the cellular level by a wide range of inducible molecules and their associated genes, and such cross-regulation may limit the effectiveness of available treatments. This chapter will now focus on the molecules and their related genes responsible for the main biochemical processes that control energy homeostasis.

Regulation of Food Energy Intake

Feeding behaviour may be regulated via inter-related short- and long-term control. Short-term control is involved with individual meal intake where hunger and satiety signals reflect the acute energy status of the body. Feeding process is affected by metabolic, neural and endocrine factors and is modified by powerful sensory, emotional and cognitive inputs.

Ultimately, all of these factors must be integrated, so that decisions to begin and end periods of feeding will result [17]. Gastrointestinal hormones play an important role in the acute stage of feeding, e.g. plasma levels of ghrelin increase during fasting and infusion of this hormone stimulates appetite, an effect mediated via agouti-related protein (AgRP) [18]. Likewise, mechanical distension of the stomach and release of gut peptides, such as cholecystokinin, gastrin-releasing peptide 29 (GRP-29) and peptide YY3–36, produce satiety signals [19,20].

Long-term signals that reflect the status of energy stores are mainly that of leptin and insulin, and they provide information to the CNS to regulate feeding behaviour and promote energy homeostasis. As a result of these feedback signals, the energy stores and body weight remain generally stable for most humans over long period of time despite the wide variations in day-to-day food intake. However, chronic excessive food intake combined with reduced energy utilisation (i.e. exercise) limits their effectiveness and leads to increased adiposity that is individually regulated within narrow margins, resulting in a constant set point to which body weight would eventually return after fat loss or gain. In addition, resetting of feedback signals to higher thresholds in hypothalamus and elsewhere results in increased levels of circulating metabolites, e.g. glucose and lipids that reflect resistance to leptin and insulin action.

The Role of Hypothalamus

Hypothalamus is critical for the regulation of homeostatic processes such as feeding, thermoregulation and reproduction. It is the primary receptor of peripheral signals that indicate food availability [21]. It receives and integrates mechanical, hormonal, metabolic and peripheral neural inputs with those from higher centres: the brainstem, cerebral cortex, olfactory areas and limbic system. Hypothalamus in response produces neurotransmitters and neurohormones resulting in behavioural, autonomic and endocrine responses that control appetite and energy balance. The central role of the hypothalamus in appetite and satiety was established from the studies where lesions in the ventromedial hypothalamus caused obesity, while, conversely, lesions in the lateral hypothalamus resulted in leanness [22].

The arcuate nucleus plays a critical role in the regulation of energy homeostasis. It occupies more than one-half of the hypothalamus and lies around the base of the third ventricle immediately above the median eminence where the capillary endothelium lacks tight junctions forming an incomplete blood–brain barrier (BBB) thus allowing larger proteins and hormones to readily access the arcuate nucleus neurons from circulation.

The hypothalamic melanocortin signalling system is also heavily regulated by circulating hormones, mainly leptin and insulin, that provide information about body energy status and reserves. Hormonal signals from leptin and insulin inhibit hypothalamic orexigenic neurons [17], while, conversely, the anorexigenic neurons are stimulated by inputs from leptin and insulin [23].

Overall, it is apparent that key neurons in the hypothalamus, most likely in the arcuate nucleus, are primed to integrate a variety of hormonal and metabolic signals in order to interpret the state of energy balance, and, in turn, mediate the necessary metabolic and behavioural responses in order to compensate for deviations from homeostasis.

The Role of Leptin

Human leptin (derived from Greek *leptos* meaning *thin*) is a protein of 167 amino acids secreted mainly by adipocytes that express the *ob* gene. The level of circulating leptin is directly proportional to the total amount of body fat [24]. Its levels are higher in women than in men as its production is stimulated by oestrogen and suppressed by testosterone. Leptin levels drop with fasting and increase with food intake, an effect mediated in part by insulin stimulation of adipocytes, and both hormones act on hypothalamus triggering similar responses [25]. Various regulatory elements have been identified within the leptin promoter, e.g. cAMP and glucocorticoid response elements, and CCATT/enhancer binding sites, suggesting a direct regulation of leptin expression through membrane and transcriptional pathways [26].

Leptin is considered the main signal that provides information to the CNS reflecting energy stores in the body. Leptin binds to receptors (Lep R) in the arcuate nucleus and other hypothalamic nuclei resulting in changes in feeding behaviour, with suppression of appetite [27], and an increase in metabolic activity and energy expenditure (thermogenesis) that on the long term regulates energy balance and maintains normal body weight, despite wide variation in daily calorie intake in most individuals.

The central effects of leptin signals stimulate the anorexigenic neuropeptides, alpha-melonocyte stimulating hormone (α-MSH) and cocaine- and amphetamine-related transcript (CART), and inhibit the orexigenic neuropeptides, neuropeptides Y (NPY) and agouti-related protein AgRP in the arcuate nucleus [27]. Leptin also inhibits melanin-concentrating hormone (MCH) and orexins in lateral hypothalamus [22]. This in turn stimulates neuroendocrine response that results in the production of thyrotropin-releasing hormone [28] and activates the sympathetic nervous system that results in increased metabolism of tissues mediated via increased metabolic rate. Deficiency of either leptin or its receptors secondary to gene mutation in *ob* and *db* (Lep R) loci causes severe obesity, the first of which can be controlled with exogenous leptin.

Direct peripheral action of leptin on many tissues, including adipose tissue, stimulates lipolysis and fatty acid oxidation [29]. Local leptin production in the stomach plays a role in short-term energy balance as its level increases under the effect of CCK and gastrin thus mediating satiety signals indicating the filling effect of a meal.

Leptin levels in most obese people are high without corresponding gene mutation indicating a disruption in the feedback signals reflecting a state of insensitivity to leptin (hyperleptinaemia) similar to that of insulin. The mechanism involved is related to reduction in leptin transport through BBB where leptin levels in the brain are lower than the plasma levels [30] (Table 3.1).

Melanocortin Signalling Pathway

Leptin acts on two distinct subtypes of adjacent neurons in the arcuate nucleus in opposing manner. The first subtype of neurons co-expresses mRNAs encoding anorexigenic peptides, CART and α-MSH (derived from POMC), and leptin induces their expression [22,25,31]. The other subtype co-expresses the orexigenic

Table 3.1 Leptin Targets of Neurotransmitters and Peptides that Affect Feeding Process

Stimulation of Feeding	Inhibition of Feeding
Neuropeptide Y	Alpha-melanocyte stimulating hormone
Agouti-related peptide	Cocaine- and amphetamine-regulated transcript
Melanin-concentrating hormone	Corticotropin-releasing hormone
Orexins	Calcitonin gene-related peptide
Ghrelin	Urocortin
Galanin	Neurotensin
Growth hormone-releasing hormone	Serotonin
Opioid peptides	Glucagon
Gamma-aminobutyric acid	Cholecystokinin
Nitric oxide	Glucagon-like peptide 1
Noreadrenaline	Bombesin/gastrin-releasing peptide

peptides NPY and AgRP, and leptin reduces their expression [17,22,25,31]. Secreted α-MSH binds to MC4R and to a lesser extent to MR3C to produce tonic inhibition with anorexic effect and increase in thermogenesis.

The axonal projections of neuropeptide α-MSH also synapse with second-order neurons in the paraventricular nucleus (PVN) and lateral hypothalamus. The PVN regulates the secretion of pituitary hormones by the release of neuropeptides such as thyrotropin- and corticotrophin-releasing hormone via the projections in the median eminence and regulates autonomic nervous system via the projections to autonomic pregangilionic neurons. Direct projection of these arcuate melanocortinergic neurons (AgRP and α-MSH) onto the neurons within the lateral hypothalamus inhibits the expression of the orexigenic neuropeptides MCH and orexins [32]. Expression of MCH stimulates food intake. Deletion of the MCH gene causes a lean phenotype [33], and transgenic overexpression promotes obesity.

Two neuropeptides orexins A and B are involved in the regulation of food intake [34]. Orexins function through G-protein-coupled receptors and their expression in the brain seems to be limited to the neurons of the lateral hypothalamus and nearby region. Central administration of orexins stimulates food intake, and production of orexin increases with fasting [34]. Recent studies indicated that the POMC-derived neuropeptides β-MSH plays a critical role in the hypothalamic control of body weight in humans [35].

Conversely, absence of leptin induced by calorie restriction or fasting activates orexigenic NPY and AgRP [36]. The activation of these orexigenic neurons leads to the inhibition of anorexigenic signalling in two ways. First, AgRP/NPY neurons synapse directly with POMC neurons, providing an inhibitory tone. Second, AgRP itself is an effective antagonist of α-MSH at MC4Rs. Release of AgRP/NPY neuropeptides stimulates food intake and reduces energy expenditure.

NPY acts via Y1, Y2 and Y5 G-protein-coupled receptors expressed in the hypothalamic neurons. The Y5 and the Y1 receptors mediate the stimulatory effects of NPY on food intake [37]. The Y2 receptor mediates an inhibitory effect of NPY at low concentrations that could be important for basal control of body weight [38]. The action of NPY on its receptors can be affected by other neurotransmitters, such as glucagon-like peptide 1, which inhibits food intake and diminishes the orexigenic effect of NPY by antagonising NPY receptors Y5 and Y1 [39]. In *ob/ob* mice model, the role of NPY in full response to leptin deficiency was determined. The results showed that NPY deficiency prevents obesity and other features of ob/ob mice. However, NPY-deficient mice feed normally and gain normal body weight [40]. Adrenal glucocorticoids potentiate the orexigenic effect of NPY by acting as endogenous antagonists of leptin and insulin thus participating in energy regulation [41]. Adrenalectomy mitigates the effect of fasting to increase both appetite and NPY expression and potentiates the anorexigenic and slimming effects of insulin and leptin; administration of glucocorticoids compensates these effects [41].

AgRP antagonises the effect of α-MSH on MC4R. Activation of MC4R by α-MSH reduces food intake, while suppression of MC4R signalling through this receptor by the endogenous antagonist AgRP increases feeding and diminishes the hypophagic response to leptin [25,42]. AgRP deficiency leads to a lean phenotype of mice with extended lifespan when the animals received a high-fat diet [43].

Gene mutation at MC4R locus affects 6% of humans with severe obesity, and most affected individuals have a single mutant allele that causes obesity through haploinsufficiency rather than a dominant-negative mechanism [44]. The obesity in several other rare human syndromes also converges on this pathway.

Targeted deletion of the MC3R produced obesity in mice [45]. However, obesity from this lesion occurs without the hyperphagia seen in MC4R mutants and may be associated with a loss of lean body mass. Thus, these two melanocortin receptors can cause obesity through distinct physiological mechanisms.

Recent studies have strongly implicated that adenosine monophosphate-activated protein kinase (AMPK) pathway plays a central role in hypothalamic control of energy homeostasis by mediating the inputs from multiple hormones, peptides, neurotransmitters and nutrients [46]. Leptin inhibition of AMPK activity in the arcuate nucleus and PVN in hypothalamus is necessary to produce its anorexic effect [47].

The melanocortin system is not the only neuropeptides system involved in weight regulation. Leptin regulates CART expression in arcuate POMC neurons, and CART axons innervate sympathetic preganglionic neurons in the thoracic spinal cord [48]. Among its central effects, leptin further regulates many other neuropeptides, including corticotropin-releasing hormone, growth hormone-releasing hormone and galanin that have been described to participate in energy regulatory pathways.

Ghrelin, a peptide expressed in stomach and brain, promotes hyperphagia and obesity, an effect mediated through NPY/AgRP neurons [49].

Noradrenalin, dopamine and serotonin (5-HT) are also known to be involved in central energy balance circuits. Serotonergic neurons within the caudal brainstem project widely within the brain, and elevated levels of 5-HT suppress food intake and reduce body weight [50]. Drugs that increase intrasynaptic 5-HT by blocking the 5-HT uptake have been used for the treatment of obesity. Mice with deletion of the 5-HT2c serotonin receptor subtype developed modest obesity [51]. Leptin increases 5-HT turnover, suggesting that these pathways can converge, but 5-HT2C-deficient mice retained an anorectic response to leptin.

In addition to neuropeptides and transmitters, the function of these neural circuits is also influenced by metabolic fuels. Neurons in the hypothalamus that respond to changes in glucose levels may be the same as, or functionally linked to those neurons that respond to leptin and express the peptides discussed earlier. Biosynthesis of long-chain fatty acids and their utilisation in hypothalamus also play an important role in the regulation of energy homeostatic responses [52] (Figure 3.1).

Control of Energy Expenditure

The hypothalamic melanocortin pathway regulates the two aspects of energy equation in a coherent way, whereby signals of excess energy result in a reduction of the intake of food energy while activating energy expenditure [21,53]. Human body utilises food

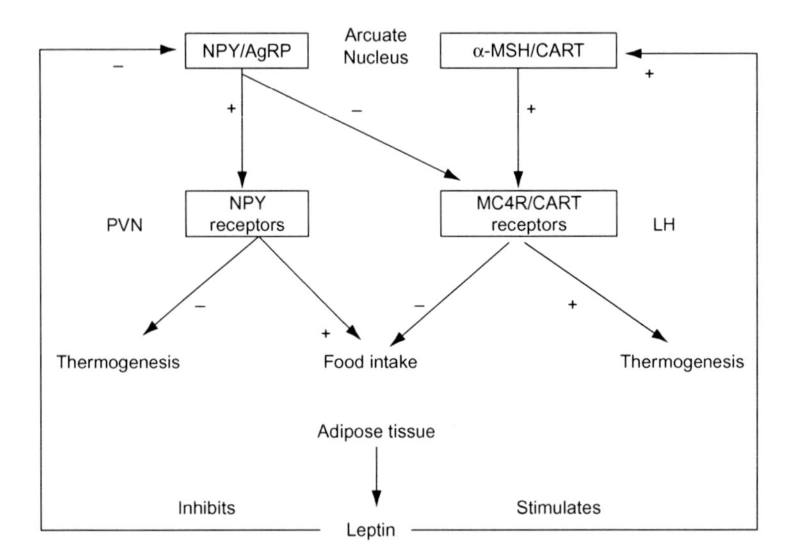

Figure 3.1 Leptin acts on two subtypes of neurons in the arcuate nucleus. Leptin stimulates the first subtype that produces α-MSH and CART. α-MSH/CART neurons project to PVN and lateral hypothalamus and they inhibit MCH and orexins thus inhibiting food intake and regulate the secretion of pituitary neuropeptides and the sympathetic nervous system to effect energy expenditure. Absence of leptin activates the second subtype of neurons that produce NPY and AgRP. AgRP antagonises α-MSH at MC4R and NPY stimulates Y5 and Y1 receptors in the PVN thus stimulating food intake and reducing energy expenditure.

energy via three forms of thermogenesis: either *obligatory* related to BMR and part of diet-induced thermogenesis, or *facultative* which is activated acutely such as through shivering or physical activity when additional heat is required or *adaptive* when loss of heat energy occurs in response to external stimuli such as exposure to cold, diet and a variety of pathogenic stimuli, including infection, inflammation, cancer, injury and stress [54].

BMR is responsible for cellular metabolism, functioning of organs and maintenance of the body's vital functions, and normally it accounts for 50–70% of total daily energy expenditure (TEE) [55]. Diet-induced thermogenesis is used to digest and absorb food after a meal. It accounts for about 10% of the TEE and does not vary greatly among individuals [56]. Physical activity accounts for 8–15% of the total energy expenditure where thermogenesis is derived from muscular activity [57].

Adaptive thermogenesis plays a role in energy regulation by heat production without performing actual work through uncoupling phosphorylation in the mitochondrion. Adaptive thermogenesis occurs mainly in the brown adipose tissue and skeletal muscles as a result of hypothalamic activation of sympathetic nervous system and hypothalamic–thyroid axis, thus stimulating energy loss in response to various inputs that indicate excess body energy.

The body produces energy by the oxidation of substrates such as fatty acids and glucose in the inner mitochondrial membrane via the electron transport system of the respiratory chain that results in the production of reducing equivalents nicotinamide adenine dinucleotide and flavin adenine dinucleotide. These are then oxidised to produce protons that are pumped to the outer surface of the inner mitochondrial membrane. Proton transport generates gradient potential that stimulates the oxidative phosphorylation of ADP to ATP via ATP synthase. However, energy coupling to ATP is not 100% efficient due to proton leak into the inner mitochondrial membrane that could contribute to resting metabolic rate by 20–50% [58]. Proton leak increases with the rise in gradient potential across the mitochondrial membrane and is enhanced by binding to uncoupling protein 1 (UCP1) resulting in respiration with heat production not linked to ATP.

UCP1 belongs to a group of proteins involved in the transport of anionic fatty acids into the mitochondria that mainly functions in the brown adipose tissue [59]. Fatty acid uncoupling enhances proton binding to the protein thus facilitating entry into the inner mitochondrial membrane resulting in thermogenesis [60]. Other UCPs have been identified in various tissues including UCP2, UCP3, UCP4 and UCP5; however, none of them is involved in thermogenesis.

Thermogenesis in brown adipose tissue, which is present in small amounts in human adults, could account for 20% of daily energy expenditure from as little as 50 g [61]. Heat production by brown adipose tissue is triggered and regulated primarily by sympathetic nervous system. Noradrenalin binds to the β3-receptors on the surfaces of brown adipocytes and stimulates lipolysis via cAMP-protein kinase A and activates the expression of PPARγ coactivator 1α (PGC-1α) that controls transcription of the UCP1 genes [62]. Fatty acids produced locally and recruited from the circulation act as substrate for oxidation and activate UCP1 with resultant thermogenesis [63]. The generation of heat is dependent on the role played by peroxisome proliferator-activated receptors (PPARs). In brown adipose tissues, all the three subtypes of PPARs are expressed but PPAR-α is in the highest concentration. PPAR-α activates lipolysis and fatty oxidation required for active thermogenesis. PPARγ is equally expressed in the both white and brown adipose tissues and involved in adipocyte differentiation.

Human skeletal muscles compose approximately 40% of the total body mass. They account for 20–30% of the total resting oxygen uptake [64] and contribute to adrenalin-induced thermogenesis by up to 50% [65]. Skeletal muscles can increase heat production in response to cold via mitochondrial uncoupling [66]. However, the mechanism differs from that in brown adipose tissues as skeletal muscles express only UCP3 which is not involved in uncoupling in the muscle [67], and the sarco/endoplasmic reticulum Ca^{2+}-ATPase is likely to be the main pathway for adaptive thermogenesis [68]. Activation of hypothalamic–thyroid axis is involved indirectly in thermogenesis in brown adipose tissue and skeletal muscle by stimulating lipolysis.

Hence, alteration in the uncoupling respiration in brown adipose tissues and skeletal muscles could significantly affect energy balance and influence accumulation of fat stores and thus becomes a target for anti-obesity treatments (Figure 3.2).

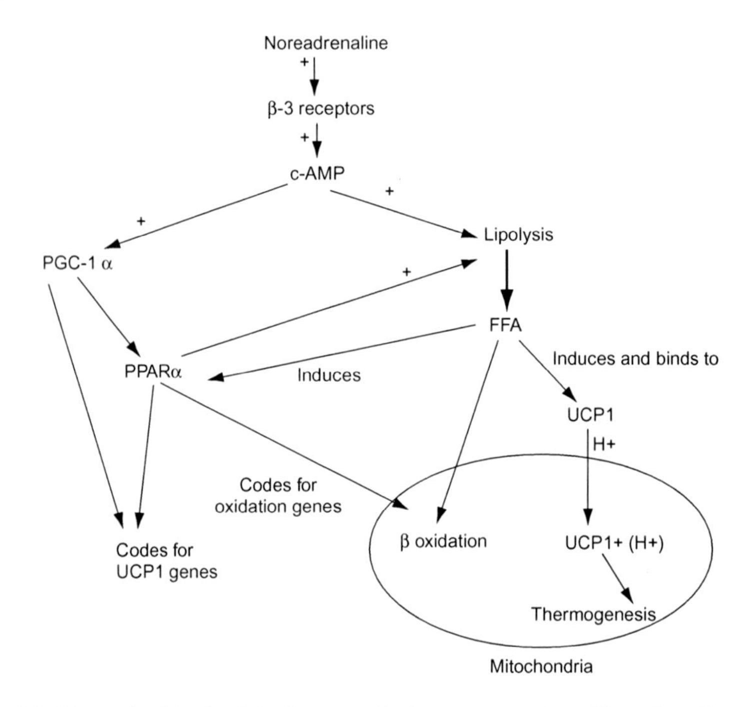

Figure 3.2 Diet and cold stimulate the sympathetic nervous system. Noreadrenaline acts on β-3 receptors and stimulates cAMP which in turn stimulates lipolysis and PGC-1α. Free fatty acids act as substrate for oxidation, induce UCP1 and form ligands that induce PPARα. PPARα stimulates lipolysis and codes for the expression of genes involved in β oxidation. PGC-1α stimulation leads to the expression of genes that code for UCP1. UCP1 binds to anionic fatty acids and facilitates their transport into mitochondria and this in turn enhances protons produced via the respiratory chain to re-enter across the mitochondrial membrane and produce heat uncoupled to ATP production.

Adipogenesis

Adipose tissue is an active endocrine and exocrine organ that responds to and produces hormonal and metabolic signals that play important role in regulation of energy balance and body weight. Thus adipocyte dysfunction could contribute to metabolic diseases in obesity.

Moderate obesity results mainly from an increase in the size of the adipocytes due to accumulation of lipids (hypertrophic obesity), while more extreme obesity or obesity in childhood also implies an increase in the number of adipose cells (hyperplasic obesity) [69]. The capacity to make new adipocytes continues throughout life and can be activated by the size, frequency and composition of meals, and by other environmental factors. Almost 10% of adipocytes undergo turnover in the human adipose tissues each year [70].

The adipocyte cell lines are derived from a pluripotent stem cell precursor, which can differentiate into various types of mesodermic cells [71]. Stem cells first develop into preadipocytes that exhibit similar morphology to stem cells; however, they are committed to the adipogenic lineage and are no longer able to transform into osteoblasts, myocytes or chondrocytes. Secondly, preadipocytes differentiate to mature fat cells with accumulation of fat.

Adipogenesis is tightly regulated at a molecular level by several transcription factors. The activity of these transcription factors is coordinated by extracellular signals regulated by related genes such as WNT signalling pathway. Peptides secreted by the WNT family have been shown to inhibit the early steps of adipogenesis, and suppression of WNT signalling by endogenous inhibitors is essential to generate adipocytes [71].

Two groups of transcription factors seem to be essential for the development of both white and brown adipose tissues – CCAAT-enhancer binding proteins (C/EBPs) and PPARs.

C/EBPβ and C/EBPδ are the key early regulators of adipogenesis; their expression is induced rapidly by insulin, and both of them are regulated by anti-adipogenic preadipocyte factor 1 [72]. In addition, C/EBPβ is regulated by proadipogenic desumoylating enzyme sentrin-specific peptidase 2 that reduces its degradation by sumoylation [73].

C/EBPβ and C/EBPδ act on the key adipogenic transcription factors C/EBPα, PPARγ and sterol regulatory element-binding protein (SREBP1) [74]. SREBP1, which specifically regulates the aspects of cholesterol and fatty acid metabolism, plays a key role in the generation of the endogenous lipid ligand that activates PPARγ [75]. PPARγ and C/EBPα promote expression of each other via a positive feedback loop. Genowide binding analyses have shown that PPARγ and C/EBPα cooperate on multiple binding sites to regulate a wide range of genes involved in developing and mature adipocytes [76]. They both induce the expression of genes involved in insulin sensitivity, lipogenesis and lypolysis including those encoding glucose transporter type 4, fatty-acid-binding protein 4 also known as adipocyte protein 2, lipoprotein lipase, *sn*-1-acylglycerol-3-phosphate acyl-transfrase 2, perilipin and the secretion of leptin and adiponectin. This transcription network is regulated by many factors such as C/EBP homologous protein 10 and members of Kruppel-like factor [74,77], in addition to an ever-expanding list of positive and negative regulators including phosphorylation which inhibits the binding of PPARγ to its target promoters [78].

PPARγ is considered a crucial regulator of adipogenesis and of many factors that influence the expansion of white adipose tissues.

Interactions Among Social, Environmental, Genetic and Behavioural Factors in Obesity

In a comparative study, 247 Punjabi immigrants in London were noted to have significantly higher BMI values, serum cholesterol and fasting blood glucose against their siblings still living in Punjab, India [79]. Similarly, Ravussin et al. [80] reported a comparative study of Pima Indians living in remote rural regions of Mexico with those living in Arizona. Compared to Pima Indians living in rural Mexico, those living in Arizona had significantly higher BMI, higher cholesterol levels and a higher incidence of type-2 diabetes (11% vs 37% in women and 6% vs 54% in men). These two studies suggest that genetic influences on the development of obesity can be mitigated by environmental conditions.

Epstein et al. [81] recently reported an association between the 'reinforcing value of food' phenotype and ad libitum energy intake moderated by the dopamine transporter gene (SLc6A3) and the dopamine 2 receptor gene (DRD2). In this study of 88 smokers of American European descent, the subjects who lacked the SLc6A4 allele consumed more total energy than the participants with SLC6A3 genotypes. Furthermore, the subjects with AI allele of DRD2 scored high on reinforcing value of food and consumed more total calories compared to participants with any other DRD2 genotype. This unique study has essentially focused on the genetics of food reward as they relate to dopamine pathways. This is an area of future research with considerable potential.

Conclusion

Obesity is a complex disorder that results from excessive energy intake without corresponding expenditure leading to accumulation of adipose tissues with serious health consequences. Adiposity is rarely inherited as a monogenic trait, but in the case of 70–80% of obese people, it is a manifestation of interaction between multiple gene loci and environmental factors that induces changes in the biological processes that control energy homeostasis and body weight. Understanding the biochemical and molecular process involved in energy regulation will facilitate the development of innovate preventive/treatment measures in susceptible individuals in a world of abundance of nutrients augmented with reduced physical activity.

It must be recognised that the onset of obesity is a developmental process that may be influenced at different genetic or environmental influences at different stages of life. Therefore, prospective studies focusing on the critical growth periods for obesity such as growth in the intra-uterine environment, adiposity in early childhood and adolescence and life experiences specific to those periods would be of immense interest.

Furthermore, studies around the genetic–environmental interactions could elucidate the nature of environmental influences, reinforcing value of food, delayed satiation, home environments and the effect of cognitive stimulation in childhood and eating in the absence of hunger. All these factors have been linked to obesity status. It will be equally interesting to study the phenotype of individuals who will seek out 'obesity-promoting environments' such as fast-food restaurants.

References

1. Great Britain Parliament House of Commons Health Committee. *Obesity – Volume I – HCP 23-I, Third Report of Session 2003–04. Report Together with Formal Minutes.* London, UK: TSO (The Stationery Office); 2004.

2. Flegal KM, Carroll MD, Kit BK, Ogden CL. Prevalence of obesity and trends in the distribution of body mass index among US adults, 1999–2010. *JAMA*. 2012;307(5): 491–497.
3. Berrington de Gonzalez A, Hartge P, Cerhan JR. Body-mass index and mortality among 1.46 million white adults. *N Engl J Med*. 2010;363:2211–2219.
4. Mokdad AH, Marks JS, Stroup DF, Gerberding JL. Actual causes of death in the United States, 2000. *JAMA*. 2004;291:1238–1245.
5. Maes HH, Neale MC, Eaves LJ. Genetic and environmental factors in relative body weight and human adiposity. *Behav Genet*. 1997;27(4):325–351.
6. Stunkard AJ, Harris JR, Pederson NL, McClearn GE. The body mass index of twins who have been reared apart. *N Engl J Med*. 1990;322(21):1483–1487.
7. Hewitt JK. The genetics of obesity: what have genetic studies told us about the environment. *Behav Genet*. 1997;27(4):353–358.
8. Perusse L, Rankinen T, Zuberi A, Chagnon YC. The human obesity gene map. The 2004 update. *Obes Res*. 2005;13(3): 381–490.
9. Clement K, Vaisse C, Lahlou N. A mutation in the human leptin receptor gene causes obesity and pituitary dysfunction. *Nature*. 1998;392:398–401.
10. Krude H, Biebermann H, Schnabel D. Obesity due to proopiomelanocortin deficiency: three new cases and treatment trials with thyroid hormone and ACTH4-10. *J Clin Endocrinol Metab*. 2003;88:4633.
11. Jackson RS, Creemers JW, Ohagi S. Obesity and impaired prohormone processing associated with mutations in the human prohormone convertase 1 gene. *Nat Genet*. 1997;16:303.
12. Montague CT, Farooqi IS, Whitehead JP. Congenital leptin deficiency is associated with severe early-onset obesity in humans. *Nature*. 1997;387:903.
13. Feng N, Young SF, Aguilera. G. Co-occurrence of two partially inactivating polymorphisms of MC3R is associated with pediatric-onset obesity. *Diabetes*. 2005;54:2663.
14. Lubrano-Berthelier C, Dubern B, Lacorte J-M, et al. Melanocortin 4 receptor mutations in a large cohort of severely obese adults: prevalence, functional classification, genotype-phenotype relationship and lack of association with binge eating. *J Clin Endocrinol Metab*. 2006;91:1811–1818.
15. den Hoed M, Ekelund U, Brage S. Genetic susceptibility to obesity and related traits in childhood and adolescence: influence of loci identified by genome-wide association studies. *Diabetes*. 2010;59:2980–2988.
16. Wu Q, Saunders RA, Szkudlarek-Mikho M, Serna Ide L, Chin K-V. The obesity-associated FTO gene is a transcriptional coactivator. *Biochem Biophys Res Commun*. 2010;401:390–395.
17. Schwartz MW, Woods SC, Porte Jr D, Seeley RJ, Baskin DG. Central nervous system control of food intake. *Nature*. 2000;404:661–671.
18. Arvat E, Maccario M, Di Vito L, et al. Endocrine activities of ghrelin, a natural growth hormone secretagogue (GHS), in humans: comparison and interactions with hexarelin, a nonnatural peptidyl GHS, and GH-releasing hormone. *J Clin Endocrinol Metab*. 2001;86:1169–1174.
19. Wright SA, Washington MC, Garcia C, Sayegh AI. Gastrin releasing peptide-29 requires vagal and splanchnic neurons to evoke satiation and satiety. *Peptides*. 2012;33:125–131.
20. Batterham RL, Ffytche DH, Rosenthal JM, et al. PYY modulation of cortical and hypothalamic brain areas predicts feeding behaviour in humans. *Nature*. 2007;450:106–109.

21. Horvath TL, Diano S, Tschöp M. Brain circuits regulating energy homeostasis. *Neuroscientist.* 2004;10:235–246.
22. Elmquist JK, Elias CF, Saper CB. From lesions to leptin: hypothalamic control of food intake and body weight. *Neuron.* 1999;22:221–232.
23. Plum L, Ma X, Hampel B, et al. Enhanced PIP3 signalling in POMC neurons causes KATP channel activation and leads to diet-sensitive obesity. *J Clin Invest.* 2006;116:1886–1901.
24. Hamilton BS, Paglia D, Kwan AYM, Deitel M. Increased obese mRNA expression in omental fat cells from massively obese humans. *Nature Med.* 1995;1:953–956.
25. Woods S, Seeley R, Porte DJ, Schwartz M. Signals that regulate food intake and energy homeostasis. *Science.* 1998;280:1378–1383.
26. He Y, Chen H, Quon MJ, Reitman M. The mouse obese gene. Genomic organization, promoter activity, and activation by CCAAT/enhancer-binding protein alpha. *J Biol Chem.* 1995;270:28887–28891.
27. Elmquist JK, Ahima RS, Maratos-Flier E, Flier JS, Saper CB. Leptin activates neurons in the ventrobasal hypothalamus and brainstem. *Endocrinology.* 1997;138:839–842.
28. Harris M, Aschkenasi C, Elias CF, Chandrankunnel A, Nillni EA, Bjoorbaek C. Transcriptional regulation of the thyrotropin-releasing hormone gene by leptin and melanocortin signaling. *J Clin Invest.* 2001;107:111–120.
29. Wang MY, Lee Y, Unger RH. Novel forms of lipolysis induced by leptin. *J Biol Chem.* 1999;274:17541–17544.
30. Banks WA, DiPalma CR, Farrell CL. Impaired transport of leptin across the blood–brain barrier in obesity. *Peptides.* 1999;20:1341–1345.
31. Shimizu H, Inoue K, Mori M. The leptin-dependent and -independent melanocortin signaling system: regulation of feeding and energy expenditure. *J Endocrinol.* 2007;193:1–9.
32. Elias CF, Aschkenasi C, Lee C, et al. Leptin differentially regulates NPY and POMC neurons projecting to the lateral hypothalamic area. *Neuron.* 1999;23:775–786.
33. Shimada M, Tritos NA, Lowell BB, Flier JS, Maratos-Flier E. Mice lacking melanin-concentrating hormone are hypophagic and lean. *Nature.* 1998;396:670–674.
34. Sakurai T, Amemiya A, Ishii M, et al. Orexins and orexin receptors: a family of hypothalamic neuropeptides and G protein-coupled receptors that regulate feeding behavior. *Cell.* 1998;92:573–585.
35. Harrold JA, Widdowson PS, Williams G. Beta-MSH: a functional ligand that regulated energy homeostasis via hypothalamic MC4R? *Peptides.* 2003;24(3):397–405.
36. Luquet S, Perez FA, Hnasko TS, Palmiter RD. NPY/AgRP neurons are essential for feeding in adult mice but can be ablated in neonates. *Science.* 2005;310:683–685.
37. Gerald C, Walker M, Criscione L, et al. A receptor subtype involved in neuropeptide Y-induced food intake. *Nature.* 1996;382:168–170.
38. Naveilhan P, Hassani H, Canals JM, et al. Normal feeding behavior, body weight and leptin response require the neuropeptide Y Y2 receptor. *Nat Med.* 1999;5:1188–1193.
39. Turton MD, O'Shea D, Gunn I, et al. A role for glucagon-like peptide-1 in the central regulation of feeding. *Nature.* 1996;379:69–72.
40. Erickson JC, Hollopeter G, Palmiter RD. Attenuation of the obesity syndrome of ob/ob mice by the loss of neuropeptide Y. *Science.* 1996;274:1704–1707.
41. Woods S, Seeley R, Porte DJ, Schwartz M. Signals that regulate food intake and energy homeostasis. *Science.* 1998;280:1378–1383.
42. Bewick GA, Gardiner JV, Dhillo WS, et al. Post-embryonic ablation of AgRP neurons in mice leads to a lean, hypophagic phenotype. *FASEB J.* 2005;19:1680–1682.

43. Wortley KE, Anderson KD, Yasenchak J, et al. Agouti-related proteindeficient mice display an age-related lean phenotype. *Cell Metab.* 2005;2:421–427.
44. Farooqi IS, Yeo GS, Keogh JM, et al. Dominant and recessive inheritance of morbid obesity associated with melanocortin 4 receptor deficiency. *J Clin Invest.* 2000;106:271–279.
45. Butler AA, Kesterson RA, Khong K, et al. A unique metabolic syndrome causes obesity in the melanocortin-3-receptor-deficient mouse. *Endocrinology.* 2000;141: 3518–3521.
46. Hardie DG, Hawley SA, Scott JW. AMP-activated protein kinase – development of the energy sensor concept. *J Physiol.* 2006;574:7–15.
47. Minokoshi Y. AMP-kinase regulates food intake by responding to hormonal and nutrient signals in the hypothalamus. *Nature.* 2004;428:569–574.
48. Elias CF. Leptin activates hypothalamic CART neurons projecting to the spinal cord. *Neuron.* 1998;21:1375–1385.
49. Tschop M, Smiley DL, Heiman ML. Ghrelin induces adiposity in rodents. *Nature.* 2000;407:908–913.
50. Leibowitz SF, Alexander JT. Hypothalamic serotonin in control of eating behavior, meal size, and body weight. *Biol Psychiatry.* 1998;44:851–864.
51. Nonogaki K, Strack AM, Dallman MF, Tecott LH. Leptin-independent hyperphagia and type 2 diabetes in mice with a mutated serotonin 5-HT2C receptor gene. *Nat Med.* 1998;4:1152–1156.
52. Obici S, Feng Z, Arduini A, Conti R, Rossetti L. Inhibition of hypothalamic carnitine palmitoyltransferase-1 decreases food intake and glucose production. *Nat Med.* 2003;9:756–761.
53. Cone RD. The central melanocortin system and energy homeostasis. *Trends Endocrinol Metab.* 1999;10:211–216.
54. Rothwell NJ. CNS regulation of thermogenesis. *Crit Rev Neurobiol.* 1994;8:1–10.
55. Levine JA. Nonexercise activity thermogenesis – liberating the life-force. *J Intern Med.* 2007;262:273–287.
56. Saito M, Okamatsu-Ogura Y, Matsushita M, Watanabe K, Yoneshiro T, Nio-Kobayassssshi J. High incidence of metabolically active brown adipose tissue in healthy adult humans: effects of cold exposure and adiposity. *Diabetes.* 2009; 58:1526–1531.
57. Ravussin E, Lillioja S, Anderson TE, Christin L, Bogardus C. Determinants of 24-hour energy expenditure in man. Methods and results using a respiratory chamber. *J Clin Invest.* 1986;78:1568–1578.
58. Rolfe DF, Brown GC. Cellular energy utilization and molecular origin of standard metabolic rate in mammals. *Physiol Rev.* 1997;77:731–758.
59. Villarroya F, Iglesial R, Giralt M. PPARs in the control of uncoupling proteins gene expression. *PPAR Res.* 2007;2007:74364 [Article ID 74364]. (published on line 2006, November 28, doi: 10.1155/2007/74364).
60. Wojtczak L, Schönfeld P. Effect of fatty acids on energy coupling processes in mitochondria. *Biochim Biophys Acta.* 1993;1183:41–57.
61. Rothwell NJ, Stock MJ. A role for brown adipose tissue in diet-induced thermogenesis. *Nature.* 1979;281:31–35.
62. Luquest S, Lopez-Soriano J, Holst D, Gaudel C, Jehl-Pietri C, Fredenrich A. Roles of peroxisome proliferator-activated receptor delta (PPARdelta) in the control of fatty acid catabolism. A new target for the treatment of metabolic syndrome. *Biochimie.* 2004;86 (11):833–837.

63. Cannon B, Nedergaard J. Brown adipose tissue: function and physiological significance. *Physiol Rev*. 2004;84:277–359.
64. Zurlo F, Larson K, Bogardus C, Ravussin E. Skeletal muscle metabolism is a major determinant of resting energy expenditure. *J Clin Invest*. 1990;86:1423–1427.
65. Astrup A, Bulow J, Madsen J, Christensen NJ. Contribution of BAT and skeletal muscle to thermogenesis induced by ephedrine in man. *Am J Physiol*. 1985;248:507–515.
66. Sander L, Wijers J, Schrauwen P, Saris WHM, van Marken Lichtenbelt D. Human skeletal muscle mitochondrial uncoupling is associated with cold induced adaptive thermogenesis. *PLoS One*. 2008;3(3):1777.
67. Azzu V, Jastroch M, Divakaruni AS, Brand MD. The regulation and turnover of mitochondrial uncoupling proteins. *Biochim Biophys Acta*. 2010;1797:785–791.
68. de Meis L. Role of the sarcoplasmatic reticulum Ca^{2+}-ATPase on heat production and thermogenesis. *Biosci Rep*. 2001;21:113–137.
69. Spiegelman BM, Flier JS. Lipogenesis and obesity: rounding out the picture. *Cell*. 1996;87:377–389.
70. Spalding KL, Arner E, Westermark PO, et al. Dynamics of fat cell turnover in humans. *Nature*. 2008;453:783–787.
71. Christodoulides C, Lagathu C, Sethi JK, VidalPuig A. Adipogenesis and WNT signaling. *Trends Endocrinol Metab*. 2009;20:16–24.
72. Wang Y, Sul HS. Pref-1 regulates mesenchymal cell commitment and differentiation through Sox9. *Cell Metab*. 2009;9:287–302.
73. Chung SS, Ahn BY, Kim M, et al. Control of adipogenesis by the SUMO specific protease SENP2. *Mol Cell Biol*. 2010;30:2135–2146.
74. White UA, Stephens JM. Transcriptional factors that promote formation of white adipose tissue. *Mol Cell Endocrinol*. 2010;318:10–14.
75. Kim JB, Wright HM, Wright M, Spiegelman BM. ADD1/SREBP1 activates PPAR gamma through the production of endogenous ligand. *Proc Natl Acad Sci U S A*. 1998;95:4333–4337.
76. Lefterova MI, Zhang Y, Steger DJ, et al. PPARgamma and C/EBP factors orchestrate adipocyte biology via adjacent binding on a genome-wide scale. *Genes Dev*. 2008;22:2941–2952.
77. Rosen ED, MacDougald OA. Adipocyte differentiation from the inside out. *Nat Rev Mol Cell Biol*. 2006;7:885–896.
78. Helenius K, Yang Y, Alasaari J, Makela TP. Mat1 inhibits peroxisome proliferator-activated receptor gamma-mediated adipocyte differentiation. *Mol Cell Biol*. 2009;29:315–323.
79. Bhatnagar D, Anand IS, Durrington PN, Patel DJ, et al. Coronary risk factors in people from Indian subcontinent living in West London and their siblings in India. *Lancet*. 1995;345(8947):405–409.
80. Ravussin E, Valencia ME, Esparza J, Bennett P, Schulz LO. Effects of a traditional lifestyle on obesity in Pima Indians. *Diabetes Care*. 1994;17(9):1067–1074.
81. Epstein LH, Wright SM, Paluch RA, Leddy JJ, et al. Relation between food reinforcement and deopamine genotypes and its effect on food intake in smokers. *Am J Clin Nutr*. 2004;80(1):82–88.

4 The Psychological Basis of Obesity

Gyöngyi Kökönyei[1], Alexander Baldacchino[2], Róbert Urbán[1] and Zsolt Demetrovics[1]

[1]Institute of Psychology, Eötvös Loránd University, Budapest, Hungary
[2]School of Medicine, Dundee University, Dundee, UK

Introduction

There are at least five perspectives in the analysis of psychological factors contributing to the formation and maintenance of obesity (which will be defined in this chapter as the unsuccessful attempt to lose weight). All five perspectives are not mutually exclusive.

1. Programs: based on the simplistic notion that obesity is the result of consuming more calories than one actually needs to use during their day-to-day activities. Therefore, these tend to encourage a combination of diet and increase in physical activities which has been shown to be ineffective in maintaining reduced weight on the long term [1].
2. Psychological distress: binge-eating behaviour is often linked to obesity [2]. Obesity is also often linked with depression [3,4] and anxiety [5]. A significant number of obese people eat as a way of coping with negative emotions [6].
3. Social issues: especially social deprivation and poverty [7].
4. Public health issues: this is permeated through a process of stigmatisation and moralisation discourse [8]. In the media, the individual's own responsibility and weak skills of self-control get a disproportionately great emphasis concerning the issue [9].
5. Cultural issues: the predetermined concepts and expectancies concerning the agreeable mean body mass is also another perspective that women (and recently more and more men) are expected to conform [10] is in connection with the superwoman ideal [11].

The interpretation of studies dealing with the psychological aspects of obesity is however hindered by the fact that overweight and obesity are often treated together. This is an especially important aspect for the reason that overweight and obesity, if examined separately, represent different types of health risks with specific health problems. For example, obesity, compared to normal weight, increases the risk of mortality due to cardiovascular diseases, while overweight does not. Overweight in turn (as well as obesity) is associated with an increased mortality due to diabetes and renal diseases [12]. Kulie et al. [13] in their evidence-based review, in addition to the above-mentioned diseases, highlighted the increased risk of infertility, unfavourable outcomes of pregnancy, low back pain and osteoarthritis of the knees in obese women. These risks are also dealt with in this book.

Obesity. DOI: http://dx.doi.org/10.1016/B978-0-12-416045-3.00004-2

Age with its related biological changes is an important factor in the complex and multi-factorial determination of obesity. It is highly relevant whether one looks at childhood or adulthood obesity separately. Therefore, in case of adulthood obesity, it is an important question whether it is a problem formed during childhood and persisting since then or excess body weight was gained during adulthood. Adolescence, monthly menstrual periods and menopause increase the risk of gaining excess weight and obesity [14].

In addition to age, gender can also moderate the roles of factors such as metabolism and the level of physical inactivity [15]. Moreover, economical and psychological factors related to poverty also influence food intake [7]. The nutrition paradox refers to the observation that among people with lower socio-economic status (SES), food insecurity might also be associated with obesity, especially in women [7,16]. One possible reason for this association might be that in families with lower income, cheap but high-calorie and nutrition-poor foods are preferred. Moreover, SES moderates the association between obesity and health-related quality of life (HRQoL). In obese people having lower SES, HRQoL is lower compared to those of the same weight in higher SES groups [17].

These results clearly raise the issue that the obese population is a heterogeneous population [18]. In line with this, many emphasise the importance of an individualised approach in unravelling the reasons of and treating obesity [15]. This needs to be contextualised in an environment that encourages sensorial and rewarding signals related to food intake rather than environment that encourages people to be more aware of the importance of satiety [19]. This is of course compounded by a society that is predominantly occupied by sedentary activities, work [20], studying, and recreation. This has been described as the obesogenic environment [21].

Eating Behaviour

The eating behaviour and attitude towards food are influenced by several factors like our innate taste preference determined by evolutional, genetic [22], socio-cultural and environmental factors, besides the knowledge acquired through behavioural modelling and conditioning as part of socialisation [23,24]. These effects are (also) present in the process when eating and food become associated with emotions.

Foods having high fat and carbohydrate content (or comfort foods) stimulate the opioid systems in the brain [25]. Food-related information experienced in the course of socialisation (e.g. prohibition of a specific food) might also increase the food's desirability [26]. Similarly, foods presented by the media to which other, emotionally loaded information is associated (e.g. social desirability and happiness) have their influence partly through emotional channels as well. In connection with this, it is to be mentioned that in the media, specific foods described as temptations that need to be resisted or classified as unhealthy are advertised a lot (e.g. chips). These effects enhance the ambivalence towards specific foods [27].

Food-related decisions (the when, what, how much and where we eat) are influenced by numerous factors. Decision-making is always carried out in a characteristic ecological environment [28]. In this environment, there are factors independent of food (e.g. atmosphere, time of the day, social interactions) and signals definitely related to food (e.g. size of the meal and way of serving) [29]. The phenomenon when we do not pay attention to and do not monitor how much we eat is called 'mindless eating' by Wansink [30].

It is characteristic of external eaters, who, instead of visceral signals (hunger), rely basically on external, food-related information in deciding how much they eat and so for this reason will have a tendency to overeat. Externality is usually treated by studies as a stable trait. However, it seems that specific visceral conditions of our body, such as hunger, influence the self-reported externality scores in a positive manner [31].

Besides mindless eating and external eating, emotional eating can also play a part in obesity. Emotional eating refers to the phenomenon when a person eats during experiencing and/or in order to cope with emotional distress (e.g. anxiety, sorrow and feeling of boredom or loneliness) without feeling hunger [6]. This can happen in order to reduce the negative emotion [6] or to escape from the unpleasant stimulus and distract their attention from that [32]. This is then further compounded by the sense of guilt producing further emotional distress resulting in further overeating behaviour [33].

Emotional eating can also be observed in normal weight as well as obese adults [34,35], adolescents [36] and preschoolers [37]. A review by Buckroyd [38] identified that as many as 50% of obese persons can be characterised by this emotional eating behaviour. Walfish [39] reported that the rate of emotional eaters is 40% among a sample of bariatric surgery patients.

Kaplan and Kaplan [40] applied classical psychodynamic approaches, in a somewhat oversimplified way [6], and emphasised the anxiety reducing function of eating. A more complex psychodynamic theory was raised by Bruch [41], who suggested that overeating connected with obesity is a pseudo-solution for deep-seated conflicts and problems. According to this hypothesis, emotional eaters are either not able to identify signals of hunger in an appropriate way and/or able to differentiate between hunger and other negative emotional states. This clearly raises the phenomenon of alexithymia. Alexithymia is conceptualised as a deficiency in understanding, processing, or describing emotions *that* is linked to a reduced capacity of cognitive processing and regulation of emotions [42–44]. Prevalence of alexithymia in obese persons is almost twice as much as in a normal-weight control group [45]. Moreover, the characteristic of difficulties in identifying feelings, which is one factor of alexithymia, was constantly observed in a group of obese women [46]. More recent studies point out that alexithymia is primarily associated with binge eating [45,47]. In addition, among women with binge-eating disorder (BED), it is also associated with emotional eating [46,47], while among non-BED obese women, emotional eating was explained as a consequence of stress and depression [47].

Early Attachment and Trauma

Difficulties of the mother in reacting in a differentiated way to the states of hunger and emotional distress are supposed to be in the background of the establishment of emotional eating [41]. Bowlby [48] in his attachment theory highlighted that food might be able to reduce emotional distress caused by the absent mother, which in turn, however, might be associated with eating disorders, and eating without the feeling of hunger, or overeating. This latter approach, unambiguously, sheds light on the importance of early attachment. Due to the insecure attachment, position eating might take the role of distress regulation.

It is also a well-documented observation that childhood traumas, e.g. sexual abuse [49], neglect [50] and interpersonal aggression [51] represent a risk factor of obesity in both childhood and adulthood. It is important to mention however that early attachment traumas are considered to be non-specific risk factors, as they have an aetiological role in several other psychological disorders (e.g. depression, anxiety disorders and conduct disorders).

Emotion Regulation

Experimental and theoretical approaches concerning emotional eating clearly draw attention to the relevance of emotional processing and emotional regulation in the background of the formation and subsistence of obesity/overweight. Suppression of emotions, for example, significantly enhances the consumption of the so-called comfort foods [52].

Moreover, emotional eating is associated with elevated calorie intake in self-threatening situations (in situations when the value of one's self is questioned) [53]. According to Heatherton and Baumeister [32], overeating in this situation prevents the state of becoming aware of the self-threat, thus suppressing aversive self-awareness, just like it is suggested by the escape theory .

According to Polivy and Herman [54], in case of restrained eaters, who wilfully restrict their food intake to control their body weight, distress-induced emotional eating is explained by the fact that eating can distract them from the true source of distress. The difference between emotional eaters and restrained eaters is that the emotional eaters in distressful situations eat in order to reduce negative emotions, while restrained eaters eat more in emotional distress situations, because emotions undermine their cognitive control of restricted eating patterns [55,56].

There are several ways to explain the association between emotional eating and distress. One is from the psychodynamic perspective described earlier, while another is from the learning perspective. From a learning perspective, eating is an operant behaviour, and the eating-induced reduction of negative emotions or distress is the negative reinforcement [57]. Supposedly different mechanisms explain the emotional eating of persons with different eating patterns (normal eaters, restrained eaters and emotional eaters) [56]. It seems logical that in cases of emotional eating, such methods are applied in the treatment of obesity or to help sustain long-term weight loss. Common applications include relaxation training [58] and emotional regulation include skill training [59].

Personality Traits

Results of both short- (e.g. 3-year study [60]) and long-term [61,62] longitudinal studies suggest that elevated scores of neuroticism and lower scores of conscientiousness are associated with higher values of BMI, higher rate of body fat and larger waist and hip sizes. Impulsivity facets of neuroticism [60,62], and order and self-discipline facets of conscientiousness were identified as significantly relevant. It is proposed that the negative association between BMI and conscientiousness, especially in facets of order and self-discipline, is due to increasing eating restraints and the practice of regular workouts. Similarly, low eating disinhibition, which is supposedly dependent of conscientiousness, is a predictor of effective long-term weight loss [63].

Depression and Anxiety

Cross-sectional studies show significant association between obesity and depression [3]. The factor responsible for this cross-sectional association is abdominal fat. If results are adjusted for abdominal fat, the association between overall body mass and depression disappears [64]. A few studies however remind us of important moderator factors. The association between major depression and obesity is significant among non-smokers, while it is not among smokers [65]. Gender differences can be considered as another moderator, since in women both overweight and obesity can increase lifetime prevalence of mood disorders, while in men this association can be observed only in case of obesity [66]. Gender differences are however also influenced by age. Among adolescent girls, being overweight is associated with higher subjective depression scores than when being obese [67]. In adults, an inverse relationship between depression and high BMI [68] is observed. This result fits well with the hypothesis of 'jolly fat'. In the 1970s and 1980s, it was found among the middle-aged (40–65 years) male individuals that obesity is associated with a lower depression and anxiety scores. Lower anxiety scores were also found in women [69,70]. Later prospective data however indicated that obesity cannot be regarded as a preventive factor for positive mental functioning [71]. The issue of temporal sequence of the two disorders however emerges. Presence of depressive and anxiety symptoms in adolescence, for example, predict later BMI values primarily in girls [72]. Furthermore, according to Liem et al. [3], depression symptoms in childhood or adolescence mean that the risk of overweight is increased from 1.90 to 3.50. Another review suggests that obesity increases the risk of depression primarily in adulthood and this association is stronger among Americans than among Europeans [4]. In a group of elderly people (70–79 years), obesity, which is described more precisely as excessive amount of visceral fat, predicted the prevalence of depression among the male cohort [73].

Several explanations have been given to the association between obesity and depression. Besides a shared genetic influence [74], according to one hypothesis, the elevated level of inflammatory agents observed in fat cells [75] might contribute to the appearance of depressed mood [76]. Studying this phenomenon from the other perspective as in depression, normal circadian rhythm of the body is disturbed, thus causing activation of the stress systems (the hypothalamic–pituitary–adrenal axis and the sympathetic axis) which increase the risk of insulin resistance and obesity as well [77]. Also, the low serotonin level associated with depression might increase appetite [78]. Other theories call attention to the relevance of lifestyle – the roles of physical inactivity and changing diet [76,78]. To compound this relationship further, the administration of antidepressants, antipsychotic drugs and mood stabilisers might also lead to weight gain [79,80].

Cross-sectional studies also highlight the association between anxiety disorders and obesity. Irrespective of gender, obesity increases the prevalence of anxiety disorders and specific phobias. Moreover, obesity is associated with the increased prevalence of social phobia among women [66]. However, there was a negative association found between obesity and anxiety [64,69,70]. These inconsistent results might have been caused partly by the methodological limitations.

Cognitive Factors

Experimental studies on food-related attentional bias and attentional allocation indicate that in overweight persons compared to normal weight individuals, initial attentional orientation, as an indicator of attentional bias, is more frequent in cases of food-related stimuli (e.g. pictures), especially in case of food with higher fat content [81,82]. There is fast attention shift from food stimuli in overweight individuals [82], while in heavily obese individuals, one observes an increased attention fixation (reflected by prolonged gaze dwell) [81]. For the resolution of this contradiction, Wethmann et al. [82] suggested that fast attention shift after the initial orientation might reflect the ambivalent attitude towards high-fat foods ('delicious, but has negative consequences if I eat that') which results in a conscious attentional avoidance. This, however, rarely has a long-run protective effect for the reason that overweight persons tend to eat more, e.g. in a tasting task.

Hunger, as a strong motivational factor, is also associated with selective attentional bias concerning food-related words [83] and food-related visual targets [84] and/or it is some kind of an attempt to overcome the temptation of the 'prohibited' food means harnessing these automatic attentional shifts and biases become highly relevant in sustaining restraint or continuing with low-calorie diet food-related intake [85].

Using the ironic process theory [86], it is proposed that planning what not to eat might also have inverse consequences. If an intention is put in a negative form (e.g. 'If I am sad, then I will not eat cake'), then this can have a cognitive and behavioural rebound effect. While if an alternative choice of thought processes are used (e.g. 'If I am sad, I eat an apple') there is no such rebound effect [87].

Goal-conflict theory of Stroebe et al. [88] offers one explanation for disinhibited eating patterns present among restrained eaters. According to the model, restrained eaters are driven by two incompatible goals: eating enjoyment and weight control. The obesogenic environment is however full of signals that pre-activate the goal of eating enjoyment, which in turn impede and subsequently prevent the fulfilment of the other conflicting goal. Palatable food-related signals spontaneously activate food-related hedonic thoughts [89] and inhibit the availability of representations linked to the conflicting goal [90].

Among overweight and obese persons without eating disorders, cognitive biases can often be observed in the forms of dysfunctional beliefs and maladaptive cognitive schemes (e.g. 'If I lose weight, I will be attractive, If I am fat, I cannot wear pretty clothes...') related to the consequences of obesity and possible weight loss [91]. Dysfunctional beliefs to shape, body weight and eating can also be found [92]. If obesity is accompanied by binge eating, these beliefs often contain as well negative overgeneralisations concerning the self [92].

In the long run, obese individuals tend to blame either external factors such as genetic and hormonal factors or internal factors such as overeating and poor exercise for their obesity [93]. Attribution to external causes and lack of internalisation to the problem generally come with worse treatment outcomes [93,94].

Summary

In our review of the psychological basis of obesity, eating behaviour (restrained and emotional eating), emotion regulation deficits, early attachment problems, specific personality traits (neuroticism and conscientiousness), depression and anxiety, and cognitive factors (e.g. attention, goals and biases) were linked to the formation and maintenance of obesity.

Although this chapter emphasised primarily the attachment/relationship experiences form the childhood environment, we should not ignore that other characteristics of the child and early environment can also be associated with the increased risk of obesity. Based on a summary of systematic reviews concerning the associations between the characteristics of early childhood and later obesity [95], maternal smoking, bottle feeding or short duration of breastfeeding, increased infant size, short sleep duration and increased television viewing further increase the risks of obesity. Among childhood characteristics, the birth weight of the child [96], BMI of the child and mother and the income of the family are all related to adulthood obesity [97].

Furthermore, we have seen that automatic and non-automatic processes, like attentional bias and implementation plans, respectively, contribute to overeating and the success of dieting and long-term weight loss. These processes need to be emphasised especially for the reason that they show us how complex the question of self-regulation is when looking at obesity. Although there are personality traits related to the risk of obesity, such as high neuroticism and low conscientiousness, these are other non-specific risk factors, which are also associated with other disorders like mood disorders and anxiety disorders.

Specific factors might as well have interactions. For example, there is an association between emotional eating and depressed mood and both have an impact on the choice of food. Emotional eating induces sweet-food ingestion while depression is associated with the reduced consumption of fruits and vegetables [98]. In this chapter, the issue of craving for specific foods is not discussed [99]; however, this can also be linked to overeating [100] and could also explain the attempts made for coping with negative mood [101].

All these aspects discussed in this chapter point towards the importance of a personalised approach in the treatment of obesity. Characteristics of eating, cognitive processes and biases, early attachment history, presence or absence of mental disorders and personality traits are all significant aetiological factors. Therefore, their exploration and the personalised responses created based on these results in the planning phase of interventions might be an important step towards successful treatment.

Acknowledgements

This work was supported by the Hungarian Scientific Research Fund (grant number: 83884) and by the European Union and co-financed by the European Social Fund (grant agreement no. TAMOP 4.2.1/B-09/1/KMR-2010-0003). Zsolt Demetrovics and Gyöngyi Kökönyei acknowledge the financial support of the János Bolyai Research Fellowship awarded by the Hungarian Academy of Science.

References

1. Garner D, Wooley S. Confronting the failure of behavioural and dietary treatments for obesity. *Clin Psychol Rev*. 1991;11:729–780.
2. de Zwann M. Binge eating disorder and obesity. *Int J Obes Relat Metab Disord*. 2001;25:S51–S55.
3. Liem ET, Sauer PJ, Oldehinkel AJ, Stolk RP. Association between depressive symptoms in childhood and adolescence and overweight in later life: review of the recent literature. *Arch Pediatr Adolesc Med*. 2008;162:981–988.
4. Luppino FS, de Wit LM, Bouvy PF, et al. Overweight, obesity, and depression: a systematic review and meta-analysis of longitudinal studies. *Arch Gen Psychiatry*. 2010;67:220–229.

5. Bodenlos JS, Lemon SC, Schneider KL, August MA, Pagoto SL. Associations of mood and anxiety disorders with obesity: comparisons by ethnicity. *J Psychosom Res.* 2011;71:319–324.
6. Ganley R. Emotion and eating in obesity: a review of the literature. *Int J Eat Disord.* 1989;8:343–361.
7. Crawford PB, Webb KL. Unraveling the paradox of concurrent food insecurity and obesity. *Am J Prev Med.* 2011;40:274–275.
8. Townend L. The moralizing of obesity: new name for an old sin? *Crit Social Policy.* 2009;29:171–190.
9. Inthorn SBT. It is disgusting how much salt you eat: television discourse of obesity, healthy and morality. *Int J Cult Stud.* 2010;13:83–100.
10. Bonafini BA, Pozzilli P. Body weight and beauty: the changing face of the ideal female body weight. *Obes Rev.* 2011;12:62–65.
11. Thornton BLR, Alberg K. Gender role typing, the superwoman ideal, and the potential for eating disorders. *Sex Roles.* 1991;25:469–484.
12. Flegal KM, Graubard BI, Williamson DF, Gail MH. Cause-specific excess deaths associated with underweight, overweight, and obesity. *JAMA.* 2007;298:2028–2037.
13. Kulie T, Slattengren A, Redmer J, Counts H, Eglash A, Schrager S. Obesity and women's health: an evidence-based review. *J Am Board Fam Med.* 2011;24:75–85.
14. Tzankoff SP, Norris AH. Longitudinal changes in basal metabolism in man. *J Appl Physiol.* 1978;45:536–539.
15. Sharma AM, Padwal R. Obesity is a sign – over-eating is a symptom: an aetiological framework for the assessment and management of obesity. *Obes Rev.* 2010;11:362–370.
16. Townsend MS, Peerson J, Love B, Achterberg C, Murphy SP. Food insecurity is positively related to overweight in women. *J Nutr.* 2001;131:1738–1745.
17. Kinge JM, Morris S. Socioeconomic variation in the impact of obesity. *Soc Sci Med.* 2010;71:1964–1971.
18. Foster GD, Kendall PC. The realistic treatment of obesity: changing the scales of success. *Clin Psychol Rev.* 1994;14:701–736.
19. Rolls ET. Sensory processing in the brain related to the control of food intake. *Proc Nutr Soc.* 2007;66:96–112.
20. Chaput JP, Tremblay A. Acute effects of knowledge-based work on feeding behavior and energy intake. *Physiol Behav.* 2007;90:66–72.
21. Swinburn B, Egger G, Raza F. Dissecting obesogenic environments: the development and application of a framework for identifying and prioritizing environmental interventions for obesity. *Prev Med.* 1999;29:563–570.
22. Krebs JR. The gourmet ape: evolution and human food preferences. *Am J Clin Nutr.* 2009;90:707S–711S.
23. Fiese BH, Tomcho TJ, Douglas M, Josephs K, Poltrock S, Baker T. A review of 50 years of research on naturally occurring family routines and rituals: cause for celebration? *J Fam Psychol.* 2002;16:381–390.
24. Giskes K, Kamphuis CB, van Lenthe FJ, Kremers S, Droomers M, Brug J. A systematic review of associations between environmental factors, energy and fat intakes among adults: is there evidence for environments that encourage obesogenic dietary intakes? *Public Health Nutr.* 2007;10:1005–1017.
25. Yanovski S. Sugar and fat: cravings and aversions. *J Nutr.* 2003;133:835S–837S.
26. Birch LL, Fischer JA. The role of experience in the development of children's eating behavior. In: Capaldi ED, ed. *Why We Eat What We Eat. The Psychology of Eating.* Washington, DC: American Psychological Association; 2004.

27. Maio GR, Haddock GG, Jarman HL. Social psychological factors in tackling obesity. *Obes Rev*. 2007;8(suppl 1):123–125.
28. Egger G, Swinburn B. An 'ecological' approach to the obesity pandemic. *BMJ*. 1997;315:477–480.
29. Wansink B, Sobal J. Mindless eating. The 200 daliy food decisions we overlook. *Environ Behav*. 2007;39:106–123.
30. Wansink B. *Mindless Eating: Why We Eat More than We Think*. New York, NY: Bantam-Dell; 2006.
31. Evers C, Stok FM, Danner UN, Salmon SJ, de Ridder DT, Adriaanse MA. The shaping role of hunger on self-reported external eating status. *Appetite*. 2011;57:318–320.
32. Heatherton TF, Baumeister RF. Binge eating as escape from self-awareness. *Psychol Bull*. 1991;110:86–108.
33. Holland S, Dallos R, Olver L. An exploration of young women's experiences of living with excess weight. *Clin Child Psychol Psychiatry*. 2011; [Epub ahead of print].
34. Macht M, Simons G. Emotions and eating in everyday life. *Appetite*. 2000;35:65–71.
35. Raspopow K, Abizaid A, Matheson K, Anisman H. Psychosocial stressor effects on cortisol and ghrelin in emotional and non-emotional eaters: influence of anger and shame. *Horm Behav*. 2010;58:677–684.
36. Nguyen-Rodriguez ST, Unger JB, Spruijt-Metz D. Psychological determinants of emotional eating in adolescence. *Eat Disord*. 2009;17:211–224.
37. Meers M. Emotional Eating in Preschoolers. MA Thesis. Ohio, USA: Bowling Green State University; 2010.
38. Buckroyd J. *Emotional eating as factor in the obesity of those with a BMI>35. In: Obesity in the UK. A Psychologicaal Perspective Obesity Working Group 2011*. Leicester: The British Psychological Society; 2011:pp. 66–78.
39. Walfish S. Self-assessed emotional factors contributing to increased weight gain in pre-surgical bariatric patients. *Obes Surg*. 2004;14:1402–1405.
40. Kaplan HI, Kaplan HS. The psychosomatic concept of obesity. *J Nerv Ment Dis*. 1957;125:181–201.
41. Bruch H. Psychological aspects of overeating and obesity. *Psychosomatics*. 1964;5:269–274.
42. Aftanas LI, Varlamov AA, Reva NV, Pavlov SV. Disruption of early event-related theta synchronization of human EEG in alexithymics viewing affective pictures. *Neurosci Lett*. 2003;340:57–60.
43. Frawley W, Smith RN. A processing theory of alexithymia. *J Cogn Syst Res*. 2001;2:189–206.
44. Taylor GJ, Bagby RM, Parker JD. The alexithymia construct. A potential paradigm for psychosomatic medicine. *Psychosomatics*. 1991;32:153–164.
45. Pinna F, Lai L, Pirarba S, et al. Obesity, alexithymia and psychopathology: a case–control study. *Eat Weight Disord*. 2011;16:e164–170.
46. Zijlstra H, van Middendorp H, Devaere L, Larsen JK, van Ramshorst B, Geenen R. Emotion processing and regulation in women with morbid obesity who apply for bariatric surgery. *Psychol Health*. 2011 [Epub ahead of print].
47. Pinaquy S, Chabrol H, Simon C, Louvet JP, Barbe P. Emotional eating, alexithymia, and binge-eating disorder in obese women. *Obes Res*. 2003;11:195–201.
48. Bowlby J. *Attachment and Loss: Attachment*. New York, NY: Basic Books; 1969.
49. Gustafson TB, Sarwer DB. Childhood sexual abuse and obesity. *Obes Rev*. 2004;5:129–135.

50. Vamosi M, Heitmann BL, Kyvik KO. The relation between an adverse psychological and social environment in childhood and the development of adult obesity: a systematic literature review. *Obes Rev*. 2010;11:177–184.
51. Midei AJ, Matthews KA. Interpersonal violence in childhood as a risk factor for obesity: a systematic review of the literature and proposed pathways. *Obes Rev*. 2011;12:e159–e172.
52. Evers C, Stok FM, de Ridder DT. Feeding your feelings: emotion regulation strategies and emotional eating. *Pers Soc Psychol Bull*. 2010;36:792–804.
53. Wallis DJ, Hetherington MM. Stress and eating: the effects of ego-threat and cognitive demand on food intake in restrained and emotional eaters. *Appetite*. 2004;43:39–46.
54. Polivy J, Herman CP. Distress and eating: why do dieters overeat? *Int J Eat Disord*. 1999;26:153–164.
55. Boon B, Stroebe W, Schut H, Ijntema R. Ironic processes in the eating behaviour of restrained eaters. *Br J Health Psychol*. 2002;7:1–10.
56. Macht M. How emotions affect eating: a five-way model. *Appetite*. 2008;50:1–11.
57. Booth DA. *Psychology of Nutrition*. London: Taylor & Francis; 1994.
58. Manzoni GM, Pagnini F, Gorini A, et al. Can relaxation training reduce emotional eating in women with obesity? An exploratory study with 3 months of follow-up. *J Am Diet Assoc*. 2009;109:1427–1432.
59. Telch CF, Agras WS, Linehan MM. Dialectical behavior therapy for binge eating disorder. *J Consult Clin Psychol*. 2001;69:1061–1065.
60. Terracciano A, Sutin AR, McCrae RR, et al. Facets of personality linked to underweight and overweight. *Psychosom Med*. 2009;71:682–689.
61. Brummett BH, Babyak MA, Williams RB, Barefoot JC, Costa PT, Siegler IC. NEO personality domains and gender predict levels and trends in body mass index over 14 years during midlife. *J Res Pers*. 2006;40:222–236.
62. Sutin AR, Ferrucci L, Zonderman AB, Terracciano A. Personality and obesity across the adult life span. *J Pers Soc Psychol*. 2011;101:579–592.
63. Legenbauer TM, de Zwaan M, Muhlhans B, Petrak F, Herpertz S. Do mental disorders and eating patterns affect long-term weight loss maintenance? *Gen Hosp Psychiatry*. 2010;32:132–140.
64. Rivenes AC, Harvey SB, Mykletun A. The relationship between abdominal fat, obesity, and common mental disorders: results from the HUNT study. *J Psychosom Res*. 2009;66:269–275.
65. Leventhal AM, Mickens L, Dunton GF, Sussman S, Riggs NR, Pentz MA. Tobacco use moderates the association between major depression and obesity. *Health Psychol*. 2010;29:521–528.
66. Barry D, Pietrzak RH, Petry NM. Gender differences in associations between body mass index and DSM-IV mood and anxiety disorders: results from the National Epidemiologic Survey on Alcohol and Related Conditions. *Ann Epidemiol*. 2008;18:458–466.
67. Revah-Levy A, Speranza M, Barry C, et al. Association between body mass index and depression: the 'fat and jolly' hypothesis for adolescents girls. *BMC Public Health*. 2011;11:649.
68. Palinkas LA, Wingard DL, Barrett-Connor E. Depressive symptoms in overweight and obese older adults: a test of the 'jolly fat' hypothesis. *J Psychosom Res*. 1996;40:59–66.
69. Crisp AH. Jolly fat: relation between obesity and psychoneurosis in general population. *BMJ*. 1975;1:7–9.

70. Crisp AH, Queenan M, Sittampaln Y, Harris G. 'Jolly fat' revisited. *J Psychosom Res.* 1980;24:233–241.
71. Roberts RE, Strawbridge WJ, Deleger S, Kaplan GA. Are the fat more jolly? *Ann Behav Med.* 2002;24:169–180.
72. Gaysina D, Hotopf M, Richards M, Colman I, Kuh D, Hardy R. Symptoms of depression and anxiety, and change in body mass index from adolescence to adulthood: results from a British birth cohort. *Psychol Med.* 2011;41:175–184.
73. Vogelzangs N, Kritchevsky SB, Beekman AT, et al. Obesity and onset of significant depressive symptoms: results from a prospective community-based cohort study of older men and women. *J Clin Psychiatry.* 2010;71:391–399.
74. Afari N, Noonan C, Goldberg J, et al. Depression and obesity: do shared genes explain the relationship? *Depress Anxiety.* 2010;27:799–806.
75. Hevener AL, Febbraio MA, Stock Conference Working Group. The 2009 stock conference report: inflammation, obesity and metabolic disease. *Obes Rev.* 2010;11:635–644.
76. Shelton RC, Miller AH. Inflammation in depression: is adiposity a cause? *Dialogues Clin Neurosci.* 2011;13:41–53.
77. Chrousos GP. Stress, chronic inflammation, and emotional and physical well-being: concurrent effects and chronic sequelae. *J Allergy Clin Immunol.* 2000;106:S275–S291.
78. Rihmer Z, Purebl G, Faludi G, Halmy L. Association of obesity and depression. *Neuropsychopharmacol Hung.* 2008;10:183–189.
79. Schwartz TL, Nihalani N, Jindal S, Virk S, Jones N. Psychiatric medication-induced obesity: a review. *Obes Rev.* 2004;5:115–121.
80. Smits JA, Rosenfield D, Mather AA, Tart CD, Henriksen C, Sareen J. Psychotropic medication use mediates the relationship between mood and anxiety disorders and obesity: findings from a nationally representative sample. *J Psychiatr Res.* 2010;44:1010–1016.
81. Castellanos EH, Charboneau E, Dietrich MS, et al. Obese adults have visual attention bias for food cue images: evidence for altered reward system function. *Int J Obes (Lond).* 2009;33:1063–1073.
82. Werthmann J, Roefs A, Nederkoorn C, Mogg K, Bradley BP, Jansen A. Can(not) take my eyes off it: attention bias for food in overweight participants. *Health Psychol.* 2011;30:561–569.
83. Mogg K, Bradley BP, Hyare H, Lee S. Selective attention to food-related stimuli in hunger: are attentional biases specific to emotional and psychopathological states, or are they also found in normal drive states? *Behav Res Ther.* 1998;36:227–237.
84. Piech RM, Pastorino MT, Zald DH. All I saw was the cake. Hunger effects on attentional capture by visual food cues. *Appetite.* 2010;54:579–582.
85. Fadardi JS, Bazzaz MM. A Combi-Stroop test for measuring food-related attentional bias. *Exp Clin Psychopharmacol.* 2011;19:371–377.
86. Wegner DM. Ironic processes of mental control. *Psychol Rev.* 1994;101:34–52.
87. Adriaanse MA, van Oosten JM, de Ridder DT, de Wit JB, Evers C. Planning what not to eat: ironic effects of implementation intentions negating unhealthy habits. *Pers Soc Psychol Bull.* 2011;37:69–81.
88. Stroebe W, Papies EK, Aarts H. From homeostatic to hedonic theories of eating: self-regulatory failure in food rich environments. *Appl Psychol Int Rev.* 2008;57:172–193.
89. Papies EK, Stroebe W, Aarts H. Pleasure in the mind: restrained eating and spontaneous hedonic thoughts about food. *J Exp Soc Psychol.* 2007;43:810–817.
90. Stroebe W, Mensink W, Aarts H, Schut H, Kruglanski AW. Why dieters fail: testing the goal conflict model of eating. *J Exp Soc Psychol.* 2008;44:26–36.

91. Werrij MQ, Jansen A, Mulkens S, Elgersma HJ, Ament AJ, Hospers HJ. Adding cognitive therapy to dietetic treatment is associated with less relapse in obesity. *J Psychosom Res.* 2009;67:315–324.
92. Nauta H, Hospers HJ, Jansen A, Kok G. Cognitions in obese binge eaters and obese non-binge eaters. *Cognit Ther Res.* 2000;24:521–531.
93. Keightley J, Chur-Hansen A, Princi R, Wittert GA. Perceptions of obesity in self and others. *Obes Res Clin Pract.* 2011;5:e341–e349.
94. Wamsteker EW, Geenen R, Iestra J, Larsen JK, Zelissen PM, van Staveren WA. Obesity-related beliefs predict weight loss after an 8-week low-calorie diet. *J Am Diet Assoc.* 2005;105:441–444.
95. Monasta L, Batty GD, Cattaneo A, et al. Early-life determinants of overweight and obesity: a review of systematic reviews. *Obes Rev.* 2010;11:695–708.
96. Yu ZB, Han SP, Zhu GZ, et al. Birth weight and subsequent risk of obesity: a systematic review and meta-analysis. *Obes Rev.* 2011;12:525–542.
97. Juonala M, Juhola J, Magnussen CG, et al. Childhood environmental and genetic predictors of adulthood obesity: the cardiovascular risk in young Finns study. *J Clin Endocrinol Metab.* 2011;96:E1542–E1549.
98. Konttinen H, Mannisto S, Sarlio-Lahteenkorva S, Silventoinen K, Haukkala A. Emotional eating, depressive symptoms and self-reported food consumption. A population-based study. *Appetite.* 2010;54:473–479.
99. Baker TB, Morse E, Sherman JE. The motivation to use drugs: a psychobiological analysis of urges. *Nebr Symp Motiv.* 1986;34:257–323.
100. Tiggemann M, Kemps E. The phenomenology of food cravings: the role of mental imagery. *Appetite.* 2005;45:305–313.
101. Corsica JA, Spring BJ. Carbohydrate craving: a double-blind, placebo-controlled test of the self-medication hypothesis. *Eat Behav.* 2008;9:447–454.

5 Obesity in Adolescence

Gail Busby[1] *and Mourad W. Seif*[2]

[1]Gynaecology Directorate, St. Mary's Hospital, Manchester, UK, [2]Academic Unit of Obstetric and Gynaecology, University of Manchester at St. Mary's Hospital, Manchester, UK

Introduction

Obesity in childhood and adolescence has major negative health impacts extending to adulthood. In addition to negative consequences that occur later in life, childhood and adolescent obesity confers increased risk of adverse outcomes including asthma, increased risk of fractures, hypertension, early markers of cardiovascular disease, insulin resistance and other endocrine abnormalities and psychological effects.

The incidence of childhood and adolescent obesity is increasing worldwide, both in developed and in developing countries. In recognition of the severity of this modern epidemic, the World Health Organization (WHO) published population-based strategies to control it [1]. Similarly, the Royal College of Obstetricians and Gynaecologists consider prevention of childhood obesity to be a priority, due to the implications of obesity on reproductive, obstetric and gynaecological health [2].

There is good evidence that adolescent obesity leads to adult obesity. A cohort of 8834 American adolescents were followed up until their adulthood and it was found that a significant proportion of obese adolescents became severely obese by their early 30s. Among the individuals who were obese in adolescence, 37.1% men and 51.3% women became severely obese adults. Severe obesity was highest among black women. In contrast, across all sex and racial/ethnic groups, less than 5% of adolescents who were at a normal weight became severely obese in adulthood [3].

Prevalence of Childhood Obesity: A Global Perspective

Worldwide, the prevalence of childhood obesity has been increasing over recent decades and increased from 4.2% in 1990 to 6.7% in 2010. This trend is expected to continue and reach a prevalence of 9.1% in 2020.

In 2010, based on the WHO criteria for weight (>2 SD above median), the estimate worldwide for preschool children aged from birth to 5 years, who were overweight and obese, was 43 million. The prevalence of overweight and obesity in developed countries is about double than that in developing countries (11.7% and

Obesity. DOI: http://dx.doi.org/10.1016/B978-0-12-416045-3.00005-4

6.1%, respectively); however, the majority of affected children live in developing countries (35 million) [4].

Although the increasing prevalence of childhood obesity is an international phenomenon, there are significant variations in prevalence throughout the world, withthe highest rates being seen in Eastern Europe (levels > 25%) and the lowest rates being found in Asia (levels < 1%).

The prevalence of childhood obesity is also higher in western and southern Europe than that in northern Europe. The prevalence rates are approximately double in Mediterranean nations than those of northern European countries [5].

At a country level, the correlation between childhood obesity and ethnicity varies. In the United States, obesity rates of both genders are highest in Mexican Americans (31%), followed by non-Hispanic Blacks (20%), non-Hispanic Whites (15%) and Asian Americans (11%) [6]. In the United Kingdom, children of Bangladeshi or Black ethnicity were significantly associated with rapid weight gain in childhood [7].

In the United Kingdom, the prevalence of childhood obesity has continued to rise as reported by the survey of National Child Measurement Programme (NCMP), which included children of both sexes between 4–5 and 10–11 years of age. This is the case for both boys and girls and across both age groups. The NCMP data suggest that mean BMI has increased by around one BMI centile from the 2007/2008 survey to the 2009/2010 survey. Obesity prevalence among children living in the most deprived areas was roughly twice than that of the children living in the least deprived areas [8].

In 2009, The Health Survey for England reported that 31% of boys and 28% of girls aged 2–15 were classed as either overweight or obese; however, the mean BMI was higher among girls than among boys (difference of 0.2 kg/m^2). This difference was greatest among older children aged 12–15 where it ranged between 0.3 and 0.9 kg/m^2 [9]. Figure 5.1 shows the aggregated prevalence of children at risk for overweight or of children classified as overweight or obese in developed countries, developing countries and globally.

Obesity and the Pubertal Transition

The age of onset of puberty in girls has decreased over the past decades. Data collected from 1940 to 1994 support the contention that thelarche and menarche are occurring earlier in the US girls. This apparent trend has coincided with the increase in prevalence of obesity. It is unknown whether the early changes of puberty in obese girls are related to neuroendocrine maturation as, for example, oestrogens from any source can result in the development of breast tissues [10]. Adipose tissues contain aromatase that can convert adrenal androgen precursors to oestrogens. There also may be an obesity-related decrease in the hepatic metabolism of oestrogens. Finally, peri-pubertal obesity is associated with insulin-induced reductions in sex-hormone-binding globulins (SHBG), thereby increasing the bioavailability of sex steroids including oestrogens.

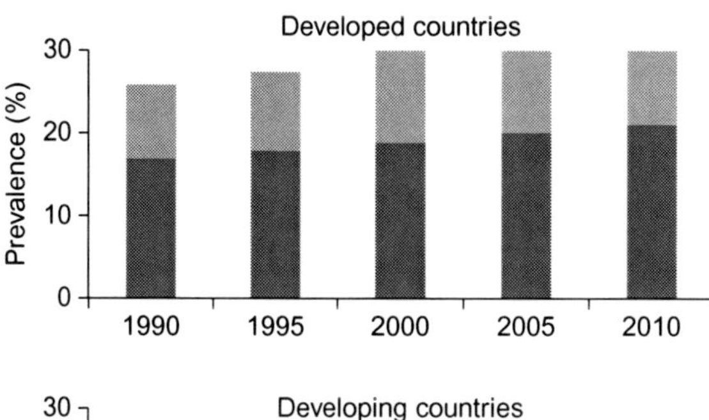

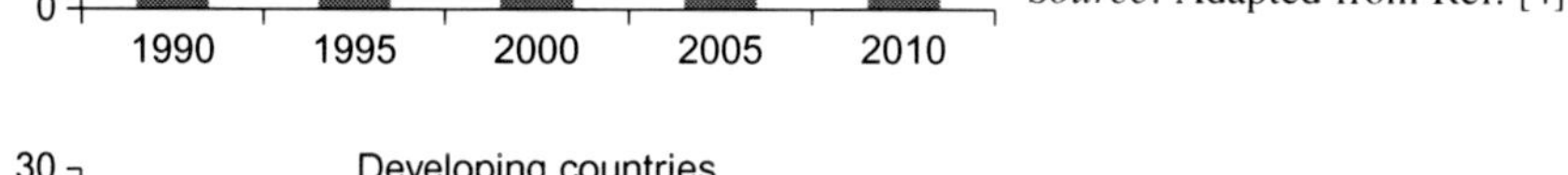

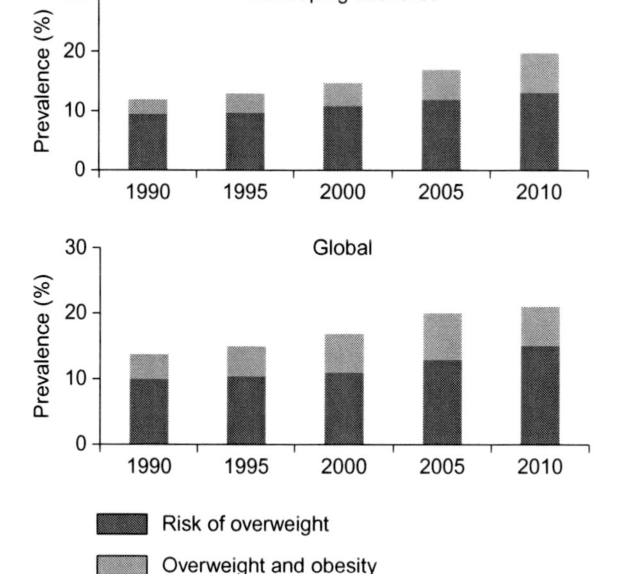

Figure 5.1 Aggregated prevalence of children at risk of overweight or of children classified as overweight or obese in developing and developed countries and globally.
Source: Adapted from Ref. [4].

Unlike their female counterparts, pubertal development in obese boys may be delayed. The reasons for this are unclear, but increased aromatisation of androgens to oestrogens in adipose tissue with feedback inhibition of gonadotrophin secretion may be involved.

Factors Affecting Childhood and Adolescent Obesity

The causes of obesity are complex and multifactorial. Increases in the amount of calorie-dense foods eaten and increases in the screen time (television, computer and video games) with a simultaneous decrease in the amount of physical activity undertaken by children have been cited as reasons for the current epidemic [5].

Weight gain in early childhood (between 3 and 5 years) has been shown to be impacted upon by biological, early life and social factors. Pre-pregnancy maternal overweight and maternal and paternal overweight status at age 3 were all independently associated with more rapid weight gain in the child. Maternal smoking during pregnancy and postnatal exposure of the child through passive smoking were also independently associated with an increased risk of more rapid weight gain. Bangladeshi or black ethnicity and lone child status were also significant [11].

Adolescent Obesity – Adverse Outcomes

General

Obese adolescents are more likely to develop pathologies such as diabetes mellitus which may be an insidious presentation. Obese adolescents have a threefold-increased risk of developing hypertension due to sodium retention, increased sympathetic tone or increased angiotensin system activity, but adolescents are usually asymptomatic and identified only via surveillance. Various pathologies are also related to childhood and adolescent obesity such as hyperlipidaemia in the pattern of increased triglycerides, low-density lipoprotein and decreased HDL.

Obstructive sleep apnoea is four to six times more common in obese children and results in hypertension, left ventricular remodelling, daytime sleepiness, hyperactivity, restlessness and inactivity.

Becoming obese significantly increases the risk of developing asthma. Overweight is significantly and independently associated with increased C-reactive protein concentration and other inflammatory indices, with repercussions on endothelial function.

Orthopaedic complications such as musculoskeletal discomfort, impaired mobility, lower extremity asymmetry and fractures are more common in childhood obesity. These problems further discourage physical activity, thereby exacerbating the underlying problem of obesity.

Obesity leads to various gastrointestinal diseases such as gastro-oesophageal reflux, non-alcoholic fat liver disease and cholelithiasis. Neurological disorders, in particular, the prevalence of benign intracranial hypertension, increases with increasing BMI [12]. Table 5.1 details the multiple adverse outcomes of childhood obesity.

Table 5.1 Adverse Outcomes of Childhood Obesity

- Cardiovascular disease
- Insulin resistance and type-2 diabetes
- PCOS
- Dyslipidaemia
- Hypertension
- Psychological and social morbidity
- Asthma
- Orthopaedic-hips, ankles
- Breathing problems and sleep apnoea
- Fatty liver disease
- Persistence of obesity into adulthood

Psychological Effects

Child/adolescent obesity is independently associated with internalising (emotional) difficulties, after adjusting for confounding variables such as gender and family income [13]. Psychological disturbances associated with childhood obesity include negative self-esteem, withdrawal from peer interaction, anxiety, depression and suicide.

Nearly half of the obese adolescents report moderate to severe depressive symptoms and one-third report anxiety. Obese children are more likely to experience psychological or psychiatric problems than non-obese children. Girls are at greater risk than boys, especially concerning self-esteem [12].

In a study of nearly 1000 adolescents aged 12–18, it was found that elevated BMI was directly associated with depression at a 1-year follow-up. Social networking mapping studies indicate that overweight children have fewer secluded relationships than their normal-weight peers who have many relationships within a central network of children.

Teasing by an overweight adolescent's peers has been established to directly correlate with the child's suicidal ideation and number of attempts at suicide [14].

The Metabolic Syndrome

The metabolic syndrome is a constellation of cardiovascular risk factors associated with insulin resistance. These include glucose intolerance, dyslipidaemia, hypertension and central obesity.

There is no consensus regarding the definition of metabolic syndrome in children and adolescents; however, the International Diabetes Federation (IDF) has proposed criteria depending on age groups [15].

The presence of the metabolic syndrome in adolescents in the United States has increased from 4.2% to 6.4% over the past two decades and is significantly higher in Hispanic and White youths. Nearly one-third of the overweight/obese adolescents meet the criteria for metabolic syndrome.

Childhood adiposity is a good predictor of metabolic syndrome in adulthood, and hence, a greater risk of cardiovascular disease, especially if there is a family history of type-2 diabetes. Metabolic syndrome is also more likely in adults who experienced a rapid increase in adipose tissue in childhood [16].

Polycystic Ovarian Syndrome in Adolescence

Polycystic ovarian syndrome (PCOS) is the most common endocrinological problem in adult women. This was often unrecognised in adolescence. Although most adolescents and adults with PCOS are obese, only 20% of obese women have PCOS [17].

PCOS may present with any, or a combination of obesity, menstrual abnormalities, hirsutism, acanthosis nigricans, acne, hair loss or premature adrenarche.

PCOS is associated with insulin resistance and the metabolic syndrome and subsequent adult morbidities may include infertility and cardiovascular disease.

In a study of 71 PCOS and 94 healthy adolescent girls, Fulghesu et al. [18] showed that the incidence of altered lipid profiles was not different between both groups of adolescents but instead was related to anthropometric characteristics (BMI, waist measurement and waist-to-hip ratio). The differences which were statistically significant between the groups were hirsutism, and grogens including total testosterone levels and hyperinsulinaemia. This suggests that PCOS confers no additional risk over obesity to dyslipidaemia in adolescence .

There have been several different criteria established for the diagnosis of PCOS, including the NIH consensus in 1990 [19], the Rotterdam criteria in 2003 [20] and the criteria by Androgen Excess and PCOS Society in 2006 [21]. However, none of the definitions fits all cases and so the exact prevalence is difficult to define precisely.

The diagnostic label of PCOS in adolescence implies an increased risk for infertility, dysfunctional uterine bleeding, endometrial cancer, obesity, type-2 diabetes, dyslipidaemia, hypertension and possible cardiovascular disease; therefore, diagnostic accuracy is important. Use of the adult diagnostic criteria in adolescence may not be appropriate, as the mean ovarian volume is higher in young women, hirsutism is uncommon and not related to PCOS and acne is similarly not related to PCOS [22]. Finally adolescents frequently have menstrual irregularity, thus making the definitive diagnosis difficult.

The diagnosis of PCOS in adolescence may be delayed for the above reasons, as well as that physiological adolescent anovulation may mimic or mask PCOS. There are, however, risk factors which may assist in the identification of adolescents at risk for PCOS [23]. Figure 5.2 shows the risk factors for adolescent PCOS.

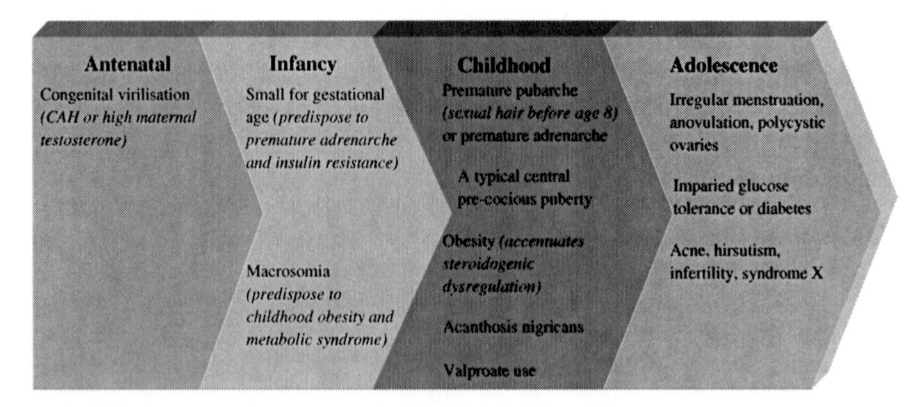

Figure 5.2 Risk factors for adolescent PCOS.
Source: Adapted from Ref. [23].

Obstetric Outcomes in Obese Adolescents

In one recent study, adolescent obesity has been shown to be independently associated with lifetime nulliparity. This study used self-reported weight and height at high school, and weight and height were measured in adulthood. The lifetime nulliparity percentage in this population of 3154 women was 16.7% in comparison with a nulliparity rate of 30.9 in women with a recalled adolescent BMI ≥ 30. The nulliparity rate increased with increasing BMI across all ranges [24].

The complications associated with teenage pregnancy include an increased incidence of very low-birthweight babies (defined as birthweight < 1500 g). Young teenagers aged 11–15 years have nearly double (4% vs 2%) the risk as women aged 20–22 years. In addition, this young age group more frequently delivered babies of birthweight between 1500 and 2500 g. In this study of 16,512 consecutive nulliparous women, the incidences of stillbirth and index values of foetal condition at birth were not significantly different between age groups [25].

Younger maternal age has also been shown to be associated with increased risk of foetal death and anaemia during pregnancy. The risk of pre-eclampsia, Caesarean section and instrumental vaginal delivery and post-parutm haemorrhage are lower in teenage pregnancy [26].

Obesity confers an increased risk of adverse foetal and pregnancy outcomes, including gestational diabetes, foetal macrosomia, delivery by Caesarean section and pre-eclampsia.

The combination of obesity and teenage pregnancy results in an increased risk of developing pre-eclampsia and eclampsia. In a retrospective cohort study of 290,807 women in Florida, USA, extremely obese (BMI ≥ 40) girls aged 16–17 years had the highest rate (13.2%) of pre-eclampsia and eclampsia compared with all the other age groups. The lowest rate of pre-eclampsia and eclampsia occurred in non-obese women aged 20–24 years (4.0%).

Extremely obese teenagers had a 71% increased risk of pre-eclampsia and eclampsia compared with extremely obese women aged 20–24 years [17].

The Impact of Childhood and Adolescent Obesity on Adult Health

Several studies have shown that childhood and adolescent overweight and obesity (based on BMI for age) have been associated with premature mortality at the adult stage of life [27]. In addition to an increased risk of premature mortality, childhood/adolescent obesity is also associated with an increased risk of later diabetes, stroke, coronary heart disease and hypertension.

A retrospective study of 230,000 Norweigan adolescents were followed up on average for 34.9 years. There were 9650 deaths within this time. Mean age of death was 40 years for men and 43 years for women. The relative risk of death from endocrine, nutritional and metabolic diseases and from diseases of the circulatory system was elevated in the two highest BMI groups of both men and women. The relative risks of death from diseases of the respiratory system and ill-defined causes were increased in the highest BMI group of both sexes [28].

In case of diseases of the circulatory system in men, ischaemic heart disease was the dominant cause of death, whereas in women, the death was due to cerebrovascular disease. In both men and women, the risk of death from ischaemic heart disease was increased in the two highest BMI categories. Similar findings were reported in a Danish study which showed a linear association between BMI in childhood (7−13 years) and both fatal and non-fatal coronary heart disease events in adulthood (age 25 or older) [29].

The risk of death from diabetes was increased in men and women in the two highest BMI categories. There was an increased risk of sudden death for both sexes and of death from chronic lower respiratory diseases in men in the highest BMI category. There was no association between BMI and mortality from mental and behavioural disorders.

Management Principles

Prevention

Prevention programmes should address parental weight status and smoking habits, both of which are modifiable risk factors. Reductions in weight pre-conception may result in an additional reduction in risk. Parents should be encouraged to integrate healthy lifestyle behaviours into the family unit, thereby decreasing the risk of obesity in their children and improving their own physical and mental health.

The long-term impact of maternal weight loss was elegantly demonstrated by the reported impressive reduction in childhood/adolescent obesity in children of mothers who underwent bariatric surgery. An accompanying improvement in cardio-metabolic risk factors also occurred which was sustained into adolescence and adulthood [30].

Obesity prevention programmes in kindergartens and schools based on exercise therapy and dietary intervention have failed in preventing childhood obesity. It is widely accepted that changing the environment (e.g. banning all sugary drinks in favour of water) is more effective in preventing obesity. This may suggest that public health restrictions on advertising and marketing of sweetened drinks are a meaningful approach to the fight against childhood obesity.

Lifestyle Interventions

A Cochrane review [31] regarding treatment in obese children included 64 randomised controlled trials with 5230 participants. Meta-analysis indicated a reduction in overweight at 6 and 12 months in response to lifestyle interventions. Dietary modification and exercise programmes are effective in the treatment of already-obese children. Parental involvement was a major predictor of success.

Recommended family lifestyle changes:

- Be physically active for 1 h/day (moderate to vigorous intensity).
- Reduce screen time (sedentary behaviour such as watching TV, using computers and playing computer games to not more than 2 h/day).

- Encourage low-energy snacks (e.g. fruit, raw vegetables and a plain biscuit).
- Avoid/cut down on high-energy foods such as crisps, chips, chocolate, sweets.
- Avoid grazing and keep food to meal times and small snacks.
- Avoid sugary juice.
- Parents should be positive about a healthy family lifestyle.

Drugs

The Cochrane review also concluded that consideration should be given to the use of drugs such as orlistat or sibutramine as an adjunct to lifestyle interventions in obese adolescents. This must, of course, be balanced with the risk of adverse events. Sibutramine is no longer available on the European market due to its side effects such as arterial and pulmonary hypertension.

Surgery

Gastric banding in obese adolescents is an effective intervention leading to substantial and durable reduction in obesity and to better health. This treatment requires long-term supportive follow-up. There is also a significant possibility of requiring revisional procedures [30].

Management of PCOS in Adolescence

In adolescents with PCOS, the target goals of therapy include protection of endometrial health, improvement in cosmetic appearance (e.g. hirsutism and acne) and reduction in weight and obesity-related metabolic complications.

Lifestyle interventions including weight loss is the first line of treatment of adolescents with PCOS. The study by Lass et al. [32] showed a significant decrease in the prevalence of metabolic syndrome and insulin resistance in adolescents with successful weight loss. Furthermore, testosterone concentrations, free testosterone index, luteinizing hormone (LH) levels and LH/follicle-stimulating hormone (FSH) ratio decreased significantly. The prevalence of amenorrhoea and/or oligomenorrhoea decreased significantly in the weight-loss group [32].

Lifestyle modifications and weight loss, while being very effective in treating the hormonal abnormalities of PCOS, are difficult to sustain. In one study of adolescents with PCOS, 30% of the subjects enrolled into a study which included intensive lifestyle modification dropped out, whereas 40% attended less than 50% of the sessions and demonstrated no weight change [33]. Therefore, in most cases, pharmacological therapy for PCOS becomes necessary.

Combined Oral Contraceptive Pills

Combined oral contraceptive pills (COCP) are among the primary treatment options for adolescents with PCOS. COCPs improve symptoms via several mechanisms. Oestrogens increase the production of SHBG, resulting in a decrease in circulating androgens, as well as their bioavailability. Progestins protect the endometrium against hyperplasia induced by unopposed oestrogen stimulation. Some progestins such as drospirenone and cyproterone acetate have been proven to have anti-androgenic effects and therefore may be of added benefit in PCOS [34]. COCPs also suppress FSH and LH, resulting in reduced ovarian stimulation and androgen production.

None of these actions, however, affects insulin resistance in PCOS, and their use may actually be associated with long-term metabolic derangements such as glucose intolerance, abnormal lipid profiles and cardiovascular diseases. A recent research showed that in adolescents with PCOS, the use of COCPs containing desogestrel or cyproterone as progestin was associated with decreased insulin sensitivity and increased total, LDL and HDL cholesterol and with variable changes in triglycerides [35].

Use of the COCP does have other benefits in this population, such as contraception in sexually active adolescents.

Insulin Sensitisers

These drugs act to reduce insulin levels (metformin) and increase insulin sensitivity (metformin and thiazolidenediones), thus treating the metabolic co-morbidities associated with PCOS and obesity.

Metformin increases insulin sensitivity by the liver, increases peripheral glucose uptake, decreases fatty acid oxidation and decreases glucose absorption from the gut. Metformin therapy in adolescents seems to be associated with several benefits including an improvement in glucose tolerance, a decrease in testosterone levels [36] and an improvement in menstrual cyclicity; some report between 90% and 100% resumption of menses in adolescents [37]. These studies were all relatively small, and, in most, significant weight loss occurred, making the data difficult to interpret the effect of metformin independent of weight loss. This has not yet been substantiated by meta-analysis or randomised controlled trials.

The dose of metformin required is approximately 1.5–2 g/day, and even larger doses up to 2.55 g/day have been reported. The dose should be titrated upwards over a period of 1 month due to gastrointestinal side effects.

Thiazolindenediones act as insulin sensitisers through their activation of the nuclear receptor peroxisome proliferator-activated receptors γ, leading to increased production of insulin-sensitive adipocytes and increased glucose uptake in these cells, increased secretion of adiponectin and decreased secretion of pro-inflammatory cytokines. The thiazolindenedione, pioglitazone, has been shown to ameliorate the signs and symptoms of PCOS in a cohort of women who failed a previous trial of metformin [38]. These medications have not been well studied in adolescence and remain off-label in this age group due to lack of evidence on efficacy and safety.

Conclusion

Obesity in childhood and adolescence is a global epidemic with health implications. There are geographic and ethnic variations in the prevalence of adolescent obesity, but it is an international problem.

All efforts should be made to control this epidemic and therefore to avoid or reduce the healthcare burden of chronic illness on healthcare systems as well as to reduce the personal cost including premature mortality.

In particular, a clear focus on lifestyle and prevention measures should be adopted for all age groups from preconception to early childhood, adolescence and adulthood.

References

1. WHO. *Population-based Prevention Strategies for Childhood Obesity: Report of a WHO Forum and Technical Meeting*. Geneva: WHO Press; 2009.
2. RCOG. *Consensus Views Arising from the 53rd Study Group: Obesity and Reproductive Health*. RCOG; 2011.
3. The NS, Suchindran C, North KE, Popkin BM, Gordon-Larsen P. Association of adolescent obesity with risk of severe obesity in adulthood. *JAMA*. 2010;304(18): 2042–2047.
4. De Onis M, Blossner M, Borghi E. Global prevalence and trends of overweight and obesity among preschool children. *Am J Clin Nutr*. 2010;92:1257–1264.
5. Ben-Sefer E, Ben-Natan M, Ehrenfeld M. Childhood obesity: current literature, policy and implications for practice. *Int Nurs Rev*. 2009;56(2):166–173.
6. Sorof JM, Lai D, Turner J, et al. Overweight, ethnicity and the prevalence of hypertension in school-aged children. *Pediatrics*. 2004;113(3):475–482.
7. Griffiths LJ, Hawkins SS, Cole TJ, Dezateux C, Millenium Cohort Study Child Health Group. Risk factors for rapid weight gain in preschool children: findings from a UK-wide prospective study. *Int J Obes*. 2010;34:624–632.
8. National Obesity Observatory. *National Child Measurement Programme. Changes in Children's Body Mass Index between 2006/7 and 2009/10*. NOO; 2011.
9. The Health and Social Care Information Centre. Statistics on Obesity, Physical Activity and Diet. England; 2011.
10. Burt Solorzano CM, McCartney CR. Obesity and the pubertal transition in girls and boys. *Reproduction*. 2010;140:399–410.
11. Stewart L. Childhood obesity. *Medicine*. 2010;39(1):42–44.
12. Bruni V, Dei M, Peruzzi E, Seravalli V. The anorectic and obese adolescent. *Best Pract Res Clin Obstet Gynaecol*. 2010;24:243–258.
13. Tiffin PA, Arnott B, Moore HJ, Summerbell CD. Modelling the relationship between obesity and mental health in children and adolescents: findings from the Health Survey for England 2007. *Child Adolesc Psychiatry Ment Health*. 2001;5:31.
14. Sinha A, Kling SA. Review of adolescent obesity: prevalence, etiology and treatment. *Obes Surg*. 2009;19:113–120.
15. Zimmet P, Alberti KG, Kaufman F, et al. The metabolic syndrome in children and adolescents – an IDF consensus report. *Pediatr Diabetes*. 2007;8:299.
16. In-Iw S, Biro FM. Adolescent women and obesity. *J Pediatr Adolesc Gynecol*. 2011;24:58–61.
17. Alihu MH, Luke S, Kristensen S, Alio A, Salihu HM. Joint effect of obesity and teenage pregnancy on the risk of preeclampsia: a population-based study. *J Adolesc Health*. 2010;46:77–82.

18. Fulghesu A, Magnini R, Portoghese E, Angioni S, Minerba L, Melis GB. Obesity-related lipid profile and altered insulin incretion in adolescents with polycystic ovary syndrome. *J Adolesc Health*. 2010;46:474–481.
19. Zawadzki JK, Dunaif A. Diagnostic criteria for polycystic ovary syndrome: towards a rational approach. In: Dunaif A, ed. *Polycystic Ovary Syndrome*. Boston, MA: Blackwell Scientific; 1995:377–384.
20. Rotterdam. ESHRE/ASRM-Sponsored PCOS Consensus Workshop Group. Revised 2003 consensus on diagnostic criteria and long-term health risks related to polycystic ovary syndrome. *Fertil Steril*. 2004;81:19–25.
21. Azziz R, Carmina E, Dewailly D, et al. The Androgen Excess and PCOS Society criteria for the polycystic ovary syndrome: the complete taskforce report. *Fertil Steril*. 2009;91:456–488.
22. Hickey M, Doherty DA, Atkinson H, et al. Clinical, ultrasound and biochemical features of polycystic ovary syndrome in adolescents: implications for diagnosis. *Hum Reprod*. 2011;26(6):1469–1477.
23. Yii MF, Lim CED, Luo X, Wong WSF, Cheng ACL, Zhan X. Polycystic ovarian syndrome in adolescence. *Gynecol Endocrinol*. 2009;25(10):634–639.
24. Polotsky AJ, Hailpern SM, Skurnick JH, et al. Association of adolescent obesity and lifetime nulliparity – The Study of Women's Health Across the Nation (SWAN). *Fertil Steril*. 2010;93(6): 2004–2011.
25. Satin AJ, Leveno KJ, Sherman ML, Reedy NJ, Lower TW, McIntire DD. Maternal youth and pregnancy outcomes: middle school versus high school age groups compared with women beyond the teen years. *Am J Obstet Gynecol*. 1994;171:184–187.
26. de Vienne CM, Creveuil C, Dreyfus M. Does young maternal age increase the risk of adverse obstetric, fetal and neonatal outcomes: a cohort study. *Eur J Obstet Gynecol Rep Bio*. 2009;147:151–156.
27. Reilly JJ, Kelly J. Long-term impact of overweight and obesity in childhood and adolescence on morbidity and premature mortality in adulthood: systematic review. *Int J Obes*. 2011;35:891–898.
28. Bjorge T, Engeland A, Tverdal A, Davey Smith G. Body mass index in adolescence in relation to cause-specific mortality: a follow-up of 230,000 Norwegian adolescents. *Am J Epidemiol*. 2008;168:30–37.
29. Baker JL, Olsen LW, Sorensen TIA. Childhood body-mass index and the risk of coronary heart disease in adulthood. *N Engl J Med*. 2007;357:2329–2337.
30. Reinehr T, Wabitsch M. Childhood obesity. *Curr Opin Lipidol*. 2011;22:21–35.
31. Oude Luttikhuis H, Baur L, Jansen H, et al. Interventions for treating obesity in children. *Cochrane Database Syst Rev*. 2009;1:CD001872.
32. Lass N, Kleber M, Winkel K, Wunsch R, Reinehr T. Effect of lifestyle intervention on features of polycystic ovarian syndrome, metabolic syndrome, and intima-media thickness in obese adolescent girls. *J Clin Endocrinol Metab*. 2011;96(11):3533–3540.
33. Hoeger K, Davidson K, Kochman L, Cherry T, Kopin L, Guzick DS. The impact of metformin, oral contraceptives, and lifestyle modification on polycystic ovary syndrome in obese adolescent women in two randomised, placebo-controlled clinical trials. *J Clin Endocrinol Metab*. 2008;93:4299–4306.
34. Franks S, Layton A, Glasier A. Cyproterone acetate/ethinyl estradiol for acne and hirsutism: time to revise prescribing policy. *Hum Reprod*. 2008;23:231–232.
35. Mastorakos G, Koliopoulos C, Deligeoroglou E, Diamanti-Kandarakis E, Creatsas G. Effects of two forms of combined oral contraceptives on carbohydrate metabolism in adolescents with polycystic ovary syndrome. *Fertil Steril*. 2006;85:420–427.

36. Arslanian SA, Lewy V, Danadian K, Saad R. Metformin therapy in obese adolescents with polycystic ovary syndrome and impaired glucose tolerance: amelioration of exaggerated adrenal response to adrenocorticotropin with reduction of insulinaemia/insulin resistance. *Clin Endocrinol Metab*. 2002;87:1555–1559.
37. De Leo V, Musacchio MC, Morgante G, Piomboni P, Petraglia F. Metformin treatment is effective in obese teenage girls with PCOS. *Hum Reprod*. 2006;21:2252–2256.
38. Gluck CJ, Moreira A, Goldenberg N, Sieve L, Wang P. Pioglitazone and metformin in obese women with polycystic ovary syndrome not optimally responsive to metformin. *Hum Reprod*. 2003;18:1618–1625.

Section 2

Obesity and Reproduction

6 Obesity and Contraception

Sujeetha Damodaran[1] and Krishnan Swaminathan[2]

[1]Iswarya Women's Hospital and Fertility Centre, Madurai, Tamil Nadu, India,
[2]Apollo Speciality Hospitals, Madurai, Tamil Nadu, India

Introduction

Obesity has reached epidemic proportions around the globe. The World Health Organization (WHO) estimates that, in 2008, 1.5 billion adults aged 20 and older were overweight. Of these, nearly 300 million women were obese[1]. Not surprisingly, the prevalence of obesity in pregnancy is rising exponentially, with 15–20% of pregnant women satisfying the criteria that would define them as 'obese' [1]. Obesity during pregnancy is associated with numerous maternal, foetal, perinatal and post-natal risks. Pregnancy in an obese woman also poses a serious challenge to the skills of the obstetrician, anaesthetist and midwives. Unintended or unplanned pregnancies, by themselves, are associated with increased risk of both maternal and infant morbidity and mortality [2]. These risks are magnified by maternal obesity. This dual combination of maternal obesity and unintended pregnancy interacts together, causing far-reaching consequences to both the mother and the foetus. Prevention of unwanted pregnancy in obese women is, therefore, a major priority for health care professionals. This is all the more relevant in the light of data suggesting that obese women are less likely to access contraceptive health care services and have more chances of unwanted pregnancy [3]. The aim of this chapter is to review the various contraceptive options available to the obese woman. Knowledge of such contraceptive options may lead to a reduction of unwanted pregnancies and support women in their quest for a well-planned pregnancy.

Risks of Obesity in Pregnancy

Maternal obesity during pregnancy is associated with several adverse maternal and foetal outcomes (Table 6.1). Adverse maternal outcomes include increased risk of miscarriage [11], gestational diabetes [12] and its consequences, pregnancy-associated hypertension [13–15], preterm birth [16,17] probably related to medical complications, post-term pregnancy [18–20], prolonged labour [21–23], higher

[1] http://www.who.int/mediacentre/factsheets/fs311/en/

Obesity. DOI: http://dx.doi.org/10.1016/B978-0-12-416045-3.00006-6

Table 6.1 Effects of Obesity on Pregnancy Outcomes

Condition	Type of Study	Effect[a]
GDM [5]	Meta-analysis	OR, 2.14 (1.82–2.53)[b] OR, 3.56 (3.05–4.21)[c] OR, 8.56 (5.07–16.04)[d]
PIH [6]	Meta-analysis	OR, 2.5 (2.1–3.0)[c] OR, 3.2 (2.6–4.0)[d]
C-section [7]	Population-based cohort study	RR, 2.6 (2.04–2.51)[c] RR, 3.38 (2.49–4.57)
Pre-eclampsia [5]	Meta-analysis	OR, 1.6 (1.1–2.25)[c] OR, 3.3 (2.4–4.5)[d]
Pre-eclampsia [8]	Retrospective cohort study	OR, 7.2 (4.7–11.2)[d]
Induction of labour [8]	Retrospective cohort study	OR, 1.8 (1.3–2.5)[d]
Postpartum haemorrhage [8]	Population-based cohort study	OR, 1.5 (1.3–1.7)[e]
Preterm delivery (<33 weeks) [8]	Population-based cohort study	OR, 2.0 (1.3–2.9)[e]
Stillbirth [9]	Systematic review and meta-analysis	OR, 1.47[b] RR, 2.07[e]
Stillbirth [10]	Population-based cohort study	OR, 2.8 (1.5–5.3)[e]
Neonatal death [10]	Population-based cohort study	OR, 2.6 (1.2–5.8)[e]

GDM, gestational diabetes mellitus; PIH, pregnancy-induced hypertension.
[a]All odds ratio (OR) and relative risk (RR) are compared to normal weight pregnant women (BMI 18–25). Values in parentheses indicate 95% CI.
[b]BMI 25–30.
[c]BMI 30–35.
[d]BMI >35.
[e]BMI >30.
Source: Reproduced by permission of Ref. [4].

rates of anaesthetic complications [24,25], higher chances of operative deliveries [26–29], wound infections [13,30–32], longer hospital stay [31] and shorter duration of breastfeeding [33–35]. Maternal obesity also contributes to adverse foetal outcomes including macrosomia [18,36], shoulder dystocia [37], perinatal mortality [38] and predisposition to obesity later in life [39].

Classification of Obesity Based on Body Mass Index

The WHO and the National Institute of Health classify obesity as given in Table 6.2.

It has to be pointed out that the arbitrary cut-offs are based on data derived from whites, as ethnic-based data are currently unavailable.

Table 6.2 Classification of Obesity

Classification	BMI (kg/m^2)
Underweight	<18.5
Normal	≥18.5–24.9
Overweight	25–29.9
Class I Obesity	30–34.9
Class II Obesity	35–39.9
Class III Obesity	≥40

Potential Concerns with Obesity and Contraception

The main concerns with contraceptive methods in obese woman are as follows:

1. Historically, overweight and obese woman have been excluded from trials in contraception, leading to lack of robust evidence regarding the safety and efficacy of contraceptive methods in such a population.
2. 'One dose fits all' has been the traditional practice with hormonal contraception. However, the effects of obesity on drug pharmacokinetics and pharmacogenetics, especially steroidal contraceptives, are poorly understood.
3. Obesity, per se, doubles the risk of venous thrombo embolism (VTE) as compared with someone with a normal body mass index (BMI) [40]. There is always a concern as to whether oral contraceptives increase the risk of a potentially life-threatening complication like VTE and other health hazards associated with obesity (diabetes, dyslipidaemia, cardiovascular disease, hepatobiliary disease and cancer).
4. As a generalisation, women tend to blame contraception for weight gain. This perceived weight gain is a leading cause of discontinuation of contraception at least in some parts of the world [41,42].
5. Finally, procedure-dependent contraceptive methods (intrauterine devices (IUDs) and sterilisation) are technically more challenging to perform in an obese woman than their normal BMI counterparts.

Obesity and Contraceptive Efficacy

Evidence (or Lack of) for Contraceptive Efficacy in Overweight or Obese Women

The current literature is inadequate to provide clear information to the overweight or obese woman as to the efficacy of contraception. There are multiple reasons for this lack of clarity. As previously mentioned, a number of studies excluded women over a certain weight/BMI cut-off, thereby greatly limiting the ability to draw reasonable conclusions. The limited numbers of studies that are available are too heterogeneous, assessing different delivery mechanisms (e.g. patch, injectables and rings).

In a retrospective cohort study of 2822 person-years of oral contraceptive use, women in the highest body quartile (≥70.5 kg) had a significantly higher risk of oral contraceptive failure compared with women of lower body weight [43]. A more rigorous case–control study from the same population found that the risk of accidental pregnancy was higher in women with a BMI of >32.2 compared with women of normal body mass [44]. However, a large cohort study of 17,032 women showed no evidence of any influence of body weight on the risk of accidental pregnancy with either combined oral contraceptive or progestogens-only contraceptive (POC) [45]. In two further studies exploring the obesity–contraceptive failure association, there was no statistically significant association between BMI/body weight and oral contraceptive failure [46,47].

In a large intervention review by the Cochrane Fertility Regulation Group [48], data from 39,531 women using different types of contraception including combined oral contraceptives (COCs), transdermal skin patches, vaginal rings, levonorgestrel (LNG) implants and subcutaneous depo medroxyprogesterone acetate (DMPA-SC) were studied. The primary purpose of this study was to look at the effectiveness of hormonal contraceptives in preventing unplanned pregnancies among women who are obese or overweight compared to women of lower weight or BMI. Only one of the three trials using BMI showed that overweight or obese women are more likely to get pregnant compared to women with normal body mass. Transdermal skin patch trials demonstrated some differences for highest weight group but pregnancies were not particularly clustered around any BMI subgroup. There was no evidence for increased risk of pregnancy in overweight or obese women using DMPA-SC. The risks from implant studies were inconsistent. Overall, there was not a consistently high risk of pregnancy in an overweight or obese woman with hormonal contraceptives, and these methods were still the most effective methods of contraception if the regime is adhered to properly.

Non-hormonal contraception such as barrier methods, IUDs and sterilisation are as effective in overweight or obese women but there are no direct comparisons to a cohort with normal BMI.

Mechanisms by Which Obesity Could Affect Contraceptive Efficacy

Obesity can have profound effects on different physiologic processes including absorption, distribution, metabolism and excretion of contraceptive drugs [49]. Higher cardiac output, increased gut perfusion, accelerated gastric emptying and alterations in enterohepatic recirculation have been reported in obese human and animal models [50–56], factors which can potentially affect absorption of contraceptive drugs. Obesity is also associated with altered body composition with an increase in fat mass, which can affect the distribution of hydrophilic and lipophilic drugs [57]. Other physiological alterations in obesity that can have a potential impact in contraceptive drug metabolism and excretion include increased sphlanchnic and renal flow, fatty infiltration of liver, inflammatory cytokines, reduction in expression of biliary canalicular transporters, increased kidney size and reduced urinary pH. The combined consequences of the above alterations include a reduction in specific cytochrome P450 activities, altered biliary metabolism and enterohepatic circulation, increased renal clearance and tubular secretion [5–9,58].

In spite of all the potential mechanisms by which obesity could affect contraceptive efficacy, there have been only three studies to date that have investigated the pharmacokinetics of contraceptive steroids in obese women. In a study conducted to determine whether increased BMI affects the pharmacokinetics of oral contraceptives, the LNG half-life in the obese subjects was twice that of normal BMI subjects and the time taken to reach a steady state was doubled as well [10]. In another study conducted to compare the pharmacokinetics of oral contraceptives in obese and normal weight women, obese women had a lower area under curve and lower maximum values for ethinyl oestradiol than normal weight women [59]. But the observed differences in pharmacokinetics did not translate into more ovarian follicular activity in obese oral contraceptive users. Finally, in a 26 week prospectively designed experimental study conducted to determine the incidence of ovulation and follicular development in different classes of obese women using DMPA-SC, median MPA were consistently lowest among class 3 obese women but above the levels needed to inhibit ovulation [60]. It is therefore, very clear that we need more clinical trials to understand the impact of obesity on drug pharmacokinetics and therapeutics, so that it is useful in counselling obese women in day-to-day practice.

Contraceptives and Weight Gain: Myth or Truth?

Potential Mechanisms by Which Contraceptives Can Cause Weight Gain

In general, weight gain is due to one of the following factors: fluid retention, fat deposition or muscle mass. Treatment with hormonal contraceptives may lead to a considerable activation of the renin–angiotensin–aldosterone system [61]. Fluid retention could, therefore, be induced by the mineralocorticoid activity of contraceptive steroids [62]. Experimentally, oestrogens increase the size and number of subcutaneous adipocytes [63] that can be associated with increased subcutaneous fat in breasts, hips and thighs [64]. In a prospective study to determine whether the use of low-dose oestrogen oral contraceptives is associated with changes in weight, body composition or fat distribution, there was no overall impact of low-dose oral contraceptive on any of the above parameters. However, when weight gain did occur, it was related to the increase in body fat and not due to fluid retention or fat distribution [65].

The Scale of the Problem

Many women and clinicians worldwide believe that an association exists between weight gain and oral contraceptives [66–70]. Young women, particularly, may be preoccupied with body image, and this fear of potential weight gain can deter already obese individuals and clinicians from initiating combination contraceptives. It can also lead to early discontinuation among users [71–73]. In a prospective nationwide study [74], 6 months after a new oral contraceptive prescription, only 68% of women were still continuing the medication. Of the women who discontinued, 46% of women did so because of side effects, predominantly weight gain. More importantly, 80% of failed to adopt another method or adopted a less effective method, putting themselves at high risk of unintended pregnancy. It is, therefore, very clear that concerns about possible weight gain limits the use of a very effective method of contraception. However, a causal relationship between weight gain and combination contraceptives has not been established.

COCs and Weight Gain

A recent review to evaluate the potential association between combination contraceptive use and changes in weight showed no significant effect, though the available evidence was insufficient to determine the effect of combination contraceptives on weight [75]. In this large review of 49 trials that included 52 distinct contraceptive pairs (or placebo), neither comparisons of different combination contraceptives nor comparisons with placebo showed any substantial differences in weight. In a recently published longitudinal study to assess the long-term effects of COCs on body weight, COC use was not found to be a predictor for weight increase in the long term [76]. Interestingly, this long-term follow-up study showed a weight increase of 10.6 kg for women between 19 and 44 years of age. There are further studies describing weight change in women with increasing age [77–79] and it may be possible that the perceived weight gain with COC may be related to the natural changes in weight from a lifetime perspective. There have been other studies which show minimal or no weight increase with COC to suggest a causal relationship [80–83]. Based on the above statements, there is little evidence for significant weight alterations in relation to COC pill use but this has to be confirmed with further long-term studies.

POCs and Effect on Weight

POCs are ideally suited for women who have contraindications to or who are unable to tolerate oestrogens. There has always been a concern of weight gain associated with progesterone preparations especially DMPA and LNG implants. In a study designed to compare women using either DMPA or an oral contraceptive pill and those who were not using hormonal contraception [84], weight gain was reported more with DMPA users (OR, 2.3). There is also some evidence that obese adolescent users of DMPA may gain more weight compared to those with normal BMI [85], and weight gain is an important reason for discontinuation of DMPA in this age group [86]. LNG implants have also been implicated in weight gain [87].

In a recent review to evaluate the potential association between POCs and alterations in body weight, the authors could find little evidence of weight gain when using POCs [88]. In this review, 1 trial examined progestogens-only pills, 10 studies examined DMPA and 4 studies addressed Norplant, comparing POCs with combination contraceptives, no hormonal contraceptives and with other POCs.

Only 4 of the 15 studies showed a significant difference in body composition change or weight change [89–92] and these differences were shown in comparison to groups using no hormonal contraceptives. The actual mean weight gain was less than 2 kg for most studies for 6–12 months. More weight gain was noted at 2 and 3 years but was similar for both groups, reinforcing our previous discussion on natural changes in weight over a period of time. Therefore, appropriate counselling regarding the degree of weight gain associated with POCs based on current evidence will help to reduce discontinuation of POCs and unintended pregnancies.

Safety of Hormonal Contraceptives in Obese Women

Obesity is associated with a diverse array of health hazards including hypertension [93,94], diabetes mellitus [95], dyslipidaemia [96], heart disease [97], stroke [98] venous thrombosis [99,100], hepatobiliary disease [101] and cancer [102,103]. It is, therefore imperative that an obese woman, who chooses a contraceptive method, does so without increasing her health risks.

Obesity, Contraception and Cardiovascular Disease

Potential mechanisms by which contraceptives can impact on cardiovascular disease involve its effects on vasculature and metabolic parameters (lipids and glucose). Current users of oral contraceptives had a moderately increased risk of hypertension, which decreased quickly with the cessation of the drug [104]. Oral contraceptives use, especially in smokers, may be associated with increased levels of fibrinogen and intravascular fibrin deposition [105], factors which have a role in arterial thrombotic diseases. In general, the oestrogen component of the oral contraceptive pill causes modest increases in triglycerides, which is offset by increases in HDL and lowering of LDL cholesterol [106]. The androgenic progestogens (norgestrel and LNG) usually increase the serum LDL and lower serum HDL concentrations but the newer progestogens such as desogestrel appear to be more favourable [107]. Oral contraceptives can also affect carbohydrate metabolism, mainly through the actions of progestogens. Studies have shown insulin resistance, glucose intolerance and increased risk of diabetes mellitus, especially with POCs [108,109]. The big question is whether these statistically significant changes in fibrinogen, lipid and glucose levels in hormonal contraceptive users are clinically relevant, especially in an obese population. Unfortunately, there are minimal data at present regarding the effects of hormonal contraception on the above parameters in the obese population, as this group has been traditionally excluded from most studies.

Some, but not all studies, report that oral contraceptives may be associated with an increased risk of myocardial infarction and stroke. In a study to assess whether current use of newer low-dose oral contraceptives increased the risk of myocardial infarction, only women who were heavy smokers (defined in this study as smoking ≥ 25 cigarettes/day) were at increased risk of myocardial infarction, with no evidence of increased risk in non-smokers or light smokers [110]. However, a meta-analysis of 10 studies suggested an overall doubling of cardiovascular mortality, mainly driven by coronary heart disease in women using low-dose COC (less than 50 mcg of ethinyl oestradiol) [111]. Low-dose oral contraceptives may also be associated with a small increase in risk of ischaemic stroke, but the absolute risk is very low [112]. Myocardial infarction and stroke are rare events in women of reproductive age group; therefore, even a doubling of this risk would still result in a very low attributable risk. With the currently available evidence, obese women who use combined hormonal contraceptives do not seem to have a higher risk of cardiovascular disease (coronary disease or stroke) compared to obese non-users, though data are limited [113]. Currently, there is no evidence for any detrimental metabolic or cardiovascular effects of non-hormonal contraception, barrier methods, copper IUDs and sterilisation.

Obesity, Contraception and VTE

Oral contraceptive use (both high- and low-dose oestrogen) has been associated with VTE [114–116]. The risk of VTE amongst COC users is twice that of non users across all brands (9–10/10,000 woman years versus 4–5/10,000 woman years) [117]. The risks of VTE associated with transdermal contraceptive patch compared to COC are not clear. In a nested case–control study, the risk for non fatal VTE with a transdermal contraceptive patch was closely similar to COC containing 35 mcg of EE and norgestimate [118]. Some, but not all studies suggest that the risk of VTE with third generation progestogens (desogestrel and gestodene but not norgestimate) is slightly higher than the second-generation progestogens (LNG) [119–121]. Even if this difference exists between different progestogens, the absolute risk appears relatively small. Currently, there are no data on the VTE risk associated with vaginal contraceptive ring.

Obesity is an independent risk factor for VTE [122] and the likelihood of VTE increases when two or more risk factors are combined. Among oral contraceptive users, obesity appears to increase the risk of VTE compared to non-obese women [40,123,124]. However, the absolute risk of VTE is still very low in healthy women, and in general, the benefits of oral contraceptives in preventing unintended pregnancy far outweighs the risks of VTE.

Obesity, Contraception and Cancer

Oral contraceptive use has been associated with increase in risk of cervical cancer [125], conflicting results with breast cancer [126–129] and decrease in risk of uterine [128], ovarian [130,131] and colorectal cancer [128]. Interestingly, the benefits for ovarian and endometrial cancers seem to persist for many years after stopping contraception [132,133]. Obesity, per se, has been associated with increased risk of cancer, especially breast, endometrial, ovarian and colorectal cancer [103,134,135]. It is, therefore, theoretically possible that oral contraceptive use in obese women may have a significant effect on incident cancer.

Data from the Royal College of General Practitioner's oral contraceptive study [128] showed statistically significant trends of decreasing risk of endometrial and ovarian malignancies but increasing risk of cervical cancer in ever users of oral contraceptive pills. The authors concluded that oral contraception was not associated with an overall increased risk of cancer and may even produce a net public health gain. While the data on this large inception cohort study were adjusted for potentially confounding factors such as age, smoking, parity and social class, there was no data on BMI or lifestyle variables. It is not clear at this point whether obese women would benefit from oral contraceptive use as cancer prevention strategies [136] but it is reassuring to see the trends in ovarian and endometrial malignancies, for which obesity is an independent risk factor.

It is widely accepted that intrauterine contraceptive devices (IUCDs) are not implicated in cancer [137,138]. There is no evidence to suggest that inert IUDs are associated with cervical cancer [139] nor has there been any difference in premalignant or malignant cervical pathologies between copper-containing IUDs and inert IUDs [140].

Contraceptive Issues After Bariatric Surgery

The demand for bariatric surgery has greatly increased in recent times, as it is believed to be the most effective treatment method for the morbidly obese. The incidence of bariatric surgery in United States increased by 800% between 1998 and 2005, predominantly accounted for by women of reproductive age group [141]. Since the chances of fertility increases after bariatric surgery [142], such women are at higher risk of unintended pregnancies. The main concerns of such an unintended pregnancy in this group of women are the risks associated with maternal and foetal outcomes due to the nutritional effects of weight loss surgery [143,144]. The general consensus is that pregnancy should be avoided for 12–24 months after bariatric surgery, coinciding with the time when significant weight loss and postoperative complications occur [141,145].

There are legitimate concerns regarding the efficacy and safety of hormonal contraception in women who have undergone bariatric surgery. Data on pharmacokinetics, efficacy and safety of oral contraceptive drugs after bariatric surgery was extremely limited to a couple of pharmacokinetic studies, some observational studies and a single case report [146]. In a study evaluating the pharmacokinetics of two commonly used progestogens (Norethisterone and LNG) in morbidly obese women after jejunoileal bypass compared to healthy controls, the mean plasma levels of both the progestogens were lower in the obese surgical patients at 1–8 h after ingestion compared to controls [147]. Whether this translates into contraceptive failure is not clear but the authors of this study recommend that low-dose progestogens-only minipills should not be used after jejunoileal bypass. This procedure is not performed nowadays, and there are no pharmacokinetic studies in more modern bypass procedures at this point of time. Three case reports on young women who had sub-dermal etonorgestrel (ENG)-releasing implant (Implanon) prior to Roux-en-Y bypass surgery showed ENG concentrations sufficient to inhibit ovulation until 8 months after insertion and may therefore be a safe contraceptive choice for women undergoing bariatric surgery [148].

In terms of efficacy, there are concerns around malabsorption of oral contraceptives, especially in women undergoing malabsorptive procedures like Roux-en-Y gastric bypass and biliopancreatic diversion with duodenal switch. In a prospective study of 40 women who underwent biliopancreatic diversion [149], 2 out of 9 patients, who post-operatively used oral contraceptives only, had their first pregnancy after surgery while still using the same contraception that they had used pre-operatively. There does not appear to be any particular concern with restrictive procedures. In a study of 215 morbidly obese women who had agreed to be on contraception for 2 years after laparoscopic adjustable gastric banding [150], 7 women had unexpected pregnancies but all these women were using unreliable methods like periodic abstinence. There were no pregnancies observed in patients using oral contraception.

In terms of safety of oral contraceptives after bariatric surgery, evidence is limited to a single case report [151] of an 18-year-old lady who developed an acute ischaemic stroke 4 months after a Roux-en-Y gastric bypass procedure. She was on oral contraceptive at the time of the event but there were other confounding factors including tobacco smoking and recreational drug use. Apart from the risk of cardiovascular disease, the other concern is the risk of VTE in post-operative users of contraceptive pills after bariatric surgery. In a survey of the members of American Society of bariatric surgery, the self-reported incidence of deep vein thrombosis and pulmonary embolism was 2.63% and 0.95%, respectively, even though routine prophylaxis was used by more than 95% of the members [152]. Finally, there is a theoretical concern to using DMPA in women who have undergone bariatric surgery. There is some evidence for increased bone turnover and reduced bone mineral density (BMD) in women who have undergone Roux-en-Y gastric bypass procedure [153,154]. DMPA use is associated with small and usually reversible changes in BMD [155] but at this point, it is not clear whether DMPA use further aggravates bone loss in women undergoing bariatric surgery.

To summarise, though there is no evidence for a significant decrease in oral contraceptive effectiveness post-bariatric surgery, there are potential concerns regarding oral contraceptive efficacy in women undergoing malabsorptive procedures, more so if they have long-term diarrhoea and/or vomiting. There does not appear to be an increased risk of cardiovascular disease in women using oral contraceptives post-bariatric surgery but there are potential concerns regarding risk of VTE, especially if there is prolonged immobilisation. The relationship between DMPA use and bone loss in post-bariatric surgery patients needs further investigation.

IUCDs in Obese Women

IUD insertion presents some challenges in the obese woman. It may be difficult to ascertain the size and direction of the uterus. In addition, visualisation of the cervix may be difficult. Simple measures like use of a larger speculum, placing a condom with the tip removed over the blades of the speculum, comfortable positioning of the patient and use of ultrasound during the device insertion can overcome the above-mentioned problems [106,156]. These difficulties are negated to a large extent by the long-term highly effective contraception with these devices. There is also evidence of some benefit with the LNG intrauterine system in selected obese women with abnormal uterine bleeding [157]. To summarise, there are no specific contraindications to hormone containing devices or copper IUDs in the obese woman, and such devices, apart from being effective contraceptives, may produce additional benefit in obese women with abnormal uterine bleeding or endometrial hyperplasia.

Sterilisation Procedures in Obese Women

Obesity can complicate tubal sterilisation procedures [158–160]. In a large prospective multicentre cohort study of 9475 women who underwent interval laparoscopic sterilisation, obesity was an independent predictor for one or more complications [159]. While there is evidence with a large international data set that obesity can be associated with higher incidence of surgical difficulties, technical failure rate and longer surgical times compared to non-obese controls, none of the above issues led to serious consequences [158]. In addition, not all studies have shown a link between obesity and poor outcomes after laparoscopic sterilisation procedures. In a retrospective study of 248 consecutive patients undergoing laparoscopic tubal sterilisation [161], there were no differences in complications, mean operating times or blood loss between obese women and non-obese controls. However, morbidly obese women have their own set of problems including theatre requirements and hazards related to anaesthetic procedures. In such cases, vasectomy for the woman's partner seems to be the best available option.

What Do the Guidelines Say?

The UK Medical Eligibility Criteria (UKMEC) are a set of evidence-based recommendations designed to help both the clinician and the patient to select the most appropriate method of contraception for various clinical conditions without imposing unnecessary restrictions. The evidence was updated in 2009[2]. A *UK Category 1* indicates that there is no restriction for use. A *UK Category 2* indicates that the method can generally be used, but more careful follow-up may be required. A contraceptive method with a *UK Category 3* can be used; however, this may require expert clinical judgement and/or referral to a specialist contraceptive provider, since use of the method is not usually recommended unless other methods are not available or not acceptable. A *UK Category 4* indicates that use poses an unacceptable health risk. Apart from BMI categories, we have summarised selected clinical conditions that may have a particular relevance to obesity (Table 6.3).

[2] http://www.fsrh.org/pdfs/UKMEC2009.pdf

Table 6.3 UKMEC Criteria (2009) for Contraceptive Use – Obesity and Selected Clinical Conditions that Are of Particular Relevance to Obese Women

Condition	CHC	POP	DMPA/ NET-EN	IMP	Cu-IUD	LNG-IUD
BMI[a] ≥ 30–34	2	1	1	1	1	1
BMI ≥ 35	3	1	1	1	1	1
MRF[b]	3/4	2	3	2	1	2
h/o of VTE[c]	4	2	2	2	1	2
Current VTE on anticoagulants	4	2	2	2	1	2
Controlled hypertension	3	1	2	1	1	1
SBP[d] 140–159 or DBP[e] 90–94	3	1	1	1	1	1
SBP ≥ 160 or DBP ≥ 95	4	1	2	1	1	1
Vascular disease	4	2	3	2	1	2
Major surgery with prolonged immobility	4	2	2	2	1	2
History of gestational diabetes	1	1	1	1	1	1

CHC, combined hormonal contraception including COC, combined transdermal patch and vaginal ring; POP, progestogen-only pill; DMPA/NET-EN, depo medroxyprogesterone acetate/Norethisterone enanthate; IMP, implant (progestogen only); Cu-IUD, copper IUD; LNG-IUD, levonorgestrel-releasing IUD.
[a]BMI – body mass index.
[b]MRF – multiple risk factors for cardiovascular disease (older age, diabetes, hypertension, obesity, smoking).
[c]h/o of VTE – history of VTE.
[d]SBP – systolic blood pressure.
[e]DBP – diastolic blood pressure.

Table 6.4 Selected Categories of Relevance to Obesity and Contraception Based on the US Medical Eligibility Criteria for Contraceptive Use (2010)

Condition	COC	POP	DMPA	Implant	Cu-IUD	LNG-IUD
BMI ≥30	2	1	1	1	1	1
Menarche to <18 years and BMI ≥30	2	1	2	1	1	1
Bariatric surgery, restrictive procedure[a]	1	1	1	1	1	1
Bariatric surgery, malabsorptive procedure[b,c]	3	3	1	1	1	1

[a]Vertical-banded gastroplasty, laparoscopic adjustable gastric band, laparoscopic sleeve gastrectomy.
[b]Roux-en-Y gastric bypass, biliopancreatic diversion.
[c]Rating Category 1 for combined hormonal patch or ring.

The Centers for Disease Control and Prevention created the US Medical Eligibility criteria for contraceptive use [113] from guidelines developed from the WHO. The indications and categories are largely similar to the UKMEC with a few exceptions. The salient features that are different or not mentioned in UKMEC guidelines are summarised in Table 6.4.

Conclusion

Unintended pregnancies in obese women have significant consequences for both maternal and foetal health. Unfortunately, there are too many misconceptions and too few well-designed trials to guide the clinician and the patient towards an appropriate contraceptive choice in obese women. Recently updated guidelines such as the UK and US Medical Eligibility Criteria for Contraceptive Use give useful information for the counselling clinician. Future studies assessing contraceptive efficacy and safety should include women of all BMI categories, reflecting current trends in obesity.

References

1. Norman JE, Reynolds R. The consequences of obesity and excess weight gain in pregnancy. *Proc Nutr Soc Nov.* 2011;70(4):450–456.
2. Dehlendorf C, Rodriguez MI, Levy K, Borrero S, Steinauer J. Disparities in family planning. *Am J Obstet Gynecol Mar.* 2010;202(3):214–220.
3. Bajos N, Wellings K, Laborde C, Moreau C, CSF Group. Sexuality and obesity a gender perspective: results from French national random probability survey of sexual behaviours. *BMJ.* 2010;340:c2573. doi:10.1136/bmj.c2573.
4. Kulie T, Slattengren A, Redmer J, Counts H, Eglash A, Schrager S. Obesity and women's health: an evidence-based review. *J Am Board Fam Med.* 2011;24(1):75–85.
5. Wisse BE. The inflammatory syndrome: the role of adipose tissue cytokines in metabolic disorders linked to obesity. *J Am Soc Nephrol.* 2004;15(11):2792–2800.
6. Irizar A, Barnett CR, Flatt PR, Loannides C. Defective expression of cytochrome P450 proteins in the liver of the genetically obese Zucker rat. *Eur J Pharmacol.* 1995;293(4): 385–393.
7. Cheng Q, Aleksunes LM, Manautou JE, et al. Drug-metabolizing enzyme and transporter expression in a mouse model of diabetes and obesity. *Mol Pharm.* 2008;5(1): 77–91.
8. Henegar JR, Bigler SA, Henegar LK, Tyagi SC, Hall JE. Functional and structural changes in the kidney in the early stages of obesity. *J Am Soc Nephrol.* 2001;12(6): 1211–1217.
9. Li WM, Chou YH, Li CC, et al. Association of body mass index and urine pH in patients with urolithiasis. *Urol Res.* 2009;37(4):193–196.
10. Edelman AB, Carlson NE, Cherala G, et al. Impact of obesity on oral contraceptive pharmacokinetics and hypothalamic–pituitary–ovarian activity. *Contraception.* 2009; 80(2):119–127.
11. Metwally M, Ong KJ, Ledger WL, Li TC. Does high body mass index increase the risk of miscarriage after spontaneous and assisted conception? A meta-analysis of the evidence. *Fertil Steril.* 2009;90(3):714.
12. Ehrenberg HM, Dierker L, Milluzzi C, Mercer BM. Prevalence of maternal obesity in an urban center. *Am J Obstet Gynecol.* 2002;187(5):1189.
13. Robinson HE, O'Connell CM, Joseph KS, McLeod NL. Maternal outcomes in pregnancies complicated by obesity. *Obstet Gynecol.* 2005;106(6):1357.

14. Sibai BM, Gordon T, Thom E, Caritis SN, Klebanoff M, Mc Nellis D, et al. Risk factors for preeclampsia in healthy nulliparous women: a prospective multicenter study. The National Institute of Child Health and Human Development Network of Maternal–Fetal Medicine Units. *Am J Obstet Gynecol.* 1995;172(2 Pt 1):642.
15. O'Brien TE, Ray JG, Chan WS. Maternal body mass index and the risk of preeclampsia: a systematic overview. *Epidemiology.* 2003;14(3):368.
16. Cnattingius S, Bergström R, Lipworth L, Kramer MS. Prepregnancy weight and the risk of adverse pregnancy outcomes. *N Engl J Med.* 1998;338(3):147.
17. Ehrenberg HM, Iams JD, Goldenberg RL, et al. Eunice Kennedy Shriver National Institute of Child Health and Human Development (NICHD) Maternal–Fetal Medicine Units Network (MFMU). Maternal obesity, uterine activity, and the risk of spontaneous preterm birth. *Obstet Gynecol.* 2009;113(1):48.
18. Johnson JW, Longmate JA, Frentzen B. Excessive maternal weight and pregnancy outcome. *Am J Obstet Gynecol.* 1992;167(2):353.
19. Stotland NE, Washington AE, Caughey AB. Prepregnancy body mass index and the length of gestation at term. *Am J Obstet Gynecol.* 2007;197(4):378.e1.
20. Denison FC, Price J, Graham C, Wild S, Liston WA. Maternal obesity, length of gestation, risk of postdates pregnancy and spontaneous onset of labour at term. *BJOG.* 2008;115(6):720.
21. Nuthalapaty FS, Rouse DJ, Owen J. The association of maternal weight with cesarean risk, labor duration, and cervical dilation rate during labor induction. *Obstet Gynecol.* 2004;103(3):452.
22. Vahratian A, Zhang J, Troendle JF, Savitz DA, Siega-Riz AM. Maternal prepregnancy overweight and obesity and the pattern of labor progression in term nulliparous women. *Obstet Gynecol.* 2004;104(5 Pt 1):943.
23. Kominiarek MA, Zhang J, Vanveldhuisen P, Troendle J, Beaver J, Hibbard JU. Contemporary labor patterns: the impact of maternal body mass index. *Am J Obstet Gynecol.* 2011;205(3):244.e1–244.e8.
24. Hood DD, Dewan DM. Anesthetic and obstetric outcome in morbidly obese parturients. *Anesthesiology.* 1993;79(6):1210.
25. Soens MA, Birnbach DJ, Ranasinghe JS, van Zundert A. Obstetric anesthesia for the obese and morbidly obese patient: an ounce of prevention is worth more than a pound of treatment. *Acta Anaesthesiol Scand.* 2008;52(1):6.
26. Poobalan AS, Aucott LS, Gurung T, Smith WC, Bhattacharya S. Obesity as an independent risk factor for elective and emergency caesarean delivery in nulliparous women – systematic review and meta-analysis of cohort studies. *Obes Res.* 2009;10:28.
27. Jensen DM, Damm P, Sørensen B, et al. Pregnancy outcome and prepregnancy body mass index in 2459 glucose-tolerant Danish women. *Am J Obstet Gynecol.* 2003;189(1):239.
28. Kaiser PS, Kirby RS. Obesity as a risk factor for cesarean in a low-risk population. *Obstet Gynecol.* 2001;97(1):39.
29. Witter FR, Caulfield LE, Stoltzfus RJ. Influence of maternal anthropometric status and birth weight on the risk of cesarean delivery. *Obstet Gynecol.* 1995;85(6):947.
30. Sebire NJ, Jolly M, Harris JP, et al. Maternal obesity and pregnancy outcome: a study of 287,213 pregnancies in London. *Int J Obes Relat Metab Disord.* 2001;25(8): 1175.
31. Perlow JH, Morgan MA. Massive maternal obesity and perioperative cesarean morbidity. *Am J Obstet Gynecol.* 1994;170(2):560.
32. Myles TD, Gooch J, Santolaya J. Obesity as an independent risk factor for infectious morbidity in patients who undergo cesarean delivery. *Obstet Gynecol.* 2002;100(5 Pt 1):959.

33. Hilson JA, Rasmussen KM, Kjolhede CL. Maternal obesity and breast-feeding success in a rural population of white women. *Am J Clin Nutr*. 1997;66(6):1371.
34. Li R, Jewell S, Grummer-Strawn L. Maternal obesity and breast-feeding practices. *Am J Clin Nutr*. 2003;77(4):931.
35. Rasmussen KM, Hilson JA, Kjolhede CL. Obesity as a risk factor for failure to initiate and sustain lactation. *Adv Exp Med Biol*. 2002;503:217.
36. Frentzen BH, Dimperio DL, Cruz AC. Maternal weight gain: effect on infant birth weight among overweight and average-weight low-income women. *Am J Obstet Gynecol*. 1988; 159(5):1114.
37. Hope P, Breslin S, Lamont L, et al. Fatal shoulder dystocia: a review of 56 cases reported to the Confidential Enquiry into Stillbirths and Deaths in Infancy. *Br J Obstet Gynaecol*. 1998;105(12):1256.
38. Chu SY, Kim SY, Lau J, et al. Maternal obesity and risk of stillbirth: a metaanalysis. *Am J Obstet Gynecol*. 2007;197(3):223.
39. Hull HR, Dinger MK, Knehans AW, Thompson DM, Fields DA. Impact of maternal body mass index on neonate birthweight and body composition. *Am J Obstet Gynecol*. 2008;198(4):416.e1.
40. Abdollahi M, Cushman M, Rosendaal FR. Obesity: risk of venous thrombosis and the interaction with coagulation factor levels and oral contraceptive use. *Thromb Haemost*. 2003;89(3):493–498.
41. Picardo CM, Nichols M, Edelman A, Jensen JT. Women's knowledge and sources of information on the risks and benefits of oral contraception. *J Am Med Womens Assoc Spring*. 2003;58(2):112–116.
42. Rosenberg M. Weight change with oral contraceptive use and during the menstrual cycle. Results of daily measurements. *Contraception*. 1998;58(6):345–349.
43. Holt VL, Cushing-Haugen KL, Daling JR. Body weight and risk of oral contraceptive failure. *Obstet Gynecol*. 2002;99(5 Pt 1):820–827.
44. Holt VL, Scholes D, Wicklund KG, Cushing-Haugen KL, Daling JR. Body mass index, weight, and oral contraceptive failure risk. *Obstet Gynecol*. 2005;105(1):46–52.
45. Vessey M. Oral contraceptive failures and body weight: findings in a large cohort study. *J Fam Plann Reprod Health Care*. 2001;27(2):90–91.
46. Brunner LR, Hogue CJ. The role of body weight in oral contraceptive failure: results from the 1995 national survey of family growth. *Ann Epidemiol*. 2005;15(7):492–499.
47. Brunner Huber LR, Toth JL. Obesity and oral contraceptive failure: findings from the 2002 National Survey of Family Growth. *Am J Epidemiol*. 2007;166(11):1306–1311.
48. Lopez LM, Grimes DA, Chen-Mok M, Westhoff C, Edelman A, Helmerhorst FM. Hormonal contraceptives for contraception in overweight or obese women. *Cochrane Database Syst Rev*. 2010;7(7):CD008452
49. Edelman AB, Cherala G, Stanczyk FZ. Metabolism and pharmacokinetics of contraceptive steroids in obese women: a review. *Contraception*. 2010;82(4):314–323.
50. Wisén O, Hellström PM. Gastrointestinal motility in obesity. *J Intern Med*. 1995;237(4): 411–418.
51. Dubois A. Obesity and Gastric emptying. *Gastroenterology*. 1983;84(4):875–876.
52. Tosetti C, Corinaldesi R, Stanghellini V, et al. Gastric emptying of solids in morbid obesity. *Int J Obes Relat Metab Disord*. 1996;20(3):200–205.
53. Wisén O, Johansson C. Gastrointestinal function in obesity: motility, secretion, and absorption following a liquid test meal. *Metabolism*. 1992;41(4):390–395.
54. Bolt HM. Interactions between clinically used drugs and oral contraceptives. *Environ Health Perspect*. 1994;102(suppl 9):35–38.

55. Dobrinska MR. Enterohepatic circulation of drugs. *J Clin Pharmacol.* 1989;29(7):577–580.
56. Geier A, Dietrich CG, Grote T, et al. Characterization of organic anion transporter regulation, glutathione metabolism and bile formation in the obese Zucker rat. *J Hepatol.* 2005;43(6):1021–1030.
57. Forbes GB, Welle SL. Lean body mass in obesity. *Int J Obes.* 1983;7(2):99–107.
58. Chagnac A, Weinstein T, Korzets A, Ramadan E, Hirsch J, Gafter U. Glomerular hemodynamics in severe obesity. *Am J Physiol Renal Physiol.* 2000;278(5):F817–F822.
59. Westhoff CL, Torgal AH, Mayeda ER, Pike MC, Stanczyk. Pharmacokinetics of a combined oral contraceptive in obese and normal-weight women. *Contraception.* 2010;81(6): 474–480.
60. Segall-Gutierrez P, Taylor D, Liu X, Stanzcyk F, Azen S, Mishell Jr DR. Follicular development and ovulation in extremely obese women receiving depo-medroxyprogesterone acetate subcutaneously. *Contraception.* 2010;81(6):487–495.
61. Kaulhausen H, Klingsiek L, Breuer H. Changes of the renin–angiotensin–aldosterone system under contraceptive steroids. Contribution to the etiology of hypertension under hormonal contraceptives. *Fortschr Med.* 1976;94(33):1925–1930.
62. Corvol P, Elkik F, Feneant M, Oblin ME, Michaud A, Claire M. Effect of progesterone and progestogens on water and salt metabolism. In: Bardin CW, Milgrom E, Mauvais-Jarvis P, eds. *Progesterone and Progestogens.* New York, NY: Raven Press; 1983.
63. Mattsson C, Olsson T. Estrogens and glucocorticoid hormones in adipose tissue metabolism. *Curr Med Chem.* 2007;14(27):2918–2924.
64. Nelson AL. Combined oral contraceptives. In: Hatcher RA, Trussell J, Nelson AL, Cates W, Stewart FH, Kowal D, et al. eds. *Contraceptive Technology.* 19th ed. New York, NY: Ardent Media, Inc; 2007:193–270.
65. Reubinoff BE, Grubstein A, Meirow D, Berry E, Schenker JG, Brzezinski A. Effects of low-dose estrogen oral contraceptives on weight, body composition, and fat distribution in young women. *Fertil Steril.* 1995;63(3):516.
66. Turner R. Most British women use reliable contraceptive methods, but many fear health risks from use. *Fam Plann Perspect.* 1994;26:183–184.
67. Gaudet LM, Kives S, Hahn PM, Reid RL. What women believe about oral contraceptives and the effect of counseling. *Contraception.* 2004;69:31–36.
68. Emans SJ, Grace E, Woods ER, Smith DE, Kelin K, Merola J. Adolescents' compliance with the use of oral contraceptives. *JAMA.* 1987;257:3377–3381.
69. Oddens BJ. Women's satisfaction with birth control: a population survey of physical and psychological effects of oral contraceptives, intrauterine devices, condoms, natural family planning, and sterilization among 1466 women. *Contraception.* 1999;59:277–286.
70. Le MG, Laveissiere MN, Pelissier C. Factors associated with weight gain in women using oral contraceptives: results of a French 2001 opinion poll survey conducted on 1665 women. *Gynecol Obstet Fertil.* 2003;31:230–239.
71. Rosenberg M. Weight change with oral contraceptive use and during the menstrual cycle. Results of daily measurements. *Contraception.* 1998;58:345–349.
72. Wysocki S. A survey of American women regarding the use of oral contraceptives and weight gain [abstract].. *Int J Gynecol Obstet.* 2000;70:114.
73. Rosenberg MJ, Waugh MS, Meehan TE. Use and misuse of oral contraceptives: risk indicators for poor pill taking and discontinuation. *Contraception.* 1995;51:283–288.
74. Rosenberg MJ, Waugh MS. Oral contraceptive discontinuation: a prospective evaluation of frequency and reasons. *Am J Obstet Gynecol.* 1998;179(3 Pt 1):577–582.
75. Gallo MF, Lopez LM, Grimes DA, Schulz KF, Helmerhorst FM. Combination contraceptives: effects on weight. *Cochrane Database Syst Rev.* 2011;7:9:CD003987.

76. Lindh I, Ellström AA, Milsom I. The long-term influence of combined oral contraceptives on body weight. *Hum Reprod.* 2011;26(7):1917–1924.
77. Sheehan TJ, DuBrava S, DeChello LM, Fang Z. Rates of weight change for black and white Americans over a twenty year period. *Int J Obes Relat Metab Disord.* 2003;27(4): 498–504.
78. Nooyens AC, Visscher TL, Verschuren WM, et al. Age, period and cohort effects on body weight and body mass index in adults: The Doetinchem Cohort Study. *Public Health Nutr.* 2009;12(6):862–870.
79. Flegal KM. Obesity, overweight, hypertension, and high blood cholesterol: the importance of age. *Obes Res.* 2000;8(9):676–677.
80. Gupta S. Weight gain on the combined pill – is it real? *Hum Reprod Update.* 2000;6(5): 427–431.
81. Milsom I, Lete I, Bjertnaes A, et al. Effects on cycle control and bodyweight of the combined contraceptive ring, NuvaRing, versus an oral contraceptive containing 30 microg ethinyl estradiol and 3 mg drospirenone. *Hum Reprod.* 2006;21(9):2304–2311.
82. Berenson AB, Rahman M. Changes in weight, total fat, percent body fat, and central-to-peripheral fat ratio associated with injectable and oral contraceptive use. *Am J Obstet Gynecol.* 2009;200(3):329.e1–329.e8.
83. Beksinska ME, Smit JA, Kleinschmidt I, Milford C, Farley TM. Prospective study of weight change in new adolescent users of DMPA, NET-EN, COCs, nonusers and discontinuers of hormonal contraception. *Contraception.* 2010;81(1):30–34.
84. Berenson AB, Odom SD, Breitkopf CR, Rahman M. Physiologic and psychologic symptoms associated with use of injectable contraception and 20 μg oral contraceptive pills. *Am J Obstet Gynecol.* 2008;199(4):351e1–351e12.
85. Curtis KM, Ravi A, Gaffield ML. Progestogen-only contraceptive use in obese women. *Contraception.* 2009;80(4):346–354.
86. Bonny AE, Britto MT, Huang B, Succop P, Slap GB. Weight gain, adiposity, and eating behaviors among adolescent females on depo medroxyprogesterone acetate (DMPA). *J Pediatr Adolesc Gynecol.* 2004;17(2):109–115.
87. Sivin I. Risks and benefits, advantages and disadvantages of levonorgestrel-releasing contraceptive implants. *Drug Saf.* 2003;26(5):303–335.
88. Lopez LM, Edelman A, Chen-Mok M, Trussell J, Helmerhorst FM. Progestogen-only contraceptives: effects on weight. *Cochrane Database Syst Rev.* 2011;13(4):CD008815.
89. Salem HT, Abdullah KA, Shaaban MM. Norplant and lactation. In: Shaaban MM, ed. *The Norplant Subdermal Contraceptive System. Proceedings of the Symposium on Longterm Subdermal Contraceptive Implants.* Assiut, Egypt: Assiut University; 1984:139–149.
90. Sule S, Shittu O. Weight changes in clients on hormonal contraceptives in Saria, Nigeria. *Afr J Reprod Health.* 2005;9(2):92–100.
91. Bonny AE, Secic M, Cromer BA. A longitudinal comparison of body composition changes in adolescent girls receiving hormonal contraception. *J Adolesc Health.* 2009; 45(4):423–425.
92. Pantoja M, Medeiros T, Baccarin MC, Morais SS, Bahamondes L, Fernandes AM. Variations in body mass index of users of depo-medroxyprogesterone acetate as a contraceptive. *Contraception.* 2010;81(2):107–111.
93. Alpert MA, Hashimi MW. Obesity and the heart. *Am J Med Sci.* 1993;306(2):117–123.
94. Huang Z, Willett WC, Manson JE, et al. Body weight, weight change, and risk for hypertension in women. *Ann Intern Med.* 1998;128(2):81–88.
95. Colditz GA, Willett WC, Rotnitzky A, Manson JE. Weight gain as a risk factor for clinical diabetes mellitus in women. *Ann Intern Med.* 1995;122(7):481–486.

96. Grundy SM, Barnett JP. Metabolic and health complications of obesity. *Dis Mon.* 1990;36(12):641–731.
97. Yusuf S, Hawken S, Ounpuu S, et al. Effect of potentially modifiable risk factors associated with myocardial infarction in 52 countries (the INTERHEART study): case–control study. *Lancet.* 2004;364(9438):937–952.
98. Rexrode KM, Hennekens CH, Willett WC, et al. A prospective study of body mass index, weight change, and risk of stroke in women. *JAMA.* 1997;277(19):1539–1545.
99. Ageno W, Becattini C, Brighton T, Selby R, Kamphuisen PW. Cardiovascular risk factors and venous thromboembolism: a meta-analysis. *Circulation.* 2008;117(1):93–102.
100. Holst AG, Jensen G, Prescott E. Risk factors for venous thromboembolism: results from the Copenhagen City Heart Study. *Circulation.* 2010;121(17):1896–1903.
101. Stampfer MJ, Maclure KM, Colditz GA, Manson JE, Willett WC. Risk of symptomatic gallstones in women with severe obesity. *Am J Clin Nutr.* 1992;55(3):652–658.
102. Deslypere JP. Obesity and cancer. *Metabolism.* 1995;44(9 suppl 3):24–27.
103. Schapira DV, Clark RA, Wolff PA, Jarrett AR, Kumar NB, Aziz NM. Visceral obesity and breast cancer risk. *Cancer.* 1994;74(2):632–639.
104. Chasan-Taber L, Willett WC, Manson JE, et al. Prospective study of oral contraceptives and hypertension among women in the United States. *Circulation.* 1996;94(3):483.
105. Scarabin PY, Vissac AM, Kirzin JM, et al. Elevated plasma fibrinogen and increased fibrin turnover among healthy women who both smoke and use low-dose oral contraceptives – a preliminary report. *Thromb Haemost.* 1999;82(3):1112.
106. Kelsey B. Contraceptive for obese women: considerations. *Nurse Pract.* 2010;35(3):24–31.
107. Lobo RA, Skinner JB, Lippman JS, Cirillo SJ. Plasma lipids and desogestrel and ethinyl estradiol: a meta-analysis. *Fertil Steril.* 1996;65(6):1100.
108. Krauss RM, Burkman Jr RT. The metabolic impact of oral contraceptives. *Am J Obstet Gynecol.* 1992;167(4 Pt 2):1177.
109. Kjos SL, Peters RK, Xiang A, Thomas D, Schaefer U, Buchanan TA. Contraception and the risk of type 2 diabetes mellitus in Latina women with prior gestational diabetes mellitus. *AMA.* 1998;280(6):533.
110. Rosenberg L, Palmer JR, Rao RS, Shapiro S. Low-dose oral contraceptive use and the risk of myocardial infarction. *Arch Intern Med.* 2001;161(8):1065.
111. Baillargeon JP, McClish DK, Essah PA, Nestler JE. Association between the current use of low-dose oral contraceptives and cardiovascular arterial disease: a meta-analysis. *J Clin Endocrinol Metab.* 2005;90(7):3863–3870.
112. Gillum LA, Mamidipudi SK, Johnston SC. Ischemic stroke risk with oral contraceptives: a meta-analysis. *AMA.* 2000;284(1):72.
113. Centers for Disease Control and Prevention (CDC). US Medical Eligibility Criteria for contraceptive use. *MMWR Recomm Rep.* 2010;59(RR-4):1–86.
114. Vandenbroucke JP, Rosing J, Bloemenkamp KW, et al. Oral contraceptives and the risk of venous thrombosis. *N Engl J Med.* 2001;344(20):1527.
115. Douketis JD, Ginsberg JS, Holbrook A, Crowther M, Duku EK, Burrows RF. A reevaluation of the risk for venous thromboembolism with the use of oral contraceptives and hormone replacement therapy. *Arch Intern Med.* 1997;157(14):1522.
116. Bloemenkamp KW, Rosendaal FR, Büller HR, Helmerhorst FM, Colly LP, Vandenbroucke JP. Risk of venous thrombosis with use of current low-dose oral contraceptives is not explained by diagnostic suspicion and referral bias. *Arch Intern Med.* 1999;159(1):65.

117. Dinger JC, Heinemann LA, Kühl-Habich D. The safety of a drospirenone-containing oral contraceptive: final results from the European Active Surveillance Study on oral contraceptives based on 142,475 women-years of observation. *Contraception*. 2007;75(5): 344–354.
118. Jick S, Kaye JA, Li L, Jick H. Further results on the risk of nonfatal venous thromboembolism in users of the contraceptive transdermal patch compared to users of oral contraceptive containing norgestimate and 35 microg of ethinyl estradiol. *Contraception*. 2007;76(1):4–7.
119. Spitzer WO, Lewis MA, Heinemann LA, Thorogood M, MacRae KD. Third generation oral contraceptives and risk of venous thromboembolic disorders: an international case–control study. Transnational Research Group on Oral Contraceptives and the Health of Young Women. *BMJ*. 1996;312(7023):83.
120. Jick H, Jick SS, Gurewich V, Myers MW, Vasilakis C. Risk of idiopathic cardiovascular death and nonfatal venous thromboembolism in women using oral contraceptives with differing progestogen components. *Lancet*. 1995;346(8990):1589.
121. Lidegaard Ø, Løkkegaard E, Svendsen AL, Agger C. Hormonal contraception and risk of venous thromboembolism: national follow-up study. *BMJ*. 2009;339:b2890.
122. Burkman R, Darney P. Contraception for women with risk factors for venous and arterial thrombosis. *Dialogues Contracept*. 2005;9(4):4–6.
123. Sidney S, Petitti DB, Soff GA, Cundiff DL, Tolan KK, Quesenberry Jr CP. Venous thromboembolic disease in users of low-estrogen combined estrogen–progestogen oral contraceptives. *Contraception*. 2004;70(1):3.
124. Pomp ER, le Cessie S, Rosendaal FR, Doggen CJ. Risk of venous thrombosis: obesity and its joint effect with oral contraceptive use and prothrombotic mutations. *Br J Haematol*. 2007;139(2):289.
125. Smith JS, Green J, Berrington de Gonzalez A, et al. Cervical cancer and use of hormonal contraceptives: a systematic review. *Lancet*. 2003;361(9364):1159.
126. Hankinson SE, Colditz GA, Manson JE, et al. Prospective study of oral contraceptive use and risk of breast cancer (Nurses' Health Study, United States). *Cancer Causes Control*. 1997;8(1):65–72.
127. Marchbanks PA, McDonald JA, Wilson HG, et al. Oral contraceptives and the risk of breast cancer. *N Engl J Med*. 2002;346(26):2025.
128. Hannaford PC, Selvaraj S, Elliott AM, Angus V, Iversen L, Lee AJ. Cancer risk among users of oral contraceptives: cohort data from the Royal College of General Practitioner's oral contraception study. *BMJ*. 2007;335(7621):651.
129. Grabrick DM, Hartmann LC, Cerhan JR, et al. Risk of breast cancer with oral contraceptive use in women with a family history of breast cancer. *JAMA*. 2000;284(14):1791.
130. Collaborative Group on Epidemiological Studies of Ovarian Cancer, Beral V, Doll R, Hermon C, Peto R, Reeves G. Ovarian cancer and oral contraceptives: collaborative reanalysis of data from 45 epidemiological studies including 23,257 women with ovarian cancer and 87,303 controls. *Lancet*. 2008;371(9609):303.
131. Vessey M, Painter R. Oral contraceptive use and cancer. Findings in a large cohort study, 1968–2004. *Br J Cancer*. 2006;95(3):385.
132. Cogliano V, Grosse Y, Baan R, Straif K, Secretan B, El Ghissassi F. WHO International Agency for Research on Cancer. Carcinogenicity of combined oestrogen–progestogen contraceptives and menopausal treatment. *Lancet Oncol*. 2005;6(8):552–553.
133. La Vecchia C, Altieri A, Franceschi S, Tavani A. Oral contraceptives and cancer: an update. *Drug Saf*. 2001;24(10):741–754.

134. Pan SY, Johnson KC, Ugnat AM, Wen SW, Mao Y, Canadian Cancer Registries Epidemiology Research Group. Association of obesity and cancer risk in Canada. *Am J Epidemiol.* 2004;159(3):259.
135. Rapp K, Schroeder J, Klenk J, et al. Obesity and incidence of cancer: a large cohort study of over 145,000 adults in Austria. *Br J Cancer.* 2005;93(9):1062.
136. Kwon JS, Lu KH. Cost-effectiveness analysis of endometrial cancer prevention strategies for obese women. *Obstet Gynecol.* 2008;112(1):56–63.
137. Misra JS, Engineer AD, Tandon P. Cytopathological changes in human cervix and endometrium following prolonged retention of copper-bearing intrauterine contraceptive devices. *Diagn Cytopathol.* 1989;5:237–242.
138. Rivera R, Best K. Current opinion: consensus statement on intrauterine contraception. *Contraception.* 2002;65:385–388.
139. Batar I. The Szontagh IUD and cervical carcinoma (results of a 10-year follow up study). *Orv Hetil.* 1990;131:1871–1874 [in Hungarian].
140. Ganacharya S, Bhattoa HP, Batár I. Pre-malignant and malignant cervical pathologies among inert and copper-bearing intrauterine contraceptive device users: a 10-year follow-up study. *Eur J Contracept Reprod Health Care.* 2006;11(2): 89–97.
141. Maggard MA, Yermilov I, Li Z, et al. Pregnancy and fertility following bariatric surgery: a systematic review. *JAMA.* 2008;300:2286–2296.
142. Marceau P, Kaufman D, Biron S, et al. Outcome of pregnancies after biliopancreatic diversion. *Obes Surg.* 2004;14:318–324.
143. Merhi ZO. Challenging oral contraception after weight loss by bariatric surgery. *Gynecol Obstet Invest.* 2007;64:100–102.
144. Gosman GG, King WC, Schrope B, et al. Reproductive health of women electing bariatric surgery. *Fertil Steril.* 2010;94:1426–1431.
145. Patel JA, Colella JJ, Esaka E, Patel NA, Thomas RL. Improvement in infertility and pregnancy outcomes after weight loss surgery. *Med Clin North Am.* 2007;91:515–528.
146. Paulen ME, Zapata LB, Cansino C, Curtis KM, Jamieson DJ. Contraceptive use among women with a history of bariatric surgery: a systematic review. *Contraception.* 2010; 82(1):86–94.
147. Victor A, Odlind V, Kral JG. Oral contraceptive absorption and sex hormone binding globulins in obese women: effects of jejunoileal bypass. *Gastroenterol Clin North Am.* 1987;16:483–491.
148. Ciangura C, Corigliano N, Basdevant A, et al. Etonorgestrel concentrations in morbidly obese women following Roux-en-Y gastric bypass surgery: three case reports. *Contraception.* 2011;84(6):649–651.
149. Gerrits EG, Ceulemans R, van Hee R, Hendrickx L, Totté E. Contraceptive treatment after biliopancreatic diversion needs consensus. *Obes Surg.* 2003;13:378–382.
150. Weiss HG, Nehoda H, Labeck B, Hourmont K, Marth C, Aigner F. Pregnancies after adjustable gastric banding. *Obes Surg.* 2001;11:303–306.
151. Choi JY, Scarborough TK. Stroke and seizure following a recent laparoscopic Roux-en-Y gastric bypass. *Obes Surg.* 2004;14:857–860.
152. Wu EC, Barba CA. Current practices in the prophylaxis of venous thromboembolism in bariatric surgery. *Obes Surg.* 2000;10(1):7–13.
153. Wucher H, Ciangura C, Poitou C, Czernichow S. Effects of weight loss on bone status after bariatric surgery: association between adipokines and bone markers. *Obes Surg.* 2008;18:58–65.

154. Coates PS, Fernstrom JD, Fernstrom MH, Schauer PR, Greenspan SL. Gastric bypass surgery for morbid obesity leads to an increase in bone turnover and a decrease in bone mass. *J Clin Endocrinol Metab.* 2004;89:1061–1065.
155. Curtis KM, Martins SL. Progestogen-only contraception and bone mineral density: a systematic review. *Contraception.* 2006;73:470–487.
156. Grimes DA, Shields WC. Family planning for obese women: challenges and opportunities. *Contraception.* 2005;72(1):1–4.
157. Vilos GA, Marks J, Tureanu V, Abu-Rafea B, Vilos AG. The levonorgestrel intrauterine system is an effective treatment in selected obese women with abnormal uterine bleeding. *J Minim Invasive Gynecol.* 2011;18(1):75–80.
158. Chi IC, Wilkens L. Interval tubal sterilization in obese women – an assessment of risks. *Am J Obstet Gynecol.* 1985;152(3):292–297.
159. Jamieson DJ, Hillis SD, Duerr A, Marchbanks PA, Costello C, Peterson HB. Complications of interval laparoscopic tubal sterilization: findings from the United States Collaborative Review of Sterilization. *Obstet Gynecol.* 2000;96(6):997–1002.
160. Chi IC, Wilkens LR, Reid SE. Prolonged hospital stay after laparoscopic sterilization. *IPPF Med Bull.* 1984;18(4):3–4.
161. Singh KB, Huddleston HT, Nandy I. Laparoscopic tubal sterilization in obese women: experience from a teaching institution. *South Med J.* 1996;89(1):56–59.

7 Sexual Health and Obesity

Sharon Cameron[1,2]

[1]Consultant Gynaecologist, Chalmers Sexual and Reproductive Health Service, NHS Lothian, 2a Chalmers Street, Edinburgh EH3 9ES, [2]Department of Reproductive and Developmental Sciences, University of Edinburgh, 51 Little France Crescent, Edinburgh EH16 5SU

Introduction

Data from national surveys from the United States have shown a steady rise in the prevalence of obesity. In 2009/2010, more than one-third of adults in the United States were obese [1]. Corresponding data from the United Kingdom show that in 2009 almost one-quarter (22% of men and 24% of women) of the adults in England were classified as obese, and more than one-quarter (27%) of the adults in Scotland were considered obese in 2010 [2,3]. Data from England has shown clearly that obesity is more prevalent in women from poorer socio-economic groups, although this does not appear to be the case for men [2]. Data from the United States shows a similar finding of tendency for increasing prevalence of obesity among women (but not men) with lower incomes [1]. The association between obesity and physical illnesses are well recognised. In addition, there is now a growing realisation of the negative impact that obesity may have on sexual health, including sexual behaviour and sexual health outcomes. There is also growing evidence that obese individuals may experience a reduced quality of their sexual life.

Obesity and Sexual Behaviour

Probably, the most important recent contribution to information to date regarding body mass index (BMI) and sexual behaviour comes from the French national survey of sexual behaviours, which was conducted in 2005/2006 (Contexte de la Sexualite en France) [4]. This was a population-based survey of over 10,000 men and women, selected at random, who underwent lengthy telephone interviews (average duration 49 min) about their sexual practices, contraception and history of sexually transmitted infections (STI). Respondents also provided data on their height and weight. Interestingly, in this study, only one-half of the female respondents who were actually obese considered themselves to be so. This lack of awareness of realising oneself as being 'very overweight' was even more marked among men, with only one-quarter of obese men considering themselves to be in that

Obesity. DOI: http://dx.doi.org/10.1016/B978-0-12-416045-3.00007-8

category. Compared to normal-weight respondents, those who were obese were less likely to have had more than one sexual partner in the previous year, and women who were obese (but not men) were more likely not to have had any sexual partner. However, obese individuals who did have sexual partners did not differ from others, in terms of frequency of sexual intercourse and the proportion who considered themselves to be 'very satisfied' with their sexual life [4].

Women who were obese tended to have sexual partners who were also obese; a tendency that was less marked for obese men. This French study also showed that obese women were more likely to have met a sexual partner through the Internet than normal-weight women. The authors of the paper suggested that women who are obese might find it more difficult to attract a sexual partner and/or that they can establish a rapport with a potential partner while at the same time concealing their weight. This study also showed that in terms of the importance placed on sexuality for ones 'personal life balance', there was a significant trend for women of higher BMI to downgrade the importance of sexuality for their well being. However, this was not the case for men. The authors also suggested that the gender differences observed in sexual activity of obese men and women may be due to one or a combination of psychological factors, such as low self-esteem, social factors such as the stigma attached to being overweight being greater for women than for men or physical factors associated with obesity.

In contrast to the findings of this French study, data from a nationally representative database from the United States from 2002 (National Survey of Family Growth), that surveyed women of reproductive age about sexual behaviour, showed no difference in objective measures of sexual behaviour such as the number of lifetime male partners or the frequency of sexual intercourse between obese and normal-weight women [5]. One possibility this might suggest is that being an obese woman in the United States is not associated with the same stigma as in France.

In general, however, in Western countries, a slender physique is associated with physical attractiveness in women. Studies have shown that obese individuals perceive their obesity as a serious handicap. Low self-esteem due to obesity could lead to difficultly in initiating a sexual relationship. In one small study of morbidly obese individuals who lost weight, most subjects stated that they would prefer to be normal weight and have a major physical handicap (such as being deaf, dyslexic or blind) rather than be obese again [6].

Physical conditions associated with obesity that may limit sexual intercourse for women include musculoskeletal problems, urinary incontinence and menorrhagia. In addition, women who are obese may also have hyperandrogenism [7]. Thus, the associated acne or hirsuitism may also contribute to lowering a women's confidence about her physical attractiveness.

It should be remembered, however, that most of the studies to date about sexual health outcomes of interest including both height and weight have been self-reported. Previous research studies have noted that with reporting of weight there is a tendency for respondents to underestimate weight [5]. Since this appears to affect individuals of all BMIs equally, it is likely that this would lead to an underestimation of the association of obesity and the outcome of interest. Furthermore, weight is a dynamic variable and so BMI at the time of survey may not be an accurate reflection of BMI at the time of the event of interest. There is also evidence that men and women may under-report sexual behaviour. Using data from the National Health and Nutrition Examination Survey from the United States that was conducted using computer-assisted self interview (widely used to gather information on sensitive topics), 18% of the men and 28% of the women respondents who reported no lifetime sex partners ever tested positive for antibodies to Herpes simplex virus type 2 (HSV-2), which was used as a serological marker of sexual exposure [8]. In this study, obese and overweight individuals reported fewer sex partners than individuals of normal weight, although this was not reflected in their HSV-2 status [8]. In addition to under-reporting of sexual partners, it is also possible that in surveys there may be an over-reporting of the number of sexual partners or frequency of intercourse, possibly due to social pressure and this may be more likely to occur among men [5].

Obesity and Sexual Dysfunction

There is good evidence that men who are obese are more likely to experience sexual dysfunction. In a survey of over 3000 men aged 40–79 from across Europe, impaired sexual function was reported more frequently by obese than non-obese men [9]. In the French survey of national behaviours, obese men were more than twice as likely to have experienced erectile dysfunction in the preceding 12 months than normal-weight men [4]. This may be due to penile vascular impairment, as a consequence of atherosclerosis associated with obesity [10,11]. Psychological considerations such as low self-esteem due to obesity may also play a role. There is evidence, however, that erectile dysfunction can improve with weight reduction. In one study, approximately one-third of obese men with erectile dysfunction experienced improved sexual function after weight loss and lifestyle changes [12]. In older men, obesity has been associated with lower sexual satisfaction [13].

Regarding sexual dysfunction among obese women, there is a growing concern that female sexual dysfunction, in general, is being medicalised by the pharmaceutical industry [14]. Experts have commented that female sexual dysfunction is a more complex condition than for males because it has its origins in a number of factors including relationship, emotion, psychology, medicine, hormones, reproduction and ageing [15]. Some studies have reported, however, that women who are obese may suffer from a lack of libido and reduced satisfaction with sexual life [16]. A study that examined self-reported quality of sexual life among 500 individuals in the United States who were obese, using a validated measure of weight related quality of life, showed that obesity was associated with a high prevalence of lack of sexual enjoyment and sexual desire, difficulty with sexual performance and avoidance of sexual encounters [16]. Furthermore, in this study, the impairment in quality of sexual life was more marked for obese women than for obese men. Another study that used a validated sexual functioning questionnaire among obese men and women showed that scores for obese women were lower than those of obese men. In addition, scores for obese women were generally lower than scores reported for cancer survivors [17].

Obesity and Sexual Health Outcomes

Obesity may impact upon sexual health outcomes in several ways. Firstly, girls who are heavier will attain secondary sexual characteristics and the menarche earlier than their normal-weight peers, potentially allowing for more reproductive years [18]. However, survey data from a cross-sectional database of women of reproductive age in the United States showed that there was no association between the age at first intercourse and BMI [5]. Data from the French national survey of sexual behaviours [4] showed that obese women under 30 years of age were four times more likely than normal-weight women to report an unintended pregnancy or an abortion. This higher unintended pregnancy rate might seem surprising since obesity is linked to anovulation and thus reduced fertility. However, the French study showed that obese women were less likely to use oral contraceptive pills and to attend contraceptive services, in general, and more likely to rely on less-effective methods such as 'withdrawal', which may partly explain the higher risk of unintended pregnancies. It is also possible that clinicians may be reluctant to prescribe combined hormonal contraception to obese women due to concerns about higher risk of venous thromboembolism [19]. It is also possible that reliance on less-effective methods may reflect a difficulty in negotiating condom use with a sex partner, greater sexual risk taking or misconceptions regarding one's fertility. However, data of a study in the United States from 2002 (National survey of Family Growth) that surveyed more than 7600 women of reproductive age showed no association between BMI and self-reported history of unintended pregnancy in the past 5 years [20]. In addition, there was no significant difference between women of different BMI groups and current method of contraception nor on women's perceived fertility [20]. The authors of this study did point out, however, that woman probably under-report a history of unintended pregnancy, just as history of abortion tends to be under-reported in surveys [5].

In contrast, there is some evidence that risk-taking behaviours may be different between adolescent women of normal weight and those with extreme obesity. Data from a US survey of adolescents showed that young women with extreme obesity were less likely to have had sex but those who had were more likely to report having taken alcohol or drugs before the last sexual encounter [21].

One US study of post-partum women reported that among women who had been using contraception at the time of conception, obese and overweight women had almost twice the rate of unintended pregnancy compared to women of normal weight [22]. While an association between higher weight and contraceptive failure has been reported [23,24], it is hard to distinguish between method failure and user failure, and so the effect of weight on failure rates has been suggested to be a reflection of issues of compliance [25]. Recently, a study that examined risk factors for failure of oral emergency contraception found that failure rates were significantly higher among obese women [26]. This would tend to point towards an interaction of obesity upon mode of action. In addition, a recent pharmacokinetic study in which normal weight or obese women were treated with a combined oral contraceptive pill containing levonorgestrel (LNG) demonstrated an increase in the time taken to achieve a steady state of LNG concentrations in relation to obesity [27]. The authors suggested that this could be a possible mechanism for increasing failure rates of hormonal contraception in obese women.

An important finding from the French national survey of sexual behaviour was that among men in their late teens and twenties, the odds of contracting an STI in the previous 5 years were more than 10 times greater for obese men than for men of normal weight [4]. There was no difference, however, between obese and normal-weight women in history of STI. Evidence for higher sexual risk taking among obese men compared to normal-weight men has also been demonstrated for men who have sex with men (MSM), with obese MSM being more likely to have unprotected anal intercourse than normal-weight counterparts in one study [28]. The authors of this study suggested that this may be because obese MSM might feel less able to negotiate safe sex and may be less well placed to be selective about sexual partners.

Conclusion

In addition to the physical problems caused by obesity, there is growing evidence that obesity negatively impacts on sexual health. In particular, there is real cause for concern that obesity affects sexual risk-taking behaviour. Clearly, however, the relative contribution of social (stigma of obesity), psychological (low self-esteem) and physical consequences of obesity upon sexual health are difficult to disentangle (Table 7.1). There is evidence that obese women may be less likely to use effective contraception than normal-weight counterparts, thus placing themselves at a higher risk of unintended pregnancy. In addition to the personal distress that may be associated with an unintended pregnancy, clearly there are significantly increased obstetric and neonatal risks for a woman who is obese when she embarks upon a pregnancy. There is ongoing uncertainty as to whether or not some methods of

Table 7.1 Factors Associated with Obesity that may Impact Negatively upon Sexual Health

Factors	Contributions
Social	Stigma of obesity
Psychological	Reduced self-esteem
Physical	Musculoskeletal
	Menorrhagia (women)
	Urinary incontinence (women)
	Acne, hirstuitism (women)
	Erectile dysfunction (men)
Behaviour	Less likely to access contraceptive services (women)
	Less-effective contraception (women)
	Less condom use (men)

hormonal contraception may be less effective in obese compared to non-obese women and a concern that clinicians may be reluctant to prescribe combined hormonal methods for obese women due to concerns about relative contraindications. Another area of concern as of the recent finding from the French study is that obese men are less likely to have protected sex and have a higher rate of STI than men of normal BMI. In addition, for both sexes, there is growing evidence that obesity may affect the quality of sexual life and in men, can give rise to sexual dysfunction via erectile difficulties.

From a public health point of view, clearly we need to continue to pursue effective strategies (prevention and cure) to tackle obesity. Health promotion services should highlight how obesity affects not only ones physical health but can also damage ones sexual life. Sexual health providers need to be aware that the needs of obese individuals are no less than those of normal-weight men and women. Where possible, the most effective methods of contraception should be provided for obese women and advice on prevention of STI reinforced for both sexes at all opportunities. Health professionals working in sexual health services should also be encouraged to have sensitive discussions around weight loss and to signpost individuals to appropriate services for weight reduction since these may ultimately prove effective and result in an improved quality of sexual life for the patient.

References

1. *Centers for Disease Control and Prevention.* US obesity trends. <http://www.cdc.gov/obesity/data/trends.HTML/>. Accessed 9th July 2012.
2. *Statistics on Obesity, Physical Activity and Diet: England.* <www.aso.org.uk/wp-content/plugins/download.../download.php?id/>; 2011. Accessed 9th July 2012.
3. *Health of Scotlands Population: Healthy Weight.* <http://www.scotland.gov.uk/Topics/Statistics/Browse/Health/TrendObesity/>. Accessed 9th July 2012.
4. Bajos N, Wellings K, Laborde C, Moreau C, CSF Group. Sexuality and obesity, a gender perspective: results from French national random probability survey of sexual behaviours. *BMJ.* 2010;340(June):c2573.
5. Kaneshiro B, Jensen JT, Carlson NE, Harvey SM, Nichols MD, Edelman AB. Body mass index and sexual behaviour. *Obstet Gynecol.* 2008;112(3):586–592.
6. Rand CS, Macgregor AMC. Successful weight loss following obesity surgery and the perceived liability of morbid obesity. *Int J Obes.* 1991;15:577–579.
7. Barber TM, McCarthy MI, Wass JA, Franks S. Obesity and polycystic ovary syndrome. *Clin Endocrinol (Oxf).* 2006;65(2):137–145.
8. Nagelkerke NJ, Bernsen RM, Sgaier SK, Jha P. Body mass index, sexual behaviour, and sexually transmitted infections: an analysis using the NHANES 1999–2000 data. *BMC Public Health.* 2006;6(August):199.
9. Han TS, Tajar A, O'Neill TW, et al. Impaired quality of life and sexual function in overweight and obese men: the European male ageing study. *Eur J Endocrinol.* 2011;164(6):1003–1011.
10. Derby CA, Mohr BA, Goldstein I, Feldman HA, Johannes CB, mcKinlay JB. Modifiable risk factors and erectile dysfunction: can lifestyle modify risk? *Urology.* 2000;56:302–306.

11. Chung WS, Sohn JH, Park YY. Is obesity an underlying factor in erectile dysfunction? *Eur Urol.* 1999;36:68–70.
12. Esposito K, Giugliano F, Di Palo C, et al. Effect of lifestyle changes on erectile dysfunction in obese men: a randomized controlled trial.. *JAMA.* 2004;291(24):2978–2984.
13. Adolfsson B, Elofsson S, Rössner S, Undén AL. Are sexual dissatisfaction and sexual abuse associated with obesity? A population-based study. *Obes Res.* 2004;12(10): 1702–1709.
14. Moynihan R. Merging of marketing and medical science: female sexual dysfunction. *BMJ.* 2010;30:341.
15. Goldbeck-Wood S. Commentary: female sexual dysfunction is a real but complex problem. *BMJ.* 2010;30:341.
16. Kolotkin RL, Binks M, Crosby RD, Østbye T, Gress RE, Adams TD. Obesity and sexual quality of life. *Obesity (Silver Spring).* 2006;14(3):472–479.
17. Ostbye T, Kolotkin RL, He H, et al. Sexual functioning in obese adults enrolling in a weight loss study. *J Sex Marital Ther.* 2011;37(3):224–235.
18. Cooper C, Kuh D, Egger P, Wadsworth M, Barker D. Childhood growth and age at menarche. *Br J Obstet Gynecol.* 1996;103:814–817.
19. *UK Medical Eligibility Criteria for Contraceptive Use.* Faculty of sexual and reproductive healthcare UK. <www.fsrh.org.uk/>; 2009. Accessed 9th July 2012.
20. Kaneshiro B, Edelman A, Carlson N, Nichols M, Jensen J. The relationship between body mass index and unintended pregnancy: results from the 2002 National Survey of Family Growth. *Contraception.* 2008;77(4):234–238.
21. Ratcliff MB, Jenkins TM, Reiter-Purtill J, Noll JG, Zeller MH. Risk-taking behaviors of adolescents with extreme obesity. *Paediatrics.* 2011;127(5):827–834.
22. Brunner Huber LR, Hogue CJ. The association between body weight, unintended pregnancy resulting in a livebirth, and contraception at the time of conception. *Matern Child Health J.* 2005;9(4):413–420.
23. Vessey MP, Lawless M, Yeates D, McPherson K. Progestogen-only oral contraception. Findings in a large prospective study with special reference to effectiveness. *Br J Fam Plann.* 1985;10:117–121.
24. Brunner Huber LR, Toth JL. Obesity and oral contraceptive failure: findings from the 2002 National Survey of Family Growth. *Am J Epidemiol.* 2007;166:1306–1311.
25. Westhoff CL, Torgal AH, Mayeda ER, et al. Ovarian suppression in normal-weight and obese women during oral contraceptive use. *Obstet Gynecol.* 2010;116:275–283.
26. Glasier A, Cameron ST, Blithe D, et al. Can we identify women at risk of pregnancy despite using emergency contraception? Data from randomized trials of ulipristal acetate and levonorgestrel. *Contraception.* 2011;84(4):363–367.
27. Edelman AB, Carlson NE, Cherala G, et al. Impact of obesity on oral contraceptive pharmacokinetics and hypothalamic–pituitary–ovarian activity. *Contraception.* 2009;80:119–127.
28. Moskowitz DA, Seal DW. Revisiting obesity and condom use in men who have sex with men. *Arch Sex Behav.* 2010;39(3):761–765.

8 Obesity in PCOS and Infertility

Ioannis E. Messinis, Christina I. Messini and Konstantinos Dafopoulos

Department of Obstetrics and Gynaecology, Medical School, University of Thessalia, Larissa, Greece

Introduction

The National Institutes of Health (NIH) diagnostic criteria [1] for the polycystic ovary syndrome (PCOS) are the presence of hyperandrogenism and chronic oligo-anovulation, with the exclusion of other causes of hyperandrogenism such as adult-onset congenital adrenal hyperplasia, hyperprolactinaemia and androgen-secreting neoplasms. The European Society of Human Reproduction and Embryology/American Society for Reproductive Medicine (ESHRE/ASRM)-sponsored PCOS consensus workshop in Rotterdam [2] concluded that ultrasound morphology of the ovaries should be included in the diagnostic criteria and that at least two of the following criteria are sufficient for the diagnosis: oligo-anovulation, clinical and/or biochemical signs of hyperandrogenism and polycystic ovaries on ultrasound, while other causes of hyperandrogenism should be excluded. The PCOS is the most common endocrinopathy in women. According to the NIH criteria the prevalence of PCOS ranges between 6% and 10%, and with utilisation of the ESHRE/ASRM consensus criteria it is as high as 15% [3].

The main reason of infertility in women with PCOS is anovulation. The PCOS represents 75% of all anovulatory disorders causing infertility, and 90% of women with oligomenorrhoea [4].On the other hand, on average 79% of women with PCOS have oligomenorrhoea [5]. Women with PCOS are commonly (35–80%) overweight (body mass index, BMI, above 25 kg/m^2) or obese (BMI above 30 kg/m^2) [6–9]. The range depends on the setting of the study and the ethnic characteristics of the patients. Women in the United States with PCOS have a higher BMI than their European counterparts. This may explain the increase in the incidence of the PCOS in the US population, which parallels the increase in obesity [10]. It has been also shown that women with PCOS may be more likely to exhibit an abdominal and/or visceral pattern of fat distribution [11], although the distribution of fat to abdominal regions is not probably related to insulin resistance in such women [12]. On the other hand, PCOS (defined by NIH criteria) was found in nearly 35% of morbidly obese women from Spain undergoing bariatric surgery [13], compared with only 6.5% in unselected female blood donors from

Obesity. DOI: http://dx.doi.org/10.1016/B978-0-12-416045-3.00008-X

Spain, and studied by the same investigators [14]. However, obesity is not included in the diagnostic criteria for PCOS.

Obesity may intensify the severity of the phenotypic characteristics of the PCOS, including disturbed menstrual cycle [15]. In particular, a higher prevalence of anovulation has been found in obese as compared to non-obese women with PCOS [16]. Obesity seems to have a negative effect on spontaneous and induced ovulation in PCOS.

Obesity and Infertility – Possible Mechanisms

Hyperandrogenism

Ovarian hyperandrogenism in PCOS may arrest the development of antral follicles. Increased androgen secretion originates from intrinsic amplified steroidogenetic capacity of theca cells, due to 17α-hydroxylase/17,20-lyase (CYP17a1), HSD3B2 and side-chain cleavage enzyme (CYP11A1) activity. Besides, endocrine mechanisms may contribute, including luteinizing hormone (LH) hypersecretion which stimulates thecal androgen secretion, relative follicle-stimulating hormone (FSH) insufficiency resulting in reduced aromatase activity and hyperinsulinaemia synergizing with LH for thecal androgen production. Intra-ovarian mechanisms, such as anti-Müllerian hormone (AMH) inhibition of FSH and subsequent inhibition of aromatase activity, may further deteriorate hyperandrogenism. Finally, ovarian hyperandrogenism in PCOS may arrest folliculogenesis through inhibition of granulosa cells proliferation and maturation, secretion of oestrogen and progesterone, action of aromatase and increase of 5α-reductase activity [17–19].

The severity of hyperandrogenism seems to be amplified in obese women with PCOS. It has been shown that obese women with PCOS have higher total and free T levels as compared to non-obese PCOS [16,20]. In fact, the increase of body weight and fat tissue, especially in the form of abdominal obesity, is associated with abnormality of sex steroid balance. This is mainly due to the reduction of sex hormone–binding globulin (SHBG) levels in circulation, resulting to an increased fraction of free androgens in blood. Reduced SHBG synthesis in liver originates from hyperinsulinaemia that compensates for insulin resistance associated with obesity. Although hyperinsulinaemia is associated with PCOS, it is clear that obese women with PCOS exhibit a higher degree of insulin resistance and hyperinsulinaemia [21]. Finally, increased androgens in obese PCOS further contribute to inhibition of SHBG secretion. It is obvious that obesity may deteriorate hyperandrogenism in women with PCOS, a mechanism that is involved in anovulatory infertility.

Hypersecretion of LH

In anovulatory women with PCOS, 75% have high LH levels, while in up to 94% an elevated LH/FSH ratio may be found [22]. Anovulation and lack of progesterone [23] and hyperandrogenaemia followed by progesterone negative feedback effect attenuation [24] are the main mechanisms for inappropriate LH secretion. This is characterised by accelerated LH pulse frequency and amplitude and elevated LH response to gonadotrophin releasing hormone (GnRH). However, obesity in PCOS is associated with blunted LH secretion, acting at the pituitary and not at the hypothalamic level. The overall quantity of GnRH secreted and the LH pulse frequency is not affected by BMI [25]. The mechanisms that may mediate the negative effect of BMI on LH secretion may be hyperinsulinaemia, since insulin infusions decrease basal and GnRH-induced LH secretion [26]. Leptin may also be important as data have shown an inverse correlation between leptin and LH levels and LH pulse amplitude; furthermore, a decrease in leptin levels and an increase in LH pulse amplitude were found following short-term caloric restriction [27–28]. According to the ceiling hypothesis, high LH levels in circulation may luteinise prematurely and inhibit the proliferation of granulosa cells resulting in anovulation [29]. Although, in lean women with PCOS, elevated LH is a significant mechanism leading to hyperandrogenaemia and anovulation, in obese women with PCOS probably this is not the case.

Hyperinsulinaemia

As reported above, in obese women with PCOS, insulin resistance and hyperinsulinaemia are higher than in lean women with PCOS. High insulin levels in circulation may be mainly related to anovulation in women with PCOS. Hyperinsulinaemia may cause premature maturation of granulosa cells, because they respond prematurely to LH (small follicles of 4 mm), which is in contrast to the normal response that occurs when follicles reach the 10 mm diameter [30]. Premature exposure of granulosa cells to LH inhibits their proliferation and further development. In normal folliculogenesis, when granulosa cells can respond to LH at about 10 mm, they undergo two more cell divisions to reach the pre-ovulatory size (20–25 mm). In the anovulatory PCOS, granulosa cells responsive to LH in follicles as small as 4 mm in diameter undergo two more divisions reaching a maximum size of around 8–10 mm [31]. Furthermore, high insulin levels amplify LH-stimulated androgen secretion from theca cells through LH receptor's up-regulation [32]. The role of hyperinsulinaemia is significant in deteriorating fertility in obese women with PCOS since when such women lose weight and subsequently become ovulatory, they also have a reduction in insulin resistance and central adiposity [33].

Adipokines

In obesity, many genes were dysregulated in adipocytes of obese compared with non-obese individuals [34]. In omental fat of obese women with PCOS there was a different expression pattern in genes compared to obese non-PCOS women [35]. Adipokines secreted by adipose tissue may mediate the deleterious effects of obesity upon fertility in women with PCOS. These substances include leptin, adiponectin, interleukin-6 (IL-6), plasminogen activator inhibitor-1 (PAI-1), resistin and tumour necrosis factor-α (TNF-α).

Leptin

Leptin is a 16 kDa protein that is secreted almost exclusively by the adipocytes and is produced by the obese (*ob*) gene. Leptin may serve as a link between fat tissue and the brain, since by acting at the level of the hypothalamus, decreases food intake and increases energy expenditure [36]. Besides, leptin may have a role in reproductive function, exerting effects upon the hypothalamic–pituitary–ovarian (HPO) axis at central and gonadal levels. Leptin receptors have been demonstrated in the hypothalamus and pituitary as well as in theca cells, granulosa cells, oocytes, endometrial cells and pre-implantation embryos [15]. In obesity, circulating leptin levels are high due to leptin resistance. Furthermore, in women with PCOS, increased leptin levels in circulation compared to weight-matched controls have been found in some but not all studies. Leptin levels have been found to be positively correlated with insulin resistance in such women, although some contradictory findings have been reported. Similarly, in some but not all studies, following treatment with insulin-sensitising agents, leptin concentrations in blood had decreased.

Leptin may affect reproductive function at many levels. Physiologically, in women, by central action, leptin seems to be important for the hypothalamic–pituitary function and puberty. However, in obese women with PCOS as compared to obese controls abnormalities in the relationship between leptin and LH secretory characteristics have been found [37]. At the level of the ovary, leptin was found to modulate basal and FSH-stimulated steroidogenesis in cultured human lutein granulosa cells, with high concentrations suppressing the secretion of oestradiol and progesterone [38]. In vivo and in vitro experiments in animals have shown that high levels of leptin, representing hyperleptinaemia of obesity, may inhibit folliculogenesis [39–40]. Leptin may have a role in regulation of embryo implantation and endometrial receptivity, and it has been suggested that obesity-related perturbations of the leptin system can possibly interfere with embryo implantation, therefore causing infertility [15]. In conclusion, it seems that obesity may further intensify hyperleptinaemia in PCOS women, deteriorating the reproductive function further.

Adiponectin

Adiponectin, a 30 kDa protein, is the most abundant serum adipokine, secreted exclusively by the adipose tissue. Adiponectin, in contrast to leptin, is down-regulated in obesity and may have both anti-inflammatory and insulin-sensitising effects. A meta-analysis showed that women with PCOS have lower levels of adiponectin, independently of BMI [41]. Because adiponectin has direct insulin-sensitising effects, decreased levels of adiponectin in PCOS women could, in addition to obesity, contribute to systemic insulin resistance and hyperinsulinaemia thereby declining fertility. Besides, a direct role of low levels of adiponectin on folliculogenesis is possible. It was found that human theca cells express adiponectin and adiponectin receptors (AdipoR1 and AdipoR2), while granulosa cells express AdipoR1 and AdipoR2 but not adiponectin. Human recombinant adiponectin increased IGF-I-induced P and E2 production by human granulosa cells through increase in the IGF-I-induced p450 aromatase protein level [42].

Interleukin-6

IL-6 is an inflammatory cytokine, and approximately 30% of circulating levels are derived from adipose tissue. Circulating IL-6 levels increase in obesity and they are associated with increased insulin resistance. In rats, intra-cerebroventricular injection of IL-6 inhibits LH secretion [43], although in another study this effect was not replicated [44]. In vitro in rat, IL-6 has been observed to prevent LH-triggered ovulation, inhibit LH-/FSH-induced oestradiol production [45] and in human granulose tumour cells to suppress aromatase activity [46]. Furthermore, women with PCOS had elevated serum and follicular IL-6 levels when compared with non-PCOS controls stimulated all for in vitro fertilisation (IVF) [47]. It seems that IL-6, in the high levels seen in obese women, may contribute to impaired fertility in women with PCOS.

Plasminogen Activator Inhibitor Type-1

PAI-1 is a regulator of blood fibrinolytic activity and is mainly produced by white adipose tissue and visceral fat. Circulating PAI-1 levels increase in obesity and correlate with the elements of the metabolic syndrome [48]. Unlike in normal weight subjects, overweight/obese patients with PCOS had higher PAI-1 levels than BMI-matched controls [49]. PAI-1 has been associated with miscarriage in women with PCOS [50–51].

Resistin

Resistin, a 12.5 kDa polypeptide, is a member of the cysteine-rich secretary proteins called 'resistin-like molecules' (RELM) or 'found in inflammatory zone' (FIZZ) and is secreted by adipocytes. Resistin is associated with insulin resistance in mice [52]. However, there was no difference of plasma resistin levels between PCOS and control women with or without insulin resistance [53] and probably resistin may not be implicated in infertility in PCOS women.

Tumour Necrosis Factor-α

TNF-α is synthesised in adipose tissue by adipocytes and other cells in the tissue matrix [48]. Blood levels and adipocyte production of TNF-α correlate with BMI and hyperinsulinaemia, and TNF-α impairs insulin action by inhibiting insulin signalling. TNF-α may affect several levels of the reproductive axis: inhibition of gonadotrophin secretion, ovulation, steroidogenesis, corpus luteum regression and endometrial development [48]. A meta-analysis involving 9 studies of circulatory

TNF-α levels revealed no statistically significant differences between PCOS and controls [54]. Obese women with PCOS probably have an additional factor, impairing fertility at multiple levels.

Ghrelin

Ghrelin is a 28 amino acid peptide hormone produced mainly by the stomach and is the endogenous ligand for the growth hormone secretagogue receptor (GHS-R) type 1a. Ghrelin stimulates the secretion of growth hormone (GH), prolactin (PRL) and adrenocorticotrophic hormone (ACTH) from pituitary and also increases appetite, promoting food intake and regulates energy balance via hypothalamic action. Evidence has been provided that ghrelin may affect reproductive function in animals and humans. Plasma ghrelin concentrations have been shown to be lower in obese when compared with normal subjects. In obese women with PCOS, lower ghrelin levels have been found than those expected based on their obesity [55]. Furthermore, obese women with PCOS showed a negative correlation between ghrelin and insulin resistance, while regardless of the presence of PCOS, a marked negative correlation existed between ghrelin and androstenedione levels [55]. Therefore, it is possible that in obese women with PCOS, ghrelin may contribute to modification of factors such as insulin resistance and androgens mediating a negative effect on fertility.

Impact of Obesity on Infertility Treatment

Treatment of anovulatory infertility in women with PCOS involves various modalities of ovulation induction [56]. Although the basis of such treatment is the administration of different drugs, in obese women diet and lifestyle changes are considered the first-line approach [57]. In case of non-compliance, various treatments or interventions including clomiphene citrate, gonadotrophins, insulin sensitisers and laparoscopic ovarian drilling (LOD) are applied. Nevertheless, the treatment outcome during ovulation induction in PCOS may be influenced by the excessive body fat.

Diet and Life-style Changes

Lifestyle modifications are based on diet and exercise and aim at the restoration of the disturbed reproductive function. Weight loss programmes applied to obese patients with PCOS result in the improvement of the abnormal biochemical and hormonal parameters. Especially, a reduction in serum-free testosterone and insulin concentrations and an increase in SHBG values have been reported, while in more than 50% of the cases regular ovulation and menstruation are re-established [57]. It has been suggested that even 5–10% decrease in body weight of overweight women with PCOS can be effective [33,58–60,]. In such cases, a 30% decrease in visceral fat has been estimated [61]. A steady decrease of intra-abdominal fat is associated with restoration of ovulation [62]. Although energy-restricted diet is the key factor, information regarding the specific type of exercise that is more effective is limited. Recent data have shown that the addition of aerobic resistance exercise to an energy-restricted diet did not further improve reproductive outcomes [63]. In contrast, the combination of hypocaloric diet and sibutramine, an oral anorexiant, showed a better effect at 6 months on the weight loss and also led to reduction in androgens levels and insulin resistance in women with PCOS than the diet alone [64]. However, sibutramine has been withdrawn from the market in the majority of the European countries and therefore it is not recommended. Evidence from studies suggests that diets with reduced glycaemic load may provide a better control of hyperinsulinaemia and the metabolic consequences as well as menstrual cyclicity [65].

It is clear that in the majority of the women a period of 3–6 months is required for losing 5–10% in body weight. This might be a hindrance for the women who are very anxious to get pregnant quickly. On the other hand, it is not known if the reduction of weight, via caloric restriction or pharmacological intervention during the peri-conceptional period, has a negative impact on the conceptus or it is more reasonable to postpone conception until the end of the effort to lose weight. However, taking into account that obesity can adversely affect human reproduction by increasing perinatal and maternal risks, it is advisable for the women to reduce their weight before attempting conception [66]. Obese women during pregnancy carry a greater risk for congenital anomalies, miscarriage, gestational diabetes and hypertension either after spontaneous or after assisted conception [67–68]. An increased miscarriage rate was seen in a retrospective analysis of data in obese women (BMI $>$ 28 kg/m^2) undergoing ovulation induction when compared to normal controls [69]. Similarly, a mixed population of obese women undergoing IVF/ICSI (intracytoplasmic sperm injection) treatment showed an increased miscarriage rate as compared to controls with normal weight [70], while a recent meta-analysis showed that women with a BMI $>$ 25 kg/m^2 had a higher miscarriage rate regardless of the mode of conception [71]. It is evident from this information that weight reduction prior to any intervention in obese women with PCOS would be an advisable approach for the improvement of the treatment outcome. National guidelines in the United Kingdom advise for a weight loss to a BMI less than 30 kg/m^2 prior to the onset of any treatment for ovulation induction [72].

Clomiphene Citrate

Clomiphene citrate is an anti-oestrogenic compound that belongs to the selective oestrogen receptor modulators. By binding to the oestrogen receptors, clomiphene blocks the negative feedback effect of oestrogens on the central nervous system and this leads to an increased secretion of GnRH and gonadotrophins from the hypothalamus and the pituitary, respectively. When this drug is given orally to women for a few days in the early follicular phase of the cycle, it creates an inter-cycle type of FSH rise that leads to follicle recruitment selection [73]. The selected dominant follicle then secretes oestrogens, while the sequence of hormonal events is similar to that in the normal menstrual cycle resulting in the occurrence of an endogenous LH surge at mid-cycle.

Clomiphene is used as first-line treatment in anovulatory infertile women with PCOS [74]. The protocol for ovulation induction involves the administration of clomiphene at a starting dose of 50 mg/day immediately after a spontaneous period or withdrawal bleeding induced by the administration of progesterone. Detailed analysis of the literature shows that clomiphene treatment leads to ovulation in 70–86% of the women, while the pregnancy rate is lower (34–43%) [75]. Nevertheless, in properly selected patients, cumulative pregnancy rates as high as 63% at 6 months and 97% at 10 months have been reported [76–77]. In cases of clomiphene failure or clomiphene resistance, a second-line treatment is used.

Clomiphene resistance is attributed to several hormonal and clinical characteristics of the women. For example, free androgen index, BMI, age and cycle abnormalities play an important role [78–79]. In a multivariate prediction model, it was shown that decreased insulin sensitivity, hyperandrogenaemia and obesity are associated with reduced response to clomiphene treatment in PCOS [80]. Especially, women with less-reduced insulin sensitivity had a higher possibility to ovulate on clomiphene treatment [81], while obesity had a negative impact on the treatment outcome with clomiphene [82]. Obese women with PCOS respond less well to clomiphene, and the chance of ovulation is reduced particularly in women with amenorrhoea as compared to those with oligomenorrhoea [79]. Such women may need higher dosages of clomiphene even up to 250 mg/day, although the evidence is limited to retrospective data [83]. According to a recent study, patients who are expected to ovulate and become pregnant on clomiphene have significantly lower BMI than those who remain anovulatory during treatment [81]. It has been proposed that leptin produced by the adipose tissue is a more direct index of ovarian dysfunction in PCOS which can predict women remaining anovulatory during treatment with clomiphene citrate [80].

Aromatase Inhibitors

Aromatase inhibitors are drugs that are given orally to women suffering from breast cancer. These compounds inhibit the action of the enzyme aromatase, which converts androgens into oestrogens. Consequently, the reduced production of oestrogens and the reduced circulating levels of these steroids lead to the attenuation of the negative feedback and the increase in the secretion of gonadotrophins from the pituitary. For these reasons, aromatase inhibitors can be used for ovulation induction in PCOS. Letrozole is one of the third-generation aromatase inhibitors used more extensively than others for the treatment of infertility. Based on evidence derived from a meta-analysis, letrozole is equally effective compared with clomiphene in inducing ovulation in women with PCOS [84]. However, letrozole is still considered an 'off-label' medication for infertility treatment due to possible teratogenic effects in pregnancy, although this has been debated [74,85]. It is advisable therefore to discuss with the patients the possible risks and benefits.

As yet there are no data in the literature regarding the role of BMI in the treatment outcome during ovulation induction in PCOS with aromatase inhibitors.

Low-Dose FSH

In early 1980s, low-dose protocols of human menopausal gonadotrophin (HMG) or FSH were introduced for ovulation induction in PCOS as a second-line treatment in women with clomiphene failure or resistance. Low dosages were adopted in order to stimulate single follicle maturation and thus to prevent multiple pregnancies and the ovarian hyperstimulation syndrome (OHSS). Two protocols, the step-up and the step-down, are in use. In the *step-up protocol*, the starting dose was initially 75 IU FSH/day, but it was subsequently reduced to 50 IU [86]. A long period of 2 weeks on the starting dose has been adopted with an increase by 25 IU every week if there is no ovarian response. In the *step-down protocol*, the starting dose was initially 150 IU, but it is now 100 IU, a few days later, following the selection of a dominant follicle, reducing down to 75 IU and then to 50 IU [87–88]. With the step-up protocol, monofollicular development is achieved in about 70% of the cycles, while there is a low rate of multiple pregnancies (~6%) and the OHSS (<1%) [89]. Both protocols are equally effective in terms of pregnancy rate, although with the step-up protocol higher monofollicular development and lower hyperstimulation are achieved [88]. The treatment is monitored only by ultrasound scans of the ovaries, while oestradiol measurement is not required. When clomiphene and FSH were considered consecutive treatments in women with no other cause of infertility, a cumulative pregnancy rate at 12 months of 91% and a live-birth rate at 24 months of 71% were reported [90–91].

The effectiveness of the treatment highly depends on various parameters including BMI. An earlier study in PCOS women with clomiphene resistance has shown that during ovulation induction with low-dose gonadotrophins, moderate obesity was associated with a lower ovulation rate and a higher miscarriage rate, although the proportion of women who had at least one pregnancy was similar with that of women with normal BMI [69]. In the same study, significantly higher doses of gonadotrophins were required in the group of the obese than in the group of the lean women, a finding that was confirmed in a subsequent study [92]. A more recent study, including women with PCOS and increased BMI but less than 35 kg/m^2, showed that obesity was associated with higher insulin resistance and free androgen index and a higher number of immature follicles [93]. Although a higher FSH threshold for ovarian stimulation was noted and a greater total dose of gonadotrophins with a longer duration of stimulation was needed, careful monitoring was necessary due to the increased risk of over-response [93]. In the context of the step-up protocol, efforts have been made to calculate the individual FSH response dose based on various screening characteristics including BMI [94]. However, due to a rather complicated prediction model based on a mathematical equation, such an approach has not been proven reliably effective. Similarly, the prediction of the individual effective dose of FSH using several screening characteristics such as BMI, clomiphene resistance or failure, free IGF-I and FSH has been also attempted in the context of the step-down protocol [95]. Despite the initial optimism with both protocols, it was later shown that in the step-up protocol, the predicted FSH dose was higher than the observed response dose [96].

Laparoscopic Ovarian Drilling

LOD is used as a second-line treatment competing with FSH in women with PCOS and clomiphene resistance. Retrospective data have shown high ovulation and pregnancy rates [97]. A recent meta-analysis demonstrated no advantage of LOD over FSH regarding clinical pregnancy, live-birth and miscarriage rates except for a significantly lower multiple pregnancy rate [98]. A recent study has compared prospectively LOD with clomiphene as a first-line treatment in PCOS but showed no difference between the two treatment modalities regarding the cumulative pregnancy rate at 12 months [99]. There are now certain indications for LOD, such as in clomiphene-resistant patients particularly those with persistently elevated LH, in case of laparoscopic assessment of the pelvis for infertility problems and if the patients are unable to visit regularly the hospital. It should be emphasised, however, that LOD must not be used for non-fertility indications. Regarding the influence of obesity in the effectiveness of LOD, a retrospective study including 200 patients with PCOS, who were treated unsuccessfully with clomiphene, showed that LOD applied to women with a BMI $\geq 35\ kg/m^2$ induced significantly lower ovulation and pregnancy rates as compared to moderately overweight and normal weight women [100]. Nevertheless, once ovulation was achieved, BMI had no influence on the conception rates in agreement with previous reports [97,100–101]. Several studies have shown a negative correlation between BMI and the response to various medical treatments including clomiphene [83,102] and gonadotrophins [103]. However, LOD has been shown to sensitise clomiphene-resistant patients to this drug.

Insulin Sensitisers

To alleviate insulin resistance in women with PCOS, insulin-sensitising drugs are currently used. Metformin is the main representative and has been used more extensively than other compounds with similar actions. Although metformin is superior to placebo in inducing ovulation, it is not considered an ovulation-inducing agent, since it only increases the number of spontaneous ovulations in PCOS patients with oligomenorrhoea from 1 ovulation to 2 ovulations every 5 months [104]. Recent prospective randomised trials have shown that metformin used for the treatment of anovulatory infertility in PCOS is inferior to clomiphene regarding live-birth rate, while the addition of metformin to clomiphene has no advantages over clomiphene alone [105–106]. Two recent meta-analyses have demonstrated that metformin in combination with clomiphene might be useful only in cases of clomiphene resistance before moving to the second-line treatment of low-dose FSH protocols [107–108].

It is known that about 40% of patients with PCOS are obese. Insulin resistance is one of the main characteristics of these women. Treatment with metformin would be expected to improve insulin sensitivity and the metabolic and reproductive functions. Several studies have demonstrated that obese but not lean women with PCOS may benefit from the treatment with metformin [109–112]. Nevertheless, metformin administration over a period of 6 months in the context of a diet and lifestyle changes programme was not better than placebo regarding bodyweight reduction [113]. A recent meta-analysis confirmed the initial observations that metformin treatment alone resulted in a greater reduction in BMI than placebo, but when added to a diet programme, it did not show any advantage over placebo [114]. A more recent study, however, has shown that with metformin the weight loss was greater than with lifestyle changes, although there was a high dropout rate [115].

It is clear that when PCOS patients are treated in the context of an ovulation induction programme, body weight and BMI are important. Although it would be expected metformin to be useful in obese women, it was recently shown that women with a lower BMI as compared to those with a higher BMI had a higher possibility to become pregnant on this drug [116]. In that respect, metformin could be used as a first-line treatment in women with a BMI $<32\ kg/m^2$, since both the clinical pregnancy and live-birth rates did not differ from those achieved with clomiphene alone [117].

In Vitro Fertilisation

When the above treatment modalities fail to induce a pregnancy, women with PCOS are treated in the context of an IVF programme. Although the number of oocytes retrieved is usually higher in PCOS as compared to controls, there is no significant difference in fertilisation, pregnancy and live-birth rates [118]. Furthermore, in IVF/ICSI cycles, obesity and PCOS were found to independently decrease size of the oocytes [119]. The ideal protocol for ovarian stimulation in PCOS for IVF has not been identified. However, obese women with PCOS require higher amounts of FSH for ovarian stimulation [104]. On the other hand, the addition of metformin at any stage of the procedure had no benefits regarding pregnancy and live-birth rates [120–121], except for a significant reduction in the risk for the OHSS [121].

References

1. Zawadski JK, Dunaif A. Diagnostic criteria for polycystic ovary syndrome; towards a rational approach. In: Dunaif A, Givens JR, Haseltine F, eds. *Polycystic Ovary Syndrome*. Boston, MA: Blackwell Scientific; 1992:377–384.
2. The Rotterdam ESHRE/ASRM-Sponsored PCOS Consensus Workshop Group. Revised 2003 consensus on diagnostic criteria and long-term health risks related to polycystic ovary syndrome (PCOS). *Hum Reprod*. 2004;19:41–47.

3. The Amsterdam ESHRE/ASRM-Sponsored Third PCOS Consensus Workshop Group. Consensus on women's health aspects of polycystic ovary syndrome (PCOS). *Hum Reprod.* 2012;27:14–24.
4. Homburg R. Polycystic ovary syndrome. *Best Pract Res Clin Obstet Gynaecol.* 2008;22:261–274.
5. Azziz R, Carmina E, Dewailly D, Diamanti-Kandarakis E, Escobar-Morreale HF, Futterweit W, et al. The Androgen Excess and PCOS Society criteria for the polycystic ovary syndrome: the complete task force report. *Fertil Steril.* 2009;91:456–488.
6. Franks S. Polycystic ovary syndrome. *Trends Endocrinol Metab.* 1989;1:60–63.
7. Azziz R, Sanchez LA, Knochenhauer ES, Moran C, Lazenby J, Stephens KC, et al. Androgen excess in women: experience with over 1000 consecutive patients. *J Clin Endocrinol Metab.* 2004;89:453–462.
8. Hahn S, Tan S, Sack S, Kimmig R, Quadbeck B, Mann K, et al. Prevalence of the metabolic syndrome in German women with polycystic ovary syndrome. *Exp Clin Endocrinol Diabetes.* 2007;115:130–135.
9. Cupisti S, Kajaia N, Dittrich R, Duezenli H, Beckmann MW, Mueller A. Body mass index and ovarian function are associated with endocrine and metabolic abnormalities in women with hyperandrogenic syndrome. *Eur J Endocrinol.* 2008;158:711–719.
10. Mokdad AH, Ford ES, Bowman BA, Dietz WH, Vinicor F, Bales VS, et al. Prevalence of obesity, diabetes, and obesity-related health risk factors, 2001. *JAMA.* 2003;289:76–79.
11. Carmina E, Bucchieri. S, Esposito A, Del Puente A, Mansueto P, Orio F, et al. Abdominal fat quantity and distribution in women with polycystic ovary syndrome and extent of its relation to insulin resistance. *J Clin Endocrinol Metab.* 2007;92:2500–2505.
12. Mannerås-Holm L, Leonhardt H, Kullberg J, Jennische E, Odén A, Holm G, et al. Adipose tissue has aberrant morphology and function in PCOS: enlarged adipocytes and low serum adiponectin, but not circulating sex steroids, are strongly associated with insulin resistance. *J Clin Endocrinol Metab.* 2011;96:E304–E311.
13. Escobar-Morreale HF, Botella-Carretero JI, Alvarez-Blasco F, Sancho J, San Millán JL. The polycystic ovary syndrome associated with morbid obesity may resolve after weight loss induced by bariatric surgery. *J Clin Endocrinol Metab.* 2005;90:6364–6369.
14. Asunción M, Calvo RM, San Millán JL, Sancho J, Avila S, Escobar-Morreale HF. A prospective study of the prevalence of the polycystic ovary syndrome in unselected Caucasian women from Spain. *J Clin Endocrinol Metab.* 2000;85:2434–2438.
15. Brewer CJ, Balen AH. The adverse effects of obesity on conception and implantation. *Reproduction.* 2010;140:347–364.
16. Kiddy DS, Sharp PS, White DM, Scanlon MF, Mason HD, Bray CS, et al. Differences in clinical and endocrine features between obese and non-obese subjects with polycystic ovary syndrome: an analysis of 263 consecutive cases. *Clin Endocrinol (Oxf).* 1990;32:213–220.
17. Pradeep PK, Li X, Peegel H, Menon KM. Dihydrotestosterone inhibits granulosa cell proliferation by decreasing the cyclin D2 mRNA expression and cell cycle arrest at G1 phase. *Endocrinology.* 2002;143:2930–2935.
18. Greisen S, Ledet T, Ovesen P. Effects of androstenedione, insulin and luteinizing hormone on steroidogenesis in human granulosa luteal cells. *Hum Reprod.* 2001;16:2061–2065.
19. Jakimiuk AJ, Weitsman SR, Magoffin DA. 5alpha-reductase activity in women with polycystic ovary syndrome. *J Clin Endocrinol Metab.* 1999;84:2414–2418.
20. Holte J, Bergh T, Gennarelli G, Wide L. The independent effects of polycystic ovary syndrome and obesity on serum concentrations of gonadotrophins and sex steroids in premenopausal women. *Clin Endocrinol (Oxf).* 1994;41:473–481.

21. Gambineri A, Pelusi C, Vicennati V, Pagotto U, Pasquali R. Obesity and the polycystic ovary syndrome. *Int J Obes Relat Metab Disord*. 2002;26:883–896.
22. Taylor AE, McCourt B, Martin KA, Anderson EJ, Adams JM, Schoenfeld D, et al. Determinants of abnormal gonadotropin secretion in clinically defined women with polycystic ovary syndrome. *J Clin Endocrinol Metab*. 1997;82:2248–2256.
23. Dafopoulos K, Venetis C, Pournaras S, Kallitsaris A, Messinis IE. Ovarian control of pituitary sensitivity of luteinizing hormone secretion to gonadotropin-releasing hormone in women with the polycystic ovary syndrome. *Fertil Steril*. 2009;92:1378–1380.
24. Eagleson CA, Gingrich MB, Pastor CL, Arora TK, Burt CM, Evans WS, et al. Polycystic ovarian syndrome: evidence that flutamide restores sensitivity of the gonadotropin-releasing hormone pulse generator to inhibition by estradiol and progesterone. *J Clin Endocrinol Metab*. 2000;85:4047–4052.
25. Pagán YL, Srouji SS, Jimenez Y, Emerson A, Gill S, Hall JE. Inverse relationship between luteinizing hormone and body mass index in polycystic ovarian syndrome: investigation of hypothalamic and pituitary contributions. *J Clin Endocrinol Metab*. 2006;91:1309–1316.
26. Lawson MA, Jain S, Sun S, Patel K, Malcolm PJ, Chang RJ. Evidence for insulin suppression of baseline luteinizing hormone in women with polycystic ovarian syndrome and normal women. *J Clin Endocrinol Metab*. 2008;93:2089–2096.
27. Laughlin GA, Morales AJ, Yen SS. Serum leptin levels in women with polycystic ovary syndrome: the role of insulin resistance/hyperinsulinemia. *J Clin Endocrinol Metab*. 1997;82:1692–1696.
28. Van Dam EW, Roelfsema F, Veldhuis JD, Helmerhorst FM, Frölich M, Meinders AE, et al. Increase in daily LH secretion in response to short-term calorie restriction in obese women with PCOS. *Am J Physiol Endocrinol Metab*. 2002;282:E865–E872.
29. Hillier SG. Current concepts of the roles of follicle stimulating hormone and luteinizing hormone in folliculogenesis. *Hum Reprod*. 1994;9:188–191.
30. Willis DS, Watson H, Mason HD, Galea R, Brincat M, Franks S. Premature response to luteinizing hormone of granulosa cells from anovulatory women with polycystic ovary syndrome: relevance to mechanism of anovulation. *J Clin Endocrinol Metab*. 1998;83:3984–3991.
31. McNatty KP, Smith DM, Makris A, Osathanondh R, Ryan KJ. The microenvironment of the human antral follicle: interrelationships among the steroid levels in antral fluid, the population of granulosa cells, and the status of the oocyte in vivo and in vitro. *J Clin Endocrinol Metab*. 1979;49:851–860.
32. Poretsky L, Cataldo NA, Rosenwaks Z, Giudice LC. The insulin-related ovarian regulatory system in health and disease. *Endocr Rev*. 1999;20:535–582.
33. Clark AM, Ledger W, Galletly C, Tomlinson L, Blaney F, Wang X, et al. Weight loss results in significant improvement in pregnancy and ovulation rates in anovulatory obese women. *Hum Reprod*. 1995;10:2705–2712.
34. Lee YH, Nair S, Rousseau E, Allison DB, Page GP, Tataranni PA, et al. Microarray profiling of isolated abdominal subcutaneous adipocytes from obese vs non-obese Pima Indians: increased expression of inflammation-related genes. *Diabetologia*. 2005;48:1776–1783.
35. Cortón M, Botella-Carretero JI, Benguría A, Villuendas G, Zaballos A, San Millán JL, et al. Differential gene expression profile in omental adipose tissue in women with polycystic ovary syndrome. *J Clin Endocrinol Metab*. 2007;92:328–337.
36. Messinis IE, Milingos SD. Leptin in human reproduction. *Hum Reprod Update*. 1999;5:52–63.

37. Roelfsema F, Kok P, Veldhuis JD, Pijl H. Altered multihormone synchrony in obese patients with polycystic ovary syndrome. *Metabolism*. 2011;60:1227–1233.
38. Karamouti M, Kollia P, Kallitsaris A, Vamvakopoulos N, Kollios G, Messinis IE. Modulating effect of leptin on basal and follicle stimulating hormone stimulated steroidogenesis in cultured human lutein granulosa cells. *J Endocrinol Invest*. 2009;32:415–419.
39. Duggal PS, Van Der Hoek KH, Milner CR, Ryan NK, Armstrong DT, Magoffin DA, et al. The in vivo and in vitro effects of exogenous leptin on ovulation in the rat. *Endocrinology*. 2000;141:1971–1976.
40. Srivastava PK, Krishna A. Increased circulating leptin level inhibits folliculogenesis in vespertilionid bat, *Scotophilus heathii*. *Mol Cell Endocrinol*. 2011;337:24–35.
41. Toulis KA, Goulis DG, Farmakiotis D, Georgopoulos NA, Katsikis I, Tarlatzis BC, et al. Adiponectin levels in women with polycystic ovary syndrome: a systematic review and a meta-analysis. *Hum Reprod Update*. 2009;15:297–307.
42. Chabrolle C, Tosca L, Ramé C, Lecomte P, Royère D, Dupont J. Adiponectin increases insulin-like growth factor I-induced progesterone and estradiol secretion in human granulosa cells. *Fertil Steril*. 2009;92:1988–1996.
43. Rivier C, Vale W. Cytokines act within the brain to inhibit luteinizing hormone secretion and ovulation in the rat. *Endocrinology*. 1990;127:849–856.
44. Watanobe H, Hayakawa Y. Hypothalamic interleukin-1 beta and tumor necrosis factor-alpha, but not interleukin-6, mediate the endotoxin-induced suppression of the reproductive axis in rats. *Endocrinology*. 2003;144:4868–4875.
45. Mikuni M. Effect of interleukin-2 and interleukin-6 on ovary in the ovulatory period – establishment of the new ovarian perfusion system and influence of interleukins on ovulation rate and steroid secretion. *Hokkaido Igaku Zasshi*. 1995;70:561–572.
46. Deura I, Harada T, Taniguchi F, Iwabe T, Izawa M, Terakawa N. Reduction of estrogen production by interleukin-6 in a human granulosa tumor cell line may have implications for endometriosis-associated infertility. *Fertil Steril*. 2005;83(suppl 1):1086–1092.
47. Amato G, Conte M, Mazziotti G, Lalli E, Vitolo G, Tucker AT, et al. Serum and follicular fluid cytokines in polycystic ovary syndrome during stimulated cycles. *Obstet Gynecol*. 2003;101:1177–1182.
48. Gosman GG, Katcher HI, Legro RS. Obesity and the role of gut and adipose hormones in female reproduction. *Hum Reprod Update*. 2006;12:585–601.
49. Koiou E, Tziomalos K, Dinas K, Katsikis I, Kandaraki EA, Tsourdi E, et al. Plasma plasminogen activator inhibitor-1 levels in the different phenotypes of the polycystic ovary syndrome. *Endocr J*. 2012;59:21–29.
50. Glueck CJ, Wang P, Fontaine RN, Sieve-Smith L, Tracy T, Moore SK. Plasminogen activator inhibitor activity: an independent risk factor for the high miscarriage rate during pregnancy in women with polycystic ovary syndrome. *Metabolism*. 1999;48:1589–1595.
51. Glueck CJ, Sieve L, Zhu B, Wang P. Plasminogen activator inhibitor activity, 4G5G polymorphism of the plasminogen activator inhibitor 1 gene, and first-trimester miscarriage in women with polycystic ovary syndrome. *Metabolism*. 2006;55:345–352.
52. Steppan CM, Bailey ST, Bhat S, Brown EJ, Banerjee RR, Wright CM, et al. The hormone resistin links obesity to diabetes. *Nature*. 2001;409:307–312.
53. Zhang J, Zhou L, Tang L, Xu L. The plasma level and gene expression of resistin in polycystic ovary syndrome. *Gynecol Endocrinol*. 2011;27:982–987.
54. Escobar-Morreale HF, Luque-Ramírez M, González F. Circulating inflammatory markers in polycystic ovary syndrome: a systematic review and metaanalysis. *Fertil Steril*. 2011;95:1048–1058.

55. Pagotto U, Gambineri A, Vicennati V, Heiman ML, Tschöp M, Pasquali R. Plasma ghrelin, obesity, and the polycystic ovary syndrome: correlation with insulin resistance and androgen levels. *J Clin Endocrinol Metab*. 2002;87:5625–5629.
56. Messinis IE. Ovulation induction: a mini review. *Hum Reprod*. 2005;20:2688–2697.
57. Hoeger KM. Role of lifestyle modification in the management of polycystic ovary syndrome. *Best Pract Res Clin Endocrinol Metab*. 2006;20:293–310.
58. Kiddy DS, Hamilton-Fairley D, Seppälä M, Koistinen R, James VH, Reed MJ, et al. Diet-induced changes in sex hormone binding globulin and free testosterone in women with normal or polycystic ovaries: correlation with serum insulin and insulin-like growth factor-I. *Clin Endocrinol (Oxf)*. 1989;31:757–763.
59. Clark AM, Thornley B, Tomlinson L, Galletley C, Norman RJ. Weight loss in obese infertile women results in improvement in reproductive outcome for all forms of fertility treatment. *Hum Reprod*. 1998;13:1502–1505.
60. van Dam EW, Roelfsema F, Veldhuis JD, Hogendoorn S, Westenberg J, Helmerhorst FM, et al. Retention of estradiol negative feedback relationship to LH predicts ovulation in response to caloric restriction and weight loss in obese patients with polycystic ovary syndrome. *Am J Physiol Endocrinol Metab*. 2004;286:E615–E620.
61. Després JP, Lemieux I, Prud'homme D. Treatment of obesity: need to focus on high risk abdominally obese patients. *BMJ*. 2001;322:716–720.
62. Kuchenbecker WK, Groen H, van Asselt SJ, Bolster JH, Zwerver J, Slart RH, et al. In women with polycystic ovary syndrome and obesity, loss of intra-abdominal fat is associated with resumption of ovulation.. *Hum Reprod*. 2011;26:2505–2512.
63. Thomson RL, Buckley JD, Noakes M, Clifton PM, Norman RJ, Brinkworth GD. The effect of a hypocaloric diet with and without exercise training on body composition, cardiometabolic risk profile, and reproductive function in overweight and obese women with polycystic ovary syndrome. *J Clin Endocrinol Metab*. 2008;93:3373–3380.
64. Florakis D, Diamanti-Kandarakis E, Katsikis I, Nassis GP, Karkanaki A, Georgopoulos N, et al. Effect of hypocaloric diet plus sibutramine treatment on hormonal and metabolic features in overweight and obese women with polycystic ovary syndrome: a randomized, 24-week study. *Int J Obes (Lond)*. 2008;32:692–699.
65. Marsh KA, Steinbeck KS, Atkinson FS, Petocz P, Brand-Miller JC. Effect of a low glycemic index compared with a conventional healthy diet on polycystic ovary syndrome. *Am J Clin Nutr*. 2010;92:83–92.
66. Nelson SM, Fleming RF. The preconceptual contraception paradigm: obesity and infertility. *Hum Reprod*. 2007;22:912–915.
67. Cedergren MI. Maternal morbid obesity and the risk of adverse pregnancy outcome. *Obstet Gynecol*. 2004;103:219–224.
68. Linné Y. Effects of obesity on women's reproduction and complications during pregnancy. *Obes Rev*. 2004;5:137–143.
69. Hamilton-Fairley D, Kiddy D, Watson H, Paterson C, Franks S. Association of moderate obesity with a poor pregnancy outcome in women with polycystic ovary syndrome treated with low dose gonadotrophin. *Br J Obstet Gynaecol*. 1992;99:128–131.
70. Fedorcsák P, Dale PO, Storeng R, Ertzeid G, Bjercke S, Oldereid N, et al. Impact of overweight and underweight on assisted reproduction treatment. *Hum Reprod*. 2004;19:2523–2528.
71. Metwally M, Ong KJ, Ledger WL, Li TC. Does high body mass index increase the risk of miscarriage after spontaneous and assisted conception? A meta-analysis of the evidence. *Fertil Steril*. 2008;90:714–726.

72. National Institute for Clinical Excellence (NICE)/National Collaborating Centre for Women's and Children's Health. *Fertility: Assessment and Treatment for People with Fertility Problems*. London: RCOG Press; 2004.
73. Adashi EY. Clomiphene citrate-initiated ovulation: a clinical update. *Semin Reprod Endocrinol*. 1986;4:255–276.
74. Thessaloniki ESHRE/ASRM-Sponsored PCOS Consensus Workshop Group. Consensus on infertility treatment related to polycystic ovary syndrome. *Hum Reprod*. 2008;23:462–477.
75. Messinis IE. Clomiphene citrate. In: Tarlatzis B, ed. *Ovulation Induction*. Paris: Elsevier; 2002:87–97.
76. Hammond MG, Halme JK, Talbert LM. Factors affecting the pregnancy rate in clomiphene citrate induction of ovulation. *Obstet Gynecol*. 1983;62:196–202.
77. Kousta E, White DM, Franks S. Modern use of clomiphene citrate in induction of ovulation. *Hum Reprod Update*. 1997;3:359–365.
78. Imani B, Eijkemans MJ, te Velde ER, Habbema JD, Fauser BC. Predictors of patients remaining anovulatory during clomiphene citrate induction of ovulation in normogonadotropic oligoamenorrheic infertility. *J Clin Endocrinol Metab*. 1998;83:2361–2365.
79. Imani B, Eijkemans MJ, te Velde ER, Habbema JD, Fauser BC. A nomogram to predict the probability of live birth after clomiphene citrate induction of ovulation in normogonadotropic oligoamenorrheic infertility. *Fertil Steril*. 2002;77:91–97.
80. Imani B, Eijkemans MJ, de Jong FH, Payne NN, Bouchard P, Giudice LC, et al. Free androgen index and leptin are the most prominent endocrine predictors of ovarian response during clomiphene citrate induction of ovulation in normogonadotropic oligoamenorrheic infertility. *J Clin Endocrinol Metab*. 2000;85:676–682.
81. Palomba S, Falbo A, Orio Jr F, Tolino A, Zullo A. Efficacy predictors for metformin and clomiphene citrate treatment in anovulatory infertile patients with polycystic ovary syndrome. *Fertil Steril*. 2009;91:2557–2567.
82. Al-Azemi M, Omu FE, Omu AE. The effect of obesity on the outcome of infertility management in women with polycystic ovary syndrome. *Arch Gynecol Obstet*. 2004;270:205–210.
83. Dickey RP, Taylor SN, Curole DN, Rye PH, Lu PY, Pyrzak R. Relationship of clomiphene dose and patient weight to successful treatment. *Hum Reprod*. 1997;12:449–453.
84. Requena A, Herrero J, Landeras J, Navarro E, Neyro JL, Salvador C, et al. Use of letrozole in assisted reproduction: a systematic review and meta-analysis. *Hum Reprod Update*. 2008;14:571–582.
85. Elizur SE, Tulandi T. Drugs in infertility and fetal safety. *Fertil Steril*. 2008;89:1595–1602.
86. White DM, Polson DW, Kiddy D, Sagle P, Watson H, Gilling-Smith C, et al. Induction of ovulation with low-dose gonadotropins in polycystic ovary syndrome: an analysis of 109 pregnancies in 225 women. *J Clin Endocrinol Metab*. 1996;81:3821–3824.
87. Fauser BC, Donderwinkel P, Schoot DC. The step-down principle in gonadotrophin treatment and the role of GnRH analogues. *Baillieres Clin Obstet Gynaecol*. 1993;7:309–330.
88. Christin-Maitre S, Hugues JN, Recombinant FSH Study Group. A comparative randomized multicentric study comparing the step-up versus step-down protocol in polycystic ovary syndrome. *Hum Reprod*. 2003;18:1626–1631.
89. Homburg R, Howles CM. Low-dose FSH therapy for anovulatory infertility associated with polycystic ovary syndrome: rationale, results, reflections and refinements. *Hum Reprod Update*. 1999;5:493–499.

90. Messinis IE, Milingos SD. Current and future status of ovulation induction in polycystic ovary syndrome. *Hum Reprod Update*. 1997;3:235–253.
91. Eijkemans MJ, Imani B, Mulders AG, Habbema JD, Fauser BC. High singleton live birth rate following classical ovulation induction in normogonadotrophic anovulatory infertility (WHO 2). *Hum Reprod*. 2003;18:2357–2362.
92. Loh S, Wang JX, Matthews CD. The influence of body mass index, basal FSH and age on the response to gonadotrophin stimulation in non-polycystic ovarian syndrome patients. *Hum Reprod*. 2002;17:1207–1211.
93. Balen AH, Dresner M, Scott EM, Drife JO. Should obese women with polycystic ovary syndrome receive treatment for infertility? *BMJ*. 2006;332:434–435.
94. Imani B, Eijkemans MJ, Faessen GH, Bouchard P, Giudice LC, Fauser BC. Prediction of the individual follicle-stimulating hormone threshold for gonadotropin induction of ovulation in normogonadotropic anovulatory infertility: an approach to increase safety and efficiency. *Fertil Steril*. 2002;77:83–90.
95. van Santbrink EJ, Eijkemans MJ, Macklon NS, Fauser BC. FSH response-dose can be predicted in ovulation induction for normogonadotropic anovulatory infertility. *Eur J Endocrinol*. 2002;147:223–226.
96. van Wely M, Fauser BC, Laven JS, Eijkemans MJ, van der Veen F. Validation of a prediction model for the follicle-stimulating hormone response dose in women with polycystic ovary syndrome. *Fertil Steril*. 2006;86:1710–1715.
97. Li TC, Saravelos H, Chow MS, Chisabingo R, Cooke ID. Factors affecting the outcome of laparoscopic ovarian drilling for polycystic ovarian syndrome in women with anovulatory infertility. *Br J Obstet Gynaecol*. 1998;105:338–344.
98. Farquhar C, Lilford RJ, Marjoribanks J, Vandekerckhove P. Laparoscopic 'drilling' by diathermy or laser for ovulation induction in anovulatory polycystic ovary syndrome. *Cochrane Database Syst Rev*. 2007;3:CD001122.
99. Amer SA, Li TC, Metwally M, Emarh M, Ledger WL. Randomized controlled trial comparing laparoscopic ovarian diathermy with clomiphene citrate as a first-line method of ovulation induction in women with polycystic ovary syndrome. *Hum Reprod*. 2009;24:219–225.
100. Amer SA, Li TC, Ledger WL. Ovulation induction using laparoscopic ovarian drilling in women with polycystic ovarian syndrome: predictors of success. *Hum Reprod*. 2004;19:1719–1724.
101. Gjønnaess H. Ovarian electrocautery in the treatment of women with polycystic ovary syndrome (PCOS). Factors affecting the results. *Acta Obstet Gynecol Scand*. 1994;73:407–412.
102. Lobo RA, Paul W, March CM, Granger L, Kletzky OA. Clomiphene and dexamethasone in women unresponsive to clomiphene alone. *Obstet Gynecol*. 1982;60:497–501.
103. Fedorcsák P, Dale PO, Storeng R, Tanbo T, Abyholm T. The impact of obesity and insulin resistance on the outcome of IVF or ICSI in women with polycystic ovarian syndrome. *Hum Reprod*. 2001;16:1086–1091.
104. Harborne L, Fleming R, Lyall H, Norman J, Sattar N. Descriptive review of the evidence for the use of metformin in polycystic ovary syndrome. *Lancet*. 2003;361:1894–1901.
105. Moll E, Bossuyt PM, Korevaar JC, Lambalk CB, van der Veen F. Effect of clomifene citrate plus metformin and clomifene citrate plus placebo on induction of ovulation in women with newly diagnosed polycystic ovary syndrome: randomised double blind clinical trial. *BMJ*. 2006;332:1485.

106. Legro RS, Barnhart HX, Schlaff WD, Carr BR, Diamond MP, Carson SA, et al. Clomiphene, metformin, or both for infertility in the polycystic ovary syndrome. *N Engl J Med*. 2007;356:551–566.
107. Moll E, van der Veen F, van Wely M. The role of metformin in polycystic ovary syndrome: a systematic review. *Hum Reprod Update*. 2007;13:523–527.
108. Creanga AA, Bradley HM, McCormick C, Witkop CT. Use of metformin in polycystic ovary syndrome: a meta-analysis. *Obstet Gynecol*. 2008;111:959–968.
109. Pasquali R, Gambineri A, Aiscotti D, Vicennati V, Gagliardi L, Colitta D, et al. Effect of long-term treatment with metformin added to hypocaloric diet on body composition, fat distribution, and androgen and insulin levels in abdominally obese women with and without the polycystic ovary syndrome. *J Clin Endocrinol Metab*. 2000;85:2767–2774.
110. Chou KH, von Eye Corleta H, Capp E, Spritzer PM. Clinical, metabolic and endocrine parameters in response to metformin in obese women with polycystic ovary syndrome: a randomized, double-blind and placebo-controlled trial. *Horm Metab Res*. 2003;35:86–91.
111. Hoeger KM, Kochman L, Wixom N, Craig K, Miller RK, Guzick DS. A randomized, 48-week, placebo-controlled trial of intensive lifestyle modification and/or metformin therapy in overweight women with polycystic ovary syndrome: a pilot study. *Fertil Steril*. 2004;82:421–429.
112. Trolle B, Flyvbjerg A, Kesmodel U, Lauszus FF. Efficacy of metformin in obese and non-obese women with polycystic ovary syndrome: a randomized, double-blinded, placebo-controlled cross-over trial. *Hum Reprod*. 2007;22:2967–2973.
113. Tang T, Glanville J, Hayden CJ, White D, Barth JH, Balen AH. Combined lifestyle modification and metformin in obese patients with polycystic ovary syndrome. A randomized, placebo-controlled, double-blind multicentre study. *Hum Reprod*. 2006;21:80–89.
114. Nieuwenhuis-Ruifrok AE, Kuchenbecker WK, Hoek A, Middleton P, Norman RJ. Insulin sensitizing drugs for weight loss in women of reproductive age who are overweight or obese: systematic review and meta-analysis. *Hum Reprod Update*. 2009;15:57–68.
115. Ladson G, Dodson WC, Sweet SD, Archibong AE, Kunselman AR, Demers LM, et al. The effects of metformin with lifestyle therapy in polycystic ovary syndrome: a randomized double-blind study. *Fertil Steril*. 2011;95:1059–1066.
116. Johnson NP, Bontekoe S, Stewart AW. Analysis of factors predicting success of metformin and clomiphene treatment for women with infertility owing to PCOS-related ovulation dysfunction in a randomised controlled trial. *Aust N Z J Obstet Gynaecol*. 2011;51:252–256.
117. Johnson N. Metformin is a reasonable first-line treatment option for non-obese women with infertility related to anovulatory polycystic ovary syndrome – a meta-analysis of randomised trials. *Aust N Z J Obstet Gynaecol*. 2011;51:125–129.
118. Heijnen EM, Eijkemans MJ, Hughes EG, Laven JS, Macklon NS, Fauser BC. A meta-analysis of outcomes of conventional IVF in women with polycystic ovary syndrome. *Hum Reprod Update*. 2006;12:13–21.
119. Marquard KL, Stephens SM, Jungheim ES, Ratts VS, Odem RR, Lanzendorf S, et al. Polycystic ovary syndrome and maternal obesity affect oocyte size in in vitro fertilization/intracytoplasmic sperm injection cycles. *Fertil Steril*. 2011;95:2146–2149.
120. Tang T, Glanville J, Orsi N, Barth JH, Balen AH. The use of metformin for women with PCOS undergoing IVF treatment. *Hum Reprod*. 2006;21:1416–1425.
121. Tso LO, Costello MF, Albuquerque LE, Andriolo RB, Freitas V. Metformin treatment before and during IVF or ICSI in women with polycystic ovary syndrome. *Cochrane Database Syst Rev*. 2009;15:CD006105.

9 Obesity and Recurrent Miscarriage

Harish Malappa Bhandari[1,2] *and Siobhan Quenby*[2,3]

[1]Centre for Reproductive Medicine, UHCW NHS Trust, Coventry, CV2 2DX, UK, [2]Division of Reproductive Health, Warwick Medical School, University of Warwick, Coventry, CV2 2DX, UK, [3]Department of Obstetrics and Gynaecology, UHCW NHS Trust, Coventry, CV2 2DX, UK

Introduction

Recurrent miscarriage (RM) is defined as loss of three or more consecutive pregnancies and affects approximately 1% of fertile women trying to conceive [1]. The risk of RM, after three miscarriages, that occurs by chance alone is 0.34%, taking a loss rate of 15% for each clinically recognised pregnancy [1]. It is a heterogenous condition with a number of recognised aetiological associations such as antiphospholipid syndrome, parental and embryonic chromosomal abnormalities, congenital uterine malformations, cervical weakness, some inherited thrombophilic defects, endocrinological disorders such as polycystic ovarian syndrome (PCOS), poorly controlled diabetes mellitus and untreated thyroid dysfunction, bacterial vaginosis and immune factors [2]. Miscarriage is a highly distressing event and RM blights the lives of many women. The heterogeneity of the condition and the existence of conflicting evidence in the treatment of underlying associated aetiologies contribute to the challenge in the management of RM [3]. Adding to the distress associated with RM, in approximately 50% of couples, the underlying cause remains unidentified despite wide range of investigations [4–6].

The widely accepted WHO classification system defines a person with a body mass index (BMI) $>30\,kg/m^2$ as obese. Obesity results when the energy intake exceeds energy expenditure and the resulting excess energy is stored as fat. There is great concern at the high prevalence of and the increasing trend to obesity world-wide. A combination of factors to include increased energy intake, decreased regular exercise activities, changes in eating habits by increased consumption of convenience foods and changes in the dietary composition are responsible for the global trend of increasing obesity. Many ethnic groups who either migrate to Western societies or adopt a Western lifestyle are prone to obesity in their changed environment [7], and there is continuing debate on the validity of current definitions of obesity across different non-white ethnic groups as some are at a higher risk of ill health at lower BMI than the European population [8]. This is because the central adiposity (android fat mass) is particularly important in

Obesity. DOI: http://dx.doi.org/10.1016/B978-0-12-416045-3.00009-1

clinical sequelae associated with an increased BMI and the fat distribution differs among different ethnic groups. BMI may not correspond to the same degree of fatness in different populations due, in part, to different body proportions. Obesity is associated with increased morbidity and mortality as a result of increased risks of cardiovascular disease, type-II diabetes mellitus, malignancy and gastrointestinal diseases. Inwomen of reproductive age group, being overweight contributes to menstrual disorders, infertility, miscarriage, poor pregnancy outcome and impaired foetal well-being.

Obesity and Miscarriage

There is a much debate on the possibility of an association between obesity and miscarriage, and there are many studies that have investigated the risk of miscarriage in obese and overweight women, both in general population and in women undergoing assisted conception. Although most of the studies have suggested a possible link between obesity and increased risk of miscarriage, there are studies that have not agreed with this observation specifically during in vitro fertilisation (IVF) treatment [9]. A meta-analysis of the evidence in 2006 that included 16 studies concluded that women with BMI $\geq$ 25 kg/m^2 had significantly higher odds of miscarriage regardless of the method of conception. Subgroup analysis from a limited number of studies in the same meta-analysis suggested that this group of women may also have significantly higher odds of miscarriage after oocyte donation and ovulation induction, but there was no evidence for increased odds of miscarriage after IVF or IVF–ICSI (intra-cytoplasmic sperm injection) [10]. A systematic review of 11 studies by Maheshwari et al. in 2007 [11] concluded that overweight women with BMI $\geq$ 25 kg/m^2 face a lower likelihood of pregnancy and an increased risk of miscarriage after IVF. They also have reduced oocytes retrieved despite requiring higher doses of gonadotrophins. The risk of miscarriage in obese women may be as high as 25–37% before the first live-born child [12].

Obesity and Recurrent Miscarriage

With sufficient evidence to suggest that obesity increases the risk of miscarriage, obesity was also blamed of being a major contributing factor to RM, and evidences from two retrospective studies [13,14] suggests an increased risk of RM in obese women. The first one by Lashen et al. [13] was a matched case–control study of a large obstetric population that included 1644 obese women and 3288 normal-weight controls and showed a significantly higher odds of RM in obese patients (OR 3.51; 95% CI 1.03–12.01). The latter study by Metwally et al. [14] showed a small but significantly higher odds of miscarriage in subsequent pregnancy (OR 1.71; 95% CI 1.05–2.8). They also demonstrated that the occurrence of miscarriage in the subsequent pregnancy could be best predicted by an increased BMI and maternal age. Though maternal age was the strongest predictor, a new observation was made suggesting that an increase in the BMI acts independent of age. The exact mechanism by which obesity increases the risk of miscarriage and RM is still unclear. Possible mechanisms for the association between obesity and RM include an adverse impact on endometrial development or a detrimental effect on ovaries affecting oocyte quality and hence embryo viability or combination of both.

Metabolic Effects of Obesity on Ovary

Polycystic Ovarian Syndrome

PCOS is the most common endocrine condition in women of reproductive age and is associated with ovarian dysfunction, symptoms of hyperandrogenism and menstrual irregularity. Obesity is present in nearly half of all the women with PCOS (35–80%) and is closely linked to the development of PCOS. A high serum concentration of androgenic hormones, such as testosterone, androstenedione and dehydroepiandrosterone sulphate, is present in women with PCOS and is also associated with peripheral insulin resistance (IR) and compensatory hyperinsulinaemia, which are amplified to certain extent by obesity.

Despite the clear association between PCOS and RM, the prevalence of PCOS in RM remains highly uncertain. With the available evidence, the prevalence ranges from 4.8% to 82%, and this wide range is a result of huge variation in diagnostic criteria for PCOS before the availability of the Rotterdam diagnostic criteria [15]. In women with PCOS, multiple endocrine and metabolic alterations may be responsible for the early pregnancy loss and RM.

1. Elevated concentrations of *leutinising hormone* (LH) in women with PCOS was thought to be responsible for RM either by causing premature maturation of the oocyte or by causing a detrimental effect on the endometrial development. Cocksedge et al. [15] in 2008 based on the available evidence concluded that hypersecretion of LH was unlikely to be responsible for RM.
2. There is evidence that local high androgen concentrations (*hyperandrogenaemia*) have a detrimental effect on follicular growth and oocyte quality. van Wely et al. [16] demonstrated the possible negative association between free androgen index, a sensitive marker for androgen excess, and oocyte quality or fertilisation in women with PCOS. The prevalence of hyperandrogenaemia in RM is 11% and that in this group of women, there is significantly an increased risk of miscarriage in a subsequent pregnancy [17].
3. *Hyperinsulinaemia and IR* in PCOS are attributed to obesity as well as IR independent of body weight [18]. The suggested prevalence of IR in RM is between 17% and 27% [15]. Hyperinsulinaemia is associated with increased levels of plasminogen activator inhibitor 1 (PAI-1) [19] that is strongly associated with an increased risk of miscarriage and RM [20,21]; hyperinsulinaemia by itself is a significant independent risk factor for miscarriage [22] and is also believed to play a key role in implantation failure by suppressing the circulating glycodelin and insulin-like growth factor binding protein 1 [23].

4. Altered secretion and action of other hormones such as *leptin, resistin, ghrelin and adiponectin*, in obese women, may be responsible for affecting follicular growth, early embryo development and implantation [24].

Ovarian Dysfunction

The quality of developing embryos and their successful implantation are dependent on the maturity and quality of the oocytes with the good-quality embryos being associated with good fertilisation rates and subsequent embryo development. Based on different observations at assisted reproductive cycles, many studies have suggested the poor reproductive performance of obese women as a result of decreased ovarian function. Compared with women with normal weight, obese women, at IVF and ICSI, are known to have increased risk of insufficient follicular development and lower oocyte count [24] and show a tendency to experience more cancellation due to poor response to gonadotrophin stimulation and lower fertilisation rates [25]. It has also been observed that intra-follicular human chorionic gonadotrophin concentration is inversely related to BMI and may be related to concurrent decrease in embryo quality and pregnancy rates [26]. Obese women tend to have significantly poorer oocyte quality compared with women of normal BMI [27]. In younger women (less than 35 years of age), obesity is associated with poorer embryo quality and have less chance of having cryopreserved embryos [28]. There is enough evidence to suggest that obesity has an effect on the number and quality of oocytes and also on the quality of the embryo. These effects on the ovarian function in conjunction with the deleterious effects of obesity on endometrium may well be the reason for increased risk of RM in obese women.

Endometrial Changes in Obesity

Implantation of the embryo and a successful pregnancy require receptive endometrium, and obesity may have its effects on the endometrium or its environment causing implantation failure and pregnancy losses. Various studies that were done to define the effects of obesity on endometrium have varied in design and have reported contradictory findings [29–31]. Wattanakumtornkul et al. [29] in 2003 studied 97 consecutive first-cycle recipients of anonymous oocyte donation and concluded that increased BMI did not impair uterine receptivity in women undergoing embryo transfer after oocyte donation. Styne-Gross et al. [30] studied 536 first-cycle recipients of donor oocytes and proposed that BMI has no adverse impact on implantation or reproductive outcomes in donor oocyte recipients and, therefore, obesity does not appear to exert a negative effect on endometrial receptivity. Bellver et al. [31] proposed that the most effective way of differentiating the action of obesity on the ovary and endometrium in humans is by studying the oocyte donation model. To analyse the effects of obesity on the endometrium, they retrospectively studied 2656 oocyte donation cycles with good-quality embryos.

They concluded that the endometrium or its environment also contributes to the poor reproductive outcomes in obese women but in a subtle manner.

The precise impact of obesity on molecular and histopathological aspects of endometrium is not fully understood. The debate continues with regards to the following effects of obesity on endometrium in causing RM.

Luteal-phase Deficiency

Luteal-phase deficiency (LPD) causing endometrial defects may be related to either decreased progesterone production by the corpus luteum or poor response of the endometrium to the available progesterone. There has been a much controversy about the role of LPD in women with RM and the exact mechanisms that cause endometrial defects. Metwally et al. [32] studied 136 women with RM (66 women with BMI > 25 kg/m^2) and suggested that there was no significant difference in the prevalence of LPD between patients with normal BMI (21.4%) and those with high BMI (24.2%). The endometrial samples in the study were precisely timed based on the LH surge to minimise the error. Some authors are of the opinion that the diagnosis of LPD requires two-timed endometrial biopsy in consecutive cycles [3].

Leukaemia Inhibitory Factor

Leukaemia inhibitory factor (LIF) is a cytokine that belongs to the interleukin 6 family, which by acting within the hypothalamus appears to play a role in energy homeostasis [33]. In mice, expression of LIF in uterine endometrial glands is important for implantation, and the blastocysts fail to implant in female mice lacking functional LIF gene [34]. In humans, LIF is maximally expressed by the luminal and glandular epithelium during the mid-secretory phase of the menstrual cycle when the uterus is receptive to an implanting blastocyst [35] and is an important regulator of decidualization [36]. In a retrospective study of women with RM, a significant negative correlation was observed between glandular concentration of LIF in the endometrium and an increasing BMI [32]. There was no difference in the endometrial oestrogen and progesterone receptors, and the authors proposed that alternative pathways may be involved. They recommended a well-powered prospective study to confirm the findings and to recommend weight loss to obese women with recurrent pregnancy failure.

Leptin

Leptin is a 167-amino acid polypeptide hormone, mainly produced by adipocytes, that aids in the regulation of body weight and food intake [37]. Leptin has various physiological functions including a possible functional role in implantation by stimulating the effect on matrix metalloproteinase expression in the cytotrophoblast [38]. Low levels of plasma leptin were found in women with RM who subsequently miscarried than those with RM who subsequently had a live birth [39]. Though weight gain is associated with a rise in circulating leptin levels that parallels the increase in the percentage of body fat [40], increased adiposity is associated with defective leptin and insulin signalling in the hypothalamus, leading to a condition termed leptin/insulin resistance [41] and this relative deficiency as a result of leptin resistance may be responsible for poor reproductive outcomes in obese women.

Endometrial Proteins

Glycodelin is a glycoprotein that is secreted by the secretory and decidualised endometrium and is found in abundance in the endometrium in the mid- to late-secretory phase. It facilitates implantation by the possible inhibition of maternal immune rejection of the foetus at foetomaternal interface [42]. Hyperinsulinaemia and IR are associated with a reduction in serum glycodelin levels. The reduced levels of glycodelin in the serum and uterine flushings have been found to be associated with women with RM [43,44].

Management

The management of obese women with idiopathic RM is challenging in the absence of sufficient good-quality evidence from randomised control trials (RCTs). There is no evidence from RCTs regarding the risk of RM in obese women so as to provide counselling for obese patients who are contemplating pregnancy after two or more miscarriages. Psychological support in a dedicated early pregnancy assessment clinics and early reassurance scans in subsequent pregnancy may increase the chance of a successful pregnancy outcome.

Role of Weight Reduction

The management of obesity has great significance because it can be done by non-invasive regimes of exercise and dietary modification and is cost effective. Alternatively, bariatric surgery is increasingly offered to obese patients with co-morbidities. Weight loss decreases body fat, reduces truncal–abdominal fat and improves metabolism and hormonal balance. Weight loss in obese women with PCOS, through a protein-rich and a very low-calorie diet, has been shown to significantly reduce serum fasting glucose and insulin, improve insulin sensitivity and decrease PAI-1 activity [45]. Weight loss also significantly decreases testosterone levels and increases sex-hormone-binding globulin levels [46]. Normalising hyperinsulinaemia, improving insulin sensitivity and reducing hyperandrogenaemia through weight reduction could be important for those women in achieving positive reproductive outcomes by improving ovarian function and endometrial receptivity.

Role of Insulin-Sensitising Agents

Insulin-sensitising agents, such as metformin, used in the treatment of obese PCOS women have shown improvements in hyperinsulinaemia and hyperandrogenism.

A case–control study by Nawaz and Rizvi [47] suggested that the spontaneous miscarriage was significantly lower in obese women with PCOS who conceived while taking metformin and continued metformin throughout pregnancy (cases) when compared to 'controls' who either conceived without metformin or metformin was stopped soon after the confirmation of pregnancy. The same study also suggested that the risk of subsequent miscarriage in women with RM was less in 'cases' when compared to 'controls' [47]. Though this study was based on Pakistani women, it has provided some evidence about the use of metformin to minimise the risk of miscarriage in those women. A large-scale RCT of weight management is desirable to provide robust evidence for the management of obese women with RM. It would also be important to consider ethnicity as it is found that women of African and Hispanic–American descent have higher IR than White women after adjusting for BMI, and women of south Asian and Asian descent have a higher risk of IR compared with those of European origin with a similar BMI.

Conclusion

A global increase in obesity among women in reproductive age group is associated with poor reproductive outcomes. Obesity increases the risk of miscarriage and RM by its possible effects on the ovary, endometrium and embryo. There is no good-quality evidence to provide counselling for obese women who are contemplating pregnancy after unexplained RM. Weight loss may be the first step and the use of metformin may be useful. A multi-centre RCT taking ethnicity and standardized classification for obesity, RM and PCOS into account would be useful in providing concrete evidence in appropriate management of those women.

References

1. Stirrat GM. Recurrent miscarriage I: definition and epidemiology. *Lancet*. 1990;336:673–675.
2. Royal College of Obstetricians and Gynaecologists. *Green-top Guideline No. 17. The Investigation and Treatment of Couples with Recurrent First-trimester and Second-Trimester Miscarriage*. London, UK: RCOG Press; 2011:1–18.
3. Tang AW, Quenby S. Recent thoughts on management and prevention of recurrent early pregnancy loss. *Curr Opin Obstet Gynecol*. 2010;22(6):446–451.
4. Yang CJ, Stone P, Stewart AW. The epidemiology of recurrent miscarriage: a descriptive study of 1214 prepregnant women with recurrent miscarriage. *Aust N Z J Obstet Gynaecol*. 2006;46(4):316–322.
5. Li TC, Makris M, Tomsu M, Tuckerman E, Laird S. Recurrent miscarriage: aetiology, management and prognosis. *Hum Reprod Update*. 2002;8(5):463–481.
6. Quenby SM, Farquharson RG. Predicting recurring miscarriage: what is important? *Obstet Gynecol*. 1993;82:132–138.

7. Norman RJ, Noakes M, Wu R, Davies MJ, Moran L, Wang JX. Improving reproductive performance in overweight/obese women with effective weight management. *Hum Reprod Update*. 2004;10(3):267–280.
8. Gatineau M, Mathrani S. Obesity and Ethnicity. Oxford: National Obesity Observatory, 2011.
9. Lashen H, Ledger W, Bernal AL, Barlow D. Extremes of body mass do not adversely affect the outcome of superovulation and in-vitro fertilization. *Hum Reprod*. 1999; 14(3):712–715.
10. Metwally M, Ong KJ, Ledger WL, Li TC. Does high body mass index increase the risk of miscarriage after spontaneous and assisted conception? A meta-analysis of the evidence. *Fertil Steril*. 2008;90(3):714–726.
11. Maheshwari A, Stofberg L, Bhattacharya S. Effect of overweight and obesity on assisted reproductive technology – a systematic review. *Hum Reprod Update*. 2007; 13(5):433–444.
12. Hamilton-Fairley D, Kiddy D, Watson H, Paterson C, Franks S. Association of moderate obesity with a poor pregnancy outcome in women with polycystic ovarian syndrome treated with low dose gonadotrophin. *Br J Obstet Gynaecol*. 1992;99:128–131.
13. Lashen H, Fear K, Sturdee DW. Obesity is associated with increased risk of first trimester and recurrent miscarriage: matched case–control study. *Human Reprod*. 2004;19:1644–1646.
14. Metwally M, Saravelos SH, Ledger WL, Li TC. Body mass index and risk of miscarriage in women with recurrent miscarriage. *Fertil Steril*. 2010;94(1):290–295.
15. Cocksedge KA, Li TC, Saravelos SH, Metwally M. A reappraisal of the role of polycystic ovary syndrome in recurrent miscarriage. *Reprod Biomed Online*. 2008; 17(1):151–160.
16. van Wely M, Bayram N, van der Veen F, Bossuyt PMM. Predicting ongoing pregnancy following ovulation induction with recombinant FSH in women with polycystic ovarian syndrome. *Hum Reprod*. 2005;20:1827–1832.
17. Cocksedge KA, Saravelos SH, Wang Q, Tuckerman E, Laird SM, Li TC. Does free androgen index predict subsequent pregnancy outcome in women with recurrent miscarriage. *Hum Reprod*. 2008;23(4):797–802.
18. Fedorcsak P, Dale PO, Storeng R, Tanbo T, Åbyholm T. The impact of obesity and insulin resistance on the outcome of IVF or ICSI in women with polycystic ovarian syndrome. *Hum Reprod*. 2001;16:1086–1091.
19. Palomba S, Orio Jr F, Falbo A. Plasminogen activator inhibitor 1 and miscarriage after metformin treatment and laparoscopic ovarian drilling in women with polycystic ovarian syndrome. *Fertil Steril*. 2005;84:761–765.
20. Glueck CJ, Wang P, Bornovali S. Polycystic ovarian syndrome, the G1691A factor V Leiden mutation and plasminogen activator inhibitor activity: associations with recurrent pregnancy loss. *Metabolism*. 2003;52:1627–1632.
21. Glueck CJ, Wang P, Fontaine RN. Plasminogen activator inhibitor activity: an independent risk factor for the high miscarriage rate during pregnancy in women with polycystic ovarian syndrome. *Metabolism*. 1999;48:1589–1595.
22. Glueck CJ, Wang P, Goldenberg N. Pregnancy outcomes among women with polycystic ovarian syndrome treated with metformin. *Hum Reprod*. 2002;17:2858–2864.
23. Jakubowicz DJ, Seppälä M, Jakubowicz S, et al. Insulin reduction with metformin increases luteal phase serum glycodelin and insulin-like growth factor-binding protein 1 concentrations and enhances uterine vascularity and blood flow in the polycystic ovary syndrome. *J Clin Endocrinol Metab*. 2001;86(3):1126–1133.

24. Fedorcsak P, Dale PO, Storeng R, et al. Impact of overweight and underweight on assisted reproduction treatment. *Hum Reprod.* 2004;19:2523–2528.
25. van Swieten van der Leeuw-Harmsen, Badings EA, van der Linden PJ. Obesity and clomiphene challenge test as predictors of outcome of in vitro fertilization and intracytoplasmic sperm injection. *Gynaecol Obstet Invest.* 2005;59(4):220–224.
26. Carrell DT, Jones KP, Peterson CM, Aoki V, Emery BR, Campbell BR. Body mass index is inversely related to intrafollicular hCG concentrations, embryo quality and IVF outcomes. *Reprod Biomed Online.* 2001;3(2):109–111.
27. Wittemer C, Ohl J, Bailly M, et al. Does body mass index of infertile women have an impact on IVF procedure and outcome? *J Assist Reprod Genet.* 2000;17:547–552.
28. Metwally M, Cutting R, Tipton A, Skull J, Ledger WL, Li TC. Effect of increased body mass index on oocyte and embryo quality in IVF patients. *Reprod Biomed Online.* 2007;15:532–538.
29. Wattanakumtornkul S, Damario MA, Stevens Hall SA, Thornhill AR, Tummon IS. Body mass index and uterine receptivity in the oocyte donation model. *Fertil Steril.* 2003;80(2):336–340.
30. Styne-Gross A, Elkind-Hirsch K, Scott Jr. RT. Obesity does not impact implantation rates or pregnancy outcome in women attempting conception through oocyte donation. *Fertil Steril.* 2005;83(6):1629–1634.
31. Bellver J, Melo MA, Bosch E, Serra V, Remohí J, Pellicer A. Obesity and poor reproductive outcome: the potential role of the endometrium. *Fertil Steril.* 2007; 88(2):446–451.
32. Metwally M, Tuckerman EM, Laird SM, Ledger WL, Li TC. Impact of high body mass index on endometrial morphology and function in the peri-implantation period in women with recurrent miscarriage. *Reprod Biomed Online.* 2007;14(3): 328–334.
33. Dozio E, Ruscica M, Galliera E, Corsi MM, Magni P. Leptin, ciliary neurotrophic factor, leukemia inhibitory factor and interleukin-6: class-I cytokines involved in the neuroendocrine regulation of the reproductive function. *Curr Protein Pept Sci.* 2009;10:577–584.
34. Stewart CL, Kaspar P, Brunet LJ, et al. Blastocyst implantation depends on maternal expression of leukaemia inhibitory factor. *Nature.* 1992;359(6390):76–79.
35. Paiva P, Menkhorst E, Salamonsen L, Dimitriadis E. Leukemia inhibitory factor and interleukin-11: critical regulators in the establishment of pregnancy. *Cytokine Growth Factor Rev.* 2009;20(4):319–328.
36. Shuya LL, Menkhorst EM, Yap J, Li P, Lane N, Dimitriadis E. Leukemia inhibitory factor enhances endometrial stromal cell decidualization in humans and mice. *PLoS ONE.* 2011;6(9):e25288. doi: 10.1371/journal.pone.0025288.
37. Henson MC, Castracane VD. Leptin in pregnancy. *Biol Reprod.* 2000;63(5):1219–1228.
38. Sagawa N, Yura S, Itoh H, et al. Possible role of placental leptin in pregnancy: a review. *Endocrine.* 2002;19(1):65–71.
39. Laird SM, Quinton ND, Anstie B, Li TC, Blakemore AI. Leptin and leptin-binding activity in women with recurrent miscarriage: correlation with pregnancy outcome. *Hum Reprod.* 2001;16(9):2008–2013.
40. Kolaczynski JW, Ohannesian JP, Considine RV, Marco CC, Caro JF. Response of leptin to short-term and prolonged overfeeding in humans. *J Clin Endocrinol Metab.* 1996;81(11):4162–4165.
41. Clarke IJ. Whatever way weight goes, inflammation shows. *Endocrinology.* 2010;151(3): 846–848.

42. Okamoto N, Uchida A, Takakura K, et al. Suppression by human placental protein 14 of natural killer cell activity. *Am J Reprod Immunol.* 1991;26(4):137–142.
43. Tulppala M, Julkunen M, Tiitinen A, Stenman U-H, Seppälä M. Habitual abortion is accompanied by low serum levels of placental protein 14 in the luteal phase of the fertile cycle. *Fertil Steril.* 1995;63:792–795.
44. Salim R, Miel J, Savvas M, Lee C, Jurkovic D. A comparative study of glycodelin concentrations in uterine flushings in women with subseptate uteri, history of unexplained recurrent miscarriage and healthy controls. *Eur J Obstet Gynecol Reprod Biol.* 2007;133(1):76–80.
45. Andersen P, Seljeflot I, Abdelnoor M, et al. Increased insulin sensitivity and fibrinolytic capacity after dietary intervention in obese women with polycystic ovary syndrome. *Metabolism.* 1995;44(5):611–616.
46. Clark AM, Ledger W, Galletly C, et al. Weight loss results in significant improvement in pregnancy and ovulation rates in anovulatory obese women. *Hum Reprod.* 1995;10:2705–2712.
47. Nawaz FH, Rizvi J. Continuation of metformin reduces early pregnancy loss in obese Pakistani women with polycystic ovarian syndrome. *Gynecol Obstet Invest.* 2010; 69(3):184–189.

10 Obesity and Assisted Reproduction

Mark Hamilton and Abha Maheshwari

Clinical Senior Lecturer and Senior Lecturer, Aberdeen Fertility Centre, University of Aberdeen, Scotland, UK

Introduction

Obesity is a major issue in western society and can have profound effects on reproductive health. There is persuasive evidence that the prevalence of obesity has increased over the past 35 years with more than half of the women in the United Kingdom either overweight or obese. Many of those are in the reproductive age group and a significant number present with infertility. There is a convincing literature which suggests that there are genuine issues of concern with respect to adverse clinical outcomes, increased health risks and expense associated with assisted reproduction treatment (ART) in women who are obese. Obstetrics data suggest that maternal and foetal risks are increased in obese women and there has been debate in recent times as to whether it is appropriate to offer access to ART for this group of patients. This chapter sets out to explore these issues highlighting the need to use limited state resources to maximum effectiveness with the safety of women and children a prime concern.

Prevalence of Obesity in the Assisted Reproduction Sector

The association of obesity with infertility is well described. However, many overweight women conceive without difficulty, though pregnancies in these circumstances may be associated with increased risks to the mother and child.

A normal body mass index (BMI) is considered to lie between 19 and 24.9 kg/m^2. Overweight is defined as BMI $\geq$ 25 kg/m^2. Obesity is subdivided into moderate (BMI 30–34.9 kg/m^2), severe (BMI 35–39.9 kg/m^2) or morbid (BMI $\geq$ 40 kg/m^2) according to the level of BMI.

It is sometimes suggested that obesity is a disease of the modern age. There is certainly some evidence to suggest that proportion of obese individuals within the population is changing. In the United States, the prevalence of obesity (BMI > 30 kg/m^2) in young adults (18–29 years) has tripled from 8% in 1971–1974 to 24% in

Obesity. DOI: http://dx.doi.org/10.1016/B978-0-12-416045-3.00010-8

2004–2006. In the United Kingdom, more than half of all women are either overweight or obese [1,2].

Recent data from the United States have suggested that, specific to users of (in vitro fertilisation) IVF services, 40% had a BMI $>$ 25 kg/m^2 while over 6% had a BMI $\geq$ 35 kg/m^2 and in other words were severely obese (Figure 10.1).

National data specific to those accessing IVF treatment in the United Kingdom is lacking, although a study analysing economic costs of IVF relevant to BMI in a UK centre showed a similar proportion (41.3%) with a BMI $\geq$ 25 kg/m^2 [3].

Evidence of Reduced Fertility in Obese Women

The association of obesity with impaired fertility has been described in many reviews. There are several mechanisms whereby the obese women may have reduced fertility potential. Psychosocial factors are likely to be important, but pathophysiological mechanisms linked to disturbed ovulation patterns as well as issues of egg, embryo and endometrial receptivity have also been implicated. A positive correlation between increasing BMI and infertility has been described with a relative risk of 2.7 at a BMI $\geq$ 30 kg/m^2 [4]. The time to conception in the overweight (BMI $\geq$ 25 kg/m^2) has also been observed to be longer [5]. Sexual dysfunction has also been described to occur with greater frequency in obese women which could be related to physical or psychological disturbances [6].

Cycle Effects

Obesity is linked to disturbances in the hypothalamic–pituitary–ovarian axis. Increased levels of serum and follicular fluid leptin are described with increasing BMI. High levels of leptin impair follicular development and reduce ovarian steroidogenesis through direct effects on theca and granulose cells [7]. There is also an

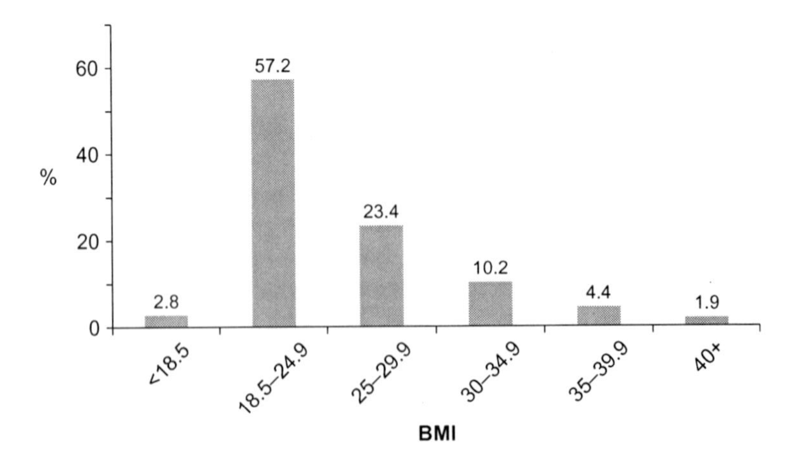

Figure 10.1 BMI of women accessing IVF treatment.
Source: Ref. [20].

inverse relationship of increasing BMI with reduced serum adiponectin levels. The low adiponectin levels are associated with elevated serum insulin levels which increase the circulating androgen levels in part linked to a reduction in the production of sex-hormone-binding globulin by the liver. Hyperandrogenism in obese women is also contributed to by insulin-like growth factor 1 (IGF-1) mediated effects on LH-induced steroidogenesis by theca cells. Enhanced androgen production causes granulosa cell apoptosis with direct consequences for follicle function. The increased availability of androgens for peripheral conversion to oestrogens in adipose tissue has pituitary effects leading to impaired production of follicle-stimulating hormone (FSH) which affects the development of ovarian follicles [8].

The clinical manifestations of the biochemical disturbances described include anovulatory cycles and sub-fertility. Ovarian dysregulation associated with hyperandrogenism, insulin resistance, menstrual irregularity and infertility is commonly found in women with polycystic ovarian syndrome many of whom are obese [2].

Specific Issues Relating to Art

ART nowadays has a therapeutic role in the management of all causes of infertility. The use of IVF may also provide some insight into the pathophysiology underlying impaired reproductive performance in obese women.

Effects on the Oocyte

A number of studies have suggested that oocyte yield after stimulation for IVF may be affected in obese women. Quantitative effects have been described where increased doses of gonadotrophins are required to elicit an ovarian response, and the ultimate yield of cumulus-oocyte complexes may be less in obese women than in normal-weight controls [9,10]. This may be linked to disturbances in leptin production or sensitivity as described earlier. Some studies have suggested that fertilisation rates of oocytes retrieved may be impaired in obese women but this observation has not been consistent. Prospective studies are needed to clarify this issue [7]. Some studies have attributed the increased risks of miscarriages in obese women after IVF to qualitative effects on oocytes leading to aberrant embryo development [11].

Effects on Embryos

As with oocytes, the literature is not consistent with respect to the effects of obesity on embryonic development. Some studies have suggested that markers of embryo quality differ in obese women. Furthermore, there may be a less availability of surplus embryos for cryostorage potentially having an impact on the cumulative pregnancy rates per episode of ovarian stimulation. Some have suggested that these observed effects are unreliable since studies may not have taken the potential confounders such as age, parity and duration of infertility into account [12]. Further work is required to inform this controversial debate.

Effects on the Endometrium

There is an increase in miscarriage in obese women both in natural conception and in that associated with infertility treatment. Specific to IVF, a 50% increased risk of miscarriage in women with a BMI $> 30\ kg/m^2$ has been described [12]. While embryo quality is an important determinant of implantation potential, studies using an egg-donation model [13] suggest that endometrial factors are likely to be involved in this phenomenon as well. The precise mechanism is not understood but ovarian steroid regulation of endometrial development, perturbations in inflammatory and coagulation pathways, perhaps linked to insulin resistance, have been suggested to be involved.

Rationale for the Use of Assisted Reproduction

The main cause of infertility in obese women relates to disturbances in ovulation. These, for the most part, can be resolved with a combined approach involving weight reduction strategies together with pharmacologically induced ovulation. Refractory dysovulation occurs with greater frequency in obese women, and for them, the use of ART has to be considered. ART, specifically IVF, will address the issues of egg and sperm availability, as well as tubal infertility for obese women, just as it does for the general infertile population. There is no definitive evidence that unexplained infertility occurs with greater frequency in obese women, though given the above remarks with respect to oocyte, embryo and endometrial factors, one might have expected this to be the case. The practical issues which arise through the use of these techniques in overweight women need to be considered carefully.

Practical Management of Obese Women Undertaking Art

Patient Selection

The selection of which patients to treat, and in whom treatment should be deferred until weight loss is achieved, should ideally depend on age, tests of ovarian reserve and the presence of co-morbidities [14].

If tests of ovarian reserve, which might include age, serum anti-mullerian hormone and/or antral follicle count, suggest a good ovarian reserve with no other co-morbidities, then it is appropriate to defer the treatment up until the desired BMI is obtained. However, if there is evidence of ovarian ageing, there is a limited time for weight loss. In these circumstances, it might be wiser to proceed with the treatment. Despite this, health care professionals have a duty of care not just to the patient but also to the potential child, and treatment should arguably not be provided if there are significant obstetric and perinatal risks such as in cases of extreme morbid obesity.

Stimulation Regimes

As alluded to above, there are observational data suggesting that the requirement of gonadotrophins is increased by at least 20% if BMI is $\geq 30\ kg/m^2$. Chong et al. [15] demonstrated that patients who have normal ±10% ideal body weight (IBW) are more likely to respond to lower doses of human menopausal gonadotrophin than patients whose weight is >10% above IBW, and, in particular, those who are >25% above their IBW. A high BMI was associated with a higher FSH threshold dose. This observation is supported by findings that the total dose of gonadotrophins needed to induce ovulation is increased in parallel with body weight [16,17]. Why heavier women may need more hormones to induce ovulation or for controlled ovarian hyperstimulation is not clear. It may be related to the greater amount of body surface, inadequate oestradiol metabolism and decreased sex-hormone-binding globulin. Also, the intramuscular absorption of the drug may be slower and incomplete in obese patients because of increased subcutaneous fat or fat infiltration of the muscle.

The effect of FSH at the ovarian level is dependent on the plasma concentrations of the hormone. This, in turn, is influenced not just by the dose administered but also by endogenous FSH secretion, metabolic clearance rate and the volume of distribution, which are individual factors, and differ from woman to woman and are influenced by BMI. Elimination of FSH is carried out largely by the kidneys and the liver. The clearance rate is dependent on filtration, secretion and reabsorption. The extent to which a drug is bound to plasma proteins also determines the fraction of the drug extracted by the eliminating organs which in turn is dependent on BMI and weight.

Pooled analysis (weighted mean differences (WMD)) from the observational studies have shown that the duration of gonadotrophin stimulation was significantly longer (WMD 0.27; 95% CI 0.26–0.28, $P < 0.00001$) and the dose was higher (WMD 406.77, 95% CI 169.26–644.2, $P = 0.0008$) in women with $BMI \geq 30\ kg/m^2$ compared to women with normal BMI [18]. However, there is no randomized controlled trial in the literature that tested the hypothesis that increasing the dose of gonadotrophins in obese women improves the live-birth rates.

There is no evidence to suggest that one regime of pituitary suppression (agonist or antagonist) is better in obese women compared to those with normal BMI.

Monitoring of Stimulation

While there are no data in the literature that quantifies the differences of monitoring those with higher BMI, it is accepted generally that the performance and interpretation of ultrasound scans can be difficult in obese women. Theoretically, where estradiol is used in monitoring the response to stimulation, the levels might be expected to differ in obese women than from those with a normal BMI. There is, however, no evidence to suggest that with overweight patients it is advantageous to use both ultrasound and estradiol in monitoring stimulated cycles [19].

Clinical Procedures

Egg Collection

Clinical staff will be sensitive to the particular challenges, presented by obese women undergoing surgical procedures. While there is no evidence from the literature, there are more problems in caring for those who are obese and this is probably because most units do not treat morbidly obese women. That said, obese women will require higher dose of sedation, due to increased surface area, which potentially may lead to side effects due to a higher risk of exposure to the drugs utilised. For the moment, in the absence of any published data in the literature, the perceived increase in risk remains theoretical.

Embryo Transfer

For the most part, the procedure of embryo transfer (ET) is simple. However, ultrasound-guided ET, which may involve the use of abdominal ultrasound, will be difficult in obese women due to poor views. Whether this leads to lower pregnancy rate remains unknown as there are no data in the literature to explore either difficulties with the procedure or lower pregnancy rates.

hCG Trigger

Theoretically, bioactive levels of hCG that triggers ovulation will be less in obese women. However, as long as more than equivalent of 1000 IU of recombinant hCG is given as the ovulatory trigger, oocyte fertilisation rates and luteal function are unlikely to be influenced by differences in bioavailable gonadotrophin. Now, most preparations used for triggering ovulation contain at least 6500 IU of hCG.

Luteal Support

Luteal support for obese women should be the same as that for women with normal BMI. This is because the vaginal pessaries are locally absorbed and bypass first-pass metabolism. There are no data comparing luteal support and outcomes in various BMI groups.

Effect of Obesity on the Results of Art

Pregnancy Rate

Several systematic reviews of observational studies [12,18] have demonstrated a detrimental impact of obesity on pregnancy rates.

There is a reduction in pregnancy rates (RR 0.87, 95% CI 0.80–0.95, $P = 0.002$) in obese women ($BMI \geq 30$ kg/m^2) when compared with those who have normal BMI (<25 kg/m^2). This reduction in pregnancy rate was also observed in women who were overweight ($BMI \geq 25$ kg/m^2) compared to those with normal BMI (RR 0.90, 95% CI 0.85–0.94, $P < 0.0001$) [18].

However, there was no difference in pregnancy rates when women with $BMI < 30$ kg/m^2 were compared with those with $BMI \geq 30$ kg/m^2, thereby indicating that there is no further detrimental effect of obesity ($BMI \geq 30$ kg/m^2) in pregnancy rates when compared to those who are overweight ($BMI \geq 25$ kg/m^2) [12].

Pooling of data from observational studies is associated with inherent bias as one cannot adjust for confounding factors such as age and number of embryos transferred. Moreover, the number of cases in these studies, where BMI was in morbid obesity range, was extremely small.

The largest single series comes from the Society of Assisted Reproduction (SART) in the United States [20]. This analysis showed that failure to achieve a clinical intrauterine gestation was significantly more likely among obese women (RR 1.22, 95% CI 1.13–1.32). This was based on 31,672 ETs from a single database, and the analysis permitted adjustment for age, parity, number of embryos transferred and the day of ET. The denominator used in this analysis was per ET rather than per woman.

Miscarriage Rate

Of those who conceive after ART, there is a 30% increased risk of miscarriage if women are overweight ($BMI \geq 25$ kg/m^2) compared to those with normal BMI (RR 1.33, 95% CI 1.06–1.68). This risk further increases to just over 50% when miscarriage rates are compared in those who are obese to those with a $BMI < 30$ kg/m^2 (OR 1.53, 95% CI 1.27–1.84) [12] (Figure 10.2).

As discussed earlier, it is uncertain whether the cause of increased miscarriages is linked to oocyte quality or other factors involved in implantation within the endometrium.

Live-Birth Rate

The SART data [20]demonstrated that the odds ratio of failure to achieve live birth was 1.27 (95% CI 1.10–1.47) in obese women ($BMI \geq 30$ kg/m^2) as compared to those with normal BMI. However, the results also indicated that there are significant differences in pregnancy and live-birth rates after ART when analysed by race and ethnicity, even within the same BMI categories. Moreover, the same data showed that there was no difference in live-birth rate based on BMI if donor oocytes were used [21] – a finding supported by others [22]. In contrast, some studies exploring the use of donor oocytes in those with high BMI have suggested a lower chance of conception [23].

Data from a recent systematic review also showed a 9% reduction in live-birth rate in overweight women (BMI 25–29.9 kg/m^2) when compared with those with

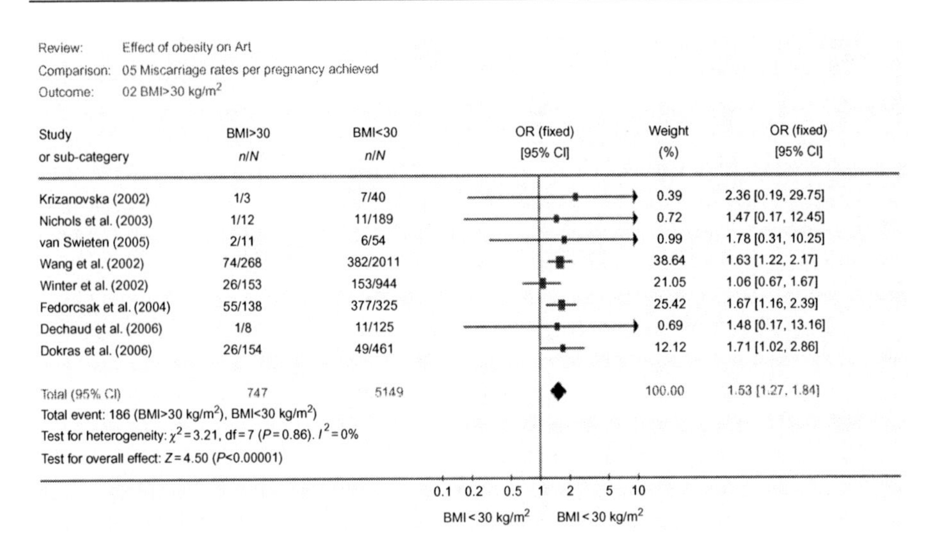

Figure 10.2 Miscarriage per pregnancy achieved after ART in obese women.[18]

Study or sub-category	BMI>30 kg/m² n/N	BMI Normal n/N	RR (fixed) [95% CI]	Weight (%)	OR (fixed) [95% CI]
Fedorcsak et al. (2004)	78/463	718/3457		31.65	0.81 [0.66, 1.00]
Dechaud et al. (2006)	99/419	1230/3930		44.23	0.75 [0.63, 0.90]
Sneed et al. (2008)	66/307	148/613		18.43	0.89 [0.69, 1.15]
Bellver et al. (2010)	6/48	76/394		3.08	0.65 [0.30, 1.41]
Zhang (2010)	7/27	582/2222		2.61	0.99 [0.52, 1.88]
Total (95% CI)	1264	10516		100.00	1.53 [1.27, 1.84]

Total events: 256 (BMI>30 kg/m²), 2754 (BMI normal)
Test for heterogeneity: $\chi^2 = 1.80$, df = 4 ($P = 0.77$). $I^2 = 0\%$
Test for overall effect: $Z = 3.72$ ($P = 0.0002$)

0.1 0.2 0.5 1 2 5 10
BMI<30 BMI normal

Figure 10.3 Live-birth rate in obese women compared to those with normal BMI.[18]

a normal BMI (OR 0.91, 95% CI 0.85–0.98). The reduction in live-birth rate in women who are obese (BMI ≥ 30 kg/m^2) was 20% compared to those with a normal BMI (RR 0.80, 95% CI 0.71–0.90). These reductions were statistically significant ($P < 0.0002$) [18] (Figure 10.3).

Safety Issues for Mothers and Offspring

Many significant health issues and chronic medical conditions for women are associated with obesity. The risk of diabetes increases with the degree and duration of being overweight. Risks of coronary artery disease increase with obesity, and weight reduction is an important adjunct to the management of those at risk. There is evidence that risks of gynaecological, particularly endometrial, and breast cancer increase in obese women, and increased weight may influence the outcome of disease.

Many adverse maternal, foetal and neonatal outcomes are known to be associated with obesity. Management of the infertile women thus poses complex questions linked to the welfare of potential mothers and their offsprings. Many pregnancy-associated complications occur with greater frequency in obese women, e.g. pregnancy-induced hypertension and gestational diabetes [24]. Need for intervention carries with the specifics of the difficulty of surgery in those who are morbidly obese together with the potential for complications such as infection, venous thrombo-embolism and anaesthetic hazards. Maternal mortality, while a rare occurrence, has associations with obesity, and, in a recent report, highlighted the fact that many maternal deaths occurred in women with pre-existing medical conditions, including obesity (BMI $\geq$ 30 kg/m^2) which seriously affected the outcome of their pregnancies [24].

Foetal risks in pregnancy are a concern with observed increased occurrence of foetal abnormality, macrosomia, low-birth weight, neonatal mortality and stillbirth. An influential report suggested that obesity is the principal modifiable risk factor for stillbirth in the developed world, greater than the increased maternal age and smoking.

Beyond these short-term outcomes, the long-term health of individuals born to obese mothers is a public health issue of concern. Children of obese women will grow up with greater risks of coronary heart disease, hypertension, glucose intolerance and diabetes as well as themselves being obese, thereby perpetuating the problem to the subsequent generation.

The management of the obese infertile women raises not only economic issues of note given increased costs associated with treatment but also those associated with the management of complicated pregnancies, particularly the need for increased surveillance, higher rates of operative delivery and the management of women with gestational diabetes and hypertension.

Ethical Issues Relevant to Access to Services

Debate within the past few years has taken place as to whether these morbidities and adverse outcomes, together with higher costs, should play a part in whether obese women should be permitted the same access to infertility services as those who are not overweight [2,3,25,26].

It could be argued, with the prevalence of obesity being at the level it is, that in fact the boundaries of what can be considered normal in the population have changed. Adverse outcomes, however, would suggest this is not the case. Charges that a restrictive policy leads to stigmatisation of obese patients have been levelled, but genuine health hazards are being identified which carry significant implications for the individuals concerned. Some have suggested that the autonomy of the individual to determine their own health would be being infringed by policies to deny access to care. On the other hand, the identification of long-term health risks could be considered as an opportunity for the empowerment of the individual to make lifestyle adjustments that may have real health benefits to themselves. Bearing in mind the issues described above, it is clear that patients have responsibilities beyond themselves, and health care professionals similarly have responsibilities to offsprings and to society at large. Scarce resources, particularly at the present time, should be used to maximum effect. Interventions to assist individuals to achieve and sustain weight loss are not always effective [27]. However, it would be anomalous for the reproductive health sector not to share with other areas of medical practice the public health responsibility of health promotion messages relevant to weight. Losing weight may of course delay the initiation of treatment and this is important particularly in those who seek assistance in later reproductive years [14]. However, in the younger patient, the amount of weight loss to make a difference may not be substantial and the time taken to achieve a target may not adversely affect the chance of treatment being successful. That said, there is no randomised trial evidence at present that weight-loss programmes prior to IVF treatment have an appreciable effect on outcomes or pregnancy-associated complications.

Conclusion

There is irrefutable evidence that fertility potential is adversely affected in obese women. The proportion of patients accessing infertility services who are obese is increasing. Natural fecundity, responses to treatment and pregnancy outcomes are sub-optimal in this group of patients. The mechanisms whereby these effects manifest are not fully understood, but it is likely that the causes are multi-factorial including endocrine, inflammatory pathways as well as effects on oocyte quality. Interventions to address sub-fertility while offering increased potential for conception raise important questions relevant to the safety of mothers and offsprings. While adverse outcomes are increased in this group of patients, the absolute risk of complications to the individual remains relatively small. Most conceptions will result in healthy live-born babies but they will have increased lifetime health risks. Ethical issues in this sphere of reproductive medicine challenge the principles of helping the individual while taking account of consequences for others, not least the potential child but, bearing in mind the costs of treatment, pregnancy care and beyond, the views of society at large.

References

1. Ogden CL, Carroll MD, Curtin LR, McDowell MA, Tabak CJ, Flegal KM. Prevalence of overweight and obesity in the United States, 1999–2004. *J Am Med Assoc.* 2006;295:1549–1555.
2. Pandey S, Pandey S, Maheshwari A, Bhattacharya S. The impact of female obesity on the outcome of fertility treatment. *J Hum Reprod Sci.* 2010;3:62–67.
3. Maheshwari A, Scotland G, Bell J, McTavish A, Hamilton M, Bhattacharya S. The direct health services costs of providing assisted reproductive services in overweight or obese women: a retrospective cross-sectional analysis. *Hum Reprod.* 2009;1:1–8.
4. Rich-Edwards JW, Goldman MB, Willett WC, et al. Adolescent body mass index and infertility caused by ovulatory disorder. *Am J Obstet Gynecol.* 1994;171:171–177.

5. Hassan MA, Killick SR. Negative lifestyle is associated with a significant reduction in fecundity. *Fertil Steril.* 2004;81:384–392.
6. Shah M. Obesity and sexuality in women. *Obstet Gynecol Clin North Am.* 2009;36:347–360.
7. Brewer CJ, Balen AH. The adverse effects of obesity on conception and implantation. *Reproduction.* 2010;140(3):347–364.
8. Metwally M, Ledger WL, Li TC. Reproductive endocrinology and clinical aspects of obesity in women. *Ann N Y Acad Sci.* 2008;1127(April):140–146.
9. Esinler I, Bozdag G, Yarali H. Impact of isolated obesity on ICSI outcome. *Reprod Biomed Online.* 2008;17:583–587.
10. Dokras A, Baredziak L, Blaine J, Syrop C, VanVoorhis BJ, Sparks A. Obstetric outcomes after in vitro fertilization in obese and morbidly obese women. *Obstet Gynecol.* 2006;108:61–69.
11. Minge CE, Bennett BD, Norman RJ, Robker RL. Peroxisome proliferator-activated receptor-gamma agonist rosiglitazone reverses the adverse effects of diet-induced obesity on oocyte quality. *Endocrinology.* 2008;149:2646–2656.
12. Maheshwari A, Stofberg L, Bhattacharya S. Effect of overweight and obesity on assisted reproductive technology a systematic review. *Hum Reprod Update.* 2007;1:433–444.
13. Bellver J, Melo MA, Bosch E, Serra V, Remohi J, Pellicer A. Obesity and poor reproductive outcome: the potential role of the endometrium. *Fertil Steril.* 2007;88:446–451.
14. Sneed ML, Uhler ML, Grotjan HE, Rapisarda JJ, Lederer KJ, Beltsos AN. Body mass index: impact on IVF success appears age-related. *Hum Reprod.* 2008;23:1835–1839.
15. Chong AP, Rafael RW, Forte CC. Influence of weight in the induction of ovulation with human menopausal gonadotropin and human chorionic gonadotropin. *Fertil Steril.* 1986;46:599–603.
16. Hamilton-Fairly D, Kiddy D, Watson H, Paterson C, Franks S. Association of moderate obesity with a poor pregnancy outcome in women with polycystic ovary syndrome treated with low dose gonadotrophin. *Br J Obstet Gynaecol.* 1992;99:128–133.
17. Balen A, Platteau P, Nyboe Andersen A, et al. The influence of body weight on response to ovulation induction with gonadotrophins in 335 women with World Health Organization group II anovulatory infertility. *Br J Obstet Gynaecol.* 2006;113:1195–1202.
18. Rittenberg V, Seshadri S, Sunkara SK, Sobaleva S, Oteng-Ntim E, El-Toukhy T. Effect of body mass index on IVF treatment outcome: an updated systematic review and meta-analysis. *Reprod Biomed Online.* 2011;23:421–439.
19. Kwan I, Bhattacharya S, McNeil A. Monitoring of stimulated cycles in assisted reproduction (IVF and ICSI). *Cochrane Database Syst Rev.* 2008;(2):CD005289. doi:10.1002/14651858.
20. Luke B, Brown MB, Stern JE, Missmer SA, Fujimoto VY, Leach R. Racial and ethnic disparities in assisted reproductive technology pregnancy and live birth rates within body mass index categories. *Fertil Steril.* 2011;95:1661–1666.
21. Luke B, Brown MB, Stern JE, et al. Female obesity adversely affects assisted reproductive technology (ART) pregnancy and live birth rates. *Hum Reprod.* 2011;26:245–252.
22. Styne-Gross A, Elkind-Hirsch K, Scott RT. Obesity does not impact implantation rates or pregnancy outcome in women attempting conception through oocyte donation. *Fertil Steril.* 2005;83:1629–1634.
23. Bellver J, Ayllón Y, Ferrando M, et al. Female obesity impairs in vitro fertilization outcome without affecting embryo quality. *Fertil Steril.* 2010;93:447–454.
24. Centre for Maternal and Child Enquiries and the Royal College of Obstetricians and Gynaecologists. *Management of Women with Obesity in Pregnancy.* London: RCOG; 2010.

25. Koning AMH, Kuchenbecker WKH, Groen H, et al. Economic consequences of overweight and obesity in infertility: a framework for evaluating the costs and outcomes of fertility care. *Hum Reprod Update*. 2010;16:246–254.
26. Dondorp W, de Wert G, Pennings G, et al. Lifestyle-related factors and access to medically assisted reproduction. *Hum Reprod*. 2010;25:578–583.
27. National Institute for Health and Clinical Excellence. *Dietary Interventions and Physical Activity Interventions for Weight Management before, during and after Pregnancy. NICE Public Health Guidance 27*. National Institute for Health and Clinical Excellence; 2010.

Section 3

Obesity and Male Reproduction

11 Obesity and Sexual Dysfunction in Men

Darius A. Paduch[1,2], Alexander Bolyakov[1,2] and Laurent Vaucher[3]

[1]Department of Urology, Weill Cornell Medical College, New York, NY, [2]Consulting Research Services, Inc, Red Bank, NJ, [3]Department of Urology, University of Lausanne, Switzerland

Prevalence of obesity has been rising over last several decades in industrialised countries and 68% of US adults are overweight and 35% are obese [1].

Obesity has been linked to increased risk of diabetes mellitus (DM), hypertension (HTN), risk of cardiovascular events and stroke. More recently low testosterone and erectile dysfunction (ED) have also been associated with obesity.

Diagnosis of obesity is based on body mass index (BMI). BMI is calculated by dividing weight by square of height. BMI is very useful and has been extensively used in epidemiological studies but should not be used as a sole measure of obesity as it does not measure per se the percentage of total body fat. Using WHO criteria for adults older than 20, normal BMI is between 18.5 and 25, subjects are overweight if their BMI is between 25 and 30, they are obese if their BMI is 30–40 and morbidly obese if BMI > 40.

Physiology of Sexual Function

The normal erectile function depends on intact vascular function of the penis – arterial blood flow and venous closing mechanism, normal sensory innervation of the penis, normal central processing and integration of sexual cues in brain and spinal cord and intact autonomous nervous system (Figure 11.1).

Sexual cues (tactile, auditory, olfactory, past experience signals) are integrated in the hypothalamic and cortical centres and descending signals sent through autonomic nervous system and spinal cord to thoracic and lumbosacral regions of sexual response (Figure 11.2). Somatosensory input from penis is carried through dorsal nerve of the penis through pudendal nerve and ends in the sacral region S2–S4. Posterocentral gyrus in cortex represents primary penile sensory activation. Scrotum and pubic area is innervated by branches of ilioinquinal nerve (L1) and femoral cutaneous nerve (L2–L3) (Figures 11.3 and 11.4). DM can impair normal conduction in pudendal nerve and has been postulated as main reason for diabetic ED [2].

Obesity. DOI: http://dx.doi.org/10.1016/B978-0-12-416045-3.00011-X

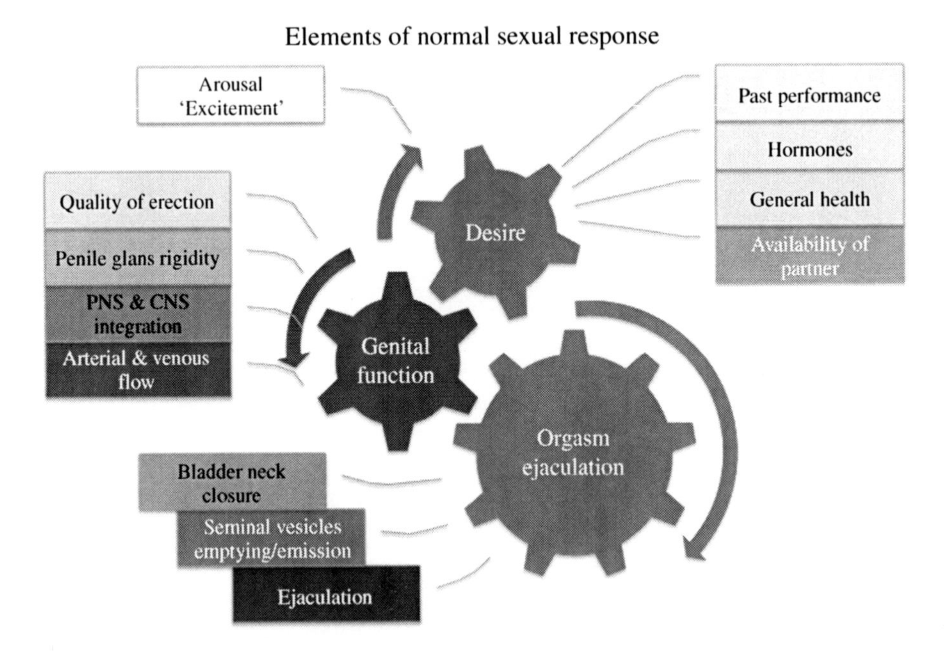

Figure 11.1 Normal sexual response in men is initiated by desire which leads to sexual act with tumescence and progressive increase in arousal leading to orgasm and ejaculation. Each component depends on complex neurobiological and hormonal environment.

The descending signals reach penile erectile tissues through cavernous nerves carrying parasympathetic and sympathetic fibres (Figure 11.4). Parasympathetic stimulation results in relaxation of smooth muscles within penis, opening of bilateral cavernosal arteries and sudden increase in blood flow through penis which results in tumescence (penile erection). At the same time subtunical veins which normally drain blood from penis have to close (Figure 11.5).

Any pathology which pushes the balance of muscle relaxation and increase in penile blood flow will result in ED.

Over last two decades most of the research has focused on ED; however, it is well known that presence of rigid erection is not sufficient for satisfactory sexual performance, thus both evaluation and therapeutic interventions need to embrace different aspects of sexuality.

Sexual dysfunction can be divided into ED (problems sustaining erections adequate for penetration), disorder of sex drive (libido), disorder of arousal (excitement), disorders of ejaculation and orgasm. The ejaculatory dysfunction can present itself as premature ejaculation (intravaginal latency time less than 2 min), delayed ejaculation (subjective prolonged time to ejaculate) and anejaculation (lack of ejaculation). Lack of ejaculation can be the result of retrograde ejaculation (semen goes back into bladder) or lack of emission. Orgasm is a subjective sensation of enhanced pleasure followed by postcoital refractory period. Orgasm is typically associated

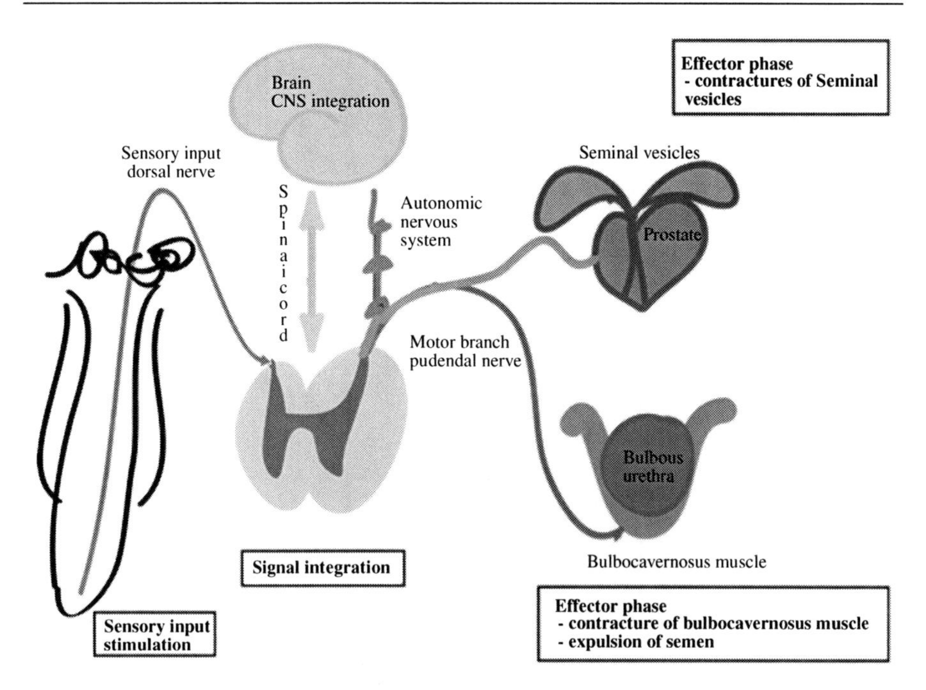

Figure 11.2 Sensory input from the penis is carried through pudendal nerve and ascending pathways to cortex. Autonomic and peripheral nervous system is involved in normal sexual response, specifically erections, emission and ejaculation.

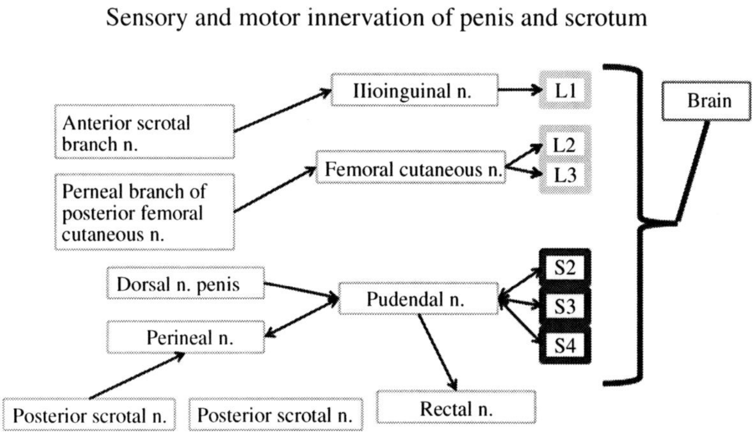

Sensory and motor innervation of penis and scrotum

Ilioinguinal n.
L1
Brain
Anterior scrotal branch n.
L2
Femoral cutaneous n.
L3
Perneal branch of posterior femoral cutaneous n.
S2
Dorsal n. penis
Pudendal n.
S3
Perineal n.
S4
Posterior scrotal n.
Posterior scrotal n.
Rectal n.

Figure 11.3 Somatosensory and motor innervation involved in sexual response in men. Pudendal nerve has both sensory and motor branches. Dorsal nerve of penis innervates most of the shaft and the glans of penis.

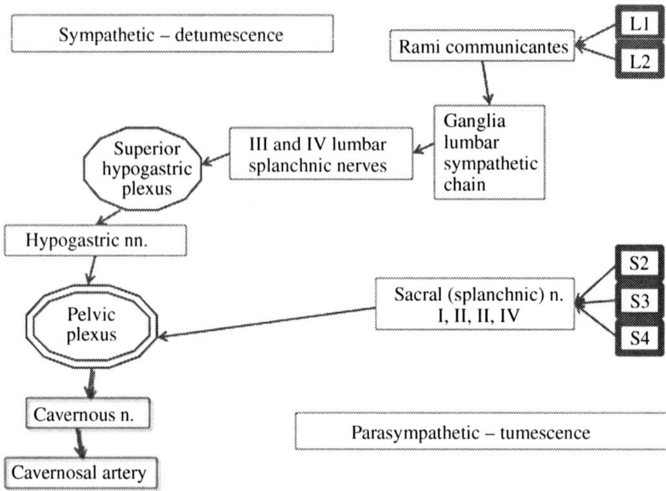

Figure 11.4 Autonomic system innervation of penis. Parasympathetic fibres through release of Ach and NO stimulate erection (green), and sympathetic fibres oppose erection (red). Cavernous nn. carry both sympathetic and parasympathetic fibres and innervated erectile tissues of penis. (For interpretation of the references to colour in this figure legend, the reader is referred to the web version of this book.)

with ejaculation in normal subjects. Orgasmic dysfunction can range from lack of orgasm through decreased sensation of orgasm (Figure 11.6).

Sex drive is a complex neuropsychological phenomenon affected by hormones, energy status, health, social norms and emotional well-being. Body image and recollection of positive sexual experience play important role in sex drive and normal sexual response in both men and women. Obesity with change in body image, decrease in functional length of penis because of pubic fat pad and limitation in physical capabilities may further erode one's confidence in sexual performance and have detrimental effect on sex drive leading to withdrawal from interpersonal relationships.

Erection (tumescence) starts with stimulation of the parasympathetic nerves with release of acetylcholine (Ach) and activation of nitric oxide synthetase within nerve endings (nNOS) and endothelial NOS (eNOS). NOS convert L-arginine into NO – a potent vasodilator which is transported by diffusion to smooth muscles. NO activates guanylate (guanylyl) cyclase converting GTP to active cGMP. cGMP activates phosphorylation of target proteins resulting in decrease in intracellular calcium, smooth muscle relaxation and erection (Figure 11.7). Sympathetic nervous system and local factors like endothelin-1 (ET-1) oppose smooth muscle relaxation and result in detumescence.

Obesity is known to affect hormonal levels, specifically increase in circulating oestradiol (E2) level and decrease in total testosterone (TT) level. In addition

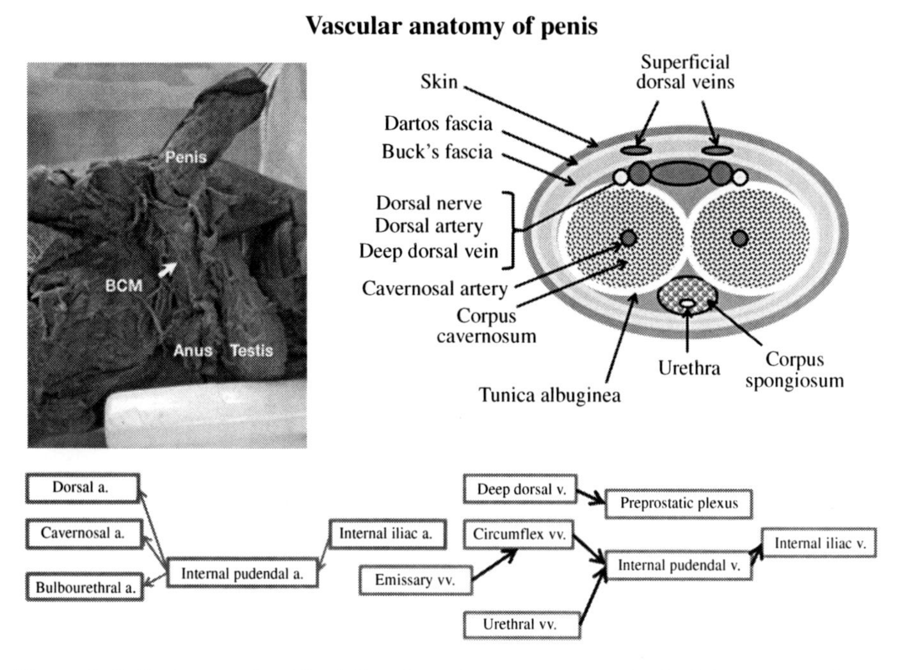

Figure 11.5 During erection blood flow through cavernosal arteries (red) increases and results in compression (closure) of emissary veins (blue). Two dorsal nerves carry sensory signals from penis. (For interpretation of the references to colour in this figure legend, the reader is referred to the web version of this book.)

obesity-related medical co-morbidities like HTN, dyslipidaemia and diabetes mellitus (DM) are linked to ED.

Exact mechanism of obesity-related sexual dysfunction is an area of intense research and multifactorial model with local impairment of penile tissue relaxation with global endocrinopathy and metabolic changes in nerve signalling seem to best describe clinically observed sexual dysfunction in obese men [3,4].

In rodents fed with high-fat diet (HFD) versus standard diet, there was clinically and statistically significant drop in cavernous strip relaxation after induction with Ach [5]. This and other studies showed decreased concentration of cGMP in penile tissue from obese mice as well as decrease in endothelial and Ach and non-adrenergic non-cholinergic nitrergic signalling thus impairing the relaxation of penis which is necessary for normal tumescence [6]. The tissue from obese penis are much more sensitive to adrenergic stimulation and ET-1 [7]. Thus obesity decreases penile ability to relax smooth muscles and increases sensitivity to stress-related signalling impairing the balance between tumescence and detumescence which is necessary for normal erection.

Hyperglycaemia decreases NO production by eNOS by O-linked glycosylation of eNOS at the Akt target S1177 [8]. Diabetes is clearly associated with diminished endothelial production of NO [9,10].

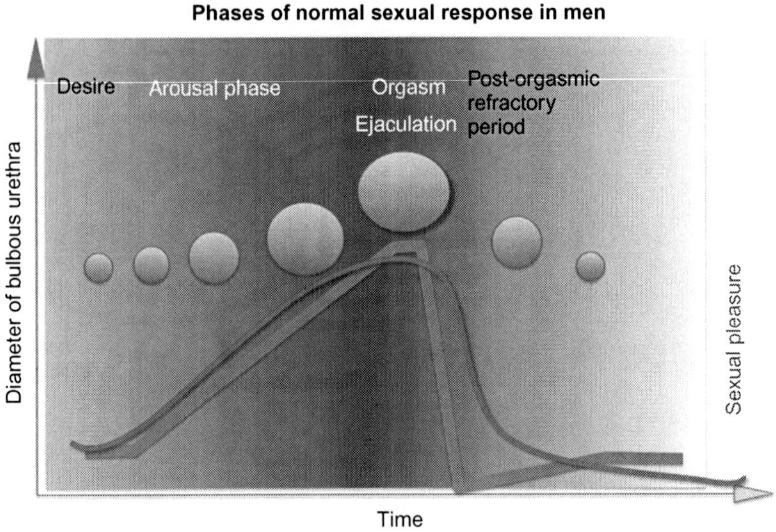

Phases of normal sexual response in men

Desire Arousal phase Orgasm Ejaculation Post-orgasmic refractory period

Diameter of bulbous urethra

Sexual pleasure

Time

Figure 11.6 During normal sexual response the arousal (excitement) has to progressively increase which is measured by increase in filling of corpus spongiosum and bulbous urethra. Impaired nerve signalling can impede arousal and result in anejaculation or anorgasmia.

Sexual response in men is initiated by desire to have sexual activity. Desire is a complex neurobiological Phenomenon which is first experienced with initiation of puberty; thus sex steroids have always been considered a critical component of normal sexual desire. Although testosterone level and desire do not have linear correlation, multiple studies in diverse ethnical groups showed similar association between low testosterone and low sexual desire [11–13].

It is well established that in obese men the sexual desire is negatively correlated with total and free testosterone [14]. Gastric bypass surgery improved testosterone levels and improved sexual desire.

The negative effects of obesity on sexual desire are most likely related to low testosterone and elevated E2, but risk of failure and changed body image may be significant contributing factors.

Testosterone is produced in testes under control of luteinising hormone (LH) and local paracrine control. Testosterone is metabolised by aromatase CYP19 to E2 (Figure 11.8). Elevated levels of E2 have been found in most obese men. E2 suppresses central release of LH but it also has negative effects on Leydig cell function and normal testosterone production [15,16].

Normal sexual function depends on adequate balance between testosterone which in men should be high and E2 which needs to be low. Treatment with aromatase inhibitors may help increasing testosterone level in some obese men with low testosterone and elevated E2 (Treatment section).

During sexual activity – progressive and linear increase in level of excitement – arousal is necessary to achieve orgasm and sustain erection. Little is known about

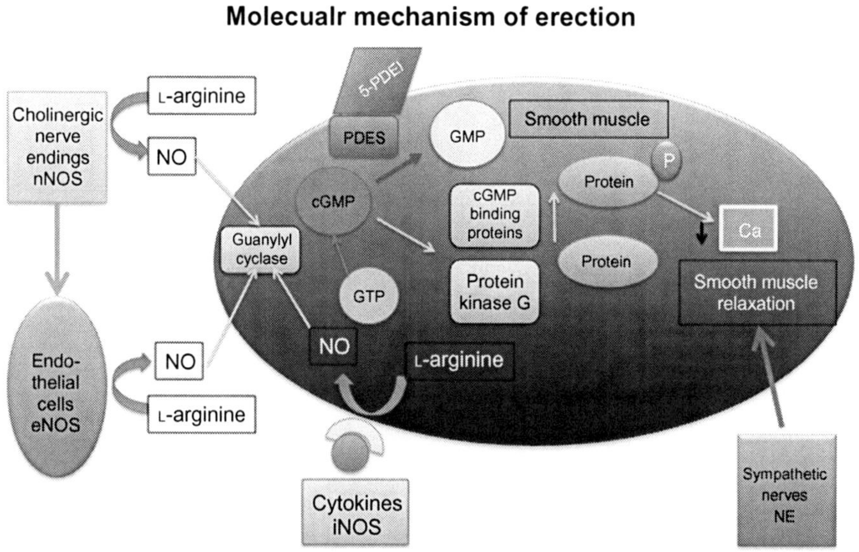

Molecualr mechanism of erection

Figure 11.7 Nitric oxide (NO) – a potent vasodilator – is produced by eNOS and nNOS from L-arginine. NO stimulates production of cGMP which triggers phosphorylation of downstream proteins and signalling cascade leading to decrease in Ca++ levels and smooth muscle relaxation. The 5-PDE inactivates cGMP to GMP and leads to detumescence. Inhibitors of 5-PDE increase cGMP and thus help in men with ED. Cytokines through inducible NOS (iNOS) as well as sympathetic nerves, ET-1 and Rho-kinase (not shown) further modulate erectile function.

arousal in men, but obesity severely impairs arousal in females [17,18]. As arousal is a central neurobiological process, it is very plausible that obese men suffer from similar issues with arousal. Here again low testosterone and hormonal abnormalities may interfere with normal processing of sexual cues necessary to achieve normal progression of arousal.

Sexual Dysfunction and Obesity-Related Co-morbidities

It is often difficult to dissect pure effects of obesity from associated obesity-related medical co-morbidities like HTN, peripheral vascular disease (PVD), hypercholesterolaemia and diabetes.

Hypercholesterolaemia

Hypercholesterolaemia – defined as elevated total cholesterol and low-density lipoprotein (LDL) – has been linked to increased risk of coronary artery disease (CAD) in Framingham Heart Study and many others [19]. Massachusetts Male

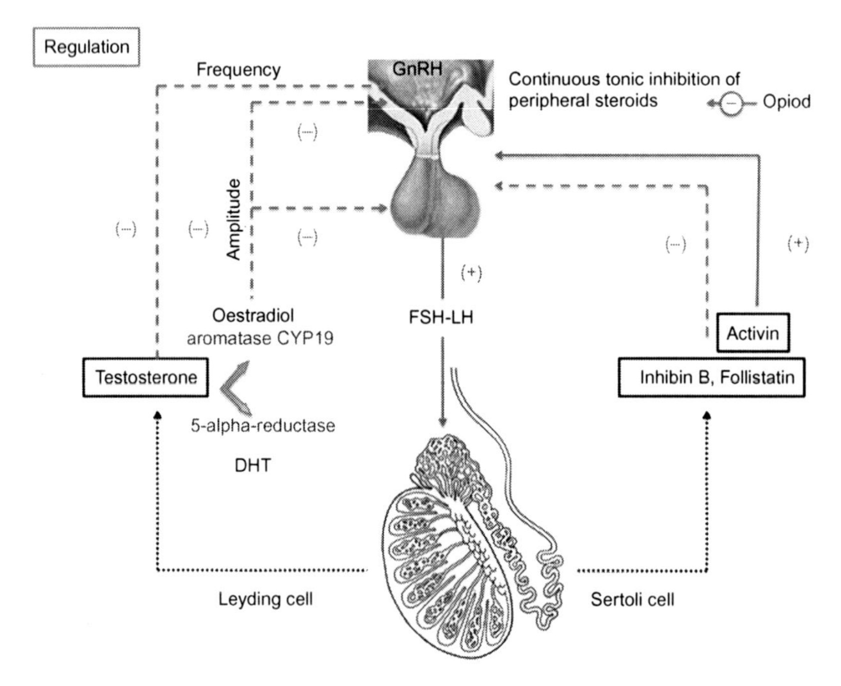

Figure 11.8 Testosterone is produced by Leydig cells under control of hypothalamic–pituitary axis. LH released from pituitary stimulates production of testosterone. Aromatase CYP19 coverts testosterone into E2. Both testosterone and E2 can inhibit release of LH through negative feedback.

Aging Study showed that ED is inversely related to baseline high-density lipoprotein and prevalence of ED doubled over 9 years in men who where obese at baseline as compared to men who were not obese at baseline or follow-up [20,21]. Obesity increases risk of vasculogenic ED as assessed by penile Doppler ultrasonography [22]. But from published literature it appears that it is a long-term obesity with obesity-related co-morbidities which correlates with ED rather than elevated BMI by itself. In Rancho Bernardo Study which followed men over 25 years the age, hypercholesterolaemia and obesity were independent predictors of severity of ED in logit model but obesity was not an independent predictor of presence of ED by itself.

Small non-randomised clinical trial showed that reduction in total cholesterol to below 200 mg/dl and LDL to less than 120 mg/dl had positive effect on ED after 3.7 months of treatment [23]. However, randomised double-blinded trial (STED TRAIL) of simvastatin failed to show improvement in erectile function as compared to placebo. It is possible that statins have role in men with ED and hypercholesterolaemia who fail to respond to sildenafil [24,25]. However, recent study of tadalafil 20 mg three times a week versus 10 mg atorvastatin showed that tadalafil was better in restoring ED than atorvastatin; however, atorvastatin showed some improvement in erectile function especially in men with hypercholesterolaemia

[26,27]. French Pharmacovigilance System Database study showed that statins may actually induce or worsen ED [28]. So far no conclusive evidence supports routine use of statins in patients with ED without clear indication for use of statins because of cardiovascular protective effects [29].

Rather disappointing results of statin treatments on improving ED in men with hypercholesterolaemia may be due to short follow-up and severity of PVD. Reduction in prevalence of cardiovascular events with statins has been showed after 5 years of therapy by most studies (CAPS, CARE, LIPID); thus it is plausible that with longer follow-up the positive effects of statin treatment on ED may be noticeable [30,31]. Another complicating factor may be lowering of TT in some men taking statins [32–34]. Thus testosterone level should be measured in men on statin therapy. In summary it is clear that dyslipidaemia is a risk factor for ED but it is not clear if obesity by itself without its sequel has similar detrimental effect on ED. Clearly further studies are needed.

Hypertension

Approximately 50–70% of men with HTN report varied degree of ED [35,36]. The underlying mechanism of ED in HTN seems to be HTN-induced PVD and severity of ED in HTN men is correlated with duration of HTN [37]. HTN impairs neurogenic induced smooth muscle relaxation and reduction in superoxide dismutase in animal models [38].

Treatment of HTN may further worsen ED and contribute to poor compliance with antihypertensive medications [39,40]. Non-selective beta-blockers, such as hydrochlorothiazide, spironolactone, angiotensin II antagonists, are known to result in ED in significant number of patients; thus angiotensin converting enzyme inhibitors and selective beta-blockers are better choice for men with HTN and pre-existing ED [41,42]. In men who require multidrug therapy for HTN, adding 5-phosphodiesterase inhibitor (5-PDEI) improves compliance with antihypertensive therapy and results in better blood pressure control [43].

Peripheral Vascular Disease

Classic epidemiological study by Blumentals showed that men with ED have increased risk of PVD by 75% [44]. Even after adjusting for the presence of other risk factors for stroke, men with ED have higher risk of stroke and lower risk of stroke-free survival over 5-year follow-up study in Taiwan [45]. Often ED is the first sign of PVD prior to claudication; thus men with ED should be screened for PVD [46]. It is believed that endothelial dysfunction combined with hypercholesterolaemia is responsible for PVD and vasculogenic ED. Large epidemiological studies showed that PVD is prevalent in obesity [47].

Coronary Artery Disease

Intracavernosal arteries in penis are less than 1 mm in diameter, thus it is no surprise that often ED is first manifestation and predates CAD [48,49]. Salem et al.

showed that presence of ED is an additional cardiovascular risk which should be considered in stratification assessment and decision-making for need for further invasive coronary evaluation [50,51]. Thus obese patients with ED should be evaluated by cardiologists to determine if they need stress test or further testing to assess their cardiovascular risks.

Diabetes Mellitus

Effects of DM on ED is multifactorial but neurogenic dysfunction and vasculopathy with impairments of NO signalling decrease in vasodilation are best understood at this point.

DM can affect both sensory signalling and result in autonomic nervous system dysfunction. In obese diabetic subjects, the peripheral neuropathy in lower extremities assessed by increased vibration is strongly associated with ED [52]. This should not be a surprise as tactile stimulation is one of most important sexual cues. Micro and macrovascular disease has been linked to DM using standardised instruments to measure sexual function [53]. Similar changes in microvascular environment with decrease in endothelial function within penis were found in animal models of DM [6].

Evaluation

Based on the high prevalence of low testosterone and sexual dysfunction in obese men, it is prudent to measure testosterone level in most obese men, especially those with loss of morning erections, decreased sex drive, fatigue and ED. Morning total and free testosterone should be obtained using the most reliable and sensitive method. In our practice we use liquid chromatography and mass spectrometry (LCMS) to measure TT. LCMS has been recommended as preferred method to measure testosterone in hypogonadal men because of its lower coefficient of variance as compared to other methods [54].

FDA uses cut-off point of 300 ng/dl to establish hypogonadism in pharmacological studies. In Europe 10.2 nmol/l is often used. One need to understand that testosterone levels change over age and 300 ng/dl cut-off point is probably most appropriate in older men >65 years of age as most of the studies on testosterone level were done in older men. Free testosterone and bioavailable testosterone is useful in men with normal or normal low TT who present with symptoms. In men who have low TT, the following values should be obtained: Follicle stimulating hormone (FSH), luteinizing hormone (LH), Prolactin (PRL), thyroid profile, E2, cortisol and baseline Prostate specific Antigen. CBC and liver function tests should be obtained at baseline. In our practice we also obtain ultrasensitive C reactive protein (CRP) and HgA1c at baseline in obese men who present with sexual dysfunction. In men with type I DM in the Diabetes Control and Complications Trial the risk of ED was correlated with HgA1c and men who had tight control of DM with insulin had significantly lower rate of ED [55]. Severity of ED increases dramatically in men whose HgA1c is above 8% and with DM type II of 6 or more years [56]. Considering that significant number of men with early diabetes are not aware of their abnormal sugars, it is critical to establish diagnosis of DM early to prevent microvascular sequels which with time may be difficult to reverse or control.

Erectile and endothelial dysfunction may have similar pathways through impaired NO activity, but in obese men the endothelial dysfunction may be further impaired through increase in interleukins (IL-6, IL-8, IL-18) and CRP. Obese men with ED but not obese men without ED had significant increase in CRP and inflammatory procytokines [57].

Elevated CRP is associated with increased risk of cardiovascular events and in many men presence of such objectivity-measured risk may aid in behavioural modifications to lose weight and to exercise.

This study also points to the fact that it is not obesity itself which result in the ED but associated vasculopathy and pro-inflammatory response. It is not known at this point why some obese men do not suffer from ED. Possible explanation may be limitation in BMI as marker of obesity and further studies with better assessment of lean body mass and per cent of body fat may help us to understand the link between obesity and ED better.

We strongly advocate active screening for sexual dysfunction as it has been shown that correction of ED with medical therapy may have positive impact on glycaemic control through better adherence to medical therapy and diet.

Depending on practice one can consider using screening questionnaires to assess ED. Questionnaires which assess broad aspects of sexuality, specifically sex drive, orgasmic and ejaculatory dysfunction in addition to ED, may be better suited for obese men considering that they often suffer from complex sexual dysfunction when ED is only one of the domains. In our practice we use Male sexual health questionnaire (MSHQ); (www.mapi-trust.org/services/questionnairelicensing) during initial visit and follow-up but each physician has to choose the instrument which is most appropriate to his or her patient population and cultural and social norms.

Although BMI is most commonly used to measure obesity, we use lean body mass and per cent body fat as well as waist circumference. BMI is very useful in large epidemiological studies but may be less useful in managing individual patients.

In clearly hypogonadal men who have elevated PRL or unexplained low or low normal LH and FSH one may consider CT or MRI of brain with attention to pituitary and hypothalamus to exclude pituitary or hypothalamic mass.

However, obesity is associated with hypogonadotropic hypogonadism because of elevated E2; thus decision to obtain additional imaging studies has to be considered on individual basis.

Typically in obese patient with ED by history and verified by questionnaires other than hormonal evaluation, no further work up is needed especially if testosterone was normal. These men may be started on one of the 5-PDEI (see treatment section) combined with diet, exercise and management of existing co-morbidities. Smoking cessation is critical.

However, in men with failed response to oral therapy, in younger men with risks for PVD or neuropathy, referral to penile Doppler ultrasound and neurosensory testing may be prudent. Often patients especially younger ones want to know why they suffer from ED.

Penile Doppler ultrasound is a minimally invasive and very well-tolerated procedure. After checking blood pressure, explaining risks like need for redosing intracavernosal injection, bruise, prolonged erection and embarrassment, patient is brought to sexual medicine laboratory when intracavernosal injection is administered in the penis to induce erection. We typically use from 5 to 10 units of Trimix and perform continuous measurement of blood flow through cavernosal arteries. Patient is allowed to self-stimulate and watch adult audiovisual aids as needed. In men with severe vasculopathy the dose of medication is escalated up to 60 units; however, risk of priapism increases with higher doses. It is critical to achieve rigid erection in the sexual medicine laboratory to exclude venous leak which is often over-diagnosed because of inadequate dosing and stimulation. We measure peak systolic velocity (PSV), end diastolic velocity and resistive index. At the same time, the length of time it takes to ejaculate or orgasm (in men with anejaculation) can be measured and amount of force to achieve tactile stimulation to sustain erection is recorded. In men with decreased sensation to the penis, penile biothesiometry is performed and vibratory thresholds measured. Typically once the subject achieves non-bending erection the study is completed, and the patient is observed at 15 min intervals to assure detumescence. No patient is allowed to leave office unless his penis is flaccid. In case of prolonged erection, the 500–1000 mcg of Neo-Synephrine is injected into the penis and the blood pressure is monitored.

In men who present with delayed ejaculation or anejaculation, the study is continued till patient achieves orgasm or at least 30 min passes to exclude lack of normal arousal as a reason for anejaculation.

In diabetic men who complain of anejaculation, post-orgasmic urine analysis is performed to determine if the subjects suffer from retrograde ejaculation.

Based on penile Doppler ultrasound diagnosis (PSV $<$ 35 cm/s or venous leak of-RI $<$ 0.75), the diagnosis of arteriogenic ED can be established. At the same time optimal dose of intracavernosal pharmacotherapy can be established. Often it is very relieving to the patient to show him that he can achieve erection even with intracavernosal therapy.

Men with venous leak will require typically much higher doses of Intracavernosal therapy and they are not likely to respond to oral medication initially, thus PDUS helps with directing therapy in individual patients. In men with venous leak the penile rings may help however they are difficult to apply in men with severe obesity and panus.

In men with neurogenic ED or ED related to low testosterone one should consider oral 5-PDE as initial therapy. Similarly in men with mild arteriogenic ED 5-PDE should be initial choice.

Treatment

Decrease in weight, improvement in hormonal profile through hormone replacement therapy, with medical and behavioural therapy to improve sexual function, and reinforcement of positive behaviour should be a goal of therapy.

Weight Management

Two approaches can be employed depending on the extent of patients' obesity, willingness to lose weight and acceptance of medical therapy. Some practitioners will defer medical therapy to improve erectile function and focus on weight loss initially. There is no question that decrease in BMI improves erectile function and sexual desire thus this approach can be tried in selective, motivated patients with mild to moderate ED; however, most of the men who present to sexual medicine specialists will present with reactive depression, poor self-esteem, marriage issues and withdrawal from sexual activities. Thus in our practice we strongly advocate combination of diet, exercise with medical therapy to help with ED and replace testosterone in men with hypogonadism.

As sexuality plays an important role in male self-esteem by improving sexual function, we often achieve improvement in mood and dedication to weight loss [58].

High protein, reduced carbohydrate, low-fat diet and low-calorie diet (1000 kcal/day) over 52 weeks have similar degree in improving sexual function, sexual desire and urinary symptoms in obese men. High-protein diet may be easier to tolerate by younger men when combined with exercise programme than restrictive caloric intake diet; however, the choice of diet should be established based on basic metabolic rate, level of daily activities and patient's own goals. Both diets reduced systemic inflammation [59]. Mediterranean diet helps reduce ED [58].

Exercise has been shown to reduce elevated blood pressure and improve ED and should be combined with diet and medical therapy [60].

Pharmacological Therapy

The main stay of therapy of ED in 2012 remains 5-PDEIs. 5-PDEI blocks the inactivation of cGMP in smooth muscles and thus improve smooth muscle relaxation. There are three drugs approved in United States in 2012: sildenafil (Viagra), vardenafil (Levitra and Staxyn) and tadalafil (Cialis). The efficacy of these agents has been established in large multinational trials in men with varied degrees of ED and broad spectrum of aetiologies.

The drugs differ in their time and duration of onset, side effects profile and effects of food on bioavailability. This is especially important issue in men with diabetes who may suffer from gastroparesis and men with frequent snacking as in obesity. Sildenafil and vardenafil bioavailability is significantly decreased by food and they should be taken on empty stomach. Bioavailability of tadalafil is not affected by food and can be taken on empty stomach as well as with food. All information about pharmacological agents described below has been extracted from latest official FDA labels for each of the medications.

Sildenafil (*Viagra*) was the first 5-PDEI approved for ED. Viagra comes in 25, 50 and 100 mg tablets. Sildenafil should be used prior to sexual activity 'on demand'. Sildenafil is rapidly absorbed with median absorption of 60 min when taken in fasting state. High-fat food delays peak level of sildenafil by 60 min and mean reduction in C_{max} of 29%. Although sildenafil's effect may last up to 4 h most of the effectiveness is observed within 2 h after oral intake. Sildenafil has been studied in men with ED and chronic stable angina limited by exercise, there was no difference in exercise-induced angina episodes between men receiving sildenafil and men on placebo. Sildenafil has dose-related impairment of colour discrimination (blue/green) without change in visual acuity or intraocular pressure. Eighty-two per cent of men taking 100 mg of sildenafil, 74% taking 50 mg of sildenafil and 63% taking 25 mg reported improvement in sexual function.

Sildenafil and all 5-PDEI are contraindicated in patients taking nitrates either regularly or intermittently. Sildenafil has systemic vasodilator properties and can cause transient decrease in supine blood pressure of 8.4/5.5 in healthy volunteers. The decrease in blood pressure may be more pronounced with aortic stenosis and severely impaired autonomic control of blood pressure. Men on antihypertensive therapy may be at increased risk of hypotensive episodes; thus in these groups the therapy should be initiated at lower doses and increased slowly. Sildenafil is contraindicated in men with retinitis pigmentosa.

Important issues to discuss with all 5-PDEI are risk of developing prolonged erection (>4 h) and priapism. Sudden loss of vision in one or both eyes may be a sign of non-arteritic anterior ischaemic optic neuropathy (NAION). Most of cases on NAION occurred in men with diabetes, HTN, CAD, hyperlipidaemia and over 50 years. Similarly sudden decrease or loss of hearing which may be accompanied by tinnitus and dizziness has been reported as well. However at this point it has not been determined that the vision and hearing loss is related to use of 5-PDEI or to other factors. Sildenafil is associated with headache 16%, flushing 10%, dyspepsia 7% and nasal congestion 4%. At 100 mg dose 17% of the patients reported dyspepsia and 11% abnormal vision.

It is critical to consider that 5-PDEI should not be used in men with angina, recent acute myocardial infarction (MI) and limited cardiac reserve as normal cardiovascular function is necessary to have sexual activities.

Diabetic patients have somehow reduced response to sildenafil, as 63% of men with type II DM and 67% of men with type I DM reported improvement in erections in pulled data from 11 trials. Eighty-six per cent of men with hyperlipidaemia reported improvement and 69% of men with HTN.

Vardenafil (*Levitra*) in oral form has been approved in 2003 by FDA. Levitra comes in 2.5, 5, 10 and 20 mg tablets. A 10 mg tablet taken 60 min prior to sexual actives is a recommended starting dose. Vardenafil, as all other 5-PDEI, is contraindicated in men taking nitrates, with unstable angina and hypotension.

Levitra is contra indicated in patients with QT syndrome, or those taking class 1A medication (quinidine, procainamide, disopyramide) or class 3 (amiodarone, sotalol, ibutilide, dofetilide) treatment. Levitra should not be used in men with unstable angina, hypotension, uncontrolled HTN (>170/110 mmHg), recent history of stroke, life-threatening arrhythmias, recent MI (<6 months) and severe cardiac failure. Levitra shares the same warnings about prolonged erections and NAION as sildenafil. Headache in 15% of subjects, flushing in 11%, rhinitis in 9% and dyspepsia in 4% are the most common side effects of Levitra. After oral intake of Levitra the peak concentration is noticed at median of 60 min in fasting state. The high-fat meals reduced in C_{max} between 18% and 50% which is of significance in obese patients who often have HFDs. Levitra shows dose-related response which in general population of men with ED is 65% for 5 mg, 75% for 10 mg and 80% for 20 mg compared to placebo response of 52%. However in men with ED and DM the response was 51% at 5 mg, 64% at 10 mg and 65% at 20 mg, thus showing a decreased response rate in patients with DM and ED by almost 15% in higher dose as compared to normal men.

Staxyn is orally disintegrable tablet of 10 mg of vardenafil. It is not interchangeable with 10 mg of Levitra as it achieves higher systemic exposure compared to 10 mg Levitra. In clinical studies the side effects profile was similar to Levitra, but the rate of per patient rate of achieving erection sufficient for penetration was 47% for 10 mg and 22% for placebo.

Tadalafil (*Cialis*) has been approved by FDA in 2003 for use in men with ED and in 2011 for men with benign prostatic hyperplasia (BPH).

Tadalafil can be used on demand at dose of 5, 10 or 20 mg or as daily 5 mg medication.

Because of its selectivity for 5-PDEI, tadalafil has much less flushing and nasal congestion 2% for 5 mg and 3% for 20 mg than sildenafil or vardenafil. Headache can occur in 6% of men on daily dose of 5 mg, and 15% of men taking highest dose of 20 mg. Dyspepsia is seen in 5% of men taking daily Cialis and 10% of men with 20 mg on-demand Cialis. Back pain can be observed in 6% of men taking 20 mg dose and 3% men (3% placebo) on daily 5 mg tadalafil. Seventy-seven per cent of men on 20 mg tadalafil versus 43% on placebo experienced improvement in ability to penetrate. Placebo patients, 64% versus 23%, reported ability to sustain erection. Fifty-seven per cent of men with DM-related ED taking 10 mg Cialis versus 30% men taking placebo reported ability to insert. Daily 5 mg tadalafil was also successful in improving ED as 67% of men on daily tadalafil versus 37% men on placebo reported ability to insert. Similar effectiveness was seen in men with DM.

Obesity increases frequency of *lower urinary tract symptoms and ED*. Tadalafil is the only medication in this group which is approved for men with BPH with or without ED; thus it represents good option in obese men who present with both. Bioavailability of tadalafil is not affected by food, hence it may be a better option in general for men with obesity; however no head-to-head study compared differences in effectiveness between three available 5-PDEI in men with delayed stomach emptying or morbid obesity.

Daily tadalafil has been now shown to improve PSV in diabetic patients with vascular disease as compared to on-demand tadalafil [61]. Considering that daily tadalafil has recently been shown to improve endothelial function which is impaired in men with obesity-related diseases, it seems prudent to initiate therapy with daily 5 mg tadalafil or fixed dose of 20 mg tadalafil thrice a week [62].

Tadalafil shares similar contraindications as sildenafil and vardenafil – as described above.

Penile Intracavernosal Injection Therapy

Subjects who fail oral therapy may benefit from *penile intracavernosal injection therapy* [63]. Self-administered penile injection therapy is well tolerated and may be more successful in diabetic men with severe ED as compared to oral agents [64]. Intra cavernous (IC) therapy can be started with premixed Caverject (prostaglandin E1). PGE1 relaxes smooth muscles in penis through E2 prostaglandin receptors [65]. *Caverject* comes in prefilled vials of 5, 10, 20 and 40 mcg to be used with self-injector or single-use syringe which can deliver from 6 to 20 mcg of medication. Side effects of Caverject are penile pain which can occur in 37% of patients, penile fibrosis in 3%, prolonged erection in 4% and injection site haematoma in 3%. Priapism has been reported in 0.4% subjects according to FDA label. Over 72% of men with diabetes responded to Caverject. Many practices including ours use multidrug combination of vasoactive agents for intracavernosal injections, which may decrease pain and improve efficacy as compared to prostaglandin E1 [66]. Most commonly used preparations are *Trimix* (papaverine 30 mg, phentolamine 1 mg and prostaglandin E1 10 mcg/1 ml) and *super-Trimix* (papaverine 30 mg, phentolamine 2 mg and prostaglandin E1 20 mcg/1 ml). Typically we use 0.05–0.1 ml of Trimix at the start. It is critical to remember that all forms of intracavernosal therapy have to be initiated and dose titrated in physician's office prior to prescribing an adequate dose.

Penile Prosthesis

Penile prosthesis can be considered in men who failed injection therapy; however, it is important that in obese men one should first consider weight loss and aggressive treatment of hypercholesterolaemia and DM prior to penile prosthesis because once placed it has to remain in the penis to prevent scarring and dramatic penis shortening. Obesity is associated with decreased satisfaction with penile prosthesis [67]. Poorly controlled diabetes is associated with increased risk of penile prosthesis infections [68]. Long-term follow-up studies showed that at 5 years 1:10 penile prosthesis has to be replaced [69].

Testosterone Replacement Therapy

It is well established that obese men suffer from hypogonadism but it is less clear if testosterone replacement therapy (TRT) should be considered in all obese men [70]. A significant amount of data exists which support combining weight loss strategy with the use of TRT [71]. TRT may decrease BMI and improve metabolic markers of cardiovascular risks [72,73].

TRT may be administered using injectable forms of therapy – like testosterone enanthate or cypionate, testosterone pellets or topical forms of replacement. Injectable testosterone is typically administered as intramuscular injection at 2 weeks intervals at 200–300 mg of depot-testosterone. Topical preparations can be applied to underarm (Axrion), shoulders (Androgel and Testim) and inner thigh (Fortesta).

Goal of therapy is to achieve normal testosterone levels. Based on recent, large data analysis of patients enrolled in MrOS study in Sweden, the cardioprotective effect of testosterone was observed if levels were above 550 ng/dl thus we typically set goal of therapy in upper normal range [54]. We use exclusively topical preparations because of significantly less risk of polycythaemia as compared to injectable agents.

TRT may have positive metabolic effect in obese men but also improve their sexual drive, ejaculatory volume and orgasmic function and improve patients' quality of life. Further studies are needed to prove the positive long term benefits of TRT in obese men. In men who failed to improve TT despite use of adequate dose of TRT, aromatase inhibitors like anastrozole 1 mg daily may be used for set amount of time (6–12 months) to overcome excessive conversion of testosterone into E2 seen in many obese men [74,75].

Summary

In summary sexual dysfunction is common in obese men. The risk of ED is related to obesity related co-morbidities like hypogonadism, PVD, hypercholesterolemia and diabetes, rather than simple increase in BMI. Reduction in BMI combined with successful treatment of sexual dysfunction improves quality of life and has positive motivator effect on compliance with control of co-morbidities. Further placebo control studies are needed to establish a role of TRT in obese men.

References

1. Flegal KM, Carroll MD, Ogden CL, Curtin LR. Prevalence and trends in obesity among US adults, 1999–2008. *JAMA*. 2010;303(3):235–241.
2. Daniels JS. Abnormal nerve conduction in impotent patients with diabetes mellitus. *Diabetes Care*. 1989;12(7):449–454.
3. Villalba N, Martinez MP, Briones AM, Sánchez A, Salaíces M, García-Sacristán A, et al. Differential structural and functional changes in penile and coronary arteries from obese Zucker rats. *Am J Physiol Heart Circ Physiol*. 2009;297(2):H696–H707.
4. Corona G, Monami M, Boddi V, Balzi D, Melani C, Federico N, et al. Is obesity a further cardiovascular risk factor in patients with erectile dysfunction? *J Sex Med*. 2010;7 (7):2538–2546.
5. Toque HA, Da Silva FH, Calixto MC, Lintomen L, Schenka AA, Saad MJ, et al. High-fat diet associated with obesity induces impairment of mouse corpus cavernosum responses. *BJU Int*. 2011;107(10):1628–1634.
6. Albersen M, Lin G, Fandel TM, Zhang H, QIu X, Lin CS, et al. Functional, metabolic, and morphologic characteristics of a novel rat model of type 2 diabetes-associated erectile dysfunction. *Urology*. 2011;78(2):476e1–476e8.
7. Contreras C, Sanchez A, Martinez P, Raposo R, Climent B, Garcia-Sacristan A, et al. Insulin resistance in penile arteries from a rat model of metabolic syndrome. *Br J Pharmacol*. 2010;161(2):350–364.
8. DU XL, Edelstein D, Dimmeler S, Ju Q, Sui C, Brownlee M. Hyperglycemia inhibits endothelial nitric oxide synthase activity by posttranslational modification at the Akt site. *J Clin Invest*. 2001;108(9):1341–1348.

9. De Angelis L, Marfella MA, Siniscalchi M, Marino L, Nappo E, Giugliano F, et al. Erectile and endothelial dysfunction in type II diabetes: a possible link. *Diabetologia.* 2001;44(9):1155–1160.
10. Akingba AG, Burnett AL. Endothelial nitric oxide synthase protein expression, localization, and activity in the penis of the alloxan-induced diabetic rat. *Mol Urol.* 2001;5(4):189–197.
11. Hofstra J, Loves S, van Wageningen B, Ruinemans-Koerts J, Janssen I, de Boer H. High prevalence of hypogonadotropic hypogonadism in men referred for obesity treatment. *Neth J Med.* 2008;66(3):103–109.
12. Kalucy RS, Crisp AH. Some psychological and social implications of massive obesity. A study of some psychosocial accompaniments of major fat loss occurring without dietary restriction in massively obese patients. *J Psychosom Res.* 1974;18(6):465–473.
13. Chao JK, Sheng Hwang IS, Ma MC, Kuo WH, Liu JH, Chen YP, et al. A survey of obesity and erectile dysfunction of men conscripted into the military in Taiwan. *J Sex Med.* 2011;8(4):1156–1163.
14. Hammoud A, Gibson M, Hunt SC, Adams TD, Carrell DT, Kolotkin RL, et al. Effect of Roux-en-Y gastric bypass surgery on the sex steroids and quality of life in obese men. *J Clin Endocrinol Metab.* 2009;94(4):1329–1332.
15. Cohen PG. Obesity in men: the hypogonadal–estrogen receptor relationship and its effect on glucose homeostasis. *Med Hypotheses.* 2008;70(2):358–360.
16. Wake DJ, Strand M, Rask E, Westerbacka J, Livinstone DEW, Soderberg S, et al. Intra-adipose sex steroid metabolism and body fat distribution in idiopathic human obesity. *Clin Endocrinol (Oxf).* 2007;66(3):440–446.
17. Ostbye T, Kolotkin RL, He H, Overcash F, Brouwer R, Binks M, et al. Sexual functioning in obese adults enrolling in a weight loss study. *J Sex Marital Ther.* 2011;37(3):224–235.
18. Bond DS, Wing RR, Vithiananthan S, Sax HC, Dean Roye G, Ryder BA, et al. Prevalence and degree of sexual dysfunction in a sample of women seeking bariatric surgery. *Surg Obes Relat Dis.* 2009;5(6):698–704.
19. Kannel WB, Castelli WP, Gordon T. Cholesterol in the prediction of atherosclerotic disease. New perspectives based on the Framingham study. *Ann Intern Med.* 1979;90(1): 85–91.
20. Feldman HA, Goldstein I, Hatzichristou DG, Krane RJ, McKinlay JB. Impotence and its medical and psychosocial correlates: results of the Massachusetts Male Aging Study. *J Urol.* 1994;151(1):54–61.
21. Derby CA, Mohr BA, Goldstein I. Modifiable risk factors and erectile dysfunction: can lifestyle changes modify risk? *Urology.* 2000;56(2):302–306.
22. Kim SC, Kim SW, Chung YJ. Men's health in South Korea. *Asian J Androl.* 2011;13(4): 519–525.
23. Saltzman EA, Guay AT, Jacobson J. Improvement in erectile function in men with organic erectile dysfunction by correction of elevated cholesterol levels: a clinical observation. *J Urol.* 2004;172(1):255–258.
24. Herrmann HC, Levine LA, Macaluso Jr J, Walsh M, Bradbury D, Schwartz S, et al. Can atorvastatin improve the response to sildenafil in men with erectile dysfunction not initially responsive to sildenafil? Hypothesis and pilot trial results. *J Sex Med.* 2006;3 (2):303–308.
25. Filippi S, Vignozzi L, Morelli A, et al. Testosterone partially ameliorates metabolic profile and erectile responsiveness to PDE5 inhibitors in an animal model of male metabolic syndrome. *J Sex Med.* 2009;6(12):3274–3288.

26. Gokce MI, Gulpinar O, Ozturk E, Gulec S, Yaman O. Effect of atorvastatin on erectile functions in comparison with regular tadalafil use. A prospective single-blind study. *Int Urol Nephrol*. 2012;44:683–687.
27. Mastalir ET, Carvalhal GF, Portal VL. The effect of simvastatin in penile erection: a randomized, double-blind, placebo-controlled clinical trial (simvastatin treatment for erectile dysfunction – STED TRIAL). *Int J Impot Res*. 2011;23(6):242–248.
28. Abdel Aziz MT, El Asmer MF, Mostafa T, Atta H, Mahfouz S, Fouad H, et al. Effects of losartan, HO-1 inducers or HO-1 inhibitors on erectile signaling in diabetic rats. *J Sex Med*. 2009;6(12):3254–3264.
29. La vignera S, Condorelli RA, Vicari E, Calogero AC. Statins and erectile dysfunction: a critical summary of current evidences. *J Androl*. 2011. doi: 10.2164/Jandrol.111.015230.
30. The Long-Term Intervention with Pravastatin in Ischaemic Disease (LIPID) Study Group. Prevention of cardiovascular events and death with pravastatin in patients with coronary heart disease and a broad range of initial cholesterol levels. *N Engl J Med*. 1998;339(19):1349–1357.
31. Lewis SJ, Moye LA, Sacks FM, Johnstone DE, Timmis O, Mitchell J, et al. Effect of pravastatin on cardiovascular events in older patients with myocardial infarction and cholesterol levels in the average range. Results of the Cholesterol and Recurrent Events (CARE) trial. *Ann Intern Med*. 1998;129(9): 68169.
32. Staworth RD, Kapoor D, Channer KS, Jones TH. Statin therapy is associated with lower total but not bioavailable or free testosterone in men with type 2 diabetes. *Diabetes Care*. 2009;32(4):541–546.
33. Dobs AS, Schrott H, Davidson MH, Bays H, Stein EA, Kush D, et al. Effects of high-dose simvastatin on adrenal and gonadal steroidogenesis in men with hypercholesterolemia. *Metabolism*. 2000;49(9):1234–1238.
34. Corona G, Boddi V, Balercia G, et al. The effect of statin therapy on testosterone levels in subjects consulting for erectile dysfunction. *J Sex Med*. 2010;7(4 Pt 1):1547–1556.
35. Kloner R. Erectile dysfunction and hypertension. *Int J Impot Res*. 2007;19(3):296–302.
36. Mittawae B, El-Nashaar AR, Fouda A, Magdy M, Shamloul R. Incidence of erectile dysfunction in 800 hypertensive patients: a multicenter Egyptian national study. *Urology*. 2006;67(3):575–578.
37. Prisant LM, Loebl Jr. DH, Waller JL. Arterial elasticity and erectile dysfunction in hypertensive men. *J Clin Hypertens (Greenwich)*. 2006;8(11):768–774.
38. Ushiyama M, Kuramochi T, Yagi S, et al. Erectile dysfunction in hypertensive rats results from impairment of the relaxation evoked by neurogenic carbon monoxide and nitric oxide. *Hypertens Res*. 2004;27(4):253–261.
39. Bener A, Al-Ansari A, Abdulla OAA, Al-Hamaq AM M. Prevalence of erectile dysfunction among hypertensive and nonhypertensive Qatari men. *Medicina (Kaunas)*. 2007;43(11):870–878.
40. Karavitakis M, Kornninos C, Theodorakis PN. Evaluation of sexual function in hypertensive men receiving treatment: a review of current guidelines recommendation. *J Sex Med*. 2011;8(9):2405–2414.
41. Engbaek M, Hjerrild M, Hallas J, Jacobsen IA. The effect of low-dose spironolactone on resistant hypertension. *J Am Soc Hypertens*. 2010;4(6):290–294.
42. Shiri R, Koskimaki J, Hakkinen J, Auvinen A, Tammela TL, Hakama M. Cardiovascular drug use and the incidence of erectile dysfunction. *Int J Impot Res*. 2007;19(2):208–212.
43. Scranton RE, Lawler E, Botteman M, et al. Effect of treating erectile dysfunction on management of systolic hypertension. *Am J Cardiol*. 2007;100(3):459–463.

44. Blumentals WA, Gomez-caminero A, Joo S, Vannappagari V. Is erectile dysfunction predictive of peripheral vascular disease? *Aging Male*. 2003;6(4):217–221.
45. Chung SD, Chen YK, Lin AC, Ching H. Increased risk of stroke among men with erectile dysfunction: a nationwide population-based study. *J Sex Med*. 2011;8(1):240–246.
46. Polonsky TS, Taillon LA, Sheth M, Min JK, Archer SL, Ward RP. The association between erectile dysfunction and peripheral arterial disease as determined by screening ankle–brachial index testing. *Atherosclerosis*. 2009;207(2):440–444.
47. Ylitalo KR, Sowers M, Heeringa S. Peripheral vascular disease and peripheral neuropathy in individuals with cardiometabolic clustering and obesity: National Health and Nutrition Examination Survey 2001–2004. *Diabetes Care*. 2011;34(7):1642–1647.
48. Fukuhara S, Tsujiimura A, Okuda H, Yamamoto K, Takao T, Miyagawa Y, et al. Vardenafil and resveratrol synergistically enhance the nitric oxide/cyclic guanosine monophosphate pathway in corpus cavernosal smooth muscle cells and its therapeutic potential for erectile dysfunction in the streptozotocin-induced diabetic rat: preliminary findings. *J Sex Med*. 2011;8(4):1061–1071.
49. Jackson G, Boon N, Eardley I. Erectile dysfunction and coronary artery disease prediction: evidence-based guidance and consensus. *Int J Clin Pract*. 2010;64(7):848–857.
50. Salem S, Abdi S, Mehrsai A, Saboury B, Saraji A, Shokohideh V, et al. Erectile dysfunction severity as a risk predictor for coronary artery disease. *J Sex Med*. 2009 Dec;6 (12):3425–32.
51. Too much, too little sleep associated with adult weight gain. Mayo Clin Womens Healthsource. 2008;12(10):3.
52. Amano T, Imao T, Takemae K. The usefulness of vibration perception threshold as a significant indicator for erectile dysfunction in patients with diabetes mellitus at a primary diabetes mellitus clinic. *Urol Int*. 2011;87(3):336–340.
53. Fukui M, Tanaka M, Okada H, et al. Five-item version of the international index of erectile function correlated with albuminuria and subclinical atherosclerosis in men with type 2 diabetes. *J Atheroscler Thromb*. 2011;18(11):991–997.
54. Ohlsson C, Barrett-Connor E, Bhasin S, et al. High serum testosterone is associated with reduced risk of cardiovascular events in elderly men. The MrOS (Osteoporotic Fractures in Men) study in Sweden. *J Am Coll Cardiol*. 2011;58(16):1674–1681.
55. Wessells H, Penson DF, Cleary P, Brandy MS, Rutledge N. Effect of intensive glycemic therapy on erectile function in men with type 1 diabetes. *J Urol*. 2011;185(5): 1828–1834.
56. Rhoden EL, Ribeiro EP, Riedner CE, et al. Glycosylated haemoglobin levels and the severity of erectile function in diabetic men. *BJU Int*. 2005;95(4):615–617.
57. Giugliano F, Esposito K, Di Palo C, et al. Erectile dysfunction associates with endothelial dysfunction and raised proinflammatory cytokine levels in obese men. *J Endocrinol Invest*. 2004;27(7): 665–669.
58. Giugliano F, Maiorino MI, Di Palo C. Adherence to Mediterranean diet and erectile dysfunction in men with type 2 diabetes. *J Sex Med*. 2010;7(5):1911–1917.
59. Khoo J, Piantadosi C, Duncan R, Worthley SG, Jenkins A, Noakes M, et al. Comparing effects of a low-energy diet and a high-protein low-fat diet on sexual and endothelial function, urinary tract symptoms, and inflammation in obese diabetic men. *J Sex Med*. 2011;8(10):2868–2875.
60. Lamina S, Okoye CG, Dagogo TT. Managing erectile dysfunction in hypertension: the effects of a continuous training programme on biomarker of inflammation. *BJU Int*. 2009;103(9):1218–1221.

61. La Vignera S, Calogero AE, Cannizzaro MA, et al. Tadalafil and modifications in peak systolic velocity (Doppler spectrum dynamic analysis) in the cavernosal arteries of patients with type 2 diabetes after continuous tadalafil treatment. *Minerva Endocrinol.* 2006;31(4):251–261.
62. Aversa A, Greco E, Bruzziches R, Pili M, Rosano G, Spera G. Relationship between chronic tadalafil administration and improvement of endothelial function in men with erectile dysfunction: a pilot study. *Int J Impot Res.* 2007;19(2):200–207.
63. Hatzimouratidis K, Amar E, Eardley I, Quliano F, Hatzichristou D, Montorsi F, et al. Guidelines on male sexual dysfunction: erectile dysfunction and premature ejaculation. *Eur Urol.* 2010;57(5):804–814.
64. Perimenis P, Markou S, Gyftopoulos K, Athanasopoulos A, Giannitsas K, Barbalias G. Switching from long-term treatment with self-injections to oral sildenafil in diabetic patients with severe erectile dysfunction. *Eur Urol.* 2002;41(4): 387–391.
65. Angulo J, Cuevas P, La Fuente JM, Pomerol JM, Edurado I. Regulation of human penile smooth muscle tone by prostanoid receptors. *Br J Pharmacol.* 2002;136(1):23–30.
66. Montorsi F, Guazzoni G, Bergamaschi F, et al. Clinical reliability of multi-drug intracavernous vasoactive pharmacotherapy for diabetic impotence. *Acta Diabetol.* 1994;31 (1):1–5.
67. Akin-Olugbade O, Parker M, Guhring P, Mulhall J. Determinants of patient satisfaction following penile prosthesis surgery. *J Sex Med.* 2006;3(4):743–748.
68. Selph JP, Carson 3rd CC. Penile prosthesis infection: approaches to prevention and treatment. *Urol Clin North Am.* 2011;38(2):227–235.
69. Chung E, Van CT, Wilson I, Cartmill RA. Penile prosthesis implantation for the treatment for male erectile dysfunction: clinical outcomes and lessons learnt after 955 procedures. *World J Urology.* 2012, doi: 10.1007/s00345-012-0859-4.
70. Drewa T, Olszewska-Slonina D, Chlosta P. Testosterone replacement therapy in obese males. *Acta Pol Pharm.* 2011;68(5):623–627.
71. Jones TH. Effects of testosterone on type 2 diabetes and components of the metabolic syndrome. *J Diabetes.* 2010;2(3):146–156.
72. Corona G, Monami M, Rastrelli G, et al. Testosterone and metabolic syndrome: a meta-analysis study. *J Sex Med.* 2011;8(1):272–283.
73. Jones TH, Saad F. The effects of testosterone on risk factors for, and the mediators of, the atherosclerotic process. *Atherosclerosis.* 2009;207(2):318–327.
74. Cohen PG. The hypogonadal-obesity cycle: role of aromatase in modulating the testosterone–estradiol shunt – a major factor in the genesis of morbid obesity. *Med Hypotheses.* 1999;52(1):49–51.
75. Simpson ER, Mendelson CR. Effect of aging and obesity on aromatase activity of human adipose cells. *Am J Clin Nutr.* 1987;45(suppl 1):290–295.

12 Male Obesity – Impact on Semen Quality

Vanessa J. Kay and Sarah Martins da Silva

Assisted Conception Unit and Developmental Biology Group, Division of Maternal and Child Health Sciences, Ninewells Hospital, University of Dundee, Dundee, UK

Background

Male infertility is a commonest defined cause of infertility, being shown to contribute to at least 40% of all causes of infertility [1]. The treatment of male infertility has been revolutionised by the advent of intra-cytoplasmic sperm injection (ICSI), which now represents nearly half of all in vitro fertilization (IVF) cycles in the United Kingdom [1]. It is a highly successful treatment, but bypasses the aetiology of infertility, and so in many cases the reason for male infertility is not identified. However, it should be remembered that ICSI treatment has significant medical risks, is expensive, and long-term safety data is still scarce [2]. Therefore it is important to establish modifiable risk factors for male fertility, to enable both preventative and curative treatments for this condition.

The incidence of obesity is rapidly rising in almost every region of the world. Obesity is defined by the World Health Organization (WHO) as a body mass index (BMI) over 30 kg/m^2 and overweight when BMI is over 25 kg/m^2. Although obesity affects women more than men, male obesity is an issue of serious concern. In Europe, the International Obesity Task Force has indicated that obesity rates in adult men range from 10% to 27%, with this prevalence rising significantly in the last 10 years [3]. It is predicted by the WHO that this will continue to rise and that by 2015, 2.3 billion adults will be overweight and 700 million obese [4].

The adverse influence of obesity on various aspects of female reproduction and fertility has been realised for some time [5] and management guidelines are now available [6]. More recently, data regarding male obesity and infertility has been accumulating [7,8]. There are now several population-based studies showing overweight and obese men having up to 50% higher rate of sub-fertility when compared with normal-weight men [9–11]. This risk is particularly high if the female partner is also overweight or obese [10]. One could argue that the association between male obesity and infertility could be related to confounding factors such as male age, smoking and alcohol use and female partner obesity. However, once these

Obesity. DOI: http://dx.doi.org/10.1016/B978-0-12-416045-3.00012-1

factors have been excluded, for every three point increase in a man's BMI, couples were 10% more likely to be infertile [11].

Impact on Semen Quality

Semen quality is accepted as a surrogate marker for male fertility, with normal reference ranges recently established [12]. Although the majority of obese men have normal semen parameters, a correlation between poor semen quality and obesity has been recognised for some time [13,14]. A relationship between obesity and different semen parameters is emerging (see Table 12.1 for overview of data).

Sperm Concentration and Count

In a cross-sectional study of 1558 military recruits undergoing compulsory physical examination, men with normal BMIs were compared with those with BMIs over 25 kg/m^2. The sperm concentration and total sperm count per ejaculate was reduced by 22% and 24%, respectively, in men with a BMI of over 25 kg/m^2, with no change in semen volume. In addition, an associated decrease in testosterone, follicle-stimulating hormone (FSH), inhibin B and sex hormone–binding globulin (SHBG) was observed in the overweight group [15]. A recent WHO surveillance study on male partners of pregnant women confirmed that obese men had significantly lower total sperm count than non-obese men (mean 231×10^6 versus 324×10^6, respectively), although other sperm parameters were not shown to be effected [23]. This finding was supported in a review by Mah and Wittert [30] who concluded that obese men have reduced sperm concentration and total sperm count, compared to lean men, but with motility and morphology unaffected.

Clearly a simplistic link between fertility and BMI is understating the complexities of male reproduction. The effect on sperm count may be marginal, with analysis of a data base of 2139 men finding only a slightly lower total sperm count among overweight men and a non-significant change in obese men (mean total sperm count in BMI 20–25: 231×10^6, BMI 25–30: 216×10^6 and BMI > 30: 265×10^6) [19]. A study of men attending an infertility clinic showed that despite major changes in reproductive hormones, only in extreme obesity (BMI > 35 kg/m^2) total sperm counts were lower [14]. A further study [16] showed that this effect was only observed in men with BMIs of over 30 kg/m^2. Qin et al. [18] did find a correlation between BMI and sperm count, but this was only significant in the men with low BMIs. Moreover in a study of 349 men, no correlation was found between BMI and total sperm count, although it was recognised that few men with extreme BMIs were included in this study [22]. Magnusdottir et al. [13] observed an association between BMI and sperm parameters but only in subfertile men and not in fertile men.

Table 12.1 Summary of Publications on Male BMI and Semen Quality

Study and Year	Population Studied	Number of Participants	Conclusions: Relationship of Sperm Parameters with BMI
Jensen et al. (2004) [15]	Young Danish military recruits	1558	Significant negative relationship with sperm concentration and total sperm count. No significant relationship with sperm motility
Koloszar et al. (2005) [16]	Hungarian normozoospermic men attending an infertility clinic	274	Significant negative relationship for sperm concentration
Magnusdottir et al. (2005) [13]	Icelandic study of men attending an infertility clinic	72	Significant negative relationship with sperm concentration and total sperm count
Kort et al. (2006) [17]	American study of men attending infertility clinic	520	Significant negative relationship with normal motile sperm count and significant positive relationship between DNA fragmentation index
Qin et al. (2007) [18]	Chinese population-based study	990	No significant relationship with sperm concentration, total sperm count, sperm morphology and semen volume (except in underweight men, but only 1.7% men had BMI over 30)
Aggerholm et al. (2008) [19]	European population-based environmental study from five studies combined	2139	No significant relationship with sperm concentration, total sperm count or sperm motility
Chavarro et al. (2010) [14]	American study of men attending an infertility clinic	483	Significant negative relationship with total sperm count and semen volume and positive relationship with sperm DNA damage. No significant relationship with sperm concentration, motility or morphology
Hammoud et al. (2008) [20]	American study of men attending an infertility clinic	526	Significant negative relationship with sperm count, progressive motility and normal morphology
Pauli et al. (2008) [21]	American population-based study	87	No significant relationship with sperm concentration, semen

(*Continued*)

Table 12.1 (Continued)

Study and Year	Population Studied	Number of Participants	Conclusions: Relationship of Sperm Parameters with BMI
			volume, sperm motility or total motile sperm count
Nicopoulou et al. (2009) [22]	American study of men attending an andrology clinic	349	No significant relationship with total sperm count (few men with BMI over 35)
Stewart et al. (2009) [23]	Australian study of male partners of pregnant women	225	Significant negative relationship with total sperm count. No significant relationship with sperm motility or morphology
Hofny et al. (2010) [24]	Egyptian study of obese men attending an andrology clinic	122	Significant negative relationship with sperm concentration and motility, normal morphology
Martini et al. (2010) [25]	South American study of men attending an andrology or reproductive clinic	794	Significant negative relationship with sperm motility and rapid motility, with nil relationship with sperm concentration
Relwani et al. (2010) [26]	American study of men attending an andrology clinic	530	No significant relationship with sperm concentration, motility or morphology or clinical pregnancy with IVF
Sekhavat et al. (2010) [27]	Iranian study of healthy men	852	Significant negative relationship with sperm concentration, sperm motility and no relationship with morphology
Tunc et al. (2011) [28]	Australian study of healthy men	81	Significant negative relationship with total sperm count and significant positive relationship with oxidative stress
Rybar et al. (2011) [29]	Czech men attending an infertility clinic	153	No significant relationship with sperm concentration, motility or normal morphology

Sperm Motility

An observational study of 520 men showed a significantly reduced number of normal motile sperms (normal BMI 18.6×10^6, overweight 3.6×10^6 and obese 0.7×10^6) and importantly an increase in sperm DNA fragmentation in overweight and obese men [17]. The findings of this study have been confirmed in a retrospective study of 526 infertile men in which the incidence of oligozoospermia and low progressive motile sperm concentration was higher in overweight and obese men compared with normal-weight men [20] and later confirmed by Sekhavat and

Moein [27] who studied 852 healthy men. Multivariate analysis also confirms a negative association between BMI and motility, progressive motility and neutral alpha-glucosidase levels (considered an epididymal functional marker) [25]. However, other large population-based studies on humans have shown no such association [15,19,23,30].

Further convincing evidence linking male obesity and sperm motility has been provided from controlled animal studies, in which an obesity model has been created in rodents by feeding them a high-fat diet (HFD) for a number of weeks. Firstly a study on mice demonstrated that the percentage of motile sperm was significantly reduced in mice fed on a HFD from 44% to 36% [31]. A similar finding was then shown in rats, in which those on a HFD had a decrease in the percentage of sperm with progressive movement [32].

Sperm Morphology

WHO criteria (2010) [33] state that a sample is normal if 4% (or 5th centile) or more of the observed sperm have normal morphology [12]. Morphology evaluation is somewhat controversial, and continues to be influenced by the subjectiveness of the observer and the lack of objective measurement. Perhaps unsurprisingly it is this area of diagnostic semen analysis where studies have found conflicting effects of obesity. Whilst some studies do support adverse effects of BMI on sperm morphology [20,24,25,27,31,32], others show no such association [14–20,22–27,30–32]. The percentage of normal sperm morphology has been shown to be adversely effected by either high or low BMIs in a study of 1558 military conscripts, although this did not reach statistical significance [15]. In a smaller study comparing obese fertile and obese infertile men, a significant positive correlation between BMI and abnormal sperm morphology was observed, with mean sperm abnormal form being 21.4% in obese fertile men compared to 35.5% in obese infertile men [24]. A more recent study has contradicted this finding, with no relationship between self-reported BMI and sperm morphology, despite showing a significant inverse correlation between BMI and sperm count and motility [27].

Combined Semen Parameters

A few studies have looked at combined measures of semen parameters. In these studies the BMI is shown to be related to poorer sperm quality, e.g. an association with low progressively motile sperm count [20] and normal motile sperm count [17].

Sperm DNA Damage

It is possible that the traditional semen parameters described above are not the most appropriate parameters to assess the effects of obesity on sperm function. Recently there has been some enthusiasm for the assessment of sperm DNA damage, using a variety of different assays. The integrity of sperm DNA has been shown to predict fertility [34]. However, fundamental questions remain regarding the nature of sperm DNA damage and the lack of standardisation of clinical assays to assess this [35].

Kort et al. [17] demonstrated a significant negative relationship between BMI and total number of motile sperm cells. In this study obese men had a significantly higher percentage of sperm with DNA damage, using a sperm chromatin structure assay (SCSA) when compared with normal-weight men (DNA fragmentation index 19.9%, 25.8% and 27.0% in normal, overweight and obese men, respectively). This was confirmed by Chavarro et al. [14] who identified significantly more sperm with high DNA damage in obese men compared with men of normal weight (adjusted mean difference from normal BMI of cells with high DNA damage 4, 5, 7 in normal, overweight and obese men, respectively), although there was no difference with three other standard measures of sperm DNA integrity. Of note, these two studies measured different aspects of DNA integrity and therefore data should be interpreted with caution (SCSA measures susceptibility of sperm chromatin to DNA denaturation, and the comet assay measures the extent of sperm DNA fragmentation in individual sperm). A recent study by Tunc et al. [28] attempted to correlate BMI with seminal reactive oxygen species (ROS) production (nitroblue terazolium assay), sperm DNA damage (TUNEL), markers of semen inflammation (CD45, seminal plasma polymorphonuclear elastase and neopterin concentration) and routine sperm parameters, as well as reproductive hormones. The study confirmed an increase in oxidative stress with increase in BMI, primarily attributed to increase in seminal macrophage activation. However, the magnitude was small and deemed to be of only minor clinical significance as there was no associated decline in sperm DNA integrity or sperm motility. Increased BMI was, however, linked to decreased sperm concentration, as well as lower serum testosterone and increased serum oestrogen. Mice fed on HFD for 9 weeks show elevated intracellular ROS (692 ± 83 vs 409 ± 22; $P < 0.01$), as well as increase in sperm DNA damage ($1.64 \pm 0.6\%$ vs $0.17 \pm 0.06\%$; $P < 0.01$). HFD mice also had a significantly lower percentage of non-capacitated sperm, which resulted in lower fertilisation rates (25.9% vs 43.9%; $P < 0.01$) [31]; however, data is conflicting. A further study examining basic semen parameters, chromatin integrity and chromatin condensation concluded no proof of impact of BMI [29].

However, it is important to note that most obese men have been shown to have normal semen quality and fertility. In a small study of 87 men [21] no association between BMI and semen parameters was identified, although correlations with several endocrine markers were observed. A further study observed no correlation between self-reported BMI and sperm concentration, motility or morphology or pregnancy rates with IVF [26]. There are various possible reasons to explain these discrepancies, including the bias from confounding lifestyle factors, insufficient sample size, presence of genetic causes of male infertility, appropriateness of the sperm parameter assessed and whether the effect is only observed at extremes of BMIs. For example, in a study on 483 infertile men [14], only the most obese men (BMI > 35 kg/m^2) had a lower sperm count when compared with normal-weight men. A recent systematic review with meta-analysis of 6800 men concluded that there was no strong evidence for a relationship between BMI and semen parameters; however, it was accepted that their data analysis was restricted to only five studies which had comparable outcome measures, with many studies excluded from the meta-analysis even though their evidence may have been useful [36]. Another limitation recognised in this meta-analysis was that BMI is not the most accurate surrogate measure of an individual's body fat, with more accurate measures being waist circumference and bioimpedance. The relationship between sperm quality and obesity is still not clearly understood and it is likely to be complex with various factors involved.

Aetiological Theories

It has not yet been clearly established how excess weight relates to the biological changes that underlie male infertility, although there are several theories worth exploring.

Endocrine Theory

The endocrine abnormalities associated with obesity in women are well known with an increase androgen metabolism and elevated oestrogen levels [37]. Similarly, obese men are known to have relative hyperoestrogenic hypogonadotrophic hypogonadism, with Aggerholm and colleagues [19] showing that serum testosterone concentrations were 25–32% lower in obese men than in normal-weight men, whereas the oestrogen concentration was 6% higher in obese men. An association between male obesity and decreased free and total testosterone and SHBG have all been well documented, with possible mechanisms including decreased luteinising hormone, inhibitory effects of oestrogen centrally and the effects of leptin and other peptides both centrally and on Leydig cells (for review see Mah and Wittert [30]).

Another hormone that may be of relevance is inhibin, which is secreted from the Sertoli cells and suppresses FSH production by the pituitary gland. Inhibin B may be important as it is known to influence spermatogenesis, with severely obese men having reduced inhibin B levels [38].

The reason why obesity causes hypoandrogenism is thought to be multifactorial. In obesity, it is known that circulating levels of oestrogens are raised by increased aromatisation of testicular and adrenal androgens in adipose tissue. Indeed when the aromatase inhibitor letrozole was administered to obese men, testosterone levels increased and serum oestradiol levels decreased [39]. These high oestrogen levels cause inappropriate suppression of the hypothalamic–pituitary–gonadal axis, resulting in decreased testosterone production. It is also possible that the elevated oestrogen levels have a direct adverse influence on spermatogenesis, although the exact nature of this is as yet undetermined.

Another factor to consider is the secretion of hormones from adipose cells. Leptin is such a hormone, with plasma concentrations rising in parallel with fat reserves. It is primarily responsible for satiety, but it has also been shown to influence male reproduction, both at a local and hypothalamic–pituitary–gonadal axis. Leptin has been discovered in semen [40], and leptin receptors have been demonstrated on sperm plasma membranes [41], suggesting a direct endocrine effect on sperm. But what is really interesting is that human spermatozoa have been shown to produce leptin [42]. Aquila et al. [42] suggest that leptin can act as an auto-regulator in spermatozoa by managing the energy status of the cell. They have also shown that sperm have insulin receptors and secrete insulin suggesting that insulin may also act as a metabolic regulator in the spermatozoa. Whilst these results require confirmation, they suggest that the effect of any changes in insulin and leptin in overweight men may be directly on the functioning of the spermatozoa post-ejaculation and that the monitoring of traditional sperm parameters may not be sufficient to detect a subtle effect on fertility. At a hypothalamic–pituitary–gonadal level, high levels of leptin are known to decrease basal and luteinising hormone–stimulated androgen level, with obese men demonstrating a 30–40% reduced testosterone level that is inversely correlated with leptin levels [40]. In a recent study [24], this link between leptin to obesity and male infertility was confirmed, with obese oligozoospermic men having higher plasma leptin levels than obese fertile controls.

Insulin resistance is known to be associated with obesity and has been negatively correlated with testosterone levels [43]. Interestingly in a meta-analysis of 80 publications, men with type 2 diabetes had a lower level of testosterone compared to controls [44]. The spermatozoa of men with type 2 diabetes did not differ from the control group in terms of semen parameters (concentration, motility and morphology), but did have a significantly higher level of DNA fragmentation [45]. This DNA damage impairs sperm function and adversely affects male fertility [34]. Insulin has also been demonstrated to influence the level of SHBG, with several studies showing that SHBG and total testosterone are inversely correlated with BMI and insulin levels [43–47].

Another mechanism for these endocrine changes relates to sleep apnoea, which is more common in obese persons [48]. It appears to decrease the nocturnal rise of testosterone, resulting in lower morning testosterone levels [49], which can be reversed by weight loss [50].

However, whether this modest decrease in testosterone level is responsible for suppression of spermatogenesis remains to be proven. In a recent observational study, despite demonstrating relative hypogonadotrophic hypoandrogenism in obese men, semen analysis parameters were unaffected [19], whilst another study has shown obesity is associated with decreases in testosterone and total sperm count [15]. These opposing results suggest that any effect of such reductions of testosterone on male infertility may be modest and of limited clinical significance, with the endocrine control of spermatogenesis usually being maintained even in obese men. Conversely it has been suggested that defective spermatogenesis causes obesity, rather than obesity causing impaired testicular function [23]. This is supported by the observation that men administered therapies to reduce testosterone, e.g. during treatment for prostatic cancer, tend to become obese [51,52].

Genetic Theory

Could a genetic abnormality independently cause both obesity and male infertility? There are a number of rare genetic conditions in which these conditions occur together, such as Laurence–Moon and Prader–Willi syndromes. More specifically the gene encoding for leptin, which is synthesised in adipose tissue, has been shown to be associated with obesity in both humans and mice, with leptin administration in leptin-deficient mice restoring fertility [53].

Sexual Dysfunction Theory

Another hypothesis to explain why obese men are less fertile is related to reduced coital frequency. Certainly it has been demonstrated that overweight and obese men have fewer sexual partners than normal-weight men [54]. However, when men with reduced coital frequency are excluded from analysis, it has still been shown that overweight and obese men have increased incidence of sub-fertility [9]. This indicates that reduced coital frequency is not necessarily the cause of infertility in obese men.

Obese men have been observed to have a higher incidence of erectile dysfunction (penile rigidity grade 1.32 vs 1.62 in non-obese men) [55]. However, when cardiovascular risk factors (such as heart disease and diabetes) were excluded, there was no difference in erectile dysfunction between obese and non-obese men, suggesting that vascular impairment rather than obesity is responsible.

Testicular Hyperthermia Theory

Excess fat in the inner thighs and suprapubic region may result in testicular hyperthermia (over 35°C). Certainly it is known that a rise in testicular temperature of a few degrees is sufficient to hinder sperm production in rats [56], and occupational heat exposure has been shown to reduce sperm output and quality in men [57]. Moreover testicular hyperthermia related to tight-fitting underwear has been shown to adversely affect semen parameters and possibly fertility [58]. There are limited data on testicular hyperthermia and obesity. In men presenting with excess suprapubic fat due to scrotal lipomatosis, scrotal lipectomy resulted in improved semen analysis [59].

Reactive Oxygen Species Theory

It is known that obesity results in an increase in oxidative stress, resulting from an imbalance between tissue free radicals, ROS and antioxidants. This may be a major cause for many of the underlying co-morbidities of obesity, including male infertility [60]. There are two main mechanisms by which that ROS affect sperm function: altered DNA resulting in defective paternal DNA being passed onto children and sperm membrane damage resulting in decreased motility and ability to fuse with the oocyte [61]. Recently it has been recommended by the Cochrane Collaboration that infertile men undergoing fertility treatment should take an oral antioxidant to increase the chance of conceiving; however, whether this is a specific benefit to obese men has not been evaluated [62].

Treatment

Less is known about treatment of infertility in obese men than in obese women. In obese men, loss of weight in a 4-month programme has been shown to improve hormonal profile, with an increase in SHBG and testosterone and a decrease in insulin and leptin [63], as well as free testosterone [64]. There is also evidence that reduction in caloric intake corrects erectile dysfunction in men with metabolic syndrome [65] especially when combined with physical exercise [66].

Despite the lack of evidence, if the changes by which obesity causes infertility are reversible, then weight loss is likely to be an effective treatment. If this is applied, it is first important initially to identify that a man is overweight. Unfortunately, increasingly individuals fail to recognise that they are obese, and this is more so in men. A recent British survey showed that 53% of the population was overweight or obese, yet in obese men only 67% recognised themselves to be overweight or obese [67]. Therefore, it is important that height and weight are measured, and BMI calculated, in all men attending for advice on infertility. Overweight and obese men should be informed regarding the association between increased weight and infertility and strongly advised to aim for a BMI of below 25 kg/m^2 before starting fertility treatment.

The primary proven method to achieve weight loss is by reducing calorie intake and increasing physical activity. Weight loss can be improved by providing dietary advice, psychological support, exercise classes and weight-reducing agents. However, sustained weight loss in patients is very difficult to achieve. As such considerable support with active monitoring will need to be provided to men aiming for significant reductions in BMI. Other options may be open.

In morbid obesity bariatric surgery, such as gastric banding or bypass may be considered. In a systematic review, all current bariatric operations lead to major weight loss, lasting for 10 years [68]. Although surgery has been shown to improve hormonal profile [69] and improve quality of sexual life [70], initial data indicates that it may reduce sperm numbers. For example in a small study of 6 severely obese previously fertile men, azoospermia developed in all men when followed up at a mean of 17 months following Roux-en-Y gastric bypass surgery [71]. Long-term data are required to explore this worrying association further.

More specific treatments to target known abnormalities which cause infertility in obese men may also be considered. In men with hypogonadotrophic hypogonadism, treatments with aromatase inhibitors and testosterone have both been shown to improve testosterone to oestradiol ratio and increase semen parameters [72].

Studies are required to evaluate whether these treatments improve fertility in obese men. Another interesting area of research is in leptin replacement therapy. In leptin-deficient mice, the administration of leptin has been shown to restore fertility [73]. As yet this has not been evaluated as a fertility treatment in men.

In obese woman there are good reasons to deny access to infertility treatment until weight loss has been achieved, namely the increased risk of maternal and perinatal complications associated with maternal obesity. It has been recommended by Balen et al. [74] that women with polycystic ovarian syndrome should be denied fertility treatment until their BMI is less than 35 kg/m^2. Obviously obesity in men does not carry such risks. As such, it is not surprising that male BMI is not usually used to restrict access to fertility treatment. However, with the known detrimental effect of male obesity on fertility, it is disappointing that more is not known about therapeutic treatment options. This may be related to the perception that male obesity has only a modest effect on fertility, compared with female obesity. However, as the incidence of male obesity increases, even a modest effect on fertility will become more clinically relevant.

Conclusions

With the increasing incidence of obesity, it is important to be aware of adverse effects of obesity on male fertility. Observational studies have shown a correlation between male obesity and a variety of sperm parameters, including concentration, motility, morphology and DNA damage. However, the association between obesity and sperm parameters is not conclusive, with some showing this association only in a proportion of men with severe obesity. Endocrine abnormalities (including increased plasma levels of oestrogen, leptin, insulin resistance, and reduced androgens and inhibin B levels) are likely to be central in the aetiology of sperm dysfunction in obese men. However, other factors may also contribute, including genetic abnormalities, sexual dysfunction, testicular hyperthermia and oxidative stress. The primary management must be to achieve weight reduction, using a reduced calorie intake combined with an exercise programme. Such regimes are difficult for patients to follow and considerable support is required. In extreme cases surgery can be considered although at present the data indicates a high incidence of azoospermia in the short term following surgery. More specific treatments to correct endocrine abnormalities associated with obesity are being evaluated, but disappointingly no effective treatment has yet been proven.

References

1. HFEA. Fertility facts and figures. <http://www.hfea.gov.uk/docs/2010-12-08_fertility_facts_and_figures_2008_publication_pdf.pdf>; 2008 Accessed 23.12.2011.
2. Steel AJ, Sutcliff A. Long-term health implications for children conceived by IVF/ICSI. *Human Fertil.* 2009;12:21–27.

3. International Obesity Task Force. <www.iaso.org/> and <www.iotf.org/>; 2005 Accessed 23.12.2011.
4. Preventing Chronic Diseases: a Vital Investment: Geneva, World Health Organization, 2005.
5. Mahmood T. Obesity: a reproductive hurdle. *Br J Diabetes Vasc Dis*. 2009;9:3–4.
6. Balen AH, Anderson R. Impact of obesity on female reproductive health: British Fertility Society, policy and practice guidelines. *Hum Fertil*. 2007;10:195–206.
7. Hammoud AO, Gibson M, Peterson CM, Meikle AW, Carrell DT. Impact of male obesity on infertility: a critical review of the literature. *Fertil Steril*. 2008;90:897–904.
8. Kay VJ, Barratt CLR. Male obesity: impact on fertility. *Br J Diabetes Vasc Dis*. 2009;9:237–241.
9. Nguyen RHN, Wilcox AJ, Skjaerven R, Baird D. Men's body mass index and infertility. *Hum Reprod*. 2007;22:2488–2493.
10. Ramlau-Hansen C, Thulstrup A, Nohr E, Bonde J, Sorensen T, Olsen J. Subfecundity in overweight and obese couples. *Hum Reprod*. 2007;22:1634–1637.
11. Sallmen M, Sandler D, Hoppin J, Baird D. Reduced fertility among overweight and obese men. *Epidemiology*. 2006;17:520–523.
12. Cooper TG, Noonan E, Von Eckardstein S, et al. World Health Organization reference values for human semen characteristics. *Hum Fertil Update*. 2009;0:1–15.
13. Magnusdottir EV, Thorsteinsson T, Thorsteinsdottir S, Heimisdottis M, Olafsdottir K. Persistent organochlorines, sedentary occupation, obesity and human male subfertility. *Hum Reprod*. 2005;20:208–215.
14. Chavarro JE, Toth TL, Wright DL, Meeker JD, Hauser R. Body mass index in relation to semen quality, sperm DNA integrity, and serum reproductive hormone levels among men attending an infertility clinic. *Fertil Steril*. 2010;93:2222–2231.
15. Jensen TK, Andersson AM, Jorgensen N, et al. Body mass index in relation to semen quality and reproductive hormone among 1558 Danish men. *Fertil Steril*. 2004;82:863–870.
16. Koloszár S, Fejes I, Závaczki Z, Daru J, Szöllosi J, Pál A. Effect of body weight on sperm concentration in normozoospermic males. *Arch Androl*. 2005;51:299–304.
17. Kort HI, Massey JB, Elsner CW, et al. Impact of body mass values on sperm quantity and quality. *J Androl*. 2006;27:450–452.
18. Qin D, Yuan W, Zhou W, Cui Y, Wu J, Gao E. Do reproductive hormones explain the association between body mass index and semen quality? *Asian J Androl*. 2007;9:827–834.
19. Aggerholm AS, Thulstrup AM, Toft G, Ramlau-Hansen CH, Bonde JP. Is overweight a risk for reduced semen quality and altered serum sex hormone profile? *Fertil Steril*. 2008;90:619–626.
20. Hammoud AO, Wilde N, Gibson M, et al. Male obesity and alteration in sperm parameter. *Fertil Steril*. 2008;90:2222–2225.
21. Pauli EM, Legro RS, Demers LM, Kunselman AR, Dodson WC, Lee PA. Diminished paternity and gonadal function with increasing obesity in men. *Fertil Steril*. 2008;90:346–351.
22. Nicopoulou SC, Alexiou M, Michalakis K, et al. Body mass index vis-a-vis total sperm count in attendees of a single andrology clinic. *Fertil Steril*. 2009;92:1016–1017.
23. Stewart TM, Lui DY, Garrett C, JØrgensen N, Brown EH, Baker HWG. Associations between andrological measures, hormones and semen quality in fertile Australian men: inverse relationship between obesity and sperm output. *Hum Reprod*. 2009;1:1–8.
24. Hofny ERM, Ali ME, Abdel-Hafez HZ. Semen parameters and hormonal profile in obese fertile and infertile men. *Fertil Steril*. 2010;94:581–583.

25. Martini AC, Tissera A, Estofán D, et al. Overweight and seminal quality: a study of 794 patients. *Fertil Steril*. 2010;94:1739–1743.
26. Relwani R, Berger D, Santoro N, et al. Semen parameters are unrelated to BMI but vary with SSRI use and prior urological surgery. *Reprod Sci*. 2011;4:391–397.
27. Sekhavat L, Moein MR. The effect of male body mass index on sperm parameters. *Aging Male*. 2010;13:155–158.
28. Tunc O, Bakos HW, Tremellen K. Impact of body mass index on seminal oxidative stress. *Andrologia*. 2011;43:121–128.
29. Rybar R, Kopecka V, Prinosilova P, Markova P, Rubes J. Male obesity and age in relationship to semen parameters and sperm chromatin integrity. *Andrologia*. 2011;43:286–291.
30. Mah PM, Wittert GA. Obesity and testicular function. *Mol Cell Endocrinol*. 2010;316:180–186.
31. Bakos HW, Mitchell M, Setchell BP, Lane M. The effect of paternal diet-induced obesity on sperm function and fertilization in a mouse model. *Int J Androl*. 2010;34:402–410.
32. Fernandez CD, Bellentani FF, Fernandes GS, et al. Diet-induced obesity in rats leads to a decrease in sperm motility. *Reprod Biol Endocrinol*. 2010;11:9–32.
33. World Health Organization. *WHO Laboratory Manual for the examination and processing of human semen*. Cambridge: Cambridge University Press; 2010.
34. Span M, Bonde JP, Hjollund HI, Kolstad HA, Cordelli E, Leter G. Sperm chromatin damage impairs human fertility. *Fertil Steril*. 2000;73:43–50.
35. Barratt CLR, Aitken J, Björndahl L, et al. Sperm DNA: organization, protection and vulnerability: from basic science to clinical applications – a position report. *Human Reprod*. 2010;0:1–15.
36. MacDonald AA, Herbison GP, Showell M, Farquhar CM. The impact of body mass index on semen parameters and reproductive hormones in human males: a systematic review with meta-analysis. *Human Reprod Update*. 2010;16:293–311.
37. Azziz R. Reproductive endocrinologic alteration in female asymptomatic obesity. *Fertil Steril*. 1989;52:702–725.
38. Globerman H, Shen-Orr Z, Karnieli Y, Charuzi I. Inhibin B in men with severely morbid obesity after weight reduction following gastroplasty. *Endocr Res*. 2005;31:16–17.
39. De Boer H, Verschoor L, Ruinemans-Koerts J, Jansen M. Letrozole normalizes serum testosterone in severely obese men with hypogonadotrophic hypogonadism. *Diabetes Obes Metab*. 2005;7:211–215.
40. Isidori AM, Caprio M, Strollo F, et al. Leptin and androgens in male obesity: evidence for leptin contribution to reduced androgen levels. *J Clin Endocrinol Metab*. 1999;84:3673–3680.
41. Jope T, Lammert A, Kratzsch J, Paasch U, Glander H. Leptin and leptin receptor in human receptor in human seminal plasma and human spermatozoa. *Int J Androl*. 2003;26:335–341.
42. Aquila S, Gentile M, Middea E, et al. Leptin secretion by human ejaculated spermatozoa. *J Clin Endocrinol Metab*. 2005;90:4753–4761.
43. Tsai EC, Matsumoto AM, Fujimoto WY, Boyko EJ. Association of bioavailable, free, and total testosterone with insulin resistance, influence of sex hormone-binding globulin and body fat. *Diabetes Care*. 2004;27:861–868.
44. Ding E, Song Y, Malik V, Lui S. Sex differences of endogenous sex hormones and risk of type 2 diabetes: a systemic review and meta-analysis. *JAMA*. 2006;295:1288–1299.

45. Agbaje IM, Rogers DA, McVicar CM, et al. Insulin dependant diabetes mellitus: implications for male reproductive function. *Hum Reprod*. 2007;22:1871 −1877.
46. Vermeulen A, Kaufman JM, Giagulli VA. Influence of some biological indexes on sex hormone-binding globulin and androgen levels in aging or obese males. *J Clin Endocrinol Metab*. 1996;81:1821−1826.
47. Osun JA, Gómez-Pérez CR, Arata-Bellabarba G, Villaroel V. Relationship between BMI, total testosterone, sex hormone-binding-globulin, leptin, insulin and insulin resistance in obese men. *Syst Biol Reprod Med*. 2006;52:355−361.
48. Wittels EH. Obesity and hormonal factors in sleep and sleep apnea. *Med Clin North Am*. 1985;69:1265−1280.
49. Luboshitzky R, Lavie L, Shen-Orr Z, Herer P. Altered luteinizing hormone and testosterone secretion in middle-aged obese men with obstructive sleep apnea. *Obes Res*. 2005;13:780−786.
50. Semple PA, Graham A, Malcolm Y, Beastall GH, Watson WS. Hypoxia, depression of testosterone, and impotence in Pickwickian syndrome reversed by weight reduction. *Br Med J*. 1984;289:801−802.
51. Smith MR. Changes in fat and lean body mass during androgen-deprivation therapy for prostate cancer. *Urology*. 2004;63:724−745.
52. Chen Z, Maricic M, Nguyen P, Ahmann FR, Bruhn R, Dalkin BL. Low bone density and high percentage of body fat among men who were treated with androgen deprivation therapy for prostatic cancer. *Cancer*. 2002;95:2136−2144.
53. Chehab FF, Lim ME, Lu RH. Correction of the sterility defect in homozygous obese? female mice is treatment with the human recombinant leptin. *Nat Genet*. 1996; 12:318−320.
54. Nagelkerke NJ, Bernsen RM, Sgaier SK, Jha P. Body mass index, sexual behaviour, and sexually transmitted infection: an analysis using the NHANES 1999−2000 data. *BMC Public Health*. 2006;6:199.
55. Chung WS, Sohn J, Park YY. Is obesity an underlying factor in erectile dysfunction? *Eur Urol*. 1999;36:68 −70.
56. Chowdhury AK, Steinberger E. Early changes in the germinal epithelium of rat testes following exposure to heat. *J Reprod Fertil*. 1970;22:205−212.
57. Thonneau P, Bujan L, Multigner L, Mieusset R. Occupational heat exposure and male infertility: a review. *Hum Reprod*. 1998;13:2122−2125.
58. Tiemessen CH, Evers JL, Bots RS. Tight-fitting underwear and sperm quality. *Lancet*. 1996;347:1844−1845.
59. Shafik A, Olfat S. Lipectomy in the treatment of scrotal lipomatosis. *Br J Urol*. 1981;53:55−61.
60. Vincent HK, Innes KE, Vincent KR. Oxidative stress and potential interventions to reduce oxidative stress in overweight and obesity. *Diab Obesity Met*. 2007;9:813−839.
61. Tremellen K. Oxidative stress and male infertility − a clinical perspective. *Hum Reprod Update*. 2008;14:243−258.
62. Showell MG, Brown J, Yazdani A, Stankiewicz MT, Hart RJ. Antioxidants for male subfertility (review). *Cochrane Database of Syst Rev*. 2011;1:CD007411.
63. Kaukau J, Pekkarinin T, Sane T, Mustajoki P. Sex hormones and sexual function in obese men loosing weight. *Obes Res*. 2003;11:689−694.
64. Niskanen L, Laaksonen DE, Punnonen K, Mustajoki P, Kaukau J, Rissanen A. Changes in sex hormone-binding globulin during weight loss and weight maintenance in abdominally obese men with metabolic syndrome. *Diabetes Obese Metab*. 2004;6:208−215.

65. Hannah JL, Maio MT, Komolova M, Adams MA. Beneficial impact of exercise and obesity interventions on erectile function and its risk factors. *J Sex Med.* 2009;6:254–261.
66. Esposito K, Giugliano F, Di Palo C, et al. Effect of lifestyle changes on erectile dysfunction in obese men. A randomized controlled trial. *JAMA.* 2004;291:2978–2984.
67. Johnson F, Cooke L, Croker H, Wardle J. Changing perceptions of weigh in Great Britain: comparison of two population studies. *BMJ.* 2008;337:270–272.
68. O'Brien PE, Brown WA, Dixon JB. Clinical update. Obesity, weight loss and bariatric surgery. *MJA.* 2005;183:310–314.
69. Bastounis EA, Karayiannakisa AJ, Syrigosb K, Zbarc A, Makria GG, Alexioua D. Sex hormone changes in morbidly obese patients after vertical banded gastroplasty. *Eur Sur Res.* 1998;30:43–47.
70. Hammoud A, Gibson M, Hunt SC, et al. Effect of Roux-en-Y gastric bypass on sex steroids and quality of life in obese men. *J Clin Endocrinol.* 2009;94:1329–1332.
71. Di Frega AS, Dale B, Di Matteo L, Wilding M. Secondary male factor infertility after? Roux-en-Y gastric bypass for morbid obesity: case report. *Hum Reprod.* 2005;20: 997–998.
72. Raman JD, Schlegel PN. Aromatase inhibitors for male infertility. *J Urol.* 2002; 167:624–629.
73. Chehab FF, Lim ME, Lu RH. Correction of the sterility defect in homozygous obese female mice is treatment with the human recombinant leptin. *Nat Genet.* 1996; 12:318–320.
74. Balen AH, Dresner M, Scott EM, Drife JO. Should obese women with polycystic ovarian syndrome receive treatment for infertility? *BMJ.* 2006;332:434–435.

13 Obesity, Bariatric Surgery and Male Reproductive Function

Jyothis George and Richard Anderson

MRC Centre for Reproductive Health, The Queen's Medical Research Institute, University of Edinburgh, Edinburgh, UK

Introduction

Obesity can be viewed in a reductionist manner as the biological consequence that naturally occurs when energy consumption in an individual is persistently in excess of energy expenditure but is, in fact, a complex pathophysiological state arising from an interplay of genetic, economic, environmental and lifestyle factors [1]. The age of onset and the progression of weight gain differ between individuals with sex, ethnicity and baseline body mass index (BMI) being predictors of progressive weight gain [2,3]. Spontaneous regression of weight gain is uncommon [2,3]. Interventions ranging from lifestyle and behavioural education to surgical methods are, therefore, required to effect weight loss in obese individuals. The combination of dietary modification and exercise regime effects weight loss and improves cardio-metabolic risk markers such as blood pressure, glucose and triglycerides, but the magnitude of weight loss observed in clinical trials is modest – with a mean weight loss of 1.5 kg (95% CI −2.3 to −0.7) – reported following intensive exercise [4]. Surgical interventions, on the other hand, have much more pronounced effects on weight loss than conventional approaches [5,6]. Therefore, when weight loss in excess of 10% baseline weight is a therapeutic goal, surgical approaches are the only reliable therapeutic options available.

In obese men, there is evidence to suggest that all the three testicular cell populations – Leydig cells, Sertoli cells and germ cells – appear to be functionally impaired resulting in diminished serum testosterone, serum inhibin B and in decreased sperm count and quality. In men, weight loss seems to have an impact on most of these markers of gonadal function as well as on sexual function, although the impact on testosterone secretion has been studied the most. In this chapter, we describe bariatric surgical techniques, the effects of obesity on male reproductive function and current evidence for the value of surgical approaches to improve male fertility.

Obesity. DOI: http://dx.doi.org/10.1016/B978-0-12-416045-3.00013-3

Bariatric Surgical Techniques

Bariatric surgical procedures aim to decrease the intake of food or to induce partial malabsorption of ingested food. Ideally introduced along with an effective patient education and preparation programme involving dietary and lifestyle modification, these interventions seek to modify the intake of food – promoting slow consumption of smaller quantities of food [7]. Several surgical techniques have been described, with gastric bypass, gastric banding, biliopancreatic diversion and sleeve gastrectomy in current use [5], of which gastric bypass and banding are more commonly employed than the others [7]. As most of these procedures are increasingly undertaken laparoscopically, perioperative morbidity and post-operative length of stay in hospital have decreased.

Gastric Banding

Gastric banding is a purely restrictive procedure where a ring (band) is placed around the fundus of the stomach to limit food intake. Unlike earlier versions, gastric bands in current use are adjustable, with provision for the addition or removal of saline through a subcutaneous access port to manipulate the size of the stoma. While the surgery, often carried out laparoscopically, is relatively easy to perform, the long-term failure and complication rates are as high as 20–30% [7,8].

Gastric Bypass

Surgical gastric bypass procedures aim to be both restrictive and malabsorptive, and involve the creation of a small gastric pouch (using surgical staples) and a gastroenterostomy stoma that bypasses the duodenum. There are several variations of the technique, a detailed discussion of which is beyond the scope of this chapter. The Roux-en-Y technique, designed to limit biliary reflux that may arise from loop gastroenterostomy, is commonly employed. Moreover, a prosthetic band is sometimes placed at the junction of the gastric pouch and the small intestine to stabilise the stoma [7]. Acute complications include anastomotic leakage and the blockage of efferent limb, while on the long term, the stoma may get narrower causing persistent vomiting. Moreover, consumption of refined sugar after gastric bypass can evoke rapid heart rate, nausea, faintness and diarrhoea – a condition termed as dumping syndrome [5]. It is technically feasible to reverse a gastric bypass.

Biliopancreatic Diversion

Biliopancreatic diversion is mainly a malabsorptive procedure and involves the removal of part of the stomach and the bypass of part of the small intestine. The resultant gastric pouch is much larger than that created in the typical Roux-en-Y technique allowing the patient to consume larger meals. However, the procedure is only partially reversible and carries higher perioperative morbidity and mortality [5].

Sleeve Gastrectomy

Sleeve gastrectomy irreversibly divides the stomach vertically, effecting an approximately 25% reduction in size. Gastric function and digestion are unaltered, as the pyloric valve remains intact. Over time, the stomach may become distended, negating some of the restrictive effects.

Future Directions

While commonly employed bariatric surgery involves laparoscopic or open surgery, recent studies have also investigated the endoscopic placement of a duodenal–jejunal bypass sleeve. In a multi-centre randomised study, where initial mean BMI was 48.9 and 47.4 kg/m^2 for the device and control patients respectively, mean excess weight loss after 3 months was 19.0% for duodenal–jejunal bypass sleeve patients versus 6.9% for controls ($P < 0.002$). Absolute changes in BMI at 3 months were 5.5 and 1.9 kg/m^2, respectively [9].

Pathophysiology of the Hypothalamic–Pituitary–Gonadal Axis in Obesity

Endocrine regulation of the human hypothalamic–pituitary–gonadal axis has been studied extensively in the last three decades since the Nobel-Prize winning experiments involving the extraction of gonadotropin-releasing hormone (GnRH) from porcine hypothalami [10,11]. It is now well established that two hormones secreted by the pituitary, luteinising hormone (LH) and follicular-stimulating hormone (FSH), regulate the functions of Leydig cells and germ cells, respectively [12]. Secretion of these pituitary hormones, in turn, is regulated by the modulation of frequency and/or amplitude of pulsatile hypothalamic secretion of GnRH [13,14]. A range of environmental, metabolic and endocrine factors regulates hypothalamic–pituitary function. In particular, testosterone [15] and oestrogen [16] (formed by the aromatization of testosterone in adipose tissue) exert negative feedback inhibition on the reproductive endocrine axis at the level of the hypothalamus and the pituitary. Inhibin B, secreted by the Sertoli cells of the testes, also plays an inhibitory role in gonadotropin secretion [17].

In obese individuals, there is evidence to suggest that all the three testicular cell populations – Leydig cells, Sertoli cells and germ cells – are functionally impaired resulting in diminished serum testosterone and inhibin B and in decreased sperm count and quality. In men, weight loss seems to have an impact on most these markers of gonadal function as well as on sexual function, although the impact on testosterone secretion has been studied the most.

Obesity and Leydig Cell Function

A reduction in the concentrations of circulating serum testosterone in obese men is now well established [18–21]. Serum concentrations of LH are not elevated in obese men [22], unlike in primary gonadal failure, suggesting that hypothalamic and/or pituitary dysfunction plays a part in the causation of hypogonadism in obese individuals. Additionally, in hypogonadal obese men, testosterone response to exogenous human chorionic gonadotrophin has been demonstrated to be normal [18–23] suggesting that Leydig cells per se are not affected by obesity. LH pulse frequency in severely obese men is preserved, while LH pulse amplitude is decreased [22–24]. This would suggest a decrease in pituitary sensitivity and/or a decrease in GnRH pulse amplitude as the central pathophysiological feature characterising hypogonadism in obesity. In this context, it has to be noted that obese men have higher concentrations of serum oestradiol [25] reflecting increased aromatase activity in the larger fat mass, which also exerts inhibitory effects on the hypothalamic—pituitary unit.

Obstructive sleep apnoea (OSA) frequently coexists with obesity and has also been implicated in the causation of hypogonadism in obese men. Studies have observed the presence of severe sleep apnoea in one-third of men with BMI > 40 kg/m^2, with some degree of disordered sleeping observed in around 98% of the patients studied [26]. In the presence of OSA, nocturnal secretion of LH and testosterone is impaired, being negatively correlated with respiratory distress index [27]. A host of other factors such as systemic inflammation, central leptin resistance, hyperglycaemia and insulin resistance have also been described to play at least a part in mediating central hypogonadism in obesity, and it is likely that the hypothalamic kisspeptin system plays a central role in relaying many of these signals to the hypothalamic GnRH neuronal population [28]. While a detailed review of the mechanisms underpinning hypogonadotropic hypogonadism in obese men is beyond the scope of this chapter, Figure 13.1 shows the current understanding of the neuroendocrine pathological processes involved.

Weight loss mediated by Roux-en-Y gastric bypass surgery has been demonstrated to double both total and free testosterone serum concentrations and reduce serum oestradiol as well [29]. In a cohort of 64 obese men, the gastric bypass surgery group ($n = 22$, baseline BMI 46.2 ± 0.9 kg/m^2) had a significantly greater decrease in BMI (-16.6 ± 1.2 vs -0.46 ± 0.51 kg/m^2) than obese controls ($n = 42$). This weight loss was associated with an increase in total testosterone (310.8 ± 47.6 vs 14.2 ± 15.3 ng/dl) and free testosterone (45.2 ± 5.1 vs -0.4 ± 3.0 pg/ml) and a decrease in serum oestradiol (-8.1 ± 2.4 vs 1.6 ± 1.4 pg/ml). Quality of sexual life also showed improvement after gastric bypass surgery [29].

Significant increases in total testosterone have also been observed in clinical studies involving the use of very low-calorie diet [30,31]. In a randomised controlled trial of a 4-month weight-loss programme including 10 weeks on a very low-energy diet (VLED), it was found that the mean weight loss in the intervention group at 8-months follow-up was 17 kg. Increases in sex-hormone-binding globulin (SHBG), testosterone and high-density lipoprotein–cholesterol were observed at the end of the follow-up and was found to be associated with decreases in insulin and leptin concentrations. During rapid weight loss observed during the VLED regime, there was further transient improvement in metabolic and reproductive endocrine markers [30]. In another observational study of 58 abdominally obese

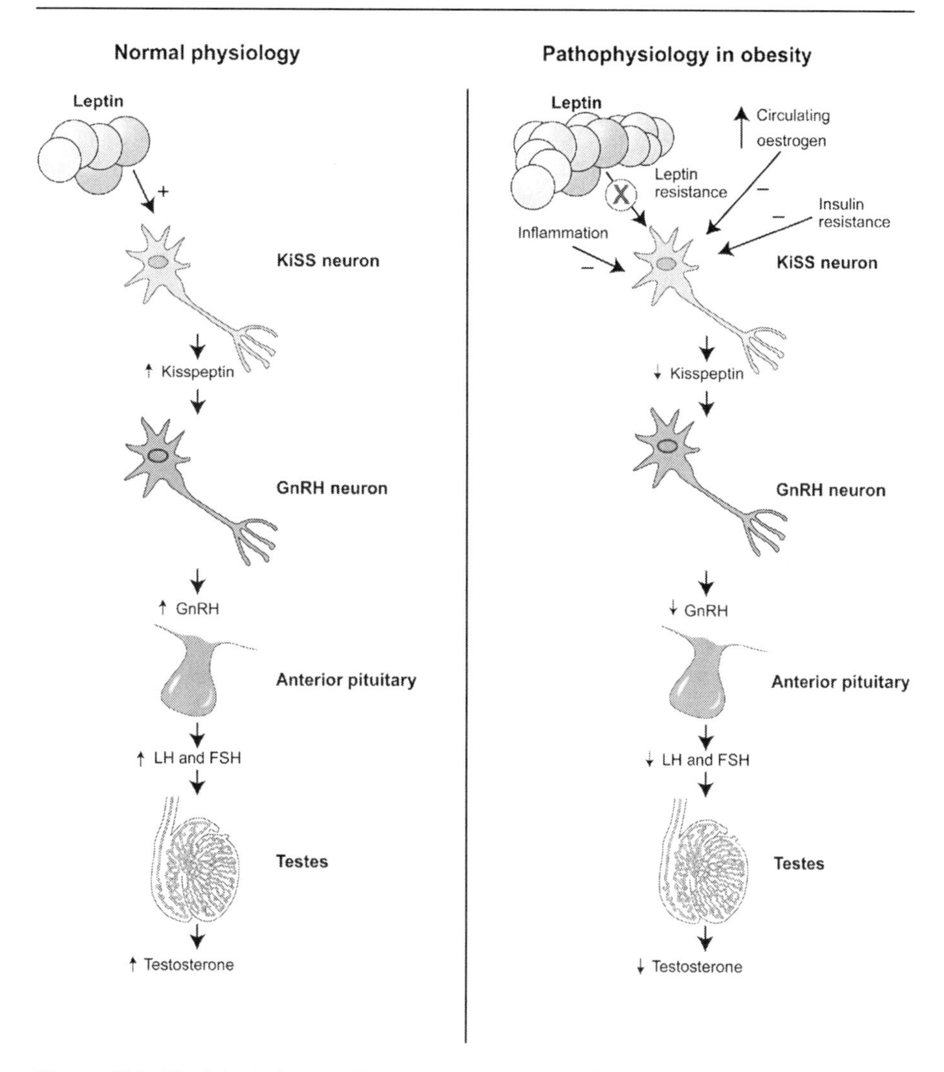

Figure 13.1 Physiological regulation of testosterone secretion and putative mechanisms by which pathological processes associated with obesity lead to impaired testosterone secretion. *Source*: Reproduced with kind permission from Ref. [28].

men (age 46.3 ± 7.5 years; BMI 36.1 ± 3.8 kg/m^2; waist girth 121 ± 10 cm) with metabolic syndrome, 9-week VLED therapy achieved weight loss of 14.3 ± 9.1 kg after a 12-month maintenance period [31]. Both SHBG and free testosterone increased significantly with the weight loss observed in this study [31].

It is, therefore, clear that obesity is associated with decreased testosterone concentrations, with negative effects on overall health that translate into increased mortality [32,33] and that this appears to be potentially reversible.

Obesity and Spermatogenesis

A reduction in sperm concentration and total sperm count in overweight men was first reported in a cohort of men drafted for military service in Denmark [34]. This study categorised men into three groups based on BMI: <20, 20–25 and >25 kg/m^2. Men with a BMI < 20 kg/m^2 had a 28.1% (95% CI 8.3–47.9) reduction in sperm concentration and a 36.4% reduction in total sperm count (95% CI 14.6–58.3) compared to those with a BMI of 20–25 kg/m^2. Men with BMI > 25 kg/m^2 also had a reduction in sperm concentration and total sperm count of 21.6% (95% CI 4.0–39.4) and 23.9% (95% CI 4.7–43.2), respectively. Percentages of normal spermatozoa were reduced, although not significantly, among men with high or low BMI. Semen volume and percentage of motile spermatozoa were not affected [34]. It has to be noted, however, that the median BMI in this study was 22.4 kg/m^2 and that the sample was drawn from 18-year-old, presumably fit, men. The extent to which the observations in this population are generalizable to wider population is open to debate, but it would seem likely that greater rather than lesser effects would be apparent in the general population.

Several other studies have reported alterations in sperm count and/or quality as well as quantity. In a study of 81 men, markers of obesity were shown to be inversely associated with sperm count and total motile sperm cell number [35]. This study showed associations between waist and hip circumferences versus sperm count, total motile sperm cell number and rapid progressive motile sperm count; between weight versus total sperm count and total motile sperm cell number and between waist circumference and waist/hip ratio versus semen volume [35].

A larger study of 520 male partners, who were overweight and obese, of couples presenting for infertility also showed a dramatic decrease in total motile sperm count in comparison to men with normal BMI [36]. Men in this study were grouped based on BMI values (normal: 20–24 kg/m^2, overweight: 25–30 kg/m^2, obese: >30 kg/m^2). The number of normal motile sperm cells in each of these groups was as follows: normal, 18.6×10^6; overweight, 3.6×10^6 and obese, 0.7×10^6. Linear regression revealed a significant negative relationship between BMI and the total number of normal motile sperm cells. Men with BMI greater than 25 kg/m^2 also had fewer chromatin-intact normal motile sperm cells per ejaculate [36]; increased sperm DNA damage is now clearly associated with reduced chance of successful pregnancy in assisted conception and with increased risk of miscarriage [37,38].

Moreover, studies in 72 men with infertility have shown that poor semen quality is three times more likely to be observed in obese men than in men with normal weight [39]. A significant negative correlation between semen quality parameters and BMI among men with normal semen quality was also observed in this study [39]. Consistent with these observations, a 3 kg/m^2 increase in male BMI was noted to be associated with infertility (OR 1.12; 95% CI 1.01–1.25; $n = 1329$) in men enrolled in the Agricultural Health Study in the United States [40]. The BMI effect was stronger when the data were limited to couples with the most complete data of highest quality. Moreover, the association between BMI and infertility was similar for older and younger men [40], and no association could be explained between age-related decline in testosterone secretion and increased prevalence of erectile dysfunction in older men.

However, it has to be noted that not all the studies have demonstrated this association between obesity and qualitative or quantitative changes in spermatogenesis [41–43]. The resultant heterogeneity of data has led the authors of a recent meta-analysis to conclude that there is a lack of association between increased BMI and semen parameters [44]. The mechanistic pathways effecting these changes in semen parameters are also unclear, as the decrease in circulating serum testosterone in obese men has not been shown to be correlated with marked decrease in intra-testicular testosterone. Adverse biochemical effects of obesity on oocyte function are now being elucidated [45], with a reduced oocyte mitochondrial membrane potential in overfed animals; it is possible or perhaps even likely that similar effects are also of relevance to male gametogenesis.

Following a lifestyle modification programme that achieved median weight loss of 15% (range 3.5–25.4%) decline from baseline, the degree of weight loss was positively correlated with an increase in total sperm count and semen volume [46]. Baseline BMI was inversely associated with sperm concentration, total sperm count, sperm morphology and motile sperm as well as total testosterone and inhibin B and positively associated with oestradiol. Weight loss, achieved with a 14-week residential weight loss programme, was associated with an increase in total sperm count, semen volume, total testosterone and SHBG. The group with the largest weight loss had a statistically significant increase in total sperm count [193 million (95% CI 45; $n = 341$)] and normal sperm morphology [46].

Serial data on sperm quality in obese men undergoing bariatric procedures are yet to be reported.

Obesity and Sertoli Cell Function

Inhibin B is a testicular peptide secreted almost exclusively by Sertoli cells which negatively regulates FSH secretion [47], acting as a readout of the spermatogenic activity of seminiferous tubules [48]. Serum concentrations of inhibin B are correlated to total sperm count, testicular volume and testicular biopsy score [49]. In this study, the receiver operating characteristic analysis showed a diagnostic accuracy of 95% compared to a value of 80% for FSH, suggesting that inhibin B is a useful endocrine marker of spermatogenesis in sub-fertile men [49].

In keeping with the studies described above linking obesity with reduced spermatogenesis, inhibin B concentrations have also been demonstrated to be lower in overweight and obese men [34–43] and to increase with weight loss [46]. However, mechanistic studies to explain these findings are yet to be reported.

Obesity and Erectile Function

Erectile dysfunction is associated with infertility [50]. Nearly four-fifths of all men reporting erectile dysfunction are found to be overweight or obese. It is unclear whether obesity per se or other confounding variables (such as smoking, dyslipidaemia, atherosclerosis, hypertension and type-2 diabetes) are the prime movers in the pathogenesis of erectile dysfunction in obese men.

In a randomised trial of obese men who achieved 10% weight loss through modification of diet and exercise regimes, one-third of men demonstrated improvement in sexual function [51]. BMI and physical activity were independently associated with improvements in International Index of Erectile Function scores [51]. However, these findings have not been replicated in other weight loss studies, where despite improvements in testosterone secretion, scores of sexual function remained unchanged [30]. Weight loss after gastric bypass surgery has also been demonstrated to be associated with an improvement in the quality of sexual life in all aspects, as assessed using the Impact of Weight on the Quality of Life-Lite questionnaire [29].

Summary and Practical Considerations

The majority of evidence, although not all, supports a link between obesity and decreased reproductive activity in men with adverse effects on endocrine, spermatogenic, and erectile function. Most studies assessing male fertility have used spermatogenic surrogates rather than fertility itself. While there are robust data linking weight loss with increases in serum testosterone, the data to recommend aggressive weight loss, with or without bariatric surgery, to men with sub-normal sperm count seeking fertility remains relatively weak. However, it has to be noted that modest weight loss, as low as 5% of baseline body weight, has been shown to reduce or eliminate disorders associated with obesity [52], and it could be argued that health care professionals should seize all opportunities to encourage weight loss.

There is also a paucity of randomised clinical trials comparing bariatric interventions with diet and exercise-based regimes. Factors including treatment allocation bias, motivation to achieve weight loss and compliance with dietary regimes are potential confounders in current studies of bariatric surgery.

While this chapter has focused on obesity and sub-fertility in individuals with no known genetic or other causes underlying obesity, it has to be emphasised that there are several endocrine conditions (e.g. Klienfelter's syndrome and Prader–Willi syndrome) that may present with obesity and hypogonadism to the clinician.

In conclusion, clinicians caring for obese men with infertility should recognise that it is likely that the obesity contributes to the reproductive dysfunction. At present, the paucity of robust data for weight loss to resolve the immediate concern means that individualisation of advice remains the appropriate clinical standpoint. The value of bariatric surgery to improve male fertility remains very uncertain.

References

1. Wilding JPH. Pathophysiology and aetiology of obesity. *Medicine*. 2011;39(1):6–10.
2. McTigue KM, Garrett JM, Popkin BM. The natural history of the development of obesity in a cohort of young U.S. adults between 1981 and 1998. *Ann Intern Med*. 2002;136(12): 857–864.

3. Kahn HS, Cheng YJ. Longitudinal changes in BMI and in an index estimating excess lipids among white and black adults in the United States. *Int J Obes*. 2007;32(1): 136–143.
4. Shaw K, Gennat H, O'Rourke P, Del Mar C. Exercise for overweight or obesity. *Cochrane Database Syst Rev*. 2006;(4):CD003817.
5. Colquitt JL, Picot J, Loveman E, Clegg AJ. Surgery for obesity. *Cochrane Database Syst Rev*. 2009;(2):CD003641.
6. Gortmaker SL, Swinburn BA, Levy D, et al. Changing the future of obesity: science, policy, and action. *Lancet*. 2011;378(9793):838–847.
7. Picot J, Jones J, Colquitt JL, et al. The clinical effectiveness and cost-effectiveness of bariatric (weight loss) surgery for obesity: a systematic review and economic evaluation. *Health Technol Assess*. 2009;13(41):1–190:215–357, iii–iv.
8. Suter M, Calmes J, Paroz A, Giusti VA. 10-year experience with laparoscopic gastric banding for morbid obesity: high long-term complication and failure rates. *Obes Surg*. 2006;16(7):829–835.
9. Schouten R, Rijs CS, Bouvy ND, et al. A multicenter, randomized efficacy study of the endobarrier gastrointestinal liner for presurgical weight loss prior to bariatric surgery. *Ann Surg*. 2010;251(2):236–243.
10. Schally AV, Arimura A, Kastin AJ, et al. Gonadotropin-releasing hormone: one polypeptide regulates secretion of luteinizing and follicle-stimulating hormones. *Science*. 1971;173 (4001):1036–1038.
11. Baba Y, Matsuo H, Schally AV. Structure of porcine LH-Releasing and FSH-Releasing hormone 2. Confirmation of proposed structure by conventional sequential analyses. *Biochem Biophys Res Commun*. 1971;44(2):459–463.
12. Millar RP, Lu Z-L, Pawson AJ, Flanagan CA, Morgan K, Maudsley SR. Gonadotropin-releasing hormone receptors. *Endocr Rev*. 2004;25(2):235–275.
13. Haisenleder DJ, Dalkin AC, Ortolan GA, Marshall JC, Shupnik MA. A pulsatile gonadotropin-releasing hormone stimulus is required to increase transcription of the gonadotropin subunit genes: evidence for differential regulation of transcription by pulse frequency in vivo. *Endocrinology*. 1991;128(1):509–517.
14. McCartney CR, Gingrich MB, Hu Y, Evans WS, Marshall JC. Hypothalamic regulation of cyclic ovulation: evidence that the increase in gonadotropin-releasing hormone pulse frequency during the follicular phase reflects the gradual loss of the restraining effects of progesterone. *J Clin Endocrinol Metab*. 2002;87(5): 2194–2200.
15. Pitteloud N, Dwyer AA, DeCruz S, et al. Inhibition of luteinizing hormone secretion by testosterone in men requires aromatization for its pituitary but not its hypothalamic effects: evidence from the tandem study of normal and gonadotropin-releasing hormone-deficient men. *J Clin Endocrinol Metab*. 2008;93(3): 784–791.
16. Veldhuis JD, Dufau ML. Estradiol modulates the pulsatile secretion of biologically active luteinizing hormone in man. *J Clin Invest*. 1987;80(3):631–638.
17. de Kretser DM, Buzzard JJ, Okuma Y, et al. The role of activin, follistatin and inhibin in testicular physiology. *Mol Cell Endocrinol*. 2004;225(1-2):57–64.
18. Glass AR, Swerdloff RS, Bray GA, Dahms WT, Atkinson RL. Low serum testosterone and sex-hormone-binding-globulin in massively obese men. *J Clin Endocrinol Metab*. 1977;45(6):1211–1219.
19. Strain GW, Zumoff B, Kream J, et al. Mild hypogonadotropic hypogonadism in obese men. *Metabolism*. 1982;31(9):871–875.

20. Zumoff B, Strain GW, Miller LK, et al. Plasma free and non-sex-hormone-binding-globulin bound testosterone are decreased in obese men in proportion to their degree of obesity. *J Clin Endocrinol Metab*. 1990;71(4):929–931.
21. Tsai EC, Matsumoto AM, Fujimoto WY, Boyko EJ. Association of bioavailable, free, and total testosterone with insulin resistance. *Diabetes Care*. 2004;27(4):861–868.
22. Giagulli VA, Kaufman JM, Vermeulen A. Pathogenesis of the decreased androgen levels in obese men. *J Clin Endocrinol Metab*. 1994;79(4):997–1000.
23. Amatruda JM, Hochstein M, Hsu TH, Lockwood DH. Hypothalamic and pituitary dysfunction in obese males. *Int J Obes*. 1982;6(2):183–189.
24. Vermeulen A, Kaufman J, Deslypere J, Thomas G. Attenuated luteinizing hormone (LH) pulse amplitude but normal LH pulse frequency, and its relation to plasma androgens in hypogonadism of obese men. *J Clin Endocrinol Metab*. 1993;76(5): 1140–1146.
25. Schneider G, Kirschner MA, Berkowitz R, Ertel NH. Increased estrogen production in obese men. *J Clin Endocrinol Metab*. 1979;48(4):633–638.
26. Valencia-Flores M, Orea A, Castano VA, et al. Prevalence of sleep apnea and electrocardiographic disturbances in morbidly obese patients. *Obesity*. 2000;8(3): 262–269.
27. Luboshitzky R, Aviv A, Hefetz A, et al. Decreased pituitary–gonadal secretion in men with obstructive sleep apnea. *J Clin Endocrinol Metab*. 2002;87(7):3394–3398.
28. George JT, Millar RP, Anderson RA. Hypothesis: kisspeptin mediates male hypogonadism in obesity and type 2 diabetes. *Neuroendocrinology*. 2010;91(4):302–307.
29. Hammoud A, Gibson M, Hunt SC, et al. Effect of Roux-en-Y gastric bypass surgery on the sex steroids and quality of life in obese men. *J Clin Endocrinol Metab*. 2009;94(4): 1329–1332.
30. Kaukua J, Pekkarinen T, Sane T, Mustajoki P. Sex hormones and sexual function in obese men losing weight. *Obes Res*. 2003;11(6):689–694.
31. Niskanen L, Laaksonen DE, Punnonen K, Mustajoki P, Kaukua J, Rissanen A. Changes in sex hormone-binding globulin and testosterone during weight loss and weight maintenance in abdominally obese men with the metabolic syndrome. *Diabetes Obes Metab*. 2004;6(3):208–215.
32. Khaw K-T, Dowsett M, Folkerd E, et al. Endogenous testosterone and mortality due to all causes, cardiovascular disease, and cancer in men. *Circulation*. 2007;116(23): 2694–2701.
33. Araujo AB, Kupelian V, Page ST, Handelsman DJ, Bremner WJ, McKinlay JB. Sex steroids and all-cause and cause-specific mortality in men. *Arch Intern Med*. 2007;167(12): 1252–1260.
34. Jensen TK, Andersson A-M, Jørgensen N, et al. Body mass index in relation to semen quality and reproductive hormonesamong 1,558 Danish men. *Fertil Steril*. 2004;82(4): 863–870.
35. Fejes I, Koloszár S, Szöllo˝si J, Závaczki Z, Pál A. Is semen quality affected by male body fat distribution? *Andrologia*. 2005;37(5):155–159.
36. Kort HI, Massey JB, Elsner CW, et al. Impact of body mass index values on sperm quantity and quality. *J Androl*. 2006;27(3):450–452.
37. Zini A, Boman JM, Belzile E, Ciampi A. Sperm DNA damage is associated with an increased risk of pregnancy loss after IVF and ICSI: systematic review and meta-analysis. *Hum Reprod*. 2008;23(12):2663–2668.
38. Aitken RJ, De Iuliis GN, McLachlan RI. Biological and clinical significance of DNA damage in the male germ line. *Int J Androl*. 2009;32(1):46–56.

39. Magnusdottir EV, Thorsteinsson T, Thorsteinsdottir S, Heimisdottir M, Olafsdottir K. Persistent organochlorines, sedentary occupation, obesity and human male subfertility. *Hum Reprod.* 2005;20(1):208–215.
40. Sallmén M, Sandler DP, Hoppin JA, Blair A, Baird DD. Reduced fertility among overweight and obese men. *Epidemiology.* 2006;17(5):520–523.
41. Teerds KJ, de Rooij DG, Keijer J. Functional relationship between obesity and male reproduction: from humans to animal models. *Hum Reprod Update.* 2011;17(5): 667–683.
42. Hammoud AO, Gibson M, Peterson CM, Hamilton BD, Carrell DT. Obesity and male reproductive potential. *J Androl.* 2006;27(5):619–626.
43. Aggerholm AS, Thulstrup AM, Toft G, Ramlau-Hansen CH, Bonde JP. Is overweight a risk factor for reduced semen quality and altered serum sex hormone profile? *Fertil Steril.* 2008;90(3):619–626.
44. MacDonald AA, Herbison GP, Showell M, Farquhar CM. The impact of body mass index on semen parameters and reproductive hormones in human males: a systematic review with meta-analysis. *Hum Reprod Update.* 2010;16(3):293–311.
45. Wu LL-Y, Dunning KR, Yang X, et al. High-Fat diet causes lipotoxicity responses in cumulus, oocyte complexes and decreased fertilization rates. *Endocrinology.* 2010;151(11): 5438–5445.
46. Hakonsen L, Thulstrup A, Aggerholm A, et al. Does weight loss improve semen quality and reproductive hormones? Results from a cohort of severely obese men. *Reprod Health.* 2011;8(1):24.
47. Anawalt BD, Bebb RA, Matsumoto AM, et al. Serum inhibin B levels reflect Sertoli cell function in normal men and men with testicular dysfunction. *J Clin Endocrinol Metab.* 1996;81(9):3341–3345.
48. Anderson RA, Wallace EM, Groome NP, Bellis AJ, Wu FC. Physiological relationships between inhibin B, follicle stimulating hormone secretion and spermatogenesis in normal men and response to gonadotrophin suppression by exogenous testosterone. *Hum Reprod.* 1997;12(4):746–751.
49. Pierik FH, Vreeburg JTM, Stijnen T, de Jong FH, Weber RFA. Serum inhibin B as a marker of spermatogenesis. *J Clin Endocrinol Metab.* 1998;83(9):3110–3114.
50. O'Brien JH, Lazarou S, Deane L, Jarvi K, Zini A. Erectile dysfunction and andropause symptoms in infertile men. *J Urol.* 2005;174(5):1932–1934.
51. Esposito K, Giugliano F, Di Palo C, et al. Effect of lifestyle changes on erectile dysfunction in obese men. *JAMA.* 2004;291(24):2978–2984.
52. Blackburn G. Effect of degree of weight loss on health benefits. *Obes Res.* 1995;3(suppl 2):211s–216s.

Section 4

Pregnancy and Obesity

14 Maternal Obesity and Developmental Priming of Risk of Later Disease

R.C.W. Ma[1], Peter D. Gluckman[2] and Mark A. Hanson[3]

[1]Department of Medicine and Therapeutics, Chinese University of Hong Kong and Hong Kong Institute of Diabetes and Obesity, Hong Kong, People's Republic of China, [2]Liggins Institute, University of Auckland, Auckland, New Zealand and Singapore Institute of Clinical Sciences, Singapore, [3]Institute of Developmental Sciences, University of Southampton, Southampton, UK

Introduction

The prevalence of obesity has reached alarming proportions globally, and continues to rise in both developed and developing countries. It was estimated that 23% of the world's adult population was overweight in 2005 (24% in men and 22% in women), with 9.8% being obese (7.7% in men and 11.9% in women), giving an estimated total number of overweight and obese adults as 937 million and 396 million, respectively [1]. If the current secular trend continues unabated, the number of overweight and obese adults is projected to increase to 2.16 billion and 1.12 billion, respectively, by 2030. In developing countries such as in Asia, the rate has increased dramatically. For example, the increase in prevalence of overweight and obesity in China over the last 20 years was 400%, compared with 20% in Australia [2]. Given the impact of obesity on morbidity and mortality, and its association with cardiovascular disease, diabetes and cancer, the escalating epidemic of obesity will lead to significant increases in health care burden [3–5]. The problem is further aggravated by the observation that for any given measure of body fat, some Asian populations have a greater risk of developing diabetes than do Caucasians [6].

In addition to increasing prevalence of obesity in the general population, the proportion of children, adolescents, young women and pregnant women who are overweight or obese has been increasing. A 40-year-old female non-smoker is estimated to lose 3.3 years of life expectancy if overweight, and 7.1 years if obese [7]. Furthermore, obesity in females is associated with a variety of reproductive issues including increased risk of menstrual disorders, polycystic ovary syndrome, infertility, recurrent miscarriage during pregnancy, worse obstetric outcome, impaired

Obesity. DOI: http://dx.doi.org/10.1016/B978-0-12-416045-3.00014-5

foetal well-being and an increased risk of gestational diabetes mellitus (GDM), as well as increased risk of malignancies including breast, ovarian and endometrial cancer [8,9].

The recent increase in adult obesity is mirrored by increasing prevalence of diabetes, as well as childhood obesity [10,11]. Whilst the increase in childhood obesity has been attributed to changes in lifestyle, it is increasingly recognised that early development has long-term effects on risk of obesity and other NCD in adults [12,13]. In this context, emerging evidence suggests that maternal obesity and maternal diabetes both have harmful long-term effects which may contribute to the rising prevalence of childhood obesity and diabetes now seen in many countries, especially in developing regions [14]. It must be borne in mind that whilst obesity increases risk of non-communicable disease (NCD), they are not synonymous [15].

Epidemiological Observations on Maternal Obesity, Macrosomia and Neonatal Adiposity

A large body of evidence has emerged linking lower birthweight and maternal under-nutrition with adult risk of obesity, type 2 diabetes and cardiovascular disease, although this is an association rather than a causal link [16]. Birthweight-independent associations between foetal epigenetic state, itself influenced by maternal nutrition, and the development of body fat have been reported [17]. However, this association reflects the compelling evidence linking limitations in foetal nutrient supply or alterations in exposure to maternal stress hormones to changes in foetal development that, while having an adaptive origin, can lead to a greater risk of compromise in a later obesogenic environment. These pathways can be induced by both maternal and placental pathophysiology, by variation in maternal nutrition and by the unavoidable mechanisms of maternal constraint. These pathways are well reviewed elsewhere [12,16] and are not the focus of this chapter. But as we have pointed out before [16], they may lead the offspring to be at greater risk of developing obesity, type 2 diabetes or indeed GDM and thus must be considered alongside maternal obesity and GDM in understanding the interactions between an obese or insulin-resistant maternal state and offspring health.

Maternal obesity is increasingly recognised to be associated with increased risk of adiposity and NCDs in the offspring, and that this effect is independent of the shared genetic and environmental factors between the mother and child [18–20]. Incidentally, birthweight has increased progressively over recent years in most populations except Japan where dieting in pregnancy is common [21], and the proportion of babies born with birthweight >4000 g has increased steadily [22].

Several lines of evidence have highlighted a potential role of intrauterine environment on the transgenerational effects of maternal obesity. For example, epidemiological studies have revealed stronger correlation between offspring body mass index (BMI) and maternal BMI than with paternal BMI, suggesting that, in addition to genetic influences, the in utero environment may confer additional impact [23]. In a retrospective study and one prospective birth cohort, maternal BMI was associated with offspring adiposity at 12 and 24 months whilst maternal glycaemia was correlated with offspring adiposity at birth although not at 12 and 24 months, suggesting a link between maternal obesity and offspring adiposity [24]. In another study of 119 term infants of mothers with GDM and 143 term control infants, it was noted that maternal pre-pregnancy weight and gestational weight gain were significant predictors of infant BMI for both infants of mothers with GDM and control infants, with maternal glucose being an additional factor that contributes to macrosomia and infant adiposity [25]. Furthermore, the increase in infant adiposity was correlated with increased blood pressure [25]. In an analysis of 187 neonates and their mothers, increased maternal pre-gravid weight and estimated insulin resistance explained close to 50% of the variance in birthweight and foetal fat mass at birth [26]. Foetuses of obese mothers not only have greater percentage body fat, but have elevated cord blood leptin, interleukin-6, and are more insulin resistant at birth. A strong correlation is present between foetal adiposity, maternal insulin resistance, as well as maternal BMI, which persists despite correcting for confounding factors [27]. As overweight or obese mothers are also more likely to have diabetes and GDM, factors known to be associated with increased foetal adiposity, the magnitude of risk associated with maternal obesity alone is not yet clear. In the Hyperglycemia and Adverse Pregnancy Outcome (HAPO) study, among participants without frank hyperglycaemia higher maternal BMI was associated with increased risk of macrosomia, Caesarian section, elevated cord C-peptide and pre-eclampsia, despite adjusting for parameters of glycaemia [28].

Maternal Obesity and Childhood Obesity

Animal models have previously demonstrated a link between maternal obesity at conception and offspring obesity [29]. In a cohort of children in Southampton with non-diabetic mothers, fat mass at 9 years was greater among the children whose mothers had a higher pre-pregnancy BMI [30]. In the Growing Up Today Study, a US nationwide study of diet, activity and growth, it was noted that children born to mothers with GDM had approximately 40% increased risk of adolescent overweight 9–14 years after birth, though this was substantially attenuated after adjustment for maternal BMI [31]. This suggests that maternal obesity plays an important part in the observed link between GDM and obesity in the offspring. In line with this, in a prospective study of 89 women with either normal glucose tolerance or GDM and their offspring at birth and at a mean age of 8.8 years, maternal pre-gravid BMI has been reported to be the strongest predictor of childhood obesity, independent of maternal glucose status or birthweight [32]. In another cohort, maternal obesity, greater gestational weight gain and parity were all predictors of offspring adiposity [20].

Although the discussion so far has focused on the impact of maternal obesity on offspring adiposity, given the link between childhood obesity and metabolic abnormalities, offspring of obese mother are at increased risk of metabolic disorders.

Children who are large for gestational age and exposed to either maternal obesity or GDM are at increased risk of developing metabolic syndrome in childhood [33].

Maternal Pregnancy Weight Gain and Childhood Obesity

In addition to increased risk in offspring of obese mothers, recent studies suggest that children born to mothers who had excessive weight gain also have increased adiposity [34]. In the Southampton Women's Survey, children born to mothers who had excessive weight gain, as defined by the Institute of Medicine recommendations, had greater fat mass in the neonatal period, which persisted at 4 years and 6 years of age [35]. In another study which included 3457 mother–offspring pairs from the Avon Longitudinal Study of Parents and Children, women who exceeded the Institute of Medicine recommendations for gestational weight gain were more likely to have children with greater BMI, waist circumference, leptin, systolic blood pressure, elevated inflammatory factors including C-reactive protein (CRP) and interleukin-6 and lower high-density lipoprotein (HDL) cholesterol [36]. These studies of excess maternal weight gain provide an opportunity to explore the long-term effects of over-nutrition as opposed to maternal obesity per se.

Maternal Diabetes, GDM and Glucose Effects

It was first postulated by Pederson in the 1950s that intrauterine exposure of the foetus to hyperglycaemia due to maternal diabetes may cause permanent changes including foetal malformations, increased birthweight and increased risk of diabetes and obesity later in life, which he termed fuel-mediated teratogenesis [37]. An increased risk of glucose intolerance and diabetes in offspring of mothers with diabetes was first highlighted by studies carried out in the Pima Indians, which revealed that they had a significantly higher risk of diabetes than offspring of pre-diabetic or non-diabetic mothers [38]. Subsequent studies in the same population found that, when compared to siblings born before the onset of diabetes in the mother, children born after the mother has been diagnosed with diabetes have significantly higher risk of diabetes, suggesting that the excess risk associated with maternal diabetes is likely to reflect exposure to intrauterine environment associated with maternal hyperglycaemia [39]. Offspring of mothers with type 1 diabetes have also been found to be at increased risk of glucose intolerance and impaired insulin secretion, suggesting that it is the exposure to intrauterine hyperglycaemia or other associated metabolic derangement that is responsible for this transgenerational effect of maternal diabetes [40]. In fact, excess maternal compared with paternal transmission of diabetes has been noted in several cohorts, emphasising the important role of the intrauterine environment [41,42].

GDM has also been found to have similar long-term transgenerational effects. Earlier studies have shown that compared with large for gestational age (LGA) offspring of mothers with normal glucose tolerance, LGA offspring of mothers with GDM have increased fat mass and decreased lean body mass [43]. GDM was twice as common in subjects with a diabetic mother compared with those with a diabetic father, again supporting maternal transmission and transgenerational effects [44]. In the multi-ethnic Exploring Perinatal Outcomes Among Children (EPOCH) study, exposure to maternal GDM was associated with increased overall and abdominal obesity, and a more central fat distribution at 6–13 years old, which persisted despite adjustment for multiple factors including socio-economic factors, birthweight, gestational age, maternal smoking during pregnancy and current diet and physical activity. Of note, the associated risk was attenuated after adjustment for maternal BMI [45].

In a systematic review and meta-analysis of studies on maternal diabetes and offspring obesity as defined by BMI, the relationship between maternal diabetes and offspring obesity was no longer present after adjustment for maternal BMI, again highlighting the important contribution of maternal obesity to the transgenerational effects of maternal diabetes and obesity [46]. In the SEARCH case–control study, which included 79 youths with type 2 diabetes and 190 normal youths, exposure to maternal obesity and maternal diabetes were both independently associated with type 2 diabetes in youth, whereby offspring of a mother who is both diabetic and obese have markedly (19-fold) elevated risk of type 2 diabetes [47].

Relationship Between Maternal Hyperglycaemia and Neonatal Adiposity

Several earlier studies have suggested a linear relationship between the level of maternal glycaemia during pregnancy and offspring adiposity. A retrospective study conducted on 143 infants of non-diabetic mothers showed a linear relationship between glucose level during the screening glucose challenge tests and infant BMI [25]. In a consecutive series of 6854 women screened for GDM, of which 6390 did not have GDM, occurrence of macrosomia was associated with higher glucose concentrations at screening [48]. Likewise, in a study conducted in South Asia, in Mysore, maternal fasting glucose at 30-weeks gestation was associated with several indices of foetal growth [49].

The HAPO study was specifically designed to test the hypothesis that maternal glucose levels during pregnancy may be related to adverse pregnancy outcomes. In this landmark international study, which included more than 20,000 pregnant women, increase in fasting, 1 and 2 h glucose at 24–32 weeks gestation was found to be associated with approximately 1- to 1.5-fold increased risk for neonatal macrosomia (birthweight above the 90th centile) and neonatal hyperinsulinaemia (cord blood C-peptide >90th centile) [50]. This large study demonstrated clearly a linear relationship between maternal glycaemia and foetal growth, and has led to revised diagnostic criteria for GDM, which are now based on the risks of pregnancy complications [51]. Furthermore, in a subsequent analysis of the body composition data, it was again noted that a continuous relationship exists between maternal glucose levels at 24–32 weeks and cord C-peptide and neonatal adiposity at birth [52].

Maternal Hyperglycaemia and Childhood Obesity

In studies conducted among the Pima Indians, it was noted that offspring of mothers with diabetes had a significantly different growth pattern, with significantly higher gestational-adjusted birthweight and a greater rate of weight gain over the first few years of life [53]. The Diabetes in Pregnancy Study at Northwestern University in Chicago has also followed prospectively a cohort of mothers with diabetes and their offspring up to a mean age of 12 years [54]. It was noted that the prevalence of impaired glucose tolerance (IGT) among offspring of diabetic mothers was 5.4% at age 5–9 years and 19.3% at 10–16 years. Offspring of mothers with diabetes had a significantly higher fasting insulin and 2 h glucose during an oral glucose tolerance test compared with offspring of non-diabetic mothers. Both childhood obesity and hyperinsulinaemia in utero, as reflected by amniotic fluid insulin concentration, were identified as independent predictors of IGT in childhood [54,55]. These results suggest that at least part of the long-term consequences of maternal diabetes and obesity are related to hyperinsulinaemia in utero and its effects on adipogenesis [56].

GDM and Cardiometabolic Diseases in the Offspring

In addition to the link between maternal diabetes or GDM and childhood obesity, GDM has been associated with increased cardiometabolic risk in the offspring. In a long-term follow-up of offspring from Chinese mothers with GDM, offspring of mothers with GDM were found to have significantly higher systolic and diastolic blood pressure, as well as lower HDL cholesterol, after adjustment for age and gender. Elevated C-peptide predicted glucose intolerance in the offspring at age 8 [57], as well as glucose intolerance and metabolic syndrome at age 15 [58]. Similar results have been observed in Project Viva, in which exposure to GDM was linked to offspring adiposity and blood pressure [59].

Other studies, with more extended follow-up of offspring of mothers with diabetes or GDM, have demonstrated a sustained increase in cardiovascular risk. In one study, offspring of mothers with GDM have increased insulin resistance and central obesity 15 years later. Maternal obesity strongly predicted daughter's BMI percentile and percentage of body fat [60]. In a retrospective cohort of children aged 6–13 years, it was noted that offspring of diabetic mothers had significantly increased markers of endothelial dysfunction including increased E-selectin, vascular adhesion molecule 1 (VCAM-1), leptin, waist circumference, BMI, systolic blood pressure and decreased adiponectin levels when compared to offspring of non-diabetic pregnancies. This relationship was substantially attenuated by adjusting for maternal BMI [61]. Similar findings have been noted in offspring of mothers with type 1 diabetes, who have significantly higher cholesterol, insulin-like growth factor-1 (IGF-1) and other inflammatory cytokines including plasminogen activator inhibitor-1 (PAI-1), VCAM-1 and E-Selectin at age 5–11 compared to offspring of non-diabetic mothers [62]. In fact, an increasing number of both animal and human studies indicates that maternal obesity can impact on the offspring's cardiometabolic health and glucose–insulin homeostasis [34]. Some of the possible mechanisms that may underline this vicious cycle of 'diabetes begetting diabetes' will be explored in later sections in this chapter [63–65] (Figure 14.1).

Maternal Insulin Resistance, Hyperinsulinaemia and Foetal Growth

In general, insulin resistance increases by approximately 40–50% in most pregnancies, with the increase predominantly occurring in late gestation [66]. For women who are lean (pre-pregnancy BMI <25 kg/m^2), insulin sensitivity is unchanged initially, whilst insulin secretion increases in early pregnancy, and this helps to promote maternal lipogenesis and thus to support the rapid foetal growth of the latter part of pregnancy [67]. In the second part of pregnancy, a decline in peripheral insulin sensitivity as a consequence of placental lactogen and growth hormone secretion into the maternal circulation results in a progressive shift from lipogenesis

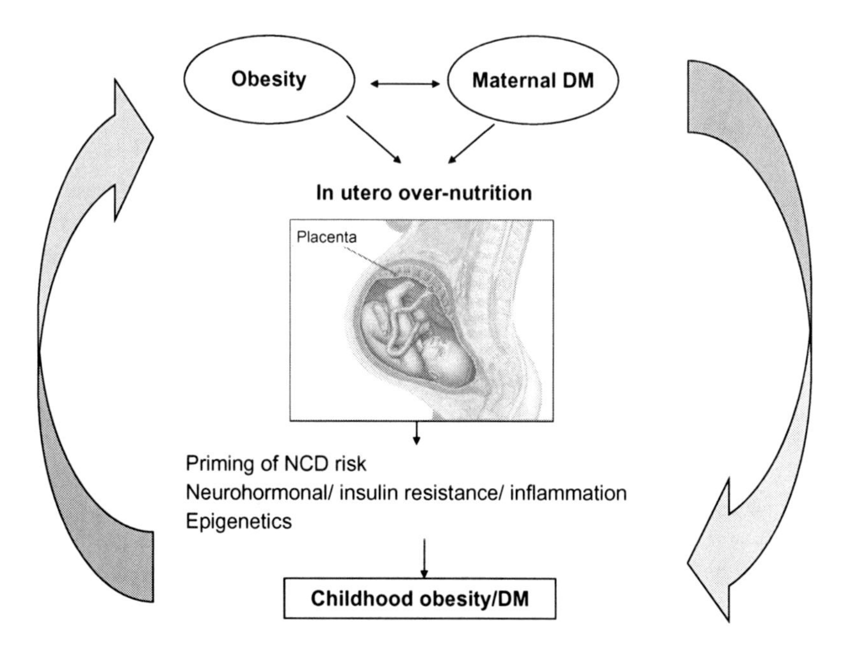

Figure 14.1 Transgenerational effects of maternal diabetes and obesity.

to lipolysis, with increased levels of free fatty acids and increased gluconeogenesis and hepatic glucose production [68]. At the same time, there is also a corresponding threefold increase in maternal insulin secretion [66].

Maternal obesity represents a different scenario whereby, in the obese mother, pre-pregnancy insulin resistance is exacerbated by the pregnant state, and such women have reduced peripheral insulin sensitivity even in early pregnancy. The increased adipose tissue in obese women also results in greater release of inflammatory markers and free fatty acids, which further exacerbates the insulin resistance and inhibits the action of insulin on suppressing lipolysis. In situations where the woman has a predisposition to impaired β-cell function, for example through inherited genetic or acquired epigenetic factors, then she may develop hyperglycaemia and GDM. Compared to women who are lean, obese women have an approximately threefold greater risk of developing GDM [69]. Women who themselves were of low birthweight and thus at greater risk of developing insulin resistance have a greater risk of developing GDM than those of normal birthweight [60].

Similar to the non-pregnant state, obesity and insulin resistance in the pregnant woman is also associated with a state of low-grade inflammation. Several studies have demonstrated that obese women have elevated markers of inflammation including elevated CRP, interleukin-6, tumour necrosis factor-α [70–72] and that these markers of inflammation are more strongly associated with pre-pregnancy BMI than other factors including GDM [72]. In a cross-sectional study in Korea, high pre-pregnancy BMI was associated with increased risk of preterm delivery, low-birthweight infants as well as macrosomia. Obese subjects had significantly lower folate status and elevated CRP [73]. This state of low-grade inflammation, with associated microvascular dysfunction, may represent one of the mechanisms by which the long-term effects of maternal obesity may be mediated, and which can occur independently of metabolic effects.

The fact that late pregnancy represents an insulin-resistant state helps to facilitate nutrient transfer to the foetus. However, in states of maternal over-nutrition, this may permit increased supply of nutrients to the foetus, resulting in increased foetal growth, especially of adipose tissue although somatic growth is also enhanced. Thus, pregnancy complicated by maternal obesity or diabetes may lead to a further increase in nutrients being transferred from the mother to the offspring.

Glucose Metabolism and Foetal Growth

Glucose transport across the placenta occurs through facilitated diffusion, aided by specific glucose transporters. The GLUT1 glucose transporter, present on both microvilli and basal membranes of the syncytial barrier, is the primary isoform involved in the transplacental movement of glucose, and is the rate-limiting step. It has been noted that, in diabetic pregnancies, there is an increase in basal GLUT1 expression and activity, which is present despite near-normal glycaemia at term, with significant consequences for the maternal–foetal flux of glucose [74,75].

Maternal Lipid Supply and Foetal Growth

In addition to maternal hyperglycaemia, over-nutrition in utero due to maternal lipid metabolism and lipotoxicity is also likely to be an important contributor to foetal priming of NCD risk seen in pregnancies complicated by maternal obesity [76]. The foetus has limited de novo lipogenic capacity, and therefore most of the precursors for foetal fat accretion are derived from the mother through transplacental transfer of substrates. During the later part of pregnancy, reduced maternal peripheral insulin sensitivity and the consequently increased lipolysis will result in increased free fatty acids, which can be transferred across the placenta. The increased nutrient supply in early pregnancy can result in increased placental and foetal somatic growth and it has been postulated that the fatty acid spillover and lipotoxicity in maternal over-nutrition may also contribute to maternal endothelial dysfunction, impaired trophoblastic invasion and altered placental metabolism and function [76].

Other lines of evidence also support a link between maternal over-nutrition and offspring obesity. For example, there is a close relationship between maternal triglyceride or fatty acid levels mid-pregnancy and offspring obesity [77–79]. The n-3 and n-6 fatty acids can only be acquired from maternal diet and placental transport. Free fatty acids are transported either by passive diffusion or through specific fatty acid–binding proteins [80,81]. Animal models, where a high-fat diet has been given to the mother, result in changes in the offspring including reduced muscle development, ectopic fat deposition, hyperinsulinaemia, hyperglycaemia, and increased triglyceride and leptin, which may be further exacerbated if the offspring is also given a high-fat diet [82,83]. This suggests that targeting maternal obesity provides an important opportunity in the life course for strategies to address the growing epidemic of diabetes and obesity [84].

Effects on Vascular Function

Long-term consequences of over-nutrition in utero have been investigated using a rat model of high-fat feeding during pregnancy [85]. It was noted that offspring developed endothelial dysfunction, which was more marked if the mother was also diabetic [85]. Subsequent studies revealed that the endothelial function was abnormal if offspring of mothers fed on a high-fat diet were given standard chow as opposed to high-fat chow during the post-natal period [86]. This demonstrated that predictive adaptive responses in the foetus exposed to maternal high-fat diet can be protective against development of cardiovascular disease, highlighting the 'mismatch' between in utero and post-natal environment as an important driver of cardiometabolic risk [87]. Recent studies suggest that hyper-responsiveness of the sympathetic nervous system in the offspring may be one of the consequences of foetal priming of NCD risk following maternal obesity, and may mediate some of the long-term effects on blood pressure regulation [88]. In a long-term follow-up of children exposed to GDM, hyperinsulinaemia in utero was associated with greater arterial stiffness in early adolescence, providing a link between the in utero environment and vascular function in the offspring [89].

Using a model of gestational protein restriction followed by rapid catch-up growth or protein restriction during lactation, it has been demonstrated that maternal low protein diet was associated with changes in antioxidant defences, and more short telomeres in the aorta, providing some mechanistic links between maternal diet with changes in vascular function [90].

More recently, it has emerged that microvascular dysfunction may precede more extensive pathological changes in blood vessels, and this may be an important site of developmental priming of risk of later disease. The evidence for this is mostly from studies of maternal under-nutrition, although it is increasingly recognised that maternal obesity may also lead to long-term cardiovascular changes that are preceded by microvascular dysfunction [91].

Animal Models Using Maternal Obesity

Several animal models have been developed to investigate the transgenerational effects of maternal obesity. For example, female mice fed on a high-fat diet for 6 weeks before mating and throughout pregnancy and lactation were found to give birth to offspring that were hyperphagic at 4–6 weeks compared with offspring of mothers given standard chow, and they also had significantly lower locomotor activity and increased adiposity by 3 months of age. Such offspring had greater central obesity at 6 months, with adipocyte hypertrophy and increased expression of 11-β-hydroxysteroid dehydrogenase (11-β-HSD) and peroxisome proliferator-activated receptor γ (PPARγ). Other notable abnormalities in the offspring include endothelial dysfunction and hypertension by 3 months, as well as reduced skeletal muscle mass, elevated fasting glucose and fasting insulin concentration [92]. This study demonstrated multiple long-term consequences of maternal obesity on cardiometabolic profile in the offspring, which may be mediated by altered appetite, physical inactivity as well as adipocyte metabolism.

To investigate some of the underlying mechanisms of maternal obesity on foetal priming of later NCD risk, a non-human primate model was established to investigate the effects of maternal high-fat diet. It was noted that chronic high-fat diet led to a threefold increase in liver triglycerides in offspring of both lean and obese mothers [93]. This was also associated with elevated foetal glycerol levels, as well as evidence of increased oxidative stress in the livers of the offspring exposed to high-fat diet. This suggests that a maternal high-fat diet is associated with development of non-alcoholic fatty liver disease in the offspring, as one of the metabolic derangements that lead to insulin resistance and other cardiometabolic abnormalities.

Earlier studies in rats have shown that exposing mothers to a high caloric diet may have long-lasting consequences. When their offspring become pregnant, even if they were given a normal chow diet they remained insulin resistant and gave birth to offspring which were hyperinsulinaemic [94,95]. As in the case of prenatal under-nutrition, maternal high-fat diets induce epigenetic changes in the offspring which may mediate the pathophysiological consequences. The offspring of rats fed on a high-fat diet during pregnancy have abnormalities of hepatic cell cycling associated with epigenetic changes in the promoter region of genes controlling cell cycle length [96].

It has also been shown that paternal diet can impact on foetal priming of NCD risk. A paternal high-fat diet was found to be associated with intergenerational transmission of impaired glucose–insulin homeostasis and β-cell dysfunction in female offspring and potential epigenetic mechanisms suggested [97,98]. This insight has implications on strategies for obesity and diabetes prevention, whereby potentially both the mother and father need to be targeted, both in view of their influence on offspring behaviour as well as from the perspective of foetal priming of risk.

Neurohormonal Changes

An increasing number of neuropeptides have been identified to be important for body weight regulation including orexigenic neuropeptides such as Neuropeptide Y (NPY), agouti-related peptide (AGRP) and anorexigenic neuropeptides such as proopiomelanocortin and α-melonocyte-stimulating hormone (MSH). Animal models have demonstrated that offspring exposed to maternal diabetes have increased hypothalamic expression of NPY and AGRP, and decreased MSH immunopositivity, which would predict increased appetite and food intake and thereby contribute to increased adiposity in them. Moreover, these changes are prevented by treatment of the maternal diabetes [99]. A rat model of maternal obesity has revealed that offspring of obese rats have central neurohormonal changes and exhibited an amplified and prolonged leptin surge, with reduced hypothalamic AGRP expression [100]. Likewise, a chronic maternal high-fat diet in non-human primates has been shown to lead to changes in hypothalamic expression of pro-inflammatory cytokines and melanocortin, with reduced AGRP levels in the offspring. These changes, if persistent, could impact on body weight homeostasis as well as cardiovascular function, consistent with the observation that offspring of mothers exposed to the high-fat diet have excess weight gain [101].

The recent demonstration that neurohormonal changes in humans in response to weight loss persist for >1 year after weight loss surgery, with increase in appetite [102], indicate that in order to achieve sustained benefit from weight loss, strategies that aim to 're-programme' neurohormonal adaptations may need to be carried out earlier in the life course, suggesting the potential opportunities associated with appropriate management of maternal diabetes or obesity.

Placenta, Foetal Priming and Foetal Development

The placenta, being at the interphase between mother and foetus, has a critical role in influencing foetal growth and development. Investigators have examined mechanisms whereby the placenta may play a role in foetal priming of later disease risk. Significant changes occur in the placenta in response to maternal diet, including changes in nutrient transporter activity, neurohormonal changes, for example in the activity of 11-β-HSD2, as well as placental oxidative stress and defence mechanisms [103]. Better understanding of the pathophysiological changes in the placenta in maternal obesity and maternal diabetes is likely to provide important insights for future intervention.

Epigenetics as the Unifying Mechanism in Foetal Priming of NCD Risk

Emerging work, in particular in relation to maternal under-nutrition, has demonstrated that changes mediated by the maternal environment can be transduced to the next generation via epigenetic changes, particularly DNA modifications other than changes in the DNA sequence itself which can be heritable [13,104]. Apart from DNA methylation such epigenetic changes include histone modifications and small non-coding RNAs [13,104]. For example, maternal exposure to a high-fat diet has been found to be associated with global as well as gene-specific alterations in DNA methylation and histone modifications [105]. The recent demonstration of consistent methylation changes in metabolic genes being closely associated with later adiposity in two prospective cohorts, independent of birthweight, highlighted the potential utility of such markers. Furthermore, it was estimated that neonatal epigenetic marks could explain a significant proportion of the variance in childhood obesity, highlighting the contribution of prenatal development towards the development of adult NCDs [17].

Implications for Public Health Intervention

The continuing epidemic of obesity and diabetes suggests that current strategies focusing on lifestyle modification in high-risk individuals or those with manifestation of early disease are unlikely to be successful in curbing the escalating burden of disease [14]. Given the emerging evidence of a strong link between maternal diabetes and obesity and risk of NCD in the offspring, a life-course perspective, focusing on maternal pre-conception and gestational health, is receiving increasing attention: it constitutes an important paradigm shift [12,63,65,106]. It has been hypothesised that interventions in the plastic phase during development may have sustained metabolic benefit compared with later interventions during the adult period [12] (Figure 14.2).

There is emerging evidence that such strategies may indeed be effective. For example, a randomised clinical trial which compared obstetric outcome in women who had intensive glucose control for GDM versus standard care found that the former was associated with less pregnancy weight gain and reduced risk of macrosomia in the foetus at term [107]. Early diagnosis and treatment of GDM will

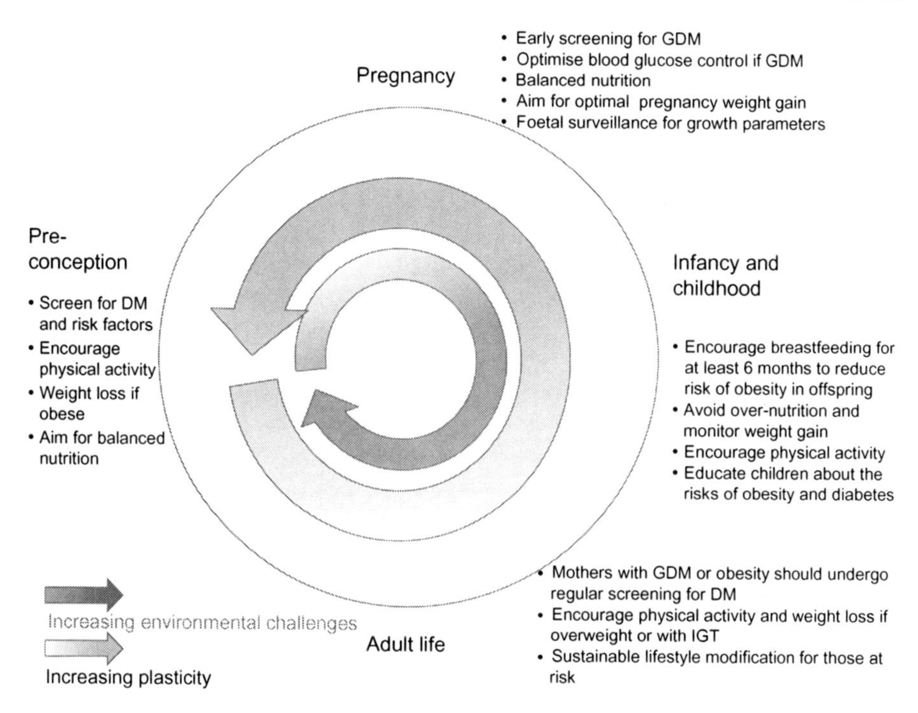

Figure 14.2 A life-course approach to the prevention of diabetes and obesity, focusing on the risks associated with unbalanced or over-nutrition rather than under-nutrition, now encountered in developed countries and increasingly in developing countries undergoing socio-economic transition.

therefore become increasingly important as the proportion of women with this condition increases on using the recently revised criteria for GDM from the International Association for Diabetes in Pregnancy Study Group [51].

A recent study which compared the cardiometabolic profile of offspring born before or after the mother underwent bariatric surgery demonstrated a significantly lower risk of obesity in offspring of those obese mothers who achieved substantial weight reduction through surgery, as well as improvement in multiple cardiometabolic risk factors, which appears sustained up to adolescence [108]. Likewise, avoiding excessive weight gain during pregnancy also appears to be associated with a more favourable outcome [109].

The Institute of Medicine has issued new guidelines for weight gain during pregnancy according to maternal pre-pregnancy BMI [110]. For women who are overweight, given the implication of excessive weight gain, there needs to be greater awareness among clinicians as well as the general public regarding the Institute of Medicine (IOM) recommendation of optimal weight gain during pregnancy [109]. A recent systematic review of randomised clinical trials of antenatal dietary or lifestyle modification for pregnant women who are overweight or obese identified only nine relatively small studies, with most involving dietary modification alone. No significant difference was observed for mean gestational weight gain or large-for-gestational-age infant [111], though it is worth noting that several larger randomised clinical trials are now underway to examine the impact of limiting gestational weight gain on pregnancy outcome [112]. On the other hand, inappropriate reduction of weight during pregnancy by excessive dieting may have adverse effects and is associated with falling birthweight in Japan [21].

Whilst the current discussion has focused on maternal obesity and diabetes, it is important to note that many other early life factors have also been identified to be associated with offspring obesity, including poor or unbalanced maternal nutrition, stress, maternal smoking, rapid infant growth, no or short duration breastfeeding, obesity in early infancy, short sleep duration, <30 min of physical activity/day and consumption of sugar-sweetened beverages [113]. Strategies focusing on maternal health and early development, as a window to target obesity and diabetes prevention, would also allow several of these additional factors to be addressed simultaneously, a holistic and contextually relevant approach which is likely to provide additional benefit [14].

Acknowledgements

RCWM acknowledges support from the Albert Renold Fellowship, the European Association for the Study of Diabetes and an EFSD/CDS/Lilly Grant from the European Association for the Study of Diabetes. PDG is supported by the National Research Centre for Growth and Development, New Zealand. MAH is supported by the British Heart Foundation.

References

1. Kelly T, Yang W, Chen CS, Reynolds K, He J. Global burden of obesity in 2005 and projections to 2030. *Int J Obes (Lond)*. 2008;32:1431–1437.
2. Asia Pacific Cohort Studies Collaboration. The burden of overweight and obesity in the Asia-Pacific region. *Obes Rev*. 2007;8:191–196.
3. Calle EE, Thun MJ, Petrelli JM, Rodriguez C, Heath Jr CW. Body-mass index and mortality in a prospective cohort of U.S. adults. *N Engl J Med*. 1999;341:1097–1105.
4. Jee SH, Sull JW, Park J, et al. Body-mass index and mortality in Korean men and women. *N Engl J Med*. 2006;355:779–787.
5. Ma R, Chan J. Metabolic complications of obesity. In: Williams G, Fruhbeck G, eds. *Obesity: Science to Practice*. Chichester: John Wiley and Sons; 2009:235–270.
6. McKeigue PM, Shah B, Marmot MG. Relation of central obesity and insulin resistance with high diabetes prevalence and cardiovascular risk in South Asians. *Lancet*. 1991;337:382–386.
7. Peeters A, Barendregt JJ, Willekens F, et al. Obesity in adulthood and its consequences for life expectancy: a life-table analysis. *Ann Intern Med*. 2003;138:24–32.
8. Ma RC, Ko GTC, Chan JCN. Health Hazards of obesity: an overview. In: Williams G, Fruhbeck G, eds. *Obesity: Science and Practice*. Chichester: John Wiley and Sons; 2009:215–236.

9. Reeves GK, Pirie K, Beral V, et al. Cancer incidence and mortality in relation to body mass index in the Million Women Study: cohort study. *BMJ*. 2007;335:1134.
10. Steinberger J, Daniels SR. Obesity, insulin resistance, diabetes, and cardiovascular risk in children: an American Heart Association scientific statement from the Atherosclerosis, Hypertension, and Obesity in the Young Committee (Council on Cardiovascular Disease in the Young) and the Diabetes Committee (Council on Nutrition, Physical Activity, and Metabolism). *Circulation*. 2003;107:1448–1453.
11. Speiser PW, Rudolf MC, Anhalt H, et al. Childhood obesity. *J Clin Endocrinol Metab*. 2005;90:1871–1887.
12. Godfrey KM, Gluckman PD, Hanson MA. Developmental origins of metabolic disease: life course and intergenerational perspectives. *Trends Endocrinol Metab*. 2010;21:199–205.
13. Gluckman PD, Hanson MA, Cooper C, Thornburg KL. Effect of in utero and early-life conditions on adult health and disease. *N Engl J Med*. 2008;359:61–73.
14. Gluckman PD, Hanson M, Zimmet P, Forrester T. Losing the war against obesity: the need for a developmental perspective. *Sci Transl Med*. 2011;3:93cm19.
15. Gluckman P, Hanson M. *Fat, Fate and Disease*. Oxford: Oxford University Press; 2012.
16. Gluckman PD, Hanson MA, Buklijas T. A conceptual framework for the developmental origins of health and disease. *J Dev Orig Health Dis*. 2010;1:6–18.
17. Godfrey KM, Sheppard A, Gluckman PD, et al. Epigenetic gene promoter methylation at birth is associated with child's later adiposity. *Diabetes*. 2011;60:1528–1534.
18. Catalano PM, Ehrenberg HM. The short- and long-term implications of maternal obesity on the mother and her offspring. *BJOG*. 2006;113:1126–1133.
19. Gluckman PD, Hanson MA. Developmental and epigenetic pathways to obesity: an evolutionary-developmental perspective. *Int J Obes (Lond)*. 2008;32(suppl 7):S62–S71.
20. Reynolds RM, Osmond C, Phillips DI, Godfrey KM. Maternal BMI, parity, and pregnancy weight gain: influences on offspring adiposity in young adulthood. *J Clin Endocrinol Metab*. 2010;95:5365–5369.
21. Gluckman PD, Seng CY, Fukuoka H, Beedle AS, Hanson MA. Low birthweight and subsequent obesity in Japan. *Lancet*. 2007;369:1081–1082.
22. Ananth CV, Wen SW. Trends in fetal growth among singleton gestations in the United States and Canada, 1985 through 1998. *Semin Perinatol*. 2002;26:260–267.
23. Danielzik S, Langnase K, Mast M, Spethmann C, Muller MJ. Impact of parental BMI on the manifestation of overweight 5–7 year old children. *Eur J Nutr*. 2002; 41:132–138.
24. Ong KK, Diderholm B, Salzano G, et al. Pregnancy insulin, glucose, and BMI contribute to birth outcomes in nondiabetic mothers. *Diabetes Care*. 2008;31:2193–2197.
25. Vohr BR, McGarvey ST, Coll CG. Effects of maternal gestational diabetes and adiposity on neonatal adiposity and blood pressure. *Diabetes Care*. 1995;18:467–475.
26. Catalano PM, Kirwan JP. Maternal factors that determine neonatal size and body fat. *Curr Diab Rep*. 2001;1:71–77.
27. Catalano PM, Presley L, Minium J, Hauguel-de Mouzon S. Fetuses of obese mothers develop insulin resistance in utero. *Diabetes Care*. 2009;32:1076–1080.
28. HAPO Study Cooperative Research Group. Hyperglycaemia and Adverse Pregnancy Outcome (HAPO) study: associations with maternal body mass index. *BJOG*. 2010;117:575–584.
29. Shankar K, Harrell A, Liu X, et al. Maternal obesity at conception programs obesity in the offspring. *Am J Physiol Regul Integr Comp Physiol*. 2008;294:R528–R538.
30. Gale CR, Javaid MK, Robinson SM, et al. Maternal size in pregnancy and body composition in children. *J Clin Endocrinol Metab*. 2007;92:3904–3911.

31. Gillman MW, Rifas-Shiman S, Berkey CS, Field AE, Colditz GA. Maternal gestational diabetes, birth weight, and adolescent obesity. *Pediatrics*. 2003;111:e221–e226.
32. Catalano PM, Farrell K, Thomas A, et al. Perinatal risk factors for childhood obesity and metabolic dysregulation. *Am J Clin Nutr*. 2009;90:1303–1313.
33. Boney CM, Verma A, Tucker R, Vohr BR. Metabolic syndrome in childhood: association with birth weight, maternal obesity, and gestational diabetes mellitus. *Pediatrics*. 2005;115:e290–e296.
34. Drake AJ, Reynolds RM. Impact of maternal obesity on offspring obesity and cardiometabolic disease risk. *Reproduction*. 2010;140:387–398.
35. Crozier SR, Inskip HM, Godfrey KM, et al. Weight gain in pregnancy and childhood body composition: findings from the Southampton Women's Survey. *Am J Clin Nutr*. 2010;91:1745–1751.
36. Fraser A, Tilling K, Macdonald-Wallis C, et al. Association of maternal weight gain in pregnancy with offspring obesity and metabolic and vascular traits in childhood. *Circulation*. 2010;121:2557–2564.
37. Freinkel N. Banting Lecture 1980. Of pregnancy and progeny. *Diabetes*. 1980;29:1023–1035.
38. Pettitt DJ, Aleck KA, Baird HR, et al. Congenital susceptibility to NIDDM. Role of intrauterine environment. *Diabetes*. 1988;37:622–628.
39. Dabelea D, Hanson RL, Lindsay RS, et al. Intrauterine exposure to diabetes conveys risks for type 2 diabetes and obesity: a study of discordant sibships. *Diabetes*. 2000;49: 2208–2211.
40. Sobngwi E, Boudou P, Mauvais-Jarvis F, et al. Effect of a diabetic environment in utero on predisposition to type 2 diabetes. *Lancet*. 2003;361:1861–1865.
41. Karter AJ, Rowell SE, Ackerson LM, et al. Excess maternal transmission of type 2 diabetes. The Northern California Kaiser Permanente Diabetes Registry. *Diabetes Care*. 1999;22:938–943.
42. Lee SC, Pu YB, Chow CC, et al. Diabetes in Hong Kong Chinese: evidence for familial clustering, maternal influence and a gender specific paternal effect. *Diabetes Care*. 2000;23:1365–1368.
43. Durnwald C, Huston-Presley L, Amini S, Catalano P. Evaluation of body composition of large-for-gestational-age infants of women with gestational diabetes mellitus compared with women with normal glucose tolerance levels. *Am J Obstet Gynecol*. 2004;191:804–808.
44. McLean M, Chipps D, Cheung NW. Mother to child transmission of diabetes mellitus: does gestational diabetes program Type 2 diabetes in the next generation? *Diabet Med*. 2006;23:1213–1215.
45. Crume TL, Ogden L, West NA, et al. Association of exposure to diabetes in utero with adiposity and fat distribution in a multiethnic population of youth: the Exploring Perinatal Outcomes among Children (EPOCH) Study. *Diabetologia*. 2011; 54:87–92.
46. Philipps LH, Santhakumaran S, Gale C, et al. The diabetic pregnancy and offspring BMI in childhood: a systematic review and meta-analysis. *Diabetologia*. 2011;54:1957–1966.
47. Dabelea D, Mayer-Davis EJ, Lamichhane AP, et al. Association of intrauterine exposure to maternal diabetes and obesity with type 2 diabetes in youth: the SEARCH Case–Control Study. *Diabetes Care*. 2008;31:1422–1426.
48. Yogev Y, Langer O, Xenakis EM, Rosenn B. The association between glucose challenge test, obesity and pregnancy outcome in 6390 non-diabetic women. *J Matern Fetal Neonatal Med*. 2005;17:29–34.

49. Hill JC, Krishnaveni GV, Annamma I, Leary SD, Fall CH. Glucose tolerance in pregnancy in South India: relationships to neonatal anthropometry. *Acta Obstet Gynecol Scand.* 2005;84:159–165.
50. Metzger BE, Lowe LP, Dyer AR, et al. Hyperglycemia and adverse pregnancy outcomes. *N Engl J Med.* 2008;358:1991–2002.
51. Metzger BE, Gabbe SG, Persson B, et al. International association of diabetes and pregnancy study groups recommendations on the diagnosis and classification of hyperglycemia in pregnancy. *Diabetes Care.* 2010;33:676–682.
52. HAPO Study Cooperative Research Group. Hyperglycemia and Adverse Pregnancy Outcome (HAPO) Study: associations with neonatal anthropometrics. *Diabetes.* 2009;58: 453–459.
53. Touger L, Looker HC, Krakoff J, et al. Early growth in offspring of diabetic mothers. *Diabetes Care.* 2005;28:585–589.
54. Silverman BL, Metzger BE, Cho NH, Loeb CA. Impaired glucose tolerance in adolescent offspring of diabetic mothers. Relationship to fetal hyperinsulinism. *Diabetes Care.* 1995;18:611–617.
55. Silverman BL, Rizzo T, Green OC, et al. Long-term prospective evaluation of offspring of diabetic mothers. *Diabetes.* 1991;40(suppl 2):121–125.
56. Simeoni U, Barker DJ. Offspring of diabetic pregnancy: long-term outcomes. *Semin Fetal Neonatal Med.* 2009;14:119–124.
57. Tam WH, Ma RC, Yang X, et al. Glucose intolerance and cardiometabolic risk in children exposed to maternal gestational diabetes mellitus in utero. *Pediatrics.* 2008; 122:1229–1234.
58. Tam WH, Ma RC, Yang X, et al. Glucose intolerance and cardiometabolic risk in adolescents exposed to maternal gestational diabetes: a 15-year follow-up study. *Diabetes Care.* 2010;33:1382–1384.
59. Wright CS, Rifas-Shiman SL, Rich-Edwards JW, et al. Intrauterine exposure to gestational diabetes, child adiposity, and blood pressure. *Am J Hypertens.* 2009;22:215–220.
60. Egeland GM, Skjaerven R, Irgens LM. Birth characteristics of women who develop gestational diabetes: population based study. *BMJ.* 2000;321:546–547.
61. West NA, Crume TL, Maligie MA, Dabelea D. Cardiovascular risk factors in children exposed to maternal diabetes in utero. *Diabetologia.* 2011;54:504–507.
62. Manderson JG, Mullan B, Patterson CC, et al. Cardiovascular and metabolic abnormalities in the offspring of diabetic pregnancy. *Diabetologia.* 2002;45:991–996.
63. Ma RC, Chan JC. Pregnancy and diabetes scenario around the world: China. *Int J Gynaecol Obstet.* 2009;104:S42–S45.
64. Yajnik CS. Nutrient-mediated teratogenesis and fuel-mediated teratogenesis: two pathways of intrauterine programming of diabetes. *Int J Gynaecol Obstet.* 2009;104(suppl 1): S27–S31.
65. Yajnik CS. Fetal programming of diabetes: still so much to learn! *Diabetes Care.* 2010;33:1146–1148.
66. Catalano PM, Tyzbir ED, Roman NM, Amini SB, Sims EA. Longitudinal changes in insulin release and insulin resistance in nonobese pregnant women. *Am J Obstet Gynecol.* 1991;165:1667–1672.
67. Lain KY, Catalano PM. Metabolic changes in pregnancy. *Clin Obstet Gynecol.* 2007;50:938–948.
68. Sivan E, Homko CJ, Chen X, Reece EA, Boden G. Effect of insulin on fat metabolism during and after normal pregnancy. *Diabetes.* 1999;48:834–838.

69. Torloni MR, Betran AP, Horta BL, et al. Prepregnancy BMI and the risk of gestational diabetes: a systematic review of the literature with meta-analysis. *Obes Rev*. 2009;10:194–203.
70. Radaelli T, Varastehpour A, Catalano P, Hauguel-de Mouzon S. Gestational diabetes induces placental genes for chronic stress and inflammatory pathways. *Diabetes*. 2003;52:2951–2958.
71. Kim C, Cheng YJ, Beckles GL. Inflammation among women with a history of gestational diabetes mellitus and diagnosed diabetes in the Third National Health and Nutrition Examination Survey. *Diabetes Care*. 2008;31:1386–1388.
72. Retnakaran R, Hanley AJ, Raif N, et al. C-reactive protein and gestational diabetes: the central role of maternal obesity. *J Clin Endocrinol Metab*. 2003;88:3507–3512.
73. Han YS, Ha EH, Park HS, Kim YJ, Lee SS. Relationships between pregnancy outcomes, biochemical markers and pre-pregnancy body mass index. *Int J Obes (Lond)*. 2011;35:570–577.
74. Gaither K, Quraishi AN, Illsley NP. Diabetes alters the expression and activity of the human placental GLUT1 glucose transporter. *J Clin Endocrinol Metab*. 1999;84:695–701.
75. Illsley NP. Glucose transporters in the human placenta. *Placenta*. 2000;21:14–22.
76. Jarvie E, Hauguel-de-Mouzon S, Nelson SM, et al. Lipotoxicity in obese pregnancy and its potential role in adverse pregnancy outcome and obesity in the offspring. *Clin Sci (Lond)*. 2010;119:123–129.
77. Nolan CJ, Riley SF, Sheedy MT, Walstab JE, Beischer NA. Maternal serum triglyceride, glucose tolerance, and neonatal birth weight ratio in pregnancy. *Diabetes Care*. 1995;18: 1550–1556.
78. Kitajima M, Oka S, Yasuhi I, et al. Maternal serum triglyceride at 24–32 weeks' gestation and newborn weight in nondiabetic women with positive diabetic screens. *Obstet Gynecol*. 2001;97:776–780.
79. Vrijkotte TG, Algera SJ, Brouwer IA, van Eijsden M, Twickler MB. Maternal triglyceride levels during early pregnancy are associated with birth weight and postnatal growth. *J Pediatr*. 2011;159:736–742.
80. Haggarty P. Placental regulation of fatty acid delivery and its effect on fetal growth – a review. *Placenta*. 2002;23(suppl A):S28–38.
81. Murphy VE, Smith R, Giles WB, Clifton VL. Endocrine regulation of human fetal growth: the role of the mother, placenta, and fetus. *Endocr Rev*. 2006;27:141–169.
82. Bayol SA, Simbi BH, Stickland NC. A maternal cafeteria diet during gestation and lactation promotes adiposity and impairs skeletal muscle development and metabolism in rat offspring at weaning. *J Physiol*. 2005;567:951–961.
83. Bayol SA, Simbi BH, Bertrand JA, Stickland NC. Offspring from mothers fed a 'junk food' diet in pregnancy and lactation exhibit exacerbated adiposity that is more pronounced in females. *J Physiol*. 2008;586:3219–3230.
84. Catalano PM. Obesity and pregnancy – the propagation of a viscous cycle? *J Clin Endocrinol Metab*. 2003;88:3505–3506.
85. Koukkou E, Ghosh P, Lowy C, Poston L. Offspring of normal and diabetic rats fed saturated fat in pregnancy demonstrate vascular dysfunction. *Circulation*. 1998;98:2899–2904.
86. Khan I, Dekou V, Hanson M, Poston L, Taylor P. Predictive adaptive responses to maternal high-fat diet prevent endothelial dysfunction but not hypertension in adult rat offspring. *Circulation*. 2004;110:1097–1102.
87. Khan IY, Dekou V, Douglas G, et al. A high-fat diet during rat pregnancy or suckling induces cardiovascular dysfunction in adult offspring. *Am J Physiol Regul Integr Comp Physiol*. 2005;288:R127–R133.

88. Samuelsson AM, Morris A, Igosheva N, et al. Evidence for sympathetic origins of hypertension in juvenile offspring of obese rats. *Hypertension*. 2010;55:76–82.
89. Tam WH, Ma RC, Yip GW, et al. The association between in utero hyperinsulinemia and adolescent arterial stiffness. *Diabetes Res Clin Pract*. 2012;95:169–175.
90. Tarry-Adkins JL, Martin-Gronert MS, Chen JH, Cripps RL, Ozanne SE. Maternal diet influences DNA damage, aortic telomere length, oxidative stress, and antioxidant defense capacity in rats. *FASEB J*. 2008;22:2037–2044.
91. Clough GF, Norman M. The microcirculation: a target for developmental priming. *Microcirculation*. 2011;18:286–297.
92. Samuelsson AM, Matthews PA, Argenton M, et al. Diet-induced obesity in female mice leads to offspring hyperphagia, adiposity, hypertension, and insulin resistance: a novel murine model of developmental programming. *Hypertension*. 2008;51:383–392.
93. McCurdy CE, Bishop JM, Williams SM, et al. Maternal high-fat diet triggers lipotoxicity in the fetal livers of nonhuman primates. *J Clin Invest*. 2009;119:323–335.
94. Srinivasan M, Aalinkeel R, Song F, Patel MS. Programming of islet functions in the progeny of hyperinsulinemic/obese rats. *Diabetes*. 2003;52:984–990.
95. Srinivasan M, Aalinkeel R, Song F, et al. Maternal hyperinsulinemia predisposes rat fetuses for hyperinsulinemia, and adult-onset obesity and maternal mild food restriction reverses this phenotype. *Am J Physiol Endocrinol Metab*. 2006;290:E129–E134.
96. Dudley KJ, Sloboda DM, Connor KL, Beltrand J, Vickers MH. Offspring of mothers fed a high fat diet display hepatic cell cycle inhibition and associated changes in gene expression and DNA methylation. *PLoS One*. 2011;6:e21662.
97. Ng SF, Lin RC, Laybutt DR, et al. Chronic high-fat diet in fathers programs beta-cell dysfunction in female rat offspring. *Nature*. 2010;467:963–966.
98. Dunn GA, Bale TL. Maternal high-fat diet effects on third-generation female body size via the paternal lineage. *Endocrinology*. 2011;152:2228–2236.
99. Franke K, Harder T, Aerts L, et al. 'Programming' of orexigenic and anorexigenic hypothalamic neurons in offspring of treated and untreated diabetic mother rats. *Brain Res*. 2005;1031:276–283.
100. Kirk SL, Samuelsson AM, Argenton M, et al. Maternal obesity induced by diet in rats permanently influences central processes regulating food intake in offspring. *PLoS One*. 2009;4:e5870.
101. Grayson BE, Levasseur PR, Williams SM, et al. Changes in melanocortin expression and inflammatory pathways in fetal offspring of nonhuman primates fed a high-fat diet. *Endocrinology*. 2010;151:1622–1632.
102. Sumithran P, Prendergast LA, Delbridge E, et al. Long-term persistence of hormonal adaptations to weight loss. *N Engl J Med*. 2011;365:1597–1604.
103. Jansson T, Powell TL. Role of the placenta in fetal programming: underlying mechanisms and potential interventional approaches. *Clin Sci (Lond)*. 2007;113:1–13.
104. Gluckman PD, Hanson MA, Buklijas T, Low FM, Beedle AS. Epigenetic mechanisms that underpin metabolic and cardiovascular diseases. *Nat Rev Endocrinol*. 2009;5:401–408.
105. Aagaard-Tillery KM, Grove K, Bishop J, et al. Developmental origins of disease and determinants of chromatin structure: maternal diet modifies the primate fetal epigenome. *J Mol Endocrinol*. 2008;41:91–102.
106. Hanson M, Gluckman P, Nutbeam D, Hearn J. Priority actions for the non-communicable disease crisis. *Lancet*. 2011;378:566–567.
107. Crowther CA, Hiller JE, Moss JR, et al. Effect of treatment of gestational diabetes mellitus on pregnancy outcomes. *N Engl J Med*. 2005;352:2477–2486.

108. Smith J, Cianflone K, Biron S, et al. Effects of maternal surgical weight loss in mothers on intergenerational transmission of obesity. *J Clin Endocrinol Metab*. 2009;94:4275–4283.
109. Vesco KK, Sharma AJ, Dietz PM, et al. Newborn size among obese women with weight gain outside the 2009 Institute of Medicine recommendation. *Obstet Gynecol*. 2011;117:812–818.
110. Subcommittee on Nutritional Status and Weight Gain During Pregnancy. *Weight Gain During Pregnancy: Reexamining the Guidelines*. Washington, DC: Institute of Medicine (US); 2009.
111. Dodd JM, Grivell RM, Crowther CA, Robinson JS. Antenatal interventions for overweight or obese pregnant women: a systematic review of randomised trials. *BJOG*. 2010;117:1316–1326.
112. Dodd JM, Turnbull DA, McPhee AJ, et al. Limiting weight gain in overweight and obese women during pregnancy to improve health outcomes: the LIMIT randomised controlled trial. *BMC Pregnancy Childbirth*. 2011;11:79.
113. Monasta L, Batty GD, Cattaneo A, et al. Early-life determinants of overweight and obesity: a review of systematic reviews. *Obes Rev*. 2010;11:695–708.

15 Foetal Ultrasound in Obese Pregnancy

Jennifer M. Walsh and Fionnuala M. McAuliffe

School of Medicine and Medical Science, University College Dublin, National Maternity Hospital, Dublin, Republic of Ireland

Introduction

Foetal health assessment is of particular relevance in overweight and obese women as these groups are at elevated risk of perinatal morbidity and mortality. Accurate prediction of those pregnancies at particular risk, however, is limited by the reduced sensitivity of imaging modalities in obese women. With the obesity epidemic likely to continue in developed countries for many years to come, antenatal assessment of foetal well being is a priority for obstetricians who manage this group of patients.

Excess Foetal Risks Associated with Obesity

Congenital Anomalies

There is good evidence that maternal obesity confers an elevated risk of congenital anomalies, in particular neural tube defects (NTDs) and congenital heart defects (CHD) (Figure 15.1).

The risk of NTDs appears to increase with increasing maternal weight in a dose–response relationship with an odds ratio of 1.2 in overweight, 1.7 in obese and 3.1 in morbidly obese women [1]. For each 10 kg increase in maternal weight, the odds ratio of NTD increases by 1.2 [2]. The risk of anencephaly appears to be particularly high in obese pregnant women [3].

The risk of CHD is also significantly greater in obese women when compared to women with a normal body mass index (BMI) with odds ratios ranging from 1.2 to 2.0 [4,5] even after excluding diabetic mothers and adjusting for potential confounders [6].

There is emerging data that a number of other congenital defects, in particular omphalocele, hypospadias, renal anomalies and hydrocephalus, may be over-represented in the obese population [4–7].

Obesity. DOI: http://dx.doi.org/10.1016/B978-0-12-416045-3.00015-7

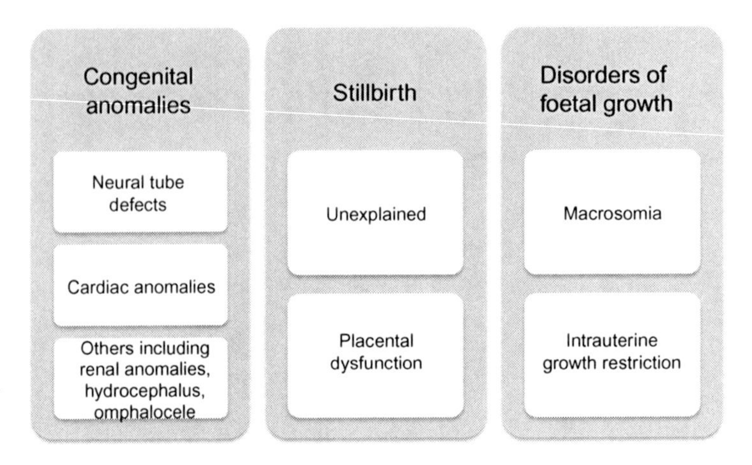

Figure 15.1 Excess foetal risks associated with obesity.

Stillbirths

Maternal obesity confers an elevated risk of intrauterine demise with the risk of foetal death, particularly late and unexplained, increasing with rising BMI [8,9].

Compared with lean mothers (BMI ≤ 19.9 kg/m^2), the odds ratios for the risk of antepartum deaths are 1.2 in normal-weight mothers (BMI 20.0–24.9 kg/m^2), 1.9 in overweight mothers (BMI 25.0–29.9 kg/m^2) and 2.1 in obese mothers (BMI ≥ 30.0 kg/m^2). For term antepartum death, the risk is even higher, with odds ratios of 1.6 for normal weight, 2.7 for overweight and 2.8 for obese women, respectively [10]. Weight gain during the interval between pregnancies also increases the risk of perinatal complications, with an increase of greater than three BMI units significantly increasing the rate of term stillbirth, independent of obesity-related diseases [11].

No single cause is particularly identified for this elevated risk; however, both unexplained intrauterine demise and placental dysfunction appear to predominate [12].

Intrauterine Growth Restriction

Though a number of studies fail to identify an association between maternal obesity and intrauterine growth restriction, this is likely to reflect the higher incidence of macrosomia in this cohort overall [13]. Obesity predisposes to diabetes, gestational hypertension and pre-eclampsia, all of which have an established link to placental insufficiency and growth restriction [14].

As mentioned above, at least some of the excess risk of stillbirth associated with obesity is attributable to placental dysfunction. Clinical assessment of foetal size is particularly limited in obese women, and it is important, therefore, that a lower threshold for referral for serial growth scans be employed in these patients.

Macrosomia

Maternal weight and maternal weight gain during pregnancy exert an important influence on the birthweight of the infant [15–17]. Increased maternal BMI confers an elevated risk of delivering a heavier infant [18], while increasing maternal weight gain during pregnancy is independently related to increasing birthweight of the infant [19–20]. Maternal weight gain of more than 11 kg is strongly associated with the birth of a large for gestational age neonate [21]. Moreover, obesity has been identified as an independent risk factor for macrosomia, even after adjustment for diabetes [14]. Even in non-diabetic pregnancies, maternal BMI exerts a significant influence on maternal glucose homeostasis, and even small variations in maternal glucose can influence both foetal growth and adiposity [22].

Ultrasound Assessment in Obesity

Ultrasound assessment in obese mother is notoriously difficult, and it is imperative, therefore, that the limitations of ultrasound assessment of foetal health are clearly outlined prior to commencement of the ultrasound scan. However, despite these limitations, the excess risks of congenital anomalies, aberrations of foetal growth and stillbirth necessitate that particular efforts are made to assess foetal well being in this cohort (Figure 15.2).

First Trimester Ultrasound

First trimester ultrasound offers a number of advantages for foetal health assessment. These include dating the pregnancy, determination of the number of foetuses, early aneuploidy screening via nuchal translucency (NT) and early detection of foetal anomalies [23].

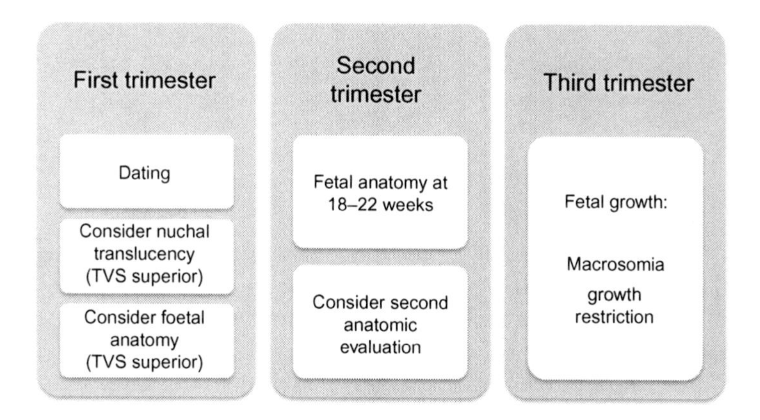

Figure 15.2 Ultrasound assessment in obese pregnant women.

All overweight and obese women should be offered routine first trimester ultrasound to accurately establish gestational age. These women have a higher incidence of irregular menstrual cycles and, therefore, the ability to calculate the gestational age from the date of last menstrual period is reduced. Moreover, obesity predisposes to dystocia and higher rates of caesarean delivery with higher operative and anaesthetic risks [24]. It is established that gestational age at onset of labour increases with increasing BMI [25]. Accurate assessment of gestational age is therefore of particular relevance in this group in order to reduce unnecessary induction of labour for post-dates.

The association of obesity with adverse maternal and foetal outcomes increases the importance of early screening for chromosomal anomalies. First trimester screening via a combination of maternal serum marker and foetal NT allows for the early identification of those at risk and guides decisions as to whether or not diagnostic tests, which may be technically more difficult in this population, are warranted. A number of authors have assessed the accuracy of first trimester ultrasound for early detection of aneuploidy in the obese population. It has been suggested that as maternal BMI increases, the time required to obtain NT measurements and the failure rate increase [26].

This may reflect the method of scanning used. Where a more frequent use of transvaginal ultrasound (TVS) is reported, it has been noted that increased BMI is in fact not associated with sub-optimal visualisation of NT, though it is associated with a longer time to perform the examination, an increased need for transvaginal examination for NT visualisation and a lower likelihood of obtaining an adequate nasal bone image [27]. Increasingly, first trimester ultrasound is used for the detection of foetal anomalies which are traditionally examined for until a second trimester scan at 18–22 weeks gestation. Again, this may have a particular role in the obese population due to the inherently increased risks of congenital anomalies and maternal complications. It also allows for the use of TVS, virtually eliminating the problem of poor visualisation due to reduced tissue penetration associated with transabdominal ultrasound (TAS).

In the general population, it is reported that a complete foetal anatomical survey can be performed in approximately two-thirds of patients by TAS and in over 80% of patients by TVS [28,29]. Compared with TAS, TVS is significantly better at visualising the cranium, spine, stomach, kidneys, bladder and upper and lower extremities [29].

Second Trimester Ultrasound

Traditionally, ultrasound at 18–22 weeks is considered the gold standard for the detection of foetal anomalies, and given the inherently increased risk in the obese population, this should at least be offered to all women. The detection of anomalous foetuses decreases with increasing BMI. In a large retrospective study of over 11,000 ultrasound examinations [30], detection rates with standard ultrasonography were 66% in those with a normal BMI, 49% in those who were overweight, falling to 48% in those with a BMI of 30–34.9 kg/m^2, 42% in those with a BMI of 35–39.9 kg/m^2 and just 25% in those with a BMI of greater than 40 kg/m^2. Detection rates were higher when targeted examinations took place, such as in known high-risk pregnancies or in those with an abnormality detected during standard ultrasonography with rates of 97% in those with a normal BMI, 91% in those who were overweight and 75% in those with a BMI of 30.0–34.9 kg/m^2. It is imperative, therefore, that counselling in the obese pregnant population should be modified to highlight not only the increased risk of foetal anomalies but also the lower detection rates of these anomalies, particularly with standard ultrasonography. Targeted examinations such as foetal echocardiography should be considered.

In particular, obese pregnant women have a higher risk of having a baby with a CHD. There is evidence that antenatal diagnosis of foetal cardiac anomalies is associated with a reduced morbidity and mortality, in particular in cases of coarctation of the aorta, hypoplastic left heart syndrome and transposition of the great arteries [31]. Women with a BMI of 30–35 kg/m^2 have 2.4 times the risk of sub-optimal ultrasound scans of foetal cardiac structures compared with women of normal weight; women with a BMI of 35–40 kg/m^2 have 5 times the risk and women with a BMI > 40 kg/m^2 have 8 times the risk [32]. Sonographic re-evaluation of the foetal heart has been shown to significantly reduce the rates of sub-optimal visualisation in obese women [33]. Given the inherently increased risk of a foetal cardiac anomaly and the high likelihood of obtaining sub-optimal views at a single traditional 20-weeks examination, a second evaluation at 24–26 weeks gestation could be considered for all obese women.

Foetal Growth Assessment

Obese pregnant woman should be considered at risk of both intrauterine growth restriction and of foetal macrosomia.

Clinical assessment of foetal growth has poor sensitivity in the general population [34] and is particularly difficult in the obese pregnant woman. Ultrasound assessment of foetal growth significantly improves the detection rates of intrauterine growth restriction [35]. A threshold of the 10th centile has better sensitivities and specificities than other commonly used centiles [35], and serial measurements are superior to single estimates in the prediction of both foetal growth restriction and poor perinatal outcome [36].

The antenatal detection of foetal macrosomia is notoriously fraught with difficulties and calls into question the policies of elective delivery based on projected estimated foetal weight cut-offs.

There is much debate in the literature as to which sonographic-estimated foetal weight formulation best predicts the macrosomic foetus. Commonly used weight formulae are associated with very large deviations when used in the macrosomic foetus [37]. This is partly because these formulae were derived from heterogeneous groups and were not specifically designed for assessing foetal weight at the upper end of the weight scale. Chahaun et al. in 2005 [38] reviewed 20 articles in the literature that calculated the sensitivity and specificity of sonographic-estimated foetal weight of >4000 g to accurately identify a macrosomic foetus. The authors found that the post-test probability of sonographic-estimated foetal weight of >4000 g to identify a macrosomic newborn varied widely from 15% to 79%. They also found that neither the type of regression equation used in the estimated foetal weight formula, the time interval between ultrasound and delivery nor the experience of the sonographer influenced the accuracy.

A number of authors have developed specific formulae for estimating the weight of the macrosomic foetus taking both foetal biometry and maternal weight into account and reported improved detection rates of macrosomia [39,40].

More recently, measurements of various markers of foetal adiposity have been assessed in terms of more accurately predicting foetal size. These include foetal upper arm or thigh subcutaneous tissue, upper arm soft tissue thickness, foetal cheek-to-cheek diameter and foetal anterior abdominal wall width (AAW). It has been reported that the use of a raised AAW measurement better predicted macrosomia than using abdominal circumference alone in both diabetic and non-diabetic populations [41,42].

However, when a comparison of five new estimation techniques that involve the measurements of soft tissue for identifying newborns with birthweights of at least 4000 g was performed, three of the five newer methods (upper arm or thigh subcutaneous tissue and ratio of thigh subcutaneous tissue to femur length) were found to be poor; the techniques that estimate either the upper arm soft tissue thickness or the cheek-to-cheek diameter were not significantly better than the clinical predictions for detecting macrosomic foetuses [43]. Clearly further, larger studies are needed for the assessment of foetal adiposity that can reliably predict subsequent birthweight.

Key Points for Clinical Practice

1. All obese pregnant women are at higher risk of congenital anomalies, stillbirth and macrosomia.
2. Ultrasound assessment is less accurate in obese pregnant women compared to the non-obese population.
3. First trimester ultrasound for calculation of gestational age is recommended.
4. First trimester evaluation of risk of aneuploidy should be considered; TVS improves visualisation rates.
5. First trimester evaluation of foetal anatomy should be considered; TVS improves visualisation rates.
6. Diagnosis of foetal anomalies, particularly cardiac, in the second trimester is improved with targeted examinations (e.g. foetal echocardiography) and second evaluations.
7. Clinical estimation of foetal size in obese women is particularly poor and ultrasound for foetal growth should be considered.
8. The sensitivity of antenatal detection of foetal macrosomia is relatively poor.

Conclusion

Obesity poses significant risks to both mother and baby during pregnancy. While it is imperative that the limitations of antenatal assessment of foetal well being be acknowledged by both caregivers and patients, every effort should be made to assess the foetal health as accurately as possible in this high-risk population and, in doing so, strive towards reducing perinatal morbidity and mortality rates.

References

1. Rasmussen SA, Chu SY, Kim SY, et al. Maternal obesity and risk of neural tube defects: a metaanalysis. *Am J Obstet Gynecol.* 2008;198:611–619.
2. Ray JG, Wyatt PR, Vermeulen MJ, Meier C, Cole DE. Greater maternal weight and the ongoing risk of neural tube defects after folic acid flour fortification. *Obstet Gynecol.* 2005;105:261–265.
3. Anderson JL, Waller DK, Canfield MA, Shaw GM, Watkins ML, Werler MM. Maternal obesity, gestational diabetes, and central nervous system birth defects. *Epidemiology.* 2005;16:87–92.
4. Watkins ML, Rasmussen SA, Honein MA, et al. Maternal obesity and risk for birth defects. *Pediatrics.* 2003;111:1152–1158.
5. Walker DR, Shaw GM, Rasmussen SA, et al. Prepregnancy obesity as a risk factor for structural birth defects. *Arch Pediatr Adolesc Med.* 2007;161:745–750.
6. Watkins ML, Botto LD. Maternal prepregnancy weight and congenital heart defects in offspring. *Epidemiology.* 2001;12:439–446.
7. Blomberg MI, Källén B. Maternal obesity and morbid obesity: the risk for birth defects in the offspring. *Birth Defects Res A Clin Mol Teratol.* 2010;88(1):35–40.
8. Froen JF, Arnestad M, Frey K, Vege A, Saugstad OD, Stray-Pedersen B. Risk factors for sudden intrauterine unexplained death: epidemiologic characteristics of singleton cases in Oslo, Norway, 1986–1995. *Am J Obstet Gynecol.* 2001;184:694–702.
9. Flenady V, Koopmans L, Middleton P, et al. Major risk factors for stillbirth in high-income countries: a systematic review and meta-analysis. *Lancet.* 2011;377(9774):1331–1340.
10. Stephansson O, Dickman PW, Johansson A, Cnattingius S. Maternal weight, pregnancy weight gain, and the risk of antepartum stillbirth. *Am J Obstet Gynecol.* 2001;184:463–469.
11. Walsh JM, Murphy DJ. Weight and pregnancy. *BMJ.* 2007;335(7612):169.
12. Kristensen J, Vestergaard M, Wisborg K, Kesmodel U, Secher NJ. Pre-pregnancy weight and the risk of stillbirth and neonatal death. *BJOG.* 2005;112:403–408.
13. Villar J, Carroli G, Wojdyla D, et al. Preeclampsia, gestational hypertension and intrauterine growth restriction, related or independent conditions? *Am J Obstet Gynecol.* 2006;194(4):921–931.
14. Arendas K, Qiu Q, Gruslin A. Obesity in pregnancy: pre-conceptional to postpartum consequences. *J Obstet Gynaecol Can.* 2008;30(6):477–488.
15. Abrams BF, Laros Jr RK. Prepregnancy weight, weight gain, and birth weight. *Am J Obstet Gynaecol.* 1986;154(3):503–509.
16. Frentzen BH, Dimperio DL, Cruz AC. Maternal weight gain: effect on infant birth weight among overweight and average-weight low-income women. *Am J Obstet Gynecol.* 1988;159(5):1114–1117.
17. Johnson JW, Longmate JA, Frentzen B. Excessive maternal weight and pregnancy outcome. *Am J Obstet Gynecol.* 1992;167(2):353–370.
18. Seidman DS, Ever-Hadani P, Gale R. The effect of maternal weight gain in pregnancy on birth weight. *Obstet Gynecol.* 1989;74(2):240–246.

19. Selvin S, Abrams B. Analysing the relationship between maternal weight gain and birthweight: exploration of four statistical issues. *Paediatr Perinat Epidemiol.* 1996;10(2): 220–234.
20. Jensen DM, Damm P, Sorensen B, et al. Pregnancy outcome and prepregnancy body mass index in 2459 glucose-tolerant Danish women. *Am J Obstet Gynecol.* 2003;189(1): 239–244.
21. Bianco AT, Smilen SW, Davis Y, Lopez S, Lapinski R, Lockwood CJ. Pregnancy outcome and weight gain recommendations for the morbidly obese woman. *Obstet Gynecol.* 1998;91(1):97–102.
22. Walsh JM, Mahony R, Byrne J, Foley M, McAuliffe FM. The association of maternal and fetal glucose homeostasis with fetal adiposity and birthweight. *Eur J Obstet Gynecol Reprod Biol.* 2011;159:338–341.
23. Fong KW, Toi A, Salem S, et al. Detection of fetal structural abnormalities with US during early pregnancy. *Radiographics.* 2004;24(1):157–174.
24. CMACE/RCOG. *CMACE/RCOG Joint Guideline: Management of Women with Obesity in Pregnancy.* London: CMACE/RCOG; 2010.
25. Walsh J, Foley M, O'Herlihy C. Dystocia correlates with body mass index in both spontaneous and induced nulliparous labors. *J Matern Fetal Neonatal Med.* 2011;24(6): 817–821.
26. Thornburg LL, Mulconry M, Post A, Carpenter A, Grace D, Pressman EK. Fetal nuchal translucency thickness evaluation in the overweight and obese gravida. *Ultrasound Obstet Gynecol.* 2009;33(6):665–669.
27. Gandhi M, Fox NS, Russo-Stieglitz K, Hanley ME, Matthews G, Rebarber A. Effect of increased body mass index on first-trimester ultrasound examination for aneuploidy risk assessment. *Obstet Gynecol.* 2009;114(4):856–859.
28. Whitlow BJ, Chatzipapas IK, Lazanakis ML, Kadir RA, Economides DL. The value of sonography in early pregnancy for the detection of fetal abnormalities in an unselected population. *Br J Obstet Gynaecol.* 1999;106(9):929–936.
29. Ebrashy A, El Kateb A, Momtaz M, et al. 13–14-week fetal anatomy scan: a 5-year prospective study. *Ultrasound Obstet Gynecol.* 2010;35(3):292–296.
30. Dashe JS, McIntire DD, Twickler DM. Effect of maternal obesity on the ultrasound detection of anomalous fetuses. *Obstet Gynecol.* 2009;113(5):1001–1007.
31. Simpson JM. Impact of fetal echocardiography. *Ann Pediatr Cardiol.* 2009;2(1): 41–50.
32. Hendler I, Blackwell SC, Bujold E, et al. The impact of maternal obesity on midtrimester sonographic visualization of fetal cardiac and craniospinal structures. *Int J Obes Relat Metab Disord.* 2004;28:1607–1611.
33. Hendler I, Blackwell SC, Bujold E, et al. Suboptimal second-trimester ultra-sonographic visualization of the fetal heart in obese women: should we repeat the examination? *J Ultrasound Med.* 2005;24:1205–1209.
34. Persson B, Stangenberg M, Lunell NO, Brodin U, Holmberg NG, Vaclavinkova V. Prediction of size of infants at birth by measurement of symphysis fundus height. *Br J Obstet Gynaecol.* 1986;93:206–211.
35. Chang TC, Robson SC, Boys RJ, Spencer JA. Prediction of the small for gestational age infant: which ultrasonic measurement is best? *Obstet Gynecol.* 1992;80:1030–1038.
36. De Jong CL, Francis A, Van Geijn HP, Gardosi J. Fetal growth rate and adverse perinatal events. *Ultrasound Obstet Gynecol.* 1999;13:86–89.

37. Melamed N, Yogev Y, Meizner I, Mashiach R, Ben-Haroush A. Sonographic prediction of fetal macrosomia: the consequences of false diagnosis. *J Ultrasound Med.* 2010;29(2): 225–230.
38. Chauhan SP, Grobman WA, Gherman RA, et al. Suspicion and treatment of the macrosomic fetus: a review. *Am J Obstet Gynecol.* 2005;193(2):332–346.
39. Hart NC, Hilbert A, Meurer B, et al. Macrosomia: a new formula for optimized fetal weight estimation. *Ultrasound Obstet Gynecol.* 2010;35(1):42–47.
40. Nahum GG, Stanislaw H. Ultrasound alone is inferior to combination methods for predicting fetal weight. *Ultrasound Obstet Gynecol.* 2007;30:913–914.
41. Higgins MF, Russell NM, Mulcahy CH, Coffey M, Foley ME, McAuliffe FM. Fetal anterior abdominal wall thickness in diabetic pregnancy. *Eur J Obstet Gynecol Reprod Biol.* 2008;140(1):43–47.
42. Petrikovsky BM, Oleschuk C, Lesser M, Gelertner N, Gross B. Prediction of fetal macrosomia using sonographically measured abdominal subcutaneous tissue thickness. *J Clin Ultrasound.* 1997;25(7):378–382.
43. Chauhan SP, West DJ, Scardo JA, Boyd JM, Joiner J, Hendrix NW. Antepartum detection of macrosomic fetus: clinical versus sonographic, including soft-tissue measurements. *Obstet Gynecol.* 2000;95(5):639–642.

16 Obesity and Prolonged Pregnancy

Mani Malarselvi[1] *and Siobhan Quenby*[2]

[1]University Hospitals Coventry and Warwickshire, Coventry, UK,
[2]Division of Reproductive Health, Clinical Science Laboratories,
Warwick Medical School, Coventry, UK

Introduction

Maternal obesity is now the commonest clinical risk factor in obstetrics. Obesity in pregnancy is defined as a body mass index (BMI) of 30 kg/m^2 or more at the first antenatal consultation in a Caucasian women but a BMI of >27 kg/m^2 is considered obese in an Asian or African women [1]. Obesity is associated with considerable maternal and foetal complications and therefore demands high on health resources. There are numerous maternal complications of obesity; however, the most common is prolonged pregnancy and induction of labour [2–8]. In the United Kingdom, obesity has reached epidemic levels so that about 1 in 5 women who are booking for antenatal care are obese [9]. The Centre for Maternal and Child Enquiries (CMACE) conducted a 3-year UK-wide Obesity in Pregnancy Project. They found that 47% of pregnant women with a BMI > 35 kg/m^2 laboured spontaneously, 33% underwent an induction of labour and 20% had Caesarean section (CS) prior to labour. The spontaneous labour and induction rate in the general maternity population is 69% and 20%, respectively. The CMACE report has suggested that in women with BMI > 35 kg/m^2, for each unit increase in BMI; there is a 3% increased risk of induction of labour [10].

Prolonged Pregnancy

Prolonged pregnancy is defined when pregnancy lasts for more than 294 days compared with term gestation which is between 260 and 293 days.

Epidemiology

Multiple epidemiological studies in many populations have found an increased risk of post-term pregnancy in obese women [2–12]. As the prevalence of obesity is increasing, women with high BMI and prolonged pregnancy are becoming increasingly common in obstetric practice. Conversely, obese women have a low

Obesity. DOI: http://dx.doi.org/10.1016/B978-0-12-416045-3.00016-9

incidence of spontaneous preterm labour, preterm delivery is associated with low pre-pregnancy BMI [4,9]. A recent Swedish Retrospective Study of 186,087 primiparous women concluded that higher maternal BMI in the first trimester was associated with longer gestation and an increased risk of post-date pregnancy. Higher maternal BMI during the first trimester was associated with decreased chance of spontaneous labour at term [4]. In another retrospective cohort study of 29,224 women with singleton pregnancies, it was found that higher maternal BMI at booking is associated with an increased risk of prolonged labour and increased rate of induction of labour [7].

Mechanisms

The mechanisms for the association between prolonged pregnancy and obesity remain unclear. There are several possible contributing factors. Some authors have suggested that the effect of obesity on the hypothalamic–pituitary–adrenal axis is contributory. Circulating levels of corticotrophin-releasing hormone (CRH), mainly synthesised by the placenta and cortisol, are considerably lower in maternal plasma at 22–24 weeks in women who deliver at term compared with those who deliver preterm, and a rise in CRH proceeds labour [13,14]. It is also known that there is an inverse linear correlation between plasma cortisol and relative weight. Hence, obese women may have lower circulating CRH and cortisol levels during pregnancy than those of normal weight thereby contributing to prolonged pregnancy [14].

Obese women have an altered metabolic status due to increased adipose tissue, which might influence the initiation of labour. This leads to another potential mechanism which is that the metabolism of oestrogen by the adipose tissue of obese women may result in an alteration in the oestrogen–progesterone ratio in maternal plasma which in turn has a role in the initiation of labour [17].

A series of publications have suggested that poor uterine contractility in obese women is an underlying patho-physiological factor in pregnancy complications in obese pregnant women [3,7,16,17,18]. However, laboratory investigations of the ex vivo contractility of myometrium from obese women have produced conflicting results. One group found that the ex vivo myometrium from obese pregnant women was contracting with less force than from normal weight women [3–18]. However, other workers did not reproduce these findings [19]. Hence, factors extrinsic to the myometrium are being sought to explain uterine quiescence in maternal obesity. Dyslipidaemia associated with obese pregnant women may be a contributory factor [20]. Ex vivo experiments found that depletion of cell membrane cholesterol increased myometrial excitability and contractility in both rats [21] and humans [18]. Furthermore, there is a reduction in tissue cholesterol synthesis at the onset of labour due to the down-regulation in cholesterol biosynthesis genes during parturition [22] and an association between single nucleotide polymorphisms in cholesterol transfer genes and gestation [23]. Cholesterol is an essential component of caveolae – flask-shaped invaginations of cellular membrane that segregates caveolin-1, large conductance potassium and L-type calcium channels essential to myometrial contractility [21]. Rats fed a high-cholesterol diet had reduced caveolin-1 and reduced markers of uterine contractility [24]. Hence, the obesity-related hypercholesterolaemia may play a direct role in uterine quiescence and prolonged pregnancy.

Obese pregnant women have increased chances of developing insulin resistance and gestational diabetes [25]. Even small increases in glucose level have been associated with an increased rate of CS [26]; however, treatment of glucose intolerance in pregnancy did not reduce this CS rate [27]. Thus, mild insulin resistance is an unlikely cause of myometrial quiescence in obese, non-diabetic, pregnant women.

Obesity in pregnancy has also been associated with pro-inflammatory cytokines [28], which have been related to spontaneous preterm labour [29]. However, obese women in our and other studies were less likely to go into spontaneous preterm labour [3–8]. Hence, obesity-related increase in pro-inflammatory cytokines is not an explanation for the failure of initiation of labour in obese women.

Management

Management of prolonged pregnancy in obese women is not straightforward. This is because it is well established that prolonged pregnancy is associated with adverse maternal and neonatal outcomes including pre-eclampsia, postpartum haemorrhage, CS delivery, stillbirth and neonatal death. These risks are all further increased by obesity. To minimise these risks in all women, the National Institute of Clinical Excellence (NICE) antenatal care guidelines recommend induction of labour between 41 and 42 weeks in all women [29]. A meta-analysis has found that induction of labour at 41 weeks has been associated with a reduced CS rate and better neonatal outcomes and is cost-effective. [30] The clinical dilemma surrounds concerns regarding the risks of emergency CS particularly if done by junior staff at night. These risks include sepsis, thromboembolism, haemorrhage, failed spinal analgesia, failed intubation and maternal death. These factors give obstetricians in antenatal clinic three options when managing obese pregnant women who are post–term:

1. induce labour to reduce the significant risk of stillbirth,
2. delivery by elective CS to reduce risk of a complicated emergency delivery,
3. offer foetal monitoring and wait for spontaneous labour, an option not recommended by NICE.

Two recent studies have provided data to help in this decision-making. An analysis of a cohort of 30,000 pregnancies found that more than 60% of obese primiparous and 90% of obese multiparous women who were induced for prolonged pregnancies achieved vaginal deliveries. Furthermore, the complications in the obese women were similar to those of normal weight with prolonged pregnancies [7]. There was no more risk of shoulder dystocia or maternal or neonatal trauma to obese women who were induced compared to those induced at a normal weight [7]. Hence, this UK cohort study [7] justified the induction of labour option for obese post-term women. The national perinatal epidemiology unit study using the UK

Obstetric Surveillance System (UKOSS) has collected data of deliveries in women with a BMI > 50 kg/m^2 in the United Kingdom and has challenged the proposition that elective CS be offered to all severely obese women [31]. They found that 70% of women with BMI > 50 kg/m^2 who had a planned vaginal delivery did indeed deliver vaginally without the expected increase in neonatal and postnatal complication rate compared to those with planned elective CS [31]. The issue of junior doctors performing out of hour's CS is illustrated by the Royal College of Obstetricians and Gynaecologists (RCOG) guideline stating 'An obstetrician and an anaesthetist at Speciality Trainee year 6 and above, or with equivalent experience in a non-training post, should be informed and available for the care of women with a BMI ≥ 40 kg/m^2 during labour and delivery'. The importance of this national UKOSS data [31] is that it was collected at a time in the United Kingdom when out of hour's consultant labour ward cover frequently involved consultants being available on call from home. Hence, it appears that the current on-call consultant cover was adequate to provide care for morbidly obese women in labour in the United Kingdom. Another important factor in these good outcomes from UKOSS was that 30% of women with extreme obesity (BMI > 50 kg/m^2) had a planned CS [31]. Hence, the favourable outcomes may be because UK obstetricians are good at assessing those women in whom planned vaginal delivery is more likely to end in complications. This suggests that an individualised decision on mode of delivery for women with extreme obesity in pregnancy is optimal.

Conclusion

Obesity has reached epidemic levels in the United Kingdom, and maternal obesity poses a significant risk for maternal and foetal health during pregnancy. The most common problem faced by the obstetrician is how to manage obese pregnant women with extreme obesity and post-term pregnancy. An individualised decision needs to be made and an informed decision should balance risks/benefits associated with both approach: induction of labour and elective CS. An adequately powered randomised clinical trial comparing both options for a selected population of extremely obese women is highly desirable assessing not only short-term complications but also to assess long-term health and economic benefits of both approaches.

References

1. WHO. *Obesity: Preventing and Managing Global Epidemic*. Geneva: WHO Publications; 2000.
2. Tennant PWG, Rankin J, Bell R. Maternal body mass index and the risk of fetal and infant death: a cohort study from the North of England. *Hum Reprod*. 2011;26(6): 1501–1511.

3. Quenby S, Zhang J, Bricker L, Wray S. Poor uterine contractility in obese women. *BJOG*. 2007;114:343–348.
4. Denison FC, Price J, Graham C, Wild S, Liston WA. Maternal obesity, length of gestation, risk of postdates pregnancy and spontaneous onset of labour at term. *BJOG*. 2008;115:720–725.
5. Stotland NE, Washington AE, Caughey AB. Prepregnancy body mass index and the length of gestation at term. *Am J Obstet Gynecol*. 2007;197:378e1–378e5.
6. Khashan AS, Kenny LC. The effects of maternal body mass index on pregnancy outcome. *Eur J Epidemiol*. 2009;24:697–705.
7. Arrowsmith S, Wray S, Quenby S. Maternal obesity and labour complications following induction of labour in postdates pregnancy. *BJOG*. 2011;118(5):578–588.
8. Caughey AB, Stotland NE, Washington AE, Escobar GJ. Who is at risk for prolonged and postterm pregnancy? *Am J Obstet Gynecol*. 2009;200:683e1–683e5.
9. Kanagalingam MG, Forouhi NG, Sattar N. Changes in booking body mass index over a decade: retrospective analysis from a Glasgow Maternity Hospital. *BJOG*. 2005;112 (10):1431–1433.
10. CMACE Release. Saving mothers' lives report – reviewing maternal deaths 2006–2008. *BJOG*. 2011;18(suppl 1):1–203.
11. Usha Kiran TS, Hemmadi S, Bethel J, Evans J. Outcome of pregnancy in a woman with an increased body mass index. *BJOG*. 2005;112:768–772.
12. Sebire NJ, Jolly M, Harris JP, et al. Maternal obesity and pregnancy outcome: a study of 287,213 pregnancies in London. *Int J Obes Relat Metab Disord*. 2001; 25:1175–1182.
13. Riley SC, Walton JC, Herlick JM, Challis JR. The localization and distribution of corticotrophin-releasing hormone in the human placenta and fetal membranes throughout gestation. *J Clin Endocrinol Metab*. 1991;72:1001–1007.
14. Mercer BM, Macpherson CA, Goldenberg RL, et al. Are women with recurrent spontaneous preterm births different from those without such history? *Am J Obstet Gynecol*. 2006;194:1176–1184.
15. Cedergren MI. Non-elective Caesarean delivery due to ineffective uterine contractility or due to obstructed labour in relation to maternal body mass index. *Eur J Obstet Gynecol Reprod Biol*. 2009;145(2):163–166.
16. Roman H, Goffinet F, Hulsey TF, Newman R, Robillard PY, Hulsey TC. Maternal body mass index at delivery and risk of Caesarean due to dystocia in low risk pregnancies. *Acta Obstet Gynecol Scand*. 2008;87(2):163–170.
17. Smith R, Mesiano S, Mc Grath S. Hormone trajectories leading to human birth. *Regul Pept*. 2002;108:159–164.
18. Zhang J, Kendrick A, Quenby S, Wray S. Contractility and calcium signaling of human myometrium are profoundly affected by cholesterol manipulation: implications for labor? *Reprod Sci*. 2007;14(5):456–466.
19. Higgins CA, Martin W, Anderson L, et al. Maternal obesity and its relationship with spontaneous and oxytocin-induced contractility of human myometrium in vitro. *Reprod Sci*. 2010;17:177–185.
20. Ramsay JE, Ferrell W, Crawford L, Wallace M, Greer IE, Sattar NJ. Maternal obesity associated with dysregulation of metabolic vascular and inflammatory pathways. *J Clin Endrocrin Metabol*. 2002;87(9):4231–4237.
21. Shmygol A, Noble K, Wray S. Depletion of membrane cholesterol eliminates the Ca2+-activated component of outward potassium current and decreases membrane capacitance in rat uterine myocytes. *J Physiol*. 2007;581(2):445–456.

22. Helguera G, Eghbali M, Sforza D, Minosyan TY, Toro L, Stefani E. Changes in global gene expression in rat myometrium in transition from late pregnancy to parturition. *Physiol Genomics*. 2009;36(2):89–97.
23. Steffen KM, Cooper ME, Shi M, et al. Maternal and fetal variation in genes of cholesterol metabolism is associated with preterm delivery. *J Perinatol*. 2007;11:672–680.
24. Elmes MJ, Tan DS, Cheng Z, Wathes DC, McMullen S. The effects of a high-fat, high-cholesterol diet on markers of uterine contractility during parturition in the rat. *Reproduction*. 2011;141:283–290.
25. Catalano PM. Obesity, insulin resistance, and pregnancy outcome. *Reproduction*. 2010;140(3):365–371.
26. Metzger BE, Lowe LP, Dyer AR, et al. Hyperglycemia and adverse pregnancy outcomes. HAPO Study Cooperative Research Group. *Engl J Med*. 2008;358(19): 1991–2002.
27. Crowther CA, Hiller JE, Moss JR, McPhee AJ, Jeffries WS, Robinson JS. Effect of treatment of gestational diabetes mellitus on pregnancy outcomes. *N Engl J Med*. 2005;352(24):2477–2486.
28. Roberts KA, Riley SC, Reynolds RM, et al. Placental structure and inflammation in pregnancies associated with obesity. *Placenta*. 2011;32(3):247–254.
29. CMACE/RCOG Joint Guideline. *Management of Women with Obesity in Pregnancy*. London: CMACE/RCOG; 2010.
30. Caughey AB, Sundaram V, Kaimal AJ, et al. Maternal and neonatal outcomes of elective induction of labor. *Evid Rep Technol Assess (Full Rep)*. 2009;(176):1–257.
31. Homer CS, Kurinczuk JJ, Spark P, Brocklehurst P, Knight M. Planned vaginal delivery or planned Caesarean delivery in women with extreme obesity. *BJOG*. 2011;118(4): 480–487.

17 Obesity and Venous Thromboembolism

Joanne Ellison and Andrew Thomson

Royal Alexandra Hospital, Paisley, UK

Introduction

Obesity is emerging as the most important risk factor for venous thromboembolism (VTE) in pregnancy [1]. In recent years, we have witnessed a reduction in maternal mortality and morbidity secondary to VTE. So much so, that in the most recent Confidential Enquiry into Maternal and Child Health (CMACE) report into maternal mortality, for the first time since the UK-wide enquiry began, VTE was no longer the leading direct cause of maternal death [1]. Nonetheless, it remains an important and largely preventable cause of maternal mortality. In the CMACE report, of the 18 women who died from VTE in the 2006–2008 triennia, 16 had risk factors for VTE. Chief amongst these was obesity, which was present in 14 of these women [1]. However, maternal mortality may be only the tip of the iceberg. Data on maternal morbidity from VTE, which include far greater numbers of women, again demonstrate a striking over-representation of women in the overweight and obese categories [2]. The exponential rise in obesity seen in the general population over recent years [3] has been mirrored in the obstetric population which has seen a greater than 50% increase in the prevalence of obesity between 1990 and 2002–2004 [4,5]. This increase shows little sign of slowing and its potential impact on the pregnant population may justifiably be likened to that of a ticking bomb.

Epidemiology, Prevalence and Trends

Maternal Mortality

The reduction in maternal mortality from VTE over the last 10–15 years could be regarded as a modern day obstetric success story. In the mid-1990s, VTE was the leading direct cause of maternal death [6]. Widespread education on the importance of thromboprophylaxis following Caesarean section coupled with the development and implementation of national guidelines resulted in a reduction in VTE in the

Obesity. DOI: http://dx.doi.org/10.1016/B978-0-12-416045-3.00017-0

following triennia [7]. Lessons learnt in subsequent triennia have led to yet further reductions [1,8,9]. Perhaps, most notable amongst these is the understanding that even women who do not undergo Caesarean delivery (including those who have uncomplicated spontaneous vaginal deliveries) may be considered at high risk for VTE when other risk factors are present [9]. Central to each reduction in maternal mortality has been an improved appreciation and quantification of risk. This includes not only vigilance in detecting isolated risk factors for VTE, but also a heightened awareness of the importance of the co-existence of multiple risk factors in the same woman (Table 17.1). For example, in the study by Jacobsen et al. [11], a body mass index (BMI) $\geq 25\ kg/m^2$ together with a history of immobilisation is associated with an odds ratio (OR) of 62.3 (95% confidence intervals (CI)

Table 17.1 Risk Assessment for VTE

Pre-existing Risk Factors	**Score**
Previous recurrent VTE	3
Previous VTE – unprovoked or oestrogen related	3
Previous VTE – provoked	2
Family history of VTE	1
Known thrombophilia	2
Medical co-morbidities	2
Age (>35 years)	1
Obesity	1/2[a]
Parity ≥ 3	1
Smoker	1
Gross varicose veins	1
Obstetric risk factors	
Pre-eclampsia	1
Dehydration/hyperemesis/OHSS	1
Multiple pregnancy or ART	1
Caesarean section in labour	2
Elective Caesarean section	1
Mid-cavity or rotational forceps	1
Prolonged labour (>24 h)	1
PPH (>1 l or transfusion)	1
Transient risk factors	
Current systemic infection	1
Immobility	1
Surgical procedure in pregnancy or ≤ 6 weeks post-partum	2

Consider thromboprophylaxis if the patient's risk factor score totals:

≥ 3 antenatally and managed as an outpatient,

≥ 2 antenatally and managed as an inpatient,

≥ 2 post-partum.

ART, assisted reproductive technologies; OHSS, ovarian hyperstimulation syndrome; PPH, post-partum haemorrhage; VTE, venous thromboembolism.

[a]Score 1 for BMI $> 30\ kg/m^2$; Score 2 for BMI $> 40\ kg/m^2$ (BMI based on booking weight).

Source: Adapted from Ref. [10].

11.5–337.6) for antenatal VTE and an OR of 40.1 (95% CI 8.0–201.5) for post-natal VTE compared with women with a BMI of <25 kg/m^2 with no associated immobility. Contrast this with an OR of 1.8 (95% CI 1.3–2.4, antenatal VTE) and 2.4 (95% CI 1.7–3.3, post-natal VTE) in women with a BMI above 25 kg/m^2 but no history of immobilisation.

In the most recent CMACE report, obesity is the most important risk factor for VTE in pregnancy [1]. Between 2006 and 2008, 18 women died from venous thrombosis giving a maternal mortality rate of 0.79 per 100,000 maternities (95% CI 0.49–1.25). Sixteen of the deaths were attributable to pulmonary embolism and two to cerebral vein thrombosis. Of the 16 of these women who had risk factors for VTE, 14 were obese: 11 women had a BMI $\geq$30 kg/m^2 and 3 women a BMI $\geq$ 40 kg/m^2 (Figure 17.1).

Maternal Morbidity

A national prospective case–control study using UK Obstetric Surveillance System (UKOSS) data identified 143 confirmed cases of antenatal pulmonary embolism between February 2005 and August 2006 [2]. This represents an incidence of 12.6 per 100,000 maternities (95% CI 10.6–14.9 per 100,000 maternities). Of these women, five died giving a case fatality rate of 3.5% (95% CI 1.1–8.0%) consistent with that reported in the CMACE enquiry. Of these women, 70% had an identifiable risk factor for VTE and 18% had at least two. Obesity was second only to multiparity as a risk factor for VTE (adjusted OR 2.65, 95% CI 1.09–6.45), again underlining its prominence.

Numerous other studies have described the strong relationship that exists between obesity and VTE in pregnancy. [2,11–15]. In particular, a Danish

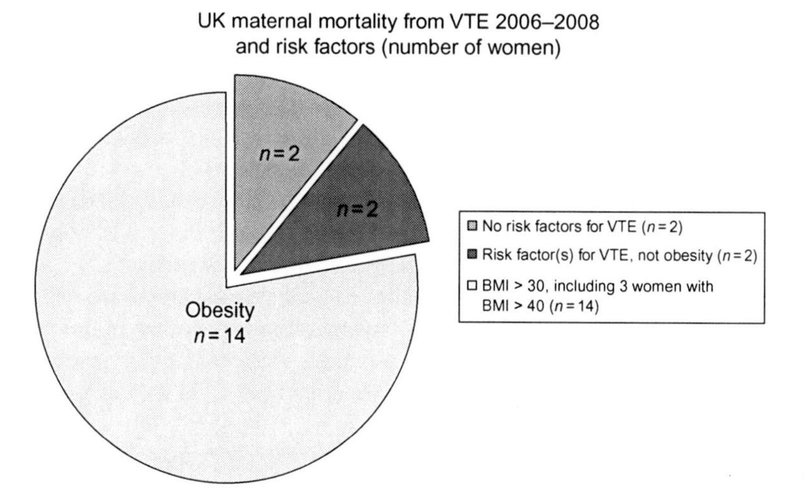

Figure 17.1 Prevalence of obesity amongst pregnant women in the United Kingdom who died from VTE in 2006–2008 [1].

retrospective case–control study involving 129 obese pregnant women identified obesity as a significant risk factor with an adjusted OR (aOR) of 5.3 (95% CI 2.1–13.5). Interestingly, the obese parturient is at greater risk of Pulmondary Embolism (PE) (aOR 14.9; 95% CI 3.0–74.8) than deep venous thrombosis (DVT) (aOR 4.4; 95% CI 1.6–11.9) [12].

Obesity and Pregnant Women

The prevalence of obesity in the general population in England has increased markedly since the early 1990s [3]. The obstetric population has undergone a parallel increase.

In a UK-based study, Heselhurst et al. compared the booking BMI of 36,821 women from 1 January 1990 to 31 December 2004 [4]. They demonstrated that the incidence of maternal obesity at the start of pregnancy is not only increasing but also accelerating. These findings were consistent with those of Kanagalingam et al. whose 2005 study in a Scottish maternity unit demonstrated a greater than twofold increase in the proportion of obese pregnant women in 2002/2004 compared to 1990 [5]. This study pre-dates the 2006–2008 triennia, yet the authors concluded somewhat prophetically that obesity-related pregnancy complications were likely to increase in the future.

Possible Underlying Mechanisms

In his eponymous triad, Virchow described the three categories of factors which contribute to thrombosis: hypercoagulability, haemodynamic changes (stasis, turbulence) and endothelial injury (Figure 17.2). Pregnancy alone impacts significantly on Virchow's triad. Pronounced increases in Factors I, V, VII, VIII, IX, X, XII, von Willebrand factor antigen and ristocetin co-factor activity engender a prothrombotic milieu [16–18]. Changes in the deep venous system also occur in normal pregnancy: a marked reduction in blood flow velocity accompanied by an increase in the diameter of the major leg veins together with the pressure of the gravid uterus lead to venous stasis [19,20]. Trauma to the venous system can occur in the course of vaginal delivery as the head passes through the pelvis. Such trauma is also a feature of operative delivery, whether abdominal or vaginal [21].

Adding obesity in to the equation further exacerbates the situation: increases in coagulation factors are exaggerated [22]. Severe adiposity further impedes venous return, worsening stasis. Additionally, in morbidly obese women, a general lack of mobility may contribute to venous stasis. Maternal obesity is also associated with endothelial injury and dysfunction [10,23].

Obesity increases the risks of a number of adverse pregnancy outcomes including operative delivery [24,25] and pre-eclampsia [26]. Indeed, data from UKOSS has shown that extreme obesity (BMI ≥ 50 kg/m^2) is associated with an increased risk of pre-eclampsia, gestational diabetes mellitus, preterm delivery, general

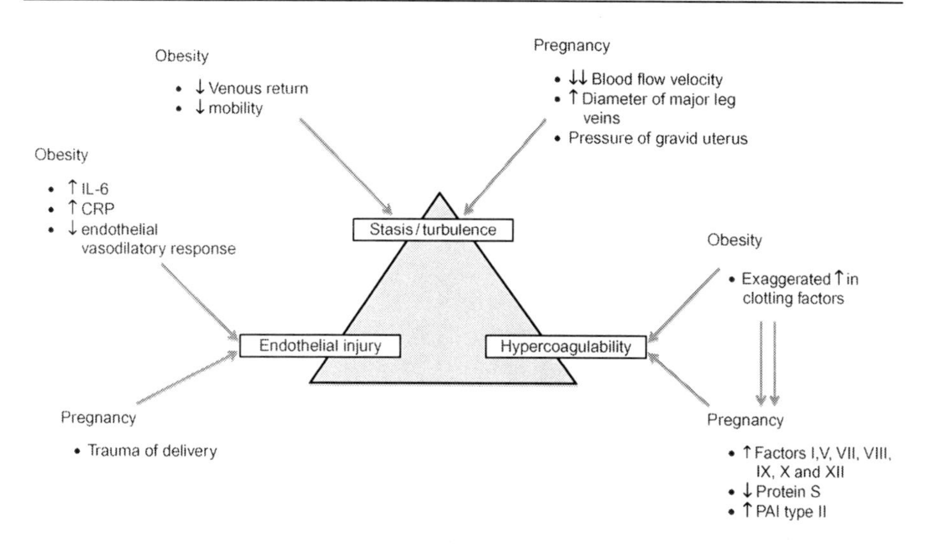

Figure 17.2 Pregnancy, obesity and Virchow's triad.

anaesthesia and admission to an intensive care unit [27]. In this study, a BMI of 50 kg/m^2 or more was also associated with a 50% Caesarean delivery rate, compared with 22% in the matched comparison group [27]. These outcomes are themselves associated with an increased risk of VTE [28]. Thus there is an element of risk amplification.

Prevention of VTE in Obese Women During Pregnancy

Recognition and Assessment of Risk

Identification of risk is pivotal. The most recent CMACE enquiry cites substandard care in 56% of the women who died from VTE [1]. This took various forms including inadequate risk assessment and thromboprophylaxis (by the standards at that time).

The Royal College of Obstetricians and Gynaecologists (RCOG) Green-Top Guideline (37a), 'Reducing the Risk of Thrombosis and Embolism During Pregnancy and the Puerperium', clearly lists and stratifies risk factors relevant to the development of VTE in pregnancy [28]. These have been reproduced in Table 17.1.

This guideline singles out obesity as a risk factor warranting particular consideration because of its high prevalence amongst the women described in maternal mortality statistics at the time of publication (Saving Mothers Lives – 2003–2005) [29]. Although obesity is considered to be only a moderate risk factor for VTE, it is particularly important because of its high prevalence within the population. In addition, the risk of VTE appears to increase further with increasing obesity [14].

Obesity was the most important risk factor for VTE in the most recent CMACE enquiry (2006–2008), and the majority of recommendations made in the chapter on thrombosis and thromboembolism relate to the care of obese pregnant women [1]. Table 17.2 summarises these recommendations and those pertinent to risk identification are discussed below. Excellent guidance is also to be found in the CMACE/RCOG Joint Guideline, Management of Women with Obesity in Pregnancy [30].

All pregnant women should have their BMI *properly* calculated at the time of the booking visit, ideally in the first trimester and clearly recorded in the case notes [30]. Height and weight should be formally measured. Correct calculation of BMI facilitates accurate risk stratification upon which basis a plan for thromboprophylaxis in the pregnancy may be formulated.

Women with a BMI $\geq$ 30 kg/m^2 should be assessed at their first antenatal visit and throughout the pregnancy for the risk of thromboembolism in accordance with CMACE/RCOG guidelines [30]. A change in circumstances such as a period of immobility, an intercurrent illness or a period of hospitalisation should automatically prompt a re-assessment of risk. Women with a BMI of 35 kg/m^2 or more are unsuitable for midwife-only care, and should be seen in pregnancy by a consultant obstetrician [1].

Risk of VTE in pregnancy exists from the very beginning of pregnancy until the end of the puerperium [1]. This has particular relevance for women who develop problems or intercurrent illnesses in early pregnancy. For example, a woman hospitalised with hyperemesis gravidarum may be dehydrated and relatively immobile, and is therefore at high risk of VTE. She should receive thromboprophylaxis despite being at an early stage in her pregnancy. The same is true for women

Table 17.2 Strategies to Reduce the Incidence of VTE in Obese Pregnant Women

BMI should be calculated at the booking visit.
Women with a booking BMI $\geq$ 30 kg/m^2 should be assessed at their first antenatal visit and throughout the pregnancy for the risk of thromboembolism.
It should be appreciated that the risk of VTE in pregnancy exists from the very beginning of pregnancy until the end of the puerperium.
Women with a booking BMI $\geq$ 30 kg/m^2 requiring pharmacological thromboprophylaxis should be prescribed a weight-specific dosage in accordance with current RCOG guidelines [10].
Women with a BMI $\geq$ 30 kg/m^2 should be encouraged to mobilise as early as practicable following childbirth.
Women with a BMI $\geq$ 40 kg/m^2 should be offered post-natal thromboprophylaxis regardless of their mode of delivery.
Vulnerable women require particular care as they may not be able to follow advice or self-administer injections.
There should be a low threshold for investigating obese women who develop chest symptoms for the first time in pregnancy.

Source: Adapted from Refs. [1,30].

undergoing surgical treatment of miscarriage, termination of pregnancy or ectopic pregnancy. For all of these women, risk analysis should be revisited and an early gestation should not be a barrier to instituting thromboprophylaxis where indicated [28,30].

Ideally, all women should have a booking visit in the first trimester. However, occasionally this visit may be delayed. Women not infrequently have no contact with maternity services until 20 weeks' gestation or beyond. These women's only point of contact with health care services may be with their General Practitioner (GP) or community health services. It is therefore important to promote awareness amongst colleagues in primary care of the risks to obese women in early pregnancy of VTE. Indeed, this promotion of awareness should encompass the wider health community including online services, pharmacists and clinicians in emergency medicine all of whom are frequently a first point of contact for pregnant women. Women who are judged to require thromboprophylaxis throughout the antenatal period should receive this from as early in the pregnancy as practical [28].

Thromboprophylaxis

Clear guidance on thromboprophylaxis in pregnancy is to be found in the RCOG guideline, 'Reducing the Risk of Thrombosis and Embolism during Pregnancy and the Puerperium' [28]. Advice more specific to the management of obese pregnant women exists in the CMACE/RCOG Joint Guideline Management of Women with Obesity in Pregnancy [30]. The points of good practice discussed below are taken from recommendations made in these sentinel guidelines.

Antenatal Thromboprophylaxis

A woman with a BMI ≥ 30 kg/m^2 who also has two or more additional risk factors for thromboembolism should be considered for prophylactic low-molecular weight heparin (LMWH) antenatally [28,30]. This should begin as early in pregnancy as practical.

Women with a booking BMI ≥ 30 kg/m^2 requiring pharmacological thromboprophylaxis should be prescribed doses appropriate for maternal weight, in accordance with the RCOG Clinical Green-Top Guideline No. 37a [28] (Table 17.3).

VTE should be discussed with women identified to be at increased risk and the reasons for individual recommendations explained [30]. The booking visit presents an ideal opportunity to do this. It also allows discussion regarding the importance of mobility, hydration and the need to urgently see a health professional should she develop de novo chest symptoms or unilateral leg swelling/pain.

Peripartum Thromboprophylaxis

Women receiving antenatal LMWH should be advised that once labour begins, they should not inject any further LMWH. They should be reassessed on admission to hospital and further doses should be prescribed by medical staff [28].

Table 17.3 Suggested Thromboprophylactic Doses for Antenatal and Post-Natal LMWH

Weight (kg)	Enoxaparin	Dalteparin	Tinzaparin
<50	20 mg daily	2500 units daily	3500 units daily
50–90	40 mg daily	5000 units daily	4500 units daily
91–130	60 mg daily[a]	7500 units daily[a]	7000 units daily[a]
131–170	80 mg daily[a]	10,000 units daily[a]	9000 units daily[a]
>170	0.6 mg/kg/day[a]	75 U/kg/day[a]	75 U/kg/day[a]

[a]May be given in two divided doses.
Source: Adapted from Ref. [10].

The pregnancy-associated prothrombotic changes in the coagulation system are maximal immediately following delivery. It is therefore desirable to continue LMWH during labour or delivery in women receiving antenatal thromboprophylaxis with LMWH. However, to allow for the use of regional anaesthesia, if requested or required, women are advised to discontinue LMWH at the onset of labour or prior to planned delivery [28].

For women receiving high prophylactic or therapeutic doses of LMWH, the dose of heparin should be reduced to its thromboprophylactic dose on the day before induction of labour and, if appropriate, continued in this dose during labour [28].

Regional anaesthesia or analgesia can be sited only after discussion with a senior anaesthetist, in keeping with local obstetric anaesthetic protocols. A large multicenter epidemiologic study involving 1863 patients has identified obesity as one factor which increases the risk of transient neurological symptoms (TNS) following spinal anaesthesia [31]. It is important to discuss the implications of treatment with LMWH for regional anaesthesia and analgesia with the women before labour or Caesarean section. This could be appropriately undertaken in an anaesthetic antenatal clinic.

In summary, to minimise or avoid the risk of epidural haematoma:

- Regional techniques should not be used until at least 12 h after the previous prophylactic dose of LMWH.
- When a woman presents while on a therapeutic regimen of LMWH, regional techniques should not be employed for at least 24 h after the last dose of LMWH.
- LMWH should not be given for 4 h after use of spinal anaesthesia or after the epidural catheter has been removed; the cannula should not be removed within 10–12 h of the most recent injection.

For delivery by elective Caesarean section in women receiving antenatal LMWH, the woman should receive a thromboprophylactic dose of LMWH on the day before delivery. On the day of delivery, any morning dose should be omitted and the operation should be performed that morning. The thromboprophylactic dose of LMWH should be given 4 h post-operatively or 4 h after removal of the epidural catheter.

There is an increased risk of wound haematoma following Caesarean section with both unfractionated heparin (UFH) and LMWH of around 2%. In obese women in particular, consideration should be given to closing the subcutaneous layer of fat to eliminate dead space to minimise this risk. Using interrupted sutures or staples for skin closure may also reduce the risk of wound haematoma. In view of the increased risk of intra-abdominal bleeding in women on heparin undergoing Caesarean delivery, consideration should be given to the use of an intraperitoneal drain at the end of the operation.

Women at high risk of haemorrhage with risk factors including major antepartum haemorrhage, coagulopathy, progressive wound haematoma, suspected intra-abdominal bleeding and post-partum haemorrhage may be more conveniently managed with UFH or graduated compression stockings (GCS) [28]. If a woman develops a haemorrhagic problem while on LMWH, the treatment should be stopped and expert haematological opinion sought [28]. It should be remembered that excess blood loss and blood transfusion is a risk factor for VTE, so thromboprophylaxis should be begun or reinstituted as soon as the immediate risk of haemorrhage is reduced [28].

Post-partum Thromboprophylaxis

The prothrombotic changes of pregnancy do not revert completely to normal until several weeks after delivery. Indeed, the time of greatest risk for VTE associated with pregnancy is the early puerperium. Although most VTE occur antenatally, the risk is greatest in the weeks immediately after delivery, with more than 50% occurring in the first post-partum week [32].

Women with a booking BMI $\geq 30\ kg/m^2$ should be encouraged to mobilise as early as practicable following childbirth to reduce the risk of thromboembolism. In addition, the importance of good hydration should be emphasised [28].

Women with a BMI $\geq 30\ kg/m^2$ who have one or more additional persisting risk factors for thromboembolism should also be considered for LMWH for 7 days after delivery [28,30].

Women with a BMI $\geq 30\ kg/m^2$ who have two or more additional persisting risk factors should be given GCS in addition to LMWH [30].

All women with a BMI $\geq 40\ kg/m^2$ should be offered post-natal thromboprophylaxis regardless of their mode of delivery and for these women thromboprophylaxis should be continued for a minimum of 1 week [28,30].

Thromboprophylaxis should be continued for 6 weeks in women at high risk (previous VTE or any woman requiring antenatal LMWH) of post-partum VTE and for 1 week in women with intermediate risk (one or more persisting risk factors besides obesity) [28,30].

In women at intermediate risk of VTE, there has been much debate as to the optimal duration of thromboprophylaxis. Clinical data suggest that the highest risk lies in the first week post-partum. Thus, a minimum of 7 days thromboprophylaxis is recommended [28]. This is in recognition of the increased risk of VTE during the first post-partum week, but also the likelihood that by this stage the majority of women will be ambulant. This is not, however, synonymous with discharge from hospital, and it is important that risk assessment is performed in each woman at least once following delivery and before discharge and arrangements is made for LMWH prescription and administration (usually by the woman herself) in the community where necessary. The most recent CMACE report highlighted the case of vulnerable women with mental illness, who may be unable to self-administer LMWH and may not present to health care services should they develop symptoms of VTE [1]. It is important that such limitations are taken account of and suitable arrangements put in place for these women prior to discharge from hospital.

The first thromboprophylactic dose of LMWH should be given as soon as possible after delivery provided that there is no post-partum haemorrhage or there has been regional analgesia, in which case LMWH should be given by 4 h after delivery or 4 h after removal of the epidural catheter, if it is removed immediately or shortly after delivery [28]. If the epidural catheter is left in place after delivery, for the purpose of post-partum analgesia, it should be removed 12 h after a dose and 4 h before the next dose of LMWH.

In the United Kingdom, approximately 20–30% of deliveries are affected by Caesarean section. Women delivered by emergency Caesarean section have double the risk of post-partum VTE compared with elective Caesarean section [28]. Women who deliver by emergency Caesarean section have a roughly fourfold increased risk of post-partum VTE compared with those who deliver vaginally [28]. These risks are higher again for obese women.

All women who have had an emergency Caesarean section (category 1–3) should be considered for thromboprophylaxis with LMWH for 7 days after delivery [28].

All women who have had an elective Caesarean section (category 4) who have one or more additional risk factors (such as age over 35 years, BMI greater than 30 kg/m^2) should be considered for thromboprophylaxis with LMWH for 7 days after delivery [28].

Thromboprophylactic Agents

Low-Molecular Weight Heparins

LMWHs are the agents of choice for antenatal thromboprophylaxis [33,34]. They are at least as effective as and safer than UFH [28], and have excellent bioavailability [35]. Their good absorption and low plasma protein binding ensure a more reliable and predictable dose response compared with UFH [35]. The risks of heparin-induced thrombocytopenia [36–38], osteoporotic fractures [36,39–42] and allergic skin reactions [36] are lower than those associated with UFH. In the systematic review of LMWH use in pregnancy by Greer and Nelson-Piercy [36], significant bleeding, usually related primarily to obstetric causes, occurred in 1.98% (95% CI 1.5–2.57). This related to both treatment and prophylactic doses of LMWH and therefore the risk of bleeding is likely to be less than 2% with prophylactic doses. A recent retrospective study of the safety and efficacy of the LMWH tinzaparin in pregnancy has provided further reassuring data [43]. LMWHs do not cross the placenta and therefore pose no risk to the developing foetus [44].

In 2009, the RCOG published weight-specific dosage guidelines for thromboprophylaxis in pregnancy (Table 17.3) [28], taking into account that the previous 'one size fits all' dosage philosophy failed women at upper extremes of the body weight range, where all too frequently the doses of LMWH administered were inadequate for effective thromboprophylaxis [1]. Doses of LMWH are based on weight, not BMI. For thromboprophylaxis, the booking weight is used to guide dosing. It is hoped that the benefit of weight-specific thromboprophylaxis guidelines will be seen when maternal morbidity and mortality are reported in subsequent triennia.

Monitoring LMWH therapy assessing anti-Xa levels provides little or no evidence on the efficacy in relation to prevention of thrombosis. Experience indicates that monitoring of anti-Xa levels is not required when LMWH is used for thromboprophylaxis, provided that the woman has normal renal function [45].

Enoxaparin, dalteparin and tinzaparin are the LMWHs most commonly used for thromboprophylaxis in pregnancy in the United Kingdom. In terms of efficacy, there is little to choose between them. Tinzaparin is more readily reversible however, as a greater proportion of its anticoagulant activity is derived from its anti-IIa action (lower anti-Xa: anti-IIa ratio) compared with dalteparin or enoxaparin [46]. This property renders tinzaparin's effect more amenable to reversal by protamine sulphate. Tinzaparin may be a more advantageous choice of LMWH in a patient where high risk for VTE and a potential bleeding problem coexist.

LMWH does not cross the breast and is safe for use by breastfeeding mothers.

Unfractionated Heparin

As previously discussed, side effects (heparin-induced thrombocytopenia, osteoporotic fractures and skin allergies) are more common with UFH than with LMWH [36–42]. In addition, the increase in coagulation factors accompanying pregnancy leads to a relative APTT-resistance, making UFH difficult to monitor and leaving potential for inadvertent over-anticoagulation. However, UFH does have a shorter half-life than LMWH, and there is more complete reversal of its activity by protamine sulphate. For this reason, UFH may occasionally be used in women at increased risk of haemorrhage.

Warfarin

Warfarin crosses the placenta leading to an increased risk of congenital abnormalities including a characteristic warfarin embryopathy in approximately 5% of foetuses exposed between 6 and 12 weeks of gestation [47]. Other reported complications associated with warfarin therapy include an increase in the risk of spontaneous miscarriage [48], stillbirth [49], neurological problems in the baby [50] and foetal and maternal haemorrhage [48]. For these reasons, the use of warfarin in pregnancy has previously been restricted to a few situations where heparin is considered unsuitable; for example in some women with mechanical heart valves. In a systematic review of anticoagulant therapy in pregnant women with prosthetic heart valves, the use of vitamin K antagonists throughout pregnancy was associated with congenital anomalies in 6.4% of live births (95% CI 4.6–8.9%) [51]. However, in a multicenter prospective cohort study of vitamin K antagonists and pregnancy outcome, there were only two cases of warfarin embryopathy in 356 live births (0.6%) suggesting that the teratogenic effect of warfarin may be less than previously thought [52]. Warfarin is not secreted into breast milk and is therefore safe for use by breastfeeding mothers.

Low-Dose Aspirin

There are no controlled trials on the use of aspirin for thromboprophylaxis in pregnancy. The American College of Chest Physicians (ACCP) guideline recommends against the use of aspirin for VTE prophylaxis in any patient group [53]. Aspirin is appropriate for women with anti-phospholipid syndrome to improve foetal outcomes.

Danaparoid

Danaparoid is a heparinoid that is mostly used in patients intolerant of heparin, either because of heparin-induced thrombocytopenia or skin allergy. A review of 91 pregnancies in 83 women concluded danaparoid is an effective and safe antithrombotic in pregnancy for women who are intolerant of heparin [54].

Other Thromboprophylactic Agents

Fondaparinux is not licensed for use in pregnancy in the United Kingdom but has been used in the setting of heparin intolerance. Lepirudin can cross the placenta and is best avoided in pregnancy unless there is no acceptable alternative. Dextran has been known to cause anaphylactic reactions and should be avoided. Dabigatran and Rivaroxaban are two new anticoagulants. They are not licensed for use in pregnancy and should also be avoided.

Graduated Elastic Compression Stockings

There are no trials to guide on the efficacy of GCS in pregnancy. We are thus largely reliant on expert opinion. The advantages and limitations of GCS and other mechanical methods of VTE prevention in the non-pregnancy setting were recently reviewed by the ACCP [53]. In the ACCP guideline on thromboprophylaxis in pregnancy, the use of properly applied GCS of appropriate strength is recommended in pregnancy and the puerperium for:

- those who are hospitalised and have a contraindication to LMWH,
- those who are hospitalised post-Caesarean section (combined with LMWH) and considered to be at particularly high risk of VTE (such as previous VTE, more than three risk factors),

outpatients with prior VTE (usually combined with LMWH),
women travelling long distance for more than 4 h.

There are few data regarding the most efficacious length of GCS to use in pregnancy. In the case of obese women it is important, but not always easy, to ensure a good fit. Ill-fitting GCS can produce a tourniquet effect promoting venous stasis and increasing the risk of VTE. Knee-high GCS that fit well would be preferred to full-length but ill-fitting stockings.

Investigation and Diagnosis of VTE in the Obese Pregnant Woman

Any woman with signs and symptoms suggestive of VTE should have objective testing performed expeditiously and treatment with LMWH until the diagnosis is excluded by objective testing, according to the RCOG Green-Top Guideline No. 37b, 'The Acute Management of Thrombosis and Embolism During Pregnancy and the Puerperium' [55]. Indeed, the CMACE report recommends that chest symptoms appearing for the first time in pregnancy or the puerperium in at-risk women need careful assessment and there should be a low threshold for investigation [1]. Obese women should be considered an 'at-risk' group and this advice should certainly be heeded in their case.

Complex duplex ultrasound is the investigation of choice for diagnosing DVT [56]. Women with suspected DVT should be commenced on LMWH at a treatment dosage until the diagnosis is excluded by objective testing [55]. If ultrasound confirms the diagnosis of DVT, anticoagulant treatment should be continued. If ultrasound is negative and a high level of clinical suspicion exists, the woman should remain anti-coagulated and the ultrasound repeated in 1 week or an alternative diagnostic test employed. If repeat testing is negative, anticoagulant treatment should then be discontinued [57].

In pregnant women, most DVT occurs in the iliofemoral segment rather than in the deep veins of the calf [21]. This area can be difficult to assess owing to the presence of the foetus and limitations of ultrasound scanning. This is particularly true for obese women where it may not be possible to achieve sufficient tissue penetration with ultrasound scanning to facilitate good visualisation. This should be borne in mind when interpreting results and on occasion it may be worth seeking the advice of a radiologist as to whether other imaging modalities may be informative.

Where there is clinical suspicion off acute pulmonary thromboembolism (PTE), a chest X-ray should be performed [55]. Compression duplex Doppler should be performed where this is normal [55]. If both tests are negative with persistent clinical suspicion of acute PTE, a ventilation–perfusion (V/Q) lung scan or a computed tomography pulmonary angiogram (CTPA) should be performed [55]. Alternative or repeat testing should be carried out where V/Q scan or CTPA and duplex Doppler are normal but the clinical suspicion of

Table 17.4 Calculation of Treatment Dosage of LMWH by Early Pregnancy Weight

	Early Pregnancy Weight (kg)			
	<50	**50–69**	**70–89**	**>90**
Enoxaparin	40 mg bd[a]	60 mg bd[a]	80 mg bd[a]	100 mg bd[a]
Dalteparin	5000 IU bd[a]	6000 IU bd[a]	8000 IU bd[a]	10,000 IU bd[a]
Tinzaparin	175 U/kg once daily (all weights)	175 U/kg once daily (all weights)	175 U/kg once daily (all weights)	175 U/kg once daily (all weights)

[a]bd = twice daily.
Source: Adapted from Ref. [44].

PTE is high [55]. Anticoagulation treatment should be continued until PTE is definitively excluded [55]. Because of the changes in the coagulation system, D-dimer testing is not informative in the diagnosis of VTE in pregnancy and is not recommended [58].

Treatment of VTE in the Obese Pregnant Woman

In clinically suspected DVT or PTE, unless treatment is strongly contraindicated, treatment with LMWH should be given until the diagnosis is excluded by objective testing [55]. Meta-analyses of randomised controlled trials indicate that LMWHs are more effective, are associated with a lower risk of haemorrhagic complications and lower mortality than UFH in the initial treatment of DVT in non-pregnant women [59,60]. LMWH should be given daily in two subcutaneous divided doses with dosage titrated against the woman's booking or most recent weight [55] (Table 17.4).

Conclusion

Obesity is currently the main risk factor for the development of VTE in pregnant women [1]. Its prevalence is increasing and accelerating. There is no doubt that further reductions in maternal mortality from VTE are possible but are largely dependent on our vigilance in recognising obesity as a risk factor and acting upon it. In seeking to achieve this, we could do little better than to look to the key recommendations in the most recent CMACE report (Table 17.2).

References

1. Confidential Enquiry into Maternal and Child Health. *Saving Mothers' Lives: Reviewing Maternal Deaths to Make Motherhood Safer: 2006–2008. The Eighth Report of the*

Confidential Enquiries into Maternal Deaths in the United Kingdom. London: CEMACH; 2011.

2. Knight M. Antenatal pulmonary embolism: risk factors, management and outcomes. *BJOG*. 2008;115:453–461.
3. Office for National Statistics. *Statistics on Obesity, Physical Activity and Diet*. London: Office for National Statistics; 2008.
4. Heslehurst N, Ells LJ, Simpson H, Batterham A, Wilkinson J, Summerbell CD. Trends in maternal obesity incidence rates, demographic predictors, and health inequalities in 36,821 women over a 15-year period. *BJOG*. 2007;114(2):187–194.
5. Kanagalingam MG, Forouhi NG, Greer IA, Sattar N. Changes in booking body mass index over a decade: retrospective analysis from a Glasgow Maternity Hospital. *BJOG*. 2005;112(**10**):1431–1433.
6. Department of Health. *Why Mothers Die. Report on Confidential Enquiries into Maternal Deaths in the United Kingdom 1994–1996*. London: Department of Health; 1998.
7. RCOG. *Why Mothers Die 1997–1999: Report on Confidential Enquiries into Maternal Deaths in the United Kingdom*. London: RCOG Press; 2001.
8. RCOG. *Why Mothers Die 2000–2002: Report on Confidential Enquiries into Maternal Deaths in the United Kingdom*. London: RCOG Press; 2004.
9. CEMACH. *Saving Mothers' Lives: Reviewing Maternal Deaths to Make Motherhood Safer – 2003 – 2005*. London: CEMACH; 2007.
10. Ramsay JE, Ferrell WR, Crawford L, Wallace AM, Greer IA, Sattar N. Maternal obesity is associated with dysregulation of metabolic, vascular, and inflammatory pathways. *J Clin Endocrinol Metab*. 2002;87:4231–4237.
11. Jacobsen AF, Skjeldestad FE, Sandset PM. Ante-natal and postnatal risk factors of venous thrombosis: a hospital-based case–control study. *J Thromb Haemost*. 2008;6: 905–912.
12. Larsen TB, Sorensen HT, Gislum M, Johnsen SP. Maternal smoking, obesity, and risk of venous thromboembolism during pregnancy and the puerperium: a population-based nested case–control study. *Thromb Res*. 2007;120:505–509.
13. James AH, Jamison MG, Brancazio LR, Myers ER. Venous thromboembolism during pregnancy and the postpartum period: incidence, risk factors, and mortality. *Am J Obstet Gynecol*. 2006;194:1311–1315.
14. Robinson HE, O'Connell CM, Joseph KS, McLeod NL. Maternal outcomes in pregnancies complicated by obesity. *Obstet Gynecol*. 2005;106:1357–1364.
15. Simpson EL, Lawrenson RA, Nightingale AL, Farmer RDT. Venous thromboembolism in pregnancy and the puerperium: incidence and additional risk factors from a London perinatal database. *Br J Obstet Gynaecol*. 2001;108:56–60.
16. Forbes CD, Greer IA. Physiology of haemostasis and the effect of pregnancy. In: Greer IA, Turpie AJJ, Forbes CD, eds. *Haemostasis and Thrombosis in Obstetrics and Gynaecology*. London: Chapman & Hall; 1992:1–25.
17. Brennand JE, Clark PA, Conkie JA, et al. Haemostatic variables in normal pregnancy. *Prenat Neonatal Med*. 1996;1(suppl):S186.
18. Greer IA. Special case of venous thromboembolism in pregnancy. In: Tooke JE, Lowe GDO, eds. *A Textbook of Vascular Medicine*. London: Arnold; 1996: 538–561.
19. Macklon NS, Greer IA. The deep venous system in the puerperium: an ultrasound study. *Br J Obstet Gynaecol*. 1997;104:198–200.
20. Macklon NS, Greer IA, Bowman AW. An ultrasound study of gestational and postural changes in the deep venous system of the leg in pregnancy. *Br J Obstet Gynaecol*. 1997;104:191–197.

21. Greer IA. Venous thrombo-embolism. In: Greer IA, ed. *Thrombo-Embolic Disease in Obstetrics and Gynaecology*. Cambridge: Ballière Tindall; 1997: 403–430.
22. Abdollahi M, Cushman M, Rosendaal FR. Obesity: risk of venous thrombosis and the interaction with coagulation factor levels and oral contraceptive use. *Thromb Haemost*. 2003;89(**3**):493–498.
23. Stewart FM, Freeman DJ, Ramsay JE, Greer IA, Caslake M, Ferrell WR. Longitudinal assessment of maternal endothelial function and markers of inflammation and placental function throughout pregnancy in lean and obese mothers. *J Clin Endocrinol Metab*. 2007;92:969–975.
24. Kabiru W, Raynor BD. Obstetric outcomes associated with increase in BMI category during pregnancy. *Am J Obstet Gynaecol*. 2004;191:928–932.
25. Sebire NJ, Jolly M, Harris JP, Wadsworth J, Joffe M, Beard RW. Maternal obesity and pregnancy outcome: a study of 287,213 pregnancies in London. *Int J Obes Relat Metab Disord*. 2001;25:1175–1182.
26. O'Brien TE, Ray JG, Chan WS. Maternal body mass index and the risk of pre-eclampsia: a systemic overview. *Epidemiology*. 2003;14:368–374.
27. Knight M, Kurinczuk JJ, Spark P, Brocklehurst P. Extreme obesity in pregnancy in the United Kingdom. *Obstet Gynaecol*. 2010;115(5):989–997.
28. Royal College of Obstetricians and Gynaecologists. *Reducing the Risk of Thrombosis and Embolism during Pregnancy and the Puerperium. Green-Top Guideline No. 37a*. London: RCOG Press; 2009.
29. Confidential Enquiry into Maternal and Child Health. *Saving Mothers' Lives: Reviewing Maternal Deaths to Make Motherhood Safer, 2003–2005. The Seventh Report of the Confidential Enquiries into Maternal Deaths in the United Kingdom*. London: CEMACH; 2007.
30. Centre for Maternal and Child Enquiries and the Royal College of Obstetricians and Gynaecologists. *CMACE/RCOG Joint Guideline: Management of Women with Obesity in Pregnancy*. London: CMACE/RCOG; 2010.
31. Freedman JM, Li D, Drasner K, Jaskela MC, Larsen B, Wi S. Transient neurologic symptoms after spinal anaesthesia. An epidemiologic study of 1863 patients. *Anaesthesiology*. 1998;89:633–641.
32. Heit JA, Kobbervig CE, James AH, Petterson TM, Bailey KR, Melton LJ. Trends in the incidence of venous thromboembolism during pregnancy or postpartum: a 30-year population-based study. *Ann Intern Med*. 2005;143(**10**): 697–706.
33. Nurmohamed MT, Rosendaal FR, Buller HR, et al. Low molecular weight heparin versus standard heparin in general and orthopaedic surgery: a meta-analysis. *Lancet*. 1992;340:152–156.
34. Weitz JI. Low-molecular-weight heparins. *N Engl J Med*. 1997;337:688–698.
35. Leizorovicz A, Haugh MC, Chapuis FR, Samama MM, Boissel JP. Low molecular weight heparin in prevention of perioperative thrombosis. *Br Med J*. 1992;305: 913–920.
36. Greer IA, Nelson-Piercy C. Low-molecular-weight heparins for thromboprophylaxis and treatment of venous thromboembolism in pregnancy: a systematic review of safety and efficacy. *Blood*. 2005;106:401–407.
37. Sanson BJ, Lensing AWA, Prins MH, et al. Safety of low-molecular-weight heparin in pregnancy: a systematic review. *Thromb Haemost*. 1999;81:668–672.
38. Warkentin TE, Levine MN, Hirsh J, et al. Heparin induced thrombocytopaenia in patients treated with low molecular weight heparin or unfractionated heparin. *N Engl J Med*. 1995;332:1330–1335.

39. Schulman S, Hellgren-Wangdahl M. Pregnancy, heparin and osteoporosis. *Thromb Haemost.* 2002;87:180–181.
40. Pettila V, Leinonen P, Markkola A, Hiilesmaa V, Kaaja R. Postpartum bone mineral density in women treated for thromboprophylaxis with unfractionated heparin or LMW heparin. *Thromb Haemost.* 2002;87:182–186.
41. Carlin AJ, Farquharson RG, Quenby SM, Topping J, Fraser WD. Prospective observational study of bone mineral density during pregnancy: low molecular weight heparin versus control. *Hum Reprod.* 2004;19:1211–1214.
42. Monreal M. Long-term treatment of venous thromboembolism with low-molecular weight heparin. *Curr Opin Pulm Med.* 2000;6:326–329.
43. Nelson-Piercy C, Powrie R, Borg JY, et al. Tinzaparin use in pregnancy: an international retrospective study of the safety and efficacy profile. *Eur J Obstet Gynecol Reprod Biol.* 2011;159:293–299.
44. Saivin S, Giroux M, Dumas JC, et al. Placental transfer of glycosaminoglycans in the human perfused cotyledon model. *Eur J Obstet Gynecol Reprod Biol.* 1991;42: 221–225.
45. Baglin T, Barrowcliffe TW, Cohen A, Greaves M, The British Committee for Standards in Haematology. Guidelines on the use and monitoring of heparin. *Br J Haematol.* 2006;6:326–329.
46. Hammerstingl C. Monitoring therapeutic anticoagulation with low molecular weight heparins: is it useful or misleading? *Cardiovasc Hematol Agents Med Chem.* 2008;6:282–286.
47. Holzgreve W, Carey JC, Hall BD. Warfarin-induced fetal abnormalities. *Lancet.* 1976;2:914–915.
48. Cotrufo M, De Feo M, De Santo LS, et al. Risk of warfarin during pregnancy with mechanical heart valve prostheses. *Obstet Gynecol.* 2002;99:35–40.
49. Nassar AH, Hobeika EM, Abd Essamad HM, Taher A, Khalil AM, Usta IM. Pregnancy outcome in women with prosthetic heart valves. *Am J Obstet Gynecol.* 2004;191: 1009–1013.
50. Wesseling J, Van Driel D, Heymans HS, et al. Coumarins during pregnancy: long-term effects on growth and development of school-age children. *Thromb Haemost.* 2001;85: 609–613.
51. Chan WS, Anad S, Ginsberg JS. Anticoagulation of pregnant women with mechanical heart valves. *Arch Intern Med.* 2000;160(**2**):191–196.
52. Schaefer C, Hannemann D, Elefant E, et al. *Thromb Haemost.* 2006;95(**6**):949–957.
53. Geerts WH, Bergqvist D, Pineo GF, et al. Prevention of venous thromboembolism. American College of Chest Physicians Evidence-Based Clinical Practice Guidelines (8th ed.). *Chest.* 2008;133:381–453.
54. Magnani HN. An analysis of clinical outcomes of 91 pregnancies in 83 women treated with danaparoid (Orgaran®). *Thromb Res.* 2010;125(4):297–302.
55. Royal College of Obstetricians and Gynaecologists. *The Acute Management of Thrombosis and Embolism During Pregnancy and the Puerperium. Green-Top Guideline No. 37b.* London: RCOG Press; 2007.
56. Scarsbrook AF, Evans AL, Owen AR, Gleeson FV. Diagnosis of suspected venous thromboembolic disease in pregnancy. *Clin Radiol.* 2006;61:1–12.
57. Thomson AJ, Greer IA. Non-haemorrhagic obstetric shock. *Ballieres Best Pract Res Clin Obstet Gynaecol.* 2000;14:19–41.
58. Francalanci I, Comeglio P, Alessandrello Liotta A, et al. D-dimer plasma levels during normal pregnancy measured by a specific ELISA. *Int J Clin Lab Res.* 1997;27:65–67.

59. Dolovich L, Ginsberg JS. Low molecular weight heparin in the treatment of venous thromboembolism: an updated mete-analysis. *Vessels*. 1997;3:4–11.
60. Gould MK, Dembitzer AD, Doyle RL, Hastie TJ, Garber AM. Low molecular weight heparins compared with unfractionated heparin for treatment of acute deep venous thrombosis. A meta-analysis of randomized controlled trials. *Ann Intern Med*. 1999;130: 800–809.

18 Obesity and pre-eclampsia

Fiona Broughton Pipkin and Pamela Loughna

Division of Obstetrics and Gynaecology, Faculty of Medicine and Health Sciences, University of Nottingham

It has been recognised since the mid-1990s that markers of cardiovascular disease (CVD) like abdominal adiposity, dyslipidaemia and hypertension in the non-pregnant state are also characteristic of, and predict, type-2 diabetes [1]. They are, in addition, pre-disposers for pre-eclampsia. Does this mean that gestational hypertension and pre-eclampsia are like gestational diabetes, in which pregnancy 'unmasks' non-pregnant diabetes? Is pregnancy simply unmasking (a tendency to) underlying CVD or are gestational hypertension and pre-eclampsia truly pregnancy-specific or is there a spectrum of 'gestational CVD' between a pure form and that driven by classic pre-disposers to CVD of non-pregnant state? In the context of this chapter, is the 'pre-eclampsia' of a woman with a booking body mass index (BMI) of 20 kg/m^2 the same disease as that in a woman with a BMI of 45 kg/m^2?

Definitions

Pre-eclampsia is defined as the occurrence of de novo hypertension (blood pressure ≥140/90 mmHg on two or more occasions at least 4 h apart) with proteinuria (>300 mg in 24 h or 2+ on urine dipstick) after 20 weeks gestation. Gestational hypertension is defined as the occurrence of de novo hypertension after 20 weeks of pregnancy without proteinuria. Pre-eclampsia is described as superimposed if it occurs in a woman known to have pre-existing hypertension or hypertension present in the first half of pregnancy (essential hypertension or secondary hypertension (i.e. with a known underlying cause)) [2].

Blood Pressure Measurement

The diagnosis of hypertension relies on the correct methodology for measuring blood pressure. It is essential that an appropriately sized cuff is used; the cuff bladder should encircle at least 80% of the arm. Given that the standard cuff is 23 cm long and the large 39 cm long, the maximum upper arm circumference for which a

Obesity. DOI: http://dx.doi.org/10.1016/B978-0-12-416045-3.00018-2

large cuff can be used is 48 cm and the maximum upper arm circumference for the standard cuff is 28 cm. As far back as 1989, the standard cuff only encircled 80% of the upper arm in 19% of male and female patients undergoing coronary angioplasty [3], and the incidence of obesity is now much greater. In a series of patients undergoing bariatric surgery with BMI of 66.7 ± 13.8 kg/m^2, the mean upper arm circumference was 48.6 ± 7.5 cm [4]. The large cuff was, therefore, too small for accurate blood pressure measurement in a significant proportion of these morbidly obese patients. It is no longer uncommon for pregnant women to be morbidly obese with an estimated prevalence of BMI of ≥ 50 kg/m^2 of 8.7/10,000 deliveries in the United Kingdom [5], which is nearly 1 in 1000. There is a direct relationship between BMI and upper arm circumference in the pregnant population [6], and, therefore, the increasing incidence of obesity will directly affect the accuracy of blood pressure measurement in pregnancy (and thus the diagnosis of pre-eclampsia and other hypertensive disorders) if the correct cuff size is not used. It may be that the current cuff used for measurement of blood pressure in the thigh (length 42 cm, maximum arm diameter 52 cm) [7] should be available in antenatal clinics, but even so it would be too small for some patients.

Urine Testing

It becomes increasingly difficult for women to collect a clean sample of urine when they are very obese and heavily pregnant. This inevitably leads to an increase in positive dip-stick testing, but over-diagnosis of proteinuria can be avoided by appropriate quantification either by protein:creatinine ratio or 24-h urine collection for protein quantification.

Changing Incidence of Obesity

There is no doubt that the incidence of obesity in both the population as a whole and in the pregnant population is increasing in the developed world. Reliable data on the BMI of women booking for antenatal care have been collected for a number of studies in the United Kingdom since 1992, mostly in the East Midlands. There has been a steady increase in the percentage of obese women, increasing from 9% in 1992 to 25% in 2008 (Figure 18.1). Our data did not record any women with a BMI of 50 kg/m^2 or more in 1992, whereas UK data obtained by the United Kingdom Obstetric Surveillance System (UKOSS) published in 2010 [5] showed an incidence of almost 1:1000, which was confirmed in the data submitted to UKOSS from Nottingham.

This increase in obesity has had a significant effect in the incidence of pregnancies complicated by both pre-existing and gestational diabetes. In the antenatal diabetic clinic in Nottingham, 26% of women in 2008 had a BMI between 25 and

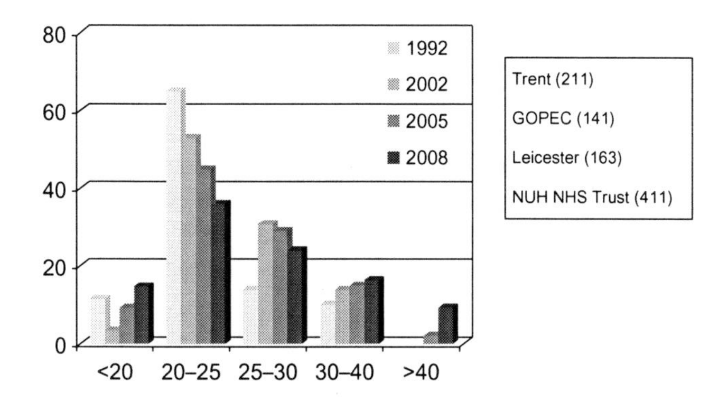

Figure 18.1 Frequency histograms of BMI (kg/m2) at antenatal booking between 1992 and 2008. There is a clear increase in obesity over this period. The emergence of an increasing proportion of young women with morbid obesity (>40 kg/m2) is particularly worrying. Trent = a health region in the East Midlands, UK, incorporating Leicestershire and Nottinghamshire; GOPEC = controls for genetics of pre-eclampsia study; NUH NHS Trust = Nottingham University Hospitals NHS Trust, Nottingham, UK.
Source: Loughna and Broughton Pipkin, unpublished data.

30 kg/m^2 with 41% being obese (BMI > 30 kg/m^2); 13% were morbidly obese (BMI > 40 kg/m^2).

The obese pregnant population has a high incidence of hypertension. A study carried out in Nottingham in 2008 showed that 47% of inpatients with a BMI of greater than 30 kg/m^2 had hypertension (blood pressure ≥ 140/90 mmHg) [8]. The rate of hypertension increased with BMI, with 8 of 11 women with a BMI of greater than 40 kg/m^2 having hypertension. This study did not differentiate between hypertension and pre-eclampsia. However, the UKOSS study on severe obesity (BMI ≥ 50 kg/m^2) demonstrated an incidence of pre-eclampsia of 9% in obese pregnant population in comparison to 2% of the non-obese controls [5]. This suggests a positive association between obesity and pre-eclampsia.

Does Obesity Predispose to Pre-Eclampsia?

A systematic review of 52 controlled studies published between 1966 and 2002 identified a raised BMI, at both pre-pregnancy and booking, as a consistently significant risk factor for pre-eclampsia. There was a clear gradation: nulliparae with a booking BMI < 20 kg/m^2 had a reduced risk (OR 0.76; 95% CI 0.62–0.92) relative to those of BMI between 20 and 24.9 kg/m^2, while those with a BMI ≥ 25 kg/m^2 had an increased risk (OR 9.3; 95% CI 2.0–48.1) [9]. A nested, case–control design that carefully studied the possible mediating factors in nulliparae at ≤20 weeks of gestation reported that about one-third of the total effect of BMI on the risk of pre-eclampsia was mediated through triglyceride levels and inflammation [10].

In order to implant, the blastocyst must erode the endometrial epithelium and replace and remodel the endothelium and spiral arteries to ensure an adequate foeto–placental blood supply. This tissue damage provokes an inflammatory response to allow the repair of the epithelium and removal of necrotic debris. The suggestion that pre-eclampsia is 'an excessive maternal inflammatory response to pregnancy' accompanied by a generalised activation of the circulating leucocytes was made more than a decade ago [11]. Obesity is a type of chronic inflammatory condition. Is an obese woman already 'primed' to have an exaggerated inflammatory response?

White adipose tissue is responsible for triglyceride storage and is much increased in obesity. The location of the fat deposits seems to be relevant with respect to long-term effects; insulin resistance and CVD are most strongly correlated with visceral white adipose tissue mass, particularly in women ([12]). During pregnancy, fat deposition is predominantly central, especially in already-obese women. These are women with an increased risk of gestational hypertension and late-onset pre-eclampsia. Visceral fat mass is strongly correlated with metabolic risk factors, such as blood pressure, insulin sensitivity, plasma lipids and hypertension, during pregnancy [13] as well as in the non-pregnant state.

Basal fat oxidation increases by at least 50% during pregnancy and there is a physiological hyperlipidaemic condition [14]. However, this is exaggerated in obese pregnant women, who have greater increases in total and very low density lipoprotein triglycerides and cholesterol and small dense low density lipoproteins (LDLs) and lower high density lipids. Early hypertriglyceridaemia is a feature of pre-eclampsia [14]. The small dense LDL particles are increased in both obesity and pre-eclampsia and are very atherogenic, promoting endothelial dysfunction and foam cell formation. It is possible that dyslipidaemia in obese women triggers the development of placental bed atherosis and pre-eclampsia. Perhaps, this dyslipidaemia, together with increased inflammation (see below), explains the strong epidemiological association of pre-eclampsia with pre-pregnancy BMI [15].

Obesity is also associated with a chronic inflammatory response, which is characterised by increased acute-phase reactants, abnormal cytokine production and activation of inflammatory signalling pathways. This is at least partly mediated through macrophage infiltration into adipose tissue [16]. These macrophages secrete a variety of cytokines, including interleukin-6 (IL-6) and tumour necrosis factor alpha (TNF-α). IL-6 is a pro-inflammatory cytokine, expressed and released into the circulation from adipose tissue, which stimulates acute-phase protein synthesis and low-grade systemic inflammation. A recent meta-analysis reported a very substantial increase in IL-6 in established pre-eclampsia [17].

Such inflammation can be assessed by measuring the serum concentration of the hepatic acute-phase C-reactive protein (CRP). Meta-analyses have shown that in healthy men and women, even concentrations of CRP well below the conventional limit of normality are associated with a two- to threefold increase in risk of peripheral arterial disease, myocardial infarction, ischaemic stroke, and coronary heart disease mortality [18]. High-sensitivity CRP (hsCRP) can be measured relatively easily and cheaply. Concentrations of hsCRP improve the prediction of risk for CVD in various populations, even when plasma lipid values are not particularly high.

CRP is a non-specific marker of inflammation. In a large ($n = 16,616$) study, it was found to be raised in 22% of men and 33% of women [18]. The gender difference was particularly striking in obesity, where obese men were twice as likely, but obese women six times more likely, to have raised concentrations of CRP than their normal BMI counterparts. Most interestingly in the context of pre-eclampsia, older (40 years and above) obese women were less likely to have elevated or clinically raised CRP levels than young obese women, an effect not seen in men. One may speculate that the raised CRP in younger women relates to the regular cycles of breakdown and repair of the endometrium since serum CRP rises by ~50% at the time of menstruation [19]. Overall, CRP rises progressively with age in women and is associated with their future risk of developing diabetes and metabolic syndrome [20]. For example, an 8-year prospective, follow-up study of nearly 15,000 women aged 45 years or more showed strong associations between the plasma concentration of CRP and the number of diagnostic components of the metabolic syndrome (upper-body obesity, hypertriglyceridaemia, decreased high density lipoprotein (HDL), hyperglycaemia and hypertension) [21]. Furthermore, the higher the CRP at diagnosis of metabolic syndrome, the higher is the risk of future cardiovascular events such as stroke, myocardial infarction or cardiovascular death. A recent systematic review reported that concentrations of hsCRP are strongly associated with the measures of abdominal adiposity even when BMI is controlled [22]. Are CRP concentrations raised in pre-eclampsia?

They certainly are in established pre-eclampsia, but is this cause or effect? A prospective study, in which CRP, TNF-α and plasminogen activator inhibitor-1 (PAI-1) were measured at 18 weeks' gestation in 71 obese women, who later developed pre-eclampsia, and in carefully matched controls reported no difference in any of these inflammatory markers. The mean BMI was only 23.4 kg/m^2 in both groups [23]. However, a subsequent prospective study, using samples taken at 13 weeks' gestation, did identify a significantly increasing trend in the risk of pre-eclampsia by tertile of CRP [24]; in this study, the effect was greatest in women whose BMI was <25 kg/m^2. Other prospective studies of women carried out between 15 and 22 weeks' gestation also reported a significant trend of increasing CRP from women who remained normotensive, through those who developed gestational hypertension to those who developed pre-eclampsia [25,26]. Thus, it seems probable that low-grade inflammation is already present in the early second trimester in women who go on to develop pre-eclampsia. What is driving this?

Adipokines

White adipose tissue is now known to be an endocrine organ in its own right. The adipocytes synthesise and release hormones (adipokines) such as leptin, PAI-1, TNF-α and adiponectin. They have many physiological actions, including effects on lipid metabolism, insulin sensitivity, haemostasis, blood pressure regulation and angiogenesis. All of these are altered during pregnancy.

Leptin is a polypeptide which is mainly synthesised in, and released from, adipose tissues. However, it is also expressed in several other tissues including the placenta. It binds to at least six leptin receptors and circulates in both bound and free forms. Clearance from the circulation is mainly through renal pathway. Interestingly, when matched for age and BMI, women have higher concentrations of leptin than men [27], which may peak in the luteal phase. Circulating concentrations of leptin are proportional to total body fat. It is one of the most important regulators of energy intake and expenditure and is synthesised in the syncytiotrophoblast as well as in adipocytes; leptin receptors are also present on the syncytiotrophoblast, where there may be cross-talk with insulin receptors. The maternal concentrations of leptin roughly double during the first trimester, and this rise is maintained until term [28]. Visceral fat is the main determinant of circulating maternal leptin in the first trimester of pregnancy [29]; therefore, leptin will be higher from early pregnancy in obese women (see above).

Leptin was originally thought to prevent obesity by inhibiting appetite, but obese individuals usually have raised concentrations of leptin. It seems likely that a state of 'leptin resistance' can develop, associated with abnormalities of signal transduction. Among other effects, this resistance decreases the activity of AMP-activated protein kinase and stimulates lipogenic enzymes such as acetyl co-A carboxylase and fatty acid synthase. These enzymes cause lipid oxidation, so there is then increased lipid build-up in muscle, liver and other tissues. Lipid oxidation is a feature of pre-eclampsia.

In human umbilical venous cells in vitro, leptin stimulates the accumulation of reactive oxygen species (ROS) through a number of pathways, including nuclear factor kappa beta (NF-κ B) [30]. The pattern of stimulation and accumulation is characteristic of changes in atherosclerotic vascular endothelial cells. Pre-eclampsia is now generally regarded as being a state of oxidative stress [31].

Adiponectin is a collagen-like adipokine which appears to be protective against vascular disease. It is positively associated with HDL cholesterol and negatively with triglycerides and LDL cholesterol. It has both anti-inflammatory and anti-oxidant properties. Concentrations of adiponectin fall in obesity and are inversely related to glucose and insulin concentrations, a mirror-image of what happens in relation to leptin. A large ($n = 1996$) study of healthy Finnish adults aged 24–39 years showed that a low serum adiponectin concentration was associated with increased carotid intima-media thickness and decreased brachial artery flow-mediated vasodilatation, even in multivariate models [32]. It has been suggested [33], on the basis of another large-scale study of young adults (33–45 years), that there may be an enhanced secretion of adiponectin in response to the metabolic environment found early in the development of vascular disease, possibly as a counter-regulatory response [33]. Concentrations of plasma adiponectin fall progressively during normal pregnancy, but there is no consensus as to whether they are increased or decreased in pre-eclampsia or related to insulin sensitivity [34].

TNF-α is part of the pathway of the systemic inflammatory response and has been linked to the development of insulin resistance, obesity and diabetes. TNF-α may contribute to insulin resistance by increasing the release of free fatty acids from adipocytes or by blocking the synthesis of adiponectin. It also activates NF-κ B, and hence increases the expression of adhesion molecules on endothelial cells and vascular smooth muscle cells. This contributes to an inflammatory state in adipose tissue, endothelial dysfunction, and, ultimately, atherogenesis. Excess macrophages in the decidua induce extravillous trophoblast apoptosis via TNF-α secretion, and this has been linked to the development of pre-eclampsia [35]. Serum concentrations of TNF-α and placental expression of the soluble TNF receptors are increased in pre-eclamptic pregnancy, although this may be cause rather than consequence [36]. Serum concentrations of TNF-α in obese women with and without pre-eclampsia were very similar [37].

Angiotensinogen (AGT) is the only precursor for angiotensin II (AngII). Although primarily synthesised in the liver, it is also synthesised by adipocytes [38] which contribute to circulating concentrations. Furthermore, these cells synthesise the other major components of the renin–angiotensin system (RAS) as well, so autocrine effects are likely. Blockade of the RAS with angiotensin type-1 receptor blockers such as losartan or irbesartan, in large-scale clinical trials (e.g. LIFE; Treat to Target), has consistently shown reductions in total cholesterol and triglycerides and increases in HDL cholesterol, the effects being significantly greater in patients with metabolic syndrome or insulin resistance [39]. AngII also regulates adiponectin secretion, which in turn regulates muscle fatty acid oxidation capacity. The chronic infusion of AngII in animal experiments leads to insulin resistance [40] via the generation of ROS. In human pregnancy, oxidative stress in pre-eclampsia is associated with the capacity for increased generation of AngI from AGT [41]. Hepatic AGT synthesis increases from early in gestation, and endometrial AGT expression is significantly increased in established pre-eclampsia [42], but the role of RAS of adipose tissue in relation to the changes in insulin resistance in pregnancy has not yet been systematically studied. In any case, RAS blockers are contraindicated in pregnancy.

Is 'Pre-Eclampsia' in the Obese Pregnant Women Really Underlying Chronic (Essential) Hypertension?

It has been exhaustively documented that obesity is associated with chronic hypertension. It is possible that some diagnoses of 'pre-eclampsia' in the obese in fact relate to the unmasking of underlying hypertension by pregnancy, in the same way that pregnancy can unmask underlying diabetes. There is some evidence to support this hypothesis. Phenotyping data from the GOPEC study [43] ($n = 994$ women with pre-eclampsia) are given in Table 18.1. Both the booking systolic and diastolic blood pressures show very clear increases ($P < 0.0001$) as the booking BMI rises from the women of normal weight to overweight to obese. The maximum blood pressures recorded before delivery are similar in the three groups, reflecting clinical practice

Table 18.1 The impact of body mass index on blood pressure, proteinuria and platelet count in women with stringently-defined pre-eclampsia

Diagnosis	BMI < 25 kg/m^2	BMI 25–29.9 kg/m^2	BMI ≥ 30 kg/m^2	Significance of Trend
n	357	343	292	
Booking BPS	112.6 ± 11.7	114.9 ± 12.5	120.6 ± 11.1	$P < 0.0001$
Booking BPD	66.8 ± 8.5	68.4 ± 8.7	72.4 ± 7.9	$P < 0.0001$
Maximum BPS	166.3 ± 17.3	166.1 ± 17.7	166.1 ± 18.0	n.s.
Maximum BPD	110.2 ± 9.5	111.4 ± 9.6	110.7 ± 9.2	n.s.
Maximum rise in BPS	53.7 ± 20.5	51.3 ± 21.0	45.4 ± 20.0	$P < 0.0001$
Maximum rise in BPD	43.5 ± 12.3	43.0 ± 12.4	38.2 ± 11.6	$P < 0.0001$
Maximum proteinuria (g/24 h)	1.5 [0.73.0] (n = 189)	1.9 [1.0–4.1] (n = 192)	1.8 [0.9–4.3] (n = 185)	n.s.
Minimum platelet count (10^9/l)	168.8 ± 64.7	177.6 ± 68.7	193.0 ± 70.5	$P < 0.0001$

There was a highly significant trend to increasing blood pressure at booking with BMI category in 994 women subsequently diagnosed with pre-eclampsia. Since the maximum recorded systolic and diastolic pressures were the same ($P > 0.9$; $P > 0.2$) across BMI category, the absolute rise in both was considerably greater in those with lower booking blood pressures ($P < 0.001$). The minimum platelet count recorded was significantly ($P < 0.001$) lower in women of lower BMI.
Maxima and minima are those routinely recorded during in-patient care. Different hospitals used different methods of recording urinary protein output; g/24 h was the most-frequently used.
Values are shown as mean ± standard deviation or median [25th–75th centiles] as appropriate. Trend was assessed using Kendall's τ.
BMI, body mass index; BPS, systolic blood pressure (mmHg); BPD, diastolic blood pressure (mmHg).
Source: The GOPEC study[43], unpublished data. FBP was CI on this study.

in the prompt delivery of women with this degree of hypertension. However, since the gestation age at delivery did not differ in the three groups, the women with lowest BMI had reached the same degree of hypertension in a shorter time, and the actual rise was, of course, much greater in those women ($P < 0.0001$). Conversely, the maximum urinary protein output (g/24 h) recorded was slightly lower in women of normal BMI ($P < 0.027$) than in women with higher BMI; obesity and the metabolic syndrome are associated with the progressive development of nephropathy [44]. The minimum platelet count recorded was significantly ($P < 0.0001$) lower in women of lower BMI. Thus 'pre-eclampsia' in this large sample is associated with subtle differences in presentation and, presumably, underlying disease. Furthermore, the incidence of pre-hypertension at a mean 11 weeks after delivery was clearly strongly related to build (Figure 18.2; $P < 0.0001$).

The women in the GOPEC study had stringently-defined pre-eclampsia. In an ongoing study in Nottingham, we are following-up women who have had carefully-differentiated gestational hypertension or pre-eclampsia. Data are currently available from 170 Caucasian women who had developed pre-eclampsia and 122 who had developed gestational hypertension. As given in Table 18.2, there was again a significant increase in booking systolic and diastolic pressures from normal weight to overweight to obese women, which was not seen in those who developed gestational hypertension, who had significantly higher systolic and diastolic pressures in each BMI category.

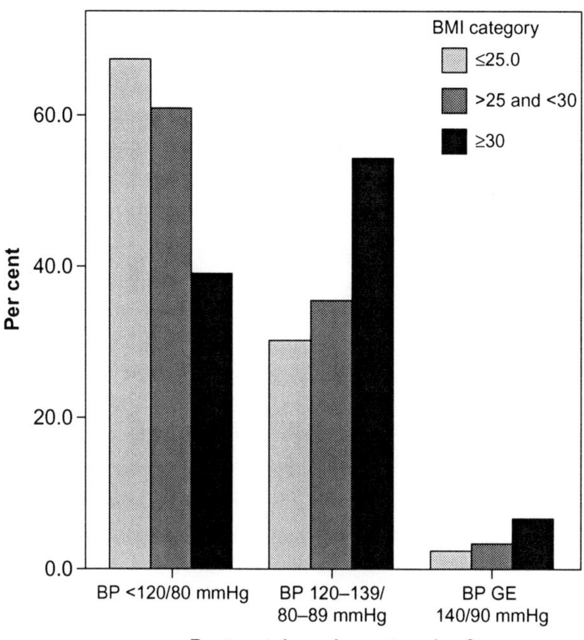

Figure 18.2 The interaction of build and the incidence of pre-hypertension 3 months after a pre-eclamptic pregnancy. $\chi^2 = 42.4$, $P < 0.0001$. BMI < 25 kg/m^2 – $n = 241$; 25–29.9 kg/m^2 – $n = 218$; ≥ 30 kg/m^2 – $n = 241$.
Source: The GOPEC study [43], unpublished data. FBP was CI on this study.

Table 18.2 The impact of body mass index on booking blood pressure in women who went on to develop either gestational hypertension or pre-eclampsia

Diagnosis	BMI <25 kg/m^2	BMI 25–29.9 kg/m^2	BMI ≥30 kg/m^2	Significance of Trend
Booking BPS–GH	127.9 ± 10.8*	129.9 ± 10.6*	127.6 ± 11.1	$P > 0.6$
Booking BPS–PE	117.9 ± 11.3	122.7 ± 13.9	124.7 ± 10.5	$P = 0.002$
Booking BPD–GH	77.7 ± 8.6	78.7 ± 7.7**	79.5 ± 7.9**	$P > 0.3$
Booking BPD–PE	71.9 ± 9.9	75.1 ± 8.4	77.6 ± 8.8	$P = 0.001$

There were significant trends to increasing blood pressure at booking with BMI category in 170 women who went on to develop pre-eclampsia, but not in 122 who developed gestational hypertension.
The number of normal weight, overweight and obese women with gestational hypertension was 39, 38, 45, respectively, while n was 75, 49, 44 in those who developed pre-eclampsia.
Values are shown as mean ± standard deviation. Trend was assessed using Kendall's τ.
BMI, body mass index; BPS, systolic blood pressure (mmHg); BPD diastolic blood pressure (mmHg); PE, pre-eclampsia; GH, gestational hypertension.
Green, A, Loughna, PV, Broughton Pipkin, F, Work in progress.
*$P < 0.005$ for difference between GH and PE.
**$P < 0.05$ for difference between GH and PE.

Final Considerations

Miscarriage is associated with over-expression of various pro-inflammatory cytokines, including IL-6 and TNF-α [45]. A recent systematic review [46] of obesity in relation to miscarriage of spontaneously-conceived pregnancy in 28,538 women concluded that there is also an increased risk of both miscarriage and recurrent miscarriage. This topic is covered in more detail in Chapter 9. The seeds of pre-eclampsia are sown in the first trimester, and women who suffer miscarriage in their first pregnancy are three times more likely to develop pre-eclampsia in their next continuing pregnancy than are those who carried a first pregnancy to term [47]. An exaggerated inflammatory response to very early pregnancy may underlie both miscarriage and pre-eclampsia, and be enhanced in obese women, who are already exposed to a hyper-inflammatory environment.

It is increasingly well documented that pre-eclampsia is associated with an increased risk of cardiovascular, renal and metabolic diseases in later life. It is perhaps less widely recognised that being born of a pre-eclamptic pregnancy is also associated with a greater risk of metabolic and CVD throughout life, so there is an impact on the baby as well. It remains to be seen whether obesity acts as an amplifier of these risks, and if so, whether it is additive or multiplicative. In any case, obese women who develop hypertension during pregnancy should be followed-up and counselled regarding weight loss and also about the importance of not allowing their babies to become obese. The post-natal check provides an excellent opportunity to do so.

References

1. Stern MP. Diabetes and cardiovascular disease. The 'common soil' hypothesis. *Diabetes*. 1995;44(4):369–374.
2. Brown MA, Lindheimer MD, de Swiet M, Van Assche A, Moutquin J-M. The classification of the hypertensive disorders of pregnancy: statement from the International Society for the Study of Hypertension in Pregnancy (ISSHP). *Hypertens Pregnancy*. 2001;20(1): ix–xiv.
3. Russell AE, Wing LM, Smith SA, et al. Optimal size of cuff bladder for indirect measurement of arterial pressure in adults. *J Hypertens*. 1989;7(8):607–613.
4. Hager H, Mandadi G, Pulley D, et al. A comparison of noninvasive blood pressure measurement on the wrist with invasive arterial blood pressure monitoring in patients undergoing bariatric surgery. *Obes Surg*. 2009;19(6):717–724.
5. Knight M, Kurinczuk JJ, Spark P, Brocklehurst P, UK Obstetric Surveillance System. Extreme obesity in pregnancy in the United Kingdom. *Obstet Gynecol*. 2010; 115(5):989–997.
6. Cooley SM, Donnelly JC, Walsh T, et al. The relationship between body mass index and mid-arm circumference in a pregnant population. *J Obset Gynaecol*. 2011; 31(7):594–596.

7. Pickering TG, Hall JE, Appel LJ, et al. Recommendations for blood pressure measurement in humans and experimental animals: part 1: blood pressure measurement in humans: a statement for professionals from the Subcommittee of Professional and Public Education of the American Heart Association Council on High Blood Pressure Research. *Hypertension*. 2005;45:142–161.
8. Shantikatara A. *The Incidence of Obesity in the Pregnant Population*. B med Sci Honours Project. Nottingham: University of Nottingham; 2008.
9. Duckitt K, Harrington D. Risk factors for pre-eclampsia at antenatal booking: systematic review of controlled studies. *BMJ*. 2005;330(7491):565.
10. Bodnar LM, et al. Inflammation and triglycerides partially mediate the effect of prepregnancy body mass index on the risk of preeclampsia. *Am J Epidemiol*. 2005; 162(12):1198–1206.
11. Redman C, Sacks G, Sargent I. Preeclampsia: an excessive maternal inflammatory response to pregnancy. *Am J Obstet Gynecol*. 1999;180(2 Pt 1):499–506.
12. Hanley A, Wagenknecht LE, Norris JM, et al. Insulin resistance, beta cell dysfunction and visceral adiposity as predictors of incident diabetes: the Insulin Resistance Atherosclerosis Study (IRAS) Family Study. *Diabetologia*. 2009;52(10):2079–2086.
13. Bartha JL, et al. Ultrasound evaluation of visceral fat and metabolic risk factors during early pregnancy. *Obesity*. 2007;15(9):2233–2239.
14. Nelson SM, Matthews P, Poston L. Maternal metabolism and obesity: modifiable determinants of pregnancy outcome. *Hum Reprod Update*. 2010;16(3):255–275.
15. Ovesen P, Rasmussen S, Kesmodel U. Effect of prepregnancy maternal overweight and obesity on pregnancy outcome. *Obstet Gynecol*. 2011;118(2 (Pt 1)):305–312.
16. Challier JC, et al. Obesity in pregnancy stimulates macrophage accumulation and inflammation in the placenta. *Placenta*. 2008;29(3):274–281.
17. Xie C, et al. A meta-analysis of tumor necrosis factor-alpha, interleukin-6, and interleukin-10 in preeclampsia. *Cytokine*. 2011;56(3):550–559.
18. Visser M, et al. Elevated C-reactive protein levels in overweight and obese adults. *JAMA*. 1999;282(22):2131–2135.
19. Gaskins AJ, et al. Whole grains are associated with serum concentrations of high sensitivity C-reactive protein among premenopausal women. *J Nutr*. 2010;140(9):1669–1676.
20. Onat A, Can G, Hergenç G. Serum C-reactive protein is an independent risk factor predicting cardiometabolic risk. *Metabolism*. 2008;57(2):207–214.
21. Ridker PM, et al. C-reactive protein, the metabolic syndrome, and risk of incident cardiovascular events. *Circulation*. 2003;107(3):391–397.
22. Brooks GC, Blaha MJ, Blumenthal RS. Relation of C-reactive protein to abdominal adiposity. *Am J Cardiol*. 2010;106(1):56–61.
23. Djurovic S, et al. Absence of enhanced systemic inflammatory response at 18 weeks of gestation in women with subsequent pre-eclampsia. *BJOG*. 2002;109(7):759–764.
24. Qiu C, et al. A prospective study of maternal serum C-reactive protein concentrations and risk of preeclampsia. *Am J Hypertens*. 2004;17(2):154–160.
25. Garciea RG, et al. Raised C-reactive protein and impaired flow-mediated vasodilation precede the development of preeclampsia. *Am J Hypertens*. 2007;20(1):98–103.
26. Thilaganathan B, et al. Early-pregnancy multiple serum markers and second-trimester uterine artery Doppler in predicting preeclampsia. *Obstet Gynecol*. 2010;115(6):1233–1238.
27. Mantzoros CS, Moschos SJ. Leptin: in search of role(s) in human physiology and pathophysiology. *Clin Endocrinol*. 1998;49(5):551–567.
28. Hauguel-de Mouzon S, Lepercq J, Catalano P. The known and unknown of leptin in pregnancy. *Am J Obstet Gynecol*. 2006;194(6):1537–1545.

29. Fattah C, et al. Maternal leptin and body composition in the first trimester of pregnancy. *Gynecol Endocrinol.* 2011;27(4):263–266.
30. Bouloumié A, et al. Leptin induces oxidative stress in human endothelial cells. *FASEB J.* 1999;13(10):1231–1238.
31. Cindrova-Davies T. Gabor Than Award Lecture 2008: pre-eclampsia – from placental oxidative stress to maternal endothelial dysfunction. *Placenta.* 2009;30(suppl A):55–65.
32. Saarikoski LA, et al. Adiponectin is related with carotid artery intima-media thickness and brachial flow-mediated dilatation in young adults – The Cardiovascular Risk in Young Finns Study. *Ann Med.* 2010;42(8):603–611.
33. Steffes MW, et al. Adiponectin, visceral fat, oxidative stress, and early macrovascular disease: the coronary artery risk development in young adults study. *Obesity.* 2006; 14(2):319–326.
34. Miehle K, Stepan H, Fasshauer M. Leptin, adiponectin and other adipokines in gestational diabetes mellitus and pre-eclampsia. *Clin Endocrinol.* 2012;76(1):2–11.
35. Reister F, et al. Macrophage-induced apoptosis limits endovascular trophoblast invasion in the uterine wall of preeclamptic women. *Lab Invest.* 2001;81(8):1143–1152.
36. Haider S, Knöfler M. Human tumour necrosis factor: physiological and pathological roles in placenta and endometrium. *Placenta.* 2009;30(2):111–123.
37. Founds SA, et al. A comparison of circulating TNF-α in obese and lean women with and without preeclampsia. *Hypertens Pregnancy.* 2008;27(1):39–48.
38. Engeli S, et al. The adipose-tissue renin–angiotensin–aldosterone system: role in the metabolic syndrome? *Int J Biochem Cell Biol.* 2003;35(6):807–825.
39. Putnam K, et al. The renin angiotensin system: a target of and contributor to dyslipidemias, altered glucose homeostasis and hypertension of the metabolic syndrome. *Am J Physiol Heart Circ Physiol.* 2012;302(6):H1219–30.
40. Olivares-Reyes JA, Arellano-Plancarte A, Castillo-Hernandez JR. Angiotensin II and the development of insulin resistance: implications for diabetes. *Mol Cell Endocrinol.* 2009;302(2):128–139.
41. Zhou A, et al. A redox switch in angiotensinogen modulates angiotensin release. *Nature.* 2010;468(7320):108–111.
42. Anton L, et al. The uterine placental bed renin–angiotensin system in normal and preeclamptic pregnancy. *Endocrinology.* 2009;150(9):4316–4325.
43. GOPEC Consortium. Disentangling fetal and maternal susceptibility for pre-eclampsia: a British multicenter candidate-gene study. *Am J Hum Genet.* 2005;77(1):127–131.
44. Griffin KA, Kramer H, Bidani AK. Adverse renal consequences of obesity. *Am J Physiol Ren Physiol.* 2008;294(4):F685–F696.
45. Calleja-Agius J, et al. Investigation of systemic inflammatory response in first trimester pregnancy failure. *Hum Reprod.* 2012;27(2):349–357.
46. Boots C, Stephenson MD. Does obesity increase the risk of miscarriage in spontaneous conception: a systematic review. *Semin Reprod Med.* 2011;29(06):507–513.
47. Bhattacharya S, et al. Does miscarriage in an initial pregnancy lead to adverse obstetric and perinatal outcomes in the next continuing pregnancy? *BJOG.* 2008; 115(13):1623–1629.

19 Obesity, Insulin Resistance and Placental Dysfunction – Foetal Growth

Anjum Doshani[1] *and Justin C. Konje*[2]

[1]University Hospitals of Leicester NHS trust, Leicester, UK,
[2]Reproductive Sciences Section, Robert Kilpatrick Clinical Sciences Building, Leicester Royal Infirmary, Leicester, UK

Introduction

Rates of maternal obesity have risen exponentially over the last 20 years, and it is currently estimated that more than one in five pregnant women are obese in the United Kingdom [1].

Pre-pregnancy obesity is associated with an increase in morbidity and mortality for both mother and offspring with pre-conceptual, antenatal, intra-partum and post-partum complications, e.g. gestational diabetes mellitus, pre-eclampsia, foetal macrosomia and late stillbirth [2].

The offsprings of obese women are two to four times more likely to be large for gestational age than the offsprings of their lean counterparts. They have a higher percentage of body fat rather than simply an increase in lean body mass [3].

There is also a sub-population of obese women in whom pregnancies fail to achieve a higher growth potential, resulting in foetal growth restriction, as shown by a Danish Birth Cohort study of stillbirths which identified a reduction in median birth weight of stillborn babies compared with live births in the obese women [4].

The reasons for aberrant foetal growth in both directions in offsprings of women with a high pre-pregnancy BMI are not well defined. Abnormal foetal growth is difficult to be accurately identified in obese women as current tools for assessing foetal size are based on populations prior to the obesity epidemic.

Obesity is associated with a significant increase in adipose tissue, with a disproportionately smaller expansion of other tissue types within the body. Fat is not just a storage organ, but it is a highly active tissue metabolically, comprising one of the largest endocrine organs in the body. In obesity, expansion of adipose tissue mass is associated with increasing inflammation of adipose tissue, and there is emerging evidence that this inflammation is causally linked both to insulin resistance and to other obesity-related morbidities such as cardiovascular disease.

Obesity. DOI: http://dx.doi.org/10.1016/B978-0-12-416045-3.00019-4

Table 19.1 Adipokines in Maternal Obesity Compared with Lean Non-Pregnant State

Adipokine	Lean Normal Pregnancy	Gestational Diabetes	Maternal Obese Pregnancy
Adiponectin	Decrease with gestation	↓	↓
Leptin	Increase to third trimester, then decrease at parturition	Unclear	↑
Insulin	Equal in first trimester, doubling by third trimester	↑↑	↑
Free IGF-1	Increase with gestation	↑	↑
Resistin	Increase with gestation	Unclear	Unclear
TNFα	Increase with gestation	↔↑	↔↑
IL-6	Increase with gestation	↑	↔↑

↑ = increased compared with lean pregnant women, ↔ = no change compared with lean pregnant women, ↓ = reduced compared with lean pregnant women.
Source: Adapted from Ref. [6].

Obesity and the Endocrine Environment

The hormonal milieu produced by white adipose tissue (WAT) (WAT volume is higher in obese compared to lean pregnancy women) includes higher circulating levels of leptin, tumour necrosis factor (TNF)-alpha, interleukins (IL) 1, 6 and 8 and reduced levels of adiponectin during pregnancy [5].

The potential consequences of these endocrine derangements in the structure and function of the placenta and their sequelae for foetal growth are given in Table 19.1 [6].

Placenta and Insulin Resistance

The placenta is a complex organ that fulfils pleiotropic roles during foetal growth. Because of its unique position (exposure to both maternal and foetal circulation), it is exposed to the regulatory influence of hormones, cytokines, growth factors and substrates present in both the foetal and maternal circulations and, hence, may be affected by changes in any of these. In turn, it can produce molecules that will affect the mother and the foetus independently.

The placenta is the key regulator of foetal growth due to its roles in supplying nutrients to the foetus and removing metabolic waste as well as hormone production. In foetal growth, restriction-specific placental phenotypes, size, morphology, blood flow and oxygen and nutrient transport mechanisms have been described [7].

The human placenta expresses virtually all known cytokines including TNF, resistin and leptin, which are also produced by adipose cells. The discovery that some of these adipokines are key players in the regulation of insulin action suggests possible novel interactions between the placenta and adipose tissue in understanding the pregnancy-induced insulin resistance.

The current view is that the abnormal maternal metabolic environment may generate stimuli within the adipose tissue and the placental cells, resulting in an increased production of inflammatory cytokines whose expression is minimal in normal pregnancy. One leading hypothesis is that changes in circulating TNF–adiponectin, leptin and resistin link inflammation to metabolic changes by enhancing insulin resistance in the mother. Declining adiponectin levels are observed in maternal obesity, leading to an increase in insulin resistance [8]. Concentrations of leptin, an adipocyte-derived molecule, were more than twofold higher in the obese women [9].

Mohamed-Ali et al. [10] have shown high concentrations of IL-6 and C-reactive protein (CRP) in obese pregnant women implicating adiposity as a key factor in low-grade chronic inflammation. They found that IL-6 was 50% higher, and CRP was almost double, in the obese pregnant women, relative to the lean group. In non-pregnant populations, high CRP correlates with endothelial dysfunction and impaired insulin sensitivity and predicts the risk for type-2 diabetes [11]. Thus, obesity-driven inflammation in pregnancy could be implicated in the pathogenesis of pre-eclampsia and gestational diabetes [12,13].

Adipokines also modulate placental function. For example, leptin has been shown to regulate placental angiogenesis [14,15], protein synthesis [16] and growth, and cause immunomodulation [17]. Thus, it is possible that the increase in local and/or circulating levels of leptin in maternal obesity may modulate placental inflammation and function.

A degree of relative leptin and insulin resistance develops in normal pregnancy; this is thought to promote constant nutrient delivery to the foetus; however, in obese women, a higher circulating concentration of leptin and insulin enhances the physiological adaptation of pregnancy leading to the development of impaired glucose tolerance or gestational diabetes.

Chronic hyperstimulation of leptin receptors, as found in obesity, leads to leptin resistance. This may affect the leptin-dependent processes of the syncytiotrophoblast and possibly foetal tissues, resulting in increased nutrient intake due to leptin insensitivity [9].

Free Insulin like growth factor 1 (IGF-1) concentrations are increased in obesity as a result of decreased binding protein; these together with insulin are important mediators of foetal growth and have been implicated in foetal overgrowth [5].

Placental Size

Observational studies have noted that as BMI increases, the average placental weight increases [18].

However, there are no robust data of foetal/placental weight ratios, and further research is needed to establish the placental exchange surface and the effect of cytokines on this and its effect on permeability, exchange and apoptosis.

Hypo- or hypercoiling of the umbilical cord, associated with foetal demise and vascular thrombosis [19], is also more common in pregnancies complicated by obesity, gestational diabetes and pre-eclampsia [20]. Although all these pregnancy complications are associated with increased inflammation, a causal link between inflammation and cord-coiling abnormalities is yet to be established.

Placental Transporter System

Maternal obesity may also affect placental transport and substrate availability.

At the microscopic level, there are no detailed studies examining the effect of maternal obesity on placental structure. However, changes in the activity of transporters on membranes such as microvillous and basal plasma membranes of the syncytiotrophoblast have been reported in women with diabetes and associated macrosomia [21].

Although placentae from gestational diabetics have a characteristic morphology (placental immaturity and oedema, chorangiosis and vascular anomalies) [22], it is plausible that obesity may also affect placental structure. A study by Challier et al. [23] demonstrated a two- to threefold increase in placental macrophages in obese women compared with non-obese women. The macrophage population was

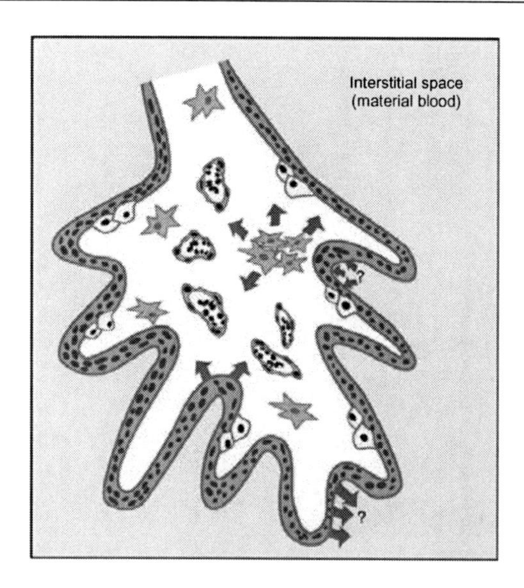

Figure 19.1 Inflammatory events within the placenta of obese pregnant women. Increased levels of placental macrophages (Hofbauer cells) in conjunction with the cytotrophoblast and syncytiotrophoblast cells are thought to contribute to a heightened inflammatory state within the placenta through the secretion of proinflammatory cytokines such as IL-1, TNF and IL-6 (as shown by arrows). An interplay between the trophoblast and immune cells may exist which could further contribute to the heightened inflammatory state of the placenta. It is unclear exactly which proinflammatory cytokines pass through the placenta barrier. However, transfer of proinflammatory cytokines from the maternal circulation across to the placenta could potentially add to the increased local inflammatory state within the placenta. Similarly, proinflammatory cytokines from the placenta may be released into the maternal circulation and contribute to the systemic increase in inflammatory mediators, which is observed in maternal obesity.
Source: From Denison FC, et al. [24].

characterised by the increased expression of the proinflammatory cytokines IL-1, TNF and IL-6 (Figure 19.1) [24].

In neonates, there is an increase in the expression of the phospholipase A2 (PLA2) genes *PLA2G2A* and *PLA2G5* (the main placenta phospholipases), leptin and TNF (body fat $> 16\%$) compared with lean neonates (body fat $< 8\%$) [25].

Amino acid transport may also be affected by maternal obesity. Physiological concentrations of the proinflammatory cytokines IL-6 and TNF stimulate the activity of the amino acid transporter system A [26]. In maternal obesity, increased levels of IL-6 and TNF in the placenta could systemically stimulate the system A transporter further, thus increasing the amino acid transport to the foetus.

Robust studies are now needed to investigate whether there is evidence of altered placental transporter expression and/or activity in obese pregnant women and whether such change correlates with foetal size.

Placenta and Inflammatory Response

Although the association between obesity, low-grade inflammation and insulin resistance is propagated in part by adipokines generated by adipocytes and cells of the stromal vascular fraction, an influx of inflammatory cells into adipose tissue is also implicated [27].

Originally, it was thought that cells of the innate immune system were the major players in this event, but increasing evidence implicates that T cells and possibly other leucocytes may be involved [28].

However, the predominant leucocytes in adipose tissues are macrophages, which are present in the stromal vascular fraction, and their density is in direct proportion to the levels of obesity [29].

Indeed, there is a significant functional overlap between adipose tissue and macrophage biology, both being phagocytic in nature, with some evidence that pre-adipocytes can be 'converted' to macrophages [27].

Prothrombotic State

Obesity is also associated with prothrombotic state with plasma concentrations of prothrombotic factors including von Willebrand factor, fibrinogen and factor VII being higher in obese controls compared with lean controls [30].

Excess adipose tissue contributes directly to the prothrombotic state by (i) impairing platelet function via low-grade inflammation and increasing in circulating leptin, (ii) impairing fibrinolysis by production of plasminogen activator inhibitor-1 and possibly thrombin-activatable fibrinolysis inhibitor, (iii) impairing coagulation by release of tissue factor and (iv) affecting hepatic synthesis of the coagulation factors fibrinogen, factor VII, factor VIII and tissue factor, by releasing free fatty acids and proinflammatory cytokines (IL-1B, TNF and IL-6) into the portal circulation and by inducing hepatic insulin resistance [30].

Diabetes and the Placenta

In pregnancy, obesity and particularly abdominal obesity is associated with glucose intolerance and insulin resistance [31] with the result that gestational diabetes is more common in obese pregnant women.

Diabetic insults at the beginning of gestation, as in many pre-gestational diabetic pregnancies, may have long-term effects on placental development. These adaptive responses of the placenta to the diabetic environment, such as buffering excess maternal glucose or increased vascular resistance, may help limit foetal growth within a normal range. If the duration or extent of the diabetic insult, including maternal hyperglycaemia, hyperinsulinaemia or dyslipidaemia, exceeds the placental capacity to mount adequate responses, then excessive foetal growth may ensue.

Diabetic insult at the later stages of gestation, such as that in gestational diabetes, will foremost lead to short-term changes in a variety of molecules involved in key functions including gene expression [32].

Insulin stimulates foetal aerobic glucose metabolism and hence increases the oxygen demand of the foetus. If adequate supply is not available because of the reduced oxygen delivery to the intervillous space, as a result of the higher oxygen affinity of glycated haemoglobin [32], thickening of the placental basement membrane [33] and reduced utero-placental or foetoplacental blood flow [34], then foetal hypoxaemia will ensue. Hypoxia is a potent stimulator of hypoxia-sensitive transcription factors which will, therefore, lead to the stimulated expression and synthesis of a variety of molecules, some of which are key players, especially in angiogenesis [35].

In women with type-1 diabetes, there is an upregulation of placental glycosylation and acylation pathways and a trend towards increased placental weight [32].

The Intrauterine Growth-Restricted Foetus

Despite the higher circulating concentrations of nutrients leading to enhanced placental transfer, not all pregnancies in obese women result in macrosomic babies. Indeed, a small number of idiopathic foetal growth-restricted babies are observed in obese pregnant women.

Theories why this may occur are varied but include the fact that obese women may lack one or more major- or micronutrients essential for foetal growth, despite overnutrition. Obese women may also limit their calorie intake; thus, the consumed nutrient may be directed to supporting maternal metabolism rather than to foetal growth [36].

Lappas and co-workers [37] have suggested a post-receptor insulin signalling defect, which may reduce the insulin-induced stimulation of placental transport systems, such as system A, resulting in reduced nutrient transfer and foetal growth restriction.

In obese individuals, local and systemic vascular and endothelial functions are significantly impaired.

In foetal growth restriction, failure of trophoblastic invasion and the resultant inadequate spiral artery transformation occur thus limiting the maternal blood flow to the placenta. However, there are no published studies that examined the process of trophoblast invasion in relation to foetal growth in obese pregnant women. In theory, foetal overgrowth may be associated with highly successful trophoblast invasion, increasing substrate transfer by overcoming flow limitations and providing substrates for placental transport processes. Trophoblast invasion may be disturbed in that subset of obese women delivering growth-restricted babies.

Blood vessel structure is altered in obesity with an increase in vessel diameter, basement membrane thickness, vascular permeability and vessel stiffness [38].

With disease progression, microvascular vessel walls start to atrophy, vessel diameter narrows and progressive microvascular rarefaction develops, increasing the risk of local tissue ischaemia.

Adipose tissue, which surrounds blood vessels (perivascular adipose tissue), also indirectly affects vascular structure and tone via the release of vasoactive inflammatory mediators including adipokines, angiotensin and endothelin-1 [39].

At a functional level, the vasodilatory response to endothelium-dependent vasodilators, such as acetylcholine, is attenuated [40], whereas the response to endothelium-independent vasodilators, such as sodium nitroprusside, remains intact.

Vasomotor responses are further blunted by an increase in sensitivity to vasoconstrictive agonists including prostanoids, endothelin-1 and a hyperactivity of the sympathetic nervous system [41].

Conclusions

Over the last decade, significant advances in our understanding of placental biology have established that placental function is dynamic and influenced by maternal health and has an important regulatory role in maternal well-being during pregnancy. Given that maternal obesity is accompanied by significant dysregulation of normal physiology, it is, therefore, plausible that placental structure and function may be altered as a consequence of maternal obesity, and equally, that the placenta may modulate maternal physiology by releasing inflammatory cytokines.

References

1. Heslehurst N, Ells LJ, Simpson H, Batterham A, Wilkinson J, Summerbell CD. Trends in maternal obesity incidence rates, demographic predictors, and health inequalities in 36,821 women over a 15 year period. *BJOG.* 2007;114(2):187–194.
2. Sebire NJ, Jolly M, Harris JP, et al. Maternal obesity and pregnancy outcome: a study of 287,213 pregnancies in London. *Int J Obes Relat Metab Disord.* 2001;25(8):1175–1182.
3. Sewell MF, Huston-Presley L, Super DM, Catalano P. Increased neonatal fat mass, not lean body mass is associated with maternal obesity. *Am J Obstet Gynaecol.* 2006; 195(4):1100–1103.
4. Nohr EA, Bech BH, Davies MJ, Frydenberg M, Henriksen TB, Olsen J. Pre-pregnancy obesity and fetal death: a study within the Danish National Birth Cohort. *Obstet Gynaecol.* 2005;106(2):250–259.
5. Jansson N, Nilsfelt A, Gellerstedt M, et al. Maternal hormones linking maternal body mass index and dietary intake to birth weight. *Am J Clin Nutr.* 2008;87(6):1743–1749.
6. Higgins L, Greenwood SL, Wareing M, Sibley CP, Mills TA. Obesity and the placenta: a consideration of nutrient exchange mechanisms in relation to aberrant fetal growth. *Placenta.* 2011;32(1):1–7.
7. Sibley CP, Turner MA, Cetin I, et al. Placental phenotypes of intrauterine growth. *Paediatr Res.* 2005;58(5):827–832.

8. Ziemke F, Mantzoros CS. Adiponectin in insulin resistance: lessons from translational research. *Am J Clin Nutr*. 2010;91(1):258S–261S.
9. McConway MG, Johnson D, Kelly A, Griffen D, Smith J, Wallace AM. Differences in circulating concentrations of total, free and bound leptin relate to gender and body composition in adult humans. *Ann Clin Biochem*. 2000;37:717–723.
10. Mohamed-Ali V, Goodrick S, Rawesh A, et al. Subcutaneous adipose tissue releases interleukin-6, but not tumor necrosis factor-alpha, in vivo. *J Clin Endocrinol Metab*. 1997;82:4196–4200.
11. Festa A, D'Agostino Jr R, Howard G, Mykkanen L, Tracy RP, Haffner SM. Chronic subclinical inflammation as part of the insulin resistance syndrome: the Insulin Resistance Atherosclerosis Study (IRAS). *Circulation*. 2000;102:42–47.
12. Pradhan AD, Manson JE, Rifai N, Buring JE, Ridker PM. C-reactive protein, interleukin 6, and risk of developing type 2 diabetes mellitus. *JAMA*. 2001;286: 327–334.
13. Austgulen R, Lien E, Vince G, Redman CW. Increased maternal plasma levels of soluble adhesion molecules (ICAM-1, VCAM-1, E-selectin) in preeclampsia. *Eur J Obstet Gynecol Reprod Biol*. 1997;71:53–58.
14. Zavalza-Gomez AB, Anaya-Prado R, Rincon-Sanchez AR, Mora-Martinez JM. Adipokines and insulin resistance during pregnancy. *Diabetes Res Clin Pract*. 2008; 80(1):8–15.
15. Islami D, Bischof P, Chardonnens D. Modulation of placental vascular endothelial growth factor by leptin and hCG. *Mol Hum Reprod*. 2003;9:395–398.
16. Perez-Perez A, Maymo J, Gambino Y, et al. Leptin stimulates protein synthesis-activating translation machinery in human trophoblastic cells. *Biol Reprod*. 2009; 81:826–832.
17. Fietta P. Focus on leptin, a pleiotropic hormone. *Minerva Med*. 2005;96:65–75.
18. Swanson LD, Bewtra C. Increase in normal placental weights related to increase in maternal body mass index. *J Matern Fetal Neonatal Med*. 2008;21:111–113.
19. Sebire NJ. Pathophysiological significance of abnormal umbilical cord coiling index. *Ultrasound Obstet Gynecol*. 2007;30:804–806.
20. de Laat MW, Franx A, van Alderen ED, Nikkels PG, Visser GH. The umbilical coiling index: a review of the literature. *J Matern Fetal Neonatal Med*. 2005;17:93–100.
21. Jansson T, Cetin I, Powell TL, et al. Placental transport and metabolism in fetal overgrowth – a workshop report. *Placenta*. 2006;27(suppl A):S109–S113.
22. Madazli R, Tuten A, Calay Z, Uzun H, Uludag S, Ocak V. The incidence of placental abnormalities, maternal and cord plasma malondialdehyde and vascular endothelial growth factor levels in women with gestational diabetes mellitus and nondiabetic controls. *Gynecol Obstet Invest*. 2008;65:227–232.
23. Challier JC, Basu S, Bintein T, et al. Obesity in pregnancy stimulates macrophage accumulation and inflammation in the placenta. *Placenta*. 2008;29:274–281.
24. Denison FC, Roberts KA, Barr SM, Norman JE. Obesity, pregnancy, inflammation and vascular function. *Reproduction*. 2010;140:373–385.
25. Varastehpour A, Radaelli T, Minium J, et al. Activation of phospholipase A2 is associated with generation of placental lipid signals and fetal obesity. *J Clin Endocrinol Metab*. 2006;91:248–255.
26. Jones HN, Jansson T, Powell TL. IL-6 stimulates system A amino acid transporter activity in trophoblast cells through STAT3 and increased expression of SNAT2. *Am J Physiol Cell Physiol*. 2009;297:C1228–C1235.
27. Hotamisligil GS. Inflammation and metabolic disorders. *Nature*. 2006;444:860–867.

28. Lumeng CN, Maillard I, Saltiel AR. T-ing up inflammation in fat. *Nat Med.* 2009;15:846–847.
29. Weisberg SP, McCann D, Desai M, Rosenbaum M, Leibel RL, Ferrante Jr AW. Obesity is associated with macrophage accumulation in adipose tissue. *Eur J Clin Invest.* 2003;112:1796–1808.
30. Faber DR, de Groot PG, Visseren FL. Role of adipose tissue in haemostasis, coagulation and fibrinolysis. *Obes Rev.* 2009;10:554–563.
31. Martin AM, Berger H, Nisenbaum R, et al. Abdominal visceral adiposity in the first trimester predicts glucose intolerance in later pregnancy. *Diabetes Care.* 2009;32:1308–1310.
32. Radaelli T, Varastehpour A, Catalano P, Hauguel-de Mouzon S. Gestational diabetes induces placental genes for chronic stress and inflammatory pathways. *Diabetes.* 2003;52:2951–2958.
33. Al-Okail MS, al-Attas OS. Histological changes in placental syncytiotrophoblasts of poorly controlled gestational diabetic patients. *Endocr J.* 1994;41:355–360.
34. Fadda GM, D'Antona D, Ambrosini G, et al. Placental and fetal pulsatility indices in gestational diabetes mellitus. *J Reprod Med.* 2001;46:365–370.
35. Lolmede K, Durand de Saint Front V, Galitzky J, Lafontan M, Bouloumie A. Effects of hypoxia on the expression of proangiogenic factors in differentiated 3T3–F442A adipocytes. *Int J Obes Relat Metab Disord.* 2003;27:1187–1195.
36. Kiel DW, Dodson EA, Artal R, Boehmer TK, Leet TL. Gestational weight gain and pregnancy outcomes in obese women: how much is enough? *Obstet Gynecol.* 2007; 110(4):752–758.
37. Colomiere M, Permezel M, Riley C, Desoye G, Lappas M. Defective insulin signaling in placenta from pregnancies complicated by gestational diabetes mellitus. *Eur J Endocrinol.* 2009;160(4):567–578.
38. Zebekakis PE, Nawrot T, Thijs L, et al. Obesity is associated with increased arterial stiffness from adolescence until old age. *J Hypertens.* 2005;23:1839–1846.
39. Zhang H, Zhang C. Regulation of microvascular function by adipose tissue in obesity and type 2 diabetes: evidence of an adipose-vascular loop. *Am J Biomed Sci.* 2009;1:133–142.
40. Steinberg HO, Chaker H, Leaming R, Johnson A, Brechtel G, Baron AD. Obesity/insulin resistance is associated with endothelial dysfunction. Implications for the syndrome of insulin resistance. *J Clin Invest.* 1996;97:2601–2610.
41. Traupe T, Lang M, Goettsch W, et al. Obesity increases prostanoid-mediated vasoconstriction and vascular thromboxane receptor gene expression. *J Hypertens.* 2002;20:2239–2245.

Section 5

Obesity and Gestational Diabetes

20 Screening for Gestational Diabetes

Peter Hornnes and Jeannet Lauenborg

Department of Obstetrics and Gynaecology, Hvidovre University Hospital, Copenhagen, Denmark

Introduction

Gestational diabetes mellitus (GDM) is defined as carbohydrate intolerance resulting in hyperglycaemia of variable severity with onset or first recognition during pregnancy [1]. Thus, GDM includes women who developed diabetes mellitus before pregnancy but only diagnosed during pregnancy and women who developed diabetes mellitus during pregnancy.

Previously, it was customary to differentiate between severity grades of GDM, the less severe grades being called impaired glucose tolerance and the more severe being called GDM. Recent studies have, however, demonstrated a linear correlation between glucose levels in pregnancy and the risk of complications, indicating that no specific critical threshold levels exist [2–4]. In 2010, the International Association of Diabetes and Pregnancy Study Groups (IADPSG) extended the definition of GDM to also include a definition for overt diabetes diagnosed in pregnancy [5].

GDM is estimated to affect up to 14% of pregnancies depending on the criteria used for screening and the population screened [6]. GDM is an asymptomatic condition and is associated with adverse outcome for both the mother and the child. In the majority of affected women, abnormal glucose metabolism disappears soon after delivery [7]. The diagnosis labels the mother and her child as high-risk subjects for developing diabetes and cardiovascular disease later in life. A history of GDM is one of the strongest predictors for later development of type-2 diabetes [8]. A strategy for screening for GDM has also long-term implications, as treatment during pregnancy and lifestyle intervention after pregnancy can delay or even prevent the onset of disease [6].

Overweight as a Risk Factor for GDM

Obesity is one of the most frequent risk factors for GDM. However, obesity has, as a single risk factor, a low predictive value [3]. Yet, the number of women with

Obesity. DOI: http://dx.doi.org/10.1016/B978-0-12-416045-3.00020-0

GDM is increasing throughout the world in the wake of the increase in the incidence of obesity also among pregnant women [9]. In a meta-analysis, Chu et al. [10] found an unadjusted odds ratio (OR) of 2.1 for GDM in overweight, 3.6 in obese and 8.6 in severely obese women.

A cohort study carried out among 370,000 Danish women, who gave birth from 2004 to 2010, found that 20.9% were overweight (BMI 25–29.9 kg/m^2), 7.7% were obese (BMI 30–35 kg/m^2) and 4% were severely obese (BMI > 35 kg/m^2). The risk of GDM increased with body weight; compared to the normal-weight women, the adjusted OR was 3.42 (3.23–3.63) for overweight women, 7.54 (7.09–8.03) for obese women and 10.83 (10.10–11.61) for severely obese women [11]. The absolute risk for GDM was only 0.9% for normal-weight women, 3.1% for overweight women, 6.7% for obese women and 9.3% for severely obese women. In addition, the increasing BMI conferred an increased risk of giving birth to a macrosomic infant >4500 g, OR for overweight women being 1.6, obese women 2.2 and severely obese women 2.7.

Pathophysiology

Even normal pregnancy is associated with insulin resistance developing during the second trimester and becoming progressively more pronounced towards term. Therefore, maternal post-prandial glucose levels are higher in normal pregnancy than in non-pregnant women. In contrast to insulin, glucose is readily transported across the placenta, and in the foetus, glucose stimulates the production of insulin, which acts as an influential growth factor. In GDM, however, insulin resistance is excessively pronounced resulting in further elevated maternal blood glucose levels [12,13]. In the foetus, the elevated glucose levels result in a further increased insulin production, and the ensuing foetal hyperinsulinaemia is responsible for excessive foetal growth particularly of adipose tissue. During pregnancy, the foetus has an increased risk of intrauterine death and polycytaemia, and during delivery, the overweight infant of a GDM mother suffers an increased risk of complications and birth trauma such as shoulder dystocia, hypoxaemia and the need for Caesarean section [14].

GDM mimics the abnormalities in glucose metabolism in type-2 diabetes regarding defects in the insulin secretory response. The metabolic stress of pregnancy may unmask the defects in women genetically pre-disposed to type-2 diabetes resulting in GDM [7]. In fact, the majority of women developing diabetes after pregnancy will have type-2 diabetes [15].

In obese pregnant women, insulin resistance is even more pronounced as obesity itself is an insulin-resistant state. This may favour an inappropriate foetal growth in obese women compared to normal-weight pregnant women due to higher serum lipid levels in early pregnancy [16]. Obese pregnant women also differ in insulin response to an intravenous glucose challenge. In lean women with GDM, the first-phase insulin response is decreased compared with weight-matched pregnant women without GDM [17]. This could not be found in obese pregnant women with GDM, who instead demonstrated an increase in second-phase insulin response [7].

Diagnosis of GDM

The diagnosis of GDM is based on an oral glucose tolerance test (OGTT). The test should be performed in the morning after an overnight fast of 8–14 h and after at least 3 days of unrestricted diet and physical activity. Several diagnostic tests and criteria for screening for gestational diabetes have been employed. Table 20.1 presents different tests and thresholds. The tests differ regarding glucose load (75 or 100 g), duration (2 or 3 h), number of sampling times (1–4) and specimen (venous plasma or capillary whole blood) [3–5,18–21]. Thresholds for the diagnosis have been based either on pregnancy outcome [22] or on long-term consequences of the mother [23]. Several studies have documented that there is no clear threshold for having or not having GDM [18]. The association between blood glucose levels and different adverse outcomes is linear and gradual. The higher the glucose values the higher the risk for an adverse outcome. The cut-off values are, therefore, based on consensus decisions [5,6,20,21]. No test has been found superior when it comes to the improvement of maternal and foetal health [24].

Due to different tests and criteria, comparison between various screening programmes regarding the effect of screening, treatment and intervention can be difficult. With the publication of a new set of screening criteria in 2010 by the IADPSG, the organisation intended to introduce a set of internationally accepted criteria. The HAPO study, on which these criteria were based, is the largest study designed to examine the risk for adverse pregnancy outcomes at different maternal glucose levels and included 25,000 pregnant women from 15 centres in nine countries. The diagnostic glucose levels were set aiming to identify an increased risk that reached 1.75 times the estimated odds of different adverse outcomes (birth weight, per cent body fat and C-peptide above 90th percentiles) at mean glucose values.

According to IADPSG, overt diabetes is diagnosed in pregnancy if one meets the following criteria: fasting plasma glucose ≥7.0 mmol/l (126 mg/dl), HbA1c ≥ 6.5% (DCCT/UKPDS standardised) or a random plasma glucose ≥11.1 mmol/l (200 mg/ml). The latter should be confirmed by fasting plasma glucose or HbA1c. In the HAPO study, 1.7% of the participants were diagnosed with overt diabetes due to a fasting plasma glucose >5.8 mmol/l (105 mg/dl) and a 2 h OGTT value > 11.1 mmol/l (200 mg/dl) [5].

Several countries have adopted the IADPSG screening programme already. However, some resistance is encountered due to the expected increase in the number of pregnant women diagnosed with GDM. In the HAPO population, 16.1% met the screening criteria. Half of them were diagnosed because of the fasting blood glucose [2], and half were diagnosed on the glucose levels during the OGTT. An additional 1.7% was diagnosed with overt diabetes, resulting in a total of 17.8% diagnosed with diabetes in pregnancy [21]. Another argument against the IADPSG

Table 20.1 Different Diagnostic Tests and Criteria for Screening Diabetes in Pregnancy

Test	Population Screened with OGTT	Glucose Load (g)	Blood Sample	Fasting mmol/l (mg/dl)	1 h mmol/l (mg/dl)	2 h mmol/l (mg/dl)	3 h mmol/l (mg/dl)	Diagnostic Criteria
IADPSG + ADA [2,21]	Early screening for overt diabetes. All women without known diabetes screening at GA 24–28 weeks	75	Venous plasma	5.1 (92)	10.0 (180)	8.5 (153)	–	One value met or exceeded
NICE/WHO [20]	Risk factors,[a] screening at GA 24–28 weeks	75	Venous plasma	7.0 (126)	–	7.8 (140)	–	One value met or exceeded
Denmark [3]	Risk factors,[b] in GA 14–20 weeks and/or GA 27–30 weeks	75	Capillary whole blood	–	–	8.9 (160)	–	Value met or exceeded
Sweden [4]	All women in week 28, week 12 if risk factors[c]	75	Capillary whole blood	–	–	9.0 (162)	–	Repeated once if 7.8–8.9 (140–160)
NDDG [18]	Risk factors,[d] in third trimester	100	Venous plasma	5.8 (105)	10.6 (190)	9.2 (165)	8.1 (145)	Two or more values met or exceeded
Carpenter and Coustan [19]	Positive GCT	100	Venous plasma	5.3 (95)	10.0 (180)	8.6 (155)	7.8 (140)	Two or more values met or exceeded

OGTT, oral glucose tolerance test; GCT, glucose challenge test; GA, gestational age; GDM, gestational diabetes mellitus.

[a]BMI > 30 kg/m^2, a previous baby weighing > 4.5 kg, previous GDM, a first-degree relative with diabetes and family origin with a high prevalence of diabetes.

[b]Glycosuria, previous GDM, maternal BMI ≥ 27 kg/m^2, a first- or second-degree relative with diabetes, a previous baby weighing ≥ 4500 g.

[c]A first-degree relative with diabetes or GDM in a previous pregnancy.

[d]Glycosuria, a first-degree relative with diabetes, previous stillbirth, spontaneous abortion, foetal malformation or macrosomia, high maternal age, parity five or more.

criteria is the lack of evidence testifying that the treatment of the milder cases is cost effective [25]. Also, evidence is lacking regarding the long-term consequences of intervention in case of mild hyperglycaemia in pregnancy. The American Diabetes Association has recommended the IADPSG criteria in all pregnant women even though the criteria are expected to double the numbers of pregnancies diagnosed with GDM. The recommendation is based on the documented benefit of treating even mild cases of hyperglycaemia in pregnancy and because the majority of the cases can be treated with lifestyle intervention [21].

Independent of which criteria are used for diagnosing GDM, different screening strategies exist. Some countries have universal screening programmes for GDM in pregnancy such as screening all pregnant women. Others recommend selective screening based on risk factors.

In universal screening, a two-step screening programme is normally applied. The initial screening is based on either a fasting blood glucose or HbA1c or a glucose challenge test (GCT). In contrast to the OGTT, a GCT can be performed without any preparation. Blood glucose is measured any time of the day 1 h after ingestion of 50 g glucose either as glucose drink or jelly beans or a candy bar. If fasting blood glucose or HbA1c are elevated or if the GCT is positive, the next step will be an OGTT, unless the initial screening indicates overt diabetes [6].

In selective screening, a one-step approach is often used. The women with moderate or high risk have an OGTT at the end of the second trimester or at the beginning of the third trimester. Commonly applied risk factors are overweight, family history of diabetes, ethnicity, previous gestational diabetes, glycosuria, macrosomia, twin pregnancy and polycystic ovaries [6,26]. With selective screening, it is estimated that around 80% of the women with gestational diabetes will be detected, leaving 20% undetected, and therefore untreated. Further analyses of undetected cases have indicated that these women have lower blood glucoses levels and fewer adverse outcomes compared with the women detected at screening [21,27].

The OGTT is time consuming, expensive and associated with some – although mild and harmless – side effects such as nausea [28]. This favours a two-step screening strategy. In populations with a high incidence of type-2 diabetes, a two-step universal screening programme may also be preferred, as only a small proportion of the pregnant population will be of low risk and thereby avoid OGTT if the one-step strategy is employed.

Treatment of GDM

The treatment for gestational diabetes is instituted to lower blood glucose levels and thereby to reduce the risk of complications for both the mother and child. Most pregnant women diagnosed with GDM can achieve good glycaemic control on dietary regulation and home blood-glucose monitoring. Furthermore, the women should be encouraged to engage in – or increase – physical activity to enhance insulin sensitivity. If blood glucose cannot be satisfactorily controlled on diet alone, the treatment must be supplemented with insulin therapy [29]. Less frequent is the use of oral hypoglycaemics such as metformin and glyburide. There is no reported severe side effects due to oral hypoglycaemics, but at present no long-term follow-up on the children exists [30,31].

There is no international standard for the treatment of gestational diabetes, and the criterion for insulin treatment differs between centres, resulting in large variations in the proportion of women with GDM treated with insulin. Yet, there is good evidence that intervention can reduce severe adverse outcome such as shoulder dystocia and Caesarean delivery [32,33].

Implications Related to GDM

A pregnancy complicated by GDM necessitates increased surveillance due to extra hospital visits including ultrasound scans, and there is an increased risk of premature delivery, in some cases due to induction of labour [6]. In the light of an expected increase in the number of cases of GDM following implementation of the IADPSG criteria, there has been some concern regarding the medicalisation of the women [21,25]. However, in the first interventional study of pregnant women with hyperglycaemia by Crowther et al. [34], a quality of life (SF36) assessment was made post-partum. The study revealed an improved health-related quality of life among women in the intervention group compared with the standard care group.

Gestational diabetes is associated with short- and long-term consequences for both the mother and the child. Elevated blood glucose levels in the mother result in elevated glucose in the foetus, increasing the secretion of insulin. Insulin is a growth factor leading to macrosomia and increased risk for hypoglycaemia within the first hours after pregnancy. Other short-term complications are hypertensive disorders including pre-eclampsia, Caesarean delivery, birth trauma (vaginal tears, shoulder dystocia, asphyxia) and hypoglycaemia [14].

In the majority of women with GDM, hyperglycaemia disappears shortly after delivery. However, women with a history of gestational diabetes have a significantly increased risk of developing pre-diabetes and overt diabetes, especially type-2 diabetes [35]. The diabetes risk is up to 70%, depending on the time of follow-up and the population examined. One study in a predominantly Caucasian population found that 27% had pre-diabetes and 40% overt diabetes at a mean of 10 years after a pregnancy complicated by GDM [36]. The major risk factors for developing diabetes are early diagnosis of GDM in pregnancy and high glucose levels at diagnosis, abnormal glucose tolerance in the first months after pregnancy and overweight.

Women with a history of gestational diabetes should be advised to maintain a healthy lifestyle the rest of their lives and should be offered regular post-partum examination for diabetes or pre-diabetes with 1–3 year interval [21]. Children born by mothers with gestational diabetes also have an increased risk for overweight and pre-diabetes, even before adulthood [37]. No studies regarding lifestyle intervention in women with a history of GDM or in children born by mothers with GDM have been published. Until then the clinicians must resort to the results from intervention studies in subjects with pre-diabetes when taking care of women with a history of GDM [38,39].

References

1. World Health Organization. *Definition, Diagnosis and Classification of Diabetes Mellitus and its Complications, Report of a WHO consultation, Part 1: diagnosis and Classification of Diabetes Mellitus*. World Health Organization; 1999:WHO/NCD/NCS/99.2 <http://www.staff.ncd.ac.uk/philip.home/who_dmg.pdf>.
2. Metzger BE, Lowe LP, Dyer AR, et al. Hyperglycemia and adverse pregnancy outcomes. *N Engl J Med*. 2008;358:1991–2002.
3. Jensen DM, Mølsted-Pedersen L, Beck-Nielsen H, Westergaard JG, Ovesen P, Damm P. Screening for gestational diabetes mellitus by a model based on risk indicators: a prospective study. *Am J Obstet Gynecol*. 2003;189:1383–1388.
4. Anderberg E, Kallen K, Berntorp K, Frid A, Aberg A. A simplified oral glucose tolerance test in pregnancy: compliance and results. *Acta Obstet Gynecol Scand*. 2007;86:1432–1436.
5. Metzger BE, Gabbe SG, Persson B, et al. International association of diabetes and pregnancy study groups recommendations on the diagnosis and classification of hyperglycemia in pregnancy. *Diabetes Care*. 2010;33:676–682.
6. Tieu J, Middleton P, McPhee AJ, Crowther CA. Screening and subsequent management for gestational diabetes for improving maternal and infant health. *Cochrane Database Syst Rev*. 2010;5:CD007222.
7. Catalano PM, Kirwan JP, Haugel-de Mouzon S, King J. Gestational diabetes and insulin resistance: role in short- and long-term implications for mother and fetus. *J Nutr*. 2003;133:1674S–1683S.
8. Retnakaran R. Glucose tolerance status in pregnancy: a window to the future risk of diabetes and cardiovascular disease in young women. *Curr Diabetes Rev*. 2009;5:239–244.
9. Dabelea D, Snell-Bergeon JK, Hartsfield CL, Bischoff KJ, Hamman RF, McDuffie RS. Increasing prevalence of gestational diabetes mellitus (GDM) over time and by birth cohort: Kaiser Permanente of Colorado GDM Screening Program. *Diabetes Care*. 2005;28:579–584.
10. Chu SY, Callaghan WM, Kim SY, et al. Maternal obesity and risk of gestational diabetes mellitus. *Diabetes Care*. 2007;30:2070–2076.
11. Ovesen P, Rasmussen S, Kesmodel U. Effect of prepregnancy maternal overweight and obesity on pregnancy outcome. *Obstet Gynecol*. 2011;118:305–312.
12. Kühl C. Etiology and pathogenesis of gestational diabetes. *Diabetes Care*. 1998;21 (suppl 2):B19–B26.
13. Yogev Y, Ben-Haroush A, Hod M. Pathogenesis of gestational diabetes mellitus. In: Hod M, Jovanovic L, Di Renzo GC, de Leiva A, Langer O, eds. *Textbook of Diabetes and Pregnancy*. London: Martin Dunitz; 2003:39–49.
14. Kjos SL, Buchanan TA. Gestational diabetes mellitus. *N Engl J Med*. 1999;341:1749–1756.
15. Lauenborg J, Hansen T, Jensen DM, et al. Abnormal glucose tolerance after gestational diabetes mellitus – a long-term follow-up study of a Danish population. *Diabetologia*. 2002;45(suppl 2):242 [Abstract].

16. Catalano PM, Hauguel-De Mouzon S. Is it time to revisit the Pedersen hypothesis in the face of the obesity epidemic? *Am J Obstet Gynecol*. 2011;204:479–487.
17. Xiang AH, Peters RK, Trigo E, Kjos SL, Lee WP, Buchanan TA. Multiple metabolic defects during late pregnancy in women at high risk for type 2 diabetes. *Diabetes*. 1999;48:848–854.
18. National Diabetes Data Group. Classification and diagnosis of diabetes mellitus and other categories of glucose intolerance. *Diabetes*. 1979;28:1039–1057.
19. Carpenter MW, Coustan DR. Criteria for screening tests for gestational diabetes. *Am J Obstet Gynecol*. 1982;144:768–773.
20. Walker JD. NICE guidance on diabetes in pregnancy: management of diabetes and its complications from preconception to the postnatal period. NICE clinical guideline 63. London March 2008. *Diabet Med*. 2008;25:1025–1027.
21. American Diabetes Association. Diagnosis and classification of diabetes mellitus. *Diabetes Care*. 2012;35:S64–S71.
22. The HAPO Study Cooperative Research Group. Hyperglycemia and adverse pregnancy outcome (HAPO) study: associations with neonatal anthropometrics. *Diabetes*. 2009;58:453–459.
23. O'Sullivan JB. The Boston gestational diabetes studies: review and perspectives. In: Sutherland HW, Stowers JM, eds. *Carbohydrate Metabolism in Pregnancy and the Newborn*. Berlin, Heidelberg, New York: Springer-Verlag; 1979:425–435.
24. Farrar D, Duley L, Lawlor DA. Different strategies for diagnosing gestational diabetes to improve maternal and infant health. *Cochrane Database Syst Rev*. 2011;10 CD007122.
25. Ryan EA. Diagnosing gestational diabetes. *Diabetologia*. 2011;54:480–486.
26. Buhling KJ, Henrich W, Starr E, et al. Risk for gestational diabetes and hypertension for women with twin pregnancy compared to singleton pregnancy. *Arch Gynecol Obstet*. 2003;269:33–36.
27. Naylor CD, Sermer M, Chen E, Farine D. Selective screening for gestational diabetes mellitus. Toronto Trihospital Gestational Diabetes Project Investigators. *N Engl J Med*. 1997;337:1591–1596.
28. Alwan N, Tuffnell DJ, West J. Treatments for gestational diabetes. *Cochrane Database Syst Rev*. 2009;3:CD003395.
29. Langer O. Maternal glycemic criteria for insulin therapy in gestational diabetes mellitus. *Diabetes Care*. 1998;21(suppl 2):B91–B98.
30. Jovanovic L. American diabetes association's fourth international workshop – conference on gestational diabetes mellitus: summary and discussion. Therapeutic interventions. *Diabetes Care*. 1998;21(suppl 2):B131–B137.
31. Nicholson W, Baptiste-Roberts K. Oral hypoglycaemic agents during pregnancy: the evidence for effectiveness and safety. *Best Pract Res Clin Obstet Gynaecol*. 2011;25:51–63.
32. Horvath K, Koch K, Jeitler K, et al. Effects of treatment in women with gestational diabetes mellitus: systematic review and meta-analysis. *BMJ*. 2010;340: c1395.
33. Langer O, Yogev Y, Most O, Xenakis EM. Gestational diabetes: the consequences of not treating. *Am J Obstet Gynecol*. 2005;192:989–997.
34. Crowther CA, Hiller JE, Moss JR, McPhee AJ, Jeffries WS, Robinson JS. Effect of treatment of gestational diabetes mellitus on pregnancy outcomes. *N Engl J Med*. 2005;352:2477–2486.

35. Kim C, Newton KM, Knopp RH. Gestational diabetes and the incidence of type 2 diabetes: a systematic review. *Diabetes Care*. 2002;25:1862–1868.
36. Lauenborg J, Hansen T, Jensen DM, et al. Increasing incidence of diabetes after gestational diabetes mellitus – a long-term follow-up in a Danish population. *Diabetes Care*. 2004;27:1194–1199.
37. Clausen TD, Mathiesen ER, Hansen T, et al. High prevalence of type 2 diabetes and pre-diabetes in adult offspring of women with gestational diabetes mellitus or type 1 diabetes: the role of intrauterine hyperglycemia. *Diabetes Care*. 2008;31:340–346.
38. Eriksson KF, Lindgärde F. No excess 12-year mortality in men with impaired glucose tolerance who participated in the Malmo Preventive Trial with diet and exercise. *Diabetologia*. 1998;41:1010–1016.
39. Pan XR, Li GW, Hu YH, et al. Effects of diet and exercise in preventing NIDDM in people with impaired glucose tolerance. The Da Qing IGT and Diabetes Study. *Diabetes Care*. 1997;20:537–544.

21 Obesity and Metabolic Disorders During Pregnancy and Pregnancy Outcome

Gerard H.A. Visser[1] *and Yariv Yogev*[2]

[1]Department of Obstetrics, University Medical Center, Utrecht, the Netherlands, [2]Rabin Medical Center, Tel Aviv, Israel

Introduction

Pre-pregnancy overweight and obesity are associated with adverse pregnancy outcome in glucose-tolerant women. Obesity is associated with gestational diabetes mellitus (GDM), which further increases the risk of poor pregnancy outcome and which may also further increase late complications as increased risk for childhood obesity, diabetes and metabolic syndrome in their offspring. In this chapter, the – relatively scarce – data on the combination obesity and GDM are reviewed.

Obesity and Gestational Diabetes

GDM is related to many factors, such as ethnicity, previous occurrence of GDM, age, parity and history of diabetes. Obesity acts as an independent risk factor. Two population-based studies in singletons in the United Kingdom and United States have shown that the odds ratio for GDM increases from 1, in case of a normal body mass index (BMI) of 20–24.9 kg/m^2 to 1.6–1.7 kg/m^2 with overweight (BMI of 25–29.9 kg/m^2) [1,2] and the odds ratio for obesity (BMI > 30 kg/m^2) is around 3.6–4. These risks persist after accounting for other confounding demographic factors [1]. With a BMI > 40 kg/m^2 a 10-fold increase in GDM has been reported [3]. This indicates that screening for GDM is mandatory in case of maternal obesity. One may even consider an early screening or offering a definitive one-step diagnosis test rather than a two-steps diagnostic approach. During early pregnancy, screening may consist of an HbA1c measurement to identify a previously undiagnosed type 2 diabetes.

Obesity. DOI: http://dx.doi.org/10.1016/B978-0-12-416045-3.00021-2

Obesity by itself is related to an impaired outcome of pregnancy. This holds for GDM, congenital malformations, foetal death, foetal macrosomia, Caesarean section, hypertension, preterm delivery, postoperative complications and neonatal morbidity, with typical odds ratios in between 2 and 3. The same holds for GDM. It is the question, if a combination of both entities further impairs outcome. The available evidence indeed suggests so. In a series of over 6000 women without GDM, a gradual increase in macrosomia, large-for-gestational age infants and incidence of Caesarean sections (CS) has been found both in obese and non-obese women in relation to increasing glucose challenge test results [4]. Obesity was the main factor impacting foetal size. Recently, the relative importance of obesity and GDM was studied using the Hyperglycemia and Adverse Pregnancy Outcome (HAPO) study data [5]. As compared to non-GDM/non-obesity (BMI < 33 kg/m^2 at 28 weeks of gestation), untreated GDM had an odds ratio for birthweight >90th centile of 2.19, obesity an odds ratio of 1.73 and the combination of GDM and obesity an odds ratio of 3.62. For pre-eclampsia the odds ratio were 1.74, 3.91 and 5.98, respectively. Results for primary Caesarean section, cord C-peptide and newborn per cent body fat >90th centile were similar (Table 21.1). Data for shoulder dystocia/birth injury were only significant for the GDM/obese subgroup. In other words, both maternal GDM and obesity are independently associated with adverse pregnancy outcomes in a symbiotic way in which the combination has a greater effect than either one alone [5].

Data on congenital malformations are likely to be similar, since maternal obesity with a normal glucose tolerance is associated with an increased incidence of malformations, and the same holds for pre-existing diabetes; the latter also in non-obese women. It is unknown if the combination results in an even higher incidence of malformations. The same holds true for stillbirths. Both entities are associated with an increased incidence of stillbirth, but it may well be that the higher incidence in obesity only is partly due to difficulties in antenatal detection of foetal growth restriction and/or to problems regarding antenatal assessment of the foetal condition.

Table 21.1 Relationship (Odds Ratios) Between Maternal GDM, Obesity and Outcome

	Control	GDM	Obesity	GDM and Obesity
Birthweight >90th centile	1	2.19	1.73	3.62
Cord C-peptide >90th centile	1	2.49	1.77	3.61
Primary Caesarean section	1	1.25	1.51	1.71
Pre-eclampsia	1	1.74	3.91	5.98
Newborn per cent body fat >90th centile	1	1.98	1.65	3.69
Shoulder dystocia/birth injury	1	1.14	1.03	1.8

Data from the HAPO study. Control women had no GDM and were not obese. All data significant from controls, except for GDM and obesity alone in case of shoulder dystocia/foetal injury.
Source: Adapted from Ref. [5].

Impact of Obesity and/or Diabetes During Pregnancy on the Offspring

Diabetes type 1 and type 2 and GDM are related to macrosomia at birth and to overweight/obesity during childhood and adolescence. In American Pima Indians, prone to develop diabetes, impaired glucose tolerance has been found in 45% of 20–24 year offspring of women with GDM during pregnancy, as compared to 8.6% in women who developed diabetes after pregnancy. These differences persisted after taking into account paternal diabetes, age at onset of diabetes in parents and birthweight [6]. A study from Scandinavia has shown similar results: 21% of offspring aged 18–27 years old of women with GDM were found to have type 2 diabetes or impaired glucose tolerance, as compared to only 12% in offspring of women genetically predisposed to diabetes but without GDM during the index pregnancy [7]. These data clearly show the impact of an abnormal intrauterine environment on later outcome. In these studies, maternal BMI had not been taken into account or level of glycaemic control and data concerning glycaemic level during pregnancy. However, the impact of abnormal glucose levels during pregnancy on offspring was also clearly demonstrated in rats that were treated with streptozotocin and who developed mild diabetes in pregnancy. This resulted in impaired glucose tolerance and GDM in the second and third generations [8]. These data provide clear examples of the developmental origins of health and disease.

In case of pre-existing maternal diabetes, it has been suggested that also maternal weight is a factor determining outcome. In a nationwide study on outcome in case of maternal type 1 diabetes we found that birthweight >90th centile (odds ratio 4.4) and maternal pre-pregnancy weight (odds ratio 2.8) were the only two independent variables related to overweight at the age of 7 years [9]. A recent study from Finland suggests that maternal overweight (BMI > 25 kg/m^2 [10]) is a far more important predictor for overweight in 16-year-old adolescents than GDM (Table 21.2). Similar data were obtained for abdominal obesity in the 16-year-old offspring. So it seems that GDM and maternal overweight are important

Table 21.2 Overweight in 16-Year-Old Adolescents in Relation to Maternal GDM and BMI (< or >25 kg/m^2) [10].

	Overweight in Offspring (%)	
	Maternal BMI < 25 (kg/m^2)	**>25 (kg/m^2)**
Risk population		
GDM (n = 84)	8.2	40
Normal OGTT (n = 657)	13.5	27.9
Control population (n = 3427)	11.7	–

contributing factors for future overweight and impaired glucose tolerance and most likely also for metabolic syndrome in their offspring.

Interventions

Reducing weight gain in obese women prior and during pregnancy reduces pregnancy complications. In a nicely designed randomised controlled study it has also been demonstrated that it reduces deterioration in glucose metabolism during pregnancy [11]. In another study in glucose-tolerant obese women, it also appeared to have a slight but significant effect on fasting glucose levels [12]. Therefore, reducing weight gain in obese women during pregnancy seems also beneficial for glucose homeostasis. Reducing weight gain in obese women during pregnancy is also associated with a significantly lower risk of pre-eclampsia, Caesarean delivery and large-for-gestational age infant and with a higher risk of a small-for-gestational age infant. These results were similar for obesity class (30–34.9, 35–35.9 and 40.0 kg/m^2), but at different amounts of gestational weight gain [13].

Metformin decreases glucose synthesis in the liver, inhibits gluconeogenesis, delays glucose resorption from the gut and increases glucose utilisation in the muscular system. It is used in (overweight) polycystic ovary syndrome (PCOS) women to lose weight and facilitate conception. It especially improves pregnancy rates in women who are clomiphene resistant. In this context, it has been used widely in PCOS women also in the first trimester of pregnancy. It crosses the placenta, but no embryonic or foetal side effects have been reported to date. In the second half of pregnancy, it is increasingly used as an oral anti-diabetic drug to treat GDM. In a large randomised controlled trial, it was not associated with increased perinatal complications as compared to insulin [14], although failure rates (additional use of insulin) may be as high as 46% [14]. A follow-up study has shown no adverse effects at 2 years of age [15]. Metformin therefore seems to be the drug of choice in PCOS, maternal obesity and in the treatment of glucose intolerance during pregnancy. However, metformin is a remarkable drug and its use may be associated with reduced risk of cancer [16], most likely by killing tumour-initiating stem cells, with anti-angiogenetic effects, anti-inflammatory effects, growth inhibitory effects and antioxidative effects [17–20]. That appears to be good for the prevention and/or treatment of cancer, but what about a 9 months exposition to the foetus? In other words, we still do not know enough about possible adverse effects of this drug in the embryo and foetus. Therefore, animal and basic human studies seem mandatory, before it might be considered safe [21].

Regarding the offspring it is known that maternal weight, diabetes, foetal macrosomia at birth and weight gain during the first 7 years of life are important variables determining obesity/overweight. Maternal overweight is hard to tackle, and birthweight centiles in women with type 1 and type 2 diabetes are rather increasing than decreasing [22] (possibly due to better periconceptional glucose control and better early placentation [23]). Treatment options exist for

GDM, since treatment has been shown to reduce the incidence of foetal macrosomia. Moreover, lifestyle interventions may be beneficial for infants at risk for obesity and/or glucose intolerance, since in several studies it has been shown that these conditions are preceded by a rapid weight gain in between 2 and 7 years of life [24–27]. Parents should be counselled to give these children a proper diet and to provide them with sufficient exercise to prevent early overweight. In this context, it might be easier to convince parents to give their children a proper diet than to lose weight themselves. On the other hand, epigenetic changes during foetal life may have triggered genes promoting childhood obesity.

Conclusions

- The incidence of GDM increases with increasing maternal weight, up to a 10-fold higher incidence in women with a BMI > 40 kg/m^2. Screening for GDM is mandatory in overweight pregnant women (HbA1c at early gestation and oral glucose tolerance test after 20 weeks).
- GDM and obesity are independently related to pregnancy complications such as pre-eclampsia, primary Caesarean section and birthweight >90th centile.
- The same holds for overweight/obesity during childhood and adolescence and for impaired glucose tolerance in their offspring.
- Reducing maternal weight gain during pregnancy reduces pregnancy complications and improves glucose homeostasis.
- There are at present insufficient data on safety of metformin treatment during pregnancy.
- Prevention of overweight and impaired glucose tolerance in offspring of women with GDM and/or obesity may best be achieved by preventing excessive growth between 2 and 7 years of life.

References

1. Sebire NJ, Jolly M, Harris JP, et al. Maternal obesity and pregnancy outcome: a study of 287,213 pregnancies in London. *Int J Obes*. 2001;25:1175–1182.
2. Baeten JM, Bukusi EA, Lambe M. Pregnancy complications and outcomes among overweight and obese nulliparous women. *Am J Public Health*. 2001;91:436–440.
3. Kumari A. Pregnancy outcome in women with morbid obesity. *Int J Gynaecol Obstet*. 2001;73:101–107.
4. Yogev Y, Langer O, Xenakis EM, Rosenn B. The association between glucose challenge test, obesity and pregnancy outcome in 6390 non-diabetic women. *J Matern Fetal Neonatal Med*. 2005;17:29–34.
5. Catalano PM, McIntyre HD, Cruickshank JK, et al. The hyperglycemia and adverse pregnancy outcome study: associations of GDM and obesity with pregnancy outcomes. *Diabetes Care*. 2012;35:780–786.
6. Pettitt DJ, Lawrence JM, Beyer J, et al. Association between maternal diabetes in utero and age at offspring's diagnosis of type 2 diabetes. *Diabetes Care*. 2008;31: 2126–2130.

7. Clausen TD, Mathiesen ER, Hansen T, et al. High prevalence of type 2 diabetes and pre-diabetes in adult offspring of women with gestational diabetes mellitus or type 1 diabetes: the role of intrauterine hyperglycemia. *Diabetes Care.* 2008;31:340–346.
8. Aerts L, Van Assche F. Animal evidence for the transgenerational development of diabetes mellitus. *Int J Biochem Cell Biol.* 2006;38:894–903.
9. Rijpert M, Evers IM, de Vroede MA, de Valk HW, Heijnen CJ, Visser GHA. Risk factors for childhood overweight in offspring of type 1 diabetic women with adequate glycemic control during pregnancy: nationwide follow-up study in the Netherlands. *Diabetes Care.* 2009;32:2099–2104.
10. Pirkola J, Pouta A, Bloigu A, et al. Risks of overweight and abdominal obesity at age 16 years associated with prenatal exposure to maternal prepregnancy overweight and gestational diabetes mellitus. *Diabetes Care.* 2010;33:1115–1121.
11. Wolff S, Legarth J, Vangsgaard K, Toubro S, Astrup A. A randomized trial of the effects of dietary counseling on gestational weight gain and glucose metabolism in obese pregnant women. *Int J Obesity.* 2008;32:495–501.
12. Jensen DM, Ovesen P, Beck-Nielsen H, et al. Gestational weight gain and pregnancy outcomes in 481 obese glucose-tolerant women. *Diabetes Care.* 2005;28: 2118–2122.
13. Kiel DW, Dodson EA, Artal R, Boehmer TK, Leet TL. Gestational weight gain and pregnancy outcomes in obese women: how much is enough? *Obstet Gynecol.* 2007;110:752–758.
14. Rowan JA, Hague WM, Gao W, Battin MR, Moore MP. Metformin versus insulin for the treatment of gestational diabetes. *N Engl J Med.* 2008;358:2003–2015.
15. Rowan JA, Rush EC, Obolonkin V, Battin M, Wouldes T, Hague WM. Metformin in gestational diabetes: the offspring follow-up (MiG TOFU): body composition at 2 years of age. *Diabetes Care.* 2011;34:2279–2284.
16. Libby G, Donnelly LA, Donnan PT, Alessi DR, Morris AD, Evans JMM. New users of metformin are at low risk of incident cancer. *Diabetes Care.* 2009;32:1620–1625.
17. Martin-Castillo B, Vazquez-Martin A, Oliveras-Ferraros C, Menendez JA. Metformin and cancer. *Cell Cycle.* 2010;9:1057–1064.
18. Tan BK, Adya R, Chen J, Lehnert H, Saint Cassia LJ, Randeva HS. Metformin treatment exerts antiinvasive and antimetastatic effects in human endometrial carcinoma cells. *J Clin Endocrinol Metab.* 2011;96:808–816.
19. Ersoy C, Kiyici S, Budak F, et al. The effect of metformin treatment on VEGF and PAI-1 levels in obese type-2 diabetic patients. *Diab Rev Clin Pract.* 2008;81:56–60.
20. Dowling RJO, Goodwin PJ, Stambolic V. Understanding the benefit of metformin use in cancer treatment. *BMC Med.* 2011;9:33.
21. Visser GHA, de Valk HW. Gestational diabetes: screening for all, which test and which treatment? *Expert Rev Endocrinol Metab.* 2012;7:165–167.
22. Evers IM, de Valk HW, Visser GHA. Risk of complications of pregnancy in women with type 1 diabetes: nationwide prospective study in the Netherlands. *BMJ.* 2004;328:915.
23. Kuc S, Wortelboer EJ, Koster MP, de Valk HW, Schielen PC, Visser GHA. Prediction of macrosomia at birth in type-1 and 2 diabetic pregnancies with biomarkers of early placentation. *BJOG.* 2011;118:748–754.

24. Eriksson JG, Forsen TJ, Osmond C, Barker DJ. Pathways of infant and childhood growth that lead to type 2 diabetes. *Diabetes Care*. 2003;26:3006–3010.
25. Touger L, Looker HC, Krakoff J, Lindsay RS, Cook V, Knowler WC. Early growth in offspring of diabetic mothers. *Diabetes Care*. 2005;28:585–589.
26. Fall CHD, Singh Sachdev H, Osmond C, et al. Adult metabolic syndrome and impaired glucose tolerance are associated with different patterns of BMI gain during infancy. *Diabetes Care*. 2008;31:2349–2356.
27. De Valk HW, van Nieuwaal NH, Visser GHA. Pregnancy outcome in type 2 diabetes mellitus: a retrospective analysis from the Netherlands. *Rev Diabet Stud*. 2006;3: 134–142.

22 Obesity, Polycystic Ovaries and Impaired Reproductive Outcome

Jyoti Balani[1], Stephen Hyer[1], Marion Wagner[2] and Hassan Shehata[2]

[1]Department of Diabetes and Endocrinology, Epsom and St Helier University Hospitals, NHS Trust, Surrey, UK, [2]Department of Maternal Medicine, Epsom and St Helier University Hospitals, NHS Trust, Surrey, UK

People of such constitution cannot be prolific...fatness and flabbiness are to blame. The womb is unable to receive the semen and they menstruate infrequently and little.

Hippocrates Essay on the Scythians [1]

Introduction

Obesity is a complex and multi-factorial chronic condition with an increasing global prevalence especially in the western countries. According to the World Health Organization (WHO), obesity has more than doubled worldwide since 1980. The WHO defines obesity as abnormal or excessive fat accumulation that may impair health. Body mass index (BMI), calculated as weight (kg)/height (metres) [2], is used as an estimate of obesity – a BMI greater than 25 kg/m^2 is classified as overweight and a BMI equal to or greater than 30 kg/m^2 is categorised as obesity. Obesity is further subdivided into class I (BMI 30–34.9 kg/m^2), class II (BMI 35–39.9 kg/m^2) and class III (BMI $\geq$ 40 kg/m^2). Overweight and obesity rank fifth as leading risks for global deaths. In addition, 44% of the burden of diabetes, 23% of ischaemic heart disease and between 7–41% of certain cancers are attributable to overweight and obesity [2].

The fundamental problem in obesity is an energy imbalance between calories consumed and calories expended. An increased intake of energy-dense foods and a decrease in physical activity associated with increasing urbanisation are important contributory factors to the increased prevalence of obesity globally.

Obesity. DOI: http://dx.doi.org/10.1016/B978-0-12-416045-3.00022-4

Obesity and Reproductive Health

Overweight women have a higher incidence of menstrual dysfunction, anovulation and infertility compared with women of similar reproductive age. Spontaneous abortions after natural conception or after using assisted reproductive technology occur more frequently in obese women compared with women having normal BMI. Lashen et al. [3] reported a significantly higher risk for early miscarriages among obese patients compared in normal weight controls. Maheshweri et al. [4] documented similar results in obese women after assisted reproduction.

Obesity is a known risk factor for subfertility due to anovulation. A recently published study by Van der Steeg et al concluded that obesity is also associated with lower pregnancy rates in subfertile ovulatory women [5]. For every BMI unit above 29 kg/m^2, the probability of a successful pregnancy was reduced by approximately 5%, equivalent to the impact on pregnancy of being 1 year older.

Obese men have erectile dysfunction and reduced coital frequency [6] and this is thought to relate to decreased testosterone concentration and elevated pro-inflammatory cytokines which induce endothelial dysfunction through the nitric oxide pathway [7]. Obesity results in increased peripheral aromatisation of androgens resulting in low testosterone levels. Increased scrotal fat and raised testicular temperature may result in reduced spermatogenesis.

Pathogenic Mechanisms

There is new and increasing evidence that point to the importance of genetics influencing body fat mass. Approximately 20 different genes have been implicated in human monogenic obesity, but they account for only a small percentage of cases [8]. Alterations in energy balance in humans have been linked to mutations in the leptin–melanocortin pathway which helps regulate energy homeostasis acting through the satiety centre in the hypothalamus [9].

The pathogenic mechanisms responsible for obesity having a negative impact on reproductive health are uncertain. One hypothesis is that obesity affects the hypothalamic–pituitary–ovary axis. Excess free oestrogen resulting in part from increased peripheral aromatisation of androgens to oestrogen in adipose tissue, combined with decreased availability of gonodotropin-releasing hormone, could interfere with the hypothalamic–pituitary regulation of ovarian function resulting in irregular or anovulatory cycles.

Hyperinsulinaemia is another important factor implicated in fertility disorders in obesity. It may be directly responsible for the development of androgen excess through its effects in reducing sex hormone–binding globulin (SHBG) synthesis and in stimulating ovarian androgen production rates. Hyperandrogenaemia in turn leads to altered ovarian function. Preliminary results from our department show that fat is preferentially distributed centrally in obese women during pregnancy. Visceral fat area as measured by bio-impedance is elevated whilst lean mass in the lower limbs is reduced. The significance of this for pregnancy outcomes is currently being investigated.

It is now well accepted that the adipocyte is effectively an endocrine cell capable of releasing many active substances including interleukins, tumour necrosis factor, leptin, complement factors and plasminogen activator inhibitor. Leptin is thought to inhibit ovarian follicular development through both the induction of insulin resistance and a direct impairment of ovarian function. Alterations in the secretion and action of insulin and other hormones as leptin, resistin, ghrelin and adiponectin in obese women may affect follicle growth, corpus luteum function, early embryo development, trophoblast function and endometrial receptivity [10].

There is recent in vitro data indicating that leptin may exert a direct inhibitory effect on ovarian function by inhibiting human granulosa and theca cell steroidogenesis, probably by antagonising stimulatory factors, such as insulin growth factor-1, transforming growth factor-β, insulin and luteinising hormone [11] Also, in vitro and in vivo studies have demonstrated that high leptin concentrations in the ovary may interfere with the development of a dominant follicle and oocyte maturation [12]. There is also evidence that the endometrium may also have a subtle role in the detrimental effects of obesity on reproduction [13].

Psychosocial factors have been implicated in reduced fertility in obese women. Obese people do not have sexual intercourse as frequently as slimmer people even if they have a cohabiting sexual partner. This could be explained in part by decreased sex drive resulting from decreased dopamine activity and increased serotonin activity in the brain caused by overeating [14,15].

Polycystic Ovary Syndrome

Polycystic ovary syndrome (PCOS) is one of the most common disorders to cause infertility from anovulation and affects 4–7% of women. Nearly one-half of the women with PCOS are obese. A consensus conference held in Rotterdam established that at least two of the following criteria are sufficient for the diagnosis of PCOS: oligo and/or anovulation, clinical and/or biochemical signs of hyperandrogenism and polycystic ovaries at ultrasound [16].

The clinical features of PCOS are heterogeneous and may change throughout life and are largely influenced by obesity and metabolic alterations. It was originally known in its severe form as the Stein–Levinthal syndrome. PCOS is characterised by multiple small cysts in the ovary and hyperandrogenism. Excess androgen production arises from the ovaries and to a lesser extent from the adrenals. It is still unclear whether the basic defect is in the ovary, adrenal, pituitary or a more generalised metabolic defect. The androgens are normally converted to oestrogens in adipose tissue, but in PCOS, androstenedione is secreted and converted to testosterone in peripheral tissues.

Obesity is strongly associated with the PCOS; obesity is present in at least 30% of cases and in some series up to 75% of cases [17]. Levels of SHBG tend to decrease with increasing body fat leading to higher circulating free androgens delivered to target sensitive tissues [18]. SHBG is regulated by a complex of factors, including oestrogen and growth hormone as stimulating factors and androgens and insulin as inhibiting factors [19]. Hyperinsulinaemia in obesity overcomes insulin resistance and inhibits SHBG synthesis in the liver.

Leptin may exert a direct inhibitory effect on ovarian function. Higher circulating levels of leptin than those expected in relation to BMI, or normal concentrations of leptin, have been reported in obese women with PCOS [20]. However, it is presently unknown whether high leptin levels in the peripheral circulation and/or in the ovarian tissues may play a role in determining anovulation in obese women with PCOS.

The possible role of a dysfunctional endocannabinoid system in the pathophysiology of obesity-related PCOS is currently being studied. One study has also shown that PCOS and maternal obesity affect oocyte size in in vitro fertilisation/intracytoplasmic sperm injection cycles. Women with PCOS and obesity have smaller oocytes compared with control subjects and both PCOS and obesity independently influence oocyte size [21].

The therapeutic efficacy of weight loss indirectly supports the pathogenic role of obesity in PCOS. Lifestyle interventions with hypocaloric diet with or without associated increased physical activity have proved their efficacy [22]. A recent study examined the effect of a 48-week period of intensive lifestyle intervention; dietary advice and a standardised physical activity programme with or without metformin treatment showed a significant positive effect on ovulatory performance which was related to the amount of weight loss, rather than the effect of metformin [23].

Obesity in Pregnancy

Obesity in pregnancy has been identified by the Confidential Enquiry into Maternal and Child Health (CEMACH) (2008–2011) as a major health risk to mother and baby (Figure 22.1). There is substantial evidence that obesity in pregnancy contributes to increased morbidity and mortality for both mother and baby. CEMACH found that approximately

- 35% of women who died were obese [24] and
- 30% of the mothers who had a stillbirth or a neonatal death were obese.

Obese women spend an average of 4.83 more days in hospital and the increased levels of complications in pregnancy and interventions in labour represent a five-fold increase in cost of antenatal care [25]. The costs associated with babies born to obese mothers are also increased as there is a 3.5-fold increase in admission to Neonatal Intensive Care Unit (Figure 22.1).

- Antepartum risks
  - pregnancy-induced hypertension/PET
  - Miscarriages
  - gestational diabetes
  - TED
  - Ultrasound difficulties
- Intrapartum risks
  - labour induction, CS and failed VBAC, perineal trauma
  - Dystocia
  - Anaesthetic risks
  - Surgical difficulties
- Post-partum risks
  - increased rates of puerperal infection
  - decreased rates of breastfeeding initiation or continuation
  - Postnatal depression (PND)
  - Thromboembolic disease (TED)
- Offspring risks
  - higher risk for having congenital anomalies
  - Stillborn
  - Childhood and adult obesity

Figure 22.1 The risks of obesity in pregnancy.

Gestational Weight Gain in Obese Women

A population-based cohort study of 120,251 pregnant, obese women delivering full-term, live born, singleton infants investigated the risk of four pregnancy outcomes including pre-eclampsia (PET), Caesarean section (CS), small for gestational age (SGA) and large for gestational age (LGA) by obesity class and total gestational weight gain. The results showed that gestational weight gain of less than 15 lb for overweight or obese pregnant women was associated with a significantly lower risk of PET, CS, and LGA and higher risk of SGA birth. The authors concluded that limited or no weight gain during pregnancy in obese pregnant women results in a more favourable pregnancy outcome. The same study showed that in obese women the LGA incidence is 15% [26].

Weight gain in pregnancy is a complex biological phenomenon that supports the function of growth and development of the foetus. Gestational weight gain is influenced not only by changes in maternal physiology and metabolism but also by placental metabolism. The placenta functions as an endocrine organ, a barrier and a transporter of substances between maternal and foetal circulation. Changes in maternal homeostasis can modify placental structure and function and thus impact on foetal growth rate. Conversely, placental function may influence maternal metabolism through alterations in insulin sensitivity and systemic inflammation.

The Institute of Medicine (IOM) has revised its guidelines for weight gain in pregnancy in 2009 [27]. According to its new recommendations, normal weight women should gain 25–35 lb (11.5–16 kg), overweight women should gain 15–25 lb (7–11.5 kg) and obese women (BMI > 30 kg/m^2) only 11–20 lb (5–9 kg). The recommended rates of weight gain in the second and third trimesters are 0.42 kg/week for normal weight women, 0.28 kg/week for overweight women and 0.22 kg/week for obese women. These calculations assume a 0.5–2 kg weight gain in the first trimester.

The IOM committee intends that its recommendations be used in concert with good clinical judgement. If the woman's gestational weight gain is not within the proposed guidelines, clinicians should consider other modifiable factors that may cause excessive or inadequate weight gain especially the presence of fluid retention/oedema. The adequacy and consistency of foetal growth should be assessed before suggesting modified target weight gain.

Guidelines for Management for Obese Women of Reproductive Age

Full implementation of the IOM Committee guidelines to improve reproductive health in obese women would mean:

Offering pre-conception services, including dietary counselling, advice on physical activity and contraception, to all overweight or obese women to help them reach a healthy weight before conceiving. This should reduce their obstetric risk as well as improve long-term health of their offspring.

Offering lifestyle advice services to all pregnant women to help them achieve the recommended weight gain targets and thereby reduce post-partum weight retention, improve their long-term health, normalise infant birthweight and offer an additional tool to help reduce childhood obesity.

Offering lifestyle advice services to all post-partum women. This may help them to conceive again at a healthy weight as well as improve their long-term health.

Currently, there are ongoing studies in the United Kingdom to determine whether the addition of insulin sensitisers like metformin to lifestyle interventions would improve the neonatal and pregnancy outcomes in obese non-diabetic pregnant women.

Management of Obesity in Pregnancy

Dietary Approaches

Intermittent fasting during pregnancy is not recommended and is associated with a higher incidence of gestational diabetes mellitus and induction of labour [28]. However, milder caloric restrictions of 1600–1800 Kcal/day are beneficial for weight gain without the risks of ketosis to the foetus [29]. Based on a Cochrane review, severe protein and energy restriction for obese and overweight women are unlikely to be beneficial and may even be harmful [30].

The available evidence supports balanced dietary interventions in overweight women before conception. Even though the results are often conflicting, there is evidence to suggest that diets rich in protein, fat or high gastrointestinal carbohydrates as well as very low caloric diets are best avoided in pregnancy. Nevertheless, most of the studies are observational in nature, and despite some strong associations, the content of the diet cannot necessarily be implicated in the pregnancy outcomes. Additionally, methodological limitations include a large number of confounding variables, selection and recall bias.

Physical Activity

A Cochrane review on the effects of physical activity on pregnancy concluded that the available data were 'insufficient to infer important risks or benefits to the mother or infant' [31]. Guidelines from the American College of Obstetricians and Gynaecologists [32] recommended that pregnant women should exercise for 30 min or more on most days of the week and participate in moderate intensity exercise unless there were medical or obstetric complications. Whilst these recommendations have been widely adopted they are consensus rather than evidence based.

Behavioural Interventions

Claesson et al. [33] used a 'motivational' talk approach in early pregnancy followed by an invitation to an aqua-aerobic class and then followed weekly by a midwife. This programme resulted in obese women in the intervention group gaining significantly less weight compared to the control group (8.7 vs 11.3 kg) independently of socio-demographic background. The authors credited the increased frequency of contact with the health care professional for the success of their lifestyle intervention compared to previous studies.

More recently Asbee et al. [34] offered pregnant women with BMIs of less than 40 kg/m^2 a simple intervention consisting of a single contact with the dietician at the initial visit where standardised dietary counselling was provided aiming for a diet consisting of 40% carbohydrates, 30% protein, 30% fat and moderate intensity exercise. This approach reduced the weight gain in the intervention group compared to the control group (13 vs 16.1 kg). It should be noted that none of the above lifestyle interventions had any significant effect on birthweight or any other pregnancy outcomes.

Pharmacological Interventions

The only available anti-obesity medication, Orlistat, is not licensed for use in pregnancy. Metformin has been increasingly used in pregnant women with diabetes mellitus or polycystic ovarian syndrome, i.e. states of increased insulin resistance.

Limited data suggest that in combination with caloric restriction it reduces pregnancy weight gain in PCOS [35]. A trial of metformin in pregnancy is needed to clarify its potential benefits for obese pregnant women. The authors are currently undertaking such a study.

Bariatric Surgery

Two recent reviews of case control and cohort trials showed that bariatric surgery improved fertility and unlike lifestyle intervention, decreased a number of maternal and foetal/neonatal complications associated with obesity [36,37].

Patients undergoing these procedures must be managed by an experienced multidisciplinary team and monitored for nutritional deficiencies and surgical complications in any subsequent pregnancies. The safety and timing of gastric surgical procedures need to be further investigated by controlled clinical trials.

References

1. Chadwick J, Mann WN. *The Medical Works of Hippocrates*. Oxford; 1950.
2. *WHO Factsheet No. 311* – Obesity and Overweight. September 2006. Geneva. http://www.who.int/mediacentre/factsheets/fs311/en/index.html.
3. Lashen H, Fear K, Sturdee DW. Obesity is associated with increased risk of first trimester and recurrent miscarriage: matched case control study. *Human Reprod*. 2004;19(7):1644–1646.
4. Maheshwari A, Stofberg L, Bhattacharya S. Effect of overweight and obesity on assisted reproductive technology – a systematic review. *Hum Reprod Update*. 2007;13(5):433–444.
5. Van der Steeg JW, Steures P, Eijkemans MJC, et al. Obesity affects spontaneous pregnancy chances in subfertile, ovulatory women. *Human Reprod*. 2008;23(2):324–328.
6. Bacon CG, Mittleman MA, Kawachi I, et al. Sexual function in men older than 50 years of age: results from the Health Professionals Follow-Up Study. *Ann Intern Med*. 2003;139:161–168.
7. Sullivan ME, Thompson CS, Dashwood MR, et al. Nitric oxide and penile erection: is erectile dysfunction another manifestation of vascular disease? *Cardiovasc Res*. 1999;43:658–665.
8. Ranadive S, Vaisse C. Lessons from extreme human obesity: monogenic disorders. *Endocrinal Metab Clin North Am*. 2008;37(3):733–751.
9. Farooqi IS, O'Rahilly S. Genetics of obesity in humans. *Endocr Rev*. 2006;27(7):710–718.
10. Budak E, Fernandez M, Bellver J. Interactions of the hormones leptin, ghrelin, adiponectin, resistin, and PYY3-36 with the reproductive system. *Fertil Steril*. 2006;85:1563–1581.
11. Agarwal SK, Vogel K, Weitsman SR, Magoffin DA. Leptin antagonises the insulin-like growth factor-1 augmentation of steroidogenesis in granulose and theca cells of the human ovary. *J Clin Endocrinol Metab*. 1999;84:1072–1076.
12. Duggal PS, Van Der Hoek KH, Milner CR, et al. The in vivo and in vitro effects of exogenous leptin on ovulation in rat. *Endocrinology*. 2000;141:1971–1976.

13. Bellver J, Melo MA, Bosch E. Obesity and poor reproductive outcome: the potential role of the endometrium. *Fertil Steril*. 2007;88:446–451.
14. Brody S. Slimness is associated with greater intercourse and lesser masturbation frequency. *J Sex Marital Ther*. 2004;30:251–261.
15. Trischitta V. Relationship between obesity-related metabolic abnormalities and sexual function. *J Endocrinol Invest*. 2003;26:62–64.
16. The Rotterdam ESHRE/ASRM-Sponsored PCOS Consensus Workshop Group. Revised 2003 consensus on diagnostic criteria and long-term health risks related to polycystic ovary syndrome (PCOS). *Hum Reprod*. 2004;19:41–47.
17. Ehrmann DA. Polycystic ovary syndrome. *N Engl J Med*. 2005;352:1223–1236.
18. Gambineri A, Pelusi C, Vicennati V, Pagotto U, Pasquali R. Obesity and the polycystic ovary syndrome. *Int J Obes Rel Metab Disord*. 2002;26:883–896.
19. Von Shoultz B, Calstrom K. On the regulation of sex-hormone-binding globulin. A challenge of old dogma and outlines of an alternative mechanism. *J Steroid Biochem*. 1989;32:327–334.
20. Mitchell M, Armstrong DT, Robker RL, Norman RJ. Adipokines: implications for female fertility and obesity. *Reproduction*. 2005;130:583–597.
21. Marquard KL, Stephens MS, Jungheim ES, et al. Polycystic ovary syndrome and maternal obesity affect oocyte size in in vitro fertilisation/intracytoplasmic sperm injection cycles. *Fertil Steril*. 2011;95(6).
22. Pasquali R, Gambineri A. Treatment of polycystic ovary syndrome with lifestyle intervention. *Curr Opin Endocrinol Metab*. 2002;9:459–468.
23. Hoeger KM, Kochman L, Wixom N, et al. A randomised, 48-week, placebo-controlled trial of intensive lifestyle modification and/or metformin therapy in overweight women with polycystic ovary syndrome: a pilot study. *Fertil steril*. 2004;82:421–429.
24. RCOG. *Confidential Enquiry into Maternal and Child Health (CEMACH) Saving Mothers' Lives: Reviewing Maternal Deaths to Make Motherhood Safer – 2003–2005. Seventh Report of the Confidential Enquiries into Maternal Deaths in the United Kingdom, December 2007*. London: RCOG Press; 2007.
25. Galtier-Dereure F, Boegner C, Bringer J. Obesity and pregnancy: complications and cost. *Am J Clin Nutr*. 2000;71(suppl):1242S–1248S.
26. Kiel DW, Dodson EA, Artal R, Boehmer TK, Leet TL. Gestational weight gain and pregnancy outcomes in obese women: how much is enough? *Obstet Gynecol*. 2007;110 (4):743–744.
27. Rasmussen KM, Yaktine AL, eds. *Weight Gain in Pregnancy: Re-examining The Guidelines*. Washington, DC: The National Academies Press; 2009.
28. Mirghani HM, Hamud OA. The effect of maternal diet restriction on pregnancy outcome. *Am J Perinatol*. 2006;23:21–24.
29. Reader DM. Medical nutrition therapy and lifestyle interventions. *Diabetes Care*. 2007;30(suppl 2):S188–S193.
30. Kramer MS, Kakuma R. Energy and protein intake in pregnancy. *Cochrane Database Syst Rev*. 2003;(4):CD000032. doi:10.1002/14651858.
31. Kramer MS, McDonald SW. Aerobic exercise for women during pregnancy. *Cochrane Database Syst Rev*. 2006;3:CD000180.
32. ACOG committee opinion. Exercise during pregnancy and the postpartum period – Number 267. *Int J Gynaecol Obstet*. 2002;77:79–81.
33. Claesson IM, Sydsjö G, Brynhildsen J, et al. Weight gain restriction for obese pregnant women: a case–control intervention study. *BJOG*. 2008;115:44–50.

34. Asbee SM, Jenkins TR, Butler JR, White J, Elliot M, Rutledge A. Preventing excessive weight gain during pregnancy through dietary and lifestyle counseling: a randomized controlled trial. *Obstet Gynecol.* 2009;113:305–312.
35. Glueck CJ, Goldenberg N, Wang P, Loftspring M, Sherman A. Metformin during pregnancy reduces insulin, insulin resistance, insulin secretion, weight, testosterone and development of gestational diabetes: prospective longitudinal assessment of women with polycystic ovary syndrome from preconception throughout pregnancy. *Hum Reprod.* 2004;19:510–521.
36. Maggard MA, Yermilov I, Li Z, et al. Pregnancy and fertility following bariatric surgery: a systematic review. *JAMA.* 2008;300:2286–2296.
37. Guelinck I, Devlieger R, Vansant G. Reproductive outcome after bariatric surgery: a critical review. *Hum Reprod Update.* 2009;15:189–201.

23 Pregnancy and Metabolic Syndrome of Obesity

Shahzya S. Huda[1] *and Scott M. Nelson*[2]

[1]Forth Valley Royal Hospital, Larbert, UK, [2]School of Medicine, University of Glasgow, Glasgow, UK

Introduction

Obesity is associated with a broad range of medical complications including diabetes [1], cardiovascular disease [2], dyslipidaemia and hypertension [3], cancer [4] and osteoarthritis [5]. Moreover, obesity is associated with excess mortality, with a continuous linear relationship with overall mortality from a body mass index (BMI) of >22.5 kg/m^2. Most of the excess mortality is likely to be directly attributable to excess adiposity and its inherent increase of complications including vascular disease [6]. Similar to the non-pregnant situation, maternal obesity is related to metabolic complications, including the pregnancy-specific conditions of pre-eclampsia and gestational diabetes. This chapter endeavours to describe the metabolic milieu of obese women which may predispose to the development of these metabolic complications. In addition, we will examine the evidence as regards management of obesity in terms of optimising gestational weight gain and potential intervention strategies in an attempt to minimise metabolic consequences for both mother and baby.

Lipid Metabolism and Fat Deposition in Lean Pregnancy

Maternal metabolism during pregnancy adapts to benefit the growth and development of the foetus and can be broadly divided into two phases. During the initial two-thirds of gestation, when foetal energy demands are limited, maternal fat stores increase [7]. This is attributable in part to maternal behavioural change including hyperphagia [8] and to increased adipose tissue (AT) lipogenesis [9]. In early pregnancy, insulin sensitivity is normal or even slightly improved with normal peripheral sensitivity to insulin and hepatic basal glucose production [10]. This metabolic environment together with pregnancy-related endocrine changes including increasing levels of oestrogen, progesterone and cortisol favours lipogenesis and fat accumulation [11].

Obesity. DOI: http://dx.doi.org/10.1016/B978-0-12-416045-3.00023-6

During the latter stages of pregnancy, this anabolic state switches to a state of catabolism with a marked increase in lipolysis rates and a corresponding rise in maternal free fatty acids (FFAs) and glycerol [12,13]. This change is enhanced by increased production and activity of hormone-sensitive lipase, with a concomitant decrease in lipoprotein lipase activity [14]. Exaggerated catecholamine release in response to even modest maternal hypoglycaemia and the insulin-resistant state of late pregnancy contribute to this switch [15,16]. Insulin effects on lipolysis (AT) and fat oxidation (in liver and muscle) are significantly impaired during the third trimester compared to earlier in pregnancy and also post-partum [16]. Reduced expression of peroxisome proliferator-activated receptor gamma (PPAR-γ) and its target genes may also contribute to accelerated fat metabolism in late pregnancy [17]. This catabolic state corresponds to the time of maximum foetal weight gain, and by moving towards a primarily lipid-based metabolism in the mother, this inevitably increases availability of glucose and amino acids for the foetus [18].

As gestation advances, there are marked changes in plasma cholesterol and triglyceride concentrations, which rise by 25–50% and 200–400%, respectively. The increase in triglyceride is mainly due to very low-density lipoprotein (VLDL) triglyceride which shows a threefold increase from 14 weeks' gestation to late pregnancy [19]. These changes in triglyceride and VLDL are driven by increased AT lipolysis resulting in increased delivery of FFA and glycerol to the liver where they are re-esterified for the synthesis of triglycerides and incorporated into VLDL. Despite a rise in Triglycerides (TG) in normal pregnancy, high-density lipoprotein (HDL)-cholesterol levels are elevated by the 14 week and rise by a maximum of around 40% at 28 weeks' gestation mainly due to an increase in the HDL2 subfraction with a proportional fall in HDL3a and HDL3b [19,20]. The mean concentration of HDL-cholesterol is around 2 mmol/l compared to around 1.5 mmol/l in the non-pregnant [21]. With respect to LDL, during normal pregnancy it rises by around 70% [21]. Although this increase is less marked than TG, there are some important qualitative changes in the LDL composition favouring a more 'atherogenic' profile with a proportional increase in small dense LDL (LDLIII) in late pregnancy [21,22]. In keeping with the non-pregnant population, this is driven by higher plasma triglycerides where a "threshold" effect may be seen [21].

All women increase maternal fat stores in early pregnancy irrespective of pre-pregnancy adiposity to meet the foeto-placental and maternal demands of late gestation and lactation. Women of normal weight gain around 3.8 kg of fat [23] during pregnancy although there is substantial variation [24–26]. Total fat appears to increase to a peak towards the end of the second trimester before diminishing which corresponds to the period of increased lipolytic activity [27,28]. In women of normal weight, the majority of fat is accumulated centrally in the subcutaneous compartment of the trunk and upper thigh [29,30]. In later stages of pregnancy there is an increase in both the thickness of pre-peritoneal fat (visceral) and the ratio of pre-peritoneal to subcutaneous fat as measured by ultrasound [31]. This pattern may be relevant to increasing insulin resistance (IR) and lipid changes that occur as pregnancy progresses. Indeed, accumulation of hepatic fat has been shown to be an important mediator of IR during pregnancy in the rat model [32].

Lipid Metabolism and Fat Deposition in Obese Women

There are some important qualitative differences in lipid metabolism and accrual of adiposity in obese compared to lean women in pregnancy. It is increasingly recognised that the location of AT accumulation is also an important determinant of metabolic risk in addition to total volume of fat. Visceral adiposity is more closely related to adverse metabolic outcomes including IR, hyperinsulinaemia, dyslipidaemia, hypertension and the metabolic syndrome [33]. In visceral fat there is a higher turnover of lipids due to its greater sensitivity to catecholamine-induced lipolysis and decreased sensitivity to insulin. Visceral fat is in direct contact with the liver via the portal venous system. The liver is therefore exposed to chronic elevation of non essential fatty acids (NEFA), which produces alteration in liver metabolism and promotes hepatic IR – the basis for the 'portal paradigm'. In addition, visceral fat may further influence increased IR through a variety of inflammatory pathways [34]. Alternatively, rather than visceral fat being directly or fully responsible for the metabolic dysregulation of obesity, its rising volume may also simply signal saturation of 'good' fat storage capacity (Figure 23.1). Drolet et al. [35] have proposed a model whereby subcutaneous fat acts as the primary fat depot and when the storage capacity of this depot is reached 'overspill' into secondary fat depots including, amongst other tissues, visceral fat occurs. Excess subcutaneous fat appears to be metabolically favourable [36]. The much higher subcutaneous storage capacity of women compared to men and therefore less propensity to accumulate fat in the visceral compartment likely explains the lower prevalence of metabolic disturbances, and diabetes in middle-age, in women compared to men. This theory may also in part explain why certain racial groups such as South Asians are at increased susceptibility to central obesity and its metabolic consequence [37]. Similarly, in certain chronic illnesses, e.g. HIV, a loss of subcutaneous fat storage capacity may 'push' fat more centrally into key metabolic organs and instigate greater IR [38]. Interestingly, fat from 'ectopic sites' such as visceral fat is preferentially lost with modest weight loss and may explain how modest weight loss appears to provide significant metabolic and clinical benefits [39].

Interestingly, lean subjects have a greater increase in per cent of body fat in pregnancy compared with obese subjects, but there is no difference in actual total fat mass [30]. However, as obese pregnant women have more saturated subcutaneous fat stores as outlined above, they tend to accumulate fat more centrally than lean women, at least as estimated by using the skin-fold thickness technique, an observation which may reflect their more insulin-resistant state [30]. Visceral adiposity or central obesity appears to be correlated more strongly with adverse metabolic outcomes in pregnancy including gestational diabetes mellitus (GDM), gestational hypertension and pre-eclampsia [40–42]. Furthermore, visceral

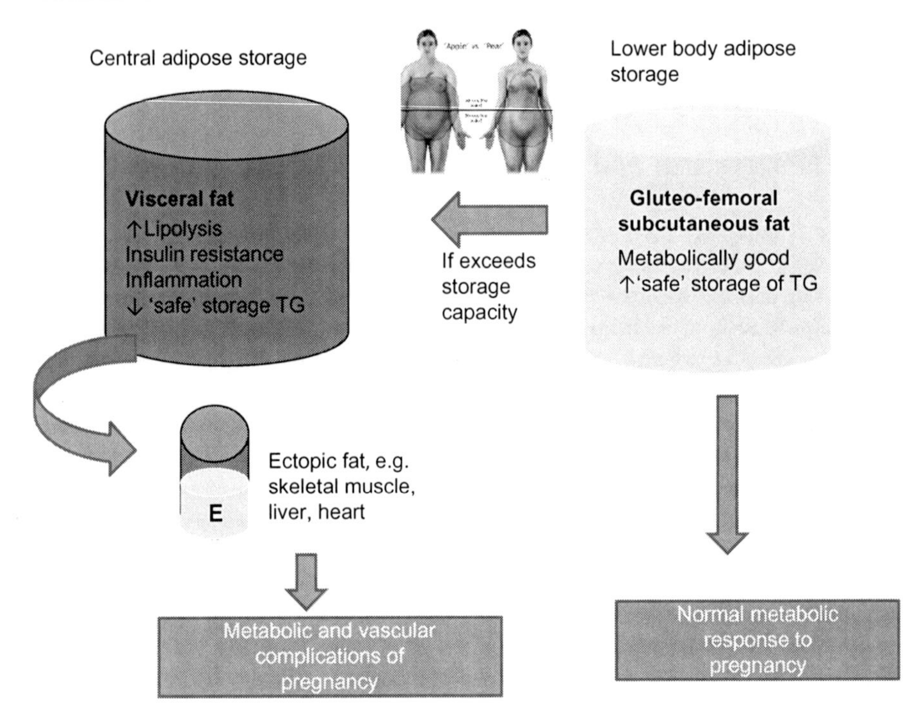

Figure 23.1 Proposed model of metabolic consequences in pregnancy related to the regional variation in storage of excess AT. Subcutaneous fat acts as the primary fat depot and when the storage capacity of this depot is reached, 'overspill' into secondary fat depots including visceral fat (VF) and ectopic (E) sites occurs, leading to increased IR and metabolic complications of pregnancy. The capacity of subcutaneous fat to store fat is determined by several factors including sex, ethnicity, genes and chronic illness. Obese women tend to accumulate more fat centrally than lean women in pregnancy. Lower body subcutaneous fat is much more insulin sensitive and is independently associated with a reduced risk of metabolic dysfunction in non-pregnant women.

adiposity in early pregnancy appears to correlate better than subcutaneous fat or BMI with metabolic risk factors such as blood pressure, IR and lipids [43]. Lower body subcutaneous (gluteo-femoral) fat is much more insulin sensitive and is independently associated with a reduced risk of metabolic dysfunction in the non-pregnant, and larger lower body fat stores are associated with more efficient storage of dietary fat [44,45] (Figure 23.1). Thus, the compartment in which fat is stored may be a critical determinant of the metabolic response to pregnancy, as opposed to the total amount of fat accrued.

Furthermore, the physiological hyperlipidaemia of pregnancy is exaggerated in obese women with higher serum triglyceride, VLDL cholesterol and FFA concentrations than those observed in lean women [46,47]. This is seen together with lower levels of the endothelial protective HDL-cholesterol, although LDL-cholesterol and total cholesterol concentrations appear similar [47]. Despite the similar concentration of LDL, its susceptibility to oxidation, a classic associate of endothelial dysfunction, atherosclerosis and cell toxicity, appears to be exacerbated by maternal obesity [48]. Collectively, this pattern of dyslipidaemia is similar to that of the metabolic syndrome in the non-pregnant population [49].

Glucose Metabolism in Lean and Obese Pregnancy

Normal pregnancy is associated with marked changes in glucose metabolism and IR to facilitate provision of fuel substrate for the foetus. In early pregnancy, insulin secretion increases, while insulin sensitivity is unchanged, or even slightly improved [50,51]. However as pregnancy progresses, insulin-mediated glucose utilisation worsens by 40–60% and insulin secretion increases several fold in order to maintain euglycaemia in the mother [10,52]. Skeletal muscle is the primary site of glucose disposal, and along with AT, becomes severely insulin resistant during the latter half of pregnancy. The ability of insulin to suppress lipolysis is also reduced during late pregnancy contributing to greater postprandial increase in FFAs, increased gluconeogenesis and IR [53].

Obesity has considerable effects on glucose metabolism in pregnancy with a loss of the reduction in fasting glucose in early pregnancy and significant enhancement of peripheral and hepatic IR, which manifests as reduced insulin-mediated glucose disposal, a large reduction in insulin-stimulated carbohydrate oxidation and a reduction in insulin suppression of endogenous glucose production [54,55]. Importantly, the overall effects of this impaired IR are not limited to glucose. In the postprandial state, this obesity-related IR exaggerates the normal circulatory increases in metabolic fuels, i.e. glucose, lipids and amino acids, exposing the foetus to increased availability of these nutrients. Indeed, these alternative nutrient pathways may independently contribute to macrosomia as higher maternal serum triglycerides and amino acid profiles have been associated with higher offspring birthweight independent of maternal glucose or pre-pregnancy BMI [56–60].

Interestingly, obese women do not show any alteration in either basal carbohydrate oxidation or non-oxidisable carbohydrate metabolism from early to late pregnancy in contrast to a 50–80% increase in basal fat oxidation [26]. This corresponds to the period of fat accumulation and reduced insulin sensitivity. In addition, women with a high BMI have the largest increase in their basal metabolic rate compared to women with normal or low BMI [61,62]. These differences lend support to the hypothesis that there is an adaptive process in obese individuals who are insulin resistant to prevent additional weight gain [63].

Not surprisingly obese women are therefore four times more likely and severely obese women almost nine times more likely to develop GDM than lean women [64]. Women who develop GDM are predisposed to developing type 2 DM in later life with the risk greatest in those who are obese [65]. The relationship between maternal obesity and diabetes is explored in more detail in Chapter 14.

Adiposity, Obesity and IR: Potential Mechanisms

The link between obesity and IR has been recognised for many years. The mechanisms for this increased risk are multi-factorial and include effects on insulin signalling similar to those seen in obese non-pregnant [17]. Subclinical inflammation and increased release of inflammatory mediators including tumour necrosis factor-α (TNFα) are also potentially implicated [66,67]. Furthermore, there is adaptation of AT function including significantly lower circulating levels of adiponectin, an insulin-sensitising hormone, and reduction in the lipogenic transcription factor PPAR-γ resulting in reduced AT insulin sensitivity and increased lipolysis [17,68,69], and this association has been attributed to disturbances in adipocyte function and metabolism. The view that AT is simply a storage organ of excess triglycerides has changed dramatically over recent years. It has been shown to secrete a diverse range of cytokines, proteins and signals which have both paracrine and endocrine actions and a wide-ranging influence on the metabolic and physiological function of other organs [70]. In particular, fat cells secrete factors involved in inflammation (TNFα, interleukin-6), haemostasis plasminogen activator inhibitor-1 (PAI-1), insulin sensitivity (adiponectin) and energy balance and control of appetite (leptin). A state of chronic positive energy balance such as obesity leads to increased storage of triglycerides which results firstly in adipocyte hypertrophy and subsequently hyperplasia through adipogenesis [71]. This in turn leads to cellular dysfunction resulting in dysregulated release of adipokines, increased release of FFAs and inflammation. Excess circulating FFA, in turn, contributes to accumulation of fat in ectopic sites including skeletal muscle, liver, heart and pancreatic β-cells [72]. Ectopic fat primarily in skeletal muscle is thought to promote peripheral IR by a reduction in insulin-mediated glucose uptake [73]. In addition, the exposure of the liver to elevated FFA leads to reduced hepatic insulin extraction leading to systemic hyperinsulinaemia and accelerated gluconeogenesis [74]. Excess FFA can also lead to decreased function and apoptosis of pancreatic β-cells contributing still further to the state of relative insulin deficiency [75].

Obesity is a state of chronic low-grade inflammation, and this is considered to provide the crucial link between obesity and IR. Inflammation which occurs in metabolically important organs such as the liver and AT is aberrant in nature and has been referred to as meta-inflammation [76]. The common pathway in which inflammation leads to IR is considered to be through modulation of insulin signalling and in particular induces serine phosphorylation of the insulin receptor substrate 1 (IRS1). Our knowledge of the mechanisms through which this occurs is becoming increasing complex [77]. One of the central pathways that links inflammation and IR is through activation of the kinase JUN N-terminal kinase (JNK) which occurs in response to a variety of stress signals including FFAs, pro-inflammatory cytokines and reactive oxygen species [78]. JNK activation has been found to be crucial in the development of IR in obese animal models – JNK1 deficient mice results in marked protection from IR and type 2 diabetes [79]. Furthermore, studies have suggested that stress on endoplasmic reticulum may provide the key link between obesity, JNK activation and IR [80].

Amino Acid Metabolism in Lean and Obese Pregnancy

In pregnancy, the mother uses the majority of amino acids for protein synthesis, with a reduction in the amount oxidised by approximately 10% [81]. Although counter-intuitively there is no increase in measured protein synthesis in the first trimester, there is an increase in the second and third trimesters of 15% and 25%, respectively [82,83]. These are greater than can simply be accounted for by highly active protein synthesis in the foetus and placenta, implying an overall increase in protein synthesis in maternal tissues including the liver, breasts and uterus. The impact of maternal protein turnover on the foetus is striking, with a greater maternal protein synthesis in the second trimester being associated with an increase in birth length and accounting for 26% of its overall variance [84].

At present, the impact of obesity on maternal and foetal amino acid metabolism is unknown. However, in non-pregnant obese women, protein synthesis is stimulated less in a hyperinsulinaemic state in comparison with lean women, with no difference in protein oxidation [85]. Obesity is also associated with a greater supply of gluconeogenic amino acids to the liver with preference of their use over glycogen for glucose production [86]. Lastly, visceral lean mass is positively correlated with maternal protein turnover [84]. Collectively, these data would suggest that the anabolic response to pregnancy may be impaired in obese women, raising the possibility that mechanisms may exist to limit what would otherwise be greater foetal growth in a hyperinsulinaemic and glucose-rich environment.

Pre-eclampsia, Obesity and Metabolic Syndrome of Pregnancy

Pregnancy-induced hypertension and pre-eclampsia are important considerations in obese pregnancy with the risk of pre-eclampsia doubling for every 5–7 kg/m^2 increase in BMI [87]. Greer and Sattar [88] proposed a model whereby pregnancy and its concomitant digression into a metabolic syndrome is a ‘stress test’ of maternal metabolic response; women who develop adverse pregnancy outcomes such as pre-eclampsia make greater excursions into metabolic disturbances during pregnancy. Obesity brings women closer to the threshold over which metabolic disease in pregnancy manifests itself. Pre-eclampsia with its observed rises in pro-inflammatory cytokines, endothelial dysfunction, IR and dyslipidaemia holds many parallels with metabolic syndrome in the non-pregnant and is generally in keeping with the metabolic and vascular phenotype of obesity. Although it is unclear to what extent either genetic or environmental factors contribute to this disease, IR is a well-recognised correlate and may explain recent rises in the incidence of pre-eclampsia due to the maternal obesity epidemic and indeed may in part explain the increased risk of cardiovascular disease in later life of women with a history of pre-eclampsia [89]. This relationship between obesity and pre-eclampsia is further explored in Chapter 18.

Intervention Strategies in Maternal Obesity and Effects on Metabolism

Due to the increasing burden of maternal obesity, there has been a plethora of interest in the development of lifestyle interventions, which may have the potential to improve adverse reproductive outcomes. As described above, we have good understanding of the changes in metabolism accompanying obesity in pregnancy and central to these is the development of IR and its metabolic sequelae. Due to the strong similarities between the risk profile for type 2 diabetes and those that characterise maternal obesity examination of the extensive literature addressing interventions in type 2 DM may provide guidance in the choice of intervention amongst obese pregnant women. Lifestyle intervention is now a critical component of the treatment strategy for diabetes, hypertension, cardiovascular disease and obesity in non-pregnant patients [90,91]. Importantly, effective lifestyle intervention strategies can prevent or at least delay the progression to type 2 diabetes in high-risk individuals [92–96]. Notably, the Finnish Diabetes Prevention Study, in addition to a 58% reduction in the incidence of diabetes incidence, also achieved within 1 year, a reduction in weight, BMI, waist circumference, fasting plasma glucose, 2 h plasma glucose, serum triglycerides and serum total cholesterol: HDL cholesterol ratio in the intervention group [92]. In the US Diabetes Prevention Program, attainment of 7% weight loss was achieved by 50% of participants at 24 weeks and 74% had achieved the physical activity targets, importantly this was also accompanied by reductions in plasma glucose [96]. These studies raise the exciting possibility that lifestyle modification in a similar form could be applied to pregnancy to prevent the onset of metabolic and obesity-related complications.

Currently, the majority of those underway in pregnancy or the preliminary studies already published focus on prevention of excessive gestational weight gain as defined by the Institute of Medicine (IOM). Observational studies show that women whose gestational weigh gain (GWG) falls within the 1990 IOM guidelines experience a better pregnancy outcome, and the new IOM 2009 guidelines provide new and more evidence-based targets for each category and, for the first time, a specific weight gain range for obese women (5–9 kg) [97] (Table 23.1). However, excessive GWG is only weakly associated with several of the primary abnormalities linked to obesity including GDM and pre-eclampsia, compared to pre-pregnancy BMI [5]. Nevertheless, if observational studies are translatable to effects of intervention, prevention of excessive GWG amongst pregnant women could reduce the risk of large for gestational age (LGA), Caesarean section, postpartum weight gain and potentially, childhood obesity.

In contrast, in view of the close association between obesity and IR and the proposed role that this plays in GDM, pre-eclampsia and macrosomia, maternal IR may be an alternative primary outcome in intervention studies. Lifestyle modification for the prevention of gestational diabetes has shown some promise in terms of controlling birthweight of newborns [98,99] and reduction of maternal fasting glucose [99,100]. These interventional strategies for primary prevention of GDM

Table 23.1 Recommendation for Total and Rate of Weight Gain During Pregnancy by Pre-pregnancy BMI

Pre-pregnancy BMI	BMI (kg/m^2) (WHO)	Total Weight Gain Range (lb)	Rates of Weight Gain II and III Trimester (Mean Range in lb/week)
Underweight	<18.5	28–40	1 (1–1.3)
Normal weight	18.5–24.9	25–35	1 (0.8–1)
Overweight	25.0–29.9	15–25	0.6 (0.5–0.7)
Obese (all classes)	≥30.0	11–20	0.5 (0.4–0.6)

Source: Adapted from Ref. [97].

do provide a template for trials in obesity. In practical terms, the lifestyle interventions which are offered to obese women to limit weight gain and to prevent IR, namely increased physical activity and dietary advice and individual counselling, may differ only slightly from those used to treat GDM, although they may start earlier in pregnancy.

Physical activity is a modifiable factor, which may reduce IR during pregnancy, and is likely to reduce GWG. In a recent study from the Project Viva cohort, mid-pregnancy walking (OR 0.92; 95% confidential interval (CI) 0.83–1.01, per 30 min/day) and vigorous physical activity (OR 0.76; 95% CI 0.60–0.97, per 30 min/day) were inversely associated with excessive GWG [101]. Dietary advice, whether it is to reduce IR or calories, involves avoidance of simple sugars and saturated fats with adherence to a 'healthy diet', but the emphasis will slightly differ. Dietary energy density is also a modifiable factor, which may assist pregnant women to manage weight gain, but this remains to be proven in adequately powered trials [101–104]. Although in some studies dietary glycaemic load has not been found to be associated with GWG [104], one study suggested that low glycaemic index diet does reduce GWG [100]. Focusing on prevention of weight gain, if misinterpreted by the pregnant women, has the disadvantage that it may increase the risk of inappropriate fasting or excessive caloric restriction leading to excessive ketonuria and ketonemia, and the potential to adversely impact upon neurocognitive and motor skill development of the offspring [105,106]. A focus on dietary quality may have beneficial effects without such risks.

Perhaps, the greatest challenge for obese women, however, is to achieve any behavioural changes in diet or physical activity. Understanding barriers to behavioural change using validated instruments in obese pregnant women is a prerequisite in development of a successful intervention. Obese women are likely to have low self-esteem, and as pre-pregnancy weight increases, so do psychosocial measures of perceived stress, trait anxiety and depressive symptoms [107]. Furthermore, women who gain excess weight in pregnancy relative to the IOM

1990 guidelines are more likely to demonstrate symptoms of depression [108]. The barriers to physical activity are also substantial, with 85% of women identifying lack of time, tiredness and the inherent physical constraints of pregnancy [109]. Socio-economic factors are also important as pregnant women who are younger, less educated, with a higher BMI and who have more children are more likely to eat a poor quality and energy-dense diet [110]. Pilot trials of complex interventions are an essential preliminary step [111], with objective measurement of dietary and physical activity before and after the intervention, to prove efficacy of the intervention prior to embarking on large randomised, controlled trials. These feasibility studies also offer insight into the practical issues which underpin success or failure [112]. Disappointingly, however, a recent systematic review of nine randomised, controlled trials comparing antenatal dietary and/or lifestyle with no treatment for overweight or obese women involving 743 women did not identify any significant difference in LGA infants, gestational weight gain or GDM [113]. This may reflect the heterogeneity of the studies and interventions and relatively small numbers of participants. This emphasises the need for larger well–designed, randomised, controlled trials with targeting of women at highest risk of metabolic complications.

Conclusion

Maternal metabolism undergoes dramatic changes during pregnancy, with further excursion into a 'metabolic syndrome' of pregnancy in obesity. Interestingly, however, not all obese women develop metabolic complications of pregnancy. Although we have a good understanding of metabolic changes in pregnancy, the identification of obese women at most risk is still to be determined. Perhaps, the key to this may be more focused on assessment of AT distribution and accrual as opposed to pre-pregnancy BMI. At present, disappointingly, intervention studies have failed to show significant differences in neonatal and maternal outcomes. Clearly, therefore, accurate targeting of interventions to those at greatest risk will be more cost-effective, with development of algorithms similar to those used for prediction of pre-eclampsia and/or GDM. Several randomised trials are planned or underway which should determine the best strategy for effective intervention.

References

1. Vazquez G, Duval S, Jacobs Jr DR, Silventoinen K. Comparison of body mass index, waist circumference, and waist/hip ratio in predicting incident diabetes: a meta-analysis. *Epidemiol Rev*. 2007;29:115–128.
2. Romero-Corral A, Montori VM, Somers VK, et al. Association of bodyweight with total mortality and with cardiovascular events in coronary artery disease: a systematic review of cohort studies. *Lancet*. 2006;368(9536):666–678.

3. Brown CD, Higgins M, Donato KA, et al. Body mass index and the prevalence of hypertension and dyslipidemia. *Obes Res*. 2000;8(9):605–619.
4. Wolk A, Gridley G, Svensson M, et al. A prospective study of obesity and cancer risk (Sweden). *Cancer Causes Control*. 2001;12(1):13–21.
5. Lievense AM, Bierma-Zeinstra SM, Verhagen AP, van Baar ME, Verhaar JA, Koes BW. Influence of obesity on the development of osteoarthritis of the hip: a systematic review. *Rheumatology (Oxford)*. 2002;41(10):1155–1162.
6. Whitlock G, Lewington S, Sherliker P, et al. Body-mass index and cause-specific mortality in 900 000 adults: collaborative analyses of 57 prospective studies. *Lancet*. 2009;373(9669):1083–1096.
7. Villar J, Cogswell M, Kestler E, Castillo P, Menendez R, Repke JT. Effect of fat and fat-free mass deposition during pregnancy on birth weight. *Am J Obstet Gynecol*. 1992;167(5):1344–1352.
8. Douglas AJ, Johnstone LE, Leng G. Neuroendocrine mechanisms of change in food intake during pregnancy: a potential role for brain oxytocin. *Physiol Behav*. 2007;91 (4):352–365.
9. Ramos MP, Crespo-Solans MD, del Campo S, Cacho J, Herrera E. Fat accumulation in the rat during early pregnancy is modulated by enhanced insulin responsiveness. *Am J Physiol Endocrinol Metab*. 2003;285(2):E318–E328.
10. Catalano PM, Tyzbir ED, Roman NM, Amini SB, Sims EA. Longitudinal changes in insulin release and insulin resistance in nonobese pregnant women. *Am J Obstet Gynecol*. 1991;165(6 Pt 1):1667–1672.
11. Ryan EA, Enns L. Role of gestational hormones in the induction of insulin resistance. *J Clin Endocrinol Metab*. 1988;67(2):341–347.
12. Diderholm B, Stridsberg M, Ewald U, Lindeberg-Norden S, Gustafsson J. Increased lipolysis in non-obese pregnant women studied in the third trimester. *BJOG*. 2005;112 (6):713–718.
13. Catalano PM, Roman-Drago NM, Amini SB, Sims EA. Longitudinal changes in body composition and energy balance in lean women with normal and abnormal glucose tolerance during pregnancy. *Am J Obstet Gynecol*. 1998;179(1):156–165.
14. Martin-Hidalgo A, Holm C, Belfrage P, Schotz MC, Herrera E. Lipoprotein lipase and hormone-sensitive lipase activity and mRNA in rat adipose tissue during pregnancy. *Am J Physiol*. 1994;266(6 Pt 1):E930–E935.
15. Herrera EM, Knopp RH, Freinkel N. Urinary excretion of epinephrine and norepinephrine during fasting in late pregnancy in the rat. *Endocrinology*. 1969;84(2): 447–450.
16. Sivan E, Homko CJ, Chen X, Reece EA, Boden G. Effect of insulin on fat metabolism during and after normal pregnancy. *Diabetes*. 1999;48(4):834–838.
17. Catalano PM, Nizielski SE, Shao J, Preston L, Qiao L, Friedman JE. Downregulated IRS-1 and PPARgamma in obese women with gestational diabetes: relationship to FFA during pregnancy. *Am J Physiol Endocrinol Metab*. 2002;282(3):E522–E533.
18. Freinkel N. Banting Lecture 1980. Of pregnancy and progeny. *Diabetes*. 1980;29(12): 1023–1035.
19. Fahraeus L, Larsson-Cohn U, Wallentin L. Plasma lipoproteins including high density lipoprotein subfractions during normal pregnancy. *Obstet Gynecol*. 1985;66(4): 468–472.
20. Alvarez JJ, Montelongo A, Iglesias A, Lasuncion MA, Herrera E. Longitudinal study on lipoprotein profile, high density lipoprotein subclass, and postheparin lipases during gestation in women. *J Lipid Res*. 1996;37(2):299–308.

21. Sattar N, Greer IA, Louden J, et al. Lipoprotein subfraction changes in normal pregnancy: threshold effect of plasma triglyceride on appearance of small, dense low density lipoprotein. *J Clin Endocrinol Metab*. 1997;82(8):2483–2491.
22. Silliman K, Shore V, Forte TM. Hypertriglyceridemia during late pregnancy is associated with the formation of small dense low-density lipoproteins and the presence of large buoyant high-density lipoproteins. *Metabolism*. 1994;43(8):1035–1041.
23. Lederman SA, Paxton A, Heymsfield SB, Wang J, Thornton J, Pierson Jr RN. Body fat and water changes during pregnancy in women with different body weight and weight gain. *Obstet Gynecol*. 1997;90(4 Pt 1):483–488.
24. Lawrence M, Coward WA, Lawrence F, Cole TJ, Whitehead RG. Fat gain during pregnancy in rural African women: the effect of season and dietary status. *Am J Clin Nutr*. 1987;45(6):1442–1450.
25. Goldberg GR, Prentice AM, Coward WA, et al. Longitudinal assessment of energy expenditure in pregnancy by the doubly labeled water method. *Am J Clin Nutr*. 1993;57 (4):494–505.
26. Okereke NC, Huston-Presley L, Amini SB, Kalhan S, Catalano PM. Longitudinal changes in energy expenditure and body composition in obese women with normal and impaired glucose tolerance. *Am J Physiol Endocrinol Metab*. 2004;287(3):E472–E479.
27. Kopp-Hoolihan LE, van Loan MD, Wong WW, King JC. Fat mass deposition during pregnancy using a four-component model. *J Appl Physiol*. 1999;87(1):196–202.
28. Pipe NG, Smith T, Halliday D, Edmonds CJ, Williams C, Coltart TM. Changes in fat, fat-free mass and body water in human normal pregnancy. *Br J Obstet Gynaecol*. 1979;86(12):929–940.
29. Sohlstrom A, Wahlund LO, Forsum E. Total body fat and its distribution during human reproduction as assessed by magnetic resonance imaging. *Basic Life Sci*. 1993;60:181–184.
30. Ehrenberg HM, Huston-Presley L, Catalano PM. The influence of obesity and gestational diabetes mellitus on accretion and the distribution of adipose tissue in pregnancy. *Am J Obstet Gynecol*. 2003;189(4):944–948.
31. Kinoshita T, Itoh M. Longitudinal variance of fat mass deposition during pregnancy evaluated by ultrasonography: the ratio of visceral fat to subcutaneous fat in the abdomen. *Gynecol Obstet Invest*. 2006;61(2):115–118.
32. Einstein FH, Fishman S, Muzumdar RH, Yang XM, Atzmon G, Barzilai N. Accretion of visceral fat and hepatic insulin resistance in pregnant rats. *Am J Physiol Endocrinol Metab*. 2008;294(2):E451–E455.
33. Bays HE. 'Sick fat', metabolic disease, and atherosclerosis. *Am J Med*. 2009;122(suppl 1): S26–S37.
34. Shah A, Mehta N, Reilly MP. Adipose inflammation, insulin resistance, and cardiovascular disease. *JPEN*. 2008;32(6):638–644.
35. Drolet R, Richard C, Sniderman AD, et al. Hypertrophy and hyperplasia of abdominal adipose tissues in women. *Int J Obes (Lond)*. 2008;32(2):283–291.
36. Fox CS, Massaro JM, Hoffmann U, et al. Abdominal visceral and subcutaneous adipose tissue compartments: association with metabolic risk factors in the Framingham Heart Study. *Circulation*. 2007;116(1):39–48.
37. Sniderman AD, Bhopal R, Prabhakaran D, Sarrafzadegan N, Tchernof A. Why might South Asians be so susceptible to central obesity and its atherogenic consequences? The adipose tissue overflow hypothesis. *Int J Epidemiol*. 2007;36(1):220–225.
38. Brown TT, Xu X, John M, et al. Fat distribution and longitudinal anthropometric changes in HIV-infected men with and without clinical evidence of lipodystrophy and

HIV-uninfected controls: a substudy of the Multicenter AIDS Cohort Study. *AIDS Res Ther*. 2009;6:8.

39. Chaston TB, Dixon JB. Factors associated with percent change in visceral versus subcutaneous abdominal fat during weight loss: findings from a systematic review. *Int J Obes (Lond)*. 2008;32(4):619–628.
40. Zhang S, Folsom AR, Flack JM, Liu K. Body fat distribution before pregnancy and gestational diabetes: findings from coronary artery risk development in young adults (CARDIA) study. *BMJ*. 1995;311(7013):1139–1140.
41. Ijuin H, Douchi T, Nakamura S, Oki T, Yamamoto S, Nagata Y. Possible association of body-fat distribution with preeclampsia. *J Obstet Gynaecol Res*. 1997;23(1):45–49.
42. Sattar N, Clark P, Holmes A, Lean ME, Walker I, Greer IA. Antenatal waist circumference and hypertension risk. *Obstet Gynecol*. 2001;97(2):268–271.
43. Bartha JL, Marin-Segura P, Gonzalez-Gonzalez NL, Wagner F, Aguilar-Diosdado M, Hervias-Vivancos B. Ultrasound evaluation of visceral fat and metabolic risk factors during early pregnancy. *Obesity*. 2007;15(9):2233–2239.
44. Snijder MB, Visser M, Dekker JM, et al. Low subcutaneous thigh fat is a risk factor for unfavourable glucose and lipid levels, independently of high abdominal fat. The Health ABC Study. *Diabetologia*. 2005;48(2):301–308.
45. Votruba SB, Mattison RS, Dumesic DA, Koutsari C, Jensen MD. Meal fatty acid uptake in visceral fat in women. *Diabetes*. 2007;56(10):2589–2597.
46. Merzouk H, Meghelli-Bouchenak M, Loukidi B, Prost J, Belleville J. Impaired serum lipids and lipoproteins in fetal macrosomia related to maternal obesity. *Biol Neonat*. 2000;77(1):17–24.
47. Ramsay JE, Ferrell WR, Crawford L, Wallace AM, Greer IA, Sattar N. Maternal obesity is associated with dysregulation of metabolic, vascular, and inflammatory pathways. *J Clin Endocrinol Metab*. 2002;87(9):4231–4237.
48. Sanchez-Vera I, Bonet B, Viana M, et al. Changes in plasma lipids and increased low-density lipoprotein susceptibility to oxidation in pregnancies complicated by gestational diabetes: consequences of obesity. *Metabolism*. 2007;56(11):1527–1533.
49. Sattar N, Tan CE, Han TS, et al. Associations of indices of adiposity with atherogenic lipoprotein subfractions. *Int J Obes Relat Metab Disord*. 1998;22(5):432–439.
50. Catalano PM, Tyzbir ED, Wolfe RR, et al. Carbohydrate metabolism during pregnancy in control subjects and women with gestational diabetes. *Am J Physiol*. 1993;264(1 Pt 1): E60–E67.
51. Catalano PM, Huston L, Amini SB, Kalhan SC. Longitudinal changes in glucose metabolism during pregnancy in obese women with normal glucose tolerance and gestational diabetes mellitus. *Am J Obstet Gynecol*. 1999;180(4):903–916.
52. Catalano PM, Tyzbir ED, Wolfe RR, Roman NM, Amini SB, Sims EA. Longitudinal changes in basal hepatic glucose production and suppression during insulin infusion in normal pregnant women. *Am J Obstet Gynecol*. 1992;167(4 Pt 1):913–919.
53. Homko CJ, Sivan E, Reece EA, Boden G. Fuel metabolism during pregnancy. *Semin Reprod Endocrinol*. 1999;17(2):119–125.
54. Mills JL, Jovanovic L, Knopp R, et al. Physiological reduction in fasting plasma glucose concentration in the first trimester of normal pregnancy: the diabetes in early pregnancy study. *Metabolism*. 1998;47(9):1140–1144.
55. Sivan E, Chen X, Homko CJ, Reece EA, Boden G. Longitudinal study of carbohydrate metabolism in healthy obese pregnant women. *Diabetes Care*. 1997;20(9):1470–1475.
56. Di Cianni G, Miccoli R, Volpe L, et al. Maternal triglyceride levels and newborn weight in pregnant women with normal glucose tolerance. *Diabet Med*. 2005;22(1):21–25.

57. Kalkhoff RK, Kandaraki E, Morrow PG, Mitchell TH, Kelber S, Borkowf HI. Relationship between neonatal birth weight and maternal plasma amino acid profiles in lean and obese nondiabetic women and in type I diabetic pregnant women. *Metabolism.* 1988;37(3):234–239.
58. Nolan CJ, Riley SF, Sheedy MT, Walstab JE, Beischer NA. Maternal serum triglyceride, glucose tolerance, and neonatal birth weight ratio in pregnancy. *Diabetes Care.* 1995;18(12):1550–1556.
59. Schaefer-Graf UM, Graf K, Kulbacka I, et al. Maternal lipids as strong determinants of fetal environment and growth in pregnancies with gestational diabetes mellitus. *Diabetes Care.* 2008;31(9):1858–1863.
60. H.S.C.R. Group. Hyperglycaemia and Adverse Pregnancy Outcome (HAPO) Study: associations with maternal body mass index. *BJOG.* 2010;117(5):575–584.
61. Butte NF, Wong WW, Treuth MS, Ellis KJ, O'Brian Smith E. Energy requirements during pregnancy based on total energy expenditure and energy deposition. *Am J Clin Nutr.* 2004;79(6):1078–1087.
62. Prentice AM, Goldberg GR, Davies HL, Murgatroyd PR, Scott W. Energy-sparing adaptations in human pregnancy assessed by whole-body calorimetry. *Br J Nutr.* 1989;62 (1):5–22.
63. Eckel RH. Insulin resistance: an adaptation for weight maintenance. *Lancet.* 1992;340 (8833):1452–1453.
64. Chu SY, Callaghan WM, Kim SY, et al. Maternal obesity and risk of gestational diabetes mellitus. *Diabetes Care.* 2007;30(8):2070–2076.
65. Ben-Haroush A, Yogev Y, Hod M. Epidemiology of gestational diabetes mellitus and its association with type 2 diabetes. *Diabet Med.* 2004;21(2):103–113.
66. Friedman JE, Kirwan JP, Jing M, Presley L, Catalano PM. Increased skeletal muscle tumor necrosis factor-alpha and impaired insulin signaling persist in obese women with gestational diabetes mellitus 1 year postpartum. *Diabetes.* 2008;57(3):606–613.
67. Winzer C, Wagner O, Festa A, et al. Plasma adiponectin, insulin sensitivity, and subclinical inflammation in women with prior gestational diabetes mellitus. *Diabetes Care.* 2004;27(7):1721–1727.
68. Worda C, Leipold H, Gruber C, Kautzky-Willer A, Knofler M, Bancher-Todesca D. Decreased plasma adiponectin concentrations in women with gestational diabetes mellitus. *Am J Obstet Gynecol.* 2004;191(6):2120–2124.
69. Mazaki-Tovi S, Romero R, Vaisbuch E, et al. Maternal serum adiponectin multimers in gestational diabetes. *J Perinat Med.* 2009.
70. Halberg N, Wernstedt-Asterholm I, Scherer PE. The adipocyte as an endocrine cell. *Endocrinol Metab Clin North Am.* 2008;37(3):753–768:x–xi.
71. Avram MM, Avram AS, James WD. Subcutaneous fat in normal and diseased states 3. Adipogenesis: from stem cell to fat cell. *J Am Acad Dermatol.* 2007;56(3):472–492.
72. de Ferranti S, Mozaffarian D. The perfect storm: obesity, adipocyte dysfunction, and metabolic consequences. *Clin Chem.* 2008;54(6):945–955.
73. Boden G. Fatty acid-induced inflammation and insulin resistance in skeletal muscle and liver. *Curr Diab Rep.* 2006;6(3):177–181.
74. Boden G. Role of fatty acids in the pathogenesis of insulin resistance and NIDDM. *Diabetes.* 1997;46(1):3–10.
75. Lupi R, Dotta F, Marselli L, et al. Prolonged exposure to free fatty acids has cytostatic and pro-apoptotic effects on human pancreatic islets: evidence that beta-cell death is caspase mediated, partially dependent on ceramide pathway, and Bcl-2 regulated. *Diabetes.* 2002;51(5):1437–1442.

76. Hotamisligil GS. Inflammation and metabolic disorders. *Nature*. 2006;444 (7121):860–867.
77. Hotamisligil GS, Erbay E. Nutrient sensing and inflammation in metabolic diseases. *Nat Rev Immunol*. 2008;8(12):923–934.
78. Hirosumi J, Tuncman G, Chang L, et al. A central role for JNK in obesity and insulin resistance. *Nature*. 2002;420(6913):333–336.
79. Tuncman G, Hirosumi J, Solinas G, Chang L, Karin M, Hotamisligil GS. Functional in vivo interactions between JNK1 and JNK2 isoforms in obesity and insulin resistance. *Proc Natl Acad Sci U S A*. 2006;103(28):10741–10746.
80. Ozcan U, Cao Q, Yilmaz E, et al. Endoplasmic reticulum stress links obesity, insulin action, and type 2 diabetes. *Science*. 2004;306(5695):457–461.
81. Duggleby SL, Jackson AA. Protein, amino acid and nitrogen metabolism during pregnancy: how might the mother meet the needs of her fetus? *Curr Opin Clin Nutr Metab Care*. 2002;5(5):503–509.
82. de Benoist B, Jackson AA, Hall JS, Persaud C. Whole-body protein turnover in Jamaican women during normal pregnancy. *Hum Nutr Clin Nutr*. 1985;39(3):167–179.
83. Jackson AA. Measurement of protein turnover during pregnancy. *Hum Nutr Clin Nutr*. 1987;41(6):497–498.
84. Duggleby SL, Jackson AA. Relationship of maternal protein turnover and lean body mass during pregnancy and birth length. *Clin Sci (Lond)*. 2001;101(1):65–72.
85. Chevalier S, Marliss EB, Morais JA, Lamarche M, Gougeon R. Whole-body protein anabolic response is resistant to the action of insulin in obese women. *Am J Clin Nutr*. 2005;82(2):355–365.
86. Chevalier S, Burgess SC, Malloy CR, Gougeon R, Marliss EB, Morais JA. The greater contribution of gluconeogenesis to glucose production in obesity is related to increased whole-body protein catabolism. *Diabetes*. 2006;55(3):675–681.
87. O'Brien TE, Ray JG, Chan WS. Maternal body mass index and the risk of preeclampsia: a systematic overview. *Epidemiology*. 2003;14(3):368–374.
88. Sattar N, Greer IA. Pregnancy complications and maternal cardiovascular risk: opportunities for intervention and screening? *BMJ*. 2002;325(7356):157–160.
89. Bellamy L, Casas JP, Hingorani AD, Williams DJ. Pre-eclampsia and risk of cardiovascular disease and cancer in later life: systematic review and meta-analysis. *BMJ*. 2007;335(7627):974.
90. Graham I, Atar D, Borch-Johnsen K, et al. European guidelines on cardiovascular disease prevention in clinical practice: executive summary. *Eur Heart J*. 2007;28 (19):2375–2414.
91. Authors/Task Force, Ryden L, Standl E, et al. Guidelines on diabetes, pre-diabetes, and cardiovascular diseases: executive summary: The Task Force on Diabetes and Cardiovascular Diseases of the European Society of Cardiology (ESC) and of the European Association for the Study of Diabetes (EASD). *Eur Heart J*. 2007;28 (1):88–136.
92. Lindstrom J, Louheranta A, Mannelin M, et al. The Finnish Diabetes Prevention Study (DPS). *Diabetes Care*. 2003;26(12):3230–3236.
93. Eriksson KF, Lindgarde F. Prevention of type 2 (non-insulin-dependent) diabetes mellitus by diet and physical exercise. The 6-year Malmo feasibility study. *Diabetologia*. 1991;34(12):891–898.
94. Pan XR, Li GW, Hu YH, et al. Effects of diet and exercise in preventing NIDDM in people with impaired glucose tolerance. The Da Qing IGT and Diabetes Study. *Diabetes Care*. 1997;20(4):537–544.

95. Ramachandran A, Snehalatha C, Mary S, Mukesh B, Bhaskar AD, Vijay V. The Indian Diabetes Prevention Programme shows that lifestyle modification and metformin prevent type 2 diabetes in Asian Indian subjects with impaired glucose tolerance (IDPP-1). *Diabetologia*. 2006;49(2):289–297.
96. Diabetes Prevention Program Research Group. Reduction in the incidence of type 2 diabetes with lifestyle intervention or metformin. *N Engl J Med*. 2002;346 (6):393–403.
97. Rasmussen KM, Yaktine AL, eds. *Committee to Reexamine IOM Pregnancy Weight Guidelines*. Washington, DC: Institute of Medicine, National Research Council; 2009.
98. Luoto R, Kinnunen TI, Aittasalo M, et al. Primary prevention of gestational diabetes mellitus and large-for-gestational-age newborns by lifestyle counseling: a cluster-randomized controlled trial. *PLoS Med*. 2011;8(5):e1001036.
99. Moses RG, Luebcke M, Davis WS, et al. Effect of a low-glycemic-index diet during pregnancy on obstetric outcomes. *Am J Clin Nutr*. 2006;84(4):807–812.
100. Clapp 3rd JF. Maternal carbohydrate intake and pregnancy outcome. *Proc Nutr Soc*. 2002;61(1):45–50.
101. Stuebe AM, Oken E, Gillman MW. Associations of diet and physical activity during pregnancy with risk for excessive gestational weight gain. *Am J Obstet Gynecol*. 2009;:e1–e8.
102. Kramer MS, Kakuma R. Energy and protein intake in pregnancy. *Cochrane Database Syst Rev (Online)*. 2003;(4)):CD000032.
103. Olafsdottir AS, Skuladottir GV, Thorsdottir I, Hauksson A, Steingrimsdottir L. Maternal diet in early and late pregnancy in relation to weight gain. *Int J Obes (Lond)*. 2006;30(3):492–499.
104. Deierlein AL, Siega-Riz AM, Herring A. Dietary energy density but not glycemic load is associated with gestational weight gain. *Am J Clin Nutr*. 2008;88(3):693–699.
105. Stehbens JA, Baker GL, Kitchell M. Outcome at ages 1, 3, and 5 years of children born to diabetic women. *Am J Obstet Gynecol*. 1977;127(4):408–413.
106. Rizzo T, Metzger BE, Burns WJ, Burns K. Correlations between antepartum maternal metabolism and child intelligence. *N Engl J Med*. 1991;325(13):911–916.
107. Laraia BA, Siega-Riz AM, Dole N, London E. Pregravid weight is associated with prior dietary restraint and psychosocial factors during pregnancy. *Obesity*. 2009;17 (3):550–558.
108. Webb JB, Siega-Riz AM, Dole N. Psychosocial determinants of adequacy of gestational weight gain. *Obesity*. 2009;17(2):300–309.
109. Evenson KR, Moos MK, Carrier K, Siega-Riz AM. Perceived barriers to physical activity among pregnant women. *Matern Child Health J*. 2008;13:364–375.
110. Rifas-Shiman SL, Rich-Edwards JW, Kleinman KP, Oken E, Gillman MW. Dietary quality during pregnancy varies by maternal characteristics in Project Viva: a US cohort. *J Am Diet Assoc*. 2009;109(6):1004–1011.
111. Craig P, Dieppe P, Macintyre S, Michie S, Nazareth I, Petticrew M. Developing and evaluating complex interventions: the new Medical Research Council guidance. *BMJ*. 2008;337:a1655.
112. Kinnunen TI, Aittasalo M, Koponen P, et al. Feasibility of a controlled trial aiming to prevent excessive pregnancy-related weight gain in primary health care. *BMC Pregnancy Childbirth*. 2008;8:37.
113. Dodd JM, Grivell RM, Crowther CA, Robinson JS. Antenatal interventions for overweight or obese pregnant women: a systematic review of randomised trials. *BJOG*. 2010;117(11):1316–1326.

Section 6

Obesity and Labour

24 Induction of Labour in Obese Women

Carolyn Chiswick and Fiona C. Denison

MRC/University of Edinburgh Centre for Reproductive Health, The Queen's Medical Research Institute, Edinburgh, UK

Background

Induction of labour refers to the process of artificially initiating the onset of regular uterine activity in order to generate progressive cervical dilatation and delivery of the baby. It is one of the most commonly performed obstetric interventions in modern practice, occurring in around 20% of pregnancies in the United Kingdom [1] and the United States [2].

Induction of labour is an important issue to discuss in the context of maternal obesity. Most, if not all, of the common indications for induction of labour are more common in women with a high body mass index (BMI). A recent retrospective cohort study including over 3000 women undergoing induction of labour demonstrates that although the risk of Caesarean section is higher in women with a higher BMI, over 70% still achieve vaginal delivery without an increase in other adverse outcomes [3]. The authors conclude that it is reasonable to offer induction of labour to those who require it. This chapter will discuss the limited evidence available to guide management of labour induction in the ever-increasing obese pregnant population.

Indications for Induction of Labour

In general, the purpose of induction of labour is to confer benefit to the mother or foetus where the risks of the procedure are outweighed by the risks of continuing with the pregnancy [1]. There are several appropriate and important indications for induction of labour. Among the commonest of these is to prevent the post-mature pregnancy that continues beyond 42 weeks gestation. Other indications include maternal disease, such as diabetes, hypertension and pre-eclampsia. Foetal indications include suspected foetal compromise, foetal growth restriction or foetal anomaly. Controversy remains regarding other acceptable indications for induction of labour. It is not uncommon for the obstetrician to be faced with a maternal request

Obesity. DOI: http://dx.doi.org/10.1016/B978-0-12-416045-3.00024-8

for induction of labour as the discomfort of pregnancy increases towards and beyond term. Where a woman has previously delivered a large-for-gestational-age baby and/or delivery has been complicated by shoulder dystocia, the clinician may be tempted to offer, or the patient may request, early induction of labour, despite a lack of evidence to support its efficacy as a preventative measure [4,5].

Post-Dates Pregnancy

Prolonged pregnancy is associated with an increased risk of foetal and neonatal morbidity and mortality [6]. For this reason, the National Institute for Health and Clinical Excellence (NICE) in the United Kingdom recommends offering induction of labour at 41–42 weeks gestation [1].

Women with a high BMI are more at risk of having a post-mature pregnancy. In a retrospective cohort study using data from the Swedish birth register, Denison et al. [7] demonstrated an inverse linear relationship between the likelihood of establishing in spontaneous labour by 42 weeks gestation and first trimester BMI. Similarly, data from an Aberdeen cohort showed raised BMI to be a risk factor for induction of labour in primiparous women, with the risk of requiring induction of labour approaching 50% for those with a BMI greater than or equal to 35 kg/m^2 (OR 1.8; 95% CI 1.3–2.5) [8].

The factors responsible for the onset of parturition in the general population remain poorly understood. Therefore, determining whether the relationship between BMI and post-dates pregnancy is causal is challenging. Various mechanisms have been suggested. Circulating cortisol is one possible factor. Obesity is associated with activation of the hypothalamic–pituitary–adrenal axis but cortisol clearance is also increased and plasma cortisol levels are often low [9–11]. Cortisol levels are significantly lower at 22–24 weeks in women who deliver at term compared with those who deliver preterm [12]. Also, longitudinal studies demonstrate a less rapid rise in maternal corticotrophin-releasing hormone in women who deliver post-dates compared with those who deliver preterm or at term [13,14]. It would seem likely that cortisol is important in the onset of parturition and the dysregulation of this in obese women may contribute to the increased incidence of a post-dates pregnancy.

Alternatively, the alteration in the oestrogen: progesterone ratio which is known to occur prior to the onset of labour [15] may be upset by a difference in circulating oestrogen in obese women given the concentration of oestrogen in the excess adipose tissue.

The evidence is conflicting about the effect of maternal obesity on myometrial contractility. Zhang et al. [16] have demonstrated impaired myometrial contractility in obese women, and suggest that the myometrial quiescence of pregnancy may be enhanced by obesity, thus inhibiting the onset of labour [3]. However, others have shown no effect of obesity on contractility [17].

Regardless of the underlying cause, it would appear probable that obese women are at greater risk of prolonged pregnancy. Consequently, it may be helpful to counsel women regarding this risk early in pregnancy so that her expectations of when she will deliver are realistic.

Hypertensive Disorders of Pregnancy

Obese women are more likely to have pre-existing hypertension and to develop hypertensive disorders of pregnancy [8]. As with a woman of normal weight, the risks and benefits of induction should be carefully considered. Particular care should be taken in obese patients to make an accurate diagnosis of hypertension in the first instance. An appropriate size of blood pressure cuff is necessary to achieve this. Too small a cuff will overestimate blood pressure and may lead to unnecessary intervention. Too large a cuff is associated with less error [18,19].

Where a diagnosis of pre-eclampsia is made, the benefits to the mother and foetus of ending the pregnancy usually outweigh the risks of induction. Results from the HYPITAT trial [20] demonstrated induction of labour for hypertensive disease beyond 37 weeks gestation to be associated with improved maternal outcome. However, there was no subgroup analysis according to BMI and no woman in the study had a BMI greater than 36 kg/m^2.

Diabetes

Diabetes is more common in obese women. This includes both gestational diabetes and non-insulin-dependent diabetes. It is generally recommended that pregnancy in association with diabetes does not progress beyond 40 weeks gestation [21] and these women are therefore highly likely to require induction of labour. Obesity should not be a preclusion to induction of labour where indicated for diabetes. Clinical judgement should be applied on a case-by-case basis if there is suspicion of severe macrosomia or placental insufficiency.

Induction of Labour for Foetal Macrosomia

Obesity, even in the absence of diabetes, is associated with an increased incidence of foetal macrosomia and may contribute to the increase in adverse pregnancy outcomes in obese women such as shoulder dystocia, instrumental delivery, Caesarean section and post-partum haemorrhage. Macrosomia is difficult to predict antenatally. Estimation of foetal size by measurement of the symphysiofundal height can be unreliable in obese women. The alternative is to perform serial ultrasound scans to measure growth but these too may be inaccurate [22,23]. There may be a strong temptation to induce labour in obese women in whom foetal macrosomia is suspected. However, two systematic reviews [4,5] of expectant management versus induction of labour for suspected foetal macrosomia demonstrated no evidence of benefit of induction of labour versus expectant management. Furthermore, a summary of the statistics from nine observational studies included in one of the reviews suggested expectant management with spontaneous onset of labour was associated with a reduced need for Caesarean birth [5].

Stillbirth

Many studies have shown an association between obesity and an increased risk of stillbirth [24–26]. Salihu et al. [24] demonstrated that obese women were 40% more likely to suffer a late intrauterine death than non-obese women, with the risk increasing in a dose-dependent fashion with increasing BMI. Although not an indication to offer early induction of labour, these data suggest careful counselling of women who decline post-dates induction is particularly important in women with a high BMI.

'Soft' Indications for Induction of Labour

Obese pregnant women are more likely to suffer with the minor complaints of pregnancy [27]. Of particular relevance in this discussion is symphysis pubis dysfunction (SPD). This is a significant contributor to morbidity in late pregnancy and a common reason for maternal request for induction of labour. Women with severe SPD may find it difficult to move easily on and off a bed or examination couch and into an appropriate position for vaginal examination. Obesity can confound these issues and the process of induction of labour may become physically and psychologically draining for the woman. Supportive care towards the end of pregnancy may help reduce the desire for induction of labour. A clear explanation of the process of induction is essential and also the associated potential complications.

Women with a Previous Caesarean Section

Whether, and how, to induce a woman who has previously been delivered by Caesarean section remains an area of controversy. Studies that have examined success rates of vaginal birth after Caesarean section overwhelmingly demonstrate a reduced success rate in women with a high BMI compared with those of normal weight [28,29]. Compared to women undergoing elective repeat Caesarean section, morbidly obese women (BMI $>$ 40 kg/m^2) undergoing a trial of labour carried a greater than fivefold risk of uterine rupture/dehiscence (2.1% vs 0.4%), almost twofold increase in composite maternal morbidity (7.2% vs 3.8%) and fivefold risk of neonatal injury (1.1% vs 0.2%). When deciding the optimal method of delivery for obese women, the examination findings, including the abdominal palpation and the favourability of the cervix, the indication for the previous Caesarean section and the woman's own preference must all be considered. However, the balance of risk may tip in favour of Caesarean section when obesity is added to the mix.

Assessment Prior to Induction of Labour

Abdominal palpation in obese women is challenging and may not yield clinically useful information. Ultrasound can be invaluable to determine foetal presentation and assess liquor volume prior to commencing induction of labour.

Vaginal examination can also be very difficult in women with a high BMI, particularly those with morbid obesity. Maternal tissues can make access difficult.

Optimal positioning of the woman is essential and the use of lithotomy poles to support the patient's legs can be very helpful to aid access to the cervix.

Methods of Inducing Labour

Numerous methods to achieve cervical ripening for induction of labour have been employed. Broadly, they fall into two categories: mechanical and pharmacological.

Those used most commonly in modern obstetric practice include membrane sweeping or stripping, prostaglandin administration, artificial rupture of membranes (ARM) and the use of intravenous syntocinon.

NICE recommends the use membrane sweeping and vaginal prostaglandin E2 (PGE2) followed by ARM and syntocinon augmentation where necessary [1].

No studies to date have specifically examined the suitability of these various methods in relation to maternal BMI. Mechanical methods may be technically more challenging in the obese population given the potential difficulties with accessing the cervix as previously discussed. Dose and method of delivery of prostaglandin used for cervical ripening varies between centres, but again no studies have specifically looked at whether a variable dose is required depending on weight.

In a population-based study examining pregnancy outcomes of women with increased BMI, Kiran et al. [30] found that obese women were less likely to achieve vaginal delivery following membrane sweeping or ARM alone than lean women (OR 0.3; 95% CI 0.2–0.6). They were also more likely to require augmentation with oxytocin (OR 1.2; 95% CI 1–1.16) and require all three methods of induction (i.e. prostaglandin, ARM and oxytocin) (OR 1.8; 95% CI 1.1–1.9) [30].

In a secondary analysis of data from a trial comparing efficacy and safety of three different sustained-release vaginal prostaglandin inserts, Pevzner et al. [31] demonstrated obese (BMI > 30 kg/m^2) and extremely obese (BMI > 40 kg/m^2) women were more likely to require pre-delivery oxytocin for labour augmentation following prostaglandin (OR 1.46; 95% CI 1.12–1.89, $P = 0.004$ and OR 2.10; 95% CI 1.49–2.95, $P < 0.001$, respectively). The total dose of oxytocin required was also proportional to BMI [31]. The authors hypothesise the increased requirement for oxytocin may be explained by the relative increase in volume of distribution in obese women which may have a dilutional effect on both the prostaglandin ripening agent and the oxytocin. This is indeed biologically plausible, as the effect of obesity on pharmacokinetics has been previously described [32].

There is a need for further trials to establish whether a weight-related dose of prostaglandin would be beneficial without an increase in harm, particularly in women who are extremely obese.

Foetal Well-Being

Clinical examination, including abdominal palpation and vaginal examination is indicated for all women prior to induction of labour. Confirmation of the abdominal findings with ultrasound scanning is particularly useful in obese women. An assessment of liquor volume by measuring the amniotic fluid index can also be helpful in the planning of induction.

Induction of labour is considered to be an indication for continuous electronic foetal monitoring [33]. This can be difficult in obese women. The signal to the external transducer to record the tocograph may be attenuated by subcutaneous fat and the accuracy of the recording may be poor. Abdominal palpation for confirmation of uterine activity can be equally unreliable. The use of an intrauterine pressure catheter is more accurate and less susceptible to failure [34] but these are rarely employed in the United Kingdom. The use of a foetal scalp electrode to record the foetal heart rate can be very helpful in women with a high BMI [35], but their use is clearly limited to women with a sufficiently dilated cervix.

Many units now offer the initial phase of induction of labour, cervical ripening, in the outpatient setting. Obese women are considered to have a 'high risk' pregnancy and it is a matter of debate as to whether these women are suitable for outpatient management. There is no evidence to inform practice at present, and data from continued audit of outpatient management of labour induction may be helpful. The decision must be made following an informed discussion between the patient and her obstetrician and based on the balance of risk and benefit to that individual.

Summary

Maternal obesity is one of the most common co-morbidities of pregnancy. It confers an elevated risk of almost all pregnancy complications contributing to, and including, an increased risk of requiring induction of labour. This in itself may then be more complicated, and less likely to be successful, for an obese women compared to a lean woman. Thus, although induction of labour is often required and there is good evidence to support its use, further studies are required to determine the optimal method used to induce labour in women who are obese.

References

1. NICE. *Induction of Labour: Guideline CG70*. London: NICE; 2008.
2. Martin JA, Hamilton BE, Sutton PD, Ventura SJ, Mathews TJ, Osterman MJK. Births: final data for 2008. *National Vital Statist Rep*. 2010;59(1):1–72.
3. Arrowsmith S, Wray S, Quenby S. Maternal obesity and labour complications following induction of labour in prolonged pregnancy. *BJOG*. 2011;118(5):578–588.
4. Irion O, Boulvain M. Induction of labour for suspected fetal macrosomia. *Cochrane Database Syst Rev*. 2000;(2):CD000938.
5. Sanchez-Ramos L, Bernstein S, Kaunitz A. Expectant management versus labor induction for suspected fetal macrosomia: a systematic review. *Obstet Gynecol*. 2002;100 (5 Pt 1):997–1002.

6. Hilder L, Costeloe K, Thilaganathan B. Prolonged pregnancy: evaluating gestation-specific risks of fetal and infant mortality. *BJOG*. 1998;178:726–731.
7. Denison F, Price J, Graham C, Wild S, Liston W. Maternal obesity, length of gestation, risk of postdates pregnancy and spontaneous onset of labour at term. *BJOG*. 2008;115:720–725.
8. Bhattacharya S, Campbell DM, Liston WA, Bhattacharya S. Effect of body mass index on pregnancy outcomes in nulliparous women delivering singleton babies. *BMC Public Health*. 2007;7:168.
9. Ljung T, Andersson B, Bengtsson B, Bjorntorp P, Marin P. Inhibition of cortisol secretion by dexamethasone in relation to body fat distribution: a dose–response study. *Obes Res*. 1996;4(3):277–282.
10. Ljung T, Holm G, Friberg P, et al. The activity of the hypothalamic–pituitary–adrenal axis and the sympathetic nervous system in relation to waist/hip circumference ratio in men. *Obes Res*. 2000;8(7):487–495.
11. Jessop D, Dallman M, Fleming D, Lightman S. Resistance to glucocorticoid feedback in obesity. *J Clin Endocrinol Metab*. 2001;86(9):4109–4114.
12. Mercer BM, Macpherson CA, Goldenberg RL, et al. Are women with recurrent spontaneous preterm births different from those without such history? *Am J Obstet Gynecol*. 2006;194(4):1176–1184.
13. Inder W, Prickett T, Ellis M, et al. The utility of plasma CRH as a predictor of preterm delivery. *J Clin Endocrinol Metab*. 2001;86(12):5706–5710.
14. McLean M, Bisits A, Davies J, Woods R, Lowry P, Smith R. A placental clock controlling the length of human pregnancy. *Nat Med*. 1995;1(5):460–463.
15. Smith R, Mesiano S, McGrath S. Hormone trajectories leading to human birth. *Regul Pept*. 2002;108(2–3):159–164.
16. Zhang J, Bricker L, Wray S, Quenby S. Poor uterine contractility in obese women. *BJOG*. 2007;114(3):343–348.
17. Higgins C, Martin W, Anderson L, et al. Maternal obesity and its relationship with spontaneous and oxytocin-induced contractility of human myometrium in vitro. *Reprod Sci*. 2010;17(2):177–185.
18. Maxwell M, Waks A, Schroth P, Karam M, Dornfeld L. Error in blood-pressure measurement due to incorrect cuff size in obese patients. *Lancet*. 1982;2(8288):33–36.
19. Milne F, Redman C, Walker J, et al. The pre-eclampsia community guideline (PRECOG): how to screen for and detect onset of pre-eclampsia in the community. *BMJ*. 2005;330(7491):576–580.
20. Koopmans CM, Bijlenga D, Groen H, et al. Induction of labour versus expectant monitoring for gestational hypertension or mild pre-eclampsia after 36 weeks gestation (HYPITAT): a multicentre, open-label randomised controlled trial. *Lancet*. 2009;374(9694):979–988.
21. NICE. *Diabetes in Pregnancy: Guideline CG63*. London: NICE; 2008.
22. Colman A, Maharaj D, Hutton J, Tuohy J. Reliability of ultrasound estimation of fetal weight in term singleton pregnancies. *N Z Med J*. 2006;119(1241):U2146.
23. Dudley NJ. A systematic review of the ultrasound estimation of fetal weight. *Ultrasound Obstet Gynecol*. 2005;25(1):80–89.
24. Salihu HM, Dunlop AL, Hedayatzadeh M, Alio AP, Kirby RS, Alexander GR. Extreme obesity and risk of stillbirth among black and white gravidas. *Obstet Gynecol*. 2007;110(3):552–557.
25. Sebire NJ, Jolly M, Harris JP, et al. Maternal obesity and pregnancy outcome: a study of 287,213 pregnancies in London. *Int J Obes Relat Metab Disord*. 2001;25(8):1175–1182.

26. Nohr EA, Bech BH, Davies MJ, Frydenberg M, Henriksen TB, Olsen J. Prepregnancy obesity and fetal death: a study within the Danish National Birth Cohort. *Obstet Gynecol.* 2005;106(2):250–259.
27. Denison F, Norrie G, Graham B, Lynch J, Harper N, Reynolds R. Increased maternal BMI is associated with an increased risk of minor complications during pregnancy with consequent cost implications. *BJOG.* 2009;116(11):1467–1472.
28. Durnwald CP, Ehrenberg HM, Mercer BM. The impact of maternal obesity and weight gain on vaginal birth after caesarean section success. *Am J Obstet Gynecol.* 2004;191 (3):954–957.
29. Hibbard J, Gilbert S, Landon M, et al. Trial of labor or repeat caesarean delivery in women with morbid obesity and previous caesarean delivery. *Obstet Gynecol.* 2006;108 (1):125–133.
30. Usha Kiran TS, Hemmadi S, Bethel J, Evans J. Outcome of pregnancy in a woman with an increased body mass index. *BJOG.* 2005;112(6):768–772.
31. Pevzner L, Powers BL, Rayburn WF, Rumney P, Wing DA. Effects of maternal obesity on duration and outcomes of prostaglandin cervical ripening and labor induction. *Obstet Gynecol.* 2009;114(6):1315–1321.
32. Cheymol G. Effects of obesity on pharmacokinetics implications for drug therapy. *Clin Pharmacokinet.* 2000;39(3):215–231.
33. NICE. *Intrapartum Care: Guideline GC55.* London: NICE; 2008.
34. Bakker PC, Zikkenheimer M, van Geijn HP. The quality of intrapartum uterine activity monitoring. *J Perinat Med.* 2008;36(3):197–201.
35. Bakker PC, Colenbrander GJ, Verstraeten AA, Van Geijn HP. The quality of intrapartum fetal heart rate monitoring. *Eur J Obstet Gynecol Reprod Biol.* 2004;116(1):22–27.

25 Challenges of Intrapartum Care in Obese Women

Lindsay Edwards and Boon H. Lim

Department of Obstetrics and Gynaecology, Royal Hobart Hospital, Hobart, Australia

Introduction

Along with the rise in obesity in the general population across the developed world is the steady increase in the number of obese pregnant females. The past two decades have seen almost a doubling in the prevalence of obesity among adults in countries like the United States, Australia and the United Kingdom [1–3].

A number of observational studies have shown that there is a higher incidence of intrapartum complications among obese women compared to those with a normal BMI. Co-morbidities such as diabetes and hypertension (both pre-existing and pregnancy related) occur with increased frequency in obese women. These add to the increase in resources and expertise required to care for those women in labour. Due to the above-mentioned complications, the place of delivery (e.g. a tertiary institution) is an important factor to consider in the obese gravida. From a practical point of view, obese women pose important occupational health and safety concerns for staff involved in their care during labour and delivery, and extra equipment and expertise is often required [4].

Antenatal preparation on a multidisciplinary basis is vital for safe intrapartum care for the mother and baby.

Preterm and Post-term Pregnancy

Maternal pre-pregnancy BMI has been shown to have an influence on preterm birth. Iatrogenic preterm delivery is increased in obese women and is explained by the significant association of medical co-morbidities, such as hypertension and diabetes mellitus, with obesity [2]. The risk of spontaneous preterm delivery, however, is lower in obese women than in women with a normal BMI [5]. A prospective observational cohort study from the United States showed that obese and overweight women exhibit less uterine activity and were significantly less likely to experience spontaneous preterm birth before 35 weeks of gestation (8.3% compared with 21.7%, $P < 0.01$) [6].

Obesity. DOI: http://dx.doi.org/10.1016/B978-0-12-416045-3.00025-X

Obese women who do not require early elective delivery are more likely to progress to post-term pregnancy. In a retrospective study of 9336 births, Stotland et al. [7] reported that high pre-pregnancy BMI was associated with a longer gestation at delivery, with 28.5% of obese women in this study progressing beyond 41 weeks' gestation (adjusted odds ratio (AOR) of 1.69; 95% CI 1.23−2.31) compared with 21.9% of normal-weight women ($P < 0.001$). A progressive relationship between increasing BMI and prolonged gestation was also observed. Similar results have been found in other studies, confirming the increased need for induction of labour for post-dates pregnancy [8,9].

Induction of Labour

Induction of labour is required more often, due to both the strong association of medical co-morbidities with obesity, such as diabetes mellitus and hypertension, and to the increased rates of post-term pregnancies seen in obese women [7,8,10]. Even when the presence of pre-eclampsia is adjusted for, compared with normal-weight women, morbidly obese women are more likely to be induced, with an AOR of 2.38 (95% CI 2.17−2.60) [11].

Obesity is associated with higher rates of failure of induction [12,13] and should only be undertaken for obstetric and medical indications. Maternal obesity alone is not an indication for induction of labour [14].

Place of Delivery

The Centre for Maternal and Child Enquiries (CMACE) and the Royal College of Obstetricians and Gynaecologists (RCOG) Guideline on the Management of Women with Obesity in Pregnancy (2010) recommends that 'Women with a BMI greater than 35 kg/m^2 should give birth in a consultant-led obstetric unit with appropriate neonatal services' [14]. This is supported by the NICE Clinical Guideline No. 55, 'Intrapartum Care' (2007), which recommends that women with a BMI between 30.0 and 34.9 kg/m^2 have an individual risk assessment regarding planned place of delivery [15]. Access to appropriate care in labour may help to lower the risk of adverse outcomes seen in the obese obstetric population. Timely assessment by midwives, obstetricians and anaesthetists should lead to the prevention of delays in performing any necessary interventions. Understandably, most maternity caregivers would advise women with a BMI greater than 35 kg/m^2 against home birth [4]. Given the increased chance of intrapartum complications, along with the added surgical and anaesthetic risks in obese women, it is essential that delivery be conducted in a facility where senior obstetric staff are immediately available as well as where there is timely access to the operating theatre.

A policy of planning delivery in a regional or tertiary centre, rather than in small, rural maternity units, has been recommended [16], as not all facilities have the equipments that enable health professionals to care appropriately and

Table 25.1 Place of Delivery by BMI Group, England (2008)

	N(%)				
	All Maternities in England (%)	**BMI 35.0–39.9 kg/m²** $N = 2824$	**BMI 40.0–49.9 kg/m²** $N = 1852$	**BMI ≥ 50 kg/m²** $N = 193$	**Professional Judgement** $N = 154$
Obstetric unit	93.0	2695 (97.0)	1781 (98.6)	189 (99.5)	146 (98.6)
Alongside midwifery unit	3.0	46 (1.7)	13 (0.7)	0 (0.0)	0 (0.0)
Freestanding midwifery unit	2.0	15 (0.5)	0 (0.0)	0 (0.0)	0 (0.0)
Home	2.0	20 (0.7)	13 (0.7)	1 (0.5)	0 (0.0)
Other	–	3 (0.1)	3 (0.2)	0 (0.0)	2 (1.4)

Source: Adapted from Ref. [4].

safely for obese women in labour [17]. In the United Kingdom, 98% of obese women give birth in an obstetric unit (Table 25.1) [4]. In cases where inter-facility transfer is required, a referral during the antenatal period is suggested. At this consultation, delivery planning, anaesthetic review and manual-handling expertise can be sought, allowing for the acquisition of special equipment if required. Though temporary relocation may be disruptive to family life in the short term, antenatal inter-facility transfer ensures the safest outcome. Transfer is not recommended once labour has established due to the increased need for, and the difficulties with, monitoring both the maternal and foetal status, as well as the technical difficulties in arranging appropriate transport for the woman with extreme obesity.

Equipment Required

Morbid obesity puts a major strain on hospital resources [18]. The UK Obstetric Surveillance System (UKOSS) reported that more than 1 in every 1200 women giving birth in the United Kingdom has a BMI of 50 kg/m^2 or above [19,20]. The implementation of a 'bariatric protocol' to identify and mobilise the necessary equipment and resources is suggested for maternity units that care for the obese parturient [21]. CMACE (2010) reported that while the majority of obstetric units in the United Kingdom had the basic equipment to care for obese women in labour, 43% of units did not have immediate access to hoists, and up to 31% of maternity units did not have delivery beds suitable for the extremely obese woman (Figure 25.1) [4,19]. When special beds for the morbidly obese are required, staff familiarity with these is required in case a woman needs to be laid flat urgently for resuscitation [22]. Twelve per cent of units did not have immediate access to appropriate operating theatre tables in the event of an unplanned admission of a woman with a very high BMI [4].

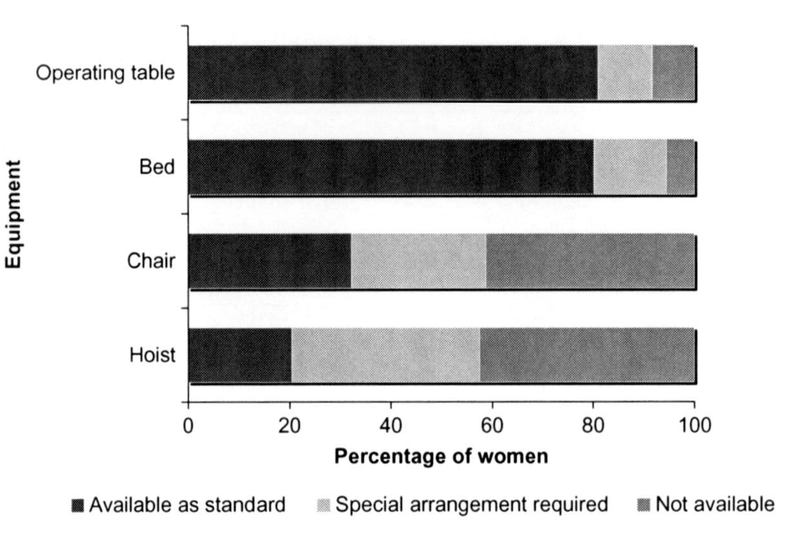

Figure 25.1 Availability of high-weight-capacity equipment at delivery of extremely obese women, United Kingdom.
Source: Reproduced with kind permission from Ref. [19].

Patient positioning and transfer are affected by obesity [16]. Weight limitations of some equipment, such as lithotomy poles, may preclude the ready use of them in some circumstances. Suitable alternatives, such as Yellofins® Stirrups and Yellofins Elite® Stirrups (Allen Medical Systems, Acton, MA), have a patient weight capacity of 159 and 227 kg, respectively, and should be readily available for this reason. Extra staff may also be required to assist in achieving the lithotomy position, both in the labour ward and in the operating theatre, especially if a regional anaesthetic is being used. An inflatable air transfer system, such as the HoverMatt® (HoverTech International, Bethlehem, PA), should be available for use in operating theatres for safe patient transfer [17].

The availability of an ultrasound machine is recognised as a good practice point in caring for the obese woman in labour for the determination of foetal presentation which can be technically very difficult. This technique can itself be difficult in cases of extreme obesity. Ultrasonography may also be required to aid venous access and is increasingly being used in the placement of epidural catheters [14]. Appropriate-sized graduated compression stockings and inflatable sequential compression devices should also be considered for use intraoperatively for thromboprophylaxis in addition to pharmacological methods [21].

Labour and Intrapartum Monitoring

During the antenatal period, all stakeholders should be involved in planning the care of the morbidly obese woman at an early stage. It is recommended that women with a BMI greater than 40 kg/m^2 be reviewed by an obstetric anaesthetist in the antenatal period, so that potential problems, such as venous access and planning for appropriate analgesia and anaesthesia, if required, can be identified before the onset of labour [4]. Anaesthetic and operating theatre staff should be notified when a woman with a BMI greater than 40 kg/m^2 is admitted to the labour ward for delivery [14].

Obese women should be admitted early in labour so that anaesthetic and intrapartum management plans devised antenatally can be adhered to. Due to the increased risk of intrapartum complications, obese women should have venous access established early in labour [14]. This may be difficult and should, therefore, be undertaken by those with the necessary experience [22]. Consideration should be given to the timing of epidural analgesia for pain relief in labour [4]. Such women can have increased difficulties with epidural placement, requiring multiple attempts, and have higher risks of inadequate analgesia [23,24].

When induction of labour or planned Caesarean section is indicated for obese women, these should be carried out when a full complement of staff (including anaesthetists) is present in the hospital [17]. Senior staff should be available at the time of assessment of progress so as to plan the delivery, as vaginal examinations and amniotomy can be more challenging to perform in obese women due to difficulties with access [17]. Lithotomy position with the legs in stirrups may help to overcome this difficulty [20].

Monitoring maternal vital signs can be challenging, with excessive adipose tissue in the upper arm causing erroneous readings, if an incorrect cuff size is used to measure blood pressure. In extreme circumstances, where the labour of a morbidly obese woman is complicated by hypertensive disorders, such as pre-eclampsia, an arterial line should be considered [21].

Monitoring contractions and assessing adequate labour progress can be challenging in obese women [21,23]. Manual palpation and/or external tocodynamometry are most commonly used, but in obese women, the distance between the skin and the uterus would render this technique inaccurate. In one study of 50 women with a BMI greater than 35 kg/m^2, difficulties in monitoring contractions was encountered in 30% of obese women compared with no such difficulties in the control group ($P < 0.001$) [20]. Intrauterine pressure monitors, while not in widespread use because of concerns over risks of infection, may need to be considered where external monitoring is impossible [21,23]. Electrohysterography has been shown in a small study to be superior to external tocodynamometry [25].

While there is no specific requirement for continuous electronic foetal monitoring in labour in an otherwise uncomplicated pregnancy, many obese parturients have other indications for continuous foetal monitoring, such as hypertension, gestational diabetes or induction of labour [17,19]. Abdominal adiposity can make external foetal monitoring through the maternal pannus more difficult. In this situation, internal monitoring, by foetal scalp electrode, should be considered if a satisfactory recording is not obtained by external monitoring [17,23]. Given the association with maternal obesity and stillbirth, including an increased risk of intrapartum stillbirth, invasive foetal monitoring should be used without hesitation, as it may be the only way of reliably monitoring the foetus [4,20].

Women with a BMI greater than 40 kg/m^2 should receive 1:1 midwifery care in established labour [14]. Two attendants should be present at a vaginal delivery with consideration given to delivery in the lithotomy position to allow adequate access should manoeuvres become necessary [20]. Senior obstetric and anaesthetic staff should be readily available to attend the birth [4] and be present for operative deliveries, due to the increased difficulty encountered in both abdominal and vaginal deliveries in the morbidly obese woman [14].

Labour Dystocia

Obese women tend to have longer labours [26–28]. Fat deposition in the maternal pelvis and foetal macrosomia may also contribute to labour dystocia [28,29]. Labour dystocia leads to an increased risk for Caesarean delivery (Table 25.2). There is overwhelming evidence that uterine contractility in obese women may be altered or impaired and that obesity is significantly associated with both elective and emergency Caesarean section [8–12,19,27–41].

In vitro studies of myometrium obtained at Caesarean section from obese women showed that they did not contract as well compared to samples from normal-weight women [40]. In a similar study, samples of myometrium, also obtained at Caesarean section, were exposed to increasing levels of leptin, a peptide secreted from adipose tissue, demonstrating a reduction in both frequency and amplitude of both spontaneous and oxytocin-induced contractions [41].

After adjusting for maternal height, pregnancy weight gain, labour induction, membrane rupture, oxytocin use, epidural analgesia and foetal size, the median duration of labour from 4 to 10 cm was 6.2 h for normal-weight women, 7.5 h for overweight women and 7.9 h for obese women. The slow progression that is seen in these women occurs mostly between 4 and 6 cm in overweight, and in active labour under 7 cm in obese women. No noticeable differences were seen after cervical dilatation of 7 cm in either group. The authors concluded that differences in labour

Table 25.2 Multiple Logistic Regression Model for Risk of Caesarean Delivery by Maternal BMI Controlling for Maternal Age, Race, Parity, Gestational Age, Weight Gain, Diabetes, Hypertension and Macrosomia

Pre-pregnancy BMI (kg/m^2)	OR (95% CI)
Normal weight (19–24.9)	Reference
Underweight (<19)	0.81 (0.76–0.86)
Overweight (25–29.9)	1.55 (1.48–1.63)
Class I obesity (30–34.9)	2.28 (2.13–2.45)
Class II obesity (35–39.9)	3.37 (3.04–3.73)
Class III obesity (>40)	4.52 (3.93–5.20)

Source: Reproduced with permission from Ref. [32].

progression seen among women with increasing BMI should be taken into account before additional interventions are performed [27]. Similar results were seen in other studies which also demonstrated that as maternal weight increased, so did oxytocin requirements, and that irrespective of parity, increasing weight is associated with a decreased rate of cervical dilatation and an increased labour duration [12,28].

The prolongation of labour duration appears limited to the first stage of labour only [27,37,42]. In a study of 5341 nulliparous women in labour at term, increased maternal BMI was not associated with a difference in second-stage duration, regardless of whether the labour was induced or spontaneous [37,42,43]. These studies support the finding that the increased risk for Caesarean section in obese women is mostly confined to the first stage of labour to which labour dystocia is a major contributor [36,42].

Increasing BMI is associated with reduced rates of spontaneous vaginal delivery [34]. The chance of a spontaneous vaginal delivery in women with a BMI over 35 kg/m^2 is 55% [4]. Data regarding instrumental delivery in the second stage is conflicting. Some studies have shown an increased odds ratio for instrumental delivery in the second stage of labour [11,26,30]. However, in the United Kingdom, CMACE reported that instrumental vaginal deliveries were seen in 7.6% of all singleton deliveries in obese women compared with 12.2% in the general maternity population [4]. Similar observations were also noted in population-based cohort studies from the United States and Canada [10,31].

Lower rates of operative vaginal delivery are thought to be due to the high rate of Caesarean sections, and perhaps added to by a reluctance of practitioners to perform instrumental vaginal deliveries in obese women [4,23,31]. Instrumental vaginal delivery can be particularly challenging in the obese woman, due to associations with foetal macrosomia, shoulder dystocia [37], perineal injury, perinatal morbidity and post-partum haemorrhage (PPH) [21,44]. A higher incidence in second-degree, but not third-degree perineal tear has been seen in primiparous obese women [15]. An attempted operative vaginal delivery in an obese woman must, therefore, take these factors into consideration, and senior obstetric staff should be present to perform or supervise the operative deliveries [14].

The risk of Caesarean delivery is more than double for obese women compared to women with a normal BMI [35], with short stature an additive risk [38]. The AOR for Caesarean section in women with a BMI greater than 35 kg/m^2 was 3.38 (95% CI 2.49–4.57) and 3.50 (95% CI 2.72–4.51) for the most morbidly obese women from the UKOSS data [19]. This is consistent with the CMACE report on maternal obesity, in which the Caesarean section rate was 37% among singleton pregnancies in women with a BMI over 35 kg/m^2 and 46% in women with a BMI over 50 kg/m^2 [4]. Similar trends were also noted by LaCoursiere et al. [32].

Caesarean Section in the Obese Woman

There is continued debate about the value of elective Caesarean section over vaginal delivery. As the rates of Caesarean section in morbidly obese women already approach 50%, morbidity from infection and haemorrhage occur with increased frequency in emergency Caesarean sections, rate of failed induction of labour higher and greater time to achieve delivery, it has been argued that planned Caesarean section allows the procedure to be carried out at a time where most personnel are available, potentially reducing the risks [21]. Conversely, a study of 591 women with BMI greater than 50 kg/m^2 demonstrated no significant differences in anaesthetic, post-natal or neonatal complications between women with a planned vaginal delivery and planned Caesarean delivery, with the exception of shoulder dystocia, seen in 3% of those in the vaginal delivery group [45]. This study does not support a routine policy of elective Caesarean for extremely obese women on the basis of concern about higher rates of perioperative complications. The CMACE (2010) [4] report on Maternal Obesity in the United Kingdom states that 'in the absence of obstetric or medical indications, labour and vaginal delivery should be encouraged for women with obesity'. The NICE Guideline (2011) states that BMI over 50 kg/m^2 should not be used as a sole indication for planned Caesarean section [46]. The decision regarding the mode of delivery should take each individual's circumstances into consideration, following a discussion with the woman and the multidisciplinary team responsible for pregnancy care. Obesity class, cervical examination, prior obstetric history and estimated foetal weight should be taken into consideration [13].

Both elective and emergency Caesarean section rates are increased in obese women [4]. Increased rates of Caesarean section are seen in obese women even after variables, such as medical complications of pregnancy and foetal macrosomia, are controlled [29,32,37,38]. The odds of Caesarean delivery increase stepwise with increasingly higher BMI categories [32]. With regard to emergency Caesarean section, increased rates occur due to an increased incidence seen in the first stage of labour, for labour dystocia most commonly, followed by foetal distress [26,27]. The rate of Caesarean section in the second stage of labour, however, does not appear to differ significantly with increasing BMI with rates of 7.1%, 9.6% and 6.9% in normal, overweight, and obese women, respectively ($P = 0.17$) [36,42].

Caesarean section is associated with increased morbidity in obese women [13,20]. In the case of emergency Caesarean section, delays from decision to delivery may occur due to longer time for patient transport and set-up, establishment of anaesthesia and longer operative time, including incision-to-delivery time [13,22,47]. Blood loss in excess of 1000 ml occurs more commonly in obese women compared with normal-weight counterparts [13,47,48]. Obese women are more likely to require general anaesthesia for Caesarean delivery, with an increased risk of difficult or failed intubation [4,19,31] and gastric aspiration [21]. Medical co-morbidities, such as diabetes mellitus and sleep apnoea, add to the perioperative risk. Postoperatively, infections and venous thromboembolism occur with increased frequency in obese women.

Surgery in the obese patient is technically more difficult. Attention to surgical preparation, technique and instrumentation help to minimise morbidity [23]. A wider skin incision, along with the use of a self-retaining retractor, such as the Alexis-O® retractor (Applied Medical, Rancho Santa Margarita, CA), may aid exposure.

Anatomical landmarks may be distorted, and the umbilicus cannot be relied upon in women with a large apron-like pannus. The pubic symphysis, however, tends to remain a reliable landmark [49]. The choice between transverse and longitudinal incision is controversial, and the type of skin incision should take the woman's body habitus into consideration. A vertical skin incision allows rapid entry into the peritoneal cavity and has the ability to be extended if required for exposure. Disadvantages include increased post-operative pain, which can lead to reduced mobility and atelectasis, and further respiratory tract complications. There is also an increased incidence of wound complications (infection, dehiscence and hernia formation) with a vertical approach [21].

Less post-operative pain and wound complications tend to be seen with a transverse skin incision with the rate of serious wound complication requiring reopening the incision reported as 12.1% [50]. Traditionally, transverse skin incisions are made in the lower abdomen two or three fingerbreadths above the pubic symphysis (Pfannenstiel and Joel-Cohen incision), but techniques have been described using a supraumbilical or higher infraumbilical approach in the obese patient [49]. In the obese woman with a voluminous pannus, a supraumbilical approach allows direct access to the lower uterine segment as the umbilicus in such women is distorted caudally, below the lower segment. This approach has the strength of a transverse repair and avoids the placement of the incision under a large pannus, which can lead to problems with wound infection. However, this approach has not been proven to be superior to the Pfannenstiel incision [29,33,50].

The National Institute of Health and Clinical Excellence (NICE) Guideline on Caesarean Section (2011) [46] recommends a transverse abdominal incision, as it is associated with less post-operative pain and an improved cosmetic effect compared with a midline incision. For the recommended suprapubic approach, extra surgical assistants, or alternatively elastoplast tape, to elevate and retract the pannus cephalad allow access for the placement of a low transverse incision (Figure 25.2) [20]. Care is required to ensure that the patient is tolerating the retraction, and that ventilation is unencumbered [21]. Regardless of the type of incision used, caution should be taken not to dissect the subcutaneous tissues excessively, so that dead space is minimised. The incision should be adequate enough to enable atraumatic delivery of a potentially macrosomic baby.

Careful handling of tissues and meticulous haemostasis is important in all cases of surgery. Mass closure, using the Smead-Jones technique incorporating the peritoneum, fascia and muscle together with a continuous non-absorbable or delayed absorbable monofilament suture, is recommended for midline incisions [21]. Stitches should be placed no further than 1 cm apart and should include more than 1 cm of fascial tissue. Closure of the subcutaneous layer in women with at least 2 cm of adipose tissue has been shown to decrease the risk of wound complications [4,51]. Meta-analysis of the use of a prophylactic subcutaneous drain after Caesarean delivery has shown no additional benefit with regard to wound disruption, infection, haematoma or seroma formation [52].

There is debate regarding the best method of skin closure in obese women following Caesarean section. A prospective pilot study from Canada comparing

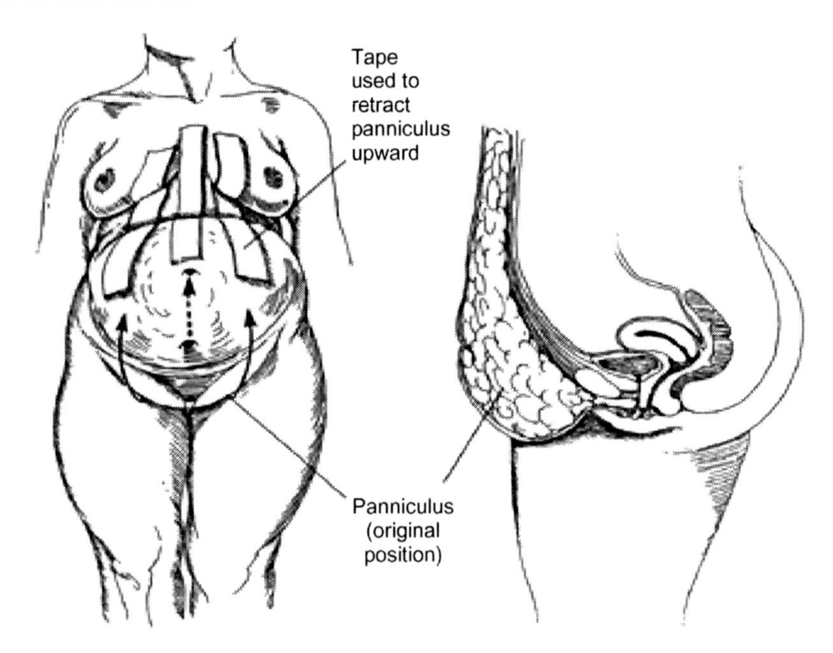

Figure 25.2 Strapping technique to retract the panniculus cephalad.
Source: Reproduced with permission from UptoDate.

staples and sub-cuticular sutures concluded that staples were superior, as operative time was shorter, pain was reduced at 6 weeks post-surgery and there was no difference noted for appearance and patient satisfaction [53]. However, this study excluded women with a BMI above 35 kg/m^2. Two recently published meta-analyses of randomised controlled trials comparing the two methods failed to demonstrate a benefit of staples over sub-cuticular sutures [54,55].

Rates of post-operative infection are higher in obese women [4,12,13,23,26,29,31,47,48,56]. The increase in infectious morbidity is also seen in elective cases, even when perioperative antibiotics are administered. A retrospective review of 287,213 deliveries reported an AOR of 2.24 (95% CI 1.91–2.64) for wound infection in obese women compared with healthy weight women [29]. In women with a BMI over 40 kg/m^2, wound infection occurs more frequently with an AOR of 3.95 (95% CI 1.77–8.82) [26]. In women with a BMI greater than 50 kg/m^2, wound infection occurred in 66% and wound dehiscence in 14% of cases [19].

Perioperative broad-spectrum antibiotics, commenced before the skin incision, reduce the risk of infections and should be given to all women regardless of BMI [4,14,23,46,57]. Of concern is that less than half of the patients (48%) with a BMI over 40 kg/m^2 achieve therapeutic tissue levels with standard perioperative antimicrobial prophylaxis [23], prompting recommendations that twice the normal dose for women with a BMI over 35 kg/m^2 be given [57].

Admission to the intensive care unit occurs more commonly among women with a BMI greater than 50 kg/m^2 compared with women of normal weight (AOR 3.86; 95% CI 1.41−10.6) [19]. Obesity and its complications are also associated with longer hospital stays and higher costs [26,34,48].

Macrosomia and Shoulder dystocia

In obese women, pre-pregnancy weight is the most important factor influencing birth weight [47]. Even after adjustment for diabetes mellitus, obese women are two to three times more likely to give birth to a large for gestational age (LGA) infant (birth weight greater than 4000 g) [4,10,21,29,30,37,48,58,59]. Almost one-third of babies born to women with a BMI over 50 kg/m^2 are LGA infants [4]. Potential complications of macrosomia include malpresentation, operative vaginal delivery, shoulder dystocia, Caesarean section, PPH, perineal trauma, low Apgar scores and admission to neonatal intensive care units [21,43].

Data regarding shoulder dystocia are conflicting with some studies showing an association with obesity and others not. In a large population-based cohort of 142,404 deliveries, shoulder dystocia was significantly increased in women with obesity (AOR 1.61; 95% CI 1.04−2.51) weighing 120 kg or more [31]. A study of 9667 women in France, in which the incidence of shoulder dystocia was 1.4%, demonstrated a 2.7-fold risk of shoulder dystocia with obesity as an independent risk factor [59]. Knight et al. [19], however, did not demonstrate the increased risk of shoulder dystocia in women with a BMI greater than 50 kg/m^2, hypothesising that this may be due to the extremely high rate of Caesarean delivery in this group. A case−control study by Robinson et al. [60] also demonstrated that maternal obesity is not an independent risk factor for shoulder dystocia, when the confounding effects of macrosomia, diabetes and midpelvic instrumental delivery are controlled for and that foetal macrosomia remained the strongest predictor (Table 25.3).

Post-partum Care

PPH is increased in women with obesity, even after accounting for the increased incidence of Caesarean section [26,29]. This may be due to the association with macrosomia, or the large volume of distribution in obese women, which may reduce the efficacy of standard doses of uterotonic drugs [29]. The CMACE report showed that [4] the incidence of PPH was almost four times higher in obese women than in the general obstetric population. After controlling for significant risk factors (Caesarean section, pre-eclampsia and birth weight over 4000 g), each BMI unit increment in women with a BMI over 35 kg/m^2 saw a 2.6% increase in the risk of PPH. Major PPH (>1000 ml) was seen in 5% [10]. Blomberg [61] reported an association between Class III obesity and atonic uterine haemorrhage over 1000 ml, following spontaneous vaginal delivery (AOR 1.23, 95% CI 1.04−1.45). This association was even more pronounced after operative vaginal delivery.

Table 25.3 Adjusted Odds of Women with and Without Diabetes Having a LGA Baby, by BMI Group

	BMI 35.0–39.9 kg/m^2 (N = 2787)	BMI 40.0–49.9 kg/m^2 (N = 1815)		BMI > 50 kg/m^2 (N = 190)	
	n (%)	*n* (%)	OR (95% CI)	*n* (%)	OR (95% CI)
Diabetes	79 (35.6)	93 (44.1)	1.43 (0.97–2.10)	18 (50.0)	1.81 (0.89–3.68)
No diabetes	358 (14.1)	324 (20.4)	**1.55 (1.32–1.83)**	43 (28.1)	**2.37 (1.64–4.43)**

N, all women within each BMI category; *n*, women with a LGA baby; percentages reflect the proportion of women within each group that have a LGA baby, after excluding missing data; an odds ratios calculated using BMI 35.0–39.9 kg/m^2 as the reference group; CI, confidence interval; bold text denotes statistically significant associations.
Source: Adapted from Ref. [4].

Perinatal Morbidity

Though the risk of intrapartum emergencies, such as foetal heart rate abnormalities, cord prolapse and placental abruption do not occur with increased frequency in obese women, should such an emergency arise, perinatal outcome may be negatively affected by the increased decision-to-delivery or incision-to-delivery intervals that can occur in the morbidly obese woman [21].

Infants born to obese mothers are more likely to require resuscitation and admission to neonatal intensive care units (NICU) due to problems including birth trauma, 5-min Apgar scores less than 7 and feeding difficulties [8,10,13,20,34]. The risk of admission to NICU increases with each increasing BMI category [4,34]. In a large population-based study, there was a more than twofold risk increase for foetal distress and low Apgar scores among infants born to morbidly obese women [11].

Risks in Subsequent Pregnancy

Caesarean delivery in the index pregnancy complicates the management of subsequent pregnancies [21]. Caesarean section is associated with increased obstetric risks related to uterine rupture, placenta praevia, placenta accreta and perioperative morbidity including injury, haemorrhage and blood transfusion. In the absence of contraindications, vaginal birth after Caesarean (VBAC) is recommended to women as an option after one previous Caesarean section. While the chance of successful VBAC has been quoted at 80% [62], several studies have shown obesity as a risk factor for reduced chance of successful VBAC [63–70]. In a study of 14,142 women who underwent a trial of labour, and 14,304 who planned repeat Caesarean delivery, the rate of unsuccessful attempt at VBAC increased as the BMI increased [63].

Morbidly obese women who attempted a trial of labour had a combined risk of rupture/dehiscence of 2.1% compared with 1.4% in the obese and 0.9% in women of normal weight [63]. Composite maternal morbidity (including one or more of endometritis, uterine rupture or dehiscence, blood transfusion, surgical injury, hysterectomy, venous thromboembolism and prolonged hospital stay) in obese women was twice as high if the trial of labour was unsuccessful. The study also demonstrated that babies born to obese women in the trial of labour group had a higher risk of neonatal injury (including fractures, brachial plexus injuries and lacerations) than those born by elective Caesarean delivery and that increasing BMI was associated with an increased risk of NICU admission and low 5-min Apgar scores [63]. The addition of risk factors such as induction of labour, no history of previous vaginal delivery and foetal macrosomia decreases the chance of successful VBAC further [68].

Women should not be prevented from attempting VBAC based on obesity alone, as even though the risk of failure is increased as BMI increases. Hibbard et al. [63] demonstrated that 60% of morbidly obese women were successful in their trial of labour. Rather, women with a BMI over 35 kg/m^2 should have an individualised decision for VBAC after detailed discussion and attention is paid to other risk factors.

Conclusion

The rising incidence of obesity in pregnancy is placing a strain on health care resources. The significant and consistent association between obesity and maternal, foetal and neonatal complications indicates the need for appropriate antenatal planning, involving a multidisciplinary approach. There should be consideration for maternity units to work on a network basis to ensure that the obese parturient is cared for in the most appropriate setting with the right equipment and expertise to ensure a safe outcome for mother and baby.

References

1. Cameron AJ, Welborn TA, Zimmet PZ, et al. Overweight and obesity in Australia: the 1999–2000 Australian diabetes, obesity and lifestyle study (AusDiab). *Med J Aust.* 2003;178:427–432.
2. Flegal KM, Carroll MD, Ogden CL, Curtin LR. Prevalence and trends in obesity among U.S. adults, 1999–2008. *JAMA.* 2010;303(3):235–241.
3. Lifestyles Statistics. *Statistics on Obesity, Physical Activity and Diet: England, 2010.* The Health and Social Care Information Centre, <www.ic.nhs.uk>, London: 2010;14.
4. Centre for Maternal and Child Enquiries (CMACE). *Maternal Obesity in the UK: Findings from a National Project.* London: CMACE; 2010.

5. Smith GC, Shah I, Pell JP, Crossley JA, Dobbie R. Maternal obesity in early pregnancy and risk of spontaneous and elective preterm deliveries: a retrospective cohort study. *Am J Public Health*. 2007;97:157–162.
6. Ehrenberg HM, Iams JD, Goldenberg RL, et al. Maternal obesity, uterine activity, and the risk of spontaneous preterm birth. *Obstet Gynecol*. 2009;113(1):48–52.
7. Stotland NE, Washington AE, Caughey AB. Prepregnancy body mass index and the length of gestation at term. *Am J Obstet Gynecol*. 2007;197:378.e1–378.e5.
8. Usha Kiran TS, Hemmadi S, Bethel J, Evans J. Outcome of pregnancy in a woman with an increased body mass index. *BJOG*. 2004;112:768–772.
9. Arrowsmith S, Wray S, Quenby S. Maternal obesity and labour complications following induction of labour in prolonged pregnancy. *BJOG*. 2011;118:578–588.
10. Dodd JM, Grivell RM, Nguyen A, Chan A, Robinson JS. Maternal and perinatal health outcomes by body mass index category. *Aust N Z J Obstet Gynaecol*. 2011;51:136–140.
11. Cedegren MI. Maternal morbid obesity and the risk of adverse pregnancy outcome. *Obstet Gynecol*. 2004;103(2):219–224.
12. Pevzner L, Powers BL, Rayburn WF, Rumney P, Wing DA. Effects of maternal obesity on duration and outcomes of prostaglandin cervical ripening and labor induction. *Obstet Gynecol*. 2009;114(6):1315–1321.
13. Wolfe KB, Rossi RA, Warshak CA. The effect of maternal obesity on the rate of failed induction of labor. *Am J Obstet Gynecol*. 2011;205:128.e, 1–7.
14. CMACE/RCOG Joint Guideline. *Management of Women with Obesity in Pregnancy*. Centre for Maternal and Child Enquiries and Royal College of Obstetricians and Gynaecologists, RCOG, London. 2010.
15. National Institute for Health and Clinical Excellence. *Intrapartum Care: Care of Healthy Women and Their Babies During Childbirth*. London: National Institute for Health and Clinical Excellence; 2007.
16. Green C, Shaker D. Impact of morbid obesity on the mode of delivery and obstetric outcome in nulliparous singleton pregnancy and the implications for rural maternity services. *Aust N Z J Obstet Gynaecol*. 2011;51:172–174.
17. Royal Women's Hospital Clinical Practice Guideline. *Obese Maternity Woman: Management*. The Royal Women's Hospital, <www.thewomens.org.au/>; Melbourne, Australia: 2006.
18. Heslehurst N, Lang R, Rankin J, Wilkinson J, Summerbell C. Obesity in pregnancy: a study of the impact of maternal obesity on NHS maternity services. *BJOG*. 2007;114:334–342.
19. Knight M, Kurinczuk JJ, Spark P, Brocklehurst P, on behalf of the UK Obstetric Surveillance System. Extreme obesity in pregnancy in the United Kingdom. *Obstet Gynecol*. 2010;**115**(5):989–997.
20. Ray A, Hildreth A, Esen UI. Morbid obesity and intrapartum care. *J Obstet Gynaecol*. 2008;28(3):301–304.
21. Gunatilake RP, Perlow JH. Obestiy and pregnancy: clinical management of the obese gravida. *Am J Obstet Gynecol*. 2011;204(2):106–119.
22. Centre for Maternal and Child Enquiries (CMACE). Saving mothers' lives: reviewing maternal deaths to make motherhood safer: 2006–2008. The eighth report on confidential enquiries into maternal deaths in the United Kingdom. *BJOG*. 2011;118 (suppl 1):1–203.
23. Davies GAL, Maxwell C, McLeod L. Obesity in pregnancy. *J Obstet Gynaecol Can*. 2010;32(2):165–173.

24. Dresner M, Brocklesby J, Bamber J. Audit of the influence of body mass index on the performance of epidural analgesia in labour and the subsequent mode of delivery. *BJOG*. 2006;113:1178–1181.
25. Euliano TY, Nguyen MT, Marossero D, Edwards RK. Monitoring contractions in obese parturients: electrohysterography compared with traditional monitoring. *Obstet Gynecol*. 2007;109(5):1136–1140.
26. Schrauwers C, Dekker G. Maternal and perinatal outcome in obese pregnant patients. *J Matern Fetal Neonatal Med*. 2009;22(3):218–226.
27. Vahratian A, Zhang J, Troendle JF, Savitz DA, Siega-Riz M. Maternal prepregnancy overweight and obesity and the pattern of labor progression in term nulliparous women. *Obstet Gynecol*. 2004;104(5):943–951.
28. Nuthalapaty FS, Rouse DJ, Owen J. The association of maternal weight with cesarean risk, labor duration, and cervical dilation rate during labor induction. *Obstet Gynecol*. 2004;103(3):452–456.
29. Sebire NJ, Jolly M, Harris JP, et al. Maternal obesity and pregnancy outcome: a study of 287 213 pregnancies in London. *Int J Obes*. 2001;25(8):1175–1182.
30. Weiss JL, Fergal DM, Emig D, et al. Obesity, obstetric complications and cesarean delivery rate – a population-based screening study. *Am J Obstet Gynecol*. 2004;190:1091–1097.
31. Robinson HE, O'Connell CM, Joseph KS, McLeod NL. Maternal outcomes in pregnancies complicated by obesity. *Obstet Gynecol*. 2005;106(6):1357–1364.
32. LaCoursiere DY, Bloebaum L, Duncan JD, Varner MW. Population-based trends and correlates of maternal overweight and obesity, Utah 1991–2001. *Am J Obstet Gynecol*. 2005;192:832–839.
33. Perlow JH, Morgan MA, Montgomery D, Towers CV, Porto M. Perinatal outcome in pregnancy complicated by massive obesity. *Am J Obstet Gynecol*. 1992;167(4):958–962.
34. Callaway LK, Prins JB, Chang AM, McIntyre HD. The prevalence and impact of overweight and obesity in an Australian obstetric population. *Med J Aust*. 2006;184(2):56–59.
35. Poobalan AS, Aucott LS, Gurung T, Smith WC, Bhattacharya S. Obesity as an independent risk factor for elective and emergency caesarean delivery in nulliparous women – systematic review and meta-analysis of cohort studies. *Obes Rev*. 2009;10(1):28–35.
36. Fyfe EM, Anderson NH, North RA, et al. Risk of first-stage and second-stage cesarean delivery by maternal body mass index among nulliparous women in labour at term. *Obstet Gynecol*. 2011;117(6):1315–1322.
37. Sheiner E, Levy A, Menes TS, Silverberg D, Katz M, Mazor M. Maternal obesity as an independent risk factor for caesarean delivery. *Paediatr Perinat Epidemiol*. 2004;18:196–201.
38. Dempsey JC, Ashiny Z, Qiu C, Miller RS, Sorensen TK, Williams MA. Maternal pre-pregnancy overweight status and obesity as risk factors for cesarean delivery. *J Matern Fetal Neonatal Med*. 2005;17(3):179–185.
39. Chu SY, Kim SY, Schmid CH, et al. Maternal obesity and risk of cesarean delivery: a meta-analysis. *Obes Rev*. 2007;8:385–394.
40. Zhang J, Bricker L, Wray S, Quenby S. Poor uterine contractility in obese women. *BJOG*. 2007;114:343–348.
41. Moynihan AT, Hehir MP, Glavey SV, Smith TJ, Morrison JJ. Inhibitory effect of leptin on human uterine contractility in vitro. *Am J Obstet Gynecol*. 2006;195:504–509.

42. Robinson BK, Mapp DC, Bloom SL, et al. Increasing maternal body mass index and characteristics of the second stage of labor. *Obstet Gynecol.* 2011;118(6):1309–1313.
43. Buhimschi CS, Buhimschi IA, Manilow AM, Weiner CP. Intrauterine pressure during the second stage of labor in obese women. *Obstet Gynecol.* 2004;103(2):225–230.
44. Zetterstrom J, Lopez A, Anzen BO, Norman M, Holmstrom BO, Mellgren A. Anal sphincter tears at vaginal delivery: risk factors and clinical outcome of primary repair. *Obstet Gynecol.* 1999;94(1):21–28.
45. Homer CS, Kurinczuk JJ, Spark P, Brocklehurst P, Knight M. Planned vaginal delivery or planned caesarean delivery in women with extreme obesity. *BJOG.* 2011;118 (4):480–487.
46. National Institute for Health and Clinical Excellence. *Caesarean Section.* London: National Institute for Health and Clinical Excellence; 2011.
47. Perlow JH, Morgan MA. Massive maternal obesity and perioperative cesarean morbidity. *Am J Obstet Gynecol.* 1994;170(2):560–565.
48. Wolfe H. High prepregnancy body-mass index – a maternal-fetal risk factor. *New Eng J Med.* 1998;338(3):147–152.
49. Tixiier H, Thouvenot S, Coulange L, et al. Cesarean section in morbidly obese women: supra or subumbilical transverse incision? *Acta Obstet Gynecol Scand.* 2009;88(9): 1049–1052.
50. Wall PD, Deucy EE, Glantz JC, Pressman EK. Vertical skin incisions and wound complications in the obese parturient. *Obstet Gynecol.* 2003;102(5):952–956.
51. Naumann RW, Hauth JC, Owen J, Hodgkins PM, Lincoln T. Subcutaneous tissue approximation in relation of wound disruption after cesarean delivery in obese women. *Obstet Gynecol.* 1995;85:412–416.
52. Hellums EK, Lin MG, Ramsey PS. Prophylactic subcutaneous drainage for prevention of wound complications after cesarean delivery – a metaanalysis. *Am J Obstet Gynecol.* 2007;197(3):229–235.
53. Rousseau J, Girard K, Turcot-Lemay L, Thomas N. A randomized study comparing skin closure in cesarean sections: staples vs subcuticular sutures. *Am J Obstet Gynecol.* 2009;200:265.e1–4.
54. Clay FSH, Walsh CA, Walsh SR. Staples vs subcuticular sutures for skin closure at cesarean delivery: a metaanalysis of randomized controlled trials. *Am J Obstet Gynecol.* 2011;204(5):378–383.
55. Tuuli MG, Rampersad RM, Carbone JF, Stamilio D, Macones GA, Odibo AO. Staples compared with subcuticular suture for skin closure after cesarean delivery: a systematic review and meta-analysis. *Obstet Gynecol.* 2011;117(3):682–690.
56. Myles TD, Gooch J, Santolaya J. Obesity as an independent risk factor for infectious morbidity in patients who undergo cesarean delivery. *Obstet Gynecol.* 2002;100 (5):959–964.
57. van Schalkwyk J, Van Eyk N. Antibiotic prophylaxis in obsetric procedures. *J Obstet Gynaecol Can.* 2010;32(9):878–884.
58. Ehrenberg HM, Mercer BM, Catalano PM. The influence of obesity and diabetes on the prevalence of macrosomia. *Am J Obstet Gynecol.* 2004;191:964–968.
59. Mazouni C, Porcu G, Cohen-Solal E, et al. Maternal and anthropomorphic risk factors for shoulder dystocia. *Acta Obstet Gynecol Scand.* 2006;85:567–570.
60. Robinson H, Tkatch S, Mayes DC, Bott N, Okun N. Is maternal obesity a predictor of shoulder dystocia? *Obstet Gynecol.* 2003;101(1):24–27.
61. Blomberg M. Maternal obesity and risk of postpartum hemorrhage. *Obstet Gynecol.* 2011;118(3):561–568.

62. Chu SY, Kim SY, Lau J, et al. Maternal obesity and risk of stillbirth: a metaanalysis. *Am J Obstet Gynecol.* 2007;197:223–228.
63. Hibbard JU, Gilbert S, Landon MB, et al. Trial of labor or repeat cesarean delivery in women with morbid obesity and previous cesarean delivery. *Obstet Gynecol.* 2006;108 (1):125–133.
64. Carrol CS, Magann EF, Chauhan SP, Klauser CK, Morrison JC. Vaginal birth after cesarean section versus elective repeat cesarean delivery: weight-based outcomes. *Am J Obstet Gynecol.* 2003;188:1516–1522.
65. Dodd J, Crowther C. Vaginal birth after caesarean versus elective repeat caesarean for women with a single prior caesarean birth: a systematic review of the literature. *Aust N Z J Obstet Gynaecol.* 2004;44:387–391.
66. Chauhan SP, Magann EF, Carroll CS, Barrilleaux PS, Scardo JA, Martin JN. Mode of delivery for the morbidly obese with prior cesarean delivery: vaginal versus repeat cesarean section. *Am J Obstet Gynecol.* 2001;185:349–354.
67. Edwards RK, Harnsberger DS, Johnson IM, Treloar RW, Cruz AC. Deciding on route of delivery for obese women with a prior cesarean delivery. *Am J Obstet Gynecol.* 2003;189(2):385–390.
68. Durnwald CP, Ehrenberg HM, Mercer BM. The impact of maternal obesity and weight gain on vaginal birth after cesarean section success. *Am J Obstet Gynecol.* 2004;191:954–957.
69. Goodall PT, Ahn JT, Chapa JB, Hibbard JU. Obesity as a risk factor for failed trial of labor in patients with previous cesarean delivery. *Am J Obstet Gynecol.* 2005;192:1423–1426.
70. Juhasz G, Gyamfi C, Gyamfi P, Tocce K, Stone JL. Effect of body mass index and excessive weight gain on success of vaginal birth after cesarean delivery. *Obstet Gynecol.* 2005;106(4):741–746.

26 The Role of the Midwife During Pregnancy, Labour and Post-partum

Yana Richens[1], Debbie M. Smith[2] and Tina Lavender[3]

[1]Institute of Womens Health, University College London Hospital, London, UK, [2]School of Psychological Sciences, The University of Manchester, Manchester, UK, [3]School of Nursing, Midwifery and Social Work, The University of Manchester, Manchester, UK

The Role of the Midwife in Pregnancy

Background

Maternal obesity is a key consideration in the provision of maternity care due to the increasing rates of women presenting with a body mass index (BMI) ≥ 30 kg/m^2 [1] and association with an increased risk of maternal co-morbidity, pregnancy-related complications and foetal morbidity and mortality [2]. Pregnancy is a 'teachable moment' [3, p. 135] due to being a time of major transition [4]. Therefore, it should be viewed as 'a window of opportunity' to promote healthy lifestyles to women and their families as a form of weight management [5, p. S50]. Maternal obesity is a public health issue and therefore requires a multi-agency approach [6]. Midwives are a key feature of this multi-agency approach as they are in an ideal position due to their relationship with women and their understanding of the pregnancy process.

The key points for midwives to remember when providing care for women with a BMI ≥ 30 kg/m^2 are summarised in Box 26.1. These points are discussed in chronological order and are a reflection of one of the author's (YR) experience of providing midwifery care for this target group in a specialist clinic, in the context of current national guidelines for the care for women with a BMI ≥ 30 kg/m^2 [7]. An increase in the proportion of women presenting with a raised BMI has meant that clinicians have had to develop relevant care protocols. However, the lack of national guidance has seen most hospital trusts devise their own local protocols and care pathways for the care of these women; slight variations to such care pathways exist. Often, many of these have not been subject to full evaluation and thus cannot

Obesity. DOI: http://dx.doi.org/10.1016/B978-0-12-416045-3.00026-1

Box 26.1 The Key Points in Chronological Order When Providing Midwifery Care for Women with a BMI ≥ 30 kg/m^2

Pregnancy

- Women's height and weight should be accurately recorded at the initial visit. If repeat measurements are taken, where possible, the same scales should be used for these measurements.
- A detailed plan of care, including discussions with the woman, should be documented.
- Women with raised BMI should be identified and, where possible, continuity of care should be encouraged with all appointments being with the same midwife.
- Weight must be discussed openly – advice on healthy eating provided either from the midwife or through a referral to a suitable health professional (dietician – where available).
- Women with a BMI ≥ 35 kg/m^2 with additional risk factor for hypertensive disease, as highlighted by the RCOG, should be prescribed aspirin (75 mg/day) from 12 weeks gestation.
- Women with a BMI ≥ 30 kg/m^2 wishing to become pregnant should take folic acid (5 mg/day) during the first trimester.
- Possible risks and complications associated with their raised BMI must be explained to the women.
- A schedule of antenatal care must be provided that reflects high risk as it is recommended that this group of women are not suitable for midwifery-led care.
- Women with a BMI ≥ 40 kg/m^2 should have an anaesthetic assessment at 22 weeks – this takes place the same day as the consultant obstetric review.
- Women with a BMI ≥ 30 kg/m^2 are advised to give birth in a consultant unit.

Labour

- Discuss birth plan with the woman and her partner.
- One-to-one care by a senior midwife should be provided.
- The registrar and anaesthetist must be informed if any problems are anticipated for women with a BMI ≥ 40 kg/m^2.
- Consider and discuss effective ways of foetal monitoring.
- Consider active management of third stage.
- Active attention to pressure areas and ensure anti-embolic stockings.
- Attention must be given to maintenance of fluid balance and bladder care.
- Ensure that the correct blood pressure cuff is used.
- Ensure that the correct size bed and equipment, including hoist and wheelchair, are available.

Post-partum

- Be aware of possibility of post-partum haemorrhage due to prolonged labour.
- Extra support whilst breastfeeding – this may be positioning due to large breasts and flattish nipples.

be used to inform evidence-based practice. Here, women with a BMI of $\geq 30\ kg/m^2$ are focused, as suggested in the CMACE and RCOG guidance. However, this cut-off is not used in all maternity units and in some cases a cut-off of BMI $\geq 40\ kg/m^2$ is used [8].

Central to effective maternity care is an understanding of the women receiving care and their needs. Their views and beliefs are essential, and measures should be in place to ensure that each key point is communicated to them.

Weighing Women

Pregnancy, for some women, is the first time that they have contact with a health professional and it may also be the first time that they have been weighed since childhood. This is particularly true when you observe the demographics of women with a BMI $\geq 30\ kg/m^2$. These women tend to be from more deprived communities and black and minority ethnic groups, who have low levels of engagement in health care services [9]. Therefore, pregnancy could be an opportune time to discuss weight and weight gain with women and start to encourage healthy lifestyle behaviours. A recent survey further highlighted that health professionals are influential when it comes to weight loss and healthy eating [10].

At the Booking Appointment: It is recommended that all pregnant women have an assessment of height and weight at the booking visit to allow calculation of the BMI [11]. BMI calculation is a necessity if we are to provide women with the most suitable care pathway. Judgement must not be made on someone's weight based on our visual perception of obesity as this can be misleading due to different body shapes. An incorrect judgement could mean that women are not offered the suitable care pathway. In addition, self-reported measures of weight should not be used, as literature has shown that women with a BMI $\geq 30\ kg/m^2$ tend to underestimate their weight [12]. Weighing women at this stage can help midwives to initiate a full and frank discussion on weight and lifestyle.

During the Third Trimester: UK guidance recommends that women who have a BMI $\geq 30\ kg/m^2$ are weighed again in the third trimester so that '...appropriate plans for equipment and personnel during labour and delivery...' [7, p. 8] can be made. The midwife's role is to undertake a full health and social care assessment and to identify any risk factors. The repeat weighing of women with a BMI $\geq 30\ kg/m^2$ is supported by the NICE Antenatal Care Guideline [11] as they recommend that repeated weight measurements should occur in circumstances where clinical management is likely to be influenced by weight. In the case of women with a BMI $\geq 30\ kg/m^2$, their care is likely to be influenced by their weight on several accounts (e.g. antenatal foetal surveillance and place of delivery), and thus weight measurements are essential on defining their care pathway and the provision of safe care. However, BMI has limitations as it does not account for body fat distribution. Recent research has recommended that the treatment of obesity requires a greater focus on abdominal fat [13] something that is not currently possible during pregnancy. Ethnicity must also be considered, as research has suggested that South Asian individuals are considered overweight when they have a BMI ≥ 23 kg/m^2 [14], and thus if a standard cut-off for BMI is used this could lead to women not having appropriate investigations and care.

The evidence supporting recommendations for women being re-weighed in the third trimester is based on good clinical practice in order to ensure appropriate medication dosage should it be required. Re-weighing women provides another opportunity to discuss lifestyle behaviours (e.g. diet and physical activity) with the women at a stage when a relationship with their midwife may have been built, encouraging a more honest and open discussion ([15]; see Section Advice on Weight Management).

Barriers to Weighing Women: As stated above, pregnancy might be the first time that women have been weighed for many years. Furthermore, it is likely that women have not previously been observed being weighed, by another individual. This process may trigger emotional feelings in the women that must be acknowledged by midwives. Midwives should not avoid asking women to be weighed but should ensure that it is done in a way that minimises any negative feelings on the part of the women. One practical example of minimising distress is locating the weighing scales in a private location and not in an area where women will feel like they are being watched. A further element to consider is whether suitable equipment is available, as bariatric sit down scales may be required for some women. Midwives working in a community setting may not always have access to the correct equipment required for this group of women. In this case, women should be weighed at the earliest opportunity in the hospital setting.

Maternity Care Pathways

As stated above, the assessment of a pregnant woman's BMI should be made at the booking appointment, ideally at less than 12 weeks gestation [7,11]. Once the woman's weight has been measured and reported the most suitable care pathway can be decided. Women with a BMI ≥ 30 kg/m^2 will follow the high-risk pathway. Women with a BMI ≥ 30 kg/m^2 are at higher risk of complications during pregnancy [2] and the number of antenatal visits that they require reflects this risk. The increased number of antenatal visits must be explained to women so that they understand why it is necessary for them to attend. A lack of understanding could result in them feeling stigmatised and may result in them failing to attend appointments.

Specialist Clinics for Women with a BMI ≥ 30 kg/m^2

Women with a BMI ≥ 30 kg/m^2 are classified as high risk and as a result receive consultant-led care [7,16]. However, it is unclear if all women with a BMI ≥ 30 kg/m^2, including those with no medical or identified obstetric issues, need to be reviewed by an obstetrician. Furthermore, it is questionable as to whether consultant obstetricians have the capacity to review all women with a BMI ≥ 30 kg/m^2. In practice, women often report back to the midwife that the obstetrician was unsure why they had been referred to them. This causes confusion and can make women feel embarrassed; during clinic sessions, women have stated their concern that they have wasted the doctors time. Midwives must make sure that they provide an explanation to women about their care to eliminate any possible feelings of embarrassment.

Midwives and obstetricians are currently tasked with the promotion of normality in an attempt to reduce Caesarean sections [17]. However, women with a BMI $\geq 30\ kg/m^2$ are excluded from accessing low-risk birth centres due to their increased risk due to their raised BMI as a result of the national guidance [7]. Some hospitals have adopted the model of specialised clinics for women with raised BMIs, although this is not in line with the recent recommendations which stated that 'specialist clinics are unlikely to be feasible in areas of high prevalence due to resource issues and it is important that all health care professionals providing maternity care are aware of the maternal and fetal risks' [7, p. 6].

Service evaluations need to be conducted to evaluate the possible impact of specialist clinics on the maternal and foetal outcomes as well as on the maternity experience for women with a BMI $\geq 30\ kg/m^2$ as currently their impact is not clear. The clinic set up by one of the authors (YR) was in response to the amount of time this group of women took up in a generic midwifery clinic. The women attending the clinic are seen and referred back to their community midwife for ongoing support. The visit is also part of the national midwifery pathway and not an extra visit.

Advice on Weight Management

It is a midwife's duty to inform women about the possible risks to them and their baby associated with maternal obesity. They are ideally placed to offer this advice due to their contact with women in the antenatal period and the fact that they are perceived by women as a source of support during pregnancy [18]. Also, the community setting has been suggested as the ideal location for weight-management interventions, both by women with a BMI $\geq 30\ kg/m^2$ [19] and national guidelines [20]. The key to the success of such weight-management interventions is in the collaboration between maternity and community services [6]. Community midwives, who are based in children's centres and GP practices, could play a central role in providing this care and ensuring effective collaboration between maternity and community services.

Recent UK guidelines suggested that the possible risks need to be discussed with women during the early stages of their pregnancy [7]. Addressing the potential short- and long-term complications with healthy lifestyle advice (e.g. increased physical activity and healthier food choices) should be the 'focus of care' in early pregnancy [21, p. 59]. Literature suggests that women want to receive information about weight management and healthy lifestyles in pregnancy [22]. However, the provision of weight-management advice is currently low in practice with a recent Scottish study reporting that although 77% of midwives thought such advice was appropriate, only 15% of midwives offered it to women [23]. Numerous barriers exist that prevent midwives and other health professionals from discussing weight and lifestyle with women. Not mentioning weight and/or giving weight-management advice during pregnancy can be perceived by women as weight not being an issue [22]. In some cases, this led to women constructing their own ideals about weight gain which often were incorrect [24]. Alternatively, they sought other sources of advice which can be incorrect [25].

Midwives not feeling confident to approach the subject of weight with women are commonly reported in the literature. The main reasons for this lack of confidence include feeling uncomfortable due to the sensitivity of the topic [26] and a lack of knowledge of support services or resources to offer to women [27,28]. In addition, continuity of care is essential and should be encouraged where possible. Women may feel vulnerable if they see a different midwife on each visit and may feel uncomfortable discussing weight-related issues with them [29]. Nurses have been dealing with obesity as a mediating factor for illness complications in their clinical practice for slightly longer than midwives, and the literature highlights techniques used to help them with communication issues. Primary Care Nurses have employed several techniques to aid their communication with obese patients; these include being non-judgemental in their approach, being aware of any stereotypes they or the patients may have, taking a patient-centred approach to care and using motivational interviewing techniques [30]. Finally, behaviour-change techniques based on psychological theories of behaviour may be of use to midwives and have been suggested in Ref. [4] as vital for inclusion in interventions.

A lack of clarity on what advice to give to women regarding weight gain is one barrier to midwives discussing weight gain with them. In the United Kingdom, there is a lack of clear guidelines regarding optimal amount of gestational weight gain. The American Institute of Medicine (IOM) provides recommended ranges of gestational weight gain, which are dependent on the woman's start weight [31]. These recommended weight-gain ranges have been found in United States to be associated with the best outcomes of pregnancy for mothers [32,33]. There is currently little evidence exploring the suitability of the IOM weight-gain recommendations to a UK population and thus should not be followed [34]. Women in the United Kingdom are not currently routinely weighed during pregnancy, so data is sparse. Women who misperceive their pre-pregnancy weight status are more likely to gain excessive amounts of weight during their pregnancy [35]. Moreover, beliefs about recommended weight gain in pregnancy are negatively correlated with BMI; underweight women are more likely to underestimate the weight gain and obese women are more likely to overestimate weight-gain recommendations [36]. These findings highlight the need for midwives to speak to women about their weight-gain expectations and to understand their motivations and views towards their pre-pregnant weight and gestational weight gain.

In summary, more training is needed for midwives to ensure they have the knowledge regarding maternal obesity and thus feel confident when addressing an issue they deem sensitive [25,26]. Such training packages need to include education on increasing the communication skills of midwives when they feel a topic is sensitive [37]. Finally, training needs to outline the possible complications and risks associated with maternal obesity so that lifestyle and obesity advice are deemed as important priorities to midwives [38].

Diabetes

In line with the NICE antenatal care guideline [11], screening for gestational diabetes and pre-eclampsia is essential for women with a BMI $\geq 30\ kg/m^2$. Midwives must determine the following risk factors for gestational diabetes at the booking appointment for each woman with a BMI $\geq 30\ kg/m^2$:

- previous macrosomic baby weighing 4.5 kg or above,
- previous gestational diabetes,
- family history of diabetes (first-degree relative with diabetes),
- family origin with a high prevalence of diabetes (South Asian, Black Caribbean and Middle Eastern).

Women with any one of the above risk factors for gestational diabetes should be offered a random blood sugar taken at the booking appointment to determine if they have gestational diabetes [11]. This should be followed by an Oral Glucose Tolerance test at 28 weeks. The increased risk for gestational diabetes could act as another opportunity for midwives to discuss weight and healthy lifestyle changes with women (see Section Advice on Weight Management for more detail). In addition, a diagnosis of gestational diabetes will see the woman referred to a consultant obstetric clinic with dietetic input. The midwife plays an important role in ensuring that the women understand the increased risk of gestational diabetes due to their BMI, the role of the test and the relationship between diabetes and lifestyle.

Folic Acid Intake

Midwives should advise all women to take 5 mg of folic acid (daily) up to 12 weeks gestation [2]. However, higher levels of folic acid are suggested for women with a BMI $\geq 30\ kg/m^2$ due to literature outlining increased risk of neural tube defects [39–41] and nutritional deficiencies, including reduced folate levels [42].

Abdominal Palpation

Fundal height measurements may not be a reliable means of estimating foetal growth in women with a BMI $\geq 30\ kg/m^2$, as the pregnancy progresses. For this reason, further ultrasound scans may be required at 28 and 36 weeks for size and foetal growth. Midwives must inform women that after 28 weeks, if they perceive any foetal movement reduction or changes, they must contact the maternity unit. In addition, women with a BMI $\geq 40\ kg/m^2$ can experience difficulty in identifying foetal movements so this will need to be discussed with them.

The ultrasound experience can be one of distress for women as they feel humiliated when health professionals do not explain to them about the process and the possibility of not having a clear image [43]. Ultrasound in pregnancy for women with a raised BMI is not only technically challenging for sonographers but also more physically challenging. The incidence of musculo-skeletal pain and discomfort for sonographers has been reported as high as 80% [44]. Prior to undertaking an abdominal palpation, it is advisable to discuss with the woman the possibility of not being able to hear the foetal heart beat with the soncaid due to the amount of adipose tissue. Also, at this examination one may find women who have thrush infections as they are unable to wash and dry their groin area as gestation increases. Practical advice on how to manage this should be offered in these instances, such as personal hygiene and asking for help when drying the area. Finally, midwives must remember that this may be one of the only times that these women expose their bodies to anyone other than their partners. Therefore, sensitive communication is essential.

Blood Pressure

A blood pressure measurement and urinalysis for protein should be carried out at each antenatal visit to screen for pre-eclampsia in all women. More frequent blood pressure measurements need to be considered for women with a BMI ≥ 30 kg/m^2. Blood pressure should be measured following the NICE guidelines [11]:

- Remove tight clothing, ensure arm is relaxed and supported at heart level.
- Use cuff of appropriate size – the midwife may need to obtain the correct cuff size in preparation for the appointment to remove any feelings of embarrassment on the part of the woman. Further guidelines and information on equipment can be found at www.apec.org.uk.
- Inflate cuff to 20–30 mmHg above palpated systolic blood pressure.
- Lower column slowly, by 2 mmHg per second or per beat.
- Read blood pressure to the nearest 2 mmHg.
- Measure diastolic blood pressure as disappearance of sounds (phase V).

A diagnosis of hypertension (a single diastolic blood pressure of 110 mmHg or two consecutive readings of 90 mmHg at least 4 h apart and/or significant proteinuria (1+)) will require increased monitoring. If the systolic blood pressure is above 160 mmHg on two consecutive readings, at least 4 h apart, treatment should be considered [11].

Venous Thromboembolism

A history should be taken of any significant pre-existing and past medical conditions (e.g. raised blood pressure, cardiac problems, deep vein thrombosis, pulmonary embolism and known thrombophilia). Nearly half of the 31 women who died of a thromboembolic event had a BMI ≥ 30 kg/m^2 [45]. All women must be assessed and a referral made to a haematologist, if there is personal or family history of deep vein thrombosis or pulmonary embolism. Midwives should also discuss the importance of mobility with women. Clinical guidance on physical activity in pregnancy concluded that, '. . .during pregnancy, aerobic and strength conditioning exercise is considered to be safe and beneficial' [46, p. 1]. The NICE guidance on weight management also emphasises the importance of physical activity in pregnancy, such as walking [20]. Therefore, midwives should encourage women to remain active in pregnancy. They must also discuss the warning signs that women should be aware of and should stop exercising as a result (e.g. dizziness and shortness of breath). There are numerous studies that outline the psychological benefits [47] and physical benefits [48] of physical activity in pregnancy. Furthermore, weight-bearing exercise throughout pregnancy can reduce the length of labour and decrease delivery complications [49], and women who incorporate physical activity into their routine during pregnancy are more likely to continue physical activity post-partum [50].

It is essential that correct weight-adjusted doses of prophylactic and therapeutic low-molecular weight heparin are prescribed to women with a raised BMI. This was highlighted in the case of a woman who was prescribed a prophylactic dose of tinzaparin for a 90 kg woman, despite weighing 200 kg; the woman had a pulmonary embolism and died [45]. CEMACH concluded that, in two-thirds of cases of pulmonary embolism, care given to women was substandard. The main reason for this was inadequate risk assessment in early pregnancy.

The Role of the Midwife in Labour

Women with a BMI $\geq 30\ kg/m^2$ have a higher risk of complications during labour, including dysfunctional labour and post-partum haemorrhage [16]. Foetal heart rate monitoring is also more technically difficult due to the central layer of adipose tissue. For these reasons, women with a BMI $\geq 30\ kg/m^2$ should be advised to labour and give birth on the consultant obstetric-led unit, as opposed to at home or in a midwifery-led unit, as continuous foetal monitoring may be required. Women with a raised BMI are discouraged from having a water birth due to the increased risk of a post-partum haemorrhage, as there is a possibility that they would not be able to safely exit the pool. Midwives need to manage the expectations of women in the antenatal period to avoid any confusion to or upset the woman who may wish to have a pool birth. For women with a raised BMI, it is essential that they have a senior midwife look after them during labour and not a student or a junior midwife, as they are considered high risk and the risks increase with the rise in BMI. The anaesthetist and senior registrar should be notified of any woman with a significantly raised BMI, and a clear plan of care made for her. This should involve the woman and her partner and should take into consideration factors such as active management of the third stage of labour.

The Role of the Midwife in the Post-partum Period

This is discussed in detail in Chapter 25. However, the continuation of care regarding weight management is required in the post-partum period as this is when women have suggested that they want to lose weight [51] and when they hope to continue with the healthy eating habits initiated in pregnancy [52]. One possibility could be a post-natal appointment which could add to the current post-partum care plan by providing women with weight-loss information and signposting to suitable services that could provide further guidance and support. This role could be undertaken by the GP at the 6 week post-natal check. This appointment would need to first be mentioned to women during the antenatal stage by their midwife, as women's motivation is greatest at this stage and the knowledge that they will receive post-natal weight-loss advice may encourage them to start making small lifestyle changes in the antenatal period. Gestational weight gain should be measured and calculated for women with a BMI ≥ 30 kg/m^2 as gestational weight gain has been found to be the biggest predictor of post-partum weight gain [53]. Gestational weight can act as a predictor of those women most at risk for post-partum weight gain and will need to be discussed with women. As briefly mentioned, the professional who leads this appointment is debatable; a midwife may have good rapport with the woman and could offer breastfeeding support in addition [54]. However, the GP will have a long-term relationship with the woman and her family and would be able to provide detailed weight-loss advice and give details of existing services.

Conclusion

Obesity is a major risk factor in pregnancy. Midwives have a key role, in providing advice on weight management; identifying women who require additional surveillance and/or intervention and providing an environment which can offer a positive experience. To support women with a BMI ≥ 30 kg/m^2, midwives need to be adequately skilled and resourced. They should also work within a multi-professional team, drawing on the expertise of others (e.g. dietician, obstetrician), as appropriate. Currently, there is a dearth of quality evidence that can be used to inform best practice regarding the most appropriate way to provide weight-management information to women with a BMI ≥ 30 kg/m^2 and the possible role of routine weighing. However, midwives should not shy away from directly discussing weight with individual women; the possible long-term implications for them and their children should be highlighted. Practical advice on managing pregnancy and obesity must also be offered to women. Research is needed to inform clinical practice and ensure that consistent, holistic and effective care is provided to women ensuring positive outcomes.

References

1. Heslehurst N, Rankin J, Wilkinson JR, Summerbell CD. A nationally representative study of maternal obesity in England, UK: trends in incidence and demographic inequalities in 619,323 births, 1989–2007. *Int J Obes*. 2010;34(3):420–428.

2. Centre for Maternal and Child Enquiries (CMACE). *Maternal Obesity in the UK: Findings from a National Project*. London: CMACE; 2010.
3. Phelan S. Pregnancy: a 'teachable moment' for weight control and obesity prevention. *Am J Obstet Gynecol*. 2010;202:135e1–135e8.
4. National Institute for Health and Clinical Excellence (NICE). *Behaviour Change at Population, Community and Individual Levels. NICE Public Health Guidance*. London: NICE; 2007.
5. Kapur A. Pregnancy: a window of opportunity for improving current and future health. *Int J Gynecol Obstet*. 2011;115(suppl 1):S50–S51.
6. Smith SA, Heslehurst N, Ells LJ, Wilkinson JR. Community-based service provision for the prevention and management of maternal obesity in the North East of England: a qualitative study. *Public Health*. 2011;125(8):515–524.
7. Centre for Maternal and Child Enquiries and Royal College of Obstetricians and Gynaecologists. *CMACE and RCOG Joint Guidance. Management of Women with Obesity in Pregnancy*. London: CMACE and RCOG; 2010.
8. Keely A, Gunning M, Denison F. Maternal obesity: understanding the risks. *Br J Midwifery*. 2011;19(6):364–369.
9. Heslehurst N, Lang R, Rankin J, Wilkinson J, Summerbell C. Obesity in pregnancy: a study of the impact of maternal obesity on NHS maternity services. *Br J Obstet Gynaecol*. 2007;114:334–342.
10. Slimming World. Lets beat obesity together – When it comes to weight loss, women trust health professionals most. *Midwives Mag*. 2012;2.
11. NICE. *Antenatal Care: Routine Care for Healthy Pregnant Women*. London: NICE–RCOG Press; 2008:<http://www.nice.org.uk/nicemedia/live/11947/40145/40145.pdf>.
12. Fattah C, Farah N, O'Toole F, Barry S, Stuart B, Turner MJ. Body mass index (BMI) in women booking for antenatal care: comparison between self-reported and digital measurements. *Eur J Obstet Gynecol Reprod Biol*. 2009;144:32–34.
13. Despres JP, Lemieux I, Prud'homme D. Treatment of obesity: need to focus on high risk abdominally obese patients. *Br Med J*. 2001;322:716–720.
14. Deurenberg P. Universal cut-off BMI points for obesity are not appropriate. *Br J Nutr*. 2001;85:135–136.
15. Richens Y. Tackling maternal obesity: suggestions for midwives. *Br J Midwifery*. 2008;16(1):14–19.
16. Centre for Maternal and Child Enquiries (CMACE). Saving Mothers' Lives: reviewing maternal deaths to make motherhood safer: 2006–08. The Eighth Report on Confidential Enquiries into Maternal Deaths in the United Kingdom. *Br J Obstet Gynaecol*. 2011;118(suppl 1):1–203.
17. National Health Service Institute. *Pathways to Success: A Self Improvement Tool Kit – Focus on Normal Birth and Reducing Caesarean Section Rates*. London: NHS; 2009: <http://www.institute.nhs.uk/index.php?option=com_joomcart&main_page=document_product_info&products_id=334&cPath=100/>.
18. Clarke PE, Gross H. Women's behaviour, beliefs and information sources about physical exercise in pregnancy. *Midwifery*. 2004;20:133–141.
19. Atkinson L, Edmunds J. *Developing a theory-based exercise intervention to reduce perinatal obesity in Warwickshire*. Coventry: Coventry University; 2009.
20. National Institute for Health and Clinical Excellence (NICE). *Dietary Interventions and Physical Activity Interventions for Weight Management Before, During and After*

Pregnancy. London: National Institute for Health and Clinical Excellence (NICE); 2010:NICE Public Health Guidance 27.

21. Aviram A, Hod M, Yogev Y. Maternal obesity: implications for pregnancy outcome and long-term risks – a link to maternal nutrition. *Int J Gynecol Obstet*. 2011;115(suppl 1): S6–S10.
22. Olander EK, Atkinson L, Edmunds JK, French DP. The views of pre- and post-natal women and health professionals regarding gestational weight gain: an exploratory study. *Sex Reprod Healthcare*. 2011;2:43–48.
23. Macleod M, Gregor A, Barnett C, Magee E, Thompson J, Anderson AS. Provision of weight management advice for obese women during pregnancy: a survey of current practice and midwives' views on future approaches. *Matern Child Nutr*. 2012;10.1111/j.1740-8709.2011.00396.x.
24. Wiles R. The views of women of above average weight gain about appropriate weight gain in pregnancy. *Midwifery*. 1998;14:254–260.
25. Furness PJ, McSeveny K, Arden MA, Garland C, Dearden AM, Soltani H. Maternal obesity support services: a qualitative study of the perspectives of women and midwives. *BMC Pregnancy Childbirth*. 2011;11:69.
26. Schmied VA, Duff M, Dahlen HG, Mills AE, Kolt GS. 'Not waving but drowning': a study of the experiences and concerns of midwives and other health professionals caring for obese childbearing women. *Midwifery*. 2011;27(4):424–430.
27. Keenan J, Stapleton H. Bonny babies? Motherhood and nurturing in the age of obesity. *Health Risk Soc*. 2010;12(4):369–383.
28. Khazaezadeh N, Pheasant H, Bewley S, Mohiddin A, Oteng-Ntim E. Using service-users' views to design a maternal obesity intervention. *Br J Midwifery*. 2011;19(1): 49–56.
29. Merill RM, Richardson JS. Validity of self-reported height, weight and body mass index: findings from the national health and nutrition examination survey, 2001–2006. *Prev Chronic Dis*. 2009;6:1–10.
30. Brown I, Thompson J. Primary care nurses' attitudes, beliefs and own body size in relation to obesity management. *J Adv Nurs*. 2007;60(5):535–543.
31. Institute of Medicine. *Weight Gain During Pregnancy: Reassessing the Guidelines*. Washington, DC: National Academy Press; 2009.
32. Abrams B, Altman SL, Pickett KE. Pregnancy weight gain: still controversial. *Am J Clin Nutr*. 2000;71:1233S–1241SS.
33. Cedergren M. Effects of gestational weight gain and body mass index in obstetric outcome in Sweden. *Int J Gynecol Obstet*. 2006;93:269–274.
34. Campbell MK, Mottola MF. Recreational exercise and occupational activity during pregnancy and birthweight: a case–control study. *Am J Obstet Gynecol*. 2001;184(3): 403–408.
35. Herring SJ, Oken E, Haines J, et al. Misperceived pre-pregnancy body weight status predicts excessive gestational weight gain: findings from a US cohort study. *Pregnancy Childbirth*. 2008;8:54–63.
36. Stotland NE, Hass JS, Brawarsky P, Jackson RA, Fuentes-Afflick E, Esconar GJ. Body mass index, provider advice, and target gestational weight gain. *Obstet Gynaecol*. 2005;105:633–638.
37. Davis DL, Raymond JE, Clements V, Adams C, Mollart LJ, Teate AJ, Foureur MJ. Addressing obesity in pregnancy: the design and feasibility of an innovative intervention in NSW, Australia. *Women Birth*. 2011;Epub ahead of print.

38. Lee D, Haynes C, Garrod D. *Exploring Health Promotion Practice Within Maternity Services*. Stockport NHS Foundation Trust; 2010:Final report.
39. Watkins M, Ramussen S, Honein M, Botto Moore C. Maternal Obesity and risk for birth defect. *Pediatrics*. 2003;111(5):1152–1158.
40. Cedergren M. Maternal morbid obesity and the risk of adverse pregnancy outcome. *Obstet Gynecol*. 2004;103:219–224.
41. Waller DK, Shaw GM, Rasmussen SA, et al. Prepregnancy obesity as a risk factor for structural birth defects. *Arch Pediatr Adolesc Med*. 2007;161:745–750.
42. Ray JG, Wyatt PR, Vermulen MJ, et al. Greater maternal weight and ongoing risk of neural tube defects after flour fortification. *Obstet Gynecol*. 2005;105:261–265.
43. Furber C, McGowen L. A qualitative study of the experiences of women who are obese and pregnant in the UK. *Midwifery*. 2011;27(4):437–444.
44. Gregory V. Occupational Health and Safety update: Report on the results of an Australian Sonography Survey on prevalence of muscle skeletal disorders among sonographers. *Sound Effects*. 1999;December:42–43.
45. Lewis GE. *The Confidential Enquiry into Maternal and Child Health (CEMACH): Saving Mothers' Lives Reviewing Maternal Deaths to make Motherhood Safer 2003–2005. The Seventh Report on Confidential Enquires into Maternal Death in the United Kingdom*. London: CEMACH; 2007.
46. Royal College of Obstetricians and Gynaecologists. *Exercise in Pregnancy*. London: Royal College of Obstetricians and Gynaecologists; 2006.
47. Armstrong K, Edwards H. The effectiveness of a pram-walking exercise programme n reducing depressive symptomatology for postnatal women. *Int J Nurs Pract*. 2004;10:177–194.
48. Horns PN, Parcliffe LP, Legget JC, Swanson MS. Pregnancy outcomes among active and sedentary primiparous women. *J Obstet Gynecol Neonatal Nurs*. 1996;25:49–54.
49. Paisley TS, Joy EA, Price RJ. Exercise during pregnancy: a practical approach. *Curr Sports Med Rep*. 2003;2:325–330.
50. Clapp JF. The course of labour after endurance exercise during pregnancy. *Am J Obstet Gynecol*. 1990;163:1799–1805.
51. Smith D, Lavender T. The pregnancy experience for women with a body mass index >30 kg/m^2: a meta-synthesis. *Br J Obstet Gynaecol*. 2011;118:778–779.
52. Olander EK, Atkinson L, Edmunds JK, French DP. Healthy eating during and after pregnancy: what do pre- and post-natal women want from a diet intervention? *Prim Health Care Res Dev*. 2012;7:1–17.
53. Huang TT, Wang HS, Dai FT. Effect of pre-pregnancy body size on postpartum weight retention. *Midwifery*. 2010;26:222–231.
54. Keating NC. Using the postnatal period to help reduce obesity: the midwife's role. *Br J Midwifery*. 2011;19(7):418–423.

27 Anaesthetic Issues During Labour

Alistair Milne and Alistair Lee

Department of Anaesthesia, Royal Infirmary, Edinburgh, UK

Introduction

Obese patients and women in labour share many risk factors for anaesthesia and surgical intervention [1]. The morbidly obese woman in labour presents a formidable challenge to the anaesthetist.

The body mass index (BMI) of women of reproductive age has increased significantly over the last two decades and over 20% of this population are obese. Studies in the United States and the United Kingdom demonstrate that over 10% of pregnant women are classed as obese with a BMI in excess of 30. The greatest increase in the prevalence of obesity has occurred in patients with a BMI over 40, who now represent over 10% of Western women of reproductive age, and will increasingly become part of everyday practice for the obstetrician and obstetric anaesthetist [2].

Obesity is associated with increased maternal mortality, and is a specific risk factor for anaesthesia-associated deaths [3].

The Cardiovascular System

A central or android distribution of fat is associated with greater myocardial fat content than a peripheral or gynaecoid distribution. Blood volume and cardiac output are increased in the obese population. Cardiac output increases by 50 ml/min for every extra 100 g of fat [4]. This increase in cardiac workload exacerbates the increase in plasma blood volume, and cardiac output of up to 50%, that occurs in pregnancy. Basal sympathetic activity is elevated and heart rate is increased. Hypertension is common in the obese [5]. Pre-eclampsia and gestational hypertension are more common in obese parturients and in those women with significant (>9 kg) interpregnancy weight gain [6,7]. All of these factors contribute to increased strain on the left ventricle and reduced cardiac reserve. ECG recording demonstrates ischaemia in a significant number of asymptomatic obese patients and echocardiography may be used as an assessment of myocardial pump function [8,9]. Pregnancy contributes to reduced cardiovascular reserve.

Obesity. DOI: http://dx.doi.org/10.1016/B978-0-12-416045-3.00027-3

Obese women are more susceptible to supine hypotensive syndrome than is generally the case in late pregnancy and careful positioning is important. Obesity supine death syndrome has been described in the super morbidly obese. Death on lying supine probably occurs due to a combination of aortocaval compression causing a sudden reduction in venous return, coupled with hypoxaemia, resulting in cardiac arrest [10].

The Respiratory System

Airway

Several authors have reported that intubation difficulties are more likely with increasing obesity, but a low incidence of significant problems has been reported when senior experienced personnel are involved [11]. The incidence of failed intubation has been reported at approximately 1:250 in the obstetric population compared to 1:2000 in a general population [12]. Enlarged breasts and oedematous airways contribute to this problem. An incidence of difficult intubation as great as 1 in 3 has been reported when parturients are morbidly obese [13]. To compound problems, obese women in labour have an even higher risk of aspiration than non-pregnant women, and a rapid sequence induction is mandatory in these patients. Awake fibreoptic intubation is possible in pregnant obese women, but this is never a rushed emergency procedure and should only be undertaken by experienced senior personnel practised in the technique.

Oxygenation

In both pregnancy and obesity, the diaphragm is pushed in a cephalad direction. This leads to a reduction in functional residual capacity, the lung volume at the end of a normal expiration. Individuals of normal weight experience closure of small airways, the closing volume, at a lung volume significantly below their functional residual capacity. In obese individuals, the functional residual capacity is reduced to an extent that it is smaller than the closing volume. As a consequence, obese individuals experience small airway closure while breathing normally. This is associated with ventilation perfusion mismatch leading to reduced oxygen saturation [14]. Oxygen consumption is significantly greater in obesity and pregnancy. In combination, these factors mean that even with good pre-oxygenation the time to critical desaturation is reduced. Desaturation to a critical level (defined as $<90\%$ saturation, the critical inflection point) occurs in less than 3 min in the obese patient compared to longer than 6 min in a normal individual [15]. This decrease is more rapid in the obstetric population. A head up tilt during pre-oxygenation increases the time to critical desaturation by about 1 min. This is due to gravity causing an increase in functional residual capacity by allowing the abdominal contents, fat pannus and diaphragm to move caudally.

During the immediate post-operative period, there is a significant risk of hypoxaemia after general anaesthesia in the obese patient. The lung volume may be reduced even allowing for delivery. Once the patient has been extubated, the lungs will no longer contain 100% oxygen as a reserve, unless the patient is breathing very effectively through a completely sealed face mask, and the time to critical oxygen desaturation can be very short. The use of volatile agents significantly depresses respiratory drive from hypercapnia, and the response to hypoxaemia is negligible until the gases are cleared. As a consequence, airway obstruction can quickly lead to death.

The Gastrointestinal System

Hiatus hernia is common in obese patients together with increased residual gastric contents compared to the non-obese [16]. Although gastric emptying time is unchanged in pregnancy, it slows significantly in labour leading to an increased risk of a full stomach. In association with the hormonal changes of pregnancy leading to an increased likelihood of reflux, the risk of aspiration in these women is very significant.

The Endocrine System

Type II diabetes is common in the obese patient and may be part of the metabolic syndrome which is defined as having three of the following: central obesity, impaired glucose tolerance, hypertension, dyslipidaemia and a pro-thrombotic, pro-inflammatory state [17]. These patients have insulin resistance and hyperinsulinaemia. Fatty liver is very common in the obese, but the incidence of HELLP (Haemolysis, Elevated Liver enzymes, Low Platelet count) syndrome appears to be unchanged.

Anaesthetic Considerations

Equipment

Modern equipment increasingly caters for an enlarging population, but the morbidly obese (BMI $>$ 40) and the super morbidly obese (BMI $>$ 50) require specialised kit. This should be available for the obstetric population, as presentation of these patients should come as no surprise. A standard operating table has an upper weight limit of about 180 kg, but a bariatric operating table will have an upper weight limit of around 450 kg. These tables have additional side adjusters that are very robust compared to normal arm supports and still allow the surgeon to stand close to the table.

Hoists may be required to move the patient; alternatively hover mattresses are easy to use in the theatre environment to move the patient effortlessly from table to bed. The hover blanket is placed underneath the patient preoperatively and is inflated when required to float the patient. Despite the ease of use, sufficient personnel are still required to control patient movement carefully. Beds need to be able to withstand the patient's weight and have hydraulic electrical adjustment so that the patient can be placed in the upright position without effort.

A full range of kit for management of the difficult airway is normally available in the obstetric unit.

Specialised pillows, or breaking the table at the thigh–torso junction, allows the patient to be placed in a 25° back up or ramped position for airway management. This position elevates the upper part of the torso and has been shown to aid intubation in the obese [18].

Continuous positive airway pressure (CPAP) from a nasal mask may be required in the perioperative period if the patient suffers from obstructive sleep apnoea. Often the patient's own machine is most suitable.

Venous access may be problematic. Ultrasonography may permit peripheral access, and on occasion will be required to secure central venous access if obtaining peripheral access proves impossible.

Blood pressure cuffs require to be larger than normal and in the morbidly obese, blood pressure may need to be monitored invasively. Radial arterial cannulation is often quite straightforward in these patients compared to difficulties with venous access.

Calf-compression boots or leggings have to be an appropriate size. The risks of venous thrombosis are greater than average in the obese patient, compounding the increased risks already present in the pro-thrombotic parturient.

Straps to hold the patient in place on the table should be available.

Personnel

Sufficient number of staff need to be available to move morbidly obese and super morbidly obese patients safely even with appropriate lifting equipment.

Management of these patients by junior staff is inappropriate. The morbidly obese parturient is becoming increasingly commonplace for the obstetric anaesthetist, but it is recommended that some form of liaison between obstetric anaesthesia and bariatric practice is in place. Bariatric services are now widespread, have dedicated anaesthesia personnel familiar with the problems of the super obese and should ensure skills are appropriately transferred to the obstetric situation. There is no excuse for providing a lesser degree of expertise to the obstetric population.

Consideration needs to be given to appropriate staffing levels in the postoperative period. Recovery room respiratory difficulties are more frequent in the obese patient and can rapidly prove life-threatening if not managed promptly.

Drugs

A consistent finding is the variability of drug effect in obese patients, both between individuals and between studies. Some broad principles are valid. Published drug doses are usually weight based and only apply to the non-obese [19]. Drug absorption from a subcutaneous or intramuscular injection is relatively poor and more variable in the obese patient. Lipid soluble drugs will have a larger volume of distribution and may be considered to require larger doses. Water soluble drugs should behave more like they would in the non-obese subject, but it is recognised that obese subjects do not have an increase in body fat alone. They have an increase in the volume of other tissues and as a consequence drug dosing based on ideal body weight may be insufficient [20].

Drug elimination is highly complex. Increased cardiac output, renal blood flow and splanchnic and hepatic blood flow may increase drug elimination, whereas an increased volume of distribution may prolong the duration of action of some drugs.

The Anaesthetic Clinic

Women with a BMI over 40 should be seen prior to labour and delivery by an anaesthetist. This allows for timely assessment of the patient and further investigations to be requested if necessary, along with a discussion of methods of analgesia and the anaesthetic options for delivery if intervention should be required. An anaesthetic plan should be discussed and documented [21].

Assessment of exercise tolerance may highlight a diminished cardiorespiratory reserve and the need for an echocardiogram. Pulmonary function tests may help in diagnosis. An increased incidence of asthma is well recognised in the obese. A history of symptoms suggestive of sleep apnoea should prompt assessment by a respiratory physician. The use of CPAP may improve sleep, reduce the incidence of episodes of hypoxaemia and ultimately prevent right ventricular failure. The current state and management of co-morbidities such as diabetes and hypertension should be reviewed [22,23].

Early assessment of the airway is essential. In exceptional circumstances, the anaesthetic team may consider that emergency general anaesthesia will be too hazardous in the context of severe foetal distress. This can lead to a significant, but ultimately safer, change to patient management at an early stage. A different method of analgesia during labour, or even a different method of delivery, may be planned for in advance. Planning for all eventualities in advance by a multidisciplinary team may avoid dangerous emergency situations. The parturient with a difficult airway and a contraindication to regional anaesthesia may be counselled towards planned elective lower segment Caesarean section with awake fibreoptic intubation.

Examination of the expectant mother's back for the bony landmarks used during the performance of regional techniques may give an early indication of extreme difficulty. Patients with no palpable midline, iliac crests or intervertebral spaces often require significant extra time and multiple attempts to site an epidural catheter or perform spinal anaesthesia. Awareness of this possibility in advance by all team members is helpful [24].

Meeting the morbidly obese expectant mother in the clinic environment allows early discussion of the challenges posed by their weight in terms of regional analgesia, and both regional and general anaesthesia. As well as forewarning the mother of potential difficulties and planning for these patients in advance, it helps the anaesthetic staff manage the mother's expectations and potentially allows anaesthetic involvement at an early stage in labour should the mother wish this.

The Labour Ward

The anaesthetist should be informed about every morbidly obese woman admitted to the labour ward. This permits early review of the patient, multidisciplinary team planning of her care and ensures senior input from an early stage.

Airway Assessment

This should be repeated on admission. There may be a worsening of difficult airway predictors as pregnancy progresses due to weight gain and fluid retention. An airway that may have appeared straightforward in the clinic may become significantly more challenging in labour.

Intravenous Access

This can be difficult to obtain in the obese parturient due to overlying fat obscuring the veins. Ultrasonography is commonly used to identify adequate-sized veins and assist with cannulation in the morbidly obese. It is prudent to site two cannulae at an early stage when time is available. This is standard bariatric practice when there is difficulty. Re-siting a failed cannula when it is needed urgently in the emergency situation may take up time that is not readily available. Central venous access may be required in exceptional circumstances. This requires time and patient cooperation to establish and requires careful maternal and foetal monitoring during positioning for the procedure. This monitoring should be performed by someone not actively undertaking the procedure.

Entonox and Opioids

Either of these forms of pain relief increases the incidence of hypoxaemia in the morbidly obese parturient and often provide suboptimal analgesia [25,26]. Remifentanil patient-controlled analgesia is an alternative technique of opioid administration that is available in some centres. It may be useful for patients who cannot have regional analgesia or in whom regional analgesia has persistently failed. Remifentanil is a powerful opioid with a peak pharmacokinetic effect at 60−120 s and a context-sensitive half-time of approximately 180 s [27]. These characteristics make it suitable for short episodes of severe pain as in labour, but its potency may cause severe respiratory depression and apnoeic episodes, especially if the peak effect of the drug does not coincide with a contraction [28]. Extreme care is required if it is used for analgesia in morbidly obese parturients. Although the effects of remifentanil rapidly wear off, oxygen administration is mandatory for this group of patients, and the rapid development of hypoxaemia is still a concern.

Regional Analgesia

Correct placement of a needle or catheter can be especially challenging in the morbidly obese parturient and regional analgesia once sited is more likely to be suboptimal [29]. When none of the bony landmarks used for siting an epidural are palpable, placement and institution of a block may take significantly longer than usual and this should be allowed for if at all possible. Once in place, epidural catheters are at a greater risk of dislodging in the obese due to lateral movements of the skin and fat pad over the vertebrae. This can potentially lead to the catheter being pulled out of the epidural space despite there being no movement of the catheter at the skin [30]. There is a significantly increased incidence of the need to re-site an epidural catheter in the obese parturient. Ultrasonography is useful as an aid to epidural insertion. It is primarily used to identify the midline and intervertebral spaces and provide surface marking of insertion points, after which long-established techniques are used to find the epidural space. Used in this manner it has been shown to reduce the number of insertion attempts required [31]. An effective epidural provides maternal analgesia, avoidance of other forms of analgesia that may be associated with respiratory depression, and permits rapid conversion to anaesthesia should this be required.

The Operating Theatre

The appropriate management of the morbidly obese parturient in the operating theatre is greatly influenced by the degree of urgency of the procedure. The anaesthetic technique employed often depends on the management of labour until this time. The morbidly obese parturient with a well-functioning epidural catheter already in situ when surgical intervention is required is in the ideal state.

Elective Cases

Advance knowledge of the morbidly obese patient is important as there are many practical issues that need to be addressed. If the patient's weight is close to or above a standard operating table's limit of 180 kg, a bariatric table will be required. The parturient's body shape and the needs for lateral tilt are important. A woman with predominantly central obesity will be too wide for the table. Solid, strongly attached side extensions and side guards that are mechanically attached to the side of the table prevent pressure sores and allow the patient to be tilted in safety. A large ward bed, a large chair and a suitable mattress to reduce the risk of pressure sores are standard bariatric requirements. These items are mundane, but without them, management of the patient is difficult, and potentially hazardous to patient and staff. Longer spinal and epidural needles should be available, together with the wedges and head supports designed to place the morbidly obese patient in the ideal position for tracheal intubation.

Invasive arterial blood pressure monitoring may be necessary. Even an appropriately large blood pressure cuff may not be suitable if the patient has a conical shaped upper arm. Accurate and frequent blood pressure monitoring is especially important during the establishment of regional blockade and when lying the patient supine. These manoeuvres are associated with significant cardiorespiratory changes. Arterial blood gas measurement may be helpful if the patient is hypoxaemic or hypercapnic.

Cell salvage is standard practice for surgical procedures associated with significant blood loss and is recommended for these women who have an increased risk of blood loss at Caesarean section. Blood and patient warming devices should be employed. The hypothermic patient has significantly increased oxygen consumption, especially when shivering, and this can contribute to cardiovascular decompensation.

Surgical access may be very difficult and hampered by a fat pannus. Taping the pannus up to a bar sitting above the patient's chest may be helpful, but if this leads to significant cardiovascular or respiratory compromise, the pannus may need to be suspended vertically. A hook system or additional surgical assistance may be required. Additional staff members are required in any case for safe patient transfer. A second anaesthetist to help with the general management of the case and to assist with any intra-operative complications is often invaluable, while an extra member of staff may be required to operate the cell salvage system.

General anaesthesia is associated with an increased morbidity and mortality in obstetric practice, most commonly due to airway difficulties leading to hypoxaemia or aspiration. These risks are exacerbated by morbid obesity. General anaesthesia is rarely the anaesthetic of choice in the elective situation.

Regional anaesthesia is almost invariably the chosen technique for elective procedures, but all methods are significantly more challenging in the morbidly obese. Identification of bony landmarks may be impossible. The distance from skin to the epidural and intrathecal spaces is significantly greater in the obese parturient, and this distance is affected by the posture the parturient adopts for needle insertion [32]. This increased distance reduces the margin of error for insertion site and needle angulation in patients who already have landmarks that are difficult to identify.

Spinal anaesthesia is associated with an unpredictable spread of local anaesthetic in the morbidly obese. This may result in a high block, probably as a consequence of increased fat content in the epidural space associated with reduced cerebrospinal fluid volume. Spinal anaesthesia causes a decrease in cardiac sympathetic innervation, and venodilation leading to a decrease in venous return. This may cause catastrophic hypotension when compounded by aortocaval compression from the foetus and fat pannus. Blockade of the intercostal muscles, along with the increased intra-abdominal pressure on the diaphragm, may lead to hypoxaemia and an increased work of breathing as the patient's functional residual capacity decreases. In an effort to reduce these effects, and to ensure satisfactory operating conditions for a procedure of unpredictable and often significantly prolonged duration, many anaesthetists would choose to use a combined spinal epidural (CSE) technique.

A CSE technique allows a single shot spinal block to be augmented with repeated local anaesthetic top ups through an epidural catheter. This maintains anaesthesia if surgery is prolonged, and the epidural may then be used to provide post-operative analgesia. A further advantage is that there is an easy means to re-establish anaesthesia in the early post-operative period should the patient require to return to theatre.

The patient must be carefully positioned once regional anaesthesia is instituted. Aortocaval compression may not be easily relieved with the standard 'tilt'. It is uncommon to achieve the often quoted 15° left tilt unless specialised table equipment is used [33]. Safe positioning requires careful attention to potential pressure areas, especially those areas that have just been rendered insensate by the regional local anaesthetic blockade. The lithotomy position may be difficult to achieve when the parturient has large, heavy legs. Nerve damage is a particular risk at susceptible sites such as the fibular head, and ligamentous damage may occur from the increased stress placed on joints. Elevating the patient's legs may lead to significant cardiorespiratory compromise as the fat pannus moves cephalad, and diaphragmatic movement is restricted.

Urgent Cases

In morbidly obese parturients, general anaesthesia is significantly more hazardous than regional anaesthesia, but regional anaesthesia is likely to take considerably longer to perform than in the slim patient. If there is a well-functioning epidural in situ, the technique of choice is to top it up to establish anaesthesia. There are many different local anaesthetic agents and adjuncts that can be used to establish epidural analgesia, with differing timescales to full effect. A lidocaine–bicarbonate–adrenaline mix can achieve sensory analgesia to the T4 level in 7 min. This compares to 11 min for levobupivacaine [34]. A recent meta-analysis confirmed the significantly more rapid conversion of epidural analgesia to surgical anaesthesia using lidocaine–adrenaline [35]. In the patient without an epidural, the balance of risks would favour establishing regional anaesthesia in the first instance. In the knowledge that this may take some time, planned frequent review of the whole situation is vital. It is very easy to get focussed on a difficult regional procedure and fail to realise that circumstances have changed. Continuous foetal monitoring is vital during all regional procedures. Patient positioning for regional anaesthesia may result in foetal compromise. Communication with the obstetric team throughout the procedure is essential as rapid changes in foetal status may require a change to the previously planned anaesthetic management. General anaesthesia may be required if it proves impossible to perform a regional block. This will usually involve a rapid sequence induction of anaesthesia, but awake fibreoptic intubation may be considered if there is the potential for extreme airway difficulty, the risk is primarily foetal, and the situation remains under control.

Emergency Cases

A major concern for every anaesthetist is the safe management of the morbidly obese parturient in the emergency situation when time is limited. The primary objective is the safety of the mother, and this should never be compromised for expeditious delivery of the foetus. The requirement for emergency delivery creates a very stressful environment. This stress is compounded when the chosen anaesthetic technique may not appear to be the most rapid to other members of the team. Inadequate communication between team members can lead to poor decisions being made in haste with potentially disastrous consequences.

It is imperative that morbidly obese parturients are referred to the anaesthetic team on admission to the labour ward. Anaesthetic assessment early in labour will allow any potential difficulties to be discussed with the patient, the obstetrician and the midwives. The proposed conduct of labour should be understood by all concerned, and a provisional plan can be made should an emergency situation develop. Continued communication throughout labour will allow team members to work more effectively together.

Obese parturients with difficult landmarks for regional anaesthesia, or difficult airways, will often be advised to have an epidural catheter sited early in labour. If these patients require emergency delivery, the epidural block can be extended. In the morbidly obese patient with a potentially difficult airway and no maternal compromise this will invariably be the choice for provision of surgical anaesthesia, no matter what the threat to the foetus.

Decisions become more difficult when the emergency situation presents in the patient who does not have an epidural catheter in place and the clinical situation would normally warrant general anaesthesia, for example severe foetal distress or a significant ante-partum or post-partum haemorrhage. Whatever technique is chosen, a senior anaesthetist must be involved for these patients, often with additional anaesthetic assistance. It will be necessary to judge the balance of risks at the time and tailor the anaesthetic appropriately. The most appropriate course of action may change, and constant review of the planned anaesthetic procedure is essential.

The obese patient judged to have a reasonable airway may be suitable for general anaesthesia using a rapid sequence induction. This may not be the case for the super-obese patient with obvious airway risk factors, or a previously known difficult intubation.

If extreme airway difficulty is judged likely, then in the context of severe foetal distress, regional anaesthesia is almost invariably most appropriate. If there is significant maternal risk from uncontrolled haemorrhage the balance of risk may then favour emergency general anaesthesia, despite the airway difficulties.

Awake fibreoptic intubation may be appropriate to consider in an urgent situation and is regarded by many as the gold standard technique, especially when bag and mask ventilation is inappropriate, as in the obstetric patient at risk of aspiration. However, it is a time-consuming process that requires a compliant patient and is almost certainly inappropriate during maternal haemorrhage or other extreme maternal obstetric emergency situations.

General anaesthesia is associated with many increased risks in the obstetric population and these risks are all exacerbated by morbid obesity. The risks of difficult and failed intubation are greatly increased, aspiration is significantly more likely, and bag and mask ventilation necessary to maintain oxygenation in the event of a failed intubation is more difficult in the obese [36]. These problems are compounded by rapid oxygen desaturation in the morbidly obese due to the reduction in oxygen reserves in the lungs and an increased metabolic demand. Increased inspiratory pressures and increased levels of positive end expiratory pressure (PEEP) are required to ventilate the lungs and keep small airways open to reduce shunting and hypoxaemia. Increased intrathoracic pressures decrease the venous return to the heart, reducing cardiac output and blood pressure. Aortocaval compression caused by the foetus is exacerbated by the fat pannus, adding to decreased venous return and hypotension. Morbid obesity increases the risk of haemorrhage, as does the myometrial relaxation caused by volatile agents used to maintain general anaesthesia.

The Post-Operative Period

There must be a low threshold for morbidly obese parturients to be managed post-operatively in a level 2 high-dependency area. Those patients with significant co-morbidities, in particular severe cardiorespiratory disorders, should be admitted to a level 3 intensive care area experienced in the management of heart failure or respiratory problems in the obese patient. Clinicians should have a low threshold for admitting obese patients who have suffered major complications such as massive haemorrhage to an intensive care area in the first instance. Every effort should be made to allow mother and baby to be together, as far as is practicable and safe, should the mother be admitted to a general critical care area. Some obese patients may require little additional medical input compared to patients of normal weight but will represent a formidable nursing challenge, and the nursing team must be forewarned and prepared.

Supplementary oxygen may be required for several days. Post-operative atelectasis, a reduced ability to breathe deeply due to pain, administration of opioids, together with the mechanical effects of obesity leading to small airway closure contribute to hypoxaemia and hypercapnia. These are common effects for every patient after abdominal surgery but are exaggerated in the obese and may be superimposed on a background co-morbidity of obstructive sleep apnoea or the obesity hypoventilation syndrome leading to the potential for a difficult early post-operative period. Sleep patterns are altered post-operatively with a reduction in rapid eye movement (REM) sleep for the first and second nights. This is followed by a rebound increase in REM sleep for the ensuing few nights. This rebound in REM sleep may be associated with sleep disordered breathing and nocturnal hypoxaemia [37]. Morbidly obese patients with sleep apnoea requiring CPAP overnight are often best managed using their own CPAP system which they should bring into hospital for use post-partum. Regular chest physiotherapy will help basal lung expansion and possibly reduce the risk of chest infection. These patients should be nursed sitting upright, not slumped down in the bed.

If the patient has an epidural catheter in place this should be retained as a method of re-establishing anaesthesia in the early post-partum period when there is a risk of post-partum haemorrhage. It should be used as the preferred method to provide post-operative analgesia, minimising or avoiding the administration of opioids and reducing the risk of respiratory depression. A functioning epidural will normally aid effective deep breathing and coughing due to superior pain control. This group of patients have an increased risk of deep venous thrombosis and pressure sores, and these risks are likely to increase with prolonged immobility. Epidural analgesia should not be allowed to delay mobilisation and the rate of local anaesthetic administration should be reduced if there is the development of significant motor blockade. Routine pressure care must be instituted even in the awake patient. These patients may be very immobile and are at significant risk of the development of pressure sores when receiving epidural analgesia.

Continued use of mechanical compression devices may be used as part of thromboprophylaxis while the patient is relatively immobile in the post-operative period. Beds, mattresses, hoists, chairs, compression boots, compression stockings, blood pressure cuffs and gowns of an appropriate size are all required at this time.

Conclusion

Morbidly obese parturients are increasingly common. An understanding of the co-morbidities associated with morbid obesity is essential for the obstetrician and obstetric anaesthetist. Appropriate equipment, together with experienced staff in sufficient numbers are essential to safe management.

References

1. Cevik B, Ilham C, Orskiran A, Colakoglu S. Morbid obesity: a risk factor for maternal mortality. *Int J Obstet Anesth*. 2006;15:263–264.
2. Pender JR, Pories WJ. Epidemiology of obesity in the United States. *Gastroenterol Clin North Am*. 2005;34:1–7.
3. Cooper GM, McClure JH. *Sixth report on confidential enquiries into maternal deaths in the United Kingdom. Why Mothers Die, 2000–2002*. London: RCOG Press; 2004:122–133.
4. Veille JC, Hanson R. Obesity, pregnancy and left ventricular functioning during the third trimester. *Am J Obstet Gynecol*. 1994;171:980–983.
5. Peluso L, Vanek VW. Efficacy of gastric bypass in the treatment of obesity-related comorbidities. *Nutr Clin Pract*. 2007;22:22–28.
6. Weiss JL, Malone FD, Emig D, et al. Obesity, obstetric complications and caesarean delivery rate – a population-based screening study. *Am J Obstet Gynecol*. 2004;190:1091–1097.
7. Getahun D, Ananth CV, Peltier MR, Salihu HM, Scorza WE. Changes in prepregnancy body mass index between the first and second pregnancies and risk of large-for-gestational-age birth. *Am J Obstet Gynecol*. 2007;196:530e1–530e8.
8. Lean ME. Obesity and cardiovascular disease: the waisted years. *Br Cardiol J*. 1999;6: 269–273.
9. Saravanakumar K, Rao SG, Cooper GM. The challenges of obesity and obstetric anaesthesia. *Curr Opin Obstet Gynecol*. 2006;18:631–635.
10. Tsueda K, Debrand M, Zeok SS, Wright BD, Griffin WO. Obesity supine death syndrome: reports of two morbidly obese patients. *Anesth Analg*. 1979;58:345–347.
11. Bellamy M, Struys M. *Obesity and the airway. Anaesthesia for the Overweight and Obese Patient*. Oxford: Oxford University Press; 2007:75–82.

12. Barnardo PD, Jenkins JG. Failed tracheal intubation in obstetrics: a 6 year review in a UK region. *Anaesthesia*. 2000;55:685–694.
13. D'Angelo R, Dewan DD. Obesity. In: Chestnut DH, ed. *Obstetric Anesthesia: Principles and Practice*. Philadelphia, PA: Mosby; 2004:893–903.
14. Adams JP, Murphy PG. Obesity in anaesthesia and intensive care. *Br J Anaesth*. 2000;85:91–108.
15. Sirian R, Wills J. Physiology of apnoea and the benefits of preoxygenation. *Contin Educ Anaesth Crit Care Pain*. 2009;9:105–108.
16. Lotia S, Bellamy MC. Anaesthesia and morbid obesity. *Contin Educ Anaesth Crit Care Pain*. 2008;8:151–156.
17. Alberti KGMM, Zimmet P, Shaw J. Metabolic syndrome – a new world-wide definition. A consensus statement from the international diabetes federation. *Diabet Med*. 2006;23:469–480.
18. Rao SL, Kunselman AR, Schuler HG, Des Harnais S. Laryngoscopy and tracheal intubation in the head-elevated position in obese patients: a randomized, controlled, equivalence trial. *Anesth Analg*. 2008;107:1912–1918.
19. Bouillon T, Shafer SL. Does size matter? *Anesthesiology*. 1998;89:557–560.
20. Lemmens HJM, Brodsky JB. The dose of succinylcholine in morbid obesity. *Anesth Analg*. 2006;102:438–442.
21. Modder J, Fitzsimons KJ. *Management of Women with Obesity in Pregnancy – CMACE/RCOG Joint Guideline*. London: CMACE/RCOG; 2010:1–29.
22. Roofthooft E. Anesthesia for the morbidly obese parturient. *Curr Opin Anaesthesiol*. 2009;22:341–346.
23. Castro LC, Avina RL. Maternal obesity and pregnancy outcomes. *Curr Opin Obstet Gynecol*. 2002;14:601–606.
24. Perlow JH, Morgan MA. Massive maternal obesity and perioperative cesarean morbidity. *Am J Obstet Gynecol*. 1994;170:560–565.
25. Elbourne D, Wiseman RA. Types of intra-muscular opioids for maternal pain relief in labour. *Cochrane Database Syst Rev*. 2000;2:CD001237.
26. Howell CJ. Epidural versus nonepidural analgesia for pain relief in labour. *Cochrane Database Syst Rev*. 2000;2:CD000331.
27. Egan TD. Remifentanil pharmacokinetics and pharmacodynamics. A preliminary appraisal. *Clin Pharmacokinet*. 1995;29:80–94.
28. Marwah R, Hassan S, Carvalho JCA, Balki M. Remifentanil versus fentanyl for intravenous patient controlled labour analgesia: an observational study. *Can J Anaesth*. 2012; 59(3):246–254.
29. Hood D. Anesthetic and obstetric outcome in morbidly obese parturients. *Anesthesiology*. 1993;79:1210–1218.
30. Wasson C. Failed epidural in an obese patient: blame it on Pythagoras. *Anaesthesia*. 2000;56:585–610.
31. Grau T, Leipold RW, Conradi R, Martin E, Motsch J. Efficacy of ultrasound imaging in obstetric epidural anaesthesia. *J Clin Anesth*. 2002;14:169–175.
32. Hamza J, Smida H, Benhamou D, Cohen SE. Parturient's posture during epidural puncture affects the distance from skin to epidural space. *J Clin Anesth*. 1995;7:1–4.
33. Jones SJ, Kinsella SM, Donald FA. Comparison of measured and estimated angles of table tilt at caesarean section. *Br J Anaesth*. 2003;90:86–87.
34. Allam J, Malhotra S, Hemingway C, Yentis SM. Epidural lidocaine–bicarbonate–adrenaline vs levobupivacaine for emergency caesarean section: a randomised controlled trial. *Anaesthesia*. 2008;63:243–249.

35. Hillyard SG, Bate TE, Corcoran TB, Paech MJ, O'Sullivan G. Extending epidural analgesia for emergency caesarean section: a meta-analysis. *Br J Anaesth.* 2011;107: 668–678.
36. Wong DT. Airway management in the operating room for a morbidly obese patient in a 'Can't Intubate, Can't Ventilate' situation. In: Hung O, Murphy M, eds. *Management of the Difficult and Failed Airway.* New York, NY: McGraw-Hill; 2008:373–380.
37. Rosenberg J, Wildschiodtz G, Pedersen MH, Von Jessen F, Kehlet H. Late postoperative nocturnal episodic hypoxaemia and associated sleep pattern. *Br J Anaesth.* 1994;72:145–150.

28 Maternal Obesity and the Risk of Stillbirth

Mairead Black and Siladitya Bhattacharya

Division of Applied Health Sciences, University of Aberdeen, Aberdeen, UK

Background

While stillbirth may be the most feared pregnancy outcome for many women, the idea that obesity may cause it is unlikely to be one with which many women are familiar. While the effort to tackle obesity is potentially overwhelming for some women, ultimately it is a reversible condition, and, therefore, making women aware of its association with the risk of stillbirth is a worthwhile exercise. A recent meta-analysis of the five high-income countries with the highest stillbirth rates reveals overweight and obesity as the highest ranking modifiable risk factors, with a population-attributable risk of 8–18% [1]. This is of particular relevance in that stillbirth rates due to a number of other causes have fallen in these countries.

Stillbirth or the death of a foetus before birth (beyond 24 weeks gestation or a birthweight of 500 g) [2] occurs approximately in 1 in 200 pregnancies in the United Kingdom [3]. While stillbirth can result from obstetric haemorrhage, congenital abnormalities or pre-eclampsia, in over half of all cases, a cause is never identified. A significant proportion of unexplained stillbirths have evidence of restricted growth, and while placental dysfunction is apparent, this has proven to be an insufficient explanation for foetal demise. Difficulties arise in classifying the aetiology of stillbirth, as many identified pathological processes could only be considered to be the potential risk factors but are not yet proven. Also, many causes of stillbirth represent a variety of underlying pathological processes, so that addressing those causes may require a multi-faceted approach. For example, congenital abnormalities resulting in stillbirth can arise from either genetic or environmental influences or a combination of both.

While precise causation of stillbirth has proven difficult to be defined, risk factors for stillbirth have been systematically assessed. Those include socio-demographic factors such as low social class and cigarette smoking, and medical conditions including diabetes mellitus. Among the known risk factors, obesity is a growing threat. Recent evidences indicate that obese women are at least twice more likely to experience a stillbirth than those of normal weight. In fact, obesity has been

Obesity. DOI: http://dx.doi.org/10.1016/B978-0-12-416045-3.00028-5

suggested as a potential explanation for the recent rise in stillbirth rates, as the soaring levels of obesity coincide with this trend [4]. Even if obesity is only moderately associated with an increased risk of stillbirth, the impact could be enormous due to the epidemic proportions of obesity in many western countries.

In describing the relationship between maternal obesity and stillbirth, this chapter addresses the epidemiology of obesity-associated stillbirth, recognised risk factors for stillbirth and their relationship with maternal obesity.

Epidemiology

Observational studies of stillbirth associated with maternal obesity have been carried out in countries across the world including Scotland [5], Sweden [6], Finland [7], Denmark [8], New Zealand [9] and China [10].

A Scottish study of over 24,000 pregnancies found an increased risk of stillbirth in obese women [5]. A similarly sized study of Danish women revealed a more than twofold risk of stillbirth in obese women but found no increased risk in overweight women. The results did not change after adjusting for multiple potentially confounding factors [8]. An English series of 48,357 births, including 324 stillbirths, suggested a dose–response relationship between BMI and risk of stillbirth, with women who were underweight having a stillbirth risk less than one-third of that associated with women of BMI over 35 kg/m^2 [11].

Stillbirth rate by BMI category Ref. [8]

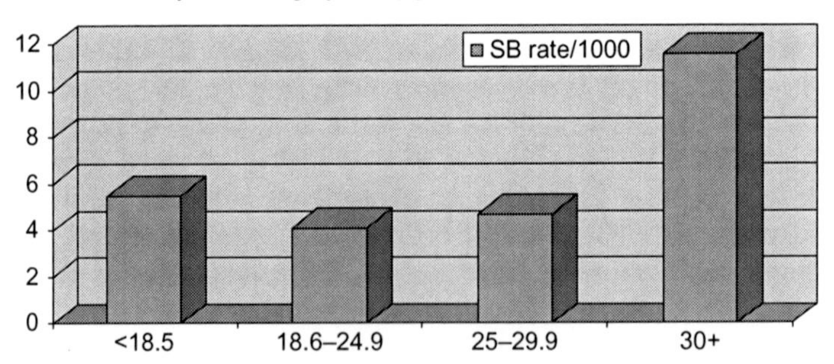

Despite variations in study design, the overwhelming message from the published literature is that the risk of stillbirth rises with increasing degree of obesity, with the lowest risk in underweight women from developed countries [6]. Several individual studies support the BMI-dependant risk relationship, and indeed many suggest the risk of stillbirth is at least doubled in obese women compared with normal-weight women [7]. These figures are supported by the findings of a meta-analysis [12].

Being relatively rare in relation to other pregnancy outcomes of interest, studies exploring causes of stillbirth incidence require a sizeable population cohort to ensure reliable results and should ideally address the potential effect of confounders including maternal age, parity, cigarette smoking, social deprivation, ethnicity, congenital abnormalities, gestational diabetes, hypertension and pre-eclampsia. Few are able to include all of these variables because few researchers have access to the full spectrum of patient data required.

Pathophysiology

A large number of risk factors for stillbirth have been identified. Of these, a significant number are recognised to occur as a direct result of obesity (direct risk factors), while others are simply associated with the condition (indirect risk factors). They are discussed here, according to their relationship with BMI and stillbirth risk (Table 28.1).

Direct Risk Factors

The incidence of many obstetric complications is increased by obesity, and a number of these are in turn independently associated with higher risk of stillbirth. Such complications include gestational diabetes, hypertension, vascular disease, foetal abnormalities and post-mature pregnancy.

Gestational Diabetes

Gestational diabetes is increasingly common as BMI rises [13]. It has been found to affect around 5.5% of obese women compared with just 0.4% of normal-weight women [8]. Insulin resistance in the mother leads to abnormally high glucose levels resulting in hyperglycaemia in the foetus who responds by increasing its insulin production. The potential impact of this includes abnormal organogenesis, increased foetal fat deposition, macrosomia, foetal death, risk of obstructed labour

Table 28.1 Factors Influencing the Relationship between High Maternal BMI and Stillbirth Risk

Direct Risk Factors	Indirect Risk Factors
Gestational diabetes	Increased maternal age
Hypertensive disease	Grandmultiparity
Foetal abnormalities	Low social class
Post-mature pregnancy	Ethnicity
Vascular disease	–
Previous pregnancy loss	–
Sleep apnoea	–
Failed antenatal screening	

and birth trauma, along with increased risk of hypoglycaemia in the neonate. Given the variation in strategies adopted to identify gestational diabetes and the cause of stillbirth, it is possible that undiagnosed gestational diabetes may form a significant proportion of those stillbirths which are classified as unexplained.

Hypertensive Disorders

Maternal obesity carries an increased likelihood that the affected woman will suffer from essential hypertension and in rare cases ischaemic heart disease. Essential hypertension increases the risk of placental dysfunction, which ultimately increases the risk of impaired foetal growth and foetal death [14]. Both hypertensive disease and gestational diabetes are more common in obese and overweight women and have been shown to increase the risk of adverse pregnancy outcomes [15]. In addition, regardless of underlying hypertension, the risk of pre-eclampsia per se is higher in obese women and is a well recognised cause of stillbirth outcome due to placental dysfunction.

Vascular Pathology and Placental Dysfunction

Sub-optimal placental function has been proposed as an underlying process predisposing to stillbirth in obese women. It has been suggested that placental perfusion is reduced by the effects of hyperlipidaemia [8]. The high lipid levels negatively affect prostacyclin secretion and increase thromboxane production, resulting in increased peroxidase which causes vasoconstriction and aggregation of platelets, potentially producing placental thrombosis, which may even occur in the absence of pre-eclampsia [6,16]. It has been suggested that the risk may increase further when hyperlipidaemia and insulin resistance co-exist, as fibrinolytic activity is reduced [17]. Recent research is indicative of the role of placental dysfunction in a sizeable proportion of stillbirths in high-income countries, with a population attributable risk of 23% in offspring small for gestational age and of 15% in placental abruption [1]. A series of 328 stillbirths with cause classified according to the ReCoDe (relative condition at death) classification revealed that only those with evidence of growth restriction were positively associated with BMI [18].

The role of 'placental dysfunction' in the pathophysiology of stillbirth depends on a specific cause and is unclear in many instances. Recognised placental dysfunction associated with stillbirth includes abruption, the processes underlying pre-eclampsia, and unrelated increased vascular resistance within the placental vasculature. There are a significant proportion of unexplained stillbirths without evidence of foetal growth restriction, where pathological evidence of acute placental dysfunction has been identified [19]. Given that birthweight centiles and foetal growth charts are often generalised (not customised), it is possible that even where biometry is accurately measured by ultrasound scan, a small for gestational age infant may not be accurately diagnosed in all women (masked growth restriction) [10]. This phenomenon is all the more likely in offspring of obese women as they may develop macrosomia with superimposed placental dysfunction, risking a 'normal' sized baby with relatively 'normal' body fat distribution as a result. The risk of such masked growth restriction could be overcome with the use of serial scans and customised growth charts specifically for obese pregnant women [18].

Metabolic Syndrome

Metabolic syndrome highlights how multiple risk factors described above may co-exist in obese women, potentially forming a synergistic effect on stillbirth risk. This syndrome comprises a cluster of clinical conditions which act as risk factors for cardiovascular disease [20]. Included within the spectrum of conditions are glucose intolerance, insulin resistance, hypertriglyceridaemia and the overt conditions polycystic ovarian syndrome, type-2 diabetes mellitus, morbid obesity and hypertension [20].

Previous Obstetric History

Previous pregnancy loss is more likely to occur in obese women and is in itself associated with an increased risk of poor obstetric outcome. In addition to the association with stillbirth, obesity is known to be associated with increased risk of miscarriage [7]. Obese women have been found to be more likely to have an interpregnancy interval of more than 6 years, which in itself is associated with an increased risk of stillbirth [7].

Foetal Abnormalities

Obese women are more likely to have a foetus with congenital anomalies such as spina bifida, omphalocoele, cardiac defects and multiple anomalies. The risk increases with rising BMI and is present in overweight women too [21]. Specifically, hyperinsulinaemia is associated with an increased risk of neural tube defects. This has given rise to targeted advice to diabetic women regarding the use of high-dose folic acid pre- and post-conception. While the mechanism for other abnormalities is less clear, the potential for gestational diabetes and obesity to work in synergy to inflate the foetal anomaly risk and stillbirth rates has been recognised.

Sleep Disorders

Obese pregnant women have been shown to be significantly more likely to suffer from snoring, sleep apnoea and oxygen desaturation [20]. Sleep apnoea involves repetitive cessation of airflow until sufficient hypoxaemia, which stimulates the patient to wake from sleep [22]. Therefore, there is a hypothetical risk of reduced oxygenation of the foetus of obese mothers and subsequently increased risk of growth restriction and stillbirth.

Failed Antenatal Screening

As ultrasound is absorbed by the overlying fat, scan detection of foetal abnormalities and growth disorders may be compromised by high maternal BMI [23] due to poor views of the foetus. This may result in undiagnosed foetal abnormalities or growth restriction. In addition, it has been hypothesised that obese women may be less perceptive of reduced foetal movement than women of normal weight [6]. This would reduce the potential contribution of foetal movement monitoring in reducing the risk of stillbirth. The routine use of symphysis–fundal height, which is used to assess the foetal growth, also has potential to be affected by maternal obesity, although a small study to date has refuted this [24].

Post-maturity

As obesity is an established risk factor for post-mature pregnancies [7], this in itself will increase the risk of stillbirth, as longer gestation carries higher risk of foetal death [25].

Indirect Risk Factors

Factors which indirectly increase the risk of stillbirth in obese women do not occur due to obesity per se but are known to occur more commonly than in normal-weight women. These factors are important when assessing the relationship between BMI and stillbirth, as they may act to inflate an independent effect of obesity or indeed may act to create a relationship which would be absent if they were accounted for in the analysis. These risk factors include increased maternal age, low socio-economic class, grandmultiparity and ethnicity.

Age

Advanced maternal age not only acts as an independent risk factor for stillbirth [1] but also coexists with multiple known risk factors including obesity and hypertension. As the average maternal age on first childbearing is increasing, this risk factor will become increasingly important over time. As developments in screening for chromosomal abnormalities has resulted in fewer stillbirths due to foetal abnormality in older women, a greater proportion are due to other reasons, including the high number of multiple pregnancies in these women [26]. In addition, very young women are at higher risk of antepartum stillbirth, as demonstrated in a large study from the United States [27]; however, these young teenagers are less likely to be obese.

Parity

Nulliparity is associated with a higher risk of stillbirth [1], with this factor estimated to contribute 15% of all stillbirths. As small family size is becoming a more popular choice, this group of women will remain a prominent cohort at risk of stillbirth, with findings of a study from Sweden suggesting that the BMI-associated risk of stillbirth is strongest in the nulliparous group [28]. There is evidence that the association between parity and stillbirth forms a J-shaped curve, with grand-multiparous women at a fourfold higher risk than those who are Para 1–3 [9].

Social Class and Dietary Habits

It is known that women of low social class are at increased risk of stillbirth, but the mechanisms are not well established [29]. Higher rates of obesity may partly explain the higher stillbirth rates associated with deprivation [6]. Poor dietary habits with sub-optimal physical activity are more common both in obese women and in those of low socio-economic class, but few adverse consequences have been recognised. Of those, increased risks of excessive gestational weight gain, pre-eclampsia and gestational diabetes have been noted [30]. Gaining more than the recommended weight during pregnancy is in itself a risk factor for poor pregnancy outcomes, but two studies from Sweden found that maternal weight gain during pregnancy did not confound the association between BMI and stillbirth [6,28].

Ethnicity

Black women are more likely to be obese and to experience a stillbirth than white women [31]. However, the precise difference in risk of stillbirth across various ethnic groups depends largely on the categories used to define obesity [32]. The realisation that the rising incidence of maternal obesity poses a significant risk factor for stillbirth between certain ethnic groups is worthy of further exploration [9].

Obesity as an Independent Risk Factor

Considering the spectrum of risk factors for stillbirth which are associated with obesity, the task of demonstrating an independent relationship between obesity and stillbirth is clearly complex, and to date, the evidence has been somewhat contradictory [8]. However, this task has been addressed in a cohort study by Kristensen et al. [8] which was designed primarily to examine the relationship between pre-pregnancy BMI and perinatal mortality. This study adjusted for smoking, age, alcohol, caffeine, height, parity, offspring gender, education level, working status and cohabitation with partner. The results confirmed a higher rate of unexplained stillbirth in the obese group, even when hypertensive and diabetic women were excluded [8]. It also found more stillbirths due to placental dysfunction than in the normal-weight group and that the stillborn offspring were smaller in the obese group, raising the possibility of subtle placental dysfunction without frank intra uterine growth restriction [33]. Given the large number of appropriate confounding factors adjusted for in this study, the findings support the role of obesity as an independent risk factor for stillbirth.

Management

In view of the potential challenges of accurate diagnostic ultrasound examination in maternal obesity, the need for a second opinion to confirm the diagnosis of stillbirth in this scenario is greater. Once confirmed, vaginal delivery should be ideally aimed. In particular, this avoids the risks associated with Caesarean section such as deep vein thrombosis, pulmonary embolism and wound infection, and importantly improves the chances of a vaginal delivery in future. Therefore, induction of labour should be offered.

Investigating the cause of stillbirth in obese women should follow the standard investigation protocols with appropriate parental consent. Post-mortem examination of the baby, external examination, X rays, maternal serum TORCH screen, Kleihauer test, rhesus antibody titres, thrombophilia screen, HbA1c, karyotype and placental examination at both macroscopic and histological level would ideally be performed [4]. Ensuring full explanation of potential benefit of post-mortem examination is crucial to the process of obtaining consent for this. Given the possibility of masked growth restriction in these infants, post-mortem examination offers the potential for the assessment of body fat distribution and organ sparing even in offspring with an apparently normal birthweight.

Counselling at the time of diagnosis should be tailored to the specific case, but may be appropriately limited to any obvious associated conditions and the immediate management, until further information is gathered. Following investigation, which may be 6–12 weeks from the time of stillbirth, a senior obstetrician should meet the couple and discuss the results, with referrals made as required. A plan for management of future pregnancy should be discussed at this time, according to the presumed cause of death. If the stillbirth is unexplained, a structured management plan should focus around the standard management of obese pregnant women, including primary prevention, plus additional care to maximise reassurance for the couple.

Prevention

Strategies for the reduction of stillbirth rates are the subject of much ongoing research. A number of routine screening and monitoring procedures are of questionable benefit. There is minimal evidence to support the use of foetal movement counting and Doppler studies in high-risk women and very little specific evidence to support the use of cardiotocographic monitoring. Circumstantial evidence of dramatic reductions in stillbirth rates associated with cardiotogography (CTG) use and Caesarean section to deliver the foetus in cases of suspected distress may suggest that their use has substantially reduced perinatal mortality [34].

Primary prevention strategies may require society level interventions to achieve positive results. Some of the principal risk factors for stillbirth can be potentially reversible, including avoiding delay in the age of first pregnancy, giving up smoking during pregnancy and lowering BMI prior to embarking on pregnancy [1]. Primary prevention of maternal obesity in the form of pre-pregnancy counselling and efforts to address obesity would appear to be the mainstay of avoidance strategies. Tackling weight issues in overweight teenagers, encouraging childbearing at a younger age, post-natal identification of women at high risk of inter-pregnancy weight gain to become obese have all been suggested [7], as has even modest weight loss to bring BMI to under 30 kg/m^2. Counselling efforts should emphasise that by avoiding obesity in pregnancy; one stands to reduce the risk not only of stillbirth but also of congenital abnormalities, gestational diabetes, high blood pressure, pre-eclampsia, difficult labour and delivery, Caesarean section, wound infection and deep vein thrombosis. Given the relatively higher frequency of many of these complications when compared to stillbirth, the incentive to avoid maternal obesity may be maximised using such a counselling approach.

Secondary prevention should take effect following the recording of the BMI at booking. However, evidence is currently limited regarding the success of any such measures. The aim of current management is to enable delivery of an antenatal programme designed to assess and address the risk of stillbirth and the remaining risks associated with obesity. Risk-scoring systems developed to identify multiple risk factors for poor obstetric outcomes have been shown to identify up to 90% of pregnancies to result in perinatal mortality [35]. Use of kick charts and uterine artery Doppler waveform analysis have been found to help predict stillbirth in high-risk pregnancies, while amniotic fluid assessment is useful in predicting stillbirth risk in both polyhydramnios and oligohydramnios [34], but their use specifically in obese women has not been reported to date. It has been suggested that a programme of intensive monitoring of pregnancies in overweight and obese women may be warranted to overcome the potentially inadequate monitoring of the foetus offered in routine care [23]. Continuous electronic foetal monitoring in labour may be advocated via foetal scalp electrode to ensure accurate recordings, but it should be remembered that evidence of a consequent reduction in stillbirths is currently lacking [36]. More frequent maternal observation to detect hypertension and gestational diabetes may be of benefit, while heightened awareness of intra-operative and post-natal risks may maximise prevention efforts.

Use of customised foetal growth charts to plot ultrasound measurement may aid the detection of foetal growth restriction, in particular, if such charts were developed to take maternal BMI into account [37].

Doppler studies carried out to identify the increased resistance within the placental vasculature of mid-trimester uterine arteries have proven beneficial in predicting adverse outcome including stillbirth [38]. However, to obtain useful positive predictive value, not to mention cost-effectiveness, such tests should be targeted to women known to be at high risk of stillbirth; so while obesity would play a role here, it would be most likely to be considered feasible when another significant risk factor exists.

Action on the part of the responsible professional to minimise the risk of complications, should they arise, would be expected to reduce the overall impact of maternal obesity on pregnancy outcome. Counselling at the outset of pregnancy and again at frequent intervals along with adequate monitoring and optimal timing of delivery is currently used to optimise outcomes. Early referral to an obstetric unit for screening of risk factor, along with careful monitoring in labour, has been promoted [7].

Future research is likely to focus on the prevention of maternal obesity. This may involve behavioural, psychological, medical or surgical therapy in obese women planning a first or subsequent pregnancy. Similar strategies to minimise weight gain during pregnancy are likely to be explored. Assessing the impact of such interventions on stillbirth rates is unlikely to be feasible in the confines of a study, but proxy outcomes may be used to assess likely impact.

Given the potential for obesity to be 'preventable', obesity-associated stillbirth may invoke significant feelings of guilt in affected women, perhaps in a similar manner to smoking, alcohol and other drug use.

As fertility treatment is often limited to women with an acceptable level of risk in relation to their BMI, the proportion of obese women becoming pregnant spontaneously should continue to outstrip those doing so with assistance. However, less regulated fertility services in certain areas of the world will risk obese women finding themselves in precarious positions when they become pregnant with one of more foetuses subsequent to fertility treatment, and with the risk of stillbirth significantly raised, outcomes may be even more devastating.

Once a stillbirth has occurred in an obese woman, addressing this as a potential risk factor in an otherwise unexplained stillbirth has been emphasised as an important part of antenatal care in a subsequent pregnancy [39].

Public Health Implication

As with the public health message regarding the dangers of smoking in pregnancy, raising the profile of the adverse effects of maternal obesity on pregnancy outcomes is likely to at least improve patient awareness, if not encourage weight loss prior to pregnancy in a number of women.

References

1. Flenady. Major risk factors for stillbirth in high-income countries: a systematic review and meta-analysis. *Obstet Gynecol Surv*. 2011;66(8):483–485.
2. WHO. *Definitions and Indicators in Family Planning Maternal and Child Health and Reproductive*. Health used in the WHO Regional Office for Europe. World Health Organization. Regional Office for Europe. 2001 Key: citeulike 4388190
3. Fleming K. *Confidential Enquiry into Maternal and Child Health. Stillbirth, Neonatal and Post-neonatal Mortality, England, Wales and Northern Ireland 2000–2002*. CEMACH; 2004. *www.cmqcc.org/resources/27/download*.
4. Smith GC, Fretts RC. Stillbirth. *Lancet*. 2007;370(9600):1715–1725.
5. Bhattacharya S, Campbell DM, Liston WA, Bhattacharya S. Effect of body mass index on pregnancy outcomes in nulliparous women delivering singleton babies. *BMC Public Health*. 2007;7(24):168.
6. Cnattingius S, Bergstrom R, Lipworth L, Kramer MS. Prepregnancy weight and the risk of adverse pregnancy outcomes. *N Engl J Med*. 1998;338(3):147–152.
7. Raatikainen K, Heiskanen N, Heinonen S. Transition from overweight to obesity worsens pregnancy outcome in a BMI-dependent manner. *Obesity (Silver Spring)*. 2006; (1):165–171.

8. Kristensen J, Vestergaard M, Wisborg K, Kesmodel U, Secher NJ. Pre-pregnancy weight and the risk of stillbirth and neonatal death. *BJOG*. 2005;112(4):403–408.
9. Stacey T, Thompson JM, Mitchell EA, Ekeroma AJ, Zuccollo JM, McCowan LM. Relationship between obesity, ethnicity and risk of late stillbirth: a case control study. *BMC Pregnancy Childbirth*. 2011;11(12):3.
10. Leung TY, Leung TN, Sahota DS, et al. Trends in maternal obesity and associated risks of adverse pregnancy outcomes in a population of Chinese women. *BJOG*. 2008; 115(12):1529–1537.
11. Francis A, Williams M, Gardosi J. Maternal obesity and perinatal mortality risk. *Am J Obstet Gynecol*. 2009;201(6):Supplement S223–224.
12. Chu SY, Kim SY, Lau J, et al. Maternal obesity and risk of stillbirth: a metaanalysis. *Am J Obstet Gynecol*. 2007;197(3):223–228.
13. Chu S, Callaghan W, et al. Maternal obesity and risk of gestational diabetes mellitus. *Diabetes Care*. 2007;30(8):2070.
14. Goldenberg RL, Kirby R, Culhane JF. Stillbirth: a review. *J Matern Fetal Neonatal Med*. 2004;16(2):79–94.
15. Garbaciak Jr JA, Richter M, Miller S, Barton JJ. Maternal weight and pregnancy complications. *Am J Obstet Gynecol*. 1985;152(2):238–245.
16. Stone JL, Lockwood CJ, Berkowitz GS, Alvarez M, Lapinski R, Berkowitz RL. Risk factors for severe preeclampsia. *Obstet Gynecol*. 1994;83(3):357–361.
17. Lindahl B, Asplund K, Eliasson M, Evrin P-. Insulin resistance syndrome and fibrinolytic activity: the northern Sweden MONICA study. *Int J Epidemiol*. 1996;25(2):291–299.
18. Gardosi J, Clausson B, Francis A. The value of customised centiles in assessing perinatal mortality risk associated with parity and maternal size. *BJOG*. 2009;116(10):1356–1363.
19. Stallmach T, Hebisch G, Meier K, Dudenhausen JW, Vogel M. Rescue by birth: defective placental maturation and late fetal mortality. *Obstet Gynecol*. 2001; 97(4):505–509.
20. Maasilta P, Bachour A, Teramo K, Polo O, Laitinen LA. Sleep-related disordered breathing during pregnancy in obese women. *Chest*. 2001;120(5):1448–1454.
21. Watkins ML, Rasmussen SA, Honein MA, Botto LD, Moore CA. Maternal obesity and risk for birth defects. *Pediatrics*. 2003;111(5 Part 2):1152–1158.
22. Packiathan I. Medical consequences of obesity. *Medicine*. 2003;31(4):5.
23. Devlieger R, Guelinckx I, Vansant M. Follow-up in obese pregnant women to prevent stillbirth. *Am J Obstet Gynecol*. 2008;199(1):e18.
24. Euans DW, Connor PD, Hahn RG, Rodney WM, Arheart KL. A comparison of manual and ultrasound measurements of fundal height. *J Fam Pract*. 1995;40(3):233–236.
25. Cnattingius S, Stephansson O. The epidemiology of stillbirth. *Semin Perinatol*. 2002; 26(1):25–30.
26. Fretts RC, Usher RH. Causes of fetal death in women of advanced maternal age. *Obstet Gynecol*. 1997;89(1):40–45.
27. Balayla. Effect of maternal age on the risk of stillbirth: a population-based cohort study on 37 million births in the United States. *Am J Perinatol*. 2011;28(8):643–650.
28. Stephansson O, Dickman PW, Johansson A, Cnattingius S. Maternal weight, pregnancy weight gain, and the risk of antepartum stillbirth. *Am J Obstet Gynecol*. 2001; 184(3):463–469.
29. Little REWC. Risk factors for antepartum and intrapartum stillbirth. *Am J Epidemiol*. 1993;137(11):1177–1189.
30. Guelinckx I, Devlieger R, Beckers K, Vansant G. Maternal obesity: pregnancy complications, gestational weight gain and nutrition. *Obes Rev*. 2008;9(2):140–150.

31. Healy AJ, Malone FD, Sullivan LM, et al. Early access to prenatal care: implications for racial disparity in perinatal mortality. *Obstet Gynecol.* 2006;107(3):625–631.
32. Rahman M, Berenson AB. Accuracy of current body mass index obesity classification for white, black, and Hispanic reproductive-age women. *Obstet Gynecol.* 2010; 115(5):982–988.
33. Gardosi J, Mul T, Mongelli M, Fagan D. Analysis of birthweight and gestational age in antepartum stillbirths. *Br J Obstet Gynaecol.* 1998;105(5):524–530.
34. Haws RA, Yakoob MY, Soomro T, Menezes EV, Darmstadt GL, Bhutta ZA. Reducing stillbirths: screening and monitoring during pregnancy and labour. *BMC Pregnancy Childbirth.* 2009;9(suppl 1):S5.
35. Pattison NS, Sadler L, Mullins P. Obstetric risk factors: can they predict fetal mortality and morbidity? *N Z Med J.* 1990;103(891):257–259.
36. Alfirevic Z, Devane D, Gyte GM. Continuous cardiotocography (CTG) as a form of electronic fetal monitoring (EFM) for fetal assessment during labour. *Cochrane Database Syst Rev.* 2006;19(3):CD006066.
37. Mongelli M, Biswas A. A fetal growth standard derived from multiple modalities. *Early Hum Dev.* 2001;60(3):171–177.
38. Lees C, Parra M, Missfelder-Lobos H, Morgans A, Fletcher O, Nicolaides KH. Individualized risk assessment for adverse pregnancy outcome by uterine artery doppler at 23 weeks. *Obstet Gynecol.* 2001;98(3):369–373.
39. Robson SJ, Leader LR. Management of subsequent pregnancy after an unexplained stillbirth. *J Perinatol.* 2011; [Publish ahead of print].

29 Obesity, Diabetes, Placental Pathology and Foetal Malformations

Margaret J. Evans

Department of Pathology, Royal Infirmary of Edinburgh, Little France Crescent, Edinburgh, UK

Introduction

Obesity (body mass index (BMI) > 30 kg/m^2) is increasing in the general population affecting all ethnic and age groups [1]. The confidential enquiry into maternal and child health (CEMACH) in the United Kingdom reported an increase of 10% in obesity between 1993 and 2002. There are few reports on the temporal changes in weight gain during pregnancy, but a study of 1200 women showed that more women were not only obese at the start of pregnancy but also gained significantly more weight during pregnancy [2]. Furthermore, it represents a significant threat to maternal health. More than half the mothers who died in the United Kingdom during the 3 year (2003–2005) were obese or overweight [3]. It has thus become increasingly important to understand the implications of this change on reproductive and maternal health, pregnancy and foetal outcome. In this chapter, we review specific features of the placenta and the foetal anomalies seen in association with diabetes, in recognition that this is the single most common presentation of the obese and overweight woman during pregnancy, and touch briefly on the underlying biochemistry.

Overweight and obese women are at an increased risk of developing insulin resistance or decreased insulin sensitivity when compared with women of average/lean weight. This combination of obesity and decreased insulin sensitivity increases the risk of these women developing metabolic syndrome with associated problems of diabetes, hypertension, hyperlipidaemia and cardiovascular disorders. In pregnancy, there is a 60% decrease in insulin sensitivity [4]; this altered metabolic state causes obese women an increased risk of metabolic dysfunction during pregnancy leading to the development of gestational diabetes, pre-eclampsia, and foetal overgrowth (macrosomia). In this chapter, we deal mainly with changes associated with diabetes in recognition that this is the single most common presentation of the obese and overweight woman during pregnancy.

Obesity. DOI: http://dx.doi.org/10.1016/B978-0-12-416045-3.00029-7

The risk of developing gestational diabetes mellitus (GDM) increases with increased BMI. Obese women are three times more likely to develop gestational diabetes than women with BMIs in the healthy range (18.9–24.9 kg/m^2). Obesity and gestational diabetes are states of insulin resistance with diverse abnormalities in oxidative stress, glycation of proteins and cellular processes. This in turn leads to impaired endothelial function, vascular inflammation and haemostasis, processes which give rise to impaired function of the microcirculation causing abnormalities in placental function leading to increased foetal morbidity and stillbirth [5].

Despite the advances made in recent years in the management of diabetes mellitus and, in particular, the management of the diabetic pregnancy, diabetes remains a significant threat during pregnancy with an increased risk of foetal malformation and stillbirth. A recent review [6] indicated that despite the well-documented relationship between morbidity and pre-gestational diabetes, modern intensive management has led to the corrected perinatal outcome in most series being equal to or better than that of the general population. This is due to a combination of strategies including improved compliance, improved glycaemic control at conception and throughout pregnancy and early foetal anomaly screening.

However, the impact of gestational diabetes is less well understood. Gugliucci et al. in 1976 [7] showed that women with fasting hyperglycaemia in pregnancy were at an increased risk of perinatal mortality and that early detection and better glycaemic control improved outcome. Recent cases have shown that placental changes and stillbirth can occur even in the face of good glycaemic control [8], and the question now is what type of intervention is appropriate, e.g. early foetal delivery, treatment with insulin or metformin use? In order to determine a better understanding of the relationship between the foetus, the placenta and the diabetic status of the mother, a careful analysis of the placental pathology is required.

As the western world is now facing an epidemic of obesity with all its associated risks including a rising incidence of diabetes, it is timely to review what is currently known about obesity, diabetes and pregnancy. It is recognised that the problems in pregnancy and early neonatal life are caused not only by high levels of glucose but also by hyperinsulinism and dyslipidaemia.

Diabetic patients may be categorised into four main groups and are as follows:

1. Insulin-dependent diabetes preceding pregnancy. Here benefit may be derived from preconceptual care.
2. Mothers with type-2 diabetes and early disease who develop hyperglycaemia in pregnancy which persists after delivery.
3. Gestational diabetes in which a disturbance of glucose occurs in pregnancy and does not persist after delivery.
4. Pre-diabetes or occult diabetes with no clinical evidence of diabetes during or after pregnancy that subsequently develops diabetes in later life. Such mothers may have late stillbirths or cherubic infants.

Categories 2, 3 and 4 may all be associated with increased BMI, and faced with these four categories, it is perhaps not surprising that placental pathology may vary. Nevertheless the typical case will show a constellation of findings, and once recognised, these changes are sufficient for an astute pathologist to raise the spectre of increased BMI or underlying diabetes mellitus.

Obesity, Diabetes and the Umbilical Cord Coiling

Macroscopically the classic 'diabetic placenta' is usually bulky and oedematous often with a thick oedematous umbilical cord. In recent years, there has been an increasing interest in cord coiling, as it is recognised that the cord represents the most vulnerable link between foetus and mother, as it lies free in the amniotic fluid and may easily be damaged. Several studies have shown that the coiling index of umbilical cords is significantly different in the obese and diabetic population with gestational diabetic mothers having either hypercoiled or non-coiled umbilical cords [9,10] (Figures 29.1 and 29.2) – both these are associated with an increase in neonatal morbidity and mortality [11].

The cord contains two umbilical arteries carrying deoxygenated blood from the foetus to the placenta, and one umbilical vein, carrying oxygenated blood from the placenta to the foetus. This blood flows at near-to-zero pressure and this together with the thin outer muscular wall renders the umbilical vein vulnerable to compression/collapse. The cord-coiling pattern is established in the first trimester and changes little thereafter.

The integrity of the umbilical cord vessels is maintained by two principal factors: spiral coiling and Wharton's jelly. It is the spiralling of the vessels which is thought to provide stability against buckling or compression. Several studies have

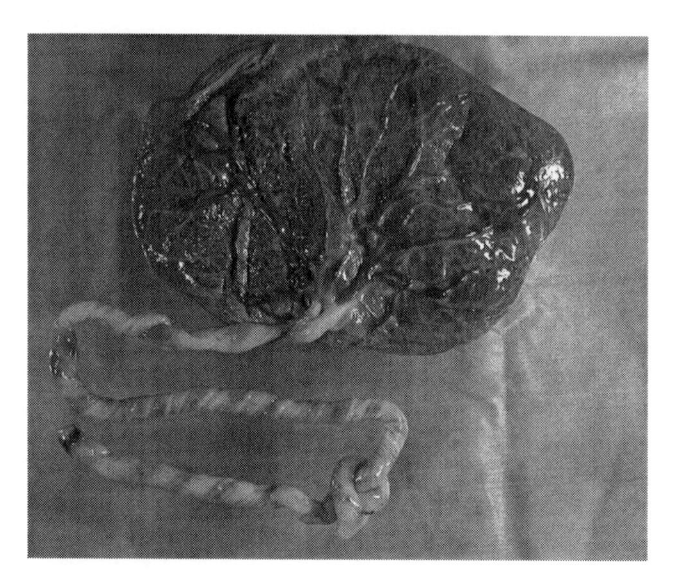

Figure 29.1 Hypercoiled cord.

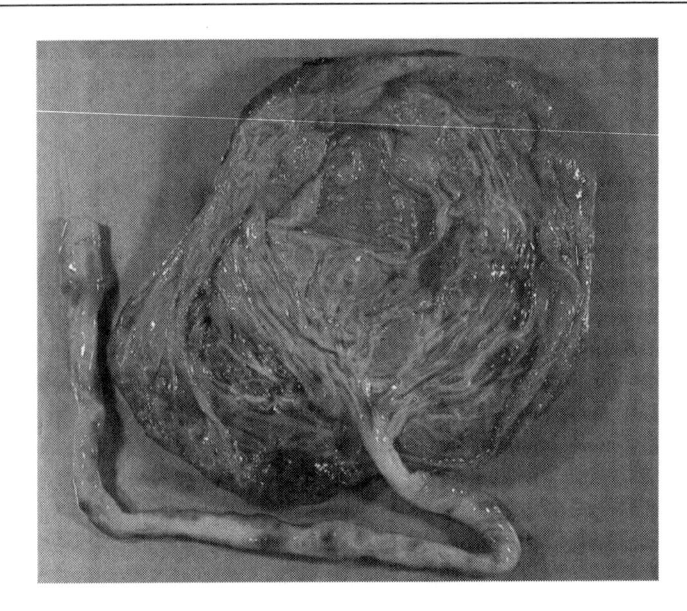

Figure 29.2 Hypocoiled cord.

shown that the normal pattern of coiling is at a rate of one coil per 5 cm, which equates to a coiling index of 0.2 cm^{-1} [12,13] Cord coiling of $<0.1\ cm^{-1}$ is considered hypocoiled while coiling greater that 0.3 cm^{-1} is hypercoiled. Hypocoiled cords lack Wharton's jelly or have Wharton's jelly at a low osmotic pregnancy. The reduced turgor of the cord fails to prevent cord kinking with compression of vessels and cessation of blood flow. On the other hand, the hypercoiled cord is susceptible to vascular obstruction die to 'pinching' of the vein.

As pregnancy progresses beyond 34 weeks of gestation, the increasing foetal demand is met by increasing the number of terminal villi and by forming vasculosyncytial membranes within these villi. De Laat et al. [14] showed that hypercoiling of the cord may affect the maturation of the placenta with a trend towards placental maturation defect in the presence of hypercoiling. Further, they confirmed the relationship between foetal death and maturation defect and hypercoiling and found an association between histological indicators of foetal hypoxia/ischaemia and placental maturation defect (mean number of vasculosyncytial membranes per terminal villous below the 10th centile). They found a significant inverse correlation between the mean number of vasculosyncytial membranes in terminal villi and umbilical cord coiling. The influence of umbilical cord coiling on perfusion pressure is unknown, but it is postulated that the reduced pressure serves as a weaker stimulus to angiogenesis with failure to form vasculosyncytial membranes. (Figures 29.3 and 29.4) The converse to this would be that there is a raised pressure in the hypo/non-coiled cords, also seen more frequently in diabetes which leads to congestion in the terminal villi and increased angiogenesis resulting in chorangiosis (Figure 29.5).

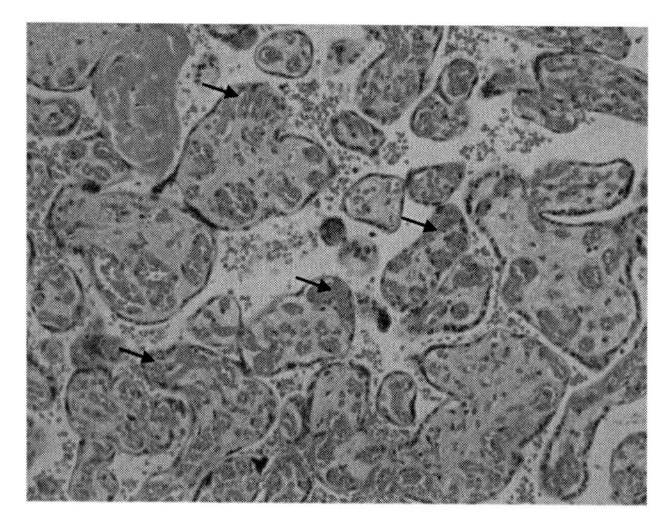

Figure 29.3 Normal development showing vasculosyncytial membranes (arrow).

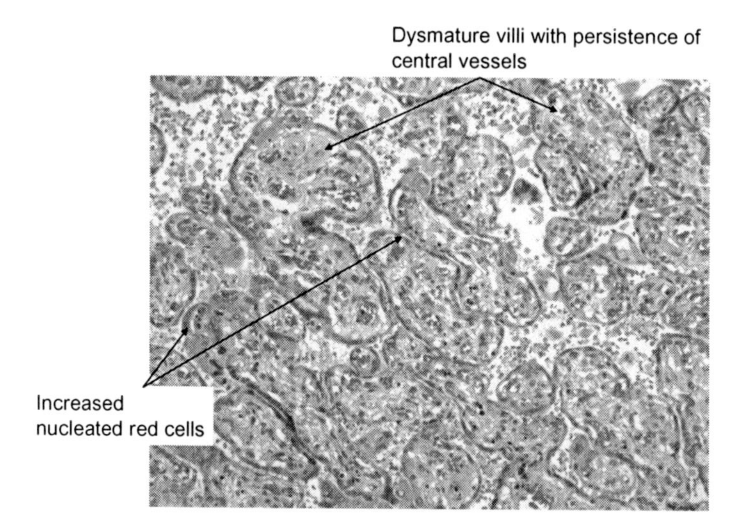

Figure 29.4 Poor formation of vasculosyncytial membranes.

The mechanism by which changes in cord coiling arise is not clear. Hyper- and non-coiling of the cord are seen in association with other disease states, and, therefore, the glycaemic index alone does not explain the changes [15]. Furthermore, there are, to date, no studies that explain how the glycaemic index early in pregnancy may relate to umbilical cord coiling. More work need to be done to delineate the underlying maternal causes of these placental changes.

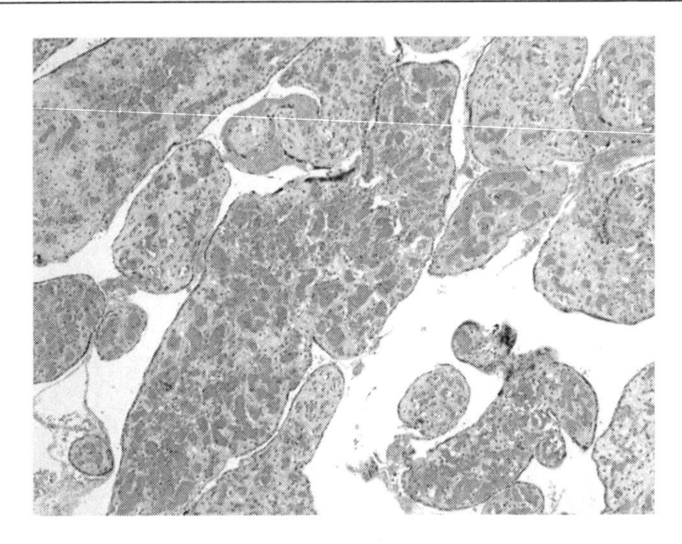

Figure 29.5 Villi showing chorangiosis.

Obesity, Diabetes and the Placenta

We have described the classical large bulky, oedematous placenta seen in association with diabetes. Microscopically the villi appear large and immature with persistence of central vessels and poor formation of vasculosyncytial membranes. This change is best seen at the centre of the cotyledons. Increased Hofbauer cells (macrophages) may also be noted. The stroma stains poorly due to the presence of increased glycosaminoglycans. Other features include thickening of the basement membrane, hyperplasia of the cytotrophoblast, chorangiosis and fibrinoid necrosis of the villi. Such changes are seen in about one-third of diabetic pregnancies, and there appears to be a relationship with these findings and poor diabetic control. They are found less frequently in gestational and occult diabetes. It is not yet clear what effect the improved glycaemic control may play in ameliorating the changes as there may be other factors involved in the angiopathy observed.

Pietryga et al. [16] noted that there was a general reduction in the vascular surface of the terminal villi with disordered vascular spaces. This was associated with foetal Intrauterine growth restriction (IUGR). However, in patients with hyperglycaemia and foetal macrosomia, there was a significant thickening of basal membranes of the trophoblast and structural abnormalities in the perivascular space with proliferation of collagen in the terminal villi. Proliferation and expansion of endothelial cells and a general decrease in vascularity of the terminal villi may account for foetal hypoxia/anoxia (Figure 29.6). These changes appeared to relate to the degree of hyperglycaemia.

In 2009, Hiden et al. [17] published a review looking at the relationship between insulin and placental vascularity. The placenta represents the interface between two circulations: that of the foetus and that of the mother. We recognise that hyperglycaemia in the mother leads to a hyperinsulinaemic state in the foetus.

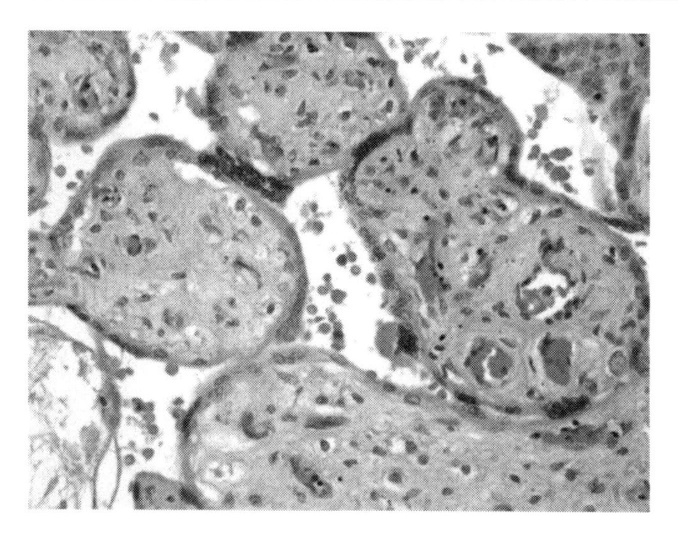

Figure 29.6 Poorly vascularised terminal villi.

Glucose-stimulated insulin secretion increases over gestation and is down-regulated by constant hyperglycaemia but enhanced by pulsatile hyperglycaemia [18]. The placental endothelium is unique in the human vasculature partly due to the need for rapid expansion in the face of foetal growth. In the third trimester, the placenta is richly endowed with insulin receptors; thus, the hyperinsulinaemic state of the foetal circulation would be expected to stimulate placental angiogenesis and overgrowth, explaining, in part, the macrosomia observed. The effect appears to be modulated through ephrin-B2 expression which is a specific arterial signalling molecule implicated in vascular sprouting. Thus, it would appear that the effect is mediated by the arterial side. This effect is seen predominantly in the placentas of type-1 diabetics. The change in gestational diabetes appears variable; the differences in vascularity may reflect different points of onset of GDM.

The idea that it is only insulin which affects the vascularity of the placenta would be an oversimplification. Diabetic pregnancies are associated with increased foetal levels of fibroblast growth factor which also stimulate angiogenesis. It appears strange that in the face of hyperglycaemia, the placenta should increase in volume thus leading to enhanced foeto-maternal transport of nutrients; this must surely reflect its prime role in protecting the foetus against hypoxic insult. Foetal insulin exerts different effects stimulating foetal aerobic glucose metabolism thus increasing oxygen demand. Glycated haemoglobin has a higher affinity for oxygen and thus there is reduced oxygen delivery to the intervillous space. This will lead to foetal hypoxia, thickening of the basement membrane and haemodynamic compromise of the foetal–placental unit. Such hypoxia, through the mediation of hypoxia inducible factor, will again lead to angiogenesis.

A better understanding of the impact which improved control may have on the pregnancy outcome along with understanding of how drugs may be exerting their effects will help in understanding the key role of the placenta in the process. Research about metformin use is currently underway and it remains to be shown what impact this might have. It may be that the benefits are related to its vascular protective effects rather than its insulin-stimulating effects in the face of insulin resistance.

Obesity and Other Placental Changes

In addition to gestational diabetes, the obese woman is at an increased risk of developing pre-eclampsia. Although the definitions vary in different studies, the association between this disease and obesity is consistent [19]. The placental changes which may be seen in association with pre-eclampsia are shown in Figures 29.7 and 29.8, with increased syncytial knots and sprouts and excess perivillous fibrin deposition denoting relative ischaemia. These changes arise in the obese pregnancy only in association with hypertension and pre-eclampsia and are not seen in the obese pregnancy in the absence of such complications [20].

Thromboembolism is also more commonly seen in the obese pregnancy due to the hypercoaguable state [21]. This may also show changes within the placenta suggestive of stem vessel occlusion (Figure 29.9).

Obesity, Diabetes and Foetal Pathology

There appears to be no increase in spontaneous abortion in patients with good glycaemic control, but the incidence of spontaneous abortion has been shown to be higher in patients with high HbA1 levels.

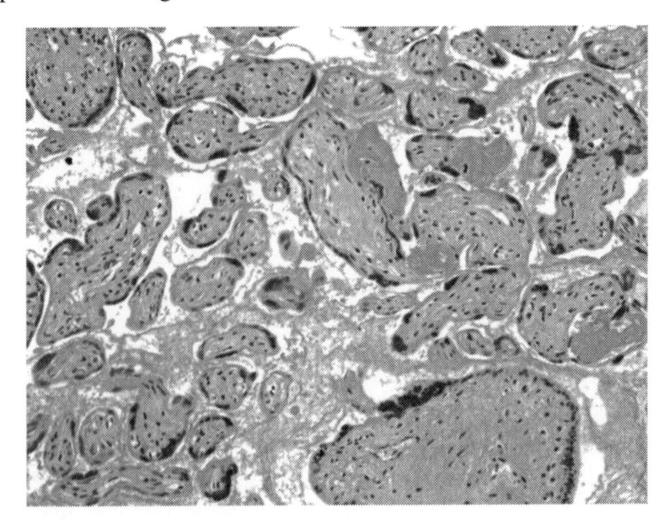

Figure 29.7 Ischaemic villi with increased syncytial sprouts and fibrin.

There is an increased incidence of major malformations in diabetic pregnancies ranging in incidence from 6% to 9% representing a threefold increase in this population. There is a correlation between elevated haemoglobin A_{1c} and major congenital abnormalities [22]. A number of systems may be affected including cardiovascular system (e.g. transposition of the great vessels, VSD and dextrocardia), central nervous system (CNS) (e.g. anencephaly, spina bifida, hydrocephaly and holoprosencephaly), genitourinary system and skeletal system.

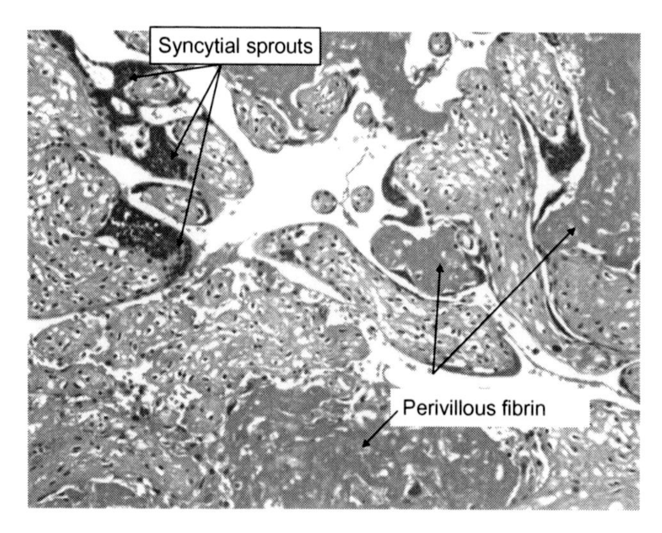

Figure 29.8 Cellular villi with increased syncytial sprouts and perivillous fibrin deposition.

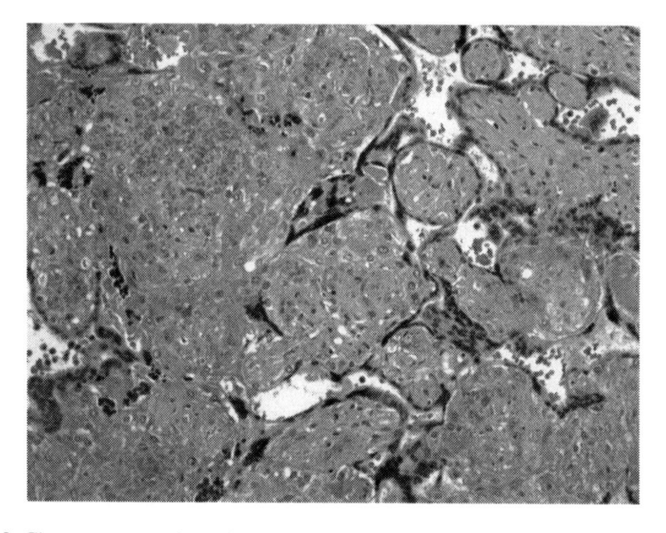

Figure 29.9 Changes suggestive of stem vessel occlusion.

The defects of the heart and CNS are likely to relate to hyperglycaemia with inhibition of myoinositol uptake which is essential for the gastrulation and neurulation stages of embryogenesis. Deficiency of myoinositol leads to problems with the phosphoinositide system which in turn leads to abnormalities in the arachidonic acid–prostaglandin pathway. Gastrulation and neurulation are also sensitive to hypoglycaemia and result in growth restriction as well as cranial and caudal Neural Tube Defects (NTDs). Obesity appears to exacerbate the effects of diabetes with an apparent rise in congenital malformations.

Recent studies by the Diabetes Control and Complications Trial Research Group have shown that some of the effects of diabetes can be ameliorated by good early glycaemic control. However, tissue damage leading to neural tube defects cannot be reversed. It has been shown that the diabetic-induced foetal anomalies are associated with disturbances in foetal metabolism including elevated superoxide dismutase activity, reduced levels of myoinositol and arachidonic acid and inhibition of the pentose phosphate shunt pathway. Moreover, the frequency of foetal malformations has been found to be reduced in patients given dietary supplements of antioxidants such as vitamins E and C and butylated hydroxytoluene. This may indicate that oxidative stress plays a significant role in foetal dysmorphogenesis [23].

Caudal regression syndrome (Figure 29.10), also known as caudal dysplasia, sacral agenesis and sirenomelia, shows a spectrum of changes ranging from agenesis of the lumbosacral spine, with absent external genitalia, anal atresia and renal agenesis, to fusion of the lower extremities with horseshoe-shaped kidney. The aetiology is not well understood, but it has a strong association with diabetes

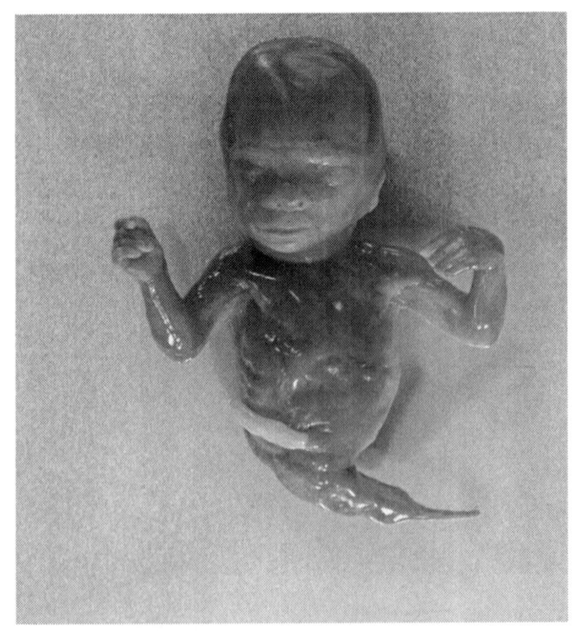

Figure 29.10 Foetus with Caudal regression syndrome showing fusion and failure of development of the lower extremities.

mellitus [24]. It is worthy of note that it is also thought to arise secondary to poor perfusion of the lower extremities. Could it be that the disturbed vascularity associated with hyperglycaemic states is responsible for the disorder? Or are other hitherto unexplained factors contributing?

A novel research approach to treatment has been undertaken using maternal immune stimulation to modulate the diabetic-induced palatal defects in mice model [25]. Using granulocyte–macrophage colony-stimulating factor and Freund's complete adjuvant prior to inducing hyperglycaemia, they were able to reduce the severity of palatal shortening. Dysregulation of cell signalling induced by hyperglycaemia is thought to play a role in foetal dysmorphogenesis; it may be that immune stimulation prevents such dysregulation. Understanding such mechanisms may allow us to better understand the anomalies which arise in diabetic pregnancies.

In women who are obese at the start of their pregnancy, there is an almost two-fold increase of being affected by neural tube defect [26], and as in the diabetic state, the risk of Cardiovascular anomalies is increased 1.3 times and cleft lip and palate are 1.23 times more common [27]. It is thought that the shared metabolic abnormalities between obese and diabetic women account for the similarities in the nature of the defects which arise. However, other studies have suggested that nutritional deficiencies such as reduced folate levels are more common in obese women and may account for the abnormalities noted [28].

It has long been recognised that diabetes mellitus has a strong association with macrosomia with its associated delivery complications. Infants of diabetic mothers are thought to grow excessively due to increase in circulating glucose and foetal hyperinsulinism [29]. Insulin acts as the primary anabolic hormone of foetal growth and development resulting in macrosomia and visceromegaly, especially affecting the heart and liver. With increased insulin, fat gets deposited towards the end of the third trimester. At the time of delivery, glucose levels fall rapidly and may lead to life-threatening hypoglycaemia. Obesity also leads to macrosomia and it appears that maternal obesity is more strongly associated with this outcome, while GDM leads more frequently to neonatal hypoglycaemia [30,31].

The elevated glucose in utero leads to an increase in the size and number of islets of Langerhans – in those who die later in pregnancy or in the first few days of life. Morphometric studies have shown that the islets in these infants are largely composed of beta cells. The most noticeable changes are seen in infants dying after 34 weeks of gestation or in those deemed large for dates. Single islet cells may stand out owing to nuclear enlargement and pleomorphism [32]. In the largest islets, the endocrine cells may be encircled by fibrous tissue containing lymphocytes and large numbers of eosinophils. Eosinophilic insulitis (Figure 29.11), hypertrophy and hyperplasia of the islets and perinsular fibrosis are pathognomonic of the diabetic pregnancy. Such changes do not occur in the obese pregnancy state in the absence of elevated glucose levels. However, the findings of eosinophilic insulitis are enough to warn of occult diabetes and it may also be seen in gestational diabetes. Its relationship with glycaemic control is not yet known. Such findings may also raise a suspicion of increased BMI and developing diabetes mellitus, though specific research in this area is lacking.

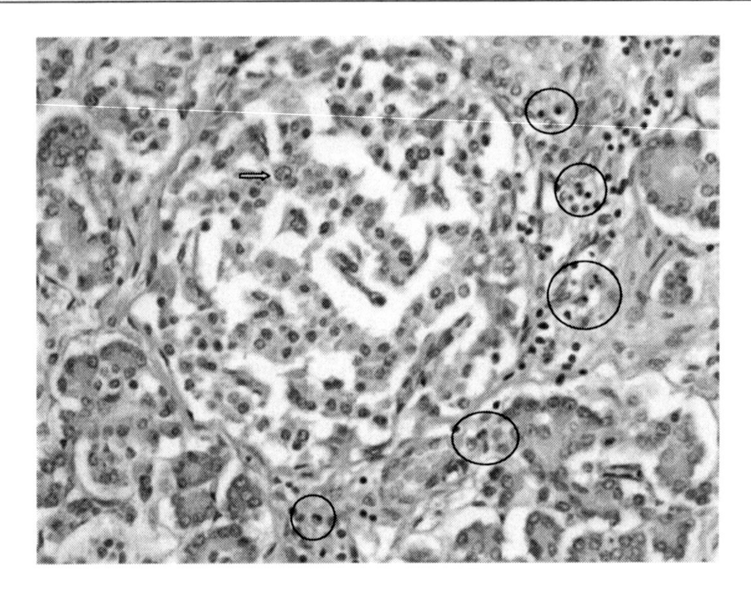

Figure 29.11 Islet of Langerhans showing nuclear pleomorphism (arrow) and eosinophils (circles).

Karmon et al. [33] showed that though improved dietary control among women with gestational diabetes may play a part in reducing perinatal mortality, with delivery up to 40 weeks showing no significant rise in the stillbirth rate among the well controlled population diabetic population, Mortality post-40 weeks gestation was significantly higher in the diabetic group with complications associated with macrosomia (e.g. shoulder dystocia) contributing to this rise. They, therefore, suggested that delivery at 40 weeks may lead to a decreased incidence of such complications.

Conclusion

Despite advances in modern medicine, diabetic pregnancy is fraught with danger and owing to the increasing obesity in the general population is one which is set to rise in the next few years. In this chapter, an overview of current knowledge has been given as it relates to the pathology of the placenta and the foetus. It is to be hoped that improved understanding of the biochemical mechanisms underpinning these morphological changes will lead to more effective therapeutic intervention.

References

1. World Health Organisation. *Obesity: Preventing and Managing the Global Epidemic.* Report of a WHO consultation, Geneva WHO; 2004.

2. Frischknecht F, Brühwiler H, Raio L, et al. Changes in pre-pregnancy weight and weight gain during pregnancy: retrospective comparison between 1986 and 2004. *Swiss Med Wkly*. 2009;24:52–55.
3. CEMACH. *Confidential Enquiry into Maternal and Child Health. Saving Mother's Lives: Reviewing Maternal Deaths to Make Motherhood Safer 2003–2005. The Seventh Report of the Confidential Enquiries into Maternal Deaths in the UK*. London: CEMACH; 2007.
4. Catalano PM, Tyzbir ED, Roman NM, Amini SB, Sims EA. Longitudinal changes in insulin release and insulin resistance in non-obese pregnant women. *Am J Obstet Gynaecol*. 1991;165:1667–1672.
5. Mondestin MAJ, Ananth CV, Smulian JC, et al. Birth weight and fetal death in the United States: the effect of maternal diabetes during pregnancy. *Am J Obstet Gynaecol*. 2002;187:922–926.
6. Lucas MJ. Diabetes complicating pregnancy. *Obstet Gynecol Clin North Am*. 2001;28 (3):513–516.
7. Gugliucci CL, O'Sullivan MJ, Opperman W, Gordon M, Stone ML. Intensive care of the pregnant diabetic. *Am J Obstet Gynecol*. 1976;125(4):435–441.
8. Campbell IW, Duncan C, Urquhart R, Evans MJ. Placental dysfunction and stillbirth in gestational diabetes mellitus. *Br J Diab Vasc Dis*. 2009;9(1):38–40.
9. Ezimokhai M, Rizk DE, Thomas L. Maternal risk factors for abnormal vascular coiling of the umbilical cord. *Am J Perinatol*. 2000;17(8):441–445.
10. Ezimokhai M, Rizk DE, Thomas L. Abnormal vascular coiling of the umbilical cord in gestational diabetes mellitus. *Arch Physiol Biochem*. 2001;109(3):209–214.
11. Machin GA, Ackerman J, Gilbert-Barness E. Abnormal umbilical cord coiling is associated with adverse perinatal outcomes. *Pediatr Dev Pathol*. 2000;3:462–471.
12. Strong TH, et al. The umbilical coiling index. *Am J Obstet Gynecol*. 1994;170:29–32.
13. De Laat MW, Franx A, Bots ML, Visser GH, Nikkels PG. Umbilical coiling index in normal and complicated pregnancies. *Obstet Gynecol*. 2006;107(5):1049–1055.
14. De Laat MW, van der Meij JJ, Visser GH, Franx A, Nikkels PG. Hypercoiling of the umbilical cord and placental maturation defect: associated pathology? *Pediatr Dev Pathol*. 2007;10(4):293–299.
15. Ezimokhai M, Rizk DE, Thomas L. Maternal risk factors for abnormal vascular coiling of the umbilical cord. *Am J Perinatal*. 2001;17(8):441–445.
16. Pietryga M, Biczysko W, Wender-Ozegowska E, et al. Ultrastructural examination of the placenta in pregnancy complicated by diabetes mellitus. *Ginekol Pol*. 2004;75 (2):111–118.
17. Hiden U, Lang I, Ghaffari-Tabrizi N, et al. Insulin action on the human placental endothelium in normal and diabetic pregnancy. *Curr Vasc Pharmacol*. 2009;Oct; 7(4): 460–466.
18. Hay Jr WW. Placental-fetal glucose exchange and fetal glucose metabolism. *Trans Am Clint Climatol Assoc*. 2006;117:321–339.
19. Baeten JM, Bukusi EA, Lambe M, et al. Pregnancy complications and outcomes among overweight and obese nulliparous women. *Am J Public Health*. 2001;91:436–440.
20. Roberts KA, Riley SC, Reynolds RM, et al. Placental structure and inflammation in pregnancies associated with obesity. *Placenta*. 2011;32(3):247–254.
21. James AH. Pregnancy-associated thrombosis. *Haematol Am Soc Haematol Educ Program*. 2009;277–285.
22. Miller E, Hare JW. Elevated maternal haemoglobin A_{1c} in early pregnancy and major congenital anomalies in infants of diabetic mothers. *N Engl J Med*. 1981;304:1331–1334.

23. Dheen ST, Tay SS, Boran J, et al. Recent studies on neural tube defects in embryos of diabetic pregnancy: an overview. *Curr Med Chem.* 2009;16(18):2345–2354.
24. Kucera J. Rate and type of congenital anomalies among offspring of diabetic women. *J Reprod Med.* 1971;Aug; 7(2):73–82.
25. Hrubec TC, Troops KA, Holladay SD. Modulation of diabetes-induced palate defects by maternal immune stimulation. *Anat Rec (Hoboken).* 2009;292(2):271–276.
26. Rasmussen SA, Chu SY, Kim SY, et al. Maternal obesity and risk of neural tube defects: a metaanalysis. *Am J Obstet Gynaecol.* 2008;198:611–619.
27. Stothard KJ, Tennant PW, Bell R, et al. Maternal overweight and obesity and the risk of congenital anomalies: a systematic review and meta-analysis. *JAMA.* 2009;301: 636–650.
28. Mojtabai R. Body mass index and serum folate in childbearing age women. *Eur J Epidemiol.* 2004;19:1029–1036.
29. North AFJ, Mazumdar S, Logrillo VW. Birthweight, gestational age and perinatal deaths in 5,471 infants of diabetic mothers. *J Pediatr.* 1977;90:444–447.
30. Maresh M, Beard RW, Bray CS, et al. Factors predisposing to and outcome of gestational diabetes. *J. Obstet Gynecol.* 1989;74:342–346.
31. Clausen T, Burski TK, Øyen N, et al. Maternal anthropometric and metabolic factors in the first half or pregnancy and risk of neonatal macrosomia in term pregnancies. A prospective study. *Eur J Endocrinol.* 2005;153:887–894.
32. Hultquist GT, Olding LB. Endocrine pathology of infants of diabetic mothers: a quantitative morphological analysis including comparison with infants of iso-immunized and non-diabetic mothers. *Acta Endocrinol Suppl (Copenh).* 1981;241:1–202.
33. Karmon A, Levy A, Holcberg G, et al. Decreased perinatal mortality among women with diet-controlled gestational diabetes mellitus. *Int J Gynaecol Obstet.* 2009;104 (3):199–202.

30 Post-Natal Care for the Recently Delivered Obese Women

Debra Bick and Sarah Beake

Florence Nightingale School of Nursing and Midwifery, King's College, London, London, UK

Introduction

The routine provision of post-natal care has played a unique role within UK public health to support birth recovery and well being of women and their infants. It is an area of clinical care which mainly falls within the sphere of practice of the midwife, who will have most contact with the woman and her infant in the first few days and weeks following birth. Midwifery post-natal care became a statutory requirement in the early twentieth century, when the midwife's main role was to undertake observations and examination to monitor a woman's physical recovery from giving birth, which included assessing her temperature, pulse, vaginal blood loss and uterine involution. The aim of this proscriptive content of care was to identify post-partum haemorrhage (PPH) and puerperal sepsis, the main causes of maternal mortality in 1902, the year the first Midwives Act was passed in England. Most contacts with the midwife took place in the woman's home, where the majority of births took place, the number and timing of contacts regulated for all women within the Midwives Rules [1]. A woman had to be visited by a midwife twice a day for the first 3 days and then daily until day 10, these timings seemingly arbitrary and not based on evidence of benefit.

The statutory provision of midwifery post-natal care as required by the 1902 Midwives Act and the pre-determined pattern of visits [1] was a reaction to the persistent high maternal mortality rate which did not start to decline until the middle of the twentieth century [2]. Minimal revision to the content and timing of midwifery post-natal care occurred during the course of the last century. In addition to routine midwifery care, all UK women are offered a consultation at 6–8 weeks with their family doctor (GP), the timing and content of this routine provision of care also lacking an evidence base. Most women will not need to see an obstetrician during the post-natal period, unless there is a specific indication.

Obesity. DOI: http://dx.doi.org/10.1016/B978-0-12-416045-3.00030-3

Midwifery post-natal care continues to be a statutory requirement in the United Kingdom, although this aspect of care is now perceived as being of lower priority for maternity service providers than the care provided during pregnancy and labour, with some indication that standards of care have declined over the last 10 years [3,4]. The most recent edition of the Midwife Rules and Standards state that the post-natal period means *the period after the labour during which the attendance of the midwife upon a woman and baby is required, being not less than 10 days and for such longer period as the midwife considers necessary* (p. 8) [5]. Findings of large UK observational studies conducted in the 1990s showed that many women experienced widespread and persistent physical and psychological health problems after giving birth, many of which did not resolve within the 6–8 week post-natal period [6,7]. The studies found that women did not report problems to their midwives or GPs, and morbidity was not identified during routine clinical contacts, indicating that the model of post-natal care was not appropriate [8]. Evidence from a large cluster randomised controlled trial (RCT) of how extended, protocol-based midwifery-led care in the community could improve women's mental health outcomes did not lead to a subsequent change in practice or revision to the definition of the post-natal period, despite the revised model of care being clinically and cost-effective [8,9].

Of concern is that the health and socio-demographic profile of women who give birth in the United Kingdom has changed dramatically [10–12], indicating an increased need for effective post-natal care in the twenty-first century. In developed countries, there is an epidemic of obesity in pregnancy, defined as body mass index (BMI) $\geq 30\ kg/m^2$ or more at the woman's first antenatal consultation [13], the impact of which on pregnancy and longer-term health outcomes is significant [14]. In the United Kingdom in 2007, it was estimated that 24% of UK women aged 16 and above were obese, an increase from 16% calculated in 1993 [15]. The UK Obstetric Surveillance System (UKOSS) conducted a study of extreme obesity during pregnancy between March 2007 and August 2008. Case definition for extreme obesity included women who weighed 140 kg or more at any time during pregnancy, had a BMI of $\geq 50\ kg/m^2$ or who were expected to fulfil these two criteria but whose weight exceeded the hospital scales. Preliminary data analyses suggested that extreme obesity affected 1 in 1000 women, and that these women experienced significantly more pregnancy, birth and post-natal complications compared with control women. The obesity epidemic is affecting obstetric populations in many other developed countries, with recent estimates suggesting that around 35% of pregnant women in Australia are overweight or obese [16].

Poor maternal health is increasing demand on a UK maternity service which is facing a shortage of midwives and an increasing birth rate. While the focus of this chapter is the post-natal care of women who are obese, all women require effective post-natal care, and other factors associated with poorer pregnancy outcomes, such as a woman's ethnicity or her exposure to social problems including domestic violence [17,18], need to be considered.

Background

Before considering what the post-natal care needs of recently delivered obese women could mean in terms of planning, content and resource use, it is important to consider how being overweight or obese during pregnancy can impact on the post-natal health of a woman and her baby. The most recent triennial report of maternal deaths in the United Kingdom [12] highlighted in stark terms why obesity is such a crucial public health issue:

> *In this Report, 47% of mothers who died from Direct causes were either overweight or obese, as were 50% of women who died from Indirect causes. This means that overall, 49% of the women who died and for whom the BMI was known were either overweight or obese. When considering obesity alone, that is a BMI of 30 or more, 30% of mothers who died from Direct causes and for whom the BMI was known were obese as were 24% of women who died from Indirect causes; 27% overall [12, p. 46].*

In the 2007 report of the Confidential Enquiry, Lewis [10] reported that maternal obesity was one of the greatest and growing overall threats to the childbearing population of the United Kingdom. Obesity remained a common factor among women who died during 2006–2008 and was associated with deaths from thromboembolism, sepsis and cardiac disease; 78% of the women who died following a thromboembolic event were overweight or obese [12]. For each woman who died as a direct or indirect consequence of pregnancy where being overweight or obese were contributing factors, it is likely that other women who experienced a potentially life-threatening event, such as a PPH, were saved due to effective and prompt emergency care.

Research into the effects of obesity in pregnancy is now highlighting the consequences of this not only for the woman, but also for the baby in utero and the health of the child during its life course. An obese woman has an increased risk of restricted foetal growth during her first pregnancy [19], a higher risk of structural foetal anomalies [20] and her child is at higher risk of developing diabetes [21]. In 2009, 10% of mothers who had a stillbirth and 10% of mothers who had a neonatal death had a BMI of $\geq$35 kg/m^2[14]. A recent systematic review and meta-analysis to identify priority areas for stillbirth prevention relevant to high-income countries where rates of stillbirth have shown little or no improvement over the last two decades included data from 96 population-based studies [22]. Maternal overweight and obesity (BMI > 25 kg/m^2) was the highest ranking modifiable risk factor, with population attributable risks (PARs) of 8–18% across five high-income countries with the highest numbers of stillbirths and where all the data required for analysis were available (Australia, Canada, the Netherlands, United Kingdom, United States). Obesity was associated with around 8000 stillbirths ($\geq$22 weeks gestation) annually across all high-income countries. Advanced maternal age (>35 years) and maternal smoking yielded PARs of 7–11% and 4–7%, contributing to more than 4200 and 2800 stillbirths annually, respectively, across all high-income countries.

Several mechanisms to account for the risk of stillbirth among overweight or obese women have been proposed, including increased risk of an obese women developing diabetes and hypertensive disorders during pregnancy [23], utero-foetal programming [24] and inability for an obese woman to feel diminished foetal movements [25]. The exact relationship between maternal obesity and stillbirth remains unknown and further research is required to assess if it is an independent effect or a consequence of co-morbidity associated with obesity during pregnancy. That important risk factors are increasing in our obstetric population means that the need to consider their interdependent effects are major challenges for the maternity services. The following risk factors for pregnancy and post-natal complications for an obese woman and her infant were included in the 2007 report of the Confidential Enquiry [10] and again in the most recent report [12].

For the Woman

Increased risks include:

- Maternal death or severe morbidity
- Cardiac disease
- Spontaneous first trimester and recurrent miscarriage
- Pre-eclampsia
- Gestational diabetes
- Thromboembolism
- Post-Caesarean wound infection
- Infection from other causes
- Post-partum haemorrhage
- Low breast-feeding rates.

For the Baby

Increased risks include:

- Stillbirth and neonatal death
- Congenital abnormalities
- Prematurity.

Centre for Maternal and Child Enquiries

Those complications more likely to influence *maternal* post-natal care will be considered further in this chapter [12, p. 47].

National statistics on the prevalence of obesity in pregnant and post-natal women are neither routinely collated in the United Kingdom, nor there are data on overall severe maternal morbidity for women in England, Wales and Northern Ireland. The UKOSS survey, launched in 2005 to collate data on severe maternal morbidity on a national population basis, is a joint initiative between the National Perinatal Epidemiology Unit and the Royal College of Obstetricians and

Gynaecologists [26]. Data are collated on a range of rare conditions in pregnancy (those with an estimated incidence of fewer than 1 in 2000 births) which can adversely affect the outcome for a woman and/or her baby, including uterine rupture, myocardial infarction (MI) and extreme obesity (BMI ≥ 50 kg/m^2). Nominated individuals at every consultant-led obstetric unit in the United Kingdom are asked to complete a return card with details of any cases of the conditions under surveillance during a defined time period in their units. In Scotland, an on-going audit of severe maternal morbidity has taken place since 2003 [27]. Data are collected for 14 categories, ranging from major PPH, pulmonary oedema, cardiac arrest and massive pulmonary embolism. During 2006–2008, the rate of severe maternal morbidity was 5.88 per 1000 births (95% CI 5.52–6.25) based on 174,430 births in Scotland. Major obstetric haemorrhage was the most frequent cause of severe morbidity.

Implications of Obesity for Planning of Post-Natal Care

That the numbers of women in the United Kingdom who are obese when they become pregnant is increasing, and the implications of this for their own and for their baby's immediate and longer-term health is not disputed, how or when to deal with the issue of obesity is less clear. The implications of being obese for a woman's post-natal health are likely to affect the content and planning of her care and frequency of contact with the multi-professional team. It also presents an opportunity to influence her future health behaviour with respect to weight management [28]. The priorities for practice should include planning post-natal care during pregnancy to prevent and minimise potential post-birth complications and assist the woman and her baby to achieve the best possible outcome. This is especially important if the woman has other risk factors in addition to being obese, such as being an older mother.

Antenatal planning may have implications for mode of birth, which in itself will influence post-natal morbidity outcomes and the level of care a woman will require, with recent evidence from UKOSS indicating that a routine policy of Caesarean section (CS) for extremely obese women is not indicated [29]. Nevertheless, with anecdotal evidence of the relatively low priority now accorded to post-natal care, obstetricians, midwives, managers and commissioners need to be aware of the challenges for practice and how planned and tailored post-natal care could make a difference. The following section highlights some of the health risks an obese woman may experience following birth. Much of the data reported are from retrospective studies which may be subject to bias due to incomplete data ascertainment, although studies consistently demonstrate the complexity of health issues faced by overweight or obese women.

Post-Natal Morbidity

Pre-Eclampsia

The precise relationship between pre-eclampsia and pregnancy-induced hypertension (PIH) remains unclear [30], but symptoms may have a pregnancy or a post-natal onset in a previously normotensive woman. One study examined the post-natal duration of hypertension among women who had already presented with PIH or pre-eclampsia. Ferrazzani et al. [31] studied 269 women with PIH ($n = 159$) or pre-eclampsia ($n = 110$) and monitored their post-partum blood pressure daily after delivery until a diastolic blood pressure of ≤ 110 mmHg was reached. The time taken for this ranged from 0 to 10 days among the women with PIH and from 0 to 23 days among those with pre-eclampsia. How long it took for women to become 'normotensive' (diastolic ≤ 80 mmHg) was not reported.

A number of retrospective studies have shown that a BMI of ≥ 30 kg/m^2 is a risk factor for the development of hypertension and onset of pre-eclampsia in women of all parities [32] as well as in primiparous women [19], a risk which appears to increase if the woman is morbidly obese (BMI ≥ 35 kg/m^2). Bhattacharya et al. [33] undertook a retrospective cohort study of women who gave birth to their first baby in one Scottish city between 1976 and 2005. In comparison with women who had a BMI of 20–24.9 kg/m^2, morbidly obese women faced the highest risk of pre-eclampsia (odds ratio (OR) 7.2, 95% CI 4.7–11.2) and under-weight women the lowest (OR 0.6, 95% 0.5–0.7). A UKOSS survey identified 214 cases of eclampsia between February 2005 and February 2006, an estimated incidence of 27.5 cases per 100,000 maternities [34]. This was almost a halving of the incidence of eclampsia since 1992, which may be attributable to the recommended use of magnesium sulphate to manage these cases [35]. In the most recent Centre for Maternal and Child Enquiries (CMACE) report [35], 19 women died of eclampsia or pre-eclampsia and three women died from acute fatty liver of pregnancy, which may be part of a spectrum of conditions associated with pre-eclampsia. Three women died following discharge from the post-natal ward. Five of the twenty-two women who died had BMI ≥ 30 kg/m^2.

Thromboembolism

Obesity is the most important risk factor for thromboembolism [35]. Eighteen women died of thrombosis/thromboembolism during 2006–2008, a mortality rate of 0.79 per 100,000 maternities [35], a lower number of deaths than reported in the previous triennial enquiry [10] and the lowest since the UK-wide Confidential Enquiry began in 1985. This was the first triennial period where thrombosis was not the leading cause of *direct* maternal death. Sixteen women died from pulmonary embolism and two from cerebral vein thrombosis. Eight deaths occurred following the birth, two after a vaginal birth and six following Caesarean birth.

One encouraging finding was the decline in deaths during pregnancy and following vaginal birth, which may reflect the routine use of thrombo-prophylaxis as recommended in national guidance [35,36]. However, of note was the high risk of thrombosis as a consequence of obesity. Fourteen of the women who died were overweight, 11 of whom had a BMI ≥ 30 kg/m^2. Substandard care was found in over half of all deaths from thrombosis, including inadequate risk assessment, inadequate thrombo-prophylaxis in line with guidance published at the time and failure to involve the multi-professional team to inform care of women with a pre-existing medical or psychiatric condition. For obese women post-birth, it is essential that content of care includes advice that women need to report symptoms of chest pain as well as adherence to RCOG guidelines on acute management of thromboembolic disease in pregnancy and the puerperium [37].

Infection

The most recent data on birth outcomes in England during 2010–2011 showed that the number of births by CS remained static from 2009 to 2010; 24.8% of women gave birth by CS [11]. Obese women are more likely to have complicated pregnancies which require medical intervention, with the rate of successful vaginal birth decreasing as maternal BMI increases [21]. A meta-analysis by Chu et al. [38] which examined maternal obesity and risk of CS based on data from 33 studies found that the odds of a CS birth were 1.46 (95% CI 1.34–1.60), 2.05 (95% CI 1.86–2.27) and 2.89 (95% CI 2.28–3.79) among overweight, obese and severely obese women, respectively, compared with pregnant women whose weight was within normal ranges. A CS birth is associated with shorter- and longer-term maternal morbidity, including surgical site infection (SSI) [39]. Despite the high numbers of Caesarean birth, the length of time women stay on inpatient post-natal wards has declined [11,40] with limited evidence that community midwifery post-natal contacts have been revised to reflect post-surgical as well as post-birth maternal needs.

SSI may affect the abdominal wound, uterus or endometrium, and infection is most likely to become evident in the community. Data on incidence of CS SSI are limited, but suggest that 10–20% of women develop SSI, based on the criteria developed by the US Centers for Disease Control. Prospective data on 715 women who had a CS birth were collected over a 35 week period during 2002/03 by Johnson et al. [41] from one maternity unit in Scotland. Eighty women developed a post-CS infection, 57 (71%) of which were detected following hospital discharge. Obese women had significanly more infections than women with a normal BMI ($P = 0.028$).

Ward et al. [42] collected data from one of the first UK prospective multicentre studies of CS infection outcomes. Data of women who had a CS birth at 11 sites in the East Midlands of England between 3 and 18 months were collected by community midwives. Of 6297 CS births during the study period, data were available for 5563 women, with a median length of follow-up of 15 days and marked inter-unit variation. A total of 745 SSIs were recorded in 738 (13.3%) of the 5563 women, 488 (65.5%) of which met study definitions. Risk factors included maternal BMI ($P < 0.0001$), emergency procedures ($P = 0.002$), SROM ($P = 0.01$), in labour at the time of surgery ($P < 0.001$), duration of procedure ($P = 0.002$) and wound closure method ($P = 0.003$). Maternal BMI was also a risk factor for SSI in the 2009–2010 All Wales CS SSI surveillance report [43], where CS surveillance was introduced as a mandatory requirement in 2006. A total of 6801 and 7611 questionnares were received for 2009 and 2010, respectively. Of these, 6237 (92%) and 7336 (96%) of forms could be analysed for determining the overall SSI rate for 2009 and 2010. During 2009, 80 inpatient SSI and 771 post-discharge SSI were detected, providing a crude overall SSI rate of 13.6%. For 2010, 48 inpatient SSI and 738 post-discharge SSI were detected, giving a crude overall SSI rate of 10.7%. Trend rates had decreased since 2007; however, lack of reliability with respect to completeness of data collection following hospital discharge and lack of application of SSI criteria remain important issues to address, given the frequency with which CS is now performed.

During 2006–2008, genital tract sepsis was the largest cause of *direct* maternal death [12]. Most deaths were associated with community acquired group A streptoccocal disease. The majority of the 26 women who died had a normal BMI or were slightly overweight; five women had a BMI greater than 30 kg/m^2. Of concern for the enquiry team was that the puerperium seemed to be regarded as a low-risk period compared with pregnancy and birth. Seven women died after a vaginal birth, following an uncomplicated pregnancy and birth and early post-natal discharge, highlighting the insidious nature of sepsis and need for vigilence among all women and increased awareness of signs of sepsis among all relevant clinical staff. A population-based case–control study from Scotland considered all cases of pregnant, intrapartum and post-natal women with sepsis or severe sepsis included in the Aberdeen Maternal and Neonatal Databank from 1986 to 2009. Dependent variables included uncomplicated sepsis or severe ('near miss') sepsis [44]. After controlling for mode of birth and demographic and clinical factors, women who were obese had twice the odds of uncomplicated sepsis (OR 2.12, 95% CI 1.14–3.89) compared with women of normal weight. Other significant predictors of sepsis included maternal age of 25 or younger and operative vaginal birth.

Women who are obese are at higher risk of experiencing other sites of infection. Heslehurst et al. [45] in a systematic review of obstetric outcomes among obese women found almost 3.5-fold increase in infections (including wound and infections of urinary tract, perineum, chest and breast) (OR 3.34, 95% CI 2.74–4.06).

Cardiac Disease

Obese women have an increased risk of cardiac disease during pregnancy. This is not the only risk factor, as those who smoke, have hypertension, diabetics, those who are older or have a family history of cardiac disease are also at risk, and a woman may present with several risk factors. Fifty-three UK women died during 2006–2008 of heart disease associated with or aggravated by pregnancy, a mortality rate of 2.31 per 100,000 maternities (95% CI 1.77–3.03) [35]. These deaths were classed as *indirect* deaths and all but two of the women died post-natally. Cardiac disease remains the most common cause overall of maternal death in the United Kingdom. Thirty (60%) of 50 women who died of cardiac disease and for whom BMI were available were overweight or obese; half had a BMI of 30 kg/m^2 or more. Of concern was that substandard care was noted in 27 of the 53 cases [35]. Concerns were raised by the CMACE assessors of the increasing number of women with ischaemic heart disease (IHD) as a cause of death during pregnancy, with the increasing poor health of women as a consequence of obesity and smoking as well as having a family history and delaying childbirth all as identified risk factors for IHD.

There is a dearth of contemporary data with respect to the number of women in the United Kingdom who suffer a cardiac arrest during pregnancy, with the current incidence estimate of 1 in 30,000 women [46]. A retrospective study from the United States conducted by Ladner et al. [47] which analysed data over a 10 year period to estimate the population incidence and pregnancy outcomes of acute MI found an incidence of 1 in 35,700 births, with 41% of events occurring during the 6 week post-natal period, 38% during pregnancy and 21% during labour. The incidence rate increased over the 10 year study period. UKOSS have been collating data on pregnancy-related MI since 2005, reporting 13 confirmed cases up until January 2008. This was lower than expected and the data collection period has been extended to enable all potential sources of data to be considered [48]. UKOSS are currently conducting a survey of all cardiac deaths in the United Kingdom for the period July 2011–June 2014 [49].

Post-partum Haemorrhage

There is no consensus on the exact amount of vaginal blood loss that constitutes a PPH. The definition of PPH is based usually on a subjective estimation of blood loss and whether symptoms occurred within the first 24 h of the birth (primary) or after the first 24 h and up to 6 weeks after the birth (secondary). The World Health Organization (WHO) defined PPH as 'vaginal bleeding in excess of 500 ml after childbirth within or following the first 24 h' [50]. Precise measurement of blood loss is subject to underestimation, and impact on maternal health and well being may also vary according to the individual's haemoglobin level (e.g. a woman with anaemia may be less tolerant of blood loss). Studies examining post-natal women's experiences of blood loss imply that there is variability in the normal range of blood loss and reports may only describe most adverse outcomes [30].

Traditionally, a blood loss greater than 500 ml is considered a valid measurement for diagnosing PPH [50], while a loss in excess from 1000 ml [51] to 1500 ml [52] is proposed to indicate severe or major obstetric haemorrhage. A survey of UK maternity units failed to establish a consistent definition for major PPH, but the majority identified blood loss $\geq$1000 ml (46%) or $\geq$1500 ml (36%) as indicators of major PPH [53]. A retrospective study of effect of BMI on pregnancy outcomes among primparous women by Bhattacharya et al. [33] found that compared with women who had a BMI of 20–24.9 kg/m^2, women who were obese had much greater risk of a PPH (OR 1.5, 95% CI 1.3, 1.7). A smaller retrospective study from South Australia which selected 100 women with normal BMI (BMI 19.1–25 kg/m^2), 100 overweight (BMI 25.1–30 kg/m^2), 110 obese (BMI 30.1–40 kg/m^2) and 60 morbidly obese women (BMI $\geq$40 kg/m^2) were identified with access to complete medical records [54]. Significantly more blood loss was recorded for women who had a BMI of 30.1–40 and $\geq$40 kg/m^2 compared with women who had a BMI of 19.1–25 kg/m^2. Heslehurst et al. [45] reported an OR of

1.24 (95% CI 1.20–1.28) of PPH among women who were obese during pregnancy in the systematic review referred to earlier.

There were nine *direct* maternal deaths from obstetric haemorrhage in the United Kingdom during 2006–2008, a decline in the number of deaths from the previous triennial enquiry [10]. Five deaths occurred post-natally. Two of the women for whom BMI data were available were overweight or obese. The numbers of women who experience an adverse obstetric event such as PPH are increasing, and attention is turning to adverse events as an indicator of the quality of maternity care in high-income countries given the rare outcome of maternal death [55]. This is discussed further in the next section.

Severe Maternal Morbidity as an Indicator of Quality of Maternity Care

Due to advances in public health and clinical management, maternal death in the United Kingdom is a relatively rare event, increasing recognition of the need to use an adverse obstetric event or severe morbidity as a 'complementary marker' of standards of care [56]. Severe morbidity is classed as a health problem that if untreated, could result in the death of the woman. The theory underlying this approach was described by Pattinson et al. [57] as:

> *'the sequence from good health to death in a pregnant woman is a clinical insult, followed by a systemic inflammatory response, organ failure and finally death. A near miss would be those women with organ dysfunction who survive'.*

Studies of severe morbidity have tended to be retrospective which may not provide robust, complete data on contributing factors or outcomes related to an event. Studies do however provide useful information on trends over time and can highlight risk factors. In one large population cohort study, it was estimated that around 1.2% of women experienced severe morbidity; in two-thirds of cases, this was caused by a massive PPH and in-one third it was caused by sepsis, uterine rupture and severe pre-eclampsia [52]. It is important that clinicians who care for obese women are aware that risk factors do not decline at birth, and in fact, women may be at greater risk of developing complications in the first few hours, days and weeks after giving birth.

Planning Post-Natal Care

Evidence of how post-natal care could be successfully revised to enhance health outcomes is limited, which is of concern given recognition of the importance of effective care over a woman's life course [58] including during and after pregnancy. The large cluster RCT by MacArthur et al.[8,9] showed that midwives who provided community-based post-natal care over an extended period of time, such as providing symptom checklists and evidence-based guidance [30], could make a significant difference to women's shorter- and longer-term psychological health. A recent study carried out to improve the outcomes of in-patient post-natal care using continuous quality improvement found statistically significant differences post-intervention in breastfeeding outcomes [59]. Further research is needed to address the post-natal needs of women with specific health needs, including those associated with obesity. From the evidence currently available, effective post-natal care could make an important contribution to a woman's future health; yet, it remains an area of low interest for many clinicians. This section discusses the evidence currently available to support the planning and content of care to enhance clinical and cost-effectiveness with reference to the care of women who are obese.

Current Guidance to Inform the Content and Timing of Post-Natal Care

In 2006, the National Institute for Health and Clinical Excellence (NICE) [60] published guidelines for the routine post-natal care of healthy women and their babies which presented recommendations for women receiving NHS maternity care in England and Wales. This is part of a suite of guidance for maternity care, which includes antenatal care [61], intrapartum care [62], CS [63,64], diabetes in pregnancy [65] and antenatal and post-natal mental health [66]. NICE also published guidelines on prevention, identification, assessment and management of obesity in adults and children in 2006 [67], SSI [68] and weight management in pregnancy and following birth [28], information from which will also be relevant for the care of the obese woman. Clinicians should also be familiar with RCOG guidelines on acute management of thromboembolic disease during pregnancy and the puerperium [36] and with joint CEMACH and RCOG guidelines on management of obese women in pregnancy [13].

All women and their infants will require elements of care as recommended in the NICE post-natal guideline [60], regardless of whether they also require more specialist care. NICE [60] recommends that planning for post-natal care for all women should commence antenatally, and recent guidance for the care of obese women during pregnancy specifically recommends that women should have risks for their pregnancy outcomes discussed with them, as well as approaches to maintaining a healthy lifestyle and infant-feeding decisions [13]. This is important as many women underestimate the impact of pregnancy and birth on their physical and psychological health even without experience of risk factors for a potentially poorer outcome [69]. Commencing the planning of care for a woman post-birth also means that relevant members of the multi-professional team can be involved to promote implementation of strategies to prevent or minimise post-natal morbidity, e.g. through advice on prophylaxis to prevent thromboembolism [67]. The need for more effective planning of maternal health needs by the multi-professional team was frequently raised by CMACE assessors in the most recent Confidential Enquiry [12]. As midwives are most likely to be the clinician women who will be seen most frequently, it is particularly important that they are aware of signs and symptoms of potentially severe morbidity and how to initiate appropriate and timely referral.

Physical Observations and Examinations

There remains a dearth of evidence to support how or when midwives should perform routine post-natal maternal physical observations and examinations. It is interesting to note that in the trial by MacArthur et al. [8], midwives providing the new model of care reported difficulties with only undertaking physical observations and examinations if indicated. Bick et al. [70] found that despite discussions with midwives to highlight that they did not need to undertake traditional observations and examinations at each post-natal contact, they continued to do so, which reflected the findings of MacArthur et al. [8]. Post-natal care documentation for the most part continues to rely on completion of a 'tick box' to indicate performance of routine observations and examinations, with limited space to record other information or promote provision of care individualised to need.

Although routine undertaking of physical observations and examinations is not necessary for all women at all contacts, women should be *asked* at each contact how they are feeling, with appropriate examination and observation if indicated based on clinical judgement or a woman's reporting of signs and symptoms of potential ill health. NICE [60] guidelines for core post-natal care includes that within 6 h of the birth, a baseline measure of a woman's blood pressure and her first void post-birth should be documented, with action to take with respect to both if concern is raised. NICE guidelines [60] also includes that at the first post-natal contact, *all* women and their partners are offered information on the signs and symptoms of major maternal morbidity including thrombosis, genital tract sepsis, PPH and pre-eclampsia/eclampsia, and if any signs or symptoms are experienced, urgent medical attention implemented. This information is essential for all women given much earlier transfer to home from hospital but especially for those at higher risk of adverse outcome.

Severe Morbidity

All relevant clinicians must be familiar with signs and symptoms of major post-natal morbidity and retain a high index of suspicion for potential adverse outcome among women who are obese. When planning care for an obese woman, the lead professional co-ordinating care should seek appropriate advice on prophylaxis. In the latest CEMACE report [12], it was again noted that clinicians failed to act on signs of symptoms of potential major morbidity and did not appreciate the seriousness of the condition leading to delays in referral and management. Of note was the failure of midwives, community midwives and doctors to recognise the signs and symptoms of sepsis or to act on findings of routine physical examinations and observations which indicated a potential life-threatening illness. As a result of reviewing cases, where care was sub-standard, a recommendation was made by CMACE [12] that women who are unwell have their physical observations and vital signs monitored using a Modified Early Obstetric Warning Scoring system (MEOWS). The introduction of MEOWS into maternity care requires further consideration, as the documentation of signs and symptoms of life-threatening illness may not prevent an adverse event, as it is likely to be the action taken by the attending clinician that will make a difference to the outcome. Given the need to tailor care to individual need, which considers physical and psychological health, research into the effectiveness of the routine use of MEOWS is urgently required.

Commonly Experienced Morbidity

Evidence of whether women who are obese are at more risk of experiencing commonly experienced health problems after giving birth, such as backache, fatigue, urinary or faecal incontinence, depression and perineal pain, is equivocal. Epidemiological studies have considered the independent effect of factors such as parity, mode of birth, maternal age, obstetric interventions and infant birthweight on maternal morbidity, but BMI has not been routinely included which may reflect lack of accurate documentation in a woman's maternity record and previous low priority as a health concern. In a population study from Cardiff [71] which included data on over 60,000 births between 1990 and 1999, adverse outcomes among women having their first baby were compared among women with a BMI of 20–30 kg/m^2 and those with a BMI $\geq$ 30 kg/m^2. With respect to immediate post-natal problems, obese women were more at risk of urinary tract infections and having their baby admitted to the neonatal care unit with feeding difficulties. Experience of health problems following hospital discharge was not a focus of this study.

NICE [60] guidance is that contacts after 24 h of birth should include asking women about their experience of common problems including perineal pain, urinary symptoms, bowel function, fatigue, headache and back pain with advice on their management and level of referral if a woman reports a problem. In addition to physical symptoms, psychological and psychiatric health problems such as onset of depression also have to be considered. Robertson et al. [72] completed a systematic review of risk factors for the development of depression and presented outcomes in order of magnitude of effect size. They were depression or anxiety during pregnancy, life events, poor social support, and previous history of depression, neuroticism and poor marital relationship. Low socio-economic status and obstetric factors had small effect sizes. The method used to assess depression in their review was not described, other than it had to have 'proven reliability'. The NICE 2007 mental health guideline [66] found eight additional studies published since the review of Robertson et al [72] and commented that these largely support the findings of the earlier studies.

Recent studies have examined links between post-natal weight retention and psychological health. In a prospective cohort study from the United States, 850 women completed the Edinburgh Post-natal Depression Scale (EPDS) at 20 weeks gestation and 6 months post-natal [73]. An EPDS score $\geq$12 indicated probable depression. Associations between antenatal and post-natal depression and substantial weight retention (at least 5 kg) were examined at 1 year after the birth. Seven-hundred and thirty-six women (87%) were not depressed during or after pregnancy, 55 (6%) experienced antenatal depression only, 22 (3%) experienced both antenatal and post-partum depression and 37 (4%) experienced post-partum depression only. At 1 year, women retained a mean weight of 0.6 kg; 12% of women retained at least 5 kg. In multivariate logistic regression analyses, new-onset post-partum depression was associated with more than a doubling of risk of retaining at least 5 kg (OR 2.54, 95% CI, 1.06, 6.09). Antenatal depression, alone or in combination with post-partum depression, was not associated with substantial weight retention.

A meta-analysis of studies which explored associations between obesity and psychological morbidity in the general population found weak evidence of an association with obesity; however, many of the 24 included studies were methodologically poor [74]. A cross-sectional study from the United States of 491 women who were overweight or obese prior to pregnancy completed the EPDS at 6 weeks post-birth, one of a series of health measures women were asked to complete, found no association with a high EPDS score (≥13) among these women [75]. A comparison of EPDS scores with women of normal BMI was not undertaken.

Current NICE post-natal guidelines [60] recommend:

> *at each postnatal contact, women should be asked about their emotional well-being, what family and social support they have and their usual coping strategies for dealing with day-to-day matters [60, p. 6].*

The NICE antenatal and post-natal mental health guideline [66] recommends women are asked three questions at their first contact with primary care, at their booking visit, at 4–6 weeks and 3–4 months. The first two questions identify possible depression:

1. During the past month, have you often been bothered by feeling down, depressed or hopeless?
2. During the past month, have you often been bothered by having little interest or pleasure in doing things?

A third question should be considered if the woman answers 'yes' to either of the above 'Is this something you feel you need or want help with' [66] (p. 13). Evidence of the effectiveness of these questions to screen pregnant and post-natal women for depression and other mental health problems is not yet available. Services for women with mental health needs remain fragmented [76], but clinicians need to be aware of signs and symptoms and implement timely and appropriate referral if this is indicated.

Care Following Caesarean Section

The current high CS birth rate has important implications for the planning and provision of maternal post-natal care; however, there is limited evidence to support the content and duration of effective care following surgery. Prophylaxis prior to or at the time of birth could minimise post-natal morbidity, e.g. use of compression stockings, low-molecular weight heparin, early mobilisation and broad-spectrum IV antibiotics, and clinicians should ensure these are implemented [12].

Few studies have addressed the content or timing of post-natal care following CS birth [64], although there is some evidence to support surgical wound closure of obese women. Women who have more than 2 cm of subcutaneous fat should have the subcutaneous tissue space sutured to reduce the risk of wound infection and separation [12]. Leddy et al. [21] also recommended closing the subcutaneous layer to minimise seroma formation and post-operative wound disruption in obese women. They also recommended that the removal of skin sutures should be delayed until at least 7 days post-birth to allow the tissue to heal completely. The evidence base to support these recommendations was not provided.

More research into CS wound management is needed, but all women should be informed of signs and symptoms of SSI and wound dehiscence (separation of the wound edges) and to immediately contact their GP or midwife if concerned. At each contact, the wound should be observed for localised pain and erythema, local oedema, exudates, pus and offensive odour and the woman should be asked about her general health and well being. For obese women, particular emphasis should be given about the need for good hygiene, to keep the wound area clean and dry, and although not based on trial evidence, it would appear to be a good practice when washing to raise the skin folds, which could occlude the wound when sitting or standing [30].

Pain relief needs should also be discussed at each contact. If the woman has been prescribed analgesia on hospital discharge which is effective, she should be encouraged to continue to take this. If this is inadequate or she has not been prescribed analgesia, paracetamol can be taken as required up to 4 g a day divided doses, or used in combination tablets with codeine based on local prescribing policy [30]. If pain relief needs continue to be unmet, appropriate referral should be made.

Diabetes

Pre-pregnancy overweight and obesity are significant risk factors for the development of type-2 diabetes (T2DM). NICE [67] guidelines recommend that women who have a BMI of >30 kg/m^2 at pregnancy booking are at risk of gestational diabetes and should be offered testing. Data from an observational cohort study of 330 Danish women with diet-treated gestational diabetes showed that 41% of women developed diabetes during a median of 10 years follow-up [77]. This was a doubling of the risk compared to an earlier cohort of 241 women with gestational diabetes followed by the same research group 10 years before (OR 2.0, 95% CI 1.1–3.4 and OR 2.6, 95% CI 1.5–4.5, respectively).

CMACE/RCOG [13] guidelines recommend that women with a pregnancy booking BMI of ≥30 kg/m^2 and gestational diabetes who have a normal glucose tolerance test after giving birth should be regularly followed up by their GP to screen for the development of T2DM. The guidelines also recommend that these women should have an annual screening for cardio-metabolic risk factors and offered lifestyle and weight management advice.

The importance of advising overweight and obese women who have type-1 diabetes (T1DM) about weight loss as part of post-natal care has recently been highlighted to prevent future perinatal complications. A prospective population-based cohort study from Sweden compared outcomes among overweight and obese pregnant women with and without T1DM (3457 with T1DM and 764498 non-diabetic pregnancies); 35% of women with T1DM were overweight and 18% obese compared with 26% and 11%, respectively, in non-diabetic pregnancies [78]. The incidences of adverse outcome increased with greater BMI category. When compared with non-diabetic normal-weight women, the adjusted OR for obese T1DM for large for gestational age infant birthweight was 13.26 (95% CI 11.27–15.59), major malformations 4.11 (95% CI 2.99–5.65) and pre-eclampsia 14.19 (95% CI 11.50–17.50). T1DM was a significant effect modifier of the association between BMI and these outcomes ($P < 0.001$). The authors concluded that the combined effect of both T1DM and overweight or obesity constituted the greatest risk.

Infant Feeding

There is emerging evidence that among other health benefits breastfeeding confers on a woman and her baby [30], it may also protect against childhood obesity [79–81], although further evidence is required to support this. 'Healthy Weight, Healthy Lives' published by the Department of Health in 2008 [82] recognised the importance of nutrition during the early years and the contribution of breastfeeding to this. However, a number of studies have highlighted that women who are overweight or obese are less likely to commence breastfeeding than women of normal weight [83,84], and if they do commence breastfeeding, they are more likely to cease earlier than women with a normal BMI [85].

In a retrospective study from Belgium to assess if pre-pregnancy BMI influenced breastfeeding outcomes, 200 women were grouped into four categories according to pre-pregnancy BMI [84]. The incidence of intention and initiation of breastfeeding was significantly lower in underweight (64%) and obese women (68%) compared with normal weight (92%) and overweight women (80%). Fifty-two per cent of underweight, 70% of normal weight and 56% of overweight women exclusively breastfed their infant during the first month of life. This incidence was significantly lower in the obese group (34%; $P = 0{\cdot}030$). Only 40% of all infants were exclusively breastfed at 3 months of age, with the lowest prevalence among obese women ($P = 0{\cdot}001$). Median duration of any breastfeeding in the obese group was significantly shorter than in the weight groups. Reasons given for ceasing breastfeeding in the obese group were maternal complications (29%), insufficient milk supply (23%), sucking problems (21%) and work resumption (21%).

In addition to social and mechanical factors, such as difficulty with latching the baby onto the breast to account for early cessation, research has also been undertaken to explore if physiological factors affect lactogenesis and prolactin levels

[86]. In one small study, researchers studied prolactin levels among 40 women who had given birth to term babies to test the hypotheses that a reduced prolactin response to suckling and higher-than-normal progesterone concentration in the first week after delivery could be the means by which maternal BMI can compromise early lactation [87]. Serum prolactin and progesterone concentrations were measured by radioimmunoassay before and 30 min after the beginning of a suckling episode at 48 h and 7 days after the birth. Prolactin levels declined between 48 h and 7 days. Women who were overweight or obese before conception had a lower prolactin response to suckling than women who were normal weight, at 48 h but not at 7 days. After conducting multi-variate analysis and adjustment for confounding by time since the birth and duration of the breastfeeding episode, only being overweight/obese remained as a significant negative predictor of prolactin response at 7 days.

In addition to these findings, and as women who are obese are more likely to have had interventions during their labour and birth, it is imperative that clinicians have appropriate skills and competencies to support women who wish to breastfeed to successfully commence and continue to exclusively breast feed for as long as possible. Following discharge, it is important that women are informed of how to access local peer groups or lactation consultants.

Weight Management Post-Birth

Excessive weight gain and persistent weight retention during the first post-natal year are strong predictors of being overweight a decade later [88]. Gradual weight loss and graded exercise programmes could impact on subsequent pregnancy outcomes, although evidence of this is lacking. One large Swedish study found that an increase in inter-pregnancy BMI was associated with a higher risk of adverse pregnancy outcome [89]. In addition to dietary advice, women should be advised to include regular physical exercise in their daily lives. Gentle exercise such as walking can be achieved by taking their baby out in the buggy, with small intervention studies showing such exercise can be safe and promote weight loss [90].

CMACE/RCOG [13] guidelines recommend that women with a BMI $\geq 30\ kg/m^2$ should receive nutritional advice post-birth from an appropriately trained professional, with a view to weight reduction. In 2010, NICE also issued guidelines on weight management before, during and after pregnancy. With respect to post-natal care, the guidelines recommends that at the 6–8 week check, women are asked about their weight, with those women who are overweight or obese asked if they would like further advice and support. Women should have a realistic expectation of how long it may take to lose weight after pregnancy and advice on healthy eating and physical activity should be tailored to the woman's lifestyle. Women with a BMI of $30\ kg/m^2$ or over who have recently given birth should have the increased risks that obesity poses to them and, if they become pregnant again, to their unborn child ensuring that this is discussed sensitively and the level of risk clearly explained. Women should be offered a structured weight-loss programme or referral to a dietician or other appropriate health professional.

Implications for Health Service Resource Use

Evidence that additional health service costs are being incurred as a consequence of the rise in numbers of women who are obese when they become pregnant is accumulating [32,91]. Chu et al. [92] analysed data on over 13,000 pregnancies in the United States between 1 January 2000 and 31 December 2004. Women were categorised as underweight (BMI < 18.5 kg/m^2), normal (BMI 18.5–24.9 kg/m^2), overweight (BMI 25.0–29.9 kg/m^2), obese (BMI 30.0–34.9 kg/m^2), very obese (BMI 35.0–39.9 kg/m^2) or extremely obese (BMI ≥ 40.0 kg/m^2). The primary outcome was mean length of hospital stay for the birth. After adjusting for age, ethnic group, level of education and parity, mean ($\pm$SE) length of hospital stay was significantly ($P < 0.05$) greater among women who were overweight (3.7 ± 0.1 days), obese (4.0 ± 0.1 days), very obese (4.1 ± 0.1 days) and extremely obese (4.4 ± 0.1 days) than among women with normal BMI (3.6 ± 0.1 days). Most of the increase in length of stay associated with higher BMI was related to increased rates of CS births and obesity-related high-risk conditions.

Heslehurst et al. [45] interviewed 33 maternity and obstetric health care professionals with personal experience of managing the care of obese pregnant women from 16 maternity units in the North East of England. Clinicians were asked for their views of the impact of maternal obesity on their services, the facilities required to care for them and what existing services were available to care for obese women. Five dominant themes relating to service delivery were identified; booking appointments, equipment, care requirements, complications and restrictions, and current and future management of care. Other issues associated with safe care of obese women in pregnancy included appropriate resources, multidisciplinary team input to manage co-existing morbidities and restricted care options and choice for women who were obese.

It is important to consider whether provision of effective post-natal care could enhance future NHS resource use, as this could determine the level of priority accorded to this aspect of routine care by maternity providers. Evidence to support cost-effective post-natal interventions for women who are obese is needed; however, there is evidence that revisions to the universal provision of midwifery community care is likely to make better use of NHS costs. MacArthur et al.[8,9] included a cost-effectiveness analysis as part of their RCT. The comparative cost of delivering post-natal care between the trial groups was close, based on women's estimates of visit frequency. Extra intervention costs (for longer visits and the post-natal check by the intervention group midwives) were balanced by the slightly lower rates of GP home visits and fewer GP post-natal checks during the 12 months following the birth. Overall costs were equivalent indicating that the intervention model was cost-effective as maternal outcomes were better.

Implications for Safety of Maternity Care

Care of pregnant and post-natal women who are overweight or obese is influencing issues that all maternity staff should be aware of with respect to compliance with the

Clinical Negligence Scheme for Trusts (CNST) maternity standards and mandatory health and safety training. This is not only to promote safe care for women but also to ensure that staffs do not implement actions which may put their own health and safety at risk. In terms of physical care, all health professionals providing maternity care should receive training in manual handling techniques and the use of specialist equipment which may be required for women who are overweight or obese [13]. In addition, maternity units should have a central list of all facilities and equipment required to provide safe care for pregnant and post-natal women with a booking BMI ≥ 30 kg/m^2 [13]. Women with a BMI ≥ 40 kg/m^2 should have an individual documented assessment in the third trimester of pregnancy by an appropriately trained professional to determine manual handling requirements for labour and birth and, importantly in terms of post-natal health, this should include tissue viability issues [93].

The Importance of Women's Views

The views of women who use the maternity services are extremely important if we are to understand how revisions to systems and processes and to the content of care could better meet their needs [94]. A small number of qualitative studies have been published of the views of women who were overweight or obese in pregnancy, including aspects of care they found supportive or negative to meet their health or support needs and how they felt that the risk of potential pregnancy complication was presented to them. Nyman et al. [95] in a small phenomenological study from Sweden interviewed 10 women with a BMI of ≥34 kg/m^2 about their experiences of pregnancy and post-birth care. Most reported negative experiences in their encounters with health professionals, with their weight forming the focus of their contacts and the perception that they were at risk of being discriminated. In another small qualitative study, this time from the north of England which recruited a purposive sample of seven midwives and six women, two overarching themes were identified. They were 'explanations for obesity and weight management, which included lack of knowledge about weight, diet and exercise during pregnancy, difficulties with maintaining motivation for a healthy lifestyle and importance of social support, and 'best care' referring to the need for weight management to include care which was consistent, supportive and non-judgemental [96].

Furber and McGowan [97] explored the views of 19 women who had a BMI of ≥35 kg/m^2 during the third trimester of pregnancy and between 3 and 9 weeks after birth. Women reported feelings of humiliation and stigma associated with being pregnant when obese, which contacts with health professionals and members of the general public reinforced. Being classed as high risk during pregnancy was also problematic for some women, with a significant source of stress experienced if difficulties were reported during an ultrasound scan with imaging of the foetus, which were not explained to the woman. Clearly, if post-natal interventions to support obese women are to be effective and lead to positive health change, women's views are of paramount importance.

References

1. Central Midwives Board. *Handbook Incorporating the Rules of the Central Midwives Board*. 1st ed. London: Central Midwives Board; 1905.
2. Loudon I. Maternal mortality in the past and its relevance to developing countries today. *Am J Clin Nutr*. 2000;72(1):241s–246s.
3. Bhavnani V, Newburn M. *Left to Your own Devices: The Postnatal Care Experiences of 1260 First-Time Mothers*. London: The National Childbirth Trust; 2010.
4. Bick D, Bastos MH. Optimising the provision and outcome of the 'Cinderella' service: why we need to prioritise postnatal care. *Eur Obstet and Gynaecol*. 2012;7(suppl 1):22–24.
5. Nursing and Midwifery Council. *Midwives Rules and Standards*. London: Nursing and Midwifery Council; 2010.
6. MacArthur C, Lewis M, Knox EG. *Health After Childbirth*. London: The Stationery Office; 1991.
7. Glazener C, Abdalla M, Stroud P, Templeton A, Russell I. Postnatal maternal morbidity: extent, causes, prevention and treatment. *Br J Obstet Gynaecol*. 1995;102: 282–287.
8. MacArthur C, Winter HR, Bick DE, et al. *Redesigning Postnatal Care: A randomised Controlled Trial of Protocol Based, Midwifery Led Care Focused on Individual Women's Physical and Psychological Health Needs*. NHS R & D, NCCHTA; 2003.
9. MacArthur C, Winter H, Bick D, et al. Effects of redesigned community postnatal care on women's health 4 months after birth: a cluster randomised controlled trial. *Lancet*. 2002;359:378–385.
10. Lewis G. *Saving mothers' lives: reviewing maternal deaths to make motherhood safer – 2003–2005*. The Seventh Report of the Confidential Enquiries into Maternal Deaths in the United Kingdom; 2007.
11. The Information Centre. *National Maternity Statistics England 2010 – 2011*. London: The Information Centre; 2011.
12. Centre for Maternal and Child Enquiries. Saving mothers' lives: reviewing maternal deaths to make motherhood safer: 2006 – 2008. The eighth report on confidential enquiries into maternal deaths in the United Kingdom. *BJOG*. 2011;118(suppl 1): 1–203.
13. CEMACH/RCOG. *Management of Obese Women in Pregnancy*. London: CEMACH/ RCOG; 2010.
14. Centre for Maternal and Child Enquiries (CMACE). *Perinatal Mortality 2009: United Kingdom*. London: CMACE; 2011.
15. Office for National Statistics. *Statistics on Obesity, Physical Activity and Diet*. England: ONS; 2008.
16. Callaway LK, Prins JB, Chang AM, MacIntyre HD. The prevalence and impact of overweight and obesity in an Australian obstetric population. *Med J Aust*. 2006; 184:56–59.
17. Mander R, Smith GD. Saving mothers' lives (formerly why mothers die): reviewing maternal deaths to make motherhood safer 2003–2005. *Midwifery*. 2008;24(1):8–12.
18. Knight M, Kurinczuk JJ, Spark P, Brocklehurst P. Inequalities in maternal health: national cohort study of ethnic variation in severe maternal morbidities. *BMJ*. 2009;338: b542. doi:10.1136/bmj.b542.
19. Rajasingam D, Seed PT, Briley AL, Shennan AH, Poston L. A prospective study of pregnancy outcome and biomarkers of oxidative stress in nulliparous obese women. *Am J Obstet Gynecol*. 2009;200:395.e1–395.e9.

20. Stothard KJ, Tennant PW, Bell R, Rankin J. Maternal overweight and obesity and the risk of congenital anomalies: a systematic review and meta-analysis. *JAMA*. 2009;301(6): 636–650.
21. Leddy MA, Power ML, Schulkin J. The impact of maternal obesity on maternal and fetal health. *Rev Obstet Gynecol*. 2008;1(4):170–178.
22. Flenady V, Koopmans L, Middleton P, et al. Major risk factors for stillbirth in high-income countries: a systematic review and meta-analysis. *Lancet*. 2011;377(9774): 1331–1340.
23. Goldenberg RL, Kirby R, Culhane JF. Stillbirth: a review. *J Matern Fetal Neonatal Med*. 2004;16:79–94.
24. de Boo HA, Harding JE. The developmental origins of adult disease (Barker) hypothesis. *Aust N Z J Obstet Gynaecol*. 2006;46:4–14.
25. Fretts RC. Etiology and prevention of stillbirth. *Am J Obstet Gynecol*. 2005;193: 1923–1935.
26. Knight M, Kurinczuk JJ, Tuffnell D, Brocklehurst P. The UK obstetric surveillance system for rare disorders of pregnancy. *BJOG*. 2005;112:263–265.
27. Lennox C. Summary of Scottish confidential audit of severe maternal morbidity Report 2008. Appendix 2b, in Centre for Maternal and Child Enquiries (CMACE). Saving mothers' lives: reviewing maternal deaths to make motherhood safer, 2006–2008. The eighth report on confidential enquiries into maternal deaths in the United Kingdom. *BJOG*. 2011;118(suppl 1):2–203.
28. National Institute for Health and Clinical Excellence. *Weight Management Before, During and After Pregnancy – Public Health Guidelines 27*. London: NICE; 2010.
29. Homer CSE, Kurinczuk J, Spark P, Brocklehurst P, Knight M. Planned vaginal delivery or planned caesarean delivery in women with extreme obesity. *Br J Obstet Gynaecol*. 2011;118:480–487.
30. Bick D, MacArthur C, Winter H. *Postnatal Care. Evidence and Guidelines for Management*. 2nd ed. London: Churchill Livingstone; 2008.
31. Ferrazzani S, De Carolis S, Pomini F, Testa AC, Mastromarino C, Caruso A. The duration of hypertension in the puerperium of pre-eclamptic women: Relationship with renal impairment and week of delivery. *Am J Obstet Gynecol*. 1994;171: 506–512.
32. Kerrigan AM, Kingdon C. Maternal obesity and pregnancy: a retrospective study. *Midwifery*. 2010;Feb;26(1):138–146. doi:10.1016/j.midw.2008.12.005.
33. Bhattacharya S, Campbell DM, Liston WA, Bhattacharya S. Effect of body mass index on pregnancy outcomes in nulliparous women delivering singleton babies. *BMC Public Health*. 2007;7:168. doi:10.1186/1471-2458-7-168.
34. Knight M., Kurinczuk J.J., Spark P. and Brocklehurst P. on behalf of UKOSS. United Kingdom Obstetric Surveillance System (UKOSS) Annual Report 2007. Oxford: National Perinatal Epidemiology Unit; 2007.
35. Neilson J. Thrombosis and thromboembolism. Centre for Maternal and Child Enquiries (CMACE). Saving mothers' lives: reviewing maternal deaths to make motherhood safer, 2006–2008. The eighth report on confidential enquiries into maternal deaths in the United Kingdom. *BJOG*. 2011;118(suppl 1):2–203.
36. Royal College of Obstetricians and Gynaecologists. *Thrombosis and Embolism during Pregnancy and the Puerperium. Reducing the Risk. Green Top Guideline No. 37a*. London: RCOG Press; 2009.
37. Royal College of Obstetricians and Gynaecologists. *Thromboembolic Disease in Pregnancy and the Puerperium: Acute Management. Green Top Guideline No. 28*. London: RCOG Press; 2007.

38. Chu SY, Kim SY, Schmid CH, et al. Maternal obesity and risk of cesarean delivery: a metaanalysis. *Obstet Rev*. 2007;8:385–394.
39. French L, Smaill F. Antibiotic regimens for endometritis after delivery. *Cochrane Database Syst Rev*. 2004;(4)CD001067. doi: 10.1002/14651858.CD001067.pub2.
40. Redshaw M, Rowe R, Hockley C, Brocklehurst P. *Recorded Delivery: National Survey of Women's Experience of Maternity Care 2006*. Oxford: National Perinatal Epidemiology Unit, University of Oxford; 2007.
41. Johnson A, Young D, Reilly J. Caesarean section surgical site infection surveillance. *J Hosp Infect*. 2006;1:1–6.
42. Ward VP, Charlett A, Fagan J, et al. Enhanced surgical site infection surveillance following caesarean section: experience ofa multicentre collaborative post-discharge system. *J Hosp Infect*. 2008;70:166–173.
43. Welsh Healthcare Associated Infections Programme. *Orthopaedic and Caesarean Section Surgical Site Infection. Annual Surveillance Report. All Wales 2009–2010*. Cardiff: Welsh Healthcare Associated Infection Programme; 2011.
44. Acosta CD, Bhattacharya S, Tuffnell D, Kurinczuk JJ, Knight M. Maternal sepsis: a Scottish population-based case–control study. *BJOG*. 2012;119(4):474–483:doi: 10.1111/j.1471-0528.2011.03239.x.
45. Heslehurst N, Lang R, Rankin J, Wilkinson JR, Summerbell CD. Obesity in pregnancy: a study of the impact of maternal obesity on NHS maternity services. *BJOG*. 2007;114(3):334–342.
46. Morris S, Stacey M. Resuscitation in pregnancy. *BMJ*. 2003;327:1277–1279.
47. Ladner HE, Danielsen B, Gilbert WM. Acute myocardial infarction in pregnancy and the puerperium: a population-based study. *Obstet Gynecol*. 2005;105(3): 480–484.
48. Knight M, Kurinczuk JJ, Spark P, Brocklehurst P, on behalf of UKOSS. United Kingdom Obstetric Surveillance System (UKOSS) Annual Report 2008. National Perinatal Epidemiology Unit, Oxford.
49. *UK Obstetric Surveillance* (*UKOSS*). Cardiac arrest in pregnancy. <www.npeu.ox.ac.uk/ukoss/current-surveillance/cap/>; Accessed 20.04.12.
50. World Health Organization. *Vaginal bleeding after childbirth. Managing Complications in Pregnancy and Childbirth: A Guide for Midwives and Doctors*. Geneva: World Health Organization; 2003.
51. Stone RW, Paterson CM, Saunders NJ. Risk factors for major obstetric haemorrhage. *Eur J Obstet Gynecol Reprod Biol*. 1993;48:15–18.
52. Waterstone M, Bewley S, Wolfe C. Incidence and predictors of severe obstetric morbidity: case–control study. *BMJ*. 2001;322:1089–1094.
53. Mousa HA, Alfirevic Z. Major postpartum hemorrhage: survey of maternity units in the United Kingdom. *Acta Obstet Gynecol Scand*. 2002;81(8):727–730.
54. Schrauwers C, Dekker G. Maternal and perinatal outcome in obese pregnant patients. *J Matern Fetal Neonatal Med*. 2009;22(3):218–226.
55. Roberts CL, Ford B, Algert C, et al. Trends in adverse maternal outcomes during childbirth: a population-based study of severe maternal morbidity. *BMC Pregnancy Childbirth*. 2009;9:7. doi:10.1186/1471-2393-9-7.
56. Penny G, Kernaghan D, Brace V. Severe maternal morbidity – the Scottish experience 2003 to 2005. In: Lewis G, ed. *Saving Mothers' Lives: Reviewing Maternal Deaths to Make Motherhood Safer – 2003–2005. The Seventh Report of the Confidential Enquiries into Maternal Deaths in the United Kingdom*; London: CEMACH, 2007:248–253.

57. Pattinson RC, Buchmann E, Mantel GD, Schoon M, Rees H. Can enquiries into severe acute maternal morbidity act as a surrogate for maternal death enquiries? *BJOG*. 2003;110:889–893.
58. Royal College of Obstetricians and Gynaecologists. *High Quality Women's Healthcare: A Proposal for Change*. London: RCOG Press; 2011.
59. Bick D, Murrells T, Rose V, Weavers A, Wray J, Beake S. Improving postnatal outcomes using continuous quality improvement: a pre and post intervention study in one English maternity unit. *BMC Pregnancy and Childbirth*. 2012;Jun 6;12(1):41 [Epub ahead of print].
60. National Institute for Health and Clinical Excellence. *Postnatal Care: Routine Postnatal Care of Women and their Babies – Clinical Guidelines 37*. London: NICE; 2006.
61. National Institute for Health and Clinical Excellence. *Antenatal Care. Routine Care for the Healthy Pregnant Woman – Clinical Guidelines 62*. 2nd ed. NICE; 2008.
62. National Institute for Health and Clinical Excellence. *Intrapartum Care. Care of Healthy Women and their Babies During Childbirth – Clinical Guidelines 55*. NICE; 2007.
63. National Institute for Health and Clinical Excellence. *Caesarean Section – Clinical Guidelines 13*. NICE; 2004.
64. National Institute for Health and Clinical Excellence. *Caesarean Section – Clinical Guidelines 132*. NICE; 2011.
65. National Institute for Health and Clinical Excellence. *Diabetes in Pregnancy: Management of Diabetes and its Complications from Pre-conception to the Postnatal Period – Clinical Guidelines 63*. NICE; 2008.
66. National Institute for Health and Clinical Excellence. *Antenatal and Postnatal Mental Health: Clinical Management and Service Guidance – Clinical Guidelines 45*. NICE; 2007.
67. National Institute for Health and Clinical Excellence. *Obesity. The Prevention, Identification, Assessment and Management of Overweight and Obesity in Adults and Children – Clinical Guidelines 43*. NICE; 2006.
68. National Institute for Health and Clinical Excellence. *Surgical Site Infection. Prevention and Treatment of Surgical Site Infection – Clinical Guidelines 74*. NICE; 2008.
69. Beake S, McCourt C, Bick D. Women's views of hospital and community-based postnatal care: the good, the bad and the indifferent. *Evid Based Midwifery*. 2005;3(2):80–86.
70. Bick D, Weavers A, Rose V, Wray J, Beake S. Improving inpatient postnatal services: midwives views and perspectives of engagement in a quality improvement initiative. *BMC Health Serv Res*. 2011;11(1):293.
71. Usha Kiran TS, Hemmadi S, Bethel J, Evans J. Outcome of pregnancy in a woman with an increased body mass index. *BJOG*. 2005;112(6):768–772.
72. Robertson E, Grace S, Wallington T, Stewart DE. Antenatal risk factors for postpartum depression: a synthesis of recent literature. *Gen Hosp Psychiatry*. 2004;26:289–295.
73. Herring SJ, Rich-Edwards JW, Oken E, Rifas-Shiman SL, Kleinman KP, Gillman MW. Association of postpartum depression with weight retention 1 year after childbirth. *Obesity*. 2008;16(6):1296–1301.
74. Atlantis E, Baker M. Obesity effects on depression: systematic review of epidemiological studies. *Int J Obes*. 2008;Jun;32(6):881–91.
75. Krause KM, Ostbye T, Swarmy GK. Occurrence and correlates of postpartum depression in overweight and obese women: results from the active mothers postpartum (AMP) study. *Matern Child Health J*. 2009;13(6):832–838.
76. Rowan C, Bick D. A survey of perinatal mental health services in two English strategic health authorities. *Evid Based Midwifery*. 2008;6(4):76–82.

77. Lauenborg J, Hansen T, Jensen DM, et al. Increasing incidence of diabetes after gestational diabetes: a long-term follow-up in a Danish population. *Diabetes Care*. 2004;27(5):1194–1199.
78. Persson M, Pasupathy D, Hanson U, Westgren M, Norman M. Pre-pregnancy body mass index and the risk of adverse outcome in type 1 diabetic pregnancies: a population-based cohort study. *BMJ Open*. 2012;2(1):e000601.
79. Arenz S, Ruckerl R, Koletzko B, et al. Breastfeeding and childhood obesity: a systematic review. *Int J Obes*. 2004;28:1247–1256.
80. Fewtrell MS. The long-term benefits of having been breast-fed. *Curr Paediatr*. 2004;14:97–103.
81. Karaolis-Danckert N, Buyken AE, Kulig M, et al. How pre- and postnatal risk factors modify the effect of rapid weight gain in infancy and early childhood on subsequent fat mass development: results from the Multicenter Allergy Study. *Am J Clin Nutr*. 2008;87(5):1356–1364.
82. Department of Health, Department for Children. *Healthy Weight, Healthy Lives: A Cross-Government Strategy for England*. London: Department of Health; Department for Children, Schools and Families; 2008.
83. Amir LH, Donath S. A systematic review of maternal obesity and breastfeeding intention, initiation and duration. *BMC Pregnancy Childbirth*. 2007;7:9.
84. Guelinckx I, Devlieger R, Bogaerts A, Pauwels S, Vansant G. The effect of pre-pregnancy BMI on intention, initiation and duration of breast-feeding. *Public Health Nutr*. 2012;15(5):840–848.
85. Baker JL, Michaelsen KF, Sørensen IA, Rasmussen KA. High pre-pregnant body mass index is associated with early termination of full and any breastfeeding in Danish women. *Am J Clin Nutr*. 2007;86:404–411.
86. Hilsson JA, Rasmussen KM, Kjolhede CL. High prepregnant body mass index is associated with poor lactation outcomes among white, rural women independent of psychosocial and demographic correlates. *J Hum Lact*. 2004;20:18–29.
87. Rasmussen KM, Kjolhede CL. Prepregnant overweight and obesity diminish the prolactin response to suckling in the first week postpartum. *Pediatrics*. 2004;113:e465–e471.
88. Rooney B, Schauberger C. Excess pregnancy weight gain and long-term obesity: one decade later. *Obstet Gynecol*. 2002;100:245–252.
89. Villamor E, Cnattingius S. Interpregnancy weight change and risk of adverse pregnancy outcomes: a population-based study. *Lancet*. 2006;368:1164–1170.
90. Oken E, Taveras EM, Folasade AP, Rich-Edwards JW, Gillman MW. Television, walking, and diet: associations with postpartum weight retention. *Am J Prev Med*. 2007;32(4):305–311.
91. Lashen H, Fear K, Sturdee DW. Obesity is associated with increased risk of first trimester and recurrent miscarriage: matched case-control study. *Hum Reprod*. 2004;19:1644–1646.
92. Chu SY, Bachman DJ, Callaghan WM, et al. Association between obesity during pregnancy and increased use of health care. *N Engl J Med*. 2008;358:1444–1453.
93. NHS Litigation Authority. *Clinical Negligence Scheme for Trusts: Maternity Clinical Risk Management Standards*. London: NHS Litigation Authority; 2012.
94. Beake S, Rose V, Bick D, Weavers A, Wray J. A qualitative study of women's experiences and expectations of in-patient postnatal care in one English maternity unit. *BMC Pregnancy Birth*. 2010;27;10:70 <http://www.biomedcentral.com/1471-2393/10/70/>.

Section 7

Interventions to Improve Care of Women During Pregnancy

31 Anti-Obesity Drugs for Obese Women Planning Pregnancy

Hang Wun Raymond Li[1], *Chin Peng Lee*[1], *Karen Siu Ling Lam*[2] *and Pak Chung Ho*[1]

[1]Department of Obstetrics and Gynaecology, The University of Hong Kong, Queen Mary Hospital, Hong Kong, People's Republic of China, [2]Department of Medicine, The University of Hong Kong, Queen Mary Hospital, Hong Kong, People's Republic of China

Introduction

Obesity is a highly prevalent health problem worldwide. For women in the reproductive age, it is one of the major concerns as discussed in other chapters. Apart from general health implications, obesity is associated with reproductive problems such as anovulatory sub-fertility, miscarriage and increased obstetric complications. Women who are planning for pregnancy are, therefore, generally advised to optimise their body weight before contemplating pregnancy.

In general, lifestyle modifications including diet control and exercise are the first-line management. However, a considerable number of obese individuals fail to lose significant weight (i.e. 10% of their initial weight) on lifestyle modifications alone. In such situations, anti-obesity drugs may be considered as an adjunctive therapy to lifestyle modifications [1]. It has been suggested in randomised controlled trials that a combination of medication and lifestyle-behavioural modifications is more effective than either intervention alone in weight reduction [2,3].

Classification of Anti-Obesity Drugs

Anti-obesity drugs can be classified into three main categories according to their mode of action [4–6] and are as follows:

1. Drugs inhibiting intestinal fat absorption

   Orlistat is the only approved anti-obesity medication which belongs to this category.

Obesity. DOI: http://dx.doi.org/10.1016/B978-0-12-416045-3.00031-5

2. Drugs suppressing food intake

   This includes medications which modulate the production of neurotransmitters or act on their receptors in the central nervous system so as to suppress appetite. Examples include the following:

   a. Noradrenergic drugs: phenylpropanolamine, amphetamine, phentermine and diethylpropion.
   b. Serotoninergic drugs: dexfenfluramine, fenfluramine.
   c. Serotoninergic and adrenergic drug: sibutramine.
   d. Selective cannabinoid type-1 (CB1) receptor antagonist: rimonabant.

3. Drugs increasing energy consumption and thermogenesis

   Ephedrine, caffeine and thyroxine are medications which have such effect. However, their use as anti-obesity agents is not recommended due to their limited evidence in weight reduction and cardiovascular side effects.

Drugs for Weight Reduction in Reproductive-Age Women: Pharmacology and Clinical Efficacy

Orlistat, sibutramine and rimonabant are the three most studied pharmacological agents for long-term treatment of obesity. Orlistat is the only currently available and approved option for long-term use, while phentermine and diethylpropion are approved for short-term use. These are elaborated in the below sections. Metformin is a commonly used insulin sensitiser which is not classically considered as an anti-obesity drug. However, it has been studied for its effect in weight reduction especially in reproductive-age women and is hence discussed here as well.

Phenylpropanolamine has been withdrawn from the market since an increased risk of haemorrhagic stroke has been demonstrated. Amphetamine is not recommended for anti-obesity treatment because of its euphoric and addictive action. Dexfenfluramine and fenfluramine have been withdrawn due to an increased risk of primary pulmonary hypertension and cardiac valvular pathology. These agents will not be further elaborated.

Orlistat (Xenical®)

Orlistat acts as a reversible inhibitor of the gastric and pancreatic lipase and hence inhibits the hydrolysis of triglycerides into free fatty acids, with the result that dietary triglycerides are excreted undigested. This reduces fat absorption by about 30% at the usual therapeutic dose. Orlistat is available in the European Union, United States and Australia as a non-prescription medication. Generic formulations are available in some countries. The recommended dose is 120 mg three times a day with meal. Higher doses do not give better clinical effects. It mainly acts locally in the gastrointestinal (GI) lumen and is predominantly excreted in faeces. Systemic absorption is minimal. The half-life of the systemically absorbed drug is about 1–2 h.

Orlistat is one of the most studied pharmacological agents for its clinical efficacy on weight reduction. In the study on the use of Xenical in the Prevention of Diabetes in Obese Subjects (XENDOS study) [7], 3305 subjects were randomised to lifestyle changes plus orlistat or placebo. Compared to the placebo (lifestyle changes only) group, subjects in the orlistat group had significantly greater weight reduction (5.8 kg vs 3.0 kg, $P < 0.001$) and lower cumulative incidence of type-2 diabetes mellitus (6.2% vs 9.0%, $P = 0.0032$) after 4 years of treatment. A meta-analysis of 29 studies on the use of orlistat demonstrated significant weight reduction by 2.59 and 2.89 kg after 6 and 12 months of treatment, respectively [8]. A more recent Cochrane review including 16 trials on long-term orlistat therapy showed a placebo-subtracted weight reduction of 2.9 kg (95% CI 2.5–3.2) and a significantly higher likelihood in achieving 5–10% weight loss in patients receiving treatment with orlistat. Subjects treated with orlistat also showed significant reductions in total, HDL and LDL cholesterol, blood pressure and incidence of diabetes mellitus [9,10]. However, there has been no large clinical trial addressing cardiovascular end-points.

The most common side effect is GI symptoms including steatorrhoea, flatus with discharge, faecal urgency, increased defaecation and faecal incontinence. These GI effects may subside in 1–4 weeks. Users are recommended to avoid diet with high fat content, which could aggravate the GI side effects. As orlistat can reduce the absorption of fat-soluble vitamins (vitamins A, D, E, K and beta-carotene), patients taking orlistat are recommended to take multi-vitamin supplements at bedtime or at least 2 h before taking orlistat.

The use of orlistat is contraindicated in patients with chronic malabsorption syndrome, cholestasis and those with known hypersensitivity to this product. Orlistat may reduce the plasma level of cyclosporin, and concurrent administration should be avoided. It can also impair the absorption of amiodarone.

Sibutramine

Sibutramine is a serotonin and noradrenaline reuptake inhibitor which was initially developed as an anti-depressant. Because of its centrally-acting effect in enhancing satiety and hence suppressing food intake and its effect in increasing basal metabolic rate and energy expenditure, it has been studied for its anti-obesity effect. It undergoes extensive first-pass metabolism in the liver and is excreted in the urine.

Sibutramine was approved by the United States Federation of Drug Administration (US FDA) as a long-term anti-obesity drug in 1997. Meta-analysis has shown that sibutramine significantly reduces body weight by 4.2% (95% CI 3.6–4.8) after 12 months of treatment [10]. It also improves plasma glucose and lipid profile [11]. In a randomised controlled trial, sibutramine alone was less effective than sibutramine combined with lifestyle modification in weight reduction (5.0 vs 12.1 kg) [2].

More common side effects include dry mouth, headache, insomnia and constipation [11]. It does not increase the release of serotonin and is hence not associated with an increased risk of valvular heart disease or pulmonary hypertension [12]. However, it may increase blood pressure and heart rate [11]. The Sibutramine Cardiovascular Outcomes (SCOUT) trial showed a significantly increased risk of major cardiovascular events by 16% (95% CI 1.03–1.31) [13]. The controversy lies in the fact that the SCOUT trial was carried out in individuals with additional cardiovascular risk factor other than obesity, and yet the all-cause mortality was not significantly higher (hazard ratio 1.04, 95% CI 0.91–1.20). The safety profile in obese patients without other co-existing cardiovascular risk factors remains uncertain. Nonetheless, despite its proven effectiveness in weight reduction, the results from the SCOUT trial led the US FDA to issue a safety announcement on 10 August 2010, with the eventual discontinuation of sibutramine in all countries subsequently.

Rimonabant

Rimonabant is the first developed drug which acts as a selective antagonist of the CB1 receptor. It acts centrally to suppress food intake and peripherally in adipose tissue, liver, muscles and the GI tract to regulate lipid and glucose metabolism. Clinical trials have shown a significant effect of rimonabant in reducing weight as well as a weight-independent effect in lowering triglyceride and raising HDL cholesterol levels. It reduces the development of metabolic syndrome. Rimonabant has never been approved by the US FDA. The European Medicines Agency approved the use of rimonabant in anti-obesity treatment in adjunct to diet and exercise in 1996, but subsequently it recommended the suspension of its use in October 2008 due to reports of significant psychiatric side effects including anxiety and depression [14].

Metformin (Glucophage®)

Metformin is a biguanide insulin-sensitising agent which potentiates insulin sensitivity in peripheral tissues. It also reduces glucose production from the liver and absorption from the intestine. It does not increase insulin production. It is used as one of the first-line pharmacological treatment for diabetes mellitus.

Metformin does not undergo systemic metabolism and is mainly excreted unchanged in the urine. The elimination half-life from the circulation is about 18 h. The recommended dose range is from 500 mg twice daily or 850 mg once daily up to 2550 mg per day in divided doses. It can be started at a low initial dose and stepped up as tolerated to achieve optimal effect. Most common adverse effects include diarrhoea, nausea, vomiting and flatulence. Hypoglycaemia is not a side effect except in hypocaloric situations or when co-administered with other hypoglycaemic agents. Lactic acidosis is a very rare but life-threatening complication which primarily occurs in diabetic patients with renal insufficiency.

Although it is not classically considered as an anti-obesity drug, there is evidence that it may contribute to modest weight loss, particularly in obese diabetic patient [15]. A recent meta-analysis [16] showed that metformin treatment resulted in a statistically significant but modest decrease in body mass index (BMI) compared to placebo with a mean difference of 0.68 kg/m^2; subgroup analysis showed significant difference only with metformin of higher dose (>1500 mg/day) but not with lower dose (≤500 mg/day) and only in those treated for more than 8 weeks.

Significant BMI reduction was not observed in those with concurrent lifestyle modification but occurred in those without. There is hence a possibility that metformin may not have additional weight reduction effect on top of lifestyle intervention, although there was vast heterogeneity in the exact form and implementation of lifestyle intervention among different studies so that such inference may not be appropriate.

Phentermine and Diethylpropion

Phentermine and diethylpropion are noradrenergic sympathomimetic agents, but unlike amphetamine, they are not dopaminergic [17]. It has been shown in a meta-analysis that both resulted in modest weight loss in combination with lifestyle modifications. The mean weight loss at 6 months was 3.6 kg (95% CI 0.6–6.0) and 3.0 kg (95% CI −1.6 to 11.5) for phentermine and diethylpropion, respectively [18]. Common side effects include tachycardia, headache, raised blood pressure, insomnia and irritability. They should be used with caution in hypertensive patients. Phentermine is approved by the US FDA as an anti-obesity drug only for short term (up to 12 weeks) [4,5]. The recommended dose for phentermine is 15–37.5 mg daily before breakfast or 10–14 h before bedtime.

Pharmacological Anti-Obesity Treatment in Women Contemplating Pregnancy

Indications of Pharmacological Treatment: A Review of Evidence

With regard to women planning for pregnancy, the use of anti-obesity drugs may be theoretically applied in the following situations:

1. Obese women who would otherwise have increased risk of poor reproductive outcome and obstetric complications;
2. Women with anovulatory sub-fertility in whom weight reduction may help to restore ovulation or improve the ovarian response to ovulation induction treatment;
3. Women requiring assisted reproduction treatment in whom optimisation of body weight before treatment may help to optimise the response to ovarian stimulation and chance of success.

Despite the ample evidence showing the adverse effects of maternal obesity on reproductive outcome, there has been no reported study on the efficacy of anti-obesity drug in reducing the rate of miscarriage or obstetric complications in obese women who are pregnant or planning conception. Hence, the use of pharmacological agents in such women who find difficulty in achieving the goal of weight reduction by lifestyle modifications alone remains entirely empirical.

There have been limited studies on the use of anti-obesity drugs in obese women suffering from anovulatory sub-fertility. An open-labelled randomised controlled trial in 40 subjects randomised to either metformin or orlistat demonstrated a statistically significant reduction in BMI and waist circumference after treatment with both agents. However, at 3 months post-treatment, the actual magnitudes of change in BMI was small (metformin group: 38.4–37.7 kg/m^2; orlistat group: 39.6–37.6 kg/m^2; $P<0.001$) and that in waist circumference was modest only (metformin group: 106.7–100.6 cm; orlistat group: 109.1–102.2 cm; $P<0.0001$). There was no significant difference in the change in BMI, waist circumference and androgen profile between the two treatment groups. The ovulation rates in the metformin and orlistat groups were 40% and 25% respectively, but probably the sample size was too small and duration of treatment too short for a sound conclusion [19]. Similar conclusions were reported in another study of similar design which focused specifically on women with polycystic ovary syndrome [20].

There is currently no data on the role of anti-obesity drugs in modifying the treatment outcomes in obese women who are undergoing assisted reproduction treatment. There is evidence that in women with the polycystic ovary syndrome, metformin co-treatment during in vitro fertilisation cycle significantly reduces the risk of ovarian hyperstimulation syndrome [21] which offers an advantage and is a recommended practice, but this is probably outside the context of weight reduction.

Safety Data in Pregnancy

Orlistat is classified by the US FDA as category B for use in pregnancy. Studies in rats and rabbits using orlistat at doses up to 800 mg/kg/day did not reveal any embryo toxicity or teratogenicity. There have been no adequate well-controlled studies for the use of orlistat in human pregnancy, and hence its use in pregnant or breastfeeding women is not recommended by the drug manufacturer, and no sound recommendation can be given to women contemplating pregnancy either [8,22]. However, due to its minimal systemic absorption and extremely low bioavailability in the circulation (<1%), its safety profile in pregnancy is theoretically favourable [22].

Metformin is assigned by the US FDA as category B for use in pregnancy. There are no adequate well-controlled studies for its use in human pregnancy. Animal studies using metformin up to 600 mg/kg/day, equivalent to about 2–6 times the body surface area-adjusted maximum recommended dose in human, showed no teratogenic effect. There is evidence from a number of observational studies that the use of metformin during pregnancy is safe and does not impose any adverse foetal effects [23]. Therefore, metformin is suggested as a possible safe treatment option for obese women planning for pregnancy [16].

Phentermine is categorised as category C by the US FDA for use in pregnancy. There have been no reported animal or human studies on its safety in pregnancy.

Timing of Intervention

Because of possible adverse foetal and neonatal effects (e.g. preterm delivery, neural tube defects and increased diabetic risk in lifetime) of acute weight loss and malnutrition during pregnancy, it is recommended that any weight loss intervention should take place before contemplating conception or fertility treatment and not during pregnancy. Contraception can be recommended during the time of active weight loss intervention [24,25].

Concluding Remarks

Lifestyle modifications including dietary control and exercise remain as the first-line treatment measures for weight reduction. In individuals who experience difficulty in reducing significant weight with lifestyle intervention alone, the use of anti-obesity drugs is a reasonable adjunctive measure. Orlistat and metformin are the options currently, and their use is probably safe for women planning for pregnancy. These weight loss interventions should be adopted before the women actually contemplate conception or fertility treatment.

References

1. National Heart, Lung, and Blood Institute (NHLBI) and the North American Association for the Study of Obesity (NAASO). *The Practical Guide: Identification, Evaluation, and Treatment of Overweight and Obesity in Adults*. Bethesda, MD: National Institutes of Health; 2000.
2. Wadden TA, Berkowitz RI, Womble LG, et al. Randomized trial of lifestyle modification and pharmacotherapy for obesity. *N Engl J Med*. 2005;353:2111–2120.
3. Poston WS, Haddock CK, Pinkston MM, et al. Evaluation of a primary care-oriented brief counseling intervention for obesity with and without orlistat. *J Intern Med*. 2006;260(4):388–398.
4. Li M, Cheung BMY. Pharmacotherapy for obesity. *Br J Clin Pharmacol*. 2009; 68(6):804–810.
5. Kaplan LM. Pharmacologic therapies for obesity. *Gastroenterol Clin North Am*. 2010;39:69–79.
6. Li MF, Cheung BMY. Rise and fall of anti-obesity drugs. *World J Diabetes*. 2011; 2(2):19–23.
7. Torgerson JS, Hauptman J, Boldrin MN, Sjostrom L. XENical in the prevention of diabetes in obese subjects (XENDOS) study: a randomized study of orlistat as an adjunct to lifestyle changes for the prevention of type 2 diabetes in obese patients. *Diabetes Care*. 2004;27:155–161.
8. Li Z, Maglione M, Tu W, et al. Meta-analysis: pharmacologic treatment of obesity. *Ann Intern Med*. 2005;142:532–546.
9. Padwal RS, Majumdar SR. Drug treatments for obesity: orlistat, sibutramine, and rimonabant. *Lancet*. 2007;369:71–77.
10. Rucker D, Padwal R, Li SK, Curioni C, Lau DC. Long-term pharmacotherapy for obesity and overweight: updated meta-analysis. *BMJ*. 2007;335:1194–1199.
11. Nisoli E, Carruba MO. A benefit-risk assessment of sibutramine in the management of obesity. *Drug Saf*. 2003;26(14):1027–1048.

12. Gardin JM, Schumacher D, Constantine G, Davis KD, Leung C, Reid CL. Valvular abnormalities and cardiovascular status following exposure to dexfenfluramine or phentermine/fenfluramine. *JAMA*. 2000;283:1703–1709.
13. James WP, Caterson ID, Coutinho W, et al. Effect of sibutramine on cardiovascular outcomes in overweight and obese subjects. *N Engl J Med*. 2010;363(10):905–917.
14. Samat A, Tomlinson B, Taheri S, Thomas GN. Rimonabant for the treatment of obesity. *Recent Pat Cardiovasc Drug Discov*. 2008;3:187–193.
15. Meneghini LF, Orozco-Beltran D, Khunti K, et al. Weight beneficial treatments for type 2 diabetes. *J Clin Endocrinol Metab*. 2011;96(11):3337–3353.
16. Nieuwenhuis-Ruifrok AE, Kuchenbecker WKH, Hoek A, Middleton P, Norman RJ. Insulin sensitizing drugs for weight loss in women of reproductive age who are overweight or obese: systematic review and meta-analysis. *Hum Reprod Update*. 2009;15(1):57–68.
17. Bray GA. Use and abuse of appetite-suppressant drugs in the treatment of obesity. *Ann Intern Med*. 1993;119:707–713.
18. Haddock CK, Poston WS, Dill PL, Foreyt JP, Ericsson M. Pharmacotherapy for obesity: a quantitative analysis of four decades of published randomized clinical trials. *Int J Obes Relat Metab Disord*. 2002;26:262–273.
19. Metwally M, Amer S, Li TC, Ledger WL. An RCT of metformin versus orlistat for the management of obese anovulatory women. *Hum Reprod*. 2009;24(4):966–975.
20. Ghandi S, Aflatoonian A, Tabibnejad N, Moghaddam MHS. The effects of metformin or orlistat on obese women with polycystic ovary syndrome: a prospective randomized open-label study. *J Assist Reprod Genet*. 2011;28:591–596.
21. Tso LO, Costello MF, Albuquerque LE, Andriolo RB, Freitas V. Metformin treatment before and during IVF or ICSI in women with polycystic ovary syndrome. *Cochrane Database Syst Rev*. 2009;(2):CD006105.
22. Metwally M, Li TC, Ledger WL. The impact of obesity on female reproductive function. *Obesity Rev*. 2007;8:515–523.
23. Feig DS, Moses RG. Metformin therapy during pregnancy: good for the goose and good for the gosling too? *Diabetes Care*. 2011;34:2329–2330.
24. Nelson SM, Fleming RF. The preconceptual contraception paradigm: obesity and infertility. *Hum Reprod*. 2007;22(4):912–915.
25. Cohen JH, Kim H. Sociodemographic and health characteristics associated with attempting weight loss during pregnancy. *Prev Chronic Dis*. 2009;6(1):A07.

32 Anti-Obesity Surgery for Women Planning Pregnancy?

Kavita Deonarine, Dilip Dan and Surujpal Teelucksingh

Departments of Clinical Medical and Surgical Sciences, University of the West Indies, St Augustine, Trinidad and Tobago, West Indies

Introduction and Background

'Man is a singular creature. He has a set of gifts which make him unique among the animals: so that, unlike them, he is not a figure in the landscape – he is the shaper of the landscape. His imagination, his reason, his emotional subtlety and toughness, make it possible for him not to accept the environment but to change it' [1]. But what if the environment changes rapidly?

A steady supply of relatively cheap, processed and reconstituted foods provides a perpetual abundance of calories. This coupled with diminishing physical exertion induced by societal and technological changes has shifted the energy equation in favour of storage. One of the consequences of excessive calorie intake is a set of metabolic aberrations, commonly referred to as the metabolic syndrome [2], with the potential to affect the life cycle in profound ways and beginning as early as intrauterine life. The animal empowered with the imagination to shape his or her environment to suit their fancy now faces a new challenge. The environment which he or she has created and in turn, his or her behavioural adaptations, or indeed maladaptation, has produced a profound change in his or her own shape. Excessive calories stored as abdominal fat produces the metabolic syndrome with 'an apple–shaped individual' with which metabolic, inflammatory and atherosclerotic consequences such as diabetes, dyslipidaemia and myocardial infarction but also cancers and a number of reproductive anomalies, most notably polycystic ovarian syndrome (PCOS) are associated.

The Global Obesity Epidemic

The obesity epidemic is spiralling out of control globally. The World Health Organization estimated that the number of people with obesity has doubled since

Obesity. DOI: http://dx.doi.org/10.1016/B978-0-12-416045-3.00032-7

the 1980s. In 2008 there were 200 million obese men and nearly 300 million obese women worldwide [3]. Surprisingly, much of this burden has been contributed by developing nations – a shift away from the traditional axis of chronic disease prevalence originally located among industrialised nations. Common to both developed and developing countries is the higher prevalence of obesity among women folk [3]. Thus, prevalence of obesity rose from 9% in the 1990s to 16% in 2000 in England with 10.9% of pregnant women recorded as obese, and obesity was present in 35% of maternal deaths [4–6].

Obesity and Reproductive Function

The impact of obesity on health is similar for both men and women with increased risk for type 2 diabetes, cardiovascular disease, degenerative arthritis and obstructive sleep apnoea. So are the profound effects on reproductive function including psychosexual dysfunction and reduced fertility. In females, however, the dysfunction can appear early in life, and precocious thelarche, pubarche and puberty are all associated with childhood obesity [7]. It is associated with PCOS, an increased risk of complicated pregnancy against a backdrop of subfertility, later onset of menopause and increased risk of some forms of gynaecologic cancers [8–12].

The Effects of Obesity on Pregnancy

Obesity affects ovulation, decreases fertility, increases obstetric risks and is associated with poorer neonatal outcomes. Table 32.1 [13] summarises the many adverse

Table 32.1 Risk of Pregnancy Complications with Obesity Expressed as Odds Ratio

Miscarriage	3.05
Spina bifida	3.5
Omphalocele	3.3
Heart defects	2.0
Gestational diabetes	3.6
Pre-eclamptic toxaemia	2.14
Macrosomia	2.36
Intrauterine death	1.4
Induction of labour	1.70
Caesarean section	1.83
Postpartum haemorrhage	1.39

effects of obesity on maternal and foetal outcomes. These risks are strikingly similar to those with poorly controlled gestational or pregestational diabetes but it should be noted that the increased risks indicated here persist even when the contribution from diabetes, per se, is excluded [14–16].

Weight loss results in significant improvement in pregnancy and ovulation rates in anovulatory women. How do we go about achieving this in the obese female?

Non-surgical measures achieve and sustain weight loss in a limited number of patients. Studies suggest that metformin has no effect on body mass index (BMI) on its own and that it should be used in conjunction with lifestyle changes to achieve weight loss in patients with PCOS [17–20]. The surgical approach to weight loss results in more significant, effective and prolonged weight loss. In the Swedish Obese Subjects study, the mean weight loss was maximal after 1–2 years (gastric bypass, 32%; vertical banded gastroplasty, 25% and banding, 20%). While those on conventional weight-loss measures only had ±2% weight change [21] (Figure 32.1).

In obese females weight loss leads to improvement in insulin insensitivity, recovery of ovulation and an increase in fertility. Table 32.2 compares the impact of metformin with bariatric surgery upon aspects of female sexual and reproductive outcomes in patients with PCOS.

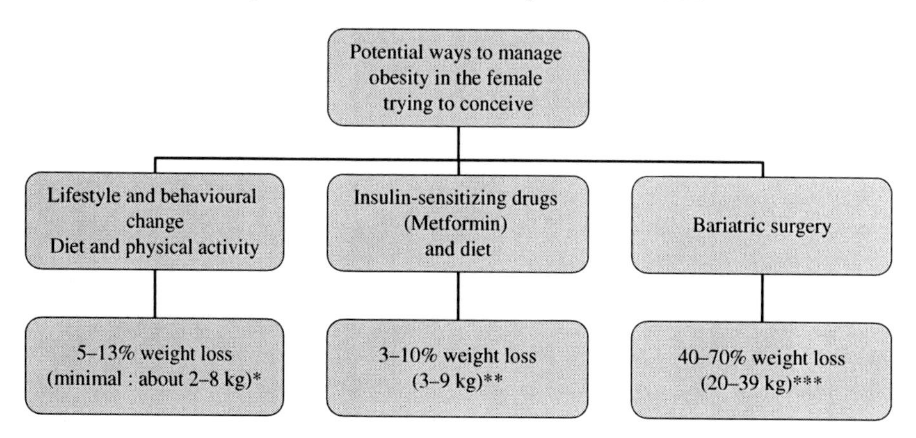

Figure 32.1 Interventions to manage obesity in the female trying to conceive and their respective weight-loss outcomes. * Refs. [22–26]; ** Refs. [17,19,27,28]; ***Refs. [21,29–31].

Table 32.2 How Metformin and Bariatric Surgery Compare in Affecting Some Aspects of Sexuality/Fertility in Patients with PCOS

	Metformin	Surgery
Psychosexual	Patients noticed increased frequency of sexual intercourse, more satisfaction with their sex life, less pain during sex, lowered impact of excessive body hair on sexuality [32]	There is a higher frequency of sexual difficulties in patients seeking bariatric surgery [33]. Higher BMI is associated with greater impairment of dysfunction. Female sexual dysfunction resolved in 68% of patients undergoing bariatric surgery [34]
PCOS	Improves the symptoms of PCOS but does not produce uniform clinical responses in patients with PCOS [35]	Almost complete improvement/ resolution of clinical and biochemical hyperandrogenism (hirsutism, decreased androgen levels, decrease in insulin resistance) [36]
Menstruation and ovulation	Achieved ovulation in 46% of patients with PCOS [37]	Recovery of regular and/or ovulation menstrual cycles (>70% resolution) [38]
Fertility and pregnancy rate	Increased fertility rates especially if clomiphene citrate is added [37]	Higher fertility rates [39], pregnancy rate in adolescents post-bariatric surgery is double the rate of the general population (12.8 vs 6.4%) [40,41]

Bariatric Surgery: Overview and Indications

Bariatric surgery is being increasingly used for the treatment of severe obesity. The success of bariatric surgery is not only reflected in weight loss but also in resolution of the metabolic disorders associated with obesity [21] and the terminology 'Metabolic Surgery' is being increasingly used. Data from United Sates have demonstrated a sixfold increase over a 7-year period (1998–2005) [39]. This is likely to be further propelled by the push factor of increasing obesity combined with the pull factors of diminishing costs and reducing complication rates associated with obesity surgery. The Department of Health and Human Services' Agency for Healthcare Research and Quality (AHRQ) found that between 2001 and 2006 the complication rate declined by 21%, re-admissions rate declined by 31% and hospital stay declined from 6 to 3.7 days with a 13% drop in cost [42,43].

NICE clinical guideline on obesity [44] recommends bariatric surgery if:

- BMI of 40 kg/m^2 or more,
- BMI between 35 and 40 kg/m^2 and other significant disease that could be improved if weight loss is present,

- Appropriate non-surgical measures are tried for at least 6 months,
- The patient receives intensive management in a specialist obesity service,
- The patient is fit for anaesthesia and surgery,
- The patient commits to the need for long-term follow-up,
- Bariatric surgery is recommended as first-line option for adults with BMI more than 50 kg/m^2.

The benefits of bariatric surgery with metabolic disorders have expanded these guidelines to cater especially for the diabetic patient with BMI > 30 kg/m^2 [45] and other features of obesity including waist circumference and waist to hip ratio. Some populations also have increased incidence of metabolic disorders at lesser degrees of obesity as judged by BMI and this has gained the attention of the Asian Consensus meeting on Metabolic Surgery in 2008 [46].

The Surgical Options

Surgery for obesity was initiated over 50 years ago and has evolved considerably over this time [47]. The options now available have widened and the older operations have been fine-tuned. The biggest change though came with the laparoscopic revolution of the 1990s resulting in significant benefits to the patient including much lower morbidity and mortality with reduced pain, better cosmetic outcome and quicker return to daily function and work [48,49]. Further advances with single-port laparoscopy and robotic surgery are rapidly evolving. Today, most procedures are done laparoscopically with over 200,000 cases annually in the United States alone [50].

Classically, bariatric procedures are divided into restrictive, mal-absorptive and combined restrictive/mal-absorptive. Restrictive includes the gastric band and sleeve gastrectomy. The vertical banded gastroplasty is an older restrictive operation that has become extinct. The classic intestinal bypass is a true mal-absorptive procedure but this has been abandoned. The duodenal switch with bilio-pancreatic diversion (BPD) and the BPD alone are classic mal-absorptive procedures, although there is some restrictive element also. They exist today but only for selected cases

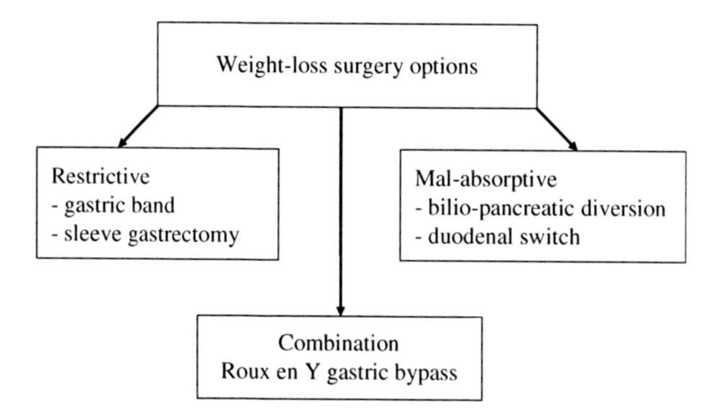

Figure 32.2 Types of weight-loss surgery.

and in selected centres. They are indicated in the super obese and for patients who have failed other procedures. The Roux en Y gastric bypass (RYGB) is considered the 'Gold Standard' operation and uses a combined mechanism (Figure 32.2).

In general, obese patients are to be regarded as high risk for surgery and should be informed and managed as such. The two main causes of mortality are pulmonary embolism (PE) and anastamotic or staple line leaks [51]. The risk of the latter is minimised by good surgical technique and the former by adequate deep vein thrombosis (DVT) prophylaxis. Compression stockings placed prior to the start of the procedure together with pre- and post-operative heparin as well as early and aggressive ambulation will minimise the risk of DVT and PE.

Laparoscopic Adjustable Gastric Band

Laparoscopic Adjustable Gastric Band (LAGB) is a silastic band placed around the proximal stomach (Figure 32.3; [52]). It is adjustable percutaneously allowing for tightening or loosening via a subcutaneous port. It is placed laparoscopically – a procedure usually completed within an hour via a pars flaccida approach, and patients are frequently discharged within 24 h. It is the most common procedure

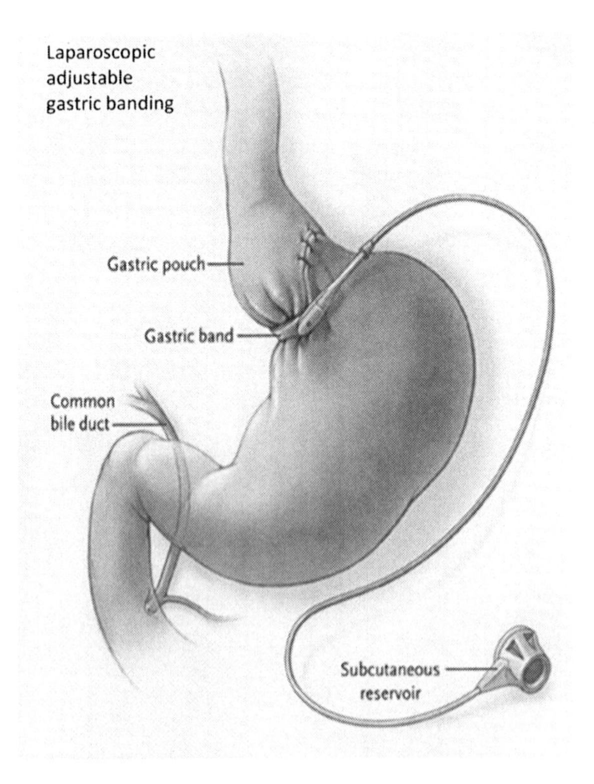

Figure 32.3 Laparoscopic adjustable gastric banding.
Source: Permission for use from Massachusetts Medical Society [52].

done in Europe. The subsequent diet is liquid, followed by purees and then solids introduced over the ensuing 4 weeks. Mineral and vitamin supplements are started immediately upon discharge. Adjustments to the band are performed monthly as an office procedure until adequate weight loss is achieved or if the band becomes completely filled. It can be released as the need arises, e.g. in pregnancy in order to balance sufficient nutrition with regain of excess weight [53,54]. Patients have to be counselled about the use of sweets and alcohol as these are readily absorbed, and overindulging can prevent maximal weight loss. The mechanism of action is restriction and appetite suppression.

Complications

Major complications include slippage (1–20%), band erosion (2–3%) and port infection (2–3%) [55]. Reflux can be made worse and this is a relative contraindication. Failure of adequate weight loss and weight regain are higher than other procedures [56].

Benefits

Expected weight loss is 40–60% excess body weight and this can take up to 2–3 years. Resolution of diabetes is 48% [57]. The procedure can theoretically be reversed but this can be quite challenging. There is also the option of conversion to another bariatric procedure if weight loss proves inadequate [58].

Sleeve Gastrectomy

This operation was initially designed as the first part of the two-part operation (BPD with duodenal switch) for patients with super-morbid obesity who were too high risk for the entire operation at a single attempt [59,60] (Figure 32.4). It was found that some of these patients never returned for the second part as the weight loss achieved was adequate. It is now considered a stand-alone operation but can still be converted to either a BPD or gastric bypass if weight loss is inadequate. It is fast becoming the preferred restrictive procedure [61], though long-term data on outcomes is not currently available. It involves removal of about 70–85% of the stomach leaving a thin remnant along the lesser curve achieved with the use of a stapling device and starting roughly 4–6 cm from the pylorus along a 38–48 Fr gastric tube towards the Angle of His. This staple line may or may not be oversewn. The greater curve is completely freed with the harmonic scalpel or equivalent. The operation is done laparoscpically and the patient discharged within 24–48 h. The post-operative diet is similar to that described above for gastric banding. The mechanism of action for weight loss through this procedure is predominantly restriction, but may also involve some element of appetite suppression, presumably mediated through mechanisms involving reduced production of Ghrelin which is produced by the stomach, and removal of the greater curve of the stomach would reduce its production.

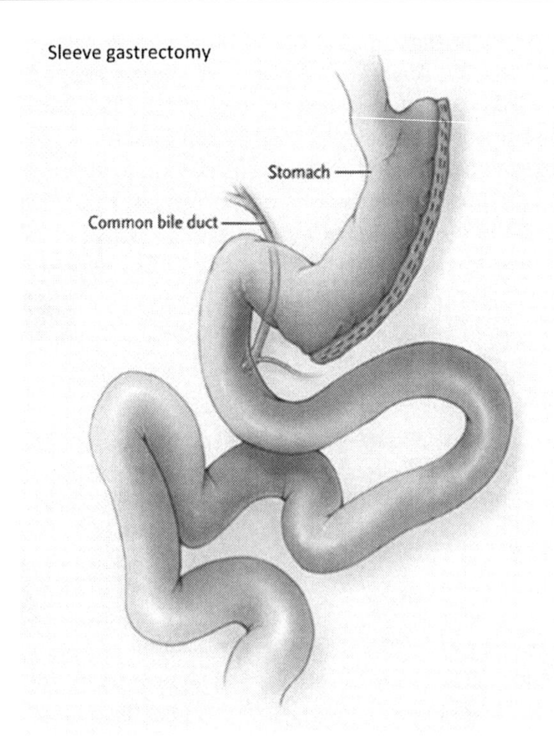

Figure 32.4 Sleeve gastrectomy.
Source: Permission for use from Massachusetts Medical Society [52].

Complications

Complications specific to the sleeve gastrectomy include bleeding, stenosis especially at the angular notch and stomach dilation with weight regain. Staple line leaks can occur especially where staple firings overlap [62]. Reflux is also increased as the tube becomes a high-pressure zone, and this procedure is not ideal for patients with gastro-oesophageal reflux disease unless the hiatus is repaired simultaneously at the time of surgery.

Benefits

Weight loss occurs over a period of 18–24 months and expected weight loss is 50–70% excess body weight. Diabetes resolves in about 70% of cases [63]. In the event of failure, conversion to either a bypass or BPD is possible.

Bilio-Pancreatic Diversion

This procedure produces mainly mal-absorption with some restriction (Figure 32.5). It can use either a sleeve or a three quarter gastrectomy combined with division of

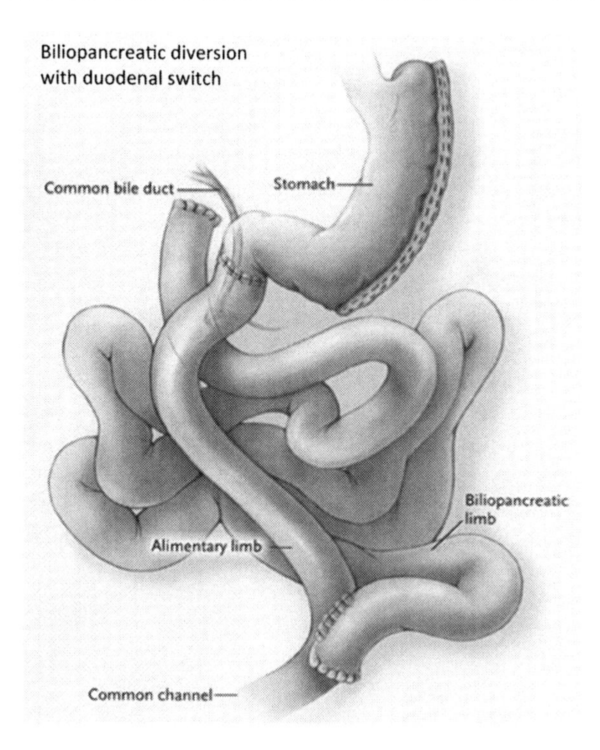

Figure 32.5 Bilioprancreatic diversion with duodenal switch.
Source: Permission for use from Massachusetts Medical Society [52].

the small bowel creating a bilio-pancreatic limb and a gastrointestinal limb. Patients can consume more than those who undergo a gastric bypass or band but this is compensated by the more significant mal-absorption associated with the procedure. This operation is, however, technically more difficult and therefore prone to more complications than the bypass, band or sleeve. The traditional open approach though still popular is being gradually replaced by laparoscopic or robotic techniques. Not all bariatric centres offer the BPD due to relatively higher complication rate and the difficulty factor.

Complications

Complications common to the BPD include diarrhoea, bloating, protein and other nutrient malnutrition and dumping [64]. Anastamotic problems (leaks, bleeding and strictures) and intestinal obstruction are also not infrequent. Mortality ranges from 1 to 1.9% and leak rates 2.7–3.75% [65].

Benefits

Weight loss and cure of diabetes are much better with up to 91% excess weight loss at 5 years and 95% resolution of diabetes [66].

Roux en Y Gastric Bypass

This is the most commonly performed surgical procedure for weight reduction (Figure 32.6). It involves creating a small gastric pouch 15–20 cc along the lesser curve of the stomach achieved with a stapling device starting between the 1st and 2nd vessel on the lesser curve angling towards the Angle of His. The duodeno–jejunal flexure is then identified and a bilio-pancreatic limb measuring 40–50 cm is created again with the use of a stapling device. A Roux limb measuring between 100 and 200 cm is then identified and a jejuno-jejunostomy created. The Roux limb is anastamosed to the gastric pouch creating an anastamotic diameter of 1.5–2 cm using either a circular or linear stapler or alternatively hand sewn. This procedure when performed laparoscopically takes 1–2 h after which patients can be discharged after 24–48 h [67]. Dietary mineral and vitamin supplements are started 2 days post-operatively. Patients are routinely followed at 1–2 weeks, 6 weeks, 3 months, 6 months and yearly thereafter. Although the proposed mechanism for weight loss is presumed to be due to restriction and mal-absorption with appetite suppression, there must be other explanations to account for control of the metabolic disorders as resolution of these occur

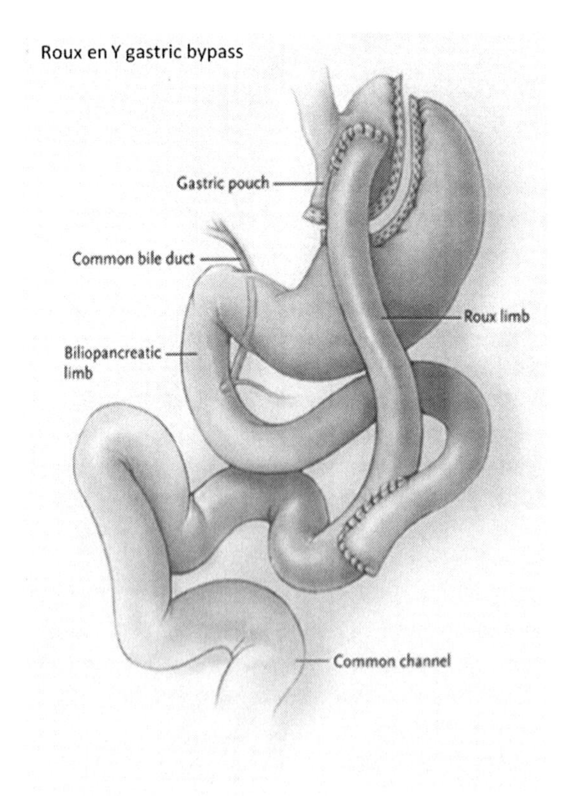

Figure 32.6 Roux en Y gastric bypass.
Source: Permission for use from Massachusetts Medical Society [52].

much before significant weight is lost [68,69]. Presumed mechanisms include the foregut and hindgut theories. The foregut theory suggests that the exclusion of a short segment of the proximal small intestine from ingested nutrients exerts a direct anti-diabetic effect. Rubino demonstrated that a duodeno–jejunal bypass in rats improved diabetic control and these results have been reproduced in human studies of duodenal–jejunal bypass showing improvement in glycemic control in the obese and non-obese [70–72].

The hindgut theory suggests that after gastric bypass or BPD, expedited delivery of ingested nutrients to the lower intestine increases the secretion of the incretin, glucagon-like peptide (GLP-1) known to have effects on insulin, and glucagon secretion beneficial to glucose homeostasis and with central effects on appetite suppression. GLP-1 is increased after the gastric bypass and BPD but not after gastric banding [73].

Complications

Major complications include DVT, PE, anastamotic leaks and bleeding which can occur in up to 4.3% of cases but mortality is generally about 0.3%. Long-term complications include internal hernias, marginal ulcers, gallstones, nutrient deficiencies and dumping with hypoglycaemia [74].

Benefits

The expected weight loss is roughly 70–80% excess weight at 5 years and diabetes resolves in 85% of cases [30]. Weight loss occurs over 12–18 months, at first rapidly then slowly thereafter. Diabetes resolves in the majority of cases and often within the first week, long before significant weight loss occurs [30,68,69]. Many components of the metabolic syndrome including PCOS improve after gastric bypass surgery [75].

Pre-Operative Assessment for Bariatric Surgery

The patient should understand the procedure and implications [50,76,77] (Table 32.3). Any unrealistic expectations should be dispelled and issues related to pre-morbid conditions should be clarified. Complications should be explained:

- 30 day mortality about 0.3–1.9% (restrictive 0.1%, gastric bypass 0.5%, for duodenal switch 1.1%) [50,78], with higher risk of mortality in with chronic diseases and super obesity, and in patients with significant co-morbidities and lower risk in centres which perform a high volume of surgeries [42,74,78]. The 6-month post-surgical death rate for patients operated on between 2005 and 2006 was 0.5% [43].
- Intra-operative injuries: liver, splenic, vascular and bowel injury.
- Early complications: Bleeding and leaks from anastamosis and/or staple line, wound infection, DVT, PE, cardiovascular and pulmonary complications [50].
- Late complications: Cholelithiasis, nutritional deficiencies including dumping and hypoglycaemia, neurological and psychiatric complications, bowel obstruction, band slippage, erosion and infection [42,43,50,56].

Table 32.3 Summary of Pre-Operative Assessment Before Weight-Loss Surgery [50,76,77]

History	Nutrition, previous efforts at weight loss, obstacles to weight management	
	Drugs: anti-depressants, oral contraceptive pill, oral hypoglycaemics	
Exam	Pulse rate, blood pressure, weight, height, BMI, waist circumference, look for clinical evidence of secondary causes of obesity (Cushing's, hypothyroidism, etc.)	
Screen for and optimise co-morbid conditions	Hypertension Chronic Renal Failure	Renal function tests
	Diabetes mellitus	Fasting blood glucose, HbA1C
	Coronary artery disease	Cardiac assessment, pre-operative beta-blocker use can be considered
	Gallbladder and liver disease	Cholelithiasis can occur after weight loss so the presence of existing gallbladder disease can be screened for. Liver function tests to assess for severity of cirrhosis and/or portal hypertension. Surgery is not recommended if Child's Class C cirrhosis is observed
	Hyperlipidaemia	Fasting lipid profile
	Sleep apnoea	Sleep studies
	Nutritional deficiencies	Ferritin, vitamin B12, 25-hydroxyvitamin D
	Pulmonary hypertension	
	Musculoskeletal disease	
	Hypothyroidism and Cushing's disease if indicated	
Psychological evaluation	Evaluate, identify and treat any disorders that may affect adherence to post-operative care requirements. Contraindicated if illicit drug abuse, schizophrenia, severe mental retardation ($IQ < 50$), heavy alcohol use and lack of knowledge about surgery	
Contraindications to surgery should be excluded	Unstable coronary artery disease, advanced liver disease with portal hypertension, major depression, psychosis, binge eating	
Consultation with a dietician or nutritionist	Counselling and detailed dietary instruction plans for the patient and caregiver. In the week prior to surgery, a high protein, low carbohydrate diet is preferable in an effort to reduce the liver size to make manipulation during surgery easier	

Women should be advised to delay pregnancy till 12–24 months after the procedure in an effort to optimise weight loss, stabilise weight and reduce the effect of nutritional deficiencies on the foetus [41]. Consideration should be given to using a second form of protection if the contraceptive pill is used or even doubling the dose of the oral pill [79].

Smokers should be encouraged to quit prior to procedure as it is a risk factor for post-operative complications and increases the risk of post-operative marginal ulceration [76].

Routine endoscopy, though recommended by some, is not mandatory. Exercise programs are started pre-operatively and can be resumed 1 week post-operatively. Weight loss prior to surgery is desirable and appears to reduce post-operative complications. A pre-operative weight loss of 5–10% of initial body weight is recommended especially for patients with a BMI ≥ 50 kg/m^2 [52,76,77].

Special Considerations in a Post-Bariatric Patient

Nutrient Replacement

After gastric bypass procedures, patients are prone to vitamin A, D, E and K (fat soluble) deficiencies and calcium deficiencies [50,76,80]. After RYGB, ingested food does not pass through the duodenum, and this may result in calcium deficiency which can be exacerbated by low intake of vitamin D in the diet with an increased risk of secondary hyperparathyroidism and subsequent osteopenia and metabolic bone disease. This can be compounded by a high foetal demand for calcium if the patient becomes pregnant post-bariatric surgery. Patients should be maintained on calcium and vitamin D supplements daily to avoid metabolic bone disease. There may be decreased production of hydrochloric acid resulting in decreased absorption secondary to altered pH. Calcium carbonate depends on acid for absorption while calcium citrate does not, so it is recommended that this be used preferentially.

Anaemia

Anaemia may develop due to inadequate iron, B12 or folate intake or absorption [50,76]. Iron should be routinely supplemented and should be combined with ascorbic acid (vitamin C) to acidify the stomach environment which facilitates absorption. B12 which requires intrinsic factor and hydrochloric acid for absorption should be routinely administered either by monthly B12 injections or daily oral supplements.

Gastrointestinal Problems in Pregnancy

There are reports of bowel obstruction, gastric ulcer, band events and staple line stricture in the pregnant post-bariatric patient [81]. They may present with non-specific symptoms ranging from abdominal pain to vomiting. Nausea and vomiting are common complaints in pregnancy and may occur early post-bariatric surgery [82]. Excessive vomiting can cause dehydration, malnutrition and an acute deficiency in thiamine. Obstetricians and bariatric surgeons should both be involved in the care of pregnant post-bariatric patients and vigilantly investigate any gastrointestinal symptoms.

Role of Bariatric Surgery to Treat Obesity-Related Female Reproductive Abnormalities

With the disappointing results from non-surgical measures together with increasing awareness, availability and reducing costs, anti-obesity surgery is being increasingly performed. As much as 80% of patients seeking bariatric procedures are women and most (50%) are in the reproductive age group 18–45 years [39,41]. Not surprisingly, reports of pregnancy and outcomes have emerged from this pool of data. Curiously, the major stated reasons for such surgical interventions are not for reproductive indications but primarily for physical appearance or psychosocial benefits [83]. Observed reproductive gains have therefore been a beneficial side effect which will have created a new indication for such procedures and which will further drive up demand. But is this indication justified on available evidence?

Data on apparent benefit to reproductive functions are derived only from observational studies. The derivation of such data and potential weaknesses are outlined in Figure 32.7.

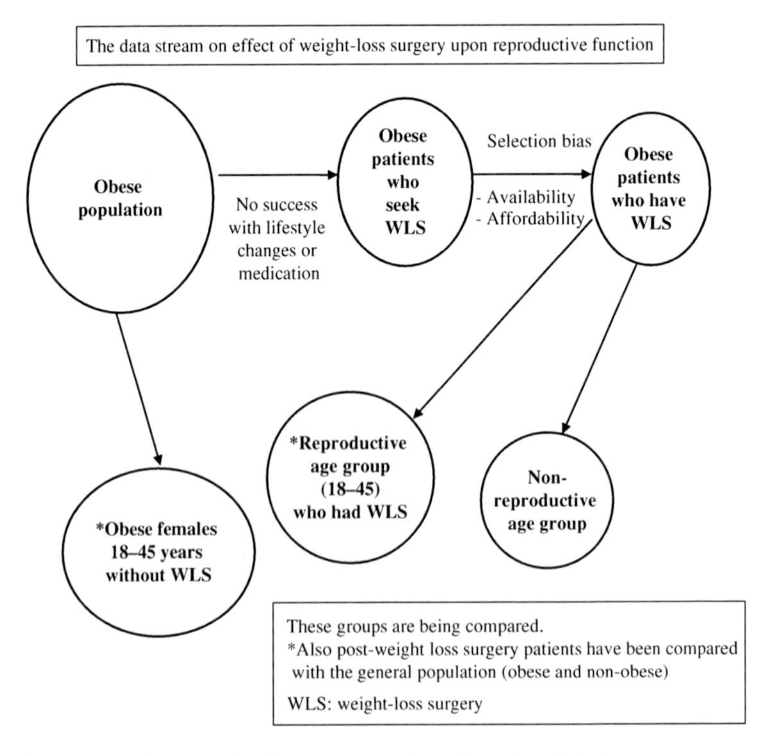

Figure 32.7 Illustrating how the data stream on the effect of weight-loss surgery upon reproduction was obtained.

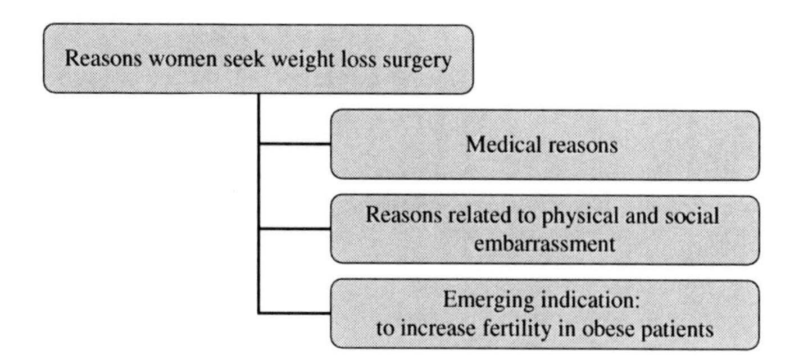

Figure 32.8 Reasons women seek weight-loss surgery.

This kind of circumstantial data may lead to a new indication for gastric bypass surgery, namely to increase fertility in the obese female (Figure 32.8).

Although bariatric surgery has been around since the 1960s, there are no high-level data to guide neither clinicians nor patients in the value, efficacy or cost-effectiveness of bariatric surgery for the treatment of infertility. Level 3 data from three expert committees exist and are summarised in the following paragraphs.

United States AHRQ

Assessment on Weight-Loss Surgery and Pregnancy

The Nationwide Inpatient Sample provided data on the incidence of bariatric procedures between 1998 and 2005 and was summarised earlier in this chapter [39]. Against this background, a review of available articles on fertility and pregnancy following bariatric surgery was undertaken. Of the screened articles, they accepted 1 case control, 12 cohort, 21 case series and 23 case reports. From this data, it was suggested that fertility improves after bariatric surgical procedures with minimal to no risk of nutritional or congenital problems if supplemental vitamins are taken and maternal nutrition is normal. However, with BPD it was observed that there is an appreciable risk for nutritional problems. There was a decreased risk for gestational diabetes, pre-eclampsia and pregnancy-induced hypertension compared with obese women who had not undergone surgery. There was no evidence that delivery complications were higher in post-surgery pregnancies [39].

Royal College of Obstetrics and Gynaecologists Scientific Advisory Committee Opinion Paper 17

The role of bariatric surgery in the management of female fertility [53]:

The Scientific Advisory Committee reviewed evidence on female fertility and bariatric surgery. Findings are summarised in the following table:

Fertility	Improved ability to conceive after LAGB and BPD	Lack of power to determine statistical significance
Miscarriage rates	Two retrospective studies were reviewed: one showed no difference in rate pre- and post-op BPD, while another recorded a reduction after BPD	Study did not comment on statistical significance
Supplementation and subsequent pregnancy	Advised on lifelong use of vitamins. Pregnancy should be delayed until surgery has been shown to be successful and until initial weight-loss phase is over as foetal growth restriction, prematurity and nutritional deficiencies can occur (mal-absorptive and mixed procedures > restrictive) and since post-surgical complications can occur in pregnancy	
Adjusting gastric band during pregnancy	Balance sufficient nutrition with regain of excess weight (may cause renewed risk of gestational diabetes)	
Diagnosis of gestational Diabetes	Delayed gastric emptying may cause misinterpretation of oral glucose tolerance test (OGTT) so diagnose with diurnal blood glucose profile instead	
Maternal and obstetric outcomes after obesity surgery	Two systemic reviews showed improved fertility outcomes and fewer maternal complications compared to obese controls and most foetal obstetric complications	Increased incidence of foetal growth restriction and caution with nutrition during pregnancy
Perinatal and obstetric complications (various studies involving heterogenous populations/samples with main findings summarised)		
Large Population Study (*comparing women who had bariatric surgery with the rest of the population*)	Similar rates of most perinatal and obstetric complications except for • Caesarean section (remained significant when controlled for possible confounders) (OR 1.4) • Premature rupture of membranes (OR 1.9) • Labour induction (OR 2.1) • Foetal macrosomia (OR 2.1)	
Case–control study (*comparing patients who had LAGB with similarly obese patients*)	Showed lower incidence of gestational diabetes, pre-eclampsia, Caesarean section, assisted vaginal deliveries, increased spontaneous delivery	
Two cohort studies (*comparing patients who had gastric bypass surgery with deliveries from non-surgical patients*) Cases were not matched for BMI so controls were less likely to be obese	No difference in gestational diabetes but higher rate of pregnancy-induced hypertension in the post-bariatric group Rates of pregnancy related complications obese > non-obese controls = post-bariatric surgery except for Caesarean section.	

The expert opinion from this committee is that bariatric surgery achieves weight loss, reverses its subsequent metabolic disturbance and thus improves infertility. After bariatric surgery maternal outcomes and morbidity are comparable with the general population *and* better than for women who are similarly obese. Cardiac and metabolic outcomes for infants are also improved. The final decision rests on the balance between the benefits and risks of surgery.

American College of Obstetrics and Gynecologists

Bariatric Surgery and Pregnancy

The ACOG [41] issued a practice bulletin to summarise the risks of obesity in pregnancy and to review the literature regarding the outcomes of pregnancy after bariatric surgery and to make recommendations for the prenatal and delivery of the post-bariatric patient.

The report is summarised in the following table.

Effect of surgery on fertility	Higher fertility rates follow surgery but it should not be used as a treatment for infertility
	Oral contraception failure may occur post-bariatric surgery and non-oral forms should be considered
	Wait 12–24 months after surgery before conceiving
Effect of surgery on maternal morbidity and mortality	No increase in congenital malformations post-surgery
	More lower mean birthweights and small for gestational age
	Decreased macrosomia after RYGB
	No increase in perinatal death
Nutritional status during pregnancy	Evaluate and correct and monitor for deficiencies at the beginning of pregnancy. Consult with nutritionist after conception
Antenatal considerations	Alternatives to the OGTT to screen for gestational diabetes due to incidence of dumping syndrome
	Medication doses: may have decreased absorption, avoid Non-Steroidal Anti-Inflammatories, therapeutic levels may need to be monitored for certain drugs
	Abdominal symptoms may be due to gastrointestinal surgical complications and should be evaluated
Considerations during labour and delivery	Higher rate of Caesarean section; however, bariatric surgery should not be considered as an indication for Caesarean section

Conclusion

Bariatric surgery has potential to ameliorate obesity-related female reproductive abnormalities and increase fertility. As the epidemic encroaches upon children greater numbers of young women are likely to present with disordered puberty and fertility as a consequence of obesity. This will mount pressure upon clinical decision-making as well as health care resources especially with the knowledge that surgery offers such hope. Data to guide such clinical or health resource allocations are currently inadequate and should include cost, cost-effectiveness, maternal and foetal outcomes as well as developmental and metabolic outcomes in later life.

References

1. Bronowski J. *The Ascent of Man*. London: BBC Books; 2011.
2. Meigs JB. Epidemiology of the metabolic syndrome, 2002. *Am J Manag Care*. 2002;8: S283–S292:quiz S293–S296.
3. World Health Organization. *Global Status Report on Noncommunicable Diseases 2010*. Geneva, Switzerland: World Health Organization; 2011.
4. Yu CKH, Teoh TG, Robinson S. Review article: obesity in pregnancy. *BJOG*. 2006;113:1117–1125.
5. Lewis G. *The Confidential Enquiry into Maternal and Child Health (CEMACH). Saving Mothers' Lives: Reviewing Maternal Deaths to Make Motherhood Safer – 2003–2005. The Seventh Report on Confidential Enquiries into Maternal Deaths in the United Kingdom. 2007 edition*. London: CEMACH; 2007.
6. Heslehurst N, Ells LJ, Simpson H, Batterham A, Wilkinson J, Summerbell CD. Trends in maternal obesity incidence rates, demographic predictors, and health inequalities in 36,821 women over a 15-year period. *BJOG*. 2007;114:187–194.
7. Burt Solorzano CM, McCartney CR. Obesity and the pubertal transition in girls and boys. *Reproduction*. 2010;140:399–410.
8. Bronstein J, Tawdekar S, Liu Y, Pawelczak M, David R, Shah B. Age of onset of polycystic ovarian syndrome in girls may be earlier than previously thought. *J Pediatr Adolesc Gynecol*. 2011;24:15–20.
9. Boots C, Stephenson MD. Does obesity increase the risk of miscarriage in spontaneous conception: a systematic review. *Semin Reprod Med*. 2011;29:507–513.
10. Metwally M, Li TC, Ledger WL. The impact of obesity on female reproductive function. *Obes Rev*. 2007;8:515–523.
11. Deslypere JP. Obesity and cancer. *Metabolism*. 1995;44:24–27.
12. Calle EE, Rodriguez C, Walker-Thurmond K, Thun MJ. Overweight, obesity, and mortality from cancer in a prospectively studied cohort of U.S. adults. *N Engl J Med*. 2003;348:1625–1638.
13. Ramsay JE, Greer I, Sattar N. ABC of obesity. Obesity and reproduction. *BMJ*. 2006;333:1159–1162.
14. Becerra JE, Khoury MJ, Cordero JF, Erickson JD. Diabetes mellitus during pregnancy and the risks for specific birth defects: a population-based case–control study. *Pediatrics*. 1990;85:1–9.
15. DCCT Research Group. Pregnancy outcomes in the diabetes control and complications trial. *Am J Obstet Gynecol*. 1996;174:1343–1353.
16. Langer O, Yogev Y, Most O, Xenakis EM. Gestational diabetes: the consequences of not treating. *Am J Obstet Gynecol*. 2005;192:989–997.
17. Nieuwenhuis-Ruifrok AE, Kuchenbecker WK, Hoek A, Middleton P, Norman RJ. Insulin sensitizing drugs for weight loss in women of reproductive age who are

overweight or obese: systematic review and meta-analysis. *Hum Reprod Update*. 2009;15:57–68.
18. Lord JM, Flight IHK, Norman RJ. Metformin in polycystic ovary syndrome: systematic review and meta-analysis. *BMJ*. 2003;327:951.
19. Haas DA, Carr BR, Attia GR. Effects of metformin on body mass index, menstrual cyclicity, and ovulation induction in women with polycystic ovary syndrome. *Fertil Steril*. 2003;79:469–481.
20. Barbieri RL. Metformin for the treatment of polycystic ovary syndrome. *Obstet Gynecol*. 2003;101:785–793.
21. Sjöström L, Narbro K, Sjöström CD, et al. Effects of bariatric surgery on mortality in swedish obese subjects. *N Engl J Med*. 2007;357:741–752.
22. Franz MJ, VanWormer JJ, Crain AL, et al. Weight-loss outcomes: a systematic review and meta-analysis of weight-loss clinical trials with a minimum 1-year follow-up. *J Am Diet Assoc*. 2007;107:1755–1767.
23. Tuah NA, Amiel C, Qureshi S, Car J, Kaur B, Majeed A. Transtheoretical model for dietary and physical exercise modification in weight loss management for overweight and obese adults. *Cochrane Database Syst Rev*. 2011;10:CD008066.
24. Yanovski SZ, Yanovski JA. Obesity. *New England J Med*. 2002;346:591–602.
25. Shaw K, Gennat H, O'Rourke P, Del Mar C. Exercise for overweight or obesity. *Cochrane Database Syst Rev*. 2006;4:CD003817.
26. Norris SL, Zhang X, Avenell A, et al. Long-term effectiveness of weight-loss interventions in adults with pre-diabetes: a review. *Am J Prev Med*. 2005;28(1): 126–139.
27. Mogul HR, Peterson SJ, Weinstein BI, Zhang S, Southren AL. Metformin and carbohydrate-modified diet: a novel obesity treatment protocol: preliminary findings from a case series of nondiabetic women with midlife weight gain and hyperinsulinemia. *Heart Dis*. 2001;3:285–292.
28. Glueck CJ, Aregawi D, Agloria M, Winiarska M, Sieve L, Wang P. Sustainability of 8% weight loss, reduction of insulin resistance, and amelioration of atherogenic-metabolic risk factors over 4 years by metformin-diet in women with polycystic ovary syndrome. *Metabolism*. 2006;55:1582–1589.
29. Maggard MA, Shugarman LR, Suttorp M, et al. Meta-analysis: surgical treatment of obesity. *Ann Intern Med*. 2005;142:547–559.
30. Buchwald H, Avidor Y, Braunwald E, et al. Bariatric surgery: a systematic review and meta-analysis. *JAMA*. 2004;292:1724–1737.
31. Colquitt J, Clegg A, Loveman E, Royle P, Sidhu MK. Surgery for morbid obesity. *Cochrane Database Syst Rev*. 2005;4:CD003641.
32. Hahn S, Benson S, Elsenbruch S, et al. Metformin treatment of polycystic ovary syndrome improves health-related quality-of-life, emotional distress and sexuality. *Hum Reprod*. 2006;21:1925–1934.
33. Kolotkin RL, Binks M, Crosby RD, Ostbye T, Gress RE, Adams TD. Obesity and sexual quality of life. *Obesity*. 2006;14:472–479.
34. Bond DS, Wing RR, Vithiananthan S, et al. Significant resolution of female sexual dysfunction after bariatric surgery. *Surg Obes Relat Dis*. 2011;7:1–7.
35. Teelucksingh S, Pinto Pereira LM. Metformin: an important tool for endocrinology in the West Indies. New tricks for an old drug. *B J Diab Vasc Dis*. 2009;9:232–236.
36. Escobar-Morreale HF, Botella-Carretero JI, Alvarez-Blasco F, Sancho J, San Millan JL. The polycystic ovary syndrome associated with morbid obesity may resolve after

weight loss induced by bariatric surgery. *J Clin Endocrinol Metab*. 2005;90: 6364–6369.
37. Lord JM, Flight IH, Norman RJ. Insulin-sensitising drugs (metformin, troglitazone, rosiglitazone, pioglitazone, D-chiro-inositol) for polycystic ovary syndrome. *Cochrane Database Syst Rev*. 2003;CD003053.
38. Teitelman M, Grotegut CA, Williams NN, Lewis JD. The impact of bariatric surgery on menstrual patterns. *Obes Surg*. 2006;16:1457–1463.
39. Maggard M, Li Z, Yermilov I, et al. *Bariatric Surgery in Women of Reproductive Age: Special Concerns for Pregnancy. Evidence Report/Technology Assessment No. 169. (Prepared by the Southern California Evidence-based Practice Center under Contract No. 290-02-003)*. Rockville, MD: Agency for Healthcare Research and Quality; 2008.
40. Roehrig HR, Xanthakos SA, Sweeney J, Zeller MH, Inge TH. Pregnancy after gastric bypass surgery in adolescents. *Obes Surg*. 2007;17:873–877.
41. American College of Obstetricians and Gynecologists. ACOG practice bulletin no. 105: bariatric surgery and pregnancy. *Obstet Gynecol*. 2009;113:1405–1413.
42. Encinosa WE, Bernard DM, Du D, Steiner CA. Recent improvements in bariatric surgery outcomes. *Med Care*. 2009;47:531–535.
43. *Complications and Costs for Obesity Surgery Declining*. Press Release, April 29, 2009. Agency for Healthcare Research and Quality, Rockville, MD. <http://www.ahrq.gov/news/press/pr2009/barsurgpr.htm>.
44. NICE. *Obesity: Guidance on the Prevention, Assessment and Management of Overweight and Obesity in Adults and Children*. London: National Institute for Health and Clinical Excellence; 2006. <http://www.nice.org.uk/CG43/>.
45. Rubino F, Kaplan LM, Schauer PR, Cummings DE, Diabetes Surgery Summit Delegates. The Diabetes Surgery Summit consensus conference: recommendations for the evaluation and use of gastrointestinal surgery to treat type 2 diabetes mellitus. *Ann Surg*. 2010;251:399–405.
46. Lakdawala M, Bhasker A, Asian Consensus Meeting on Metabolic Surgery. Report: Asian Consensus Meeting on Metabolic Surgery. Recommendations for the use of Bariatric and Gastrointestinal Metabolic Surgery for Treatment of Obesity and Type II Diabetes Mellitus in the Asian Population: August 9 and 10, 2008, Trivandrum, India. *Obes Surg*. 2010;20:929–936.
47. Mason EE. History of obesity surgery. *Surg Obes Relat Dis*. 2005;1:123–125.
48. Wittgrove A, Clark G, Schubert K. Laparoscopic gastric bypass, Roux en-Y: technique and results in 75 patients with 3–30 months follow-up. *Obes Surg*. 1996;6: 500–504.
49. Wittgrove AC, Clark GW. Laparoscopic gastric bypass, Roux-en-Y- 500 patients: technique and results, with 3–60 month follow-up. *Obes Surg*. 2000;10:233–239.
50. Mechanick JI, Kushner RF, Sugerman HJ, et al. American association of clinical endocrinologists, the obesity society, and American Society for Metabolic and Bariatric Surgery Medical Guidelines for clinical practice for the perioperative nutritional, metabolic, and nonsurgical support of the bariatric surgery patient. *Surg Obes Relat Dis*. 2008;4:S109–S184.
51. Al Harakeh AB. Complications of laparoscopic Roux-en-Y gastric bypass. *Surg Clin North Am*. 2011;91:1225–1237.
52. DeMaria EJ. Bariatric surgery for morbid obesity. *N Engl J Med*. 2007;356:2176–2183.
53. Scholtz S, Le Roux C, Balen AH. The role of bariatric surgery in the management of female fertility. *Hum Fertil (Camb)*. 2010;13(2):67–71.

54. Dixon J, Dixon M, O'Brien P. Pregnancy after lap-band surgery: management of the band to achieve healthy weight outcomes. *Obes Surg.* 2001;11:59–65.
55. Mittermair R, Aigner F, Obermüller S. High complication rate after Swedish adjustable gastric banding in younger patients ≤25 years. *Obes Surg.* 2009;19:446–450.
56. Bowne WB, Julliard K, Castro AE, Shah P, Morgenthal CB, Ferzli GS. Laparoscopic gastric bypass is superior to adjustable gastric band in super morbidly obese patients: a prospective, comparative analysis. *Arch Surg.* 2006;141:683–689.
57. Weiner R, Blanco-Engert R, Weiner S, Matkowitz R, Schaefer L, Pomhoff I. Outcome after laparoscopic adjustable gastric banding – 8 years experience. *Obes Surg.* 2003;13:427–434.
58. Mognol P, Chosidow D, Marmuse J-P. Laparoscopic conversion of laparoscopic gastric banding to Roux-en-Y Gastric bypass: a review of 70 patients. *Obes Surg.* 2004;14:1349–1353.
59. Daskalakis M, Weiner RA. Sleeve gastrectomy as a single-stage bariatric operation: indications and limitations. *Obes Facts.* 2009;2(suppl 1):8–10.
60. Rosen DJ, Dakin GF, Pomp A. Sleeve gastrectomy. *Minerva Chir.* 2009;64:285–295.
61. Abu-Jaish W, Rosenthal RJ. Sleeve gastrectomy: a new surgical approach for morbid obesity. *Expert Rev Gastroenterol Hepatol.* 2010;4:101–119.
62. Aurora AR, Khaitan L, Saber AA. Sleeve gastrectomy and the risk of leak: a systematic analysis of 4,888 patients. *Surg Endosc.* 2012;26:1509–1515.
63. Brethauer SA. Sleeve gastrectomy. *Surg Clin North Am.* 2011;91:1265–1279.
64. Laurenius A, Taha O, Maleckas A, Lonroth H, Olbers T. Laparoscopic biliopancreatic diversion/duodenal switch or laparoscopic Roux-en-Y gastric bypass for super-obesity-weight loss versus side effects. *Surg Obes Relat Dis.* 2010;6:408–414.
65. Buchwald H, Kellogg TA, Leslie DB, Ikramuddin S. Duodenal switch operative mortality and morbidity are not impacted by body mass index. *Ann Surg.* 2008;248:541–548.
66. Sudan R, Jacobs DO. Biliopancreatic diversion with duodenal switch. *Surg Clin North Am.* 2011;91:1281–1293.
67. Dan D, Harnanan D, Singh Y, Hariharan S, Naraynsingh V, Teelucksingh S. Effects of bariatric surgery on type-2 diabetes mellitus in a Caribbean setting. *Int J Surg.* 2011;9:386–391.
68. Schauer PR, Burguera B, Ikramuddin S, et al. Effect of laparoscopic Roux-en Y gastric bypass on type 2 diabetes mellitus. *Ann Surg.* 2003;238:467–484 [discussion 84–85].
69. Pories WJ, Swanson MS, MacDonald KG, et al. Who would have thought it? An operation proves to be the most effective therapy for adult-onset diabetes mellitus. *Ann Surg.* 1995;222:339–350 [discussion 350–352].
70. Rubino F. Is type 2 diabetes an operable intestinal disease? A provocative yet reasonable hypothesis. *Diabetes Care.* 2008;31(suppl 2):S290–S296.
71. Rubino F, Forgione A, Cummings DE, et al. The mechanism of diabetes control after gastrointestinal bypass surgery reveals a role of the proximal small intestine in the pathophysiology of type 2 diabetes. *Ann Surg.* 2006;244:741–749.
72. Rubino F, Schauer PR, Kaplan LM, Cummings DE. Metabolic surgery to treat type 2 diabetes: clinical outcomes and mechanisms of action. *Annu Rev Med.* 2010; 61:393–411.
73. Strader AD, Vahl TP, Jandacek RJ, Woods SC, D'Alessio DA, Seeley RJ. Weight loss through ileal transposition is accompanied by increased ileal hormone secretion and synthesis in rats. *Am J Physiol Endocrinol Metab.* 2005;288:E447–E453.

74. The Longitudinal Assessment of Bariatric Surgery (LABS) Consortium. Perioperative safety in the longitudinal assessment of bariatric surgery. *N Engl J Med.* 2009; 361:445–454.
75. Batsis JA, Romero-Corral A, Collazo-Clavell ML, Sarr MG, Somers VK, Lopez-Jimenez F. Effect of bariatric surgery on the metabolic syndrome: a population-based, long-term controlled study. *Mayo Clin Proc.* 2008;83:897–907.
76. Apovian CM, Cummings S, Anderson W, et al. Best practice updates for multi-disciplinary care in weight loss surgery. *Obesity.* 2009;17:871–879.
77. Collazo-Clavell ML, Clark MM, McAlpine DE, Jensen MD. Assessment and preparation of patients for bariatric surgery. *Mayo Clin Proc.* 2006;81:S11–S17.
78. Buchwald H, Avidor Y, Braunwald E, et al. Bariatric surgery. *JAMA.* 2004; 292:1724–1737.
79. Paulen ME, Zapata LB, Cansino C, Curtis KM, Jamieson DJ. Contraceptive use among women with a history of bariatric surgery: a systematic review. *Contraception.* 2010;82:86–94.
80. Miller AD, Smith KM. Medication and nutrient administration considerations after bariatric surgery. *Am J Health Syst Pharm.* 2006;63:1852–1857.
81. Guelinckx I, Devlieger R, Vansant G. Reproductive outcome after bariatric surgery: a critical review. *Hum Reprod Update.* 2009;15:189–201.
82. Maggard MA, Yermilov I, Li Z, et al. Pregnancy and fertility following bariatric surgery. *JAMA.* 2008;300:2286–2296.
83. Pinkney JH, Johnson AB, Gale EA. The big fat bariatric bandwagon. *Diabetologia.* 2010;53:1815–1822.

33 Evidence-Based Approach to the Management of Obese Pregnant Women

Shakila Thangaratinam and Khalid S. Khan

Women's Health Research Unit, Centre for Primary Care and Public Health, Barts and the London School of Medicine and Dentistry, Queen Mary University of London, London, UK

Background

More than half of the women who die during pregnancy, childbirth or puerperium are either obese (body mass index, BMI, 30 kg/m^2 or more) or overweight (BMI 24.9–29.9 kg/m^2). Maternal obesity has been identified as a significant and growing threat to the pregnant women in United Kingdom by the Confidential Enquiries into Maternal and Child Health (CEMACH) report [1]. Obesity in pregnancy is associated with an increase in complications to the mother and baby. The maternal risks of obesity include maternal death or severe morbidity, cardiac disease, spontaneous first trimester and recurrent miscarriage, pre-eclampsia, gestational diabetes, thromboembolism, post-Caesarean wound infection, infection from other causes such as urinary, respiratory infections, post-partum haemorrhage and low breastfeeding rates [1,2]. The foetal risks include stillbirths and neonatal deaths, macrosomia, neonatal unit admission, preterm births, congenital abnormalities and childhood obesity with associated long-term risks [3]. Maternal obesity is also a major risk factor for childhood obesity. The obesity rate is doubled in 2- and 4-year-old children born to obese mothers. Obese women spend an average of 4.83 more days in hospital with a fivefold increase in the cost of antenatal care [4].

The Royal College of Obstetricians and Gynaecologists (RCOG) has identified weight management interventions targeting mothers as an important long-term challenge.[1] Antenatal period is an ideal time to provide dietary and physical activity interventions to manage weight. Pregnant women are highly motivated to make changes and have opportunities for regular contact with health professionals [5]. There is a need to identify the optimal interventions that can be delivered in

[1] http://www.rcog.org.uk/news/cmace-release-cmace-publishes-information-obesity-pregnancy

Obesity. DOI: http://dx.doi.org/10.1016/B978-0-12-416045-3.00033-9

pregnancy targeting women who enter pregnancy as obese or non-obese women who gain significant weight in pregnancy and enter post-partum period as obese. The latter group are at further risk of retention of weight post-partum and start subsequent pregnancies as obese. The management of obese mothers include appropriate identification and referral, regular surveillance for maternal and foetal complications and interventions to minimise weight gain in pregnancy and prevent adverse outcomes in current and subsequent pregnancies. The interventions in pregnancy need to be effective, acceptable and cost-effective in improving the short-term and long-term outcomes for the mother and the baby.

Pre-Pregnancy Weight Management

Overweight and obesity have a strong impact on fertility as well as consequences on fertility treatment and pregnancy outcome. Compared to women with normal weight, overweight and obese women have live birth less often and complications more often [6]. Pre-pregnancy counselling offers a good opportunity to emphasise the need for weight control to reduce risks.

Weight Management Interventions in Pregnancy

Current recommendations from National Institute for Health and Clinical Excellence (NICE) [7], RCOG [8] and American College of Obstetricians and Gynecologists (ACOG) for the management of obesity include healthy diet and exercise in pregnancy with referral to a nutritionist if required [9]. The recent NICE guidance has recommended a 'life course approach' by focussing on pregnancy and 1 year childbirth as the crucial periods to target weight management interventions based on behavioural change, dietary and physical activity [7].

Recommended Weight Gain in Pregnancy

The target weights for gestational weight gain (GWG) were based on the recommendations provided by the Institute Of Medicine (IOM), ACOG and National Institute of Diabetes Digestive and Kidney Diseases (NIDDK) [9–12]. Obese women (BMI > 30 kg/m^2) were recommended a total weight gain of 5–9 kg in pregnancy and a mean weight gain of 0.22 kg/week (0.18–0.27 kg) in the second and third trimesters [10]. The recommendations were based on evidence from population-based cohort studies that evaluated the association between weight gain in pregnancy for women with various BMI and maternal and foetal outcomes. The risk of adverse outcomes varies with the various classes of obesity I (BMI 30–34.9 kg/m^2), II (BMI 35–39.9 kg/m^2) and III (BMI ≥ 40 kg/m^2). The risk of adverse outcomes was minimal for a GWG of 4.5–15.5 kg for obesity class I and 0–4.1 kg for obesity classes II and III. The NICE in United Kingdom refrained from providing recommended ranges for weight gain due to limitations in the evidence and concerns about the generalisability of the results to the UK population.

Effects of Interventions on Maternal and Foetal Weight

A systematic review of effects of weight management interventions in pregnant women including obese and overweight mothers identified three main interventions for weight management in pregnancy: (1) diet and nutritional advice, (2) physical activity and (3) mixed approach incorporating diet and physical activity components underpinned by behavioural approach [13]. The interventions were delivered in both primary and secondary care by trained health care professionals including dieticians, midwives and qualified fitness specialists. The sessions were either one-to-one or group sessions delivered at home, hospital or community with or without supervision. Brochures, video sessions and direct counselling sessions were employed to deliver the interventions.

Weight management interventions evaluated in 10 randomised controlled trials (RCTs) showed a reduction in weight gain by 1.8 kg with interventions compared to the control group (mean difference, MD −1.83 kg, 95% confidence interval (CI) −3.52, −0.14 kg) [13]. Interventions in obese and overweight women (7 RCTs, 1591 women) showed a reduction in the birthweight by 60 g (MD −60 g, 95% CI −110, −10 g). The risk of large-for-gestational-age fetuses in obese and overweight women was not altered with interventions (5 RCTs, 1433 women).

Diet-Based Intervention

Four randomised trials (1338 women) evaluated the effect of dietary and nutritional advice on GWG in obese and overweight women [13]. The commonest diet evaluated in the studies was a balanced calorie regime with low-fat or cholesterol and high-fibre diet. The calorie intake was pre-specified to be not more than 2200 kcal/day [14] or 2000 kcal/day [15,16] for obese women. Dietary advice was provided as a personalised plan or in a group session with a trained dietician.

Dietary intervention showed the largest reduction in GWG (−4.13 kg, 95% CI 0.28, −8.53 kg) compared to physical activity and mixed approach and (MD −0.94 kg, −2.09. 2.1 kg), respectively. For obese women (7 RCTs, 562 women), diet-based interventions were most effective in reducing GWG (MD −4.79 kg, 95% CI −11.05, 1.48 kg) compared to other methods. There was no significant difference between the two groups adherent to IOM guidelines with interventions in obese and overweight women. Diet-based interventions in pregnancy for obese and overweight mothers showed a reduction in the birthweight of the newborn (MD −80 g, 95% CI −140, 20 g) compared to control group. There was no significant difference in the risk of small-for-gestational-age fetuses.

Amongst the weight management interventions, dietary interventions showed the most benefit in obese and overweight women by demonstrating a significant reduction in pre-eclampsia (4 RCTs, 1334 women; RR 0.63, 95% CI 0.42–0.96) and a trend towards reduction in the risk of gestational diabetes (3 RCTs, 334 women; RR 0.52, 95% CI 0.27–1.03). Dietary advice resulted in a 70% decrease in the risk of gestational hypertension (RR 0.30, 95% CI 0.10–0.88) in obese women. There was a reduction in the incidence of shoulder dystocia in obese and overweight women (2 RCTs, 1052 women) with diet (RR 0.33, 95% CI 0.14–0.74). There were no differences in the rates of Caesarean section, induction of labour or vaginal delivery with diet compared to controls. Two randomised trials (1164 women) showed a reduction in preterm births with diet that was not statistically significant (RR 0.79, 95% CI 0.55–1.14) [13].

Diet-based interventions are effective in reducing GWG and maternal complications in pregnancy. A dietary advice on the principles of patient-centred care taking into account the woman's preferences and cultural background is highlighted in the NICE recommendations on 'Healthy Eating' in pregnancy [7,17]. The diet based on low-glycaemic index healthy diet includes whole grain starchy foods, fibre rich and low-fat foods, five portions of fruit and vegetables per day and avoidance of sugary drinks, fast food and high-fat food.

Physical Activity–Based Intervention

Three RCTs (124 women) evaluated the effect of physical activity in pregnancy [13]. There was a decrease in GWG that was not significant (MD −1.4 kg, 95% CI −3.50, 0.69 kg). The physical activity–based interventions involved moderate exercise including aerobic dance programme, hydrotherapy, stationary cycling and light intensity resistance training. The sessions lasted for up to 1 h with a frequency of three times a week with the intensity of exercise determined as 50–60% of the predicted maximum heart rate. The exercise programme was home-based or supervised group sessions. There was no difference in the birthweight of the newborn with physical activity. One small RCT (94 women) evaluating the effect of exercise on preterm birth showed an increase in preterm birth intervention that was not significant and with imprecise estimate (RR 2, 95% CI 0.19–21.30).

Physical activity–based interventions in obese and overweight women do not show significant benefit in minimising weight gain in pregnancy, birthweight or obstetric complications. There is paucity of evidence on the effect of exercise in obese women for major obstetric outcomes like pre-eclampsia, gestational hypertension, gestational diabetes, Caesarean section, induction of labour, preterm birth and small-for-gestational-age foetuses.

Mixed Approach Intervention

Three RCTs reported on the effects of mixed approach interventions on GWG in obese and overweight pregnant women [13]. The interventions involved a programme of dietary and/or lifestyle advice with feedback on weight gain, adherence to the intervention and counselling using behavioural modification techniques. The programs included in-depth behavioural risk assessments and tailored counselling messages, based on principles of motivational interviewing. There was no significant difference in GWG (MD −0.01 kg, 95% CI −2.85, 2.83 kg) or birthweight (2 RCTs, 134 women; MD 30 g, 95% CI −180, 130 g) with mixed approach interventions. The risk of large-for-gestational-age newborns in obese and overweight women was unchanged with mixed approach. One small RCT (49 women) reported a trend towards reduction in the risk of small-for-gestational-age newborns. There was no significant reduction in the risks of pre-eclampsia, gestational hypertension, preterm birth, Caesarean section or induction of labour.

Safety of Weight Management Interventions and Weight Loss in Pregnancy

A systematic review of 26 studies (2 RCTs, 24 observational studies) involving 468,858 women did not show any significant maternal or foetal adverse effects with diet or physical activity interventions recommended in clinical practice [13]. Most of the data on adverse effects from dietary interventions were derived from studies on extreme diet and famine. There was an increase in the rate of neural tube defects and cleft lip and palate in pregnant women with extreme forms of dieting and on high-glycaemic index diets. Starvation in pregnancy was associated with increased incidence of metabolic syndrome, dyslipidaemia, coronary artery disease and hypertension. There were no significant maternal or foetal adverse effects like cord abnormalities, threatened miscarriage, meconium stained liquor, abnormal foetal heart rate pattern, maternal sepsis or chorioamnionitis observed with physical activity in pregnancy.

A population-based retrospective cohort study involving 819,905 singleton pregnancies evaluated the effect of gestational weight loss in pregnancy on obstetric outcomes [18]. Obese women were found to be more likely to lose weight in pregnancy compared to normal-weight and underweight women. The odds ratios (ORs) of pre-eclampsia, high blood pressure and non-elective Caesarean section were significantly reduced in obese women with gestational weight loss compared with obese women who gained the recommended weight in pregnancy. In classes I (BMI 30–34.9 kg/m^2) and II (BMI 35–39.9 kg/m^2) obese women, the OR of SGA (small for gestational age) was however increased in comparison to the recommended weight gain group. There was no difference in the perinatal mortality between the two groups.

Women's Views on Weight Management Interventions in Pregnancy

A qualitative research involving pregnant and post-natal women found that pregnant women lack awareness regarding the consequences of obesity and excess GWG [19]. Further, some women were not concerned by their weight post-natally and would consider weight loss strategies after they have had another baby. Women also reported difficulty in losing weight after giving birth, and some underestimated the difficulty in losing the weight gained in pregnancy. An intervention based on healthy eating and physical activity to manage weight during pregnancy is acceptable to women when it is seen to benefit their baby. Women preferred to tailor physical activity advice according to the trimesters of pregnancy. They wanted more information on appropriate healthy eating with individually tailored advice. A lack of social support was felt identifying the need for behavioural change support. After delivery, women preferred an activity that involved spending time with their baby. Crucially, women reported that they wanted an intervention they could access when suitable for them, rather than those that required compulsory attendance. Consequently, a home-based intervention that provides the opportunities for women to engage in healthy behaviour, without demanding extensive commitment, is likely to be accepted by and participated in by both pre- and post-partum women.

Health Professionals' Views on Weight Management Interventions

In a qualitative study, health professionals, including midwives, health visitors and children's centre managers, identified the need for an intervention to manage weight during and after pregnancy [19,20]. Although the health professionals were aware of the women's lack of concern regarding weight gain in pregnancy, they felt unable to advise them on weight management due to paucities in knowledge and time. Midwives were worried about overloading the women with weight management information in addition to other antenatal advice. There was a lack of knowledge regarding appropriate physical activities to be engaged in pregnancy. They were also unsure about the optimal time to commence exercise after giving birth. Thus, it is necessary that the health professionals in contact with pregnant and post-natal women are aware of the dietary and physical activity recommendations for weight management relevant to the individual woman.

Nutritional Supplementation for Obese Women in Pregnancy

A meta analysis of 12 observational studies demonstrated that the risk of neural tube defects is higher in obese (OR 1.70, 95% CI 1.34–2.15) and severely obese women (OR 3.11, 95% CI 1.75–5.46) compared to normal-weight women [21]. Furthermore, the levels of serum folate are low in obese and overweight women after controlling for folate intake [22]. Pregnant women who are obese are therefore advised to increase their folate intake to 5 mg/day in the pre-conception period and first trimester. Obese women are at increased risk of vitamin D deficiency with associated fall in the cord blood levels of vitamin D in the newborn [23]. Vitamin D supplementation of 10 mg/day is recommended to prevent adverse outcomes associated with vitamin D deficiency.

Prophylaxis for Pre-eclampsia and Thromboembolism

A meta-analysis of six cohort studies (64,789 women) comparing women with raised and normal BMI before pregnancy showed an increased risk of pre-eclampsia (RR 2.47, 95% CI 1.66–3.67) [24]. Women with a BMI ≥ 35 kg/m^2 and one additional risk factor such as first pregnancy, age 40 years or older, more than 10 year interval between pregnancies, family history of pre-eclampsia and multiple pregnancy are advised to take 75 mg/day aspirin in pregnancy. The evidence for higher dose of aspirin is limited by the small number of participants in the studies.

There is a fivefold increase in the OR of venous thromboembolism in obese mothers compared to non-obese women (adjusted OR 5.3, 95% CI 2.1–13.5) [25]. The decision for antenatal and post-natal thromboprophylaxis in obese women will need to take into consideration other co-existent risk factors. Morbidly obese women with a BMI ≥ 40 kg/m^2 should be recommended post-natal thromboprophylaxis for at least 7 days regardless of the mode of delivery.

Screening for Gestational Diabetes

The risk of gestational diabetes increases with BMI. For every 1 kg/m^2 increase in pre-pregnancy BMI, the prevalence of Gaussian distribution moves (GDM) increased by 0.92% (95% CI 0.73–1.10). GDM poses risks to the mother in pregnancy due to associated increase in the risks of pre-eclampsia and complications in delivery due to large-for-gestational-age foetuses. In the long term, about 20–50% of women with GDM develop type 2 diabetes mellitus at 5 years [26]. A meta-analysis of 20 studies showed an increase in the OR of developing GDM in overweight (OR 2.14, 95% CI 1.82–2.53), obese (OR 3.56, 95% CI 3.05–4.21) and severely obese (OR 8.56, 95% CI 5.07–16.04) women compared with normal-weight pregnant women [27]. Screening for gestational diabetes is recommended for obese women between 24 and 28 weeks of pregnancy for early detection of gestational diabetes to minimise adverse outcomes.

Public Health Measures on Obesity

The epidemic of obesity is often explained through energy balance. Energy cannot be destroyed. It can only be gained, lost or stored. This theory emphasises excessive food intake or insufficient physical activity as the determinants of obesity. As a society afflicted by this problem gains weight, its average moves to the right. As the tail of GDM to the right, more and more individuals cross the thresholds used to define obesity. It is important to note that the definition of obesity at BMI ≥ 30 kg/m^2 is arbitrary. It follows that intervention needs to focus both on reducing the average and the tail. Current guidelines for obesity prevention and treatment tend to focus on people in the tail who are a minority. It is also important to note that the rate at which people get classified as obese is faster than the rate at which the average moves to the right. Therefore, it is important that effective prevention is targeted as population average. A large preventive effect can only be achieved through public health and policy interventions that have an effect on the average weight of the society, which stops people tipping over the thresholds that define obesity. Thus, when focussing on obesity, it is important not to ignore the need for intervention in women who are not obese by definition.

Most of the interventions currently recommended target the women who start pregnancy obese or develop obesity during pregnancy or in the post-natal period. The health benefits to the mother and the unborn child with this strategy are limited for the following reasons: the interventions usually commence when obesity is already established making it harder to reverse the trend; a significant proportion of the resources, both financial and personnel, are focussed in the management of weight and prevention of complications in secondary care placing a strain on the health care system; and in the absence of capabilities to continue the intense intervention after childbirth, the benefits may be negated for subsequent pregnancies. In addition to increasing the awareness regarding the risks of obesity itself, the specific harm to the mother and the baby need to be highlighted with an emphatic approach to change behaviour, similar to the use of alcohol, smoking and substance misuse in pregnancy. The success of public health measures on obesity risks in pregnancy can only succeed if supported by other health policy measures such as support groups, helplines, taxes on junk foods and subsidy for healthy foods and media campaigns.

Recommendations for Clinical Practice

- Risk assessment of the women early in pregnancy or pre-pregnancy and formulate appropriate management plan for pregnancy.
- Provision of evidence-based in-depth advice on the magnitude of risks of obesity in pregnancy to the mother and foetus.
- Weight management interventions to commence as early as possible, stressing the importance of diet.
- Nutritional supplementation with higher dose of folic acid in pre-pregnancy and first trimester and vitamin D throughout pregnancy.
- Close surveillance for pre-eclampsia and screening for gestational diabetes.
- Consideration of prophylactic measures like aspirin or thromboprophylaxis where indicated.

Recommendations for Research

- There is a need to evaluate the impact of adherence to IOM recommendations for obese mothers on optimal weight gain in pregnancy on maternal and foetal outcomes.
- Identification of the predictors of adverse short-term and long-term outcomes in obese pregnant women will enable commencement of appropriate preventive and treatment measures reduce complications.

- An individual patient data meta-analysis on the effects of interventions on maternal and foetal outcomes will ascertain the magnitude of incremental benefit associated with the reduction in GWG and identify the groups that benefit the most from interventions.
- Determination of the optimal weight gain in pregnancy for women stratified by BMI at the onset of pregnancy to maximise the favourable outcome to the mothers and children will provide concrete targets for weight management in pregnancy.
- Comparison of cost-effectiveness of the measures to tackle obesity in non-pregnant women of the childbearing age with those who are already pregnant will be the first step in forming health policies and measures for this problem.

References

1. Confidential Enquiry into Maternal and Child Health. *Saving Mothers' Lives: Reviewing Maternal Details to Make Motherhood Safer. The Seventh Report of the Confidential Enquiries into Maternal Deaths in the United Kingdom*. London: Confidential Enquiry into Maternal and Child Health; 2007.
2. Heslehurst N, Rankin J, et al. A nationally representative study of maternal obesity in England, UK: trends in incidence and demographic inequalities in 619,323 births, 1989–2007. *Int J Obes*. 2009;34(3):420–428.
3. Dodd JM, Crowther CA, Robinson JS. Dietary and lifestyle interventions to limit weight gain during pregnancy for obese or overweight women: a systematic review. *Acta Obst Gynecol*. 2008;87:702–706.
4. Galtier Dereure Boegner C, Bringer J. Obesity and pregnancy: complications and cost. *Am J Clin Nutr*. 2000;71:1242S–1248S.
5. Jackson RA, Stotland NE, Caughey AB, et al. Improving diet and exercise in pregnancy with video doctor counseling: a randomized trial. *Patient Educ Couns*. 2011; 83:203–209. doi:10.1016/j.pec.2010.05.019.
6. Koning AM, Kuchenbecker WK, Groen H, et al. Economic consequences of overweight and obesity in infertility: a framework for evaluating the costs and outcomes of fertility care. *Hum Reprod Update*. 2011;16:246–254.
7. National Institute for Health and Clinical Excellence. *Dietary Interventions and Physical Activity Interventions for Weight Management Before, During and After Pregnancy – NICE Public Health Guideline 27*. London: NICE; 2010.
8. Obesity and Reproductive Health RCOG Study Group Statement. *Consensus Views Arising from the 53rd Study Group: Obesity and Reproductive Health*. London: RCOG; 2007.
9. American College of Obstetricians and Gynecologists. ACOG committee opinion number 315, September 2005: obesity in pregnancy. *Obstet Gynecol*. 2005;106:671–675.
10. Rasmussen KM, Yatkine AL, eds. *Weight Gain During Pregnancy: Re-examining the Guidelines. Committee to Reexamine Institute of Medicine Pregnancy Weight Guidelines*. Washington, DC: The National Academies Press; 2009.
11. Institute of Medicine Subcommittee on Nutritional Status and Weight Gain in Pregnancy. *Nutrition During Pregnancy*. Washington, DC: National Academies Press; 1990.
12. US Department of Health and Human Services, National Institutes of Health. *Healthy Eating and Physical Activity Across Your Life Span: Fit for Two: Tips for Pregnancy: NIDDK Weight Control Information Network*. Bethesda, MD: NIH Publication; 2002: NIH Publication No. 02-5130.

13. Thangaratinam S, Rogozinska E, Jolly K, etal. Interventions to reduce or prevent obesity in pregnant women: a systematic review. *Health Technol Assess.* 16:ISSN 1366–5278.
14. Gomez TG, Delgado JG, Agudelo AA, Hurtado H. Diet effects on the perinatal result of obese pregnant patient. *Rev Colomb Obstet Ginecol.* 1994;45:313–316 [in Spanish].
15. Badrawi H, Hassanein MK, Badraoui MHH, Wafa YA, Shawky HA, Badrawi N. Pregnancy outcome in obese pregnant mothers. *J Perinat Med.* 1992;20:203.
16. Thornton YS. Preventing excessive weight gain during pregnancy through dietary and lifestyle counseling: a randomized controlled trial. *Obstet Gynecol.* 2009;114:173.
17. National Institute for Health and Clinical Excellence. *Maternal and Child Nutrition – NICE Public Health Guideline 11.* London: NICE; 2008: www.nice.org.uk/guidance/PH11/.
18. Beyerlein A, Schiessl B, Lack N, von Kries R. Associations of gestational weight loss with birth-related outcome: a retrospective cohort study. *BJOG.* 2011;118:55–61.
19. Edmunds J, French DP, Atkinson L, et al. *Exploring the Barriers and Facilitators to 'Maintaining a Healthy Weight' During Pregnancy and the Post-Natal Period – Final Report.* Nuneaton: Nuneaton and Bedworth Borough Council; 2008.
20. Atkinson L, Edmunds J. *Developing a Theory-Based Exercise Intervention to Reduce Perinatal Obesity in Warwickshire – Final Report.* Warwickshire: NHS Warwickshire; 2009.
21. Rasmussen SA, Chu SY, Kim SY, Schmid CH, Lau J. Maternal obesity and risk of neural tube defects: a metaanalysis. *Am J Obs Gynecol.* 2008;198:611–619.
22. Mojtabai R. Body mass index and serum folate in childbearing age women. *Eur J Epidemiol.* 2004;19:1029.
23. Bodnar LM, Catov JM, Roberts JM, Simhan HM. Prepregnancy obesity predicts poor vitamin D status in mothers and their neonates. *J Nutr.* 2007;137:2437–2442.
24. Duckitt K, Harrington D. Risk factors for pre-eclampsia at ante-natal booking: systematic review of controlled studies. *Br Med J.* 2005;330(7491):565.
25. Larsen TB, Sorenson HT, Gislum N, Johnsen SP. Maternal smoking, obesity, and risk of venous thromboembolism during pregnancy and the puerperium: a population-based nested case–control study. *Thromb Res.* 2007;120:505–509.
26. Kim C, Newton KM, Knopp RH. Gestational diabetes and the incidence of type 2 diabetes: a systematic review. *Diabetes Care.* 2002;25:1862–1868.
27. Chu SY, Callaghan WM, Kim SY, et al. Maternal obesity and risk of gestational diabetes mellitus. *Diabetes Care.* 2007;30:2070–2076.

34 Multimodal Framework for Reducing Obesity-Related Maternal Morbidity and Mortality

Leroy C. Edozien

Manchester Academic Health Sciences Centre, University of Manchester, St Mary's Hospital, Manchester, UK

Introduction

Although there is a dearth of robust evaluation, it is generally accepted that prenatal care was one of the success stories of the twentieth century. A simple, relatively inexpensive intervention produced marked reduction in maternal morbidity and mortality in countries where this intervention was widely available. As adverse pregnancy outcomes for healthy women diminished, women with pre-existing medical problems became the focus of attention. Here too, pregnancy outcomes (in particular, survival) improved markedly as appropriate, targeted, multidisciplinary prenatal care pathways were developed for specific high-risk groups. This was reflected in reduced mortality rates and, albeit to a lower degree, morbidity statistics for women with conditions such as diabetes, cardiac disease and renal transplant. With the emergence of obesity as a major risk factor for maternal morbidity and mortality, it can be assumed that lessons learned from the prenatal (including intrapartum) management of chronic medical conditions can be applied to obesity. Unlike the relatively homogenous medical conditions, however, obesity is multifactorial in its aetiology, co-morbidities and consequences [1,2]. It is associated with a number of interacting risks, and many of these are best addressed before pregnancy. Furthermore, it is not usually perceived as a disease condition by the patient. Indeed, treating her as someone with a disease could increase the woman's sense of stigmatisation. For these reasons, strategies over and above that currently applied to chronic medical conditions are needed. In particular, both public health and clinical interventions are required. Unfortunately, the development of strategies for meeting the challenges of maternal obesity has lagged well behind the rising prevalence (now often described colloquially but incorrectly as an 'epidemic') of the condition [3,4]. Although good practice guidelines have been produced [5,6], there is still no comprehensive framework for tackling this challenge.

Obesity. DOI: http://dx.doi.org/10.1016/B978-0-12-416045-3.00034-0

Obesity in Women of Reproductive Age

Addressing the challenge of reducing obesity-related maternal mortality and morbidity should start with recognition of the size of the problem, for unless we appreciate this we are unable to determine and provide the resources and facilities needed for reducing maternal mortality and morbidity.

The age-adjusted prevalence of obesity in adult women in the United States in 2009–2010 was 35.8%, but there are indications that the previously rising trend has levelled [7]. On the other hand, the English Health Survey [8] showed that the proportion of obese women, as defined by body mass index (BMI), increased from 16.4% in 1993 to 26.1% in 2010. The proportion of adult females with a raised waist circumference increased from 26% to 46% during this period. Similar prevalence has been reported for Scotland [9] and Wales [10].

Correspondingly, studies in different parts of the United Kingdom indicate a geometric rise in the prevalence of obesity in pregnancy (maternal obesity) since the end of the last millennium [11–13]. The UK Obstetric Surveillance System reported that nearly 1 in every 1000 women giving birth in the United Kingdom between March 2007 and August 2008 had a BMI $\geq 50\ kg/m^2$ or weighed more than 140 kg [14]. An audit conducted by the UK Centre for Maternal and Child Enquiries (CMACE) in 2009 showed that the UK prevalence of women with a BMI $\geq 35\ kg/m^2$ who give birth $\geq 24 + 0$ weeks' gestation was 4.99%, and the prevalence of morbid obesity (BMI $\geq 40\ kg/m^2$) was 2.01% [15].

The rising prevalence of obesity in women of reproductive age is not limited to the United Kingdom and United States; it is truly global. The prevalence of obesity in women has been reported to be high and/or rising in Morocco [16], Cameroon [17], Iran [18], India [19], Bolivia [20], Spain [21], Canada [22], Australia [23]; this 'epidemic' is not restricted to affluent countries [24].

Various studies across the world also show a rural/urban divide in the prevalence of obesity [8]. Statistics from the United Kingdom show that obese pregnant women are more likely to be older in pregnancy, to have a higher parity and to live in areas of high deprivation, compared with non-obese women [8]. This survey also showed that the proportion of women with a raised waist circumference was also lowest in the highest income quintile (36%) and highest in the lowest income quintile (53%). The prevalence of obesity is higher among females of Black Caribbean and Black African ethnicities, compared to the other ethnic groups and the general population. A link between obesity and social deprivation was also shown in the CMACE maternal obesity project [15].

These epidemiological observations inform the framework described later for reducing mortality and morbidity. Maternity care providers should be mindful of differences between ethnic groups when defining thresholds for obesity: a particular BMI does not necessarily connote the same degree of fatness for all, and Asian populations (including those from the Far East) have a high percentage of body fat at a relatively low BMI. For this reason, a consensus meeting concluded that the current World Health Organization BMI cut-off points 'do not provide an adequate basis for taking action on risks related to overweight and obesity in many populations in Asia' [25].

Obesity-Related Maternal Mortality

The Sixth Report of the Confidential Enquiries into Maternal Deaths in the United Kingdom [26] found that between 2000 and 2002, 29% of women who suffered a direct or indirect death were obese (compared with 23% of the population of women of reproductive age). The seventh report [27] found that between 2003 and 2005, 27% of deaths (64/231) were associated with BMI $>$ 30 kg/m^2. In this triennium, of the six maternal deaths directly related to anaesthesia, four patients were obese and two were morbidly obese. Of the eight anaesthetic-related maternal deaths reported by a survey in the state of Michigan, United States, six women were obese or morbidly obese [28].

The causes of death identified in these reviews include complications at intubation or extubation, aspiration of gastric content, post-operative respiratory failure, thromboembolism and inexperience of the attending anaesthetist [26,29].

A comprehensive framework for prevention should address the proximate and remote factors underlying these causes of death.

Obesity-Related Maternal Morbidity

The obese woman faces a range of potential complications spanning pregnancy, labour and the puerperium (Box 34.1). Studies from various countries [30–36] show that, compared to those with normal BMI, obese women are more likely to be hypertensive at the beginning of pregnancy, and more likely to develop pregnancy-induced hypertension, pre-eclampsia, gestational diabetes, preterm delivery and post-partum haemorrhage. Other problems associated with maternal obesity include increased risk of sleep apnoea [37], congenital anomaly [38], macrosomia [31], disproportionate use of Caesarean delivery [39] and difficulty in initiating and maintaining lactation [40,41]. The higher rate of Caesarean delivery has additive implications: obesity is associated with prolonged delivery interval, total operative time, major post-partum haemorrhage, multiple epidural placement failures, post-operative endometritis and prolonged hospitalisation [42]. The risk of venous thromboembolism is increased.

There are also increased risks to the newborn baby of an obese mother (Box 34.2). Obese women are approximately twice as likely to have a stillborn baby as non-obese women. [43] Other perinatal problems associated with maternal obesity include birth trauma, birth asphyxia and neonatal hypoglycaemia. These are commonly classified as foetal complications, but they contribute to both physical and psychological distress of the mother.

Box 34.1 Maternal Morbidity Associated with Obesity

Periconceptional

- Chronic hypertension
- Diabetes mellitus

Antenatal

- Miscarriage
- Sleep apnoea
- Gestational hypertension
- Pre-eclampsia
- Gestational diabetes
- Vitamin D deficiency
- Foetal macrosomia
- Thromboembolism
- Cardiac disease

Peripartum

- Induction of labour
- Delayed progress in labour
- Shoulder dystocia
- Caesarean delivery, wound infection, thrombosis
- Post-partum haemorrhage, anaemia
- Post-partum endometritis
- Thromboembolism
- Prolonged post-partum hospital stay

Anaesthetic Complications

- Difficult venepuncture/cannulation
- Difficult placement of epidural catheter
- Sub-optimal epidural analgesia
- Difficult or failed intubation

In the context of this framework, maternal morbidity refers not only to physical but also to psychological morbidity. Psychological morbidity may induce obesity or result from obesity. Pregnant obese women report feelings of humiliation and stigmatisation, and these could be unwittingly exacerbated by health professionals [44].

The Concept of a Multimodal Framework

The underlying factors, established and putative, accounting for obesity in an individual woman and for obesity 'epidemic' in the population are heterogeneous, so

Box 34.2 Fetal Risks and Complications Associated with Maternal Obesity and Potentially Contributory to Maternal Psychological Morbidity

Neural tube defect and other congenital anomalies
Macrosomia
Preterm delivery
Traumatic delivery
Low-Apgar scores
Stillbirth
Meconium aspiration
Neonatal hypoglycaemia
Jaundice
Admission to special/intensive care

Box 34.3 Constituent Elements of the Multimodal Framework for Reducing Obesity-Related Maternal Mortality and Morbidity

Coordinated primary, secondary and tertiary health care services
Implementation of primary, secondary and tertiary prevention strategies
Interventions spanning pre-pregnancy, pregnancy, post-pregnancy care
Multidisciplinary approach – input from midwives, obstetricians, anaesthetists, psychologists, endocrinologists, GP, dieticians, physiotherapists, surgeons, public health physicians and other professionals
Engagement between obstetric, midwifery and public health services
Holistic care, including provision of psychological support
Integrated approach to lifestyle issues during and beyond pregnancy – smoking, diet, alcohol, exercise, sleep
Long-term vision and life-course thinking – taking account of epigenetics, allostasis and foetal programming
Family-centred philosophy

the approach to tackling the problem should ideally be multimodal, and the various modalities should be strategically integrated into a coherent and comprehensive framework. The constituent elements of such a framework are itemised in Box 34.3.

Implementation of the framework requires the concerted efforts not only of obstetricians and midwives but also of other stakeholders (Box 34.4). It is a multi-sectoral activity, involving the three health sector tiers (primary, secondary and tertiary), and sectors such as education, social services and employment. Commissioners of women's health services have a major role to play in delivering the type of preventative care advocated in this chapter, and an evidence-based

Box 34.4 Key Stakeholders in Reducing Obesity-Related Maternal Morbidity and Mortality

Health policy leaders at national, regional and local levels
Service users
Local authority
Service commissioners
Health provider institutions at all levels and service managers
Health professionals (multidisciplinary, multiprofessional)
Third sector

guide for commissioners on weight management before, during and after pregnancy is available [45].

The multimodal framework encompasses primary, secondary and tertiary prevention programmes. Primary prevention of obesity-related maternal mortality and morbidity entails reducing the numbers of women who start pregnancy with BMI > 30 kg/m^2. This combines global efforts to reduce the prevalence of obesity in women of childbearing age with focused preconception care for individual at-risk women. Secondary prevention aims to reduce mortality and morbidity in women who are obese at the beginning of pregnancy, through interventions that prevent complications of obesity, and ensuring early detection of any such complication. Tertiary prevention targets effective management of obese women who have developed complications (such as gestational diabetes) during pregnancy, preventing the progression of disease or disability. Each of the three tiers of health service has a role in delivering all three levels of prevention.

The multimodal framework incorporates a life-course approach. With at least half of all pregnancies in the general population being unplanned, preventative services restricted to pregnant women and/or women planning a pregnancy will have sub-optimal impact. Interventions to reduce the prevalence of obesity should be targeted to all women of childbearing age and should promote in this age group an awareness of obesity-related maternal mortality and morbidity.

Linked with this is the family-centred philosophy which aims to prepare families for parenthood [46]. Obesity-related maternal mortality and morbidity afflict not just the woman but also the family. It makes sense therefore to engage partners and to view prevention of obesity-related maternal mortality and morbidity as part of preparation for parenthood. In this regard, parents need to be more aware of the impact of obesity on foetal programming (the finding that the intrauterine environment can predetermine the health of the offspring in adult life) [47]. Obese mothers tend to have big babies that are programmed to be obese in childhood, adolescence and adulthood [48].

Such a multisectoral and life-course programme of prevention may appear ambitious and aspirational, but the basis for it already exists in a raft of national policies, strategies and interventions such as *Healthy Lives, Healthy People: Our Strategy for Public Health in England* [49], *Obesity: The Prevention, Identification, Assessment and Management of Overweight and Obesity in Adults and Children* [50], *Tackling Obesities: Future Choices – Modelling Future Trends in Obesity and Their Impact on Health* [51], *Maternity and Early Years: Making a Good Start to Family Life* [46], *Healthy Child Programme: Pregnancy and the First Five Years of Life* [52] and *Healthy Lives, Brighter Futures – The Strategy for Children and Young People's Health* [53]. There are also exemplary projects at regional level which demonstrate the feasibility of an integrated approach – for example, the West Midlands Reducing Childhood Obesity project [54].

Primary Prevention of Obesity-Related Maternal Mortality and Morbidity

Despite the absence of robust health economic analysis, it can reasonably be assumed that the most cost-effective way of preventing obesity-related maternal mortality and morbidity is to reduce the number of women who start pregnancy with a high BMI. Women of childbearing age should be motivated and supported to curb excessive intake of protein, fat and simple carbohydrates, to avoid a sedentary lifestyle and to undertake regular exercise. Women who are already overweight or obese should be advised of the obstetric risks and encouraged to lose weight before becoming pregnant. Those groups identified from epidemiological studies as being at high risk should be targeted, and barriers to uptake of lifestyle interventions should be addressed. Commissioners of health services should provide for the establishment of community-based services for weight management, to help women lose weight before they become pregnant [45].

Primary prevention is not restricted to nulliparous women; women who have had babies should be encouraged to optimise their weight before the next pregnancy. Weight loss between pregnancies has been shown to reduce the risk of gestational diabetes in the subsequent pregnancy [55]. Conversely, increase in BMI from normal in the first pregnancy to overweight or obese in the second pregnancy is associated with increased risk of pre-eclampsia in the second pregnancy [56].

Taking a long-term strategic position, primary prevention of obesity-related maternal mortality and morbidity could be seen as beginning from intrauterine and early childhood life. Compared with those of normal weight, macrosomic babies and obese children are more likely to be obese in adulthood. As the childbearing population becomes taller and bigger, average birthweight increases, and it has been hypothesised that control of birth size and subsequent growth rate has the best potential for curbing the obesity epidemic. Viewed in this light, control of childhood obesity has implications for future obesity-related maternal morbidity and mortality.

Secondary Prevention of Obesity-Related Maternal Mortality and Morbidity

Primary prevention is in theory simpler and more effective than secondary prevention of obesity-related maternal mortality and morbidity but, paradoxically, is more challenging to implement. Until public health initiatives to address the problem begin to yield full dividend, secondary prevention must be fully resourced and delivered. National guidelines [5,6] for managing obesity in pregnancy are now available (efforts in this regard lagged behind the 'epidemic'), and any obese – or, in particular, morbidly obese – pregnant woman managed outwith the guidelines may be receiving substandard care which is difficult to justify in the event of mortality or morbidity.

How the obesity-in-pregnancy service is configured and delivered will vary with local circumstances, but each unit must have formal policies and procedures.

Dedicated Clinic

In some units, especially larger centres, there will be a dedicated obesity antenatal clinic. Given the high prevalence of maternal obesity, it will not usually be feasible to sequester all obese pregnant women in one clinic, and the dedicated clinic will determine its threshold for referral. The super-obese ($BMI \geq 50\ kg/m^2$) have a higher risk of mortality and morbidity than other obese women, so should be accorded priority, and a referral criterion of $BMI \geq 40\ kg/m^2$ will usually capture the higher risk women.

A dedicated obesity antenatal clinic looks good in principle, but in practice its clinical and cost-effectiveness could be limited. For an obesity antenatal clinic located in a secondary or tertiary institution to be optimally effective, it should see the patients early in pregnancy, have co-located multidisciplinary team members and offer specific interventions (over and above advice) tailored to the needs of the individual woman. It is not helpful if the woman is seen in the specialised clinic for the first time almost midway through the pregnancy and/or given detailed information on risks but offered no practical solutions and no support, or simply asked to see a dietician. The objective of the clinic should be clear (i.e. whether it is the screening and signposting women to relevant services or delivery of tailored interventions), and patients' expectations should be managed accordingly. The clinic should have written operating procedures spelling out its scope and how the service will be delivered, including not only clinical practice guidelines but also care pathways, lines of communication, staffing requirements, quality and safety, and value-for-money issues. Ideally, skill mix in a dedicated clinic should include competencies in behaviour change, diet management and exercise in pregnancy.

Whether or not the woman is cared for in a dedicated antenatal clinic, the fundamentals of care are the same and can be delivered in standard secondary care setting, in coordination with primary care.

The Booking Visit and Follow-Up Arrangements

Booking for antenatal care early in pregnancy will help reduce the risk of mortality and morbidity. A baseline risk assessment is undertaken and a plan of care devised. The assessment includes risk categorisation for venous thromboembolism and institution of appropriate prophylaxis. The blood pressure should be measured with a large cuff, and her weight should be recorded with a suitable weighing scale, rather than relying on self- or GP-reported weight.

Compared with non-obese women, obese women have lower serum folate levels [57], so should be commenced on higher dose folic acid (5 mg daily) if she is not already on this dose, as she should be periconceptionally. Obese women and their babies are also at increased risk of vitamin D deficiency [58] so the woman should be commenced on vitamin D 10 μg periconceptionally or at booking.

Screening for gestational diabetes should be arranged (this would usually but not invariably be a glucose tolerance test, GTT; see section on bariatric surgery below).

In the third trimester of pregnancy, morbidly obese women should be formally assessed for tissue viability and manual handling requirements for childbirth [5].

It is not advisable to rely on symphysio-fundal height measurements for monitoring foetal growth, as the maternal fat pad may contribute to giving the impression of a normal reading, and foetal growth restriction is missed. Serial growth scan should be performed instead, but clinicians should also be mindful of the limitations of ultrasound scan in the obese woman, particularly in relation to prenatal diagnosis [59].

The place of birth and risks of pregnancy and labour should be discussed, and all discussions and plans should be documented. Birth outside an obstetric unit is not recommended for women with BMI ≥ 35 kg/m^2. Weight management should also be discussed and referral to a dietician arranged, as appropriate.

Health professionals should be proactive in engaging the woman at all stages in pregnancy, labour and the puerperium. Comments such as expressions of difficulty in auscultating the foetal heart tones or securing good views of the baby at ultrasound scan could induce psychological distress if made insensitively.

Anaesthetic Consultation

It is essential that the woman should have a consultation with a senior anaesthetist. This affords an opportunity for her to be fully counselled regarding anaesthetics and for a perinatal care plan to be formulated. There should be a mechanism for confirming that the woman has kept the anaesthetic consultation appointment, and the consultation should be well documented.

Facilities

Obesity makes a huge call on health care resources [60,61] but the availability of appropriate facilities is central to reduction of morbidity and mortality. The required equipment include large blood pressure cuffs; weighing scales with a safe working load of 300 kg; sit-on weighing scale; large chairs without arms; large wheelchairs, ultrasound scan couches, ward and delivery beds; suitable operating theatre trolleys and tables; and lifting and lateral transfer equipment. Unfortunately, as demonstrated by a UK audit [15], the required facilities are not always available and when they are, they are not accessible (e.g. because they are in the outpatients department but not on labour ward or not accessible out of hours). The majority of the units in the United Kingdom did not (at the time of the audit in 2008) have immediate access to appropriate extra-wide wheelchairs, examination couches, trolleys or ward beds, and the minimum safe working load of 250 kg recommended by the CMACE/Royal College of Obstetricians and Gynaecologists (RCOG) guideline [5].

It is recommended that all maternity units should have a documented environmental risk assessment regarding the availability of the following facilities to care for obese pregnant women:

Circulation space
Accessibility, including doorway widths and thresholds
Safe working loads of equipment (up to 250 kg) and floors
Appropriate theatre gowns
Equipment storage
Transportation
Staffing levels
Availability of, and procurement process for, the required equipment.

Staff

Units with a dedicated obesity antenatal clinic would usually have a specialist midwife who facilitates compliance with the care pathways and protocols and coordinates the transition of the patient between the community, clinic and wards. Regardless of the existence of this coordinating professional, all obstetricians and midwives should be competent to provide the fundamentals of care for obese pregnant women. Specific training will usually be required, as it has been shown that information is often not provided adequately [62] or sensitively [44,63]. A training package [64] produced by the UK National Institute for Health and Clinical Excellence for health professionals dealing with complex issues in pregnancy can be adapted for this purpose.

Weight Control in Pregnancy

Obese women often find themselves receiving conflicting advice about weight control in pregnancy [63]. Health professionals should be confident about giving evidence-based advice. Guidelines produced by the Institute of Medicine in the United States [65] recommend a target weight gain of 5–9 kg during pregnancy for obese women, less than the 11.5–16 kg recommended for women of normal BMI. There are strong reasons for weight control in pregnancy. Women who gain excess weight in pregnancy are more likely to retain weight after the pregnancy, and this could mean that they start the next pregnancy with even more weight. In the Stockholm Pregnancy and Women's Nutrition study, women who gained more than 15.6 kg during pregnancy retained more weight 1 year and 15 years after delivery [66].

While the need to control weight during pregnancy is recognised there are, unfortunately, no interventions for achieving this that are supported by high-level evidence [67–69].

Dieting in pregnancy is not recommended, but healthy eating and exercise are recommended. The RCOG recommends that morbidly obese women undertaking exercise in pregnancy should do so under medical supervision [70]. A systematic review of weight management interventions in pregnancy showed no evidence of benefit, and this was possibly because some of the barriers to achieving healthy weight gain identified by women in qualitative studies were not addressed by the interventions evaluated [69]. The authors concluded that 'multiple types of interventions including community-based strategies are needed to address this complex health problem' [69]. Guidance in this regard is available in the United Kingdom [71]. The key point is that interventions should be individualised. It has been noted that although trials conducted in North American populations were not successful in helping women limit weight gain in pregnancy, studies in Sweden and Denmark – where the subjects received individualised attention – showed that motivated obese pregnant women were able to limit their weight gain during pregnancy to 6–7 kg [65].

Intrapartum Care

The key to reducing intrapartum mortality and morbidity in obese women is anticipatory care in labour. Obese women, particularly the morbidly obese, should have a venous cannula inserted early in labour and the theatre team should be alerted. It may be necessary to confirm foetal presentation by ultrasound scan, as the abdominal panniculus could make palpation difficult and unreliable. For the same reason, it may be difficult to obtain a good quality cardiotocographic trace using an abdominal transducer and there should be ready recourse to use of a foetal scalp electrode. There is a higher than average chance of a slow labour, and attention should be paid to pressure areas and to hydration. The third stage of labour should be actively managed.

Caesarean Section

Approximately, one in two of morbidly obese women will have a Caesarean delivery, so preparations for this should be started early [72]. Given the risks of a general anaesthetic and the likelihood of prolonged operating time, the operation is best performed under combined spinal–epidural analgesia. Obese women are more likely to require more than two attempts to achieve a regional block [73] so an experienced anaesthetist should be present and the probability of a prolonged effort to establish regional analgesia should be taken into account.

There should be a suitable bed and operating table (see Section Facilities), and the table should be capable of providing a left lateral tilt of the obese woman. A low transverse abdominal incision is usually feasible, but the panniculus will have to be retracted to allow this in the morbidly obese. Prophylactic antibiotics should be given and the subcutaneous fat layer should be repaired to reduce the risk of wound infection.

Thromboembolism Prophylaxis

As indicated above, venous thromboembolism risk should be assessed early in pregnancy, and appropriate prophylaxis commenced. The risk assessment should be revisited throughout pregnancy and labour. It is important that the woman receives the right dose of prophylactic anticoagulant therapy, based on her weight. Women assessed to be at high or moderate risk of venous thromboembolism should receive pharmacological prophylaxis for 6 weeks post-partum.

Breastfeeding

Obese women need to be supported to initiate and maintain breastfeeding. In addition to the benefits of breastfeeding that apply to all, there are a couple of benefits for the obese woman. Firstly, exclusive lactation promotes post-partum weight loss. Secondly, formula feeding increases the risk that the newborn will develop obesity.

Post-Natal Care

Post-partum pressure care is important, as is early ambulation, and she should be seen by a physiotherapist. Arrangements should be made for her to receive post-natal support for weight reduction and healthy eating, and coordination of care with the primary care tier is crucial in this regard. Unfortunately, despite policy proclamations, community services for managing post-natal obesity remain rudimentary [74]. If the woman has gestational diabetes, arrangements should be made for follow-up GTT, to exclude type 2 diabetes post-natally.

Tertiary Prevention of Obesity-Related Maternal Mortality and Morbidity

Dealing with Co-Morbidities

Diabetes and hypertension are the most common co-morbidities, and the coexistence of obesity and diabetes compounds maternal and foetal risks, with higher rates of complications such as preterm labour, pre-eclampsia, birth trauma, post-operative wound complications, macrosomia, shoulder dystocia, stillbirth, growth restriction and hypoglycaemia [75]. Increased surveillance and liaison with diabetologists and renal physicians will facilitate early detection and management of these and other co-morbidities, including mental health problems.

Vaginal Birth After Caesarean Section

Obesity is associated with a lower chance of success and higher risk of scar rupture in women undergoing a trial of vaginal birth after a prior Caesarean section (VBAC), when compared to those having a planned Caesarean delivery [76]; infectious morbidity is particularly high [76,77]. Maternal morbidity was almost doubled in morbidly obese women, and in cases where the trial was unsuccessful there was a sixfold increase in composite maternal morbidity [76]. Morbid obesity also carried a fivefold risk of neonatal injury. Although relative risks were higher, the absolute risk of morbidity was low. In other studies, however, the success rate of VBAC was as low as 15% [77]. Obese women with a prior Caesarean section (CS) should therefore be carefully counselled regarding mode of delivery. Emergency CS following failed VBAC places the woman at significant risk of morbidity (anaesthetic problems, infection, haemorrhage and thrombosis), so the risks and benefits of elective CS versus VBAC have to be carefully weighed up. Morbidly obese women with no previous vaginal delivery and/or for whom induction of labour is contemplated should be informed that an elective repeat CS offers significantly lower prospects of maternal morbidity.

Pregnancy After Bariatric Surgery

Bariatric surgery could be regarded as a means of primary prevention of maternal mortality and morbidity – as it acts to reduce the number of women who are obese at the beginning of pregnancy. Some women who have undergone bariatric surgery will, however, still be obese at the start of pregnancy. In some cases, this will be because they have not waited for the recommended 12–18 month interval between surgery and pregnancy, so have not lost enough weight. Other women may have lost weight substantially but still have a BMI over 30 kg/m^2.

The published data on pregnancy outcome following bariatric surgery suffers two limitations. First, there are no large-scale studies. Secondly, studies often fail to classify outcomes by type of surgery – gastric banding (a restrictive procedure) or gastric bypass (a malabsorptive procedure). Also, a variety of control groups have been used. In general, women who have had bariatric surgery have rates of co-morbidities (gestational diabetes, pre-eclampsia) that are lower than those of obese women who did not have surgery but higher than those of non-obese women [78,79]. There is a lower rate of macrosomia but a higher risk of intrauterine growth restriction, particularly in women who have had bypass surgery, because of maternal and foetal undernutrition.

The care of pregnant women with a history of bariatric surgery should be multidisciplinary, with input from the surgeon, nutritionist and other disciplines that may be called upon. The general principles outlined above for obese women apply.

Gastric bypass surgery can cause deficiencies of iron, folate, vitamin B12, calcium and occasionally vitamin K, due to the duodenum (site of absorption) being bypassed, and the absence of intrinsic factor produced in the stomach. The deficiencies could be aggravated by hyperemesis. Periconceptional nutritional status assessment and monitoring of iron, vitamins B1, B12, and D and red cell folate is essential, and adequate nutritional supplements should be provided, in association with the nutritionist.

Following bariatric surgery, glucose which rapidly reaches the small intestine draws fluid into the gut by osmosis, resulting in symptoms of hypovolaemia such as palpitation, tachycardia and fainting. This is known as the dumping syndrome and may also manifest as abdominal pain and diarrhoea. In pregnancy, the dumping syndrome can be precipitated by a GTT, so this should be avoided in women who have had gastric bypass (but not for gastric band cases). As an alternative to GTT, serial home monitoring of blood glucose should be done.

Pregnancy increases the risk of intestinal obstruction in women who have undergone bariatric surgery, and deaths from it have been reported, so this diagnosis should be excluded if the woman presents with abdominal pain and vomiting.

CS is not routinely indicated after bariatric surgery and labour should be managed as outlined above.

Conclusion

Growing numbers of obese women will increase the population of pregnant women who have gestational diabetes, hypertensive disease and venous thromboembolism. There is also significant psychological morbidity. The use of health care services is increased, and this includes higher CS rates. There are major intrapartum risks, particularly in relation to anaesthesia. Reduction of mortality and morbidity in this group of women calls for coordinated action before, during and after pregnancy [80]. Preventative action limited to the duration of pregnancy, whilst transiently helpful to the individual woman, will not significantly reduce the overall burden of morbidity and mortality. National and international initiatives to reduce obesity-related mortality and morbidity should employ a multimodal framework as described in this chapter.

References

1. McAllister EJ, Dhurandhar NV, Keith SW, et al. Ten putative contributors to the obesity epidemic. *Crit Rev Food Sci Nutr*. 2009;49(10):868–913.
2. Bray GA. Medical consequences of obesity. *J Clin Endocrinol Metab*. 2004;89(6): 2583–2589.

3. Krishnamoorthy U, Schram CMH, Hill SR. Maternal obesity in pregnancy: is it time for meaningful research to inform preventive and management strategies? *BJOG*. 2006; 113(10):1134–1140.
4. Bick D. Addressing the obesity epidemic: time for the maternity services to act now but what strategies should we use? *Midwifery*. 2009;25(4):337–338.
5. Centre for Maternal and Child Enquiries & Royal College of Obstetricians and Gynaecologists. *Management of Women with Obesity in Pregnancy*. London: CMACE/ RCOG; 2010.
6. *American College of Obstetricians and Gynaecologists*. Obesity in pregnancy. ACOG Committee Opinion Number 315. Reaffirmed 2008. <http://www.acog.org/ Resources_And_Publications/Committee_Opinions/Committee_on_Obstetric_Practice/ Obesity_in_Pregnancy/>; 2005 Accessed 21.03.12.
7. Flegal KM, Carroll MD, Kit BK, Ogden CL. Prevalence of obesity and trends in the distribution of body mass index among US adults, 1999–2010. *JAMA*. 2012;307(5): 491–497.
8. The NHS Information Centre, Lifestyles Statistics. *Statistics on Obesity, Physical Activity and Diet: England*. London: The Health and Social Care Information Centre; 2012.
9. *The Scottish Health Survey*. Topic Report: Obesity – Scottish Government, 2011. <http://www.scotland.gov.uk/Resource/0038/00389668.pdf>; Accessed 29 July 2012.
10. Welsh Government. *The Welsh Health Survey, 2010*. Welsh Assembly 2011. Available at http://wales.gov.uk/docs/statistics/2011/110913healthsurvey10en.pdf. Accessed 29 July 2012.
11. Heslehurst N, Rankin J, Wilkinson JR, Summerbell CD. A nationally representative study of maternal obesity in England, UK: trends in incidence and demographic inequalities in 619,323 births, 1989–2007. *Int J Obes*. 2010;34(3):420–428. doi:10.1038/ ijo.2009.250.
12. Kiran TSU, Hemmadi S, Bethal J, Evans J. Outcome of pregnancy in a woman with an increased body mass index. *Br J Obstet Gynaecol*. 2005;112(6):768–772.
13. Kanagalingam MG, Forouhi NG, Greer IA, Sattar N. Changes in booking body mass index over a decade: retrospective analysis from a Glasgow Maternity Hospital. *Br J Obstet Gynaecol*. 2005;112(10):1431–1433.
14. Knight M, Kurinczuk JJ, Spark P, Brocklehurst P, UK Obstetric Surveillance System. Extreme obesity in pregnancy in the United Kingdom. *Obstet Gynecol*. 2010;115(5): 989–997.
15. Centre for Maternal and Child Enquiries (CMACE). *Maternal Obesity in the UK: Findings from a National Project*. London: CMACE; 2010.
16. Belahsen R, Mziwira M, Fertat F. Anthropometry of women of childbearing age in Morocco: body composition and prevalence of overweight and obesity. *Public Health Nutr*. 2004;7(4):523–530.
17. Pasquet P, Temgoua LS, Melaman-Sego F, Froment A, Rikong-Adié H. Prevalence of overweight and obesity for urban adults in Cameroon. *Ann Hum Biol*. 2003;30(5): 551–562.
18. Soutoudeh G, Khosravi Sh, Khajehnasiri F, Khalkhali HR. High prevalence of overweight and obesity in women of Islamshahr, Iran. *Asia Pac J Clin Nutr*. 2005;14(2): 169–172.
19. Singh RB, Pella D, Mechirova V, et al. Prevalence of obesity, physical inactivity and undernutrition, a triple burden of diseases during transition in a developing economy. The Five City Study Group. *Acta Cardiol*. 2007;62(2):119–127.

20. Pérez-Cueto FJ, Botti AB, Verbeke W. Prevalence of overweight in Bolivia: data on women and adolescents. *Obes Rev*. 2009;10(4):373–377.
21. García-Alvarez A, Serra-Majem L, Ribas-Barba L, et al. Obesity and overweight trends in Catalonia, Spain (1992–2003): gender and socio-economic determinants. *Public Health Nutr*. 2007;10(11A):1368–1378.
22. Shields M, Carroll MD, Ogden CL. Adult obesity prevalence in Canada and the United States. *NCHS Data Brief*. 2011;56:1–8.
23. Callaway LK, Chang AM, McIntyre HD, Prins JB. The prevalence and impact of overweight and obesity in an Australian obstetric population. *Med J Aust*. 2006;184(2): 56–59.
24. Mendez MA, Monteiro CA, Popkin BM. Overweight exceeds underweight among women in most developing countries. *Am J Clin Nutr*. 2005;81(3):714–721.
25. WHO Expert Consultation. Appropriate body mass index for Asian populations and its implications for policy and intervention strategies. *Lancet*. 2004;363(9403): 157–163.
26. Lewis G. *The Confidential Enquiry into Maternal and Child Health (CEMACH). Why Mothers Die 2000–2002. The Sixth Report of Confidential Enquiry into Maternal Deaths in the United Kingdom*. London: RCOG Press; 2004.
27. Lewis G. *The Confidential Enquiry into Maternal and Child Health (CEMACH). Saving Mother's Lives: Reviewing Maternal Deaths to Make Motherhood Safer: 2003–2005. The Seventh Report on Confidential Enquiries into Maternal Deaths in the United Kingdom*. London: Confidential Enquiry into Maternal and Child Health; 2007.
28. Mhyre JM, Riesner MN, Polley LS, Naughton NN. A series of anesthesia-related maternal deaths in Michigan, 1985–2003. *Anesthesiology*. 2007;106(6):1096–1104.
29. Cooper GM, McClure JH. Anaesthesia chapter from saving mothers' lives: reviewing maternal deaths to make pregnancy safer. *Br J Anaesth*. 2008;100(1):17–22. doi:10.1093/bja/aem344.
30. Weiss JL, Malone FD, Emig D, et al. Obesity, obstetric complications and caesarean delivery rate – a population-based screening study. *Am J Obstet Gynecol*. 2004;190(4): 1091–1097.
31. Bhattacharya S, Campbell DM, Liston WA, Bhattacharya S. Effect of body mass index on pregnancy outcomes in nulliparous women delivering singleton babies. *BMC Public Health*. 2007;24(7):168.
32. Callaway LK, Prins JB, Chang AM, McIntyre HD. The prevalence and impact of overweight and obesity in an Australian obstetric population. *Med J Aust*. 2006;184(2): 56–59.
33. Chan WS. Maternal body mass index and the risk of preeclampsia: a systematic overview. *Epidemiology*. 2003;14(3):368–374.
34. Sebire NJ, Jolly M, Harris JP, et al. Maternal obesity and pregnancy outcome: a study of 287,213 pregnancies in London. *Int J Obes*. 2001;25(8):1175–1182.
35. Briese V, Voigt M, Wisser J, Borchardt U, Straube S. Risks of pregnancy and birth in obese primiparous women: an analysis of German perinatal statistics. *Arch Gynecol Obstet*. 2011;283(2):249–253.
36. Rode L, Nilas L, Wojdemann K, Tabor A. Obesity related complications in Danish single cephalic term pregnancies. *Obstet Gynecol*. 2005;105(3):537–542.
37. Maasilta P, Bachour A, Teramo K, Polo O, Laitinen. LA. Sleep-related disordered breathing during pregnancy in obese women. *Chest*. 2001;120(5):1448–1454.

38. Stothard KJ, Tennant PW, Bell R, Rankin J. Maternal overweight and obesity and the risk of congenital anomalies: a systematic review and meta-analysis. *JAMA*. 2009;301(6): 636–650.
39. Poobalan AS, Aucott LS, Gurung T, Smith WC, Bhattacharya S. Obesity as an independent risk factor for elective and emergency caesarean delivery in nulliparous women – systematic review and meta-analysis of cohort studies. *Obes Rev*. 2009;10(1):28–35.
40. Lepe M, Bacardí Gascón M, Castañeda-González LM, Pérez Morales ME, Jiménez Cruz A. Effect of maternal obesity on lactation: systematic review. *Nutr Hosp*. 2011;26(6): 1266–1269.
41. Rasmussen KM, Kjolhede CL. Prepregnant overweight and obesity diminish the prolactin response to suckling in the first week postpartum. *Pediatrics*. 2004;113(5):e465–e471.
42. Perlow JH, Morgan MA. Massive maternal obesity and perioperative caesarean morbidity. *Am J Obstet Gynecol*. 1994;170(2):560–565.
43. Chu SY, Kim SY, Lau J, et al. Maternal obesity and risk of stillbirth: a metaanalysis. *Am J Obstet Gynecol*. 2007;197(3):223–228.
44. Furber CM, McGowan L. A qualitative study of the experiences of women who are obese and pregnant in the UK. *Midwifery*. 2011;27(4):437–444.
45. National Institute for Health and Clinical Excellence. *Weight Management Before, During and After Pregnancy*. London: NICE; 2011:Commissioning Guide CMG 36
46. Department of Health. *Maternity and Early Years: Making a Good Start to Family Life*. London: Department of Health; 2010.
47. Heerwagen MJR, Miller MR, Barbou LA, Friedman JE. Maternal obesity and fetal metabolic programming: a fertile epigenetic soil. *Am J Physiol Regul Integr Comp Physiol*. 2010;299(3):R711–R722. doi:10.1152/ajpregu.00310.2010.
48. Martínez JA, Cordero P, Campión J, Milagro FI. Interplay of early-life nutritional programming on obesity, inflammation and epigenetic outcomes. *Proc Nutr Soc*. 2012;71(2): 276–283.
49. Department of Health. *Healthy Lives, Healthy People: Our Strategy for Public Health in England*. London: Department of Health; 2010. <http://www.dh.gov.uk/en/Publicationsandstatistics/Publications/PublicationsPolicyAndGuidance/DH_121941/>.
50. National Institute for Health and Clinical Excellence. *Obesity: The Prevention, Identification, Assessment and Management of Overweight and Obesity in Adults and Children*. London: NICE; 2006:NICE Clinical Guideline CG43.
51. Foresight Programme. *Tackling Obesities: Future Choices – Modelling Future Trends in Obesity and Their Impact on Health*. London: Department of Innovation Universities and Skills; 2007.
52. Shribman S, Billingham K. *Healthy Child Programme – Pregnancy and the First Five Years*. London: Department of Health and Department for Children, Schools and Families; 2009.
53. Department of Health. *Healthy Lives, Brighter Futures – The Strategy for Children and Young People's Health*. London: Department of Health; 2009.
54. *West Midlands NHS*. Investing for Health Programme. Achievement Report – Project 2b Reducing Childhood Obesity. <http://www.ifh.westmidlands.nhs.uk/InvestingforHealthKeyProjects/P02bChildhoodObesity.aspx>; 2011 Accessed 21.03.12.
55. Glazer NL, Hendrickson AF, Schellenbaum GD, Mueller BA. Weight change and the risk of gestational diabetes in obese women. *Epidemiology*. 2004;15(6):733–737.
56. Villamor E, Cnattingius S. Interpregnancy weight change and risk of adverse pregnancy outcomes: a population-based study. *Lancet*. 2006;368(9542):1164–1170.

57. Mojtabai R. Body mass index and serum folate in childbearing age women. *Eur J Epidemiol*. 2004;19(11):1029.
58. Bodnar LM, Catov JM, Roberts JM, Simhan HN. Prepregnancy obesity predicts poor vitamin D status in mothers and their neonates. *J Nutr*. 2007;137(11):2437–2442.
59. Wolfe HM, Sokol RJ, Martier SM, et al. Maternal obesity: a potential source of error in sonographic prenatal diagnosis. *Obstet Gynecol*. 1990;76(3 Pt 1):339–342.
60. Heslehurst N, Lang R, Rankin J, et al. Obesity in pregnancy: a study of the impact of maternal obesity on NHS maternity services. *Br J Obstet Gynaecol*. 2007;114(3): 334–342.
61. Chu SY, Bachman DJ, Callaghan WM, et al. Association between obesity during pregnancy and increased use of health care. *N Engl J Med*. 2008;358(14):1444–1453.
62. Mhyre JM, Greenfield ML, Polley LS. Survey of obstetric providers' views on the anesthetic risks of maternal obesity. *Int J Obstet Anesth*. 2007;16(4):316–322.
63. Furness PJ, McSeveny K, Arden MA, Garland C, Dearden AM, Soltani H. Maternal obesity support services: a qualitative study of the perspectives of women and midwives. *BMC Pregnancy Childbirth*. 2011;11:69.
64. *National Institute for Health and Clinical Excellence*. Raising sensitive issues – a training package for maternity settings. <http://www.nice.org.uk/usingguidance/implementationtools/educationaltools/raisingsensitiveissuestrainingpack.jsp/>; 2011 Accessed 23.03.12.
65. Institute of Medicine. *Weight Gain During Pregnancy: Re-Examining the Guidelines*. Washington, DC: National Academic Press; 2009.
66. Linné Y, Dye L, Barkeling B, Rössner S. Weight development over time in parous women – the SPAWN study – 15 years follow-up. *Int J Obes Relat Metab Disord*. 2003;27(12):1516–1522.
67. Birdsall KM, Vyas S, Khazaezadeh N, et al. Maternal obesity: a review of interventions. *Int J Clin Pract*. 2009;63(3):494–507.
68. Dodd JM, Crowther CA, Robinson JS. Dietary and lifestyle interventions to limit weight gain during pregnancy for obese or overweight women: a systematic review. *Acta Obstet Gynecol Scand*. 2008;87(7):702–706.
69. Campbell F, Johnson M, Messina J, Guillaume L, Goyder E. Behavioural interventions for weight management in pregnancy: a systematic review of quantitative and qualitative data. *BMC Public Health*. 2011;11:491. doi:10.1186/1471-2458-11-491.
70. Royal College of Obstetricians and Gynaecologists. *Exercise in Pregnancy*. London: RCOG; 2006.
71. National Institute for Health and Clinical Excellence. *Dietary Interventions and Physical Activity Interventions for Weight Management Before, During and After Pregnancy – NICE Public Health Guidance 27*. London: NICE; 2010.
72. Machado LS. Cesarean section in morbidly obese parturients: practical implications and complications. *N Am J Med Sci*. 2012;4(1):13–18.
73. Bamgbade OA, Khalaf WM, Sharma Ajai O, Chidambaram R, Madhavan G. Obstetric anaesthesia outcome in obese and nonobese parturients undergoing caesarean delivery: an observational study. *Int J Obstet Anesth*. 2009;18(3):221–225.
74. Smith SA, Heslehurst N, Ells LJ, Wilkinson JR. Community-based service provision for the prevention and management of maternal obesity in the North East of England: a qualitative study. *Public Health*. 2011;125(8):518–524.
75. Yogev Y, Visser GH. Obesity, gestational diabetes and pregnancy outcome. *Semin Fetal Neonatal Med*. 2009;14(2):77–84.

76. Hibbard JU, Gilbert S, Landon MB, et al. Trial of labor or repeat cesarean delivery in women with morbid obesity and previous cesarean delivery. *Obstet Gynecol*. 2006;108(1): 125–133.
77. Chauhan Suneet P, Magann Everett F, Carroll Charles S, et al. Mode of delivery for the morbidly obese with prior cesarean delivery: vaginal versus repeat cesarean section. *Am J Obstet Gynecol*. 2001;185(2):349–354.
78. Maggard MA, Yermilov I, Li Z, et al. Pregnancy and fertility following bariatric surgery: a systematic review. *JAMA*. 2008;300(19):2286–2296.
79. Shekelle PG, Newberry S, Maglione M, et al. Bariatric surgery in women of reproductive age: special concerns for pregnancy. *Evid Rep Technol Assess (Full Rep)*. 2008;(169): 1–51.
80. Davies GA, Maxwell C, McLeod L, et al. Obesity in pregnancy. *J Obstet Gynaecol Can*. 2010;32(2):165–173.

35 Developing Standards of Care for Obese Women During Pregnancy

David Churchill[1], Gemma Forbes[2] and Tahir Mahmood[3]

[1]New Cross Hospital, Wolverhampton, UK, [2]West Midlands Deanery, New Cross Hospital, Wolverhampton, UK, [3]Office of Research and Clinical Audit, Lindsay Stewart R&D Centre, Royal College of Obstetricians and Gynaecologists, London, UK

Introduction

Quality of care can only improve when the health care systems and professionals working in them provide equitable, effective, evidence-based, safe, consistent and humane care driven by high quality standards.

But what is a standard?

The term is used loosely and can refer to different aspects of clinical care depending upon the prevailing situation. Defining standards of clinical care is easy when there is a strong evidence base for a set of actions that can form a clinical guideline [1]. However, when there is uncertainty about the optimal way of managing a clinical problem, standard setting is more difficult.

In this chapter, we will examine how standards are developed, briefly overview the extent of the obesity epidemic and give examples of standards that will improve the care provided to obese women in pregnancy.

What Is a Standard in Health Care?

A standard or more commonly set of standards should underpin an overarching statement of purpose. Indeed, the overarching statement is a construct from aggregated criteria [2]. It or they should in specific terms be developed from evidence-based clinical guidelines. A standard can be a measure of outcome, e.g. a predetermined disease end point such as a pulmonary embolism or a process to which professionals and through them the health care systems must conform. In meeting the standard whether that is an outcome or a process, it informs the patient or clinician of the level of quality of health care being provided.

Obesity. DOI: http://dx.doi.org/10.1016/B978-0-12-416045-3.00035-2

Some standards are fixed and do not change, they can be likened to the Ten Commandments but most will change over time, responding to events, developments and social challenges. In this way, standards evolve to meet the prevailing scientific, social or political conditions in which health care systems are operating.

The obesity epidemic is causing significant clinical and financial stresses within health care systems [3]. The need to research new ways of treating or managing individuals who are obese is widely recognised. But setting standards of care cannot wait for the next breakthrough in medical science. A great deal can be achieved by ensuring that the current health care systems deliver the highest possible standards now. It is likely that by setting and applying high standards of care for obese people, we will have an impact on the secondary problems caused by obesity in pregnancy.

Key Point

Standards should be drawn from the highest levels of evidence available.

How to Develop Standards

The most robust standards are based upon the highest levels of evidence and should be developed from evidence-based clinical guidelines. In the absence of firm evidence, they have to rely on empirical evidence from basic science, consensus and expert opinion. These situations are far from ideal but often necessary. Thus, standard setting for equity, safety and consistency often has to rely on qualitative data and professional judgement. There is a dearth of good quality randomised trials on various interventions for obesity during pregnancy and labour, therefore drawing up standards of care for obese pregnant women requires good clinical judgement.

What Measures Should be Used When Defining Standards of Care?

Well-defined outcome measures provide the best assessment of quality for any given intervention. But they can also be problematic and comparisons between institutions or health care systems is fraught with scientific danger. These measures are often affected by other factors apart from the intervention itself, complicating the interpretation of results. An example of the problems caused can be seen in the fierce debate surrounding the usefulness of the Hospital Standardised Mortality Rate in judging the quality of care in different hospitals [4–6]. The debate has not yet been resolved but it demonstrates how an apparently simple measure is not necessarily the complete answer. Low event rates, multiple influencing co-factors known and unknown, temporal changes, disease virulence or other treatments can all have an impact on the outcome in the complex world of medicine. However when outcomes measures, confer upon an institution the status of an 'Outlier' it does mean that there is the need for further investigation to ensure that quality of care is not being truly compromised. The more extreme the deviation from the average for an outcome measure, the more likely it is to indicate a problem of some sort [7].

These difficulties can be overcome in part by choosing process measures alongside outcomes to build up a more comprehensive picture of the care being provided [8]. However, process measures have to be chosen carefully. They should represent key or critical steps in the care pathway and lead to the defined outcome or end point. While it is not necessary to measure every step in a pathway, enough steps must be examined to ensure that the process of care is being delivered in its entirety.

Therefore, a set of standards is necessary to indicate the quality of care. As well as monitoring the effectiveness of interventions treating the target disease, they should include standards that aim to mitigate the worst consequences of the main condition.

So the range of standards must be all encompassing. They must also include standards that indicate good organisational health to compliment good personal and clinical care.

Key Point

A blend of process and outcome measures will give a richer and more accurate assessment of the quality of care provided by a maternity unit.

The Obesity Epidemic

Obesity is defined as 'a condition characterised by excess of body fat, frequently resulting in a significant impairment of health and longevity'. It is defined in pregnancy as a body mass index (BMI) of 30 kg/m^2 or more at the first antenatal consultation. There are three different classes of obesity: BMI 30.0–34.9 kg/m^2 (Class I); BMI 35.0–39.9 kg/m^2 (Class 2) and BMI 40 kg/m^2 and over (Class 3 or morbid obesity) [9]. The recent report from the joint Centre for Maternal and Child Enquiries (CMACE)/Royal College of Obstetricians and Gynaecologists (RCOG) includes an additional category of 'super-morbid' obesity, defined as a BMI $\geq$ 50 kg/m^2 [10]. The addition of this category recognises the continuous relationship between increasing BMI and worsening morbidity and mortality.

In 2008 the World Health Organization (WHO) estimated that worldwide there were 500 million obese people and in the United Kingdom it estimated that 31% of people will be officially classifiable as obese by 2012 [11,12]. These figures represent a major clinical and financial burden for society. The Confidential Enquiry into Maternal and Child Health (CEMACH) report in 2007 estimated that 38,478 women giving birth in the United Kingdom each year have a BMI of $\geq$35 kg/m^2, which equates to 1 in 20 maternities [13]. This represents a very direct effect upon the maternity services in the United Kingdom.

The adverse outcomes associated with obesity in pregnancy are summarised in Table 35.1 [14,15].

The most severe outcome, maternal death in pregnancy, is a real risk for obese women. The previously mentioned CEMACH report on maternal deaths between 2003 and 2005 showed that 28% of mothers who died were obese, whereas the prevalence of obesity in the general maternity population within the same time period was 16–19%. The risk of stillbirth is also increased with an odds ratio of 2.79 (confidence interval, CI, 1.94, 4.02).

Pre-Pregnancy Care

Providing obese women with information on the risks to them and their pregnancy is the first step in partnership care. Obesity not only has an impact upon a woman's long-term health but also her reproductive capacity. In some women, the obesity itself is related to an underlying condition such as polycystic ovarian disease, which in turn results in reduced fertility.

Good health prior to pregnancy is a prerequisite for a healthy pregnancy. Pre-pregnancy counselling is recommended for women who suffer from certain medical conditions such as diabetes, but take up rate is often low. An opportunistic policy that targets young women when they access health services for other reasons, e.g. contraceptive advice, cervical screening, is likely to be more effective than a voluntary one. These approaches should be applied when encouraging obese women to accept pre-pregnancy advice.

But what advice should be given? Exhorting the woman to eat a healthy diet and take more exercise rarely has any lasting effect. Continued support and the

Table 35.1 Obesity-Related Risks in Pregnancy

Maternal
Severe morbidity
Cardiac disease
Spontaneous first trimester and recurrent miscarriage
Pre-eclampsia
Gestational diabetes
Thromboembolism
Post-Caesarean wound infection
Infection from other causes
Post-partum haemorrhage
Low-breastfeeding rates
Maternal death
Foetal
Stillbirth and neonatal death
Congenital abnormalities
Prematurity

help of specialist services may be more helpful in the long term, especially for the morbidly obese.

Women, who are having difficulty conceiving whether or not they are obese, must be investigated for an underlying cause. Fertility treatment should not only account for any medical disorder but the obesity as well. This may require advice from specialist services such as dietetics.

It is also important to give obese woman evidence-based information about the risks to and from a pregnancy. Obesity per se increases the risks of miscarriage, congenital foetal anomalies, hypertensive disease, diabetes and complications during labour and post-delivery problems such as post-partum haemorrhage and of course venous thromboembolism (VTE). Some women may choose to avoid a pregnancy until significant weight loss has occurred and the risks are reduced.

Regardless of the specific approach taken to weight reduction, obese women with a BMI > 27 kg/m^2 are less likely to receive folic acid through their diet. Therefore, they should be advised to take 5 mg of folate periconceptually and for the first trimester to reduce the risk of neural tube defects in the foetus. Likewise, obese women are more likely to be deficient in vitamin D and must be advised to take a supplement of 10 μg of vitamin D each day [16,17].

Standard

Obese women should be offered supplements of folic acid periconceptually and during the first trimester of pregnancy.

Early Pregnancy Care

At the first antenatal clinic visit, a comprehensive risk assessment must be undertaken. The BMI must be calculated accurately and the presence of co-morbid conditions accounted for in the risk assessment process. It must be clearly recorded in the care record. Once this has been completed, an individualised management plan can be put in place. It must encompass both maternal and foetal complications mitigating the risks as far as possible.

Dietary Advice

Women with a BMI > 30 kg/m^2 should be given advice about the risks in pregnancy associated with obesity and once again the opportunity should be taken to inform them about a healthy diet and appropriate levels of exercise to limit the weight gain in pregnancy to acceptable levels.

The link between a poor diet high in calories and obesity is well known. But what is not clear is how and when to intervene to produce a sustained fall in weight. Excessive gestational weight gain (GWG), in combination with pre-pregnancy obesity increases the risk of an adverse outcome further for both the mother and the baby. The Institute of Medicine has set out recommendations for weight gain during pregnancy and related it to a woman's pre-pregnancy BMI [18]. The data are shown in Table 35.2.

GWG has been used as an outcome measure to test dietary interventions during pregnancy. One randomised control trial (RCT) compared the effect of energy intake restrictions on a group of obese women, compared with three control groups (underweight, normal weight and obese) [19]. The results were promising and demonstrated that the GWG in the intervention group was halved (6.2 kg ± 2.1) compared with the other groups (underweight 12.9 kg ± 4.6; normal 11.9 kg ± 1.5 and obese 13.6 ± 5.6).

But this success was not repeated in other trials, where GWG was found to still be excessive, despite the interventions showing improved nutritional status. Further, RCTs in this field are necessary [20].

One factor influencing the nutritional interventions in obese women is the socioeconomic divide. WHO figures show that this effect is present in both developed and developing nations [11]. Because of varying resources and differing levels of deprivation, a standardised single approach to dietary intervention for obese women in pregnancy is not likely to prove effective. Nevertheless, until specific interventions become available the opportunity to advise obese women on what is a good diet should be taken. A sustained improvement in their diet will in the long term have all-round benefits.

Standards

All obese women should be offered dietary advice to enable them to limit their weight gain during pregnancy to recommended levels.

All women with a BMI in excess of 40 kg/m^2 should be offered referral to a dietician for advice and the construction of an individualised dietary plan.

Aspects of Care During Pregnancy

Health professionals caring for an obese woman need to employ a policy of total vigilance. Many complications can arise and a failure of attention to detail could put the woman at risk of serious morbidity.

Table 35.2 New Recommendations for Total and Rate of Weight Gain During Pregnancy, by Pre-pregnancy BMI

Pre-Pregnancy BMI (kg/m^2)	Ideal Total Weight Gain in Pregnancy (kg)
<18.5	12.5–18
18.5–24.9	11.5–16
25.0–29.9	7–11.5
≥30.0	6

Obstetricians need to consider the effects from other co-morbid conditions so commonly found in obese women when drawing up an individualised care plan. These range from the interventions for minor complaints such as varicose veins to major chronic diseases like diabetes.

Maintaining and Monitoring Maternal Well-Being

These women are not suitable for midwifery care alone. Women whose BMI lies between 30 and 40 kg/m^2 need a careful assessment about who are the most appropriate lead carer. A joint obstetric and midwifery decision is the ideal way forward. The woman's social situation needs to be taken into account as well as her medical status. Women with a BMI in excess of 40 kg/m^2 should be advised to deliver in a consultant obstetric unit. Complications in labour and delivery are more likely and this policy will enable them to access obstetric expertise should complications occur.

Standard

All obese women must have their BMI calculated and recorded in the pregnancy record in the first trimester. The BMI must be re-calculated in the early third trimester and similarly recorded in the notes. A further risk assessment must then be carried out.

Blood Pressure Measurement

Obese women are at greater risk of pre-eclampsia. At each antenatal visit an accurate blood pressure measurement must be taken. An arm cuff of an appropriate size must be used. The threshold arm circumference for using a large adult cuff is 32 cm. For some morbidly obese women, this size may be inadequate necessitating the use of a thigh cuff. It is important for all BP measurements that the bladder within the cuff encompasses 80% of the arm and is correctly sited. Several sequential measures, 1 min apart, will reduce the risk of the measurement errors by bias. Automated devices must be calibrated regularly to maintain their accuracy.

Obesity and Gestational Diabetes

No one needs to be reminded that obesity is linked to insulin resistance and in turn gestational diabetes mellitus (GDM). The NICE guidelines on Diabetes in Pregnancy set the threshold at which women should be screened for GDM using the 75 gm oral glucose tolerance test (OGTT) between 24 and 28 weeks gestation, at a BMI of 30 kg/m^2 [21]. Two important randomised controlled trials have shown that treatment of glucose intolerance even at mild levels improves neonatal outcomes [22,23].

It is reasonable to expect that 100% of obese women, as defined above, are offered screening for GDM. A failure to offer obese women screening for the condition should be treated as a failure in the system of care.

A positive diagnosis must then result in the patient being referred for further management in a joint diabetic/antenatal clinic with blood sugar monitoring and intervention to normalise blood sugars. The woman may need metformin and or insulin to normalise her blood sugars.

While this is probably the most straightforward standard to set for obese women, it is one of the most important given the strong evidence base for a beneficial effect from intervention.

Standard

All women with a BMI of 30 kg/m^2 or greater must be offered an OGTT between 24 and 28 weeks gestation.

Foetal Anomaly Screening and the Assessment of Foetal Growth

Several studies have shown an increased incidence of neural tube, cardiac and facial defects in association with obesity. Others have also noted limb reduction defects and ano-rectal anomalies [24]. A systematic review published in 2009, pooled data from high-quality studies to calculate the odds ratios for these anomalies [25]. A summary is shown in Table 35.3.

Among the more credible mechanisms postulated for these increased risks are undiagnosed hyperglycaemia and nutritional deficiencies.

Apart from ano-rectal anomalies, most of the specific defects are amenable to prenatal diagnosis using ultrasound. But obesity itself reduces the diagnostic effectiveness of this modality. Ultrasound is attenuated in fat and penetration is reduced. To increase the penetration and to view deeper into the body, lower transducer frequencies are needed. The trade-off though is poorer tissue differentiation and reduced image quality. Ultrasound technology is continually improving and modern machines have found many technological fixes to help these problems; nevertheless, some difficulties remain, although perhaps not to the same extent as 5 years ago.

The impact of the difficulties caused by obesity on mid-trimester ultrasound visualisation of cardiac and cranio-spinal structures has been quantified. In a study of 11,019 women comparing obese to non-obese women, the overall sub-optimal visualisation rate (SUV) was found to be 37.3–18.7%, respectively. In the specific

Table 35.3 Data from High-Quality Studies to Calculate the Odds Ratio for fetal Anomalies

	Odds Ratio	CI (95%)
Neural tube defects	1.87	1.62, 2.15
Cardiovascular anomalies	1.30	1.03, 1.47
Cleft lip and palate	1.20	1.03, 1.40
Ano-rectal anomalies	1.48	1.12, 1.97
Limb reduction defects	1.34	1.03, 1.73

instance of cranio-spinal defects, the SUV rate was 42.8% in the obese against 29.5% in the non-obese women. These differences remained after adjustment for gestational age and ultrasound machine type [26].

Clearly the situation is a real problem for the clinician. Not only does obesity cause a higher rate of congenital anomalies in the foetus, but it also reduces the ability of the clinician to detect them. Nevertheless, the service must recognise the situation and introduce measures to improve the chances of detection. Safety standards in this area must not only measure the outcomes, i.e. rates of foetal anomaly, but the processes put in place to maximise the detection rates and provide obese women with the best possible chance of having foetal anomalies identified.

Assessing foetal growth in obese women is also problematic. The symphysio-fundal height measurement is inaccurate due to the abdominal fat pad. The only solution is to offer regular ultrasound scans to assess foetal growth during the second and third trimesters. The optimal interval between the ultrasound assessments is not known but monthly scans after 20 weeks gestation, as a minimum, would seem a sensible approach to try and detect abnormal foetal growth as early as possible. Again the technical difficulties outlined above affect the foetal measurements, especially in women with morbid obesity. The obstetrician must be aware of the potential for these errors when making an assessment of the patient.

Key Point

When auditing detection rates for foetal anomalies, the data should be stratified by BMI to ensure that the detection rate in obese women reaches the highest possible rates.

Departments must have a replacement programme to refresh their machinery and take advantage of improvements in technology. The timing of the examination is also important, with the optimal time for visualisation being between 18 and 20 weeks gestation. When an abnormality is suspected, referral to for a second opinion must be prompt. But most important of all, obese women must be made aware of the limitations of ultrasound and presented with clear and accurate information.

Standards

All obese women must be offered a high quality mid-trimester foetal anomaly scan. They must be made aware of the potential difficulties in visualisation of the foetus. The ultrasound machines must be up to date and maintained to the manufacturer's requirements.

In women whose obesity renders the symphysio-fundal height measure inaccurate, should receive monthly ultrasound assessments of foetal growth in the second and third trimesters.

Thromboembolic Disease

VTE encompasses deep vein thrombosis (DVT) within the veins of the leg and pulmonary embolisms (PE) where a clot or 'thrombus' is carried to the pulmonary vessels. The latter results in haemodynamic and respiratory compromise and severe cases, death. DVTs, although less catastrophic are associated with significant morbidity including post-thrombotic syndrome.

In the last CEMACH report, pulmonary embolism was the single biggest cause of direct maternal mortality, with a rate of 1.56 per 100.000 maternities [27]. The significance of this is more acute in obese patients because of their increased risk of thromboembolic disease. In the CEMACH report of 2003–2005, of the 33 women who died, 16 were classed as overweight (BMI > 25), 12 of whom were obese (BM1 > 30). Further evidence of risk comes from the United Kingdom Obstetric Surveillance System, which reported an adjusted odds ratio of 2.65 (95% CI 1.09–6.45) for PE, in women with a BMI > 30 kg/m^2 [28].

Risk assessment for thromboembolic disease should be carried out at the earliest opportunity in pregnancy [19]. Each woman's risk can be stratified according to their BMI, and then re-adjusted taking into account other thromboembolic risk factors.

It is wise in obese women to maintain a high index of suspicion for VTE, as signs and symptoms may be masked. For example calf measurements may prove unreliable when looking for asymmetry. Investigations can be problematic too. Plain film chest X-rays may have poorer penetration. Ventilation/perfusion scanning and computerised tomography of the pulmonary arteries can overcome some of these limitations but the risks of radiation exposure to the foetus have to be balanced with the likelihood of the diagnosis. However, the X ray doses delivered nowadays are much lower and thus so are the risks from radiation exposure, making the tests useful when used wisely.

Prophylaxis and Treatment

Studies into pharmacological prophylaxis and treatment are primarily based on surgical or trauma patients, and extrapolated to pregnancy [29]. Low-molecular-weight heparin (LMWH) is often the pharmacological treatment of choice in pregnancy. There are lower incidences of osteoporosis and heparin-induced thrombocytopenia (HIT) when LMWH is compared with unfractionated heparin [30]. Long-term monitoring of the platelet count in patients taking LMWH is not thought to be necessary. However, the disadvantage of LMWH is the time taken for its effects on the clotting system to dissipate. Nevertheless, it is important to ensure that obese women at risk receive prophylaxis in the antenatal period and that it is continued for between 6 and 8 weeks after delivery. The post-natal period still remains a significantly risky for obese women, as will be seen later.

There is no clear evidence to support specific doses of LMWH in pregnancy. Twice daily dosing with LMWH is thought to provide optimal coverage in obese obstetric patients, overcoming concerns about variable absorption due to body fat ratios. Three types of LMWH are currently in use in pregnancy. They are enoxaparin, dalteparin and tinzaparin. Prophylactic and therapeutic dosing regimens are available for each drug [31]. It is important to remember that for situations where a therapeutic dose is needed; or women weighing in excess of 170 kg who require prophylaxis, the dose must be worked out on an individual basis, based upon the woman's weight in kilograms.

At lower prophylactic dosing regimens, it is not necessary to monitor the effect of the heparin on the clotting system. However, at higher prophylactic and therapeutic doses monitoring with either a heparin assay or factor Xa levels can be used. The advice of a haematologist should be sought in these situations.

In addition to heparin, other measures can also be used to reduce the risks of thromboembolic disease (TED) in the obese pregnant woman. These include properly fitted graduated compression stockings, early mobilisation after delivery, maintaining hydration, especially in labour, and mechanical leg compression boots when surgical procedures are being performed or the woman is immobile for any significant period of time.

Standards

All obese women must be formally risk assessed for venous thromboembolic disease in the first trimester and repeated in the late second trimester.

All obese women must be offered thromboprophylaxis in pregnancy based upon the schedule laid down in the RCOG guidelines.

Planning for Labour and Delivery

Planning for labour and delivery is an important aspect of the antenatal care process. A discussion must take place with obese women about the risks associated with labour and delivery. A provisional plan can be drawn up in the first half of pregnancy but this must be reviewed and finalised in the third trimester, between 34 and 36 weeks gestation. Any developments in the antenatal period such as the foetus becoming macrosomic can be included in the later assessment and discussions with the woman. Risk can be adjusted in the light of these findings and choices laid out for each individual. Once again individualising the plan is imperative and keeping the woman fully informed will give her confidence. The result is more likely to be a happy, healthy mother and baby.

Women with a BMI in excess of 40 kg/m^2 should be referred to an obstetric anaesthetist. There is the potential for problems with venous access, spinals/epidural anaesthesia and general anaesthesia. Instructions can be drawn up for all these eventualities and included in the plan of care.

Standard

All obese women must be informed of the risks of labour and delivery and an individualised care plan for labour and delivery entered into the pregnancy care record.

Care During Labour and Delivery

Intrinsically, obesity increases the risk of intra-partum problems quite apart from the risks caused by co-morbid conditions. Obese women are more likely to have a foetus that is macrosomic or that suffers from abnormal cardiotocographic findings suggesting hypoxia or requires an instrumental delivery or a Caesarean section. The values in Table 35.4 show the risk of intervention and/or adverse pregnancy outcome in the morbidly obese woman (a BMI of over 40 kg/m^2) and were taken from a study of over 900,000 pregnant women in Sweden between 1992 and 2001 [32].

Other studies have shown a similar increase in the rate of Caesarean sections. One case–control study found that nulliparous obese women had a Caesarean section rate of 33.8% compared to 20.7% in non-obese controls. For the morbidly obese, the Caesarean section rate was even higher at 47.4% [14]. The increase in the Caesarean section rate due to obesity is now widely accepted.

Key Point

When monitoring intervention rates in labour, the data should be stratified for BMI so that local data can be available to pregnant women.

But the increase in operative delivery may not be solely due to macrosomia or an increased risk of 'foetal hypoxia'. Although these factors play a part, there is experimental evidence to suggest an underlying physiological dysfunction in the calcium flux within the uterine muscle, impairing the efficacy of uterine contractility [33].

Caesarean section also carries a greater risk of complications in the obese woman. Rates of wound haematoma, infection and dehiscence are all increased. The surgery itself is technically more demanding with intra-operative complications being more difficult to deal with. It is therefore important that experienced senior operators carry out the surgery.

Research has been performed on the type of skin incision – vertical versus transverse – and the use of drains, to determine if the rates of complications can be reduced [34,35].

However, the evidence is still unclear about whether to use drains or not and which incision is better. Still, much will depend upon the skill, experience and judgement of the operator.

Table 35.4 Risk of Intervention and/or Adverse Pregnancy Outcome in the Morbidly Obese

	Odds Ratio	CI (95%)
Pre-eclampsia	4.82	4.04, 5.74
Antepartum stillbirth	2.79	1.94, 4.02
Caesarean section	2.69	2.49, 2.90
Instrumental delivery	1.34	1.16, 1.56
Shoulder dystocia	3.14	1.86, 5.31
Meconium aspiration	2.85	1.60, 5.07
Foetal distress	2.52	2.12, 2.99
Early neonatal death	3.41	2.07, 5.63
Large for gestational age	3.82	3.50, 4.16

Nevertheless, when all this is said and done it still remains true that a vaginal delivery would in most cases be the lowest risk choice all round.

Team Management on Delivery Suite

Senior members of the obstetric and anaesthetic team should be told when an obese woman with a BMI > 40 kg/m^2 is admitted to the delivery suite in labour.

An obstetrician of at least registrar level should reappraise the maternal and foetal risks and inform the consultant. Those caring for the woman must be capable of dealing with the likely complications, which means that they must have a suitable level of experience. Senior supervision and one-to-one care in labour are important in order to reduce complications and interventions.

Being prepared for all eventualities helps ensure patient safety. A woman with a BMI above 40 should have venous access secured when she is in established labour. It would also be wise to send a sample of blood to the laboratory for blood grouping in view of the greater risk of operative delivery and post-partum haemorrhage. To try to prevent the latter, active management in the third stage of labour should be recommended to all obese women.

Standard

The consultant obstetrician and anaesthetist must be informed when a woman with a BMI over 40 is in labour and a multidisciplinary approach taken to her management.

Post-Natal Care

Obese women have lower initiation and maintenance of breastfeeding. It is likely that the cause is multifactorial and includes the woman's own perceptions of breastfeeding, difficulty in positioning the baby on the breast and possibly an impaired prolactin response to suckling. The evidence shows that education and support do improve rates of breastfeeding initiation and its maintenance for longer periods of time.

The puerperium remains a period of risk for thromboembolic disease. Women with a BMI of 30 kg/m^2 or more with one or more additional risk factors should be considered for thromboprophylaxis with LMWH for the first 7 days following birth. They should also be advised to wear graduated compression stockings, provided they can be properly fitted. For women with morbid obesity, a decision on whether to continue the LMWH for longer may need to be taken. However, like all women they should be encouraged to mobilise as early as possible and to maintain adequate hydration by drinking plenty of water, especially if they are breastfeeding.

For women who suffered from gestational diabetes, post-natal testing must be arranged to exclude type 2 diabetes. This can be done with fasting blood glucose 6 weeks after delivery and an oral GTT if the fasting threshold is exceeded.

Finally, it is worth repeating the advice on lifestyle and dietary changes to reduce weight in the long term. The importance of weight loss for long-term health, including their future reproductive health, should be stressed.

Standard

All obese women must have a reassessment of their VTE risk in the first 48 hours after birth.

Organisation

Organisational Requirements

Finally, the organisation must be adequately equipped and ready to care for obese women.

The CEMACH report and the joint CMACE/RCOG guidelines recognise the need for suitable equipment to enable staff to care for obese women effectively. It is recommended that an organisational risk assessment is undertaken for obese women and that key equipment is readily available [36].

This includes assessment of:

- Adequate accessibility including doorway widths and thresholds
- Safe working loads of equipment such as delivery beds and theatre tables (up to 250 kg)
- Appropriately sized theatre gowns
- Equipment storage local to the point of need
- Transportation readily available for emergencies
- Staffing levels adequate to cope with the levels of patient dependency
- Availability of, and procurement process for, specific equipment:
  - large blood pressure cuffs
  - sit-on weighing scale
  - large chairs without arms
  - large wheelchairs
  - ultrasound scan couches
  - ward and delivery beds
  - theatre trolleys
  - strengthened operating theatre tables
  - lifting and lateral transfer equipment.

Recommendations also suggest the inclusion of other equipment such as ultrasound machines to determine foetal presentation and well-being, and longer needles for the placement of epidurals. With the ever-rising obesity rates, this equipment is essential for all delivery suites in the United Kingdom.

Staff must be made aware of the equipment requirements and receive procedural advice for handling women who are morbidly obese.

Although staff are becoming increasingly familiar with how to manage the obese patient, the organisation is responsible for ensuring that manual handling training is available and up to date.

Standard

All units should keep an inventory of equipment required to manage obese women. The equipment should be regularly tested to ensure that it is functional and the inventory reviewed on an annual basis to establish that the inventory is complete and meets current needs.

References

1. Grimshaw J, Russell I. Achieving health gain through clinical guidelines. I: developing scientifically valid guidelines. *Qual Health Care*. 1993;2:243–248.
2. Irvine D, Donaldson L. Quality and standards in health care. In: Beck JS, Bouchier IAD, Russell IT, eds. *Quality Assurance in Medical Care*. Edinburgh: Royal Society of Edinburgh, RSMED; 1993.
3. Ramsay JE, Greer I, Sattar N. ABC of obesity: obesity and reproduction. *BMJ*. 2006;333:1159–1162.
4. Black N. Editorial: assessing the quality of hospitals. *BMJ*. 2010;340:c2066.
5. Lilford R, Provnost P. Using hospital mortality rates to judge hospital performance: a bad idea that won't go away. *BMJ*. 2010;340:201–206.
6. Mohammed MA, Deekes JJ, Girling A, Rudge G, Carmalt M, Stevens AJ, Lilford RJ. Evidence of methodological bias in hospital standardised mortality ratios: retrospective database study of English Hospitals. *BMJ*. 2009;338:780.
7. Commission for Healthcare Audit and Inspection. *Investigation into Mid Staffordshire NHS Foundation Trust*. London: Healthcare Commission; 2009:www.cqc.org.uk/_db/_documents/Investigation_into_Mid_Staffordshire_NHS_Foundation_Trust.pdf
8. Lilford RJ, Brown CA, Nicholl J. Use of process measures to monitor the quality of clinical practice. *BMJ*. 2007;335:648–650.
9. Krishnamoorthy U, Schram C, Hill S. Review article: maternal obesity in pregnancy: is it time for meaningful research to inform preventive and management strategies? *BJOG*. 2006;113:1134–1140. doi:10.1111/j.1471-0528.2006.01045.x.
10. CMACE/RCOG Joint Guideline. *Management of Women with Obesity in Pregnancy*. London: CMACE/RCOG; 2010.
11. *The World Health Organization*. Obesity and overweight fact sheet (number 311). <http://www.who.int/mediacentre/factsheets/fs311/en/index.html>; May 2012.
12. *The Health and Social Care Information Centre*. <http://www.ic.nhs.uk/statistics-and-data-collections/health-and-lifestyles/obesity/statistics-on-obesity-physical-activity-and-diet-england-2011/>; 2011 Accessed 29.11.11.
13. Lewis G, ed. *The Confidential Enquiry into Maternal and Child Health (CEMACH). Saving Mothers' Lives: Reviewing Maternal Deaths to Make Motherhood Safer – 2003–2005. The Seventh Report on Confidential Enquiries into Maternal Deaths in the United Kingdom*. London: CEMACH; 2007.
14. Cedergren MI. Maternal morbid obesity and the risk of adverse pregnancy outcome. *Obstet Gynecol*. 2004;103(2):219–224.
15. National Obesity Observatory: Maternal Obesity and Maternal Health. <http://www.noo.org.uk/NOO_about_obesity/maternal_obesity/maternalhealth/>; 2010 Accessed 1.12.11.

16. Carmichael SL, Shaw GM, Schaffer DM, Laurent C, Selvin S. Dieting behaviours and risk of neural tube defects. *Am J Epidemiol*. 2003;158(12):1127–1131.
17. Bodner LM, Cator JM, Roberts JM, Simham HN. Prepregnancy obesity predicts poor vitamin D status in mothers and their neonates. *J Nutr*. 2007;137(11): 2437–2442.
18. *Committee to Reexamine IOM Pregnancy Weight Guidelines, Food and Nutrition Board, and Board on Children, Youth and Families*. Weight gain during pregnancy: reexamining the guidelines. Institute of Medicine and National Research Council.<http://www.iom.edu/Reports/2009/Weight-Gain-During-Pregnancy-Reexamining-the-Guidelines.aspx/>; 2009 Accessed 3.12.11.
19. Borberg C, Gillmer MD, Brunner EJ, Gunn PJ, Oakley NW, Beard RW. Obesity in pregnancy: the effect of dietary advice. *Diabetes Care*. 1980;3:476–481.
20. Dodd JM, Crowther CA, Robinson JS. Dietary and lifestyle interventions to limit weight gain during pregnancy for obese or overweight women: a systematic review. *Acta Obstet Gynecol Scand*. 2008;87(7):702–706.
21. National Institute of Health and Clinical Excellence. *Diabetes in Pregnancy. Management of Diabetes and its Complications from Preconception to the Postnatal Period*. London: National Institute of Health and Clinical Excellence; 2008.
22. Crowther CA, Hiller JE, Moss JR, McPhee AJ, Jeffries WS, Robinson JS. Effect of treatment of gestational diabetes on pregnancy outcomes. *N. Engl J Med*. 2005;352:2477–2486.
23. Landon MB, Spong CY, Thom E, et al. A multicentre randomised trial for treatment of gestational diabetes. *N Engl J Med*. 2009;361:1339–1348.
24. Watkins ML, Rasmussen SA, Honein MA, et al. Maternal obesity and risk for birth defects. *Pediatrics*. 2003;111:1152–1158.
25. Stothard KJ, Tennant PWG, Bell R, Rankin J. Maternal overweight and obesity and the risk of congenital anomalies. A systematic review and meta-analysis. *JAMA*. 2009;301:636–650.
26. Hendler I, Blackwell SC, Bujold E, et al. The impact of maternal obesity on midtrimester sonographic visualization of fetal cardiac and craniospinal structures. *Int J Obes*. 2004;28:1607–1611.
27. Knight M. Antenatal pulmonary embolism: risk factors, management and outcomes. *BJOG*. 2008;115(4):453–461.
28. Romero A, Alonso C, Rincon M, et al. Risk of venous thromboembolic disease in women: a qualitative systematic review. *Eur J Obstet Gynecol Reprod Biol*. 2005; 121(1):8–17. doi:10.1016/j.ejogrb.2004.11.023.
29. Geerts WH, Jay RM, Code KI. A comparison of low-dose heparin with low-molecular-weight heparin as prophylaxis against venous thromboembolism after major trauma. *N Engl J Med*. 1996;335(10):701–707.
30. Sanson BJ, Lensing AWA, Prins MH, et al. Safety of low-molecular-weight heparin in pregnancy: a systematic review. *Thromb Haemost*. 1999;81:668–672.
31. RCOG. *Reducing the Risk of Thromboembolism During Pregnancy and the Puerperium – RCOG Green Top Guideline No 37*. London: RCOG; 2009.
32. Weiss JL, Malone FD, Emig D, et al. Obesity, obstetric complications and caesarean delivery rate – a population based screening study. *Am J Obstet Gynaecol*. 2003;190:1091–1097.
33. Zhang J, Bricker L, Wray S, Quenby S. Poor uterine contractility in obese women. *BJOG*. 2007;114:343–348.

34. Ramsey PS, White AM, Guinn DA, et al. Subcutaneous tissue reapproximation, alone or in combination with drain, in obese women undergoing cesarean delivery. *Obstet Gynaecol.* 2005;105:967–973.
35. Allaire AD, Fisch J, McMahon MJ. Sub-cutaneous drain vs suture in obese women undergoing caesarean delivery. A prospective randomized trial. *J Reprod Med.* 2000;45:327–331.

Section 8

Long-Term Impact of Obesity

36 Maternal Obesity – The Road to Diabetes and Cardiovascular Risk

Ioannidis Ioannis and Grigoropoulou Pinelopi

Diabetes and Obesity Unit, 2nd Department of Internal Medicine, Konstantopoulio Hospital, Athens, Greece

Introduction

Obesity among women of childbearing age is becoming one of the most important women's health issues. A survey carried out in the United States between 2003 and 2006 reported that 32% of women aged 20–44 years were classified as obese. The problem is even bigger if we count the overweight women. According to the latest National Health and Nutrition Examination survey (1999–2002), 26% of non-pregnant women ages 20–39 years are overweight, and 29% are obese [1]. In the United Kingdom, where the prevalence of obesity in women is among the highest in Europe, the rise in obesity parallels the upward trend of obesity in the general population; almost one in five women of reproductive age are now obese (body mass index, BMI > 30 kg/m^2) [2,3]. As a result, during the last few years there has been a twofold rise in women identified as obese during pregnancy. Data from the United States estimate that nearly 23% of pregnant women are overweight, and 19% are obese. Importantly, there is also a shift towards higher gestational weight gain, which indicates excessive nutrient intake during gestation in affluent countries. Maternal obesity increases the risk of pregnancy complications, including hypertension, pre-eclampsia and gestational diabetes.

In addition to maternal obesity, an alarming trend in childhood obesity is also recorded [4]. Epidemiological studies clearly establish a strong association between maternal obesity and offspring predisposition to obesity and type 2 diabetes (T2DM) [5,6]. Published data confirm the above concept and demonstrate that obesity and insulin resistance have a foetal origin in many patients, as it is observed in offspring of parents with T2DM [7,8]. Children exposed to maternal obesity are at increased risk of developing metabolic syndrome [9]. Maternal environment especially during early foetal life seems to have a determinant role in the tendency to develop obesity, diabetes and cardiovascular risk of the progeny. In recent years, strong evidence in support of prenatal programming of offspring's obesity, diabetes and cardiovascular risk has been accrued, whereby factors mainly

Obesity. DOI: http://dx.doi.org/10.1016/B978-0-12-416045-3.00036-4

related to maternal health are believed to alter foetal development in utero in a manner that promotes excess adipose tissue deposition during pre- or post-natal life.

Most evidence suggests that children of women who begin pregnancy in an overweight or obese state are at greater risk for becoming obese themselves. Excess adiposity among women increases the risk for other metabolic health abnormalities during pregnancy (namely glucose metabolism abnormalities, diabetes and inflammation), and it is those abnormalities that appear more likely to contribute to offspring obesity risk.

Impact of Maternal Obesity on Offspring Obesity, Diabetes and Cardiovascular Disease Risk – Evidence from Experimental Studies

An increasing number of animal studies suggest that there are various time points during early development, in which maternal obesity and/or maternal/foetal overnutrition may result in programming effects in the offspring. Thus, the timing of exposure to mother's obesity and nutrition oversupply is of critical importance in determining the offspring's phenotype [10–14]. In many studies, offspring have been exposed not only to maternal obesity but also to maternal overnutrition during both pregnancy and lactation, so that the effects of maternal obesity per se cannot be adequately separated from those of overnutrition.

A number of human studies do also suggest that there may be particular developmental periods during which maternal obesity/overnutrition may have implications for offspring development. Recent animal data indicate that maternal obesity impairs oocyte quality and is associated with impaired development of the early embryo, so that programming effects in the offspring could occur as a consequence of maternal obesity even before fertilisation [15]. According to these data, the ovaries show increased apoptotic follicles, smaller oocyte size and number and delayed meiotic maturation. The pre-implantation events are also altered: insulin-like growth factor 1 receptor (IGF1R) expression is blunted in the blastocyst, a fact that correlates to increased apoptosis.

The IGF1R is critical for insulin signalling and glucose transport. Foetuses are smaller, and placental IGF1R expression is increased. These events are accompanied by metabolic changes in 3-month-old adult mice with signs that resemble those of the metabolic syndrome: increased weight in mice born from obese mothers accompanied by glucose intolerance and higher cholesterol and body fat. Thus, programming of offspring obesity as a consequence of overnutrition during pregnancy is quite possible. Another animal study also suggested that maternal obesity at conception is associated with an increased risk of obesity in the offspring even with normal maternal dietary intake during pregnancy [16].

The published data not only clearly link the effects of maternal obesity per se but also point towards the role of excessive weight gain, secondary to prevailing nutritional milieu during pregnancy, influencing the 'windows' for programming effects on the offspring.

The Link Between Maternal Obesity and Offspring Obesity, Diabetes and Cardiovascular Risk – Hypotheses and Possible Mechanisms

Although the relevance of experimental findings to humans is not clear, increased rates of obesity in mothers is associated with an increase of large-for-gestational-age infants and by an increase on obesity rates in children [4,17]. But which mechanisms mediate this road from mother's obesity to offspring's obesity, diabetes and cardiovascular risk?

The Developmental Overnutrition Hypothesis

The 'developmental overnutrition hypothesis' has been proposed as a potential pathophysiological link between maternal obesity and offspring's obesity, as well as other metabolic abnormalities (e.g. diabetes and cardiometabolic diseases). According to this *hypothesis*, high maternal glucose and high free fatty acid and amino acid plasma concentrations result in permanent changes in appetite control, neuroendocrine function and energy metabolism in the developing foetus, leading to risk of adiposity in later life [18–20]. This hypothesis is based on the concept of the association of maternal adiposity with a greater risk of insulin resistance and glucose intolerance, which result in higher concentrations of glucose and free fatty acids plasma concentrations. The offspring of these mothers would be expected to be programmed to become more obese themselves. Consequently, the obesity epidemic would be accelerated through successive generations independent of further genetic or environmental factors.

Several studies were conducted to support the developmental overnutrition hypothesis. A similar magnitude of effect of maternal and paternal adiposity on offspring adiposity would suggest that the associations are driven by factors that are just as likely to be passed from father to offspring as they are from mother to offspring, and this would reject the developmental overnutrition hypothesis. Two large studies compared the effect of maternal and paternal adiposity on offspring. In an Australian birth cohort, maternal BMI was more strongly associated with offspring BMI than was paternal BMI and this difference was impressive [21]. The Avon Longitudinal Study of Parents and Children (ALSPAC) was a longitudinal population-based birth cohort study that recruited 14,451 pregnant women resident in Avon, United Kingdom, with expected date of delivery from 1 April 1991 to 31 December 1992. In that study, the associations between parent and offspring BMI were the same in mothers and fathers [22]. Nevertheless, both of these studies used BMI as an indicator of offspring adiposity, which is not the best indicator of adiposity especially in children. When fat mass in the ALSPAC study was assessed by dual energy X-ray absorptiometry, the association of maternal BMI with offspring fat mass was stronger. Yet, paternal BMI was also positively associated with offspring fat mass. In addition, when maternal *FTO* genotype (a gene that predisposes to T2DM via an effect on BMI) was associated with offspring fat mass, there was no statistical evidence to suggest that differences in offspring fat mass were related to the maternal *FTO* genotype. This study concluded that the maternal effect on offspring BMI was not likely to be responsible for the recent epidemic obesity [18].

Despite the conflicting results, at least a part of the maternal BMI–offspring fat mass association is related to factors that are specific to the mother. A plausible maternal specific effect is overnutrition during key developmental periods (intrauterine and during breastfeeding), and hence the findings are supportive of the developmental overnutrition hypothesis [23]. Yet, the magnitude of this specific effect is relatively weak and possibly responsible for a slow and steady increase in population levels of obesity, but not for the recent sharp increase of obesity epidemic.

Predictive Adaptive Response Hypothesis

One general concept is that in response to an adverse intrauterine environment the foetus adapts its physiological development to maximise its immediate chances for survival. These adaptations may include resetting metabolic homeostasis set points, endocrine systems and down-regulation of growth, commonly manifest in an altered birth phenotype. Recently, the predictive adaptive response hypothesis proposes that the degree of mismatch between the pre- and post-natal environments is a major determinant of subsequent disease risk [24]. According to this hypothesis, whilst adaptive changes in foetal physiology may be beneficial for short-term survival in utero, they may be maladaptive in later life, contributing to adverse health outcomes when offspring are exposed to catch-up growth, diet-induced obesity and other factors.

Maternal obesity (*term which includes obesogenic nutrition, high BMI and gestational diabetes*) in combination with placental abnormalities leading to disruption in supply of nutrients to the foetus (*maternal obesity is associated with a maternal inflammatory state which induces structural and functional changes in the placenta*) and environmental factors (e.g. stress) lead to developmental adaptations. All these adaptations result mainly to increased lipid storage and decreased lipid oxidation. Interaction with post-natal environment, e.g. post-weaning high-fat nutrition, leads finally to obesity, insulin resistance and expression of different components of the so-called metabolic syndrome.

Maternal obesity may be associated with programming of altered adipocyte proliferation and differentiation capacity, increased expression of inflammatory mediators and altered lipid turnover. All these changes favour obesity, insulin resistance, and diabetes and cardiovascular diseases.

Maternal Obesity, Inflammation and Foetal Skeletal Muscle Development

Skeletal muscle and liver are the two major insulin-responsive key organs. Skeletal muscle composes of 40–50% of body mass, making it the most important tissue for glucose and fatty acid utilisation. *The foetal stage is crucial for skeletal muscle development* because there is no increase in muscle fibre numbers after birth. Poor foetal skeletal muscle development impairs glucose and fatty acid metabolism by skeletal muscle in response to insulin stimulation, and thus predisposes offspring to diabetes, obesity and other related metabolic disturbances leading to increased cardiovascular risk later in life [25,26]. Human infants who are small at birth are at greater risk for T2DM and obesity [27]. In these infants it seems that decreased muscle mass is a major factor in low birthweight [28].

Changes in foetal skeletal muscle development are likely to provide a link between maternal obesity and progeny obesity, diabetes and cardiovascular risk. In foetal muscle, there are a large number of mesenchymal stem cells (MSCs). Although the vast majority of MSCs commit to myogenesis, MSCs are also capable of differentiating into other cell types, such as adipocytes or fibroblasts [29]. A shift from myogenesis to adipogenesis or fibrogenesis will replace muscle fibres with adipocytes or fibrous tissues. This shift impairs the physiological functions of skeletal muscle, such as reduction in muscle force and oxidative capacity. Both of them might lead to decreased physical activity and capacity in adulthood, predisposing offspring to obesity [30,31]. In addition, enhanced adipogenesis within muscle leads to skeletal muscle insulin resistance, which plays a key role in the pathogenesis of T2DM [29].

Low-grade inflammation accompanies maternal obesity [32–34]. There is also evidence that offspring of obese women have increased inflammation themselves. Maternal obesity induces foetal inflammation, which changes foetal skeletal muscle development by shifting MSC differentiation from myogenesis towards adipogenesis [35]. This shift is expected to have permanent effects on offspring skeletal muscle properties.

Chronic inflammation associated with maternal obesity may also alter foetal skeletal muscle development through three major mechanisms, which include (1) down-regulation of WNT signalling (*The Wnt/b-catenin signalling pathway is required for early embryonic myogenesis. Wnt signalling enhances b-catenin nuclear translocation, but inflammation promotes the formation of b-catenin/FOXO complexes, which divert b-catenin from forming a complex with appropriate transcription factors to induce myogenesis*); (2) inhibition of AMP-activated protein kinase (AMPK) activity (*AMPK is a key serine-threonine kinase which serves as the energy status guardian within cells. AMPK is activated after ATP depletion within the cell, and responds by adjusting the rates of ATP-consuming (anabolic) and ATP-generating (catabolic) pathways in an attempt to restore and maintain cellular energy levels. AMPK switches MSCs in skeletal muscle from adipogenesis to myogenesis*) and (3) induction of epigenetic modifications (*Inflammation may induce epigenetic modifications that alter the expression of genes involved in myogenesis and adipogenesis in MSCs*).

Maternal obesity enhances adipogenesis and fibrogenesis not only in foetal but also later in offspring muscle. This action impairs the physiological function of skeletal muscle. The consequences include (1) the higher fat content in offspring muscle which impairs insulin signalling and limits the utilisation of fatty acids by peripheral tissues, leading to further accumulation of lipids; (2) enhancement of adipogenesis and fibrogenesis that limits the oxidative capacity of skeletal muscle and reduce lipid oxidation and (3) excessive lipid accumulation and obesity in offspring lead to inflammation, further deteriorating insulin signalling, forming a vicious circle. In brief, all these data show that maternal obesity has long-lasting effects on the properties of offspring muscle, which may provide a key mechanism for the foetal programming of adult metabolic diseases such as obesity and T2DM.

Epigenetics

Epigenetic processes are heritable changes in gene expression or cellular phenotype caused by mechanisms other than changes in the underlying DNA sequence [36]. The term epigenetics refers to functionally relevant modifications to the genome that do not involve a change in the nucleotide sequence. These changes may remain through cell divisions for the remainder of the cell's life and may also last for multiple generations.

Epigenetic modifications may be the principal mechanism by which exposure to an altered intrauterine milieu or metabolic perturbation may influence the phenotype of the organism much later in life [23]. There are four epigenetic modalities (*DNA methylation, non-coding RNAs, transcription factors and histone modifications*) that contribute to epigenetic memory [23,37,38]. Epigenomic changes may mediate the altered control of foetal gene expression as a consequence of maternal obesity.

In animal models, it has been shown that during pregnancy or early neonatal life epigenetic changes are taking place. In rats, high-fat maternal diet was associated with alterations in hepatic gene expression, which are associated with post-natal obese phenotype. Promoter methylation of DNA affect gene expression in pathways associated with a range of physiologic processes [39]. For example, altered promoter methylation and downstream changes in gene expression have been shown for the hepatic glucocorticoid receptor and the peroxisome proliferator-activated receptor-α, influencing carbohydrate and lipid metabolism [40,41]. It has also been shown that the promoter in the leptin gene is subject to epigenetic programming, and leptin gene expression can be modulated by DNA methylation [42,43]. Data from recent studies support the concept that impaired glucose tolerance during pregnancy leads to adaptations in leptin gene DNA methylation [44]. In addition, differential DNA methylation was observed in promoters of genes involved in glucose metabolism including GLUT4 and uncoupling protein-2, both major contributors to the development of T2DM [45,46].

Leptin Central Processes Regulating Food Intake in Offspring

There are data from few animal studies that maternal obesity induced by diet, prior to and throughout pregnancy and lactation, results in offspring with a hyperphagic and obese phenotype in adulthood [47,48]. Before the onset of the adult phenotype, these *animals show evidence of leptin resistance*. Neonates also display an amplified and prolonged surge of leptin, which is accompanied by elevated leptin mRNA expression in adipose tissue. The hypothesis is that prolonged release of abnormally high levels of leptin before weaning leads to permanently impaired leptin-signalling and a consequent reduction in leptin's neurotrophic effects, possibly due to down-regulation of leptin receptors [49]. Such effects may underlie the subsequent development of hyperphagia and increased adiposity in this experimental model.

Data from several studies suggest that leptin resistance acquired by the animals early in post-natal life play a critical role in the aetiology of the phenotype induced by maternal overnutrition (hypertension, obesity, diabetes and increased cardiovascular risk).

Maternal obesity was associated with increased levels of the orexigenic peptide, Neuropeptide Y (NPY) in the offspring, along with reduced levels of proopiomelanocortin (POMC) the major anorectic neuropeptide, α-melanocyte stimulating hormone (α-MSH). These changes may also be secondary to the development of leptin resistance in the neonate. These neurons (NPY and POMC) are located in arcuate nucleus of the hypothalamus (ARH) and send extensive projections to other parts of the hypothalamus, where they release their peptides to regulate energy balance. The critical player is likely to be leptin. Absence of leptin has been shown to permanently alter neuronal projections from the arcuate nucleus, and this can be prevented by administration of leptin during a critical period of neonatal development [50]. So, during this early post-natal life, leptin appears to act as an important neurotrophic factor that directs the formation of hypothalamic feeding circuits. Mice lacking leptin display permanent disruption of projections from the ARH to each of its target sites.

Maternal obesity is associated with elevated leptin in the neonate [49]. The high leptin, associated with elevated NPY and reduced POMC, is suggestive again of leptin resistance. It seems likely that perinatal perturbations of leptin action, through hypothalamic leptin resistance, might be the mechanism by which high perinatal nutrition could permanently alter hypothalamic circuits.

Hyperinsulinaemia, associated with maternal obesity, might also play a key role. Given its known role as a trophic factor for neuronal growth and survival, exposure of the developing foetal brain to excess insulin levels may be another important determinant of abnormal brain development (besides leptin), leading to obesity and glucose intolerance in later life [49].

Maternal Obesity and Glucose Metabolism in Mother and Offspring

T2DM is characterised by an age-related decline of β-cell mass function. This decline may be programmed at an early developmental stage. In animal studies, maternal obesity is associated with increased foetal pancreatic weight and a marked increase in the number of insulin-positive cells per unit area of the foetal pancreas, perhaps reflecting enhanced early β-cell maturation. However, such changes in early pancreatic development may result in premature post-natal β-cell loss and result in a predisposition to the development of obesity and metabolic dysfunction in adulthood [51]. There is strong evidence that T2DM is more prevalent among subjects that were in utero exposed to maternal diabetes. Maternal diabetes negatively imprints the growth of a genetically normal β-cell mass, resulting in decreased β-cell population later at adult life.

Recent studies in humans have examined the impact of maternal obesity to the offspring's glucose/insulin homeostasis. In Pima Indians, the prevalence of offspring diabetes was assessed in families in which at least one sibling was born before and one after their mother was diagnosed with T2DM. Offspring born after their mothers displayed diabetes had a fourfold higher risk of diabetes than their siblings born before the mother developed diabetes [52]. In a small study, offspring of obese mothers were more insulin resistant (using the homeostasis model from umbilical cord glucose and insulin concentrations) than offspring of lean mothers, suggesting that the foetus may have increased insulin secretion earlier in pregnancy [53]. Recent evidence from the Hyperglycaemia and Adverse Pregnancy Outcome study including 23,316 participants also reported an association between increased maternal BMI and foetal hyperinsulinaemia (assessed by cord serum C-peptide levels), even after adjustment for maternal glycaemia [54].

A recently published human study reported that obesity during pregnancy was associated with a high prevalence of overweight and obesity (overall $\sim$88%) in young adult offspring. This study cohort although had insulin resistance and hyperinsulinaemia, but there was no β-cell glucose sensitivity impairment [55]. Thus, this study confirmed in humans the data reported previously in animals that *maternal overweight at conception contributes to offspring obesity and insulin resistance in the absence of changes in birthweights.*

The existence of gestational diabetes mellitus (GDM) in the obese mother increases also the risk of offspring obesity and T2DM. This is supported from the observation that if fathers have T2DM, the disease risk for children is not as high as children born to women with DM during pregnancy. The data also showed that siblings born to the same mother only have increased disease risk if she had DM during pregnancy which supports the idea of an environmental effect rather than a genetic effect in this increased rate of obesity and diabetes in offspring of obese mothers with GDM.

Maternal DM among other detrimental effects also alters the function of foetal/neonatal endothelial progenitor cells (EPCs), ultimately effecting offspring vascular structure and function. EPCs participate in vascular repair, vessel formation and vascular remodelling. Reduced circulating EPCs correlate with increased cardiovascular disease risk. Neonatal EPCs from GDM pregnancies are hyperproliferative but yet have reduced vessel formation.

Maternal Obesity and Programming of Blood Pressure and Vascular Function

A number of animal studies have demonstrated that the offspring of mothers maintained on a high-fat diet before and during pregnancy and through lactation develop high blood pressure [13,56,57]. In animal studies, the offspring of obese mothers have endothelial dysfunction, assessed as reduced endothelium-dependent vasodilatation in both small and large vessels [58,59]. Hyperleptinaemia increases sympathetic activity (there is selective central leptin resistance not affecting sympathetic tone) and results in increased cardiovascular response to stress [60]. It seems logical that programming of autonomic function might be one mechanism underpinning the development of hypertension.

However, there are only scarce data in humans examining the association between maternal obesity and offspring blood pressure. A positive association was reported between gestational weight gain and both offspring obesity and systolic blood pressure at the age of 3 years [61]. Lastly, there are data supporting that exposure to hyperglycaemia (gestational diabetes is frequent in maternal obesity) in utero impairs nephrogenesis resulting in reduced nephron endowment in the foetus that has been associated with increased risk of developing essential hypertension and chronic renal failure [23].

Conclusions

Maternal obesity is frequently, yet not always, related to poor metabolic health. Impaired glucose tolerance, increased waist circumference, older age of obesity onset, recent weight gain and family history of metabolic diseases (e.g. diabetes) determine whether maternal obesity predispose to offspring metabolic disorders. Data from experimental and human studies consistently suggest a role for maternal obesity-related metabolic health abnormalities, rather than maternal adiposity per se, higher offspring risk for obesity, diabetes and cardiovascular risk.

Maternal metabolic health, particularly glucose intolerance, is likely to be involved in prenatal programming of obesity among neonates. The pathophysiological pathway linking the in utero environment with the foetal development is still unknown. However, there are clearly two requisites: the foetus must detect a change in environment and alter gene expression to cope with this altered environment. Cytokines, glucocorticoids, leptin, insulin or glucose are thought to act as cues that instigate altered foetal development. Epigenetic modifications such as gene methylation and histone acetylation are, most likely, the mechanism for altering foetal gene expression offspring.

What is becoming more apparent is the important role played by the maternal condition before and during gestation. Maternal health and well-being including obesity, gestational diabetes, nutritional or dietary intake are just a few of the important parameters which may need to be monitored during pregnancy especially in communities which have a higher predisposition or risk of expressing these cardiometabolic disease-prone phenotypes.

References

1. Hedley AA, Ogden CL, Johnson CL, Carroll MD, Curtin LR, Flegal KM. Prevalence of overweight and obesity among US children, adolescents, and adults, 1999–2002. *JAMA*. 2004;291(23):2847–2850.
2. Kanagalingam MG, Forouhi NG, Greer IA, Sattar N. Changes in booking body mass index over a decade: retrospective analysis from a Glasgow Maternity Hospital. *BJOG*. 2005;112(10):1431–1433.

3. Heslehurst N, Ells LJ, Simpson H, Batterham A, Wilkinson J, Summerbell CD. Trends in maternal obesity incidence rates, demographic predictors, and health inequalities in 36,821 women over a 15-year period. *BJOG*. 2007;114(2):187–194.
4. Ogden CL, Carroll MD, Curtin LR, McDowell MA, Tabak CJ, Flegal KM. Prevalence of overweight and obesity in the United States, 1999–2004. *JAMA*. 2006;295(13):1549–1555.
5. Barker DJ. Fetal programming of coronary heart disease. *Trends Endocrinol Metab*. 2002;13(9):364–368.
6. Fowden AL, Ward JW, Wooding FP, Forhead AJ, Constancia M. Programming placental nutrient transport capacity. *J Physiol*. 2006;572(Pt 1):5–15.
7. Petersen KF, Dufour S, Befroy D, Garcia R, Shulman GI. Impaired mitochondrial activity in the insulin-resistant offspring of patients with type 2 diabetes. *N Engl J Med*. 2004;350(7):664–671.
8. Warram JH, Martin BC, Krolewski AS, Soeldner JS, Kahn CR. Slow glucose removal rate and hyperinsulinemia precede the development of type II diabetes in the offspring of diabetic parents. *Ann Intern Med*. 1990;113(12):909–915.
9. Boney CM, Verma A, Tucker R, Vohr BR. Metabolic syndrome in childhood: association with birth weight, maternal obesity, and gestational diabetes mellitus. *Pediatrics*. 2005;115(3):e290–e296.
10. Guo F, Jen KL. High-fat feeding during pregnancy and lactation affects offspring metabolism in rats. *Physiol Behav*. 1995;57(4):681–686.
11. Bayol SA, Farrington SJ, Stickland NC. A maternal 'junk food' diet in pregnancy and lactation promotes an exacerbated taste for 'junk food' and a greater propensity for obesity in rat offspring. *Br J Nutr*. 2007;98(4):843–851.
12. Levin BE, Govek E. Gestational obesity accentuates obesity in obesity-prone progeny. *Am J Physiol*. 1998;275(4 Pt 2):R1374–R1379.
13. Samuelsson AM, Matthews PA, Argenton M, et al. Diet-induced obesity in female mice leads to offspring hyperphagia, adiposity, hypertension, and insulin resistance: a novel murine model of developmental programming. *Hypertension*. 2008;51(2):383–392.
14. Tamashiro KL, Terrillion CE, Hyun J, Koenig JI, Moran TH. Prenatal stress or high-fat diet increases susceptibility to diet-induced obesity in rat offspring. *Diabetes*. 2009;58 (5):1116–1125.
15. Minge CE, Bennett BD, Norman RJ, Robker RL. Peroxisome proliferator-activated receptor-gamma agonist rosiglitazone reverses the adverse effects of diet-induced obesity on oocyte quality. *Endocrinology*. 2008;149(5):2646–2656.
16. Shankar K, Harrell A, Liu X, Gilchrist JM, Ronis MJ, Badger TM. Maternal obesity at conception programs obesity in the offspring. *Am J Physiol Regul Integr Comp Physiol*. 2008;294(2):R528–R538.
17. Surkan PJ, Hsieh CC, Johansson AL, Dickman PW, Cnattingius S. Reasons for increasing trends in large for gestational age births. *Obstet Gynecol*. 2004;104(4):720–726.
18. Lawlor DA, Timpson NJ, Harbord RM, et al. Exploring the developmental overnutrition hypothesis using parental–offspring associations and FTO as an instrumental variable. *Plos Med*. 2008;5(3):e33.
19. Taylor PD, Poston L. Developmental programming of obesity in mammals. *Exp Physiol*. 2007;92(2):287–298.
20. Armitage JA, Lakasing L, Taylor PD, et al. Developmental programming of aortic and renal structure in offspring of rats fed fat-rich diets in pregnancy. *J Physiol*. 2005;565 (Pt 1):171–184.
21. Lawlor DA, Smith GD, O'Callaghan M, et al. Epidemiologic evidence for the fetal overnutrition hypothesis: findings from the mater-university study of pregnancy and its outcomes. *Am J Epidemiol*. 2007;165(4):418–424.

22. Davey Smith G, Steer C, Leary S, Ness A. Is there an intrauterine influence on obesity? Evidence from parent child associations in the Avon Longitudinal Study of Parents and Children (ALSPAC). *Arch Dis Child.* 2007;92(10):876–880.
23. Gluckman PD, Hanson MA, Cooper C, Thornburg KL. Effect of in utero and early-life conditions on adult health and disease. *N Engl J Med.* 2008;359(1):61–73.
24. Silveira PP, Portella AK, Goldani MZ, Barbieri MA. Developmental origins of health and disease (DOHaD). *J Pediatr (Rio J).* 2007;83(6):494–504.
25. Stannard SR, Johnson NA. Insulin resistance and elevated triglyceride in muscle: more important for survival than 'thrifty' genes? *J Physiol.* 2004;554(Pt 3):595–607.
26. Zambrano E, Martinez-Samayoa PM, Bautista CJ, et al. Sex differences in transgenerational alterations of growth and metabolism in progeny (F2) of female offspring (F1) of rats fed a low protein diet during pregnancy and lactation. *J Physiol.* 2005;566(Pt 1):225–236.
27. Forsen T, Eriksson J, Tuomilehto J, Reunanen A, Osmond C, Barker D. The fetal and childhood growth of persons who develop type 2 diabetes. *Ann Intern Med.* 2000;133(3):176–182.
28. Hediger ML, Overpeck MD, Kuczmarski RJ, McGlynn A, Maurer KR, Davis WW. Muscularity and fatness of infants and young children born small- or large-for-gestational-age. *Pediatrics.* 1998;102(5):E60.
29. Aguiari P, Leo S, Zavan B, et al. High glucose induces adipogenic differentiation of muscle-derived stem cells. *Proc Natl Acad Sci USA.* 2008;105(4):1226–1231.
30. Bayol SA, Macharia R, Farrington SJ, Simbi BH, Stickland NC. Evidence that a maternal 'junk food' diet during pregnancy and lactation can reduce muscle force in offspring. *Eur J Nutr.* 2009;48(1):62–65.
31. Zhu MJ, Ford SP, Means WJ, Hess BW, Nathanielsz PW, Du M. Maternal nutrient restriction affects properties of skeletal muscle in offspring. *J Physiol.* 2006;575(Pt 1):241–250.
32. Steinberg GR. Inflammation in obesity is the common link between defects in fatty acid metabolism and insulin resistance. *Cell Cycle.* 2007;6(8):888–894.
33. Greenberg AS, Obin MS. Obesity and the role of adipose tissue in inflammation and metabolism. *Am J Clin Nutr.* 2006;83(2):461S–465S.
34. Wei Y, Chen K, Whaley-Connell AT, Stump CS, Ibdah JA, Sowers JR. Skeletal muscle insulin resistance: role of inflammatory cytokines and reactive oxygen species. *Am J Physiol Regul Integr Comp Physiol.* 2008;294(3):R673–R680.
35. Du M, Yan X, Tong JF, Zhao J, Zhu MJ. Maternal obesity, inflammation, and fetal skeletal muscle development. *Biol Reprod.* 2010;82(1):4–12.
36. Egger G, Liang G, Aparicio A, Jones PA. Epigenetics in human disease and prospects for epigenetic therapy. *Nature.* 2004;429(6990):457–463.
37. Heerwagen MJ, Miller MR, Barbour LA, Friedman JE. Maternal obesity and fetal metabolic programming: a fertile epigenetic soil. *Am J Physiol Regul Integr Comp Physiol.* 2010;299(3):R711–R722.
38. Waterland RA, Michels KB. Epigenetic epidemiology of the developmental origins hypothesis. *Annu Rev Nutr.* 2007;27:363–388.
39. Vickers MH. Developmental programming of the metabolic syndrome – critical windows for intervention. *World J Diabetes.* 2011;2(9):137–148.
40. Burdge GC, Lillycrop KA, Jackson AA, Gluckman PD, Hanson MA. The nature of the growth pattern and of the metabolic response to fasting in the rat are dependent upon the dietary protein and folic acid intakes of their pregnant dams and post-weaning fat consumption. *Br J Nutr.* 2008;99(3):540–549.
41. Burdge GC, Slater-Jefferies J, Torrens C, Phillips ES, Hanson MA, Lillycrop KA. Dietary protein restriction of pregnant rats in the F0 generation induces altered methylation of hepatic gene promoters in the adult male offspring in the F1 and F2 generations. *Br J Nutr.* 2007;97(3):435–439.

42. Iliopoulos D, Malizos KN, Tsezou A. Epigenetic regulation of leptin affects MMP-13 expression in osteoarthritic chondrocytes: possible molecular target for osteoarthritis therapeutic intervention. *Ann Rheum Dis*. 2007;66(12):1616–1621.
43. Stoger R. In vivo methylation patterns of the leptin promoter in human and mouse. *Epigenetics*. 2006;1(4):155–162.
44. Bouchard L, Thibault S, Guay SP, et al. Leptin gene epigenetic adaptation to impaired glucose metabolism during pregnancy. *Diabetes Care*. 2010;33(11):2436–2441.
45. Yokomori N, Tawata M, Onaya T. DNA demethylation during the differentiation of 3T3-L1 cells affects the expression of the mouse GLUT4 gene. *Diabetes*. 1999;48(4):685–690.
46. Carretero MV, Torres L, Latasa U, et al. Transformed but not normal hepatocytes express UCP2. *FEBS Lett*. 1998;439(1–2):55–58.
47. White CL, Purpera MN, Morrison CD. Maternal obesity is necessary for programming effect of high-fat diet on offspring. *Am J Physiol Regul Integr Comp Physiol*. 2009;296 (5):R1464–R1472.
48. Rajia S, Chen H, Morris MJ. Maternal overnutrition impacts offspring adiposity and brain appetite markers-modulation by postweaning diet. *J Neuroendocrinol*. 2010;22 (8):905–914.
49. Li M, Sloboda DM, Vickers MH. Maternal obesity and developmental programming of metabolic disorders in offspring: evidence from animal models. *Exp Diabetes Res*. 2011;:592408.
50. Kirk SL, Samuelsson AM, Argenton M, et al. Maternal obesity induced by diet in rats permanently influences central processes regulating food intake in offspring. *Plos One*. 2009;4(6):e5870.
51. Drake AJ, Reynolds RM. Impact of maternal obesity on offspring obesity and cardiometabolic disease risk. *Reproduction*. 2010;140(3):387–398.
52. Dabelea D, Hanson RL, Lindsay RS, et al. Intrauterine exposure to diabetes conveys risks for type 2 diabetes and obesity: a study of discordant sibships. *Diabetes*. 2000;49 (12):2208–2211.
53. Catalano PM, Presley L, Minium J, Hauguel-de Mouzon S. Fetuses of obese mothers develop insulin resistance in utero. *Diabetes Care*. 2009;32(6):1076–1080.
54. HAPO Study Group. Hyperglycaemia and Adverse Pregnancy Outcome (HAPO) study: associations with maternal body mass index. *BJOG*. 2010;117(5):575–584.
55. Mingrone G, Manco M, Mora ME, et al. Influence of maternal obesity on insulin sensitivity and secretion in offspring. *Diabetes Care*. 2008;31(9):1872–1876.
56. Khan IY, Taylor PD, Dekou V, et al. Gender-linked hypertension in offspring of lard-fed pregnant rats. *Hypertension*. 2003;41(1):168–175.
57. Liang C, Oest ME, Prater MR. Intrauterine exposure to high saturated fat diet elevates risk of adult-onset chronic diseases in C57BL/6 mice. *Birth Defects Res B Dev Reprod Toxicol*. 2009;86(5):377–384.
58. Koukkou E, Ghosh P, Lowy C, Poston L. Offspring of normal and diabetic rats fed saturated fat in pregnancy demonstrate vascular dysfunction. *Circulation*. 1998;98 (25):2899–2904.
59. Ghosh P, Bitsanis D, Ghebremeskel K, Crawford MA, Poston L. Abnormal aortic fatty acid composition and small artery function in offspring of rats fed a high fat diet in pregnancy. *J Physiol*. 2001;533(Pt 3):815–822.
60. Samuelsson AM, Morris A, Igosheva N, et al. Evidence for sympathetic origins of hypertension in juvenile offspring of obese rats. *Hypertension*. 2010;55(1):76–82.
61. Oken E, Rifas-Shiman SL, Field AE, Frazier AL, Gillman MW. Maternal gestational weight gain and offspring weight in adolescence. *Obstet Gynecol*. 2008;112(5):999–1006.

37 Obesity and Female Malignancies

Mohamed K. Mehasseb and Mahmood I. Shafi

Department of Gynaecological Oncology, Addenbrooke's Hospital, Cambridge Biomedical Campus, Cambridge, UK

Introduction

The prevalence of obesity is rapidly increasing worldwide. The proportion of overweight and obese children and adults has steadily increased, with nearly one-third of all adults now being classified as obese [1,2]. The impact of overweight and obesity in terms of both mortality and health care costs equals or exceeds that associated with tobacco use, worldwide. There is sufficient evidence to recommend that adults and children maintain healthy weight for general health benefits, including decreasing their risk of cancer.

Overall, about 3% of all cancers are linked to obesity, while cancers linked to obesity comprise approximately 51% of newly diagnosed cases among women [3]. Obesity and overweight play an important role in female malignancies, being associated with increased risks of endometrial cancer (one and a half-fold), post-menopausal breast cancer (two-fold), and possibly cervical and ovarian cancers (Table 37.1) [4]. It is estimated that 20–30% of deaths from cancer in women could be attributed to overweight and obesity [5,6]. Furthermore, obesity has an impact on the screening, diagnosis and treatment of female malignancies.

Epidemiology

Obesity and Endometrial Carcinoma

Obesity accounts for about 40% cases of endometrial cancer in the developed world [7]. Endometrial carcinoma was the first malignancy to be recognised as being linked to obesity. A linear increase in the risk of endometrial cancer with increasing weight and BMI has been observed [8]. There is convincing and consistent evidence from both case–control and cohort studies that overweight and obesity are strongly associated with type I (oestrogen-dependent) endometrial

Obesity. DOI: http://dx.doi.org/10.1016/B978-0-12-416045-3.00037-6

Table 37.1 Relative Risk of Cancer Incidence in Relation to Body Mass Index (BMI)

Site	BMI			
	25–27.4	**27.5–29.5**	**>30**	**Overall Trend**
Endometrium	+21%	+43%	+273%	+289%
Breast (pre-menopausal)	−7%	−1%	−21%	−14%
Breast (post-menopausal)	+10%	+21%	+29%	+40%
Ovary	−1%	+13%	+12%	+14%
Cervix	−6%	−21%	+2%	+4%

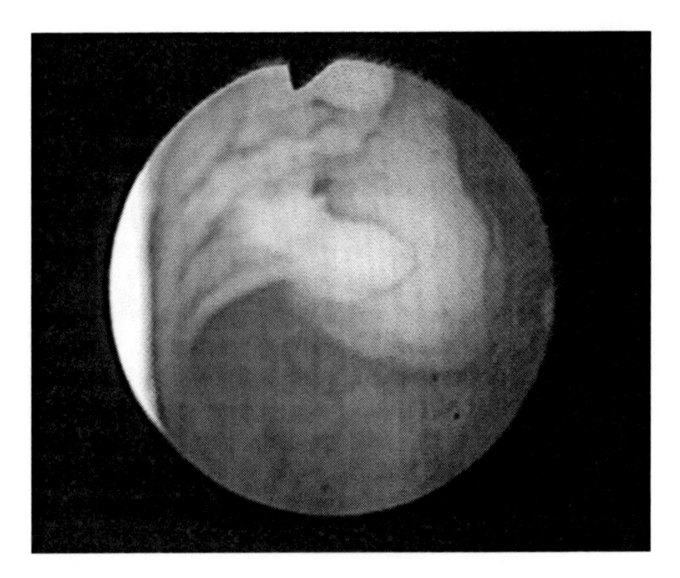

Figure 37.1 Hysteroscopy showing an endometrial polyp associated with a thickened malignant endometrium.

cancer [2,8]. Overweight and obese women have two to four times greater risk of developing endometrial cancer than do women of a healthy weight, regardless of their menopausal status [9,10]. Obesity in the menopause produces a state of excess oestrogen production. This is due to the peripheral conversion, in the adipose tissue, of androgens secreted from the adrenal glands and ovaries into oestrone, by the enzyme aromatase. Prolonged unopposed oestrogen exposure will lead to a continuous spectrum of change from proliferative endometrium through endometrial hyperplasia/polyps to endometrial carcinoma (Figures 37.1 and 37.2). Intrauterine contraceptive devices are a good contraceptive choice for obese women given their high efficacy irrespective of weight and the ability of the levonorgestrel intrauterine system (Mirena®) to prevent endometrial hyperplasia in obese anovulatory women.

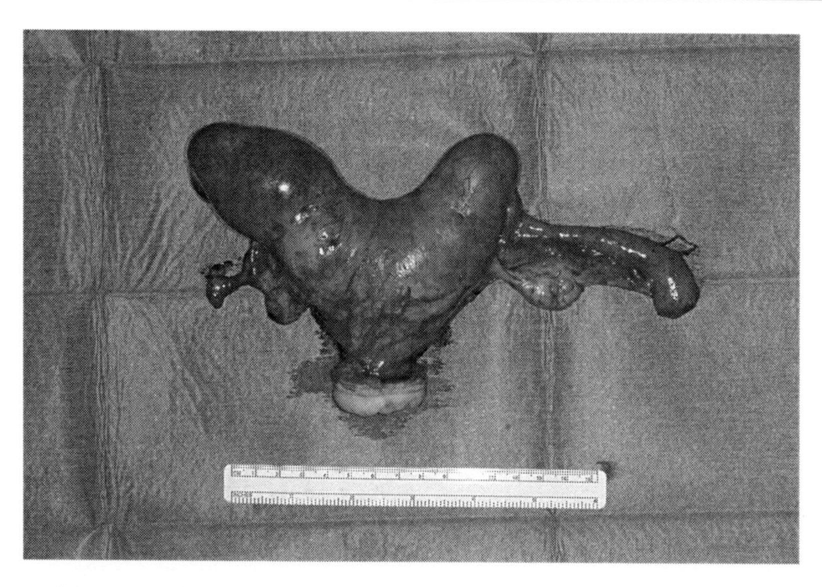

Figure 37.2 A surgical specimen following a total abdominal hysterectomy and bilateral salpingo-oophorectomy performed for the treatment of high-grade endometrial carcinoma. Note the presence of a bulky enlarged uterus with endometrial cancer filling and distending both cornua and the right fallopian tube.

Obesity and Breast Cancer

Obesity seems to increase the risk of breast cancer only among post-menopausal women who do not use hormonal replacement therapy (HRT) (30–50% increased risk) [11–13]. Obese women are also at higher risk of dying from breast cancer after the menopause [14]. Weight gain during adulthood has been found to be the most consistent and strongest predictor of post-menopausal breast cancer risk [15–17]. It is estimated that 11,000–18,000 deaths per year from breast cancer in women over age 50 might be avoided if women could maintain a BMI under 25 kg/m^2 throughout their adult lives [14]. Women with central obesity have a greater breast cancer risk than those whose fat is distributed over the hips, buttocks and lower extremities [18]. In addition, adult weight gain has been associated with a higher risk of post-menopausal breast cancer than the actual BMI level [19,20]. Among post-menopausal HRT users, there is no significant difference in breast cancer risk between obese women and women of a healthy weight [12,21]. Paradoxically, pre-menopausal obese women have a lower risk of developing breast cancer than do women of a healthy weight [12,22,23].

Both the increased risk of developing breast cancer and dying from it after the menopause seem to be due to increased levels of endogenous oestrogens in obese women [24]. Oestrogen levels in post-menopausal women are 50–100% higher among obese women, compared to lean women [25], leading to a more rapid growth of oestrogen-responsive/sensitive breast tumours. Breast cancer also seems to be detected later in obese women, leading to a poorer outcome. This is because the detection of a breast tumour is more difficult in overweight women [23].

Obesity and Ovarian Cancer

Ovarian cancer has not been consistently linked to obesity. Some studies report an increased risk among obese women, whereas others have found no association [26–34]. A report found an increased risk in women who were overweight or obese in adolescence or young adulthood; while no increased risk was found in older obese women [35]. Although endogenous hormones are believed to be involved in the aetiology of ovarian cancer [36], no solid conclusions could be drawn. It is possible that obesity increases the risk of specific histological subtypes of ovarian cancer (e.g. endometrioid), but not others. Most studies have not examined risk by histological subtype of ovarian cancer, and this might contribute to the inconsistent findings.

Obesity and Cervical Cancer

Studies of the association between BMI and cervical cancer are limited and inconclusive. While some studies reported cervical cancer to be associated with elevated BMI (two- to three-fold increased risk) [8], others found a lower relative risk [37]. No association was observed in a cohort study of Swedish women [38]. The increased risk among overweight and obese women was mainly for cervical adenocarcinoma, with a smaller increased risk for squamous-cell carcinoma [39]. Differential screening and health behaviour, where obese women might be less likely to go for screening on a regular basis than women of normal weight, could partly explain the observed increased risk.

Mechanisms Relating Obesity to Female Malignancies Risks

Obesity affects the production of peptides (e.g. insulin and insulin-like growth factor 1 (IGF1), sex hormone–binding globulin (SHBG)) and steroid hormones (i.e. oestrogen, progesterone and androgens). It is likely that prolonged exposure to high levels of oestrogen and insulin associated with obesity may contribute to the development of female malignancies [2,40,41].

BMI is directly related to circulating levels of oestrone and oestradiol, particularly in post-menopausal women [40]. Obesity in the menopause produces a state of excess oestrogen production. Adipose tissue cells express various steroid hormone–metabolising enzymes and are an important source of circulating oestrogens, especially in post-menopausal women. This is due to the peripheral conversion of androgens secreted from the adrenal glands and ovaries into oestrone, by the enzyme aromatase in the fat cells. The situation is aggravated by the fact that increased body fat is associated with decreased circulating levels of both progesterone and SHBG. With lower SHBG, there is a higher circulating level of free active oestrogens.

Excess weight, increased plasma triglyceride levels and low levels of physical activity can all raise circulating insulin levels, leading to chronic hyperinsulinaemia which has been associated with cancers of the breast [42,43] and the endometrium [2]. The carcinogenic effects of hyperinsulinaemia could be directly mediated by insulin receptors in the target cells, or might be due to related changes in endogenous female sex hormones synthesis and bioavailability. Insulin also promotes the synthesis and biological activity of IGF1. Both insulin and IGF1 can act as growth factors that promote cell proliferation and inhibit apoptosis [44,45]. Endometrial cancer risk is inversely related to blood levels of IGF-binding protein 1 and 2 (IGFBP1 and IGFBP2), which reduce the amount of bioavailable IGF1 [46]. There is an increased risk of breast cancer in women with increased serum levels of insulin or IGF1 [47–49], especially in pre-menopausal women. The increase in blood levels of insulin and IGF1 results in reduced hepatic synthesis and blood concentrations of SHBG, increasing the bioavailability of oestradiol [2,50,51]. Proteins secreted by adipose tissue (adipokines) also contribute to the regulation of immune response (leptin), inflammatory response (tumour necrosis factor-α, interleukin-6 and serum amyloid A), vasculature and stromal interactions and angiogenesis (vascular endothelial growth factor 1), as well as extracellular matrix components (type VI collagen).

Avoiding weight gain lowers the risk of endometrial and post-menopausal breast cancers [52]. However, there is limited evidence that intentional weight loss will affect cancer risk [53,54]. Physical activity among post-menopausal women at a level of walking about 30 min/day was associated with a 20% reduction in breast cancer risk, mainly among women who were of normal weight. The protective effect of physical activity was not found among overweight or obese women [55].

Effect of Obesity on Management of Female Malignancies

Obese patients have a poorer outcome compared to lean patients. For instance, obesity is associated both with reduced likelihood of survival and increased likelihood of recurrence among patients with breast cancer, regardless of menopausal status and after adjustment for stage and treatment (Table 37.2) [4,14,56]. The poorer outcomes in obese women probably reflect a true biological effect of adiposity on survival, a delayed diagnosis in heavier women, and a higher rate of

Table 37.2 Relative Risk of Cancer Mortality in Relation to BMI

Site	BMI			
	25–27.4	27.5–29.5	>30	Overall Trend
Endometrium	+9%	+21%	+228%	+246%
Breast (pre-menopausal)	+5%	−9%	−36%	−32%
Breast (post-menopausal)	+26%	+22%	+49%	+53%
Ovary	−7%	+2%	+16%	+17%
Cervix	−23%	−39%	+15%	+53%

treatment-associated complications. Heavier women are less likely to receive mammography or cervical screening [57]. For women who self-detect their breast cancers, non-localised disease is more common with a high BMI [58].

Clinical examination of the obese women with female malignancies can be difficult. Manual handling of these patients can also be challenging. Special hospital beds and operating tables should be available. The best way to ensure a safe and successful treatment is adequate pre-operative evaluation, preparation and counselling. Assistance with proper dosing and monitoring of medications should be considered. Prescriptions must take into account the concepts of total body weight as well as ideal body weight, as certain doses (e.g. corticosteroids, penicillin and cephalosporins) are calculated based on ideal body weight, while others are calculated based on total body weight (e.g. heparins).

A high BMI increases the risk of perioperative complications and mortality, particularly in the presence of co-morbidities. The risks are increased in case of morbid obesity (BMI $>$ 40). Obese women should receive careful counselling about the increased risk of complications and technical difficulties that may be encountered during surgery [59,60]. Pre-operative evaluation should include a cardiovascular and respiratory assessment. Obese patients are at much higher risk for post-operative complications given the more frequent co-morbidities, such as diabetes, hypertension, coronary artery disease, sleep apnoea, hypoventilation and osteoarthritis of the knees and hips. Thus, respiratory or cardiac failure, venous thromboembolism, aspiration, wound infection and dehiscence, and post-operative asphyxia are all more common in obese patients. Control of the airway is critical in obese surgical patients. Planned admission to high-dependency units is often advisable. Ventilation may be aided by the use of non-invasive positive pressure ventilation units, particularly if the patient has a history of sleep apnoea. Venous access can be problematic in obese patients. Doppler ultrasound scans could assist the safe placement of intravenous lines.

Women with gynaecological malignancies can be managed in a standard fashion, and in most instances there is no need to compromise surgical treatment where this is indicated [61]. The route of any surgical intervention needs to be considered, as abdominal procedures are more of an issue than vaginal surgery. The abdominal wall anatomy is distorted by the overhanging skin and fat (panniculus). Obesity is recognised as a potential limiting factor in the application of laparoscopic surgery because of a higher rate of failed entry, hindered manipulation and poor views. Obesity may not allow steep Trendelenburg because of unacceptably high peak inspiratory pressure. In addition, obesity may prevent adequate mobilisation of the small bowel out of the pelvis to allow for proper pelvic visualisation. Nevertheless, laparoscopic surgery has additional benefits for the obese: they have less post-operative ileus, fewer wound infections and they mobilise more quickly than those undergoing laparotomy [62]. Obesity presents problems with laparotomy incision placement and closure. Adequate wound antisepsis is necessary, as obese women are at increased risk of wound infection and wound failure. Possible aetiologies include decreased oxygen tension, immune impairment, and tension and secondary ischaemia along suture lines [63]. Access to the pelvis can be challenging and there is a higher incidence of intra-operative complications due to problems with access or distorted anatomy. Difficulty with haemostasis, particularly among women when removing the cervix and suturing the vaginal vault, requires experience to manage. Good assistance, retraction and lighting are essential.

The use of regional anaesthesia is to be encouraged: as well as the anaesthetic benefits, this helps with post-operative analgesia. However, in practice, regional anaesthesia may prove difficult or even impossible, but the availability of an experienced anaesthetist will minimise technical failure.

Future Directions

Further research to define the causal role of obesity in gynaecological malignancies is needed. Obesity-associated dysregulation of adipokines is likely to contribute not only to tumorigenesis and tumour progression, but also to metastatic potential. It will also be important to develop successful intervention strategies, both at the individual and community levels, for weight loss and maintenance. The tobacco control experience has taught us that policy and environmental changes are crucial to achieving changes in individual behaviour. Future trials may involve studies on the effect of dietary changes on weight gain and cancer risk, the effect of patterns of physical activity (the intensity, frequency and duration of various sorts of physical activity) in relation to weight gain and cancer risk, the combined effects of changes in diet and physical activity on obesity and female cancer risk.

References

1. Zaninotto P, Head J, Stamatakis E, Wardle H, Mindell J. Trends in obesity among adults in England from 1993 to 2004 by age and social class and projections of prevalence to 2012. *J Epidemiol Community Health*. 2009;63:140–146.
2. Kaaks R, Lukanova A, Kurzer MS. Obesity, endogenous hormones, and endometrial cancer risk: a synthetic review. *Cancer Epidemiol Biomarkers Prev*. 2002;11:1531–1543.
3. Polednak AP. Trends in incidence rates for obesity-associated cancers in the US. *Cancer Detect Prev*. 2003;27:415–421.
4. Reeves GK, Pirie K, Beral V, et al. Cancer incidence and mortality in relation to body mass index in the Million Women Study: cohort study. *BMJ*. 2007;335:1134.
5. Allison DB, Fontaine KR, Manson JE, Stevens J, VanItallie TB. Annual deaths attributable to obesity in the United States. *JAMA*. 1999;282:1530–1538.
6. Banegas JR, Lopez-Garcia E, Gutierrez-Fisac JL, Guallar-Castillon P, Rodriguez-Artalejo F. A simple estimate of mortality attributable to excess weight in the European Union. *Eur J Clin Nutr*. 2003;57:201–208.
7. Bergstrom A, Pisani P, Tenet V, Wolk A, Adami HO. Overweight as an avoidable cause of cancer in Europe. *Int J Cancer*. 2001;91:421–430.

8. Calle EE, Rodriguez C, Walker-Thurmond K, Thun MJ. Overweight, obesity, and mortality from cancer in a prospectively studied cohort of U.S. adults. *N Engl J Med.* 2003;348:1625–1638.
9. Weiderpass E, Persson I, Adami HO, Magnusson C, Lindgren A, Baron JA. Body size in different periods of life, diabetes mellitus, hypertension, and risk of postmenopausal endometrial cancer (Sweden). *Cancer Causes Control.* 2000;11:185–192.
10. Shoff SM, Newcomb PA. Diabetes, body size, and risk of endometrial cancer. *Am J Epidemiol.* 1998;148:234–240.
11. Galanis DJ, Kolonel LN, Lee J, Le Marchand L. Anthropometric predictors of breast cancer incidence and survival in a multi-ethnic cohort of female residents of Hawaii, United States. *Cancer Causes Control.* 1998;9:217–224.
12. van den Brandt PA, Spiegelman D, Yaun SS, et al. Pooled analysis of prospective cohort studies on height, weight, and breast cancer risk. *Am J Epidemiol.* 2000;152:514–527.
13. Friedenreich CM. Review of anthropometric factors and breast cancer risk. *Eur J Cancer Prev.* 2001;10:15–32.
14. Petrelli JM, Calle EE, Rodriguez C, Thun MJ. Body mass index, height, and postmenopausal breast cancer mortality in a prospective cohort of US women. *Cancer Causes Control.* 2002;13:325–332.
15. Kawai M, Minami Y, Kuriyama S, et al. Adiposity, adult weight change and breast cancer risk in postmenopausal Japanese women: the Miyagi Cohort Study. *Br J Cancer.* 2010;103:1443–1447.
16. Ahn J, Schatzkin A, Lacey Jr. JV, et al. Adiposity, adult weight change, and postmenopausal breast cancer risk. *Arch Intern Med.* 2007;167:2091–2102.
17. Eliassen AH, Colditz GA, Rosner B, Willett WC, Hankinson SE. Adult weight change and risk of postmenopausal breast cancer. *JAMA.* 2006;296:193–201.
18. Kaaks R, Van Noord PA, Den Tonkelaar I, Peeters PH, Riboli E, Grobbee DE. Breast-cancer incidence in relation to height, weight and body-fat distribution in the Dutch 'DOM' cohort. *Int J Cancer.* 1998;76:647–651.
19. Feigelson HS, Jonas CR, Teras LR, Thun MJ, Calle EE. Weight gain, body mass index, hormone replacement therapy, and postmenopausal breast cancer in a large prospective study. *Cancer Epidemiol Biomarkers Prev.* 2004;13:220–224.
20. Schairer C, Lubin J, Troisi R, Sturgeon S, Brinton L, Hoover R. Menopausal estrogen and estrogen–progestin replacement therapy and breast cancer risk. *JAMA.* 2000;283:485–491.
21. Lahmann PH, Lissner L, Gullberg B, Olsson H, Berglund G. A prospective study of adiposity and postmenopausal breast cancer risk: the Malmo Diet and Cancer Study. *Int J Cancer.* 2003;103:246–252.
22. Trentham-Dietz A, Newcomb PA, Storer BE, et al. Body size and risk of breast cancer. *Am J Epidemiol.* 1997;145:1011–1019.
23. Cui Y, Whiteman MK, Flaws JA, Langenberg P, Tkaczuk KH, Bush TL. Body mass and stage of breast cancer at diagnosis. *Int J Cancer.* 2002;98:279–283.
24. Toniolo PG, Levitz M, Zeleniuch-Jacquotte A, et al. A prospective study of endogenous estrogens and breast cancer in postmenopausal women. *J Natl Cancer Inst.* 1995;87:190–197.
25. Huang Z, Hankinson SE, Colditz GA, et al. Dual effects of weight and weight gain on breast cancer risk. *JAMA.* 1997;278:1407–1411.
26. Mori M, Nishida T, Sugiyama T, et al. Anthropometric and other risk factors for ovarian cancer in a case–control study. *Jpn J Cancer Res.* 1998;89:246–253.

27. Mink PJ, Folsom AR, Sellers TA, Kushi LH. Physical activity, waist-to-hip ratio, and other risk factors for ovarian cancer: a follow-up study of older women. *Epidemiology*. 1996;7:38–45.
28. Farrow DC, Weiss NS, Lyon JL, Daling JR. Association of obesity and ovarian cancer in a case–control study. *Am J Epidemiol*. 1989;129:1300–1304.
29. Pan SY, Johnson KC, Ugnat AM, Wen SW, Mao Y, Canadian Cancer Registries Epidemiology Research G. Association of obesity and cancer risk in Canada. *Am J Epidemiol*. 2004;159:259–268.
30. Fairfield KM, Willett WC, Rosner BA, Manson JE, Speizer FE, Hankinson SE. Obesity, weight gain, and ovarian cancer. *Obstet Gynecol*. 2002;100:288–296.
31. Peterson NB, Trentham-Dietz A, Newcomb PA, et al. Relation of anthropometric measurements to ovarian cancer risk in a population-based case–control study (United States). *Cancer Causes Control*. 2006;17:459–467.
32. Kuper H, Cramer DW, Titus-Ernstoff L. Risk of ovarian cancer in the United States in relation to anthropometric measures: does the association depend on menopausal status? *Cancer Causes Control*. 2002;13:455–463.
33. Schouten LJ, Goldbohm RA, van den Brandt PA. Height, weight, weight change, and ovarian cancer risk in the Netherlands cohort study on diet and cancer. *Am J Epidemiol*. 2003;157:424–433.
34. Hartge P, Schiffman MH, Hoover R, McGowan L, Lesher L, Norris HJ. A case–control study of epithelial ovarian cancer. *Am J Obstet Gynecol*. 1989;161:10–16.
35. Engeland A, Tretli S, Bjorge T. Height, body mass index, and ovarian cancer: a follow-up of 1.1 million Norwegian women. *J Natl Cancer Inst*. 2003;95:1244–1248.
36. Risch HA. Hormonal etiology of epithelial ovarian cancer, with a hypothesis concerning the role of androgens and progesterone. *J Natl Cancer Inst*. 1998;90:1774–1786.
37. Wolk A, Gridley G, Svensson M, et al. A prospective study of obesity and cancer risk (Sweden). *Cancer Causes Control*. 2001;12:13–21.
38. Tornberg SA, Carstensen JM. Relationship between Quetelet's index and cancer of breast and female genital tract in 47,000 women followed for 25 years. *Br J Cancer*. 1994;69:358–361.
39. Lacey Jr. JV, Swanson CA, Brinton LA, et al. Obesity as a potential risk factor for adenocarcinomas and squamous cell carcinomas of the uterine cervix. *Cancer*. 2003;98:814–821.
40. Key TJ, Appleby PN, Reeves GK, et al. Body mass index, serum sex hormones, and breast cancer risk in postmenopausal women. *J Natl Cancer Inst*. 2003;95:1218–1226.
41. Bianchini F, Kaaks R, Vainio H. Overweight, obesity, and cancer risk. *Lancet Oncol*. 2002;3:565–574.
42. Kaaks R. Nutrition, hormones, and breast cancer: is insulin the missing link? *Cancer Causes Control*. 1996;7:605–625.
43. Stoll BA. Oestrogen/insulin-like growth factor-I receptor interaction in early breast cancer: clinical implications. *Ann Oncol*. 2002;13:191–196.
44. Prisco M, Romano G, Peruzzi F, Valentinis B, Baserga R. Insulin and IGF-I receptors signaling in protection from apoptosis. *Horm Metab Res*. 1999;31:80–89.
45. Khandwala HM, McCutcheon IE, Flyvbjerg A, Friend KE. The effects of insulin-like growth factors on tumorigenesis and neoplastic growth. *Endocr Rev*. 2000;21:215–244.
46. Lukanova A, Zeleniuch-Jacquotte A, Lundin E, et al. Prediagnostic levels of C-peptide, IGF-I, IGFBP-1, -2 and -3 and risk of endometrial cancer. *Int J Cancer*. 2004;108:262–268.

47. Cust AE, Stocks T, Lukanova A, et al. The influence of overweight and insulin resistance on breast cancer risk and tumour stage at diagnosis: a prospective study. *Breast Cancer Res Treat.* 2009;113:567–576.
48. Pichard C, Plu-Bureau G, Neves ECM, Gompel A. Insulin resistance, obesity and breast cancer risk. *Maturitas.* 2008;60:19–30.
49. Schairer C, Hill D, Sturgeon SR, et al. Serum concentrations of IGF-I, IGFBP-3 and c-peptide and risk of hyperplasia and cancer of the breast in postmenopausal women. *Int J Cancer.* 2004;108:773–779.
50. Tchernof A, Despres JP. Sex steroid hormones, sex hormone–binding globulin, and obesity in men and women. *Horm Metab Res.* 2000;32:526–536.
51. Kokkoris P, Pi-Sunyer FX. Obesity and endocrine disease. *Endocrinol Metab Clin North Am.* 2003;32:895–914.
52. Vainio H, Kaaks R, Bianchini F. Weight control and physical activity in cancer prevention: international evaluation of the evidence. *Eur J Cancer Prev.* 2002;11(suppl 2): S94–100.
53. Ziegler RG, Hoover RN, Nomura AM, et al. Relative weight, weight change, height, and breast cancer risk in Asian–American women. *J Natl Cancer Inst.* 1996;88:650–660.
54. Trentham-Dietz A, Newcomb PA, Egan KM, et al. Weight change and risk of postmenopausal breast cancer (United States). *Cancer Causes Control.* 2000;11:533–542.
55. McTiernan A, Kooperberg C, White E, et al. Recreational physical activity and the risk of breast cancer in postmenopausal women: the Women's Health Initiative Cohort Study. *JAMA.* 2003;290:1331–1336.
56. Chlebowski RT, Aiello E, McTiernan A. Weight loss in breast cancer patient management. *J Clin Oncol.* 2002;20:1128–1143.
57. Wee CC, McCarthy EP, Davis RB, Phillips RS. Screening for cervical and breast cancer: is obesity an unrecognized barrier to preventive care? *Ann Intern Med.* 2000;132:697–704.
58. Reeves MJ, Newcomb PA, Remington PL, Marcus PM, MacKenzie WR. Body mass and breast cancer. Relationship between method of detection and stage of disease. *Cancer.* 1996;77:301–307.
59. Pena MM, Taveras EM. Preventing childhood obesity: wake up, it's time for sleep! *J Clin Sleep Med.* 2011;7:343–344.
60. Haslam D, Sattar N, Lean M. ABC of obesity. Obesity – time to wake up. *BMJ.* 2006;333:640–642.
61. Papadia A, Ragni N, Salom EM. The impact of obesity on surgery in gynecological oncology: a review. *Int J Gynecol Cancer.* 2006;16:944–952.
62. Lamvu G, Zolnoun D, Boggess J, Steege JF. Obesity: physiologic changes and challenges during laparoscopy. *Am J Obstet Gynecol.* 2004;191:669–674.
63. DeMaria EJ, Carmody BJ. Perioperative management of special populations: obesity. *Surg Clin North Am.* 2005;85:1283–1289.

38 Obesity and Menstrual Disorders

Hilary O.D. Critchley[1], W. Colin Duncan[1], Savita Brito-Mutunayagam[1] and Rebecca M. Reynolds[2]

[1]MRC Centre for Reproductive Health, The University of Edinburgh, The Queen's Medical Research Institute, Edinburgh, UK, [2]Endocrinology Unit, University/BHF Centre for Cardiovascular Science, The University of Edinburgh, The Queen's Medical Research Institute, Edinburgh, UK

Obesity: The Problem

An association between obesity and reproductive function has long been recognised. Hippocrates wrote: 'People of such constitution cannot be prolific...fatness and flabbiness are to blame. The womb is unable to receive the semen and they menstruate infrequently and little' [1].

The prevalence of obesity, defined as a body mass index (BMI) >30 kg/m^2, is rising worldwide and is a major health concern. Obesity increases the risk of developing type-2 diabetes, hypertension, ischaemic heart disease, cancer (particularly colon and breast), cerebrovascular disease and osteoarthritis. Obesity also increases risk of death; a recent meta-analysis including data from 57 prospective studies with 894,576 participants reported that for every increase in BMI of 5 kg/m^2, there was a 30% overall higher mortality with a 40% increase in vascular mortality, a greater than 50% increase in diabetic, renal and hepatic mortality, a 10% increase in neoplastic mortality and 20% increase in respiratory and other mortalities [2].

Data from the Health Survey for England showed that the prevalence of obesity in 2009 was 22.1% in adult men and 23.9% in adult women [3]. These rates are lower than in the United States where most recent data from 2009 to 2010 suggest that although levels of obesity appear to be stable, the prevalence is 35.5% among men and 35.8% among women [4]. The rates of obesity in the general population are paralleled among pregnant women [5,6] and currently up to one in five pregnant women in the United Kingdom have a BMI of 30 kg/m^2 or more [7,8]. This problem was highlighted in the most recent Confidential Enquiry on Maternal Mortality where 27% of women who died had a BMI >30 kg/m^2 [9]. The risks of obesity during pregnancy are reviewed in detail in Chapters 34 and 35.

Obesity. DOI: http://dx.doi.org/10.1016/B978-0-12-416045-3.00038-8

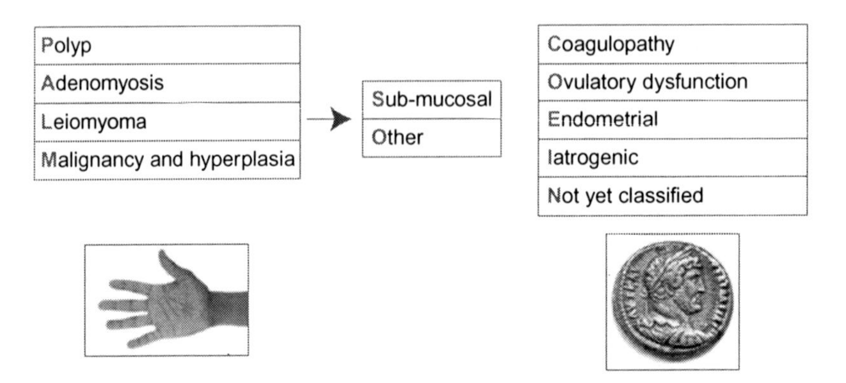

Figure 38.1 The PALM-COEIN classification of abnormal uterine bleeding.
Source: Reproduced with permission granted by FIGO, from Ref. [13].

Obesity affects virtually every organ system in the body, including the reproductive organs. Adolescent obesity is also a concern in these contemporary times [10,11].

Obesity and Abnormal Uterine Bleeding

In the women of reproductive age, obesity influences the development and progression of menstrual problems. Obese women are reported to be three times more likely to suffer from menstrual abnormalities than women of a normal weight [12]. In addition, weight loss can restore menstruation to a normal pattern [12]. Abnormal uterine bleeding may be classified using the PALM-COEIN paradigm (Figure 38.1) [13]. This useful acronym describes the aetiological basis of menstrual problems and may be used as a framework to describe the impact of obesity on the menstrual bleeding experience.

Polyps

A survey of pre-menopausal women with endometrial polyps found that 82% reported abnormal uterine bleeding [14]. In these women, obesity, particularly in combination with hypertension, was an important risk factor for polyp development [14]. In addition, in an infertility setting, BMI was an independent risk factor for the development of endometrial polyps [15]. Obese women would therefore appear to be at an increased risk of developing endometrial polyps, although the basis for this is not known.

Malignancy and Hyperplasia

Obesity also increases the risk of malignancy developing within an endometrial polyp [16]. The major risk factors for the development of endometrial cancer and endometrial hyperplasia are the same, and prominent among these is obesity [17,18]. In one study, it was reported that 86% of women with complex hyperplasia were obese [19]. Histological examination of pre-menopausal endometrial biopsies found that women with hyperplasia had a significantly higher BMI than those without hyperplasia [20]. In another study, the median BMI in the hyperplastic group was 38 kg/m^2 (95% CI 34.8–42.4) compared with 30 kg/m^2 (95% CI 29.9–33.3) in the non-hyperplastic group ($P < 0.0001$) [21]. Women with complex endometrial hyperplasia are more frequently obese (OR 2.7, 95% CI 1.5–5.0) [22]. In addition, BMI is predictive of endometrial thickness on an ultrasound scan and this is predictive of hyperplasia [20]. Obese women are thus at increased risk of developing endometrial hyperplasia.

Ovulatory Dysfunction

Several studies have described associations between obesity and irregularity of the menstrual cycle as assessed by self-reported questionnaire. A large study carried out over three decades ago, including 26,638 women aged between 20 and 40 years, reported associations between menstrual cycle irregularity and anovulation with overweight and obesity [23]. Women with evidence of anovulatory cycles, defined as irregular cycles greater than 36 days, were more than 13.6 kg heavier than women with no menstrual cycle abnormalities, after adjusting for age and height.

More contemporary studies have reported similar findings. For example, a cross-sectional study in the United States including 3941 women, who described their menstrual cycle characteristics by questionnaire, found that women with BMI $\geq$ 35 kg/m^2 had increased risk of long cycles compared to women with BMI of 22–23 kg/m^2 (OR 5.4, 95% CI 2.1–13.7) [24]. Similar findings were reported in a cross-sectional study of 726 Australian women aged 26–36 years who were not taking hormonal contraceptives and were not pregnant or breastfeeding [25]. Compared with those of normal weight, obese women had at least twofold greater odds of having an irregular cycle, defined as $\geq$15 days between the longest and shortest cycle in the last 12 months. The findings are also applicable to other ethnic groups; for example both a small study of 120 Mexican women [26] and another study of 322 Samoan women [27] found that increasing obesity was associated with increasing likelihood of amenorrhoea and oligomenorrhoea.

Generally, associations between obesity and menstrual irregularity appear to be similar when using BMI as a marker of general obesity or when using markers of central obesity such as waist circumference [25,28], waist to hip ratio [25,27] or trunk fat [29]. Importantly, weight loss in infertile women restores normal menstrual cyclicity [30,31]. Interestingly, there is evidence that the association between menstrual disorders and obesity is not only to do with current levels of obesity but is also related to a prior history of obesity. Using data from the 1958 British birth cohort study in 5799 females, obesity in early adulthood at the age of 23 years, as well as obesity in childhood at the age of 7 years, both independently increased the risk of menstrual problems by the age of 33 years (OR 1.75 and 1.59, respectively) after adjusting for other confounding factors [32]. The evidence thus strongly indicates that obese women are at increased risk of developing ovulatory dysfunction.

The Endometrium

Heavy menstrual bleeding (HMB) is defined as excessive menstrual blood loss (MBL) which interferes with a woman's physical, social, emotional and or material quality of life [33]. In clinical research, HMB has an objective definition wherein a total MBL of 80 ml or greater is considered as HMB [34].

HMB is a source of considerable morbidity among the women of reproductive age irrespective of BMI. Menstrual complaints are a significant burden to general practice as well as specialist health service resources. HMB has a significant socio-economic impact with some 800,000 women annually seeking treatment for this complaint in the United Kingdom [33]. Disorders of the menstrual cycle are one of the four commonest reasons for general practitioner consultations [35]. Although the aetiology underlying HMB of endometrial origin [13] remains to be fully defined, the body of knowledge concerning molecular and cellular mechanisms underpinning HMB of endometrial origin may lead to the development of novel therapeutic approaches [36].

A raised BMI is associated with earlier menarche and menstrual irregularities during adolescence. A question that has been raised is whether an increased BMI impacts upon the quantity of menstrual blood loss (MBL) among women who complain of heavy menstruation. A raised BMI will certainly impact on endometrial function in the context of an increased risk of endometrial hyperplasia and endometrial carcinoma [37]. Raised circulating oestrogen levels, as a consequence of the peripheral conversion of androgens by the enzyme aromatase, found in adipose tissue, have been implicated in the increased proliferative activity of endometrial cells. Circulating adipokines have also been associated with increased angiogenesis as well as cell proliferation. HMB is a common complaint among those women who are pre-menopausal and who are subsequently diagnosed with endometrial cancer. It would, therefore, not be unlikely if a raised BMI was found to impact on the volume of MBL.

A relatively small local study was carried out among some 57 subjects who attended a gynaecology clinic with menstrual problems own (unpublished data). Their MBL was measured by pictorial blood loss assessment charts and by an objective measurement (alkaline haematin technique [34]); no correlation was found

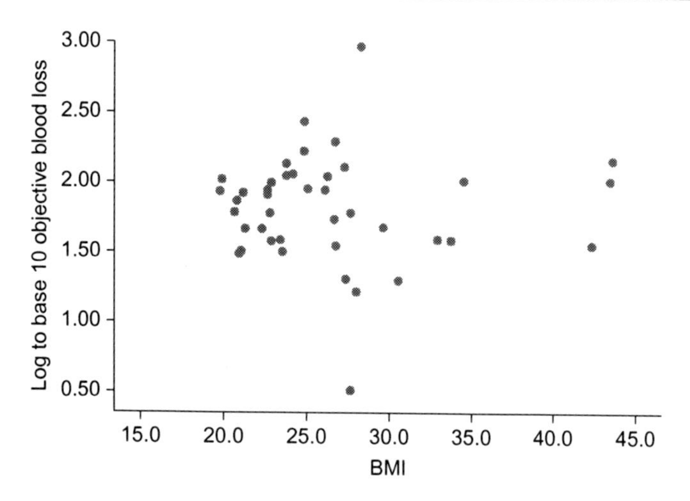

Figure 38.2 Scatter graph showing Pearson's correlation between BMI and MBL scores as determined using the alkaline haematin technique: no association with $R = 0.01$, P = non-significant (ns). Mean age (range) 42 (19–55) years.

between BMI and MBL determined by either method (Figure 38.2). Mean BMI (range) of the subjects was 26.6 kg/m^2 (19.7–43.5 kg/m^2). Adjustment for confounding factors such as age, parity, presence of fibroids, smoking and use of medications for HMB (mefenamic acid and tranexamic acid) failed to unmask an association between BMI and MBL. These data, however, are derived from a small sample of women, and the data require validation in a larger sample. A direct link between obesity and HMB, therefore, remains to be established.

One clear effect of obesity is that the management of HMB among women with a raised BMI is a challenge. The treatments available for HMB may be limited, although data showing treatment outcome in relation to BMI are lacking. Raised BMI is associated with poor efficacy of hormonal contraception [38,39] suggesting an effect of obesity on the bioavailability or action of steroids. This aspect has been covered in detail in Chapter 7.

Hysterectomy will have additional complications in the presence of a raised BMI. Data are sparse on the outcome of less invasive surgical approaches for management of HMB, i.e. endometrial ablation in obese women. A recent publication reported that patients with a BMI of greater than 34 kg/m^2 showed a trend towards failure with this intervention [40].

Some options seem to be suited to obese women. The levonorgestrel-releasing intrauterine system is considered a 'first-time' treatment option for the management of HMB [33]. It also protects against endometrial hyperplasia in ovulatory dysfunction. A recent study among adolescent women who underwent bariatric surgery showed a high acceptance rate of this method for the management of menstrual complaints [10].

Polycystic Ovary Syndrome

Therefore, it seems that the most important endometrial effects of obesity are polyps and endometrial hyperplasia. Both these are associated with ovulatory dysfunction which is clearly influenced by obesity. Anovulation leads to unopposed oestrogen stimulation of the endometrium, and the risk of hyperplasia increases with the duration of amenorrhoea [41]. The ovulatory dysfunction associated with polycystic ovary syndrome (PCOS) is a risk factor for endometrial hyperplasia. Hyperplasia has been reported in one-third of anovulatory women with PCOS [20,41], and this is associated with increased endometrial thickness on a pelvic ultrasound scan [41]. PCOS is also associated with an increased incidence of endometrial polyps [15]. Women with PCOS have ovulatory dysfunction and are at increased risk of developing endometrial polyps and hyperplasia.

The prevalence of menstrual dysfunction in women with PCOS is 75–85% [42]. In the United States, the estimated cost of hormonally treating menstrual dysfunction and abnormal uterine bleeding in women with PCOS is $1.35 billion [42]. PCOS itself is therefore a significant cause of abnormal uterine bleeding. Importantly, however, women with PCOS are commonly overweight or obese (38–66%) [12,43]. Therefore, it is difficult to separate the effects of obesity from the effects of PCOS. This is particularly problematic as obesity has a profound effect on the expression of the (PCOS) syndrome [12].

Population studies suggest that one in five women have polycystic ovaries (PCO) on a pelvic ultrasound scan [44,45], but only one-third of the women with PCO on an ultrasound scan have PCOS [46]. Obesity affects the clinical manifestations of PCOS [47] and women with PCO on ultrasound scan in the absence of PCOS are more likely to develop PCOS with increasing weight gain [48]. That means that increasing obesity will increase the prevalence of PCOS [12]. The development of the symptoms of PCOS is predicted by weight, and weight gain is attributed to the development of symptoms in more than one-third of women with PCOS [49]. As obesity rates rise, the public health significance of PCOS will therefore increase [50].

As women with PCOS gain weight, they are more likely to exhibit hyperandrogenism and menstrual cycle disturbances [12]. Thus, obesity modifies the severity of PCOS and more women will get menstrual problems as their weight increases [51]. Several studies have demonstrated that menstrual abnormalities are more prevalent in obese than non-obese women with PCOS [43]. Indeed, a history of weight gain frequently precedes the onset of oligomenorrhoea and menstrual problems [43]. An important observation is that weight loss improved menstrual function in adolescent [52], infertile women [53] and other adult women [54,55] with PCOS. Obesity will thus exaggerate PCOS and increase PCOS-related abnormal uterine bleeding.

Obesity in the Absence of PCOS

y that increasing ovulatory dysfunction associated with PCOS is the major obesity-related menstrual problems. However, obese women may suffer from perturbations of the hypothalamic–pituitary–ovarian axis, with menstrual cycle disturbance, in the absence of PCOS [12,56]. Obesity independently increases hyperandrogenism, hirsutism and infertility [50]. This may be secondary to the altered body fat-dependent hormone mileau and the development of insulin resistance associated with weight gain [12]. Both weight loss and metformin administration are reported to benefit menstrual bleeding disorders in the absence of PCOS [38]. The presence of insulin resistance predicts a thicker endometrium on ultrasound scan [57], and BMI is positively associated with the thickness of the endometrium in the absence of PCOS. In addition, obesity has been independently associated with increased uterine blood flow as measured by Doppler uterine artery pulsatility index [44]. There may therefore be a PCOS-independent effect of obesity, although this remains to be established.

Most of the studies reporting associations between menstrual cycle irregularities have been conducted among women who are not currently pregnant and so may have included women with fertility problems where PCOS is over-represented. We therefore investigated menstrual history among pregnant women who attended an antenatal metabolic clinic. Women who were severely obese ($n = 249$) (BMI 44.3 kg/m^2 (SD 4.1)) and normal weight ($n = 109$) (BMI 22.6 kg/m^2 (SD 1.6)) recorded information about their menstrual cycles prior to pregnancy. All women had conceived naturally and none had received fertility treatment.

The percentage of women with regular menstrual cycles was similar in obese and lean (76% vs 80%, $P = \text{ns}$), and there was no difference in the length of the menstrual cycle (mean (SD) 28.7 (3.4) vs 28.9 (4.6) days, $P = \text{ns}$). Likewise, there were no significant differences in the numbers of women with a history of PCOS. Thus, in this group of fertile women, severe obesity did not impact on the regularity of the menstrual cycle, suggesting there may be a subgroup of women who are less susceptible to the adverse influences of obesity. In accord with other studies, obese women had a significantly earlier age at menarche compared to normal-weight women (mean (SD) 12.5 (1.5) vs 13.0 (1.3) years, $P = 0.0002$).

Summary

As raised BMI is associated with menstrual disorders, the current epidemic of obesity is manifesting by increased rates of abnormal uterine bleeding. Further research is needed into the differential impact of obesity on endometrial and ovarian functions. Studies should address the optimal lifestyle and medical and surgical management of abnormal bleeding in obese women. As obesity is a risk factor for serious endometrial pathology, BMI as well as age should be taken into account in the decision for endometrial biopsy in women with abnormal uterine bleeding.

Acknowledgements

We thank Mrs Sheila Milne for secretarial support and Mr Ronnie Grant for assistance with the illustrations. We thank Miss N. Marielle ten Brink for collating the menstrual cycle history data. She was supported by an Erasmus studentship and Tommy's. We also acknowledge the contribution of Dr Jacqueline Maybin in the study of women attending our local gynaecology clinic with menstrual problems.

References

1. Lloyd GER, ed. *Hippocratic Writings*. London: Penguin Books; 1978.
2. Whitlock G, Lewington S, Sherliker P, et al. Body-mass index and cause-specific mortality in 900,000 adults: collaborative analyses of 57 prospective studies. *Lancet*. 2009;373(9669):1083–1096.
3. *National Health Service Information Center*. Health Survey for England, 2009: trend tables <http://www.ic.nhs.uk/statistics-and-data-collections/health-and-lifestyles-related-surveys/health-survey-for-england/health-survey-for-england–2009-trend-tables/>; 2009 Accessed 04.01.12.
4. Flegal KM, Carroll MD, Kit BK, Ogden CL. Prevalence of obesity and trends in the distribution of body mass index among US adults, 1999–2010. *JAMA*. 2012;307(5): 491–497.
5. Heslehurst N, Ells LJ, Simpson H, Batterham A, Wilkinson J, Summerbell CD. Trends in maternal obesity incidence rates, demographic predictors, and health inequalities in 36,821 women over a 15-year period. *BJOG*. 2007;114(2):187–194.
6. Heslehurst N, Rankin J, Wilkinson JR, Summerbell CD. A nationally representative study of maternal obesity in England, UK: trends in incidence and demographic inequalities in 619 323 births, 1989–2007. Int J Obes (Lond). 2010;34(3):420–428.
7. Kanagalingam MG, Forouhi NG, Greer IA, Sattar N. Changes in booking body mass index over a decade: retrospective analysis from a Glasgow maternity hospital. *BJOG*. 2005;112(10):1431–1433.
8. Denison FC, Norrie G, Graham B, Lynch J, Harper N, Reynolds RM. Increased maternal BMI is associated with an increased risk of minor complications during pregnancy with consequent cost implications. *BJOG*. 2009;116(11):1467–1472.
9. Cantwell R, Clutton-Brock T, Cooper G, et al. Saving mothers' lives: reviewing maternal deaths to make motherhood safer: 2006–2008. The eighth report of the confidential enquiries into maternal deaths in the United Kingdom. *BJOG*. 2011;118(suppl 1): 1–203.
10. Hillman JB, Miller RJ, Inge TH. Menstrual concerns and intrauterine contraception among adolescent bariatric surgery patients. J Womens Health (Larchmt). 2011;20(4): 533–538.
11. Ogden CL, Carroll MD, Flegal KM. High body mass index for age among US children and adolescents, 2003–2006. *JAMA*. 2008;299(20):2401–2405.
12. Brewer CJ, Balen AH. The adverse effects of obesity on conception and implantation. *Reproduction*. 2010;140(3):347–364.
13. Munro MG, Critchley HO, Broder MS, Fraser IS. FIGO classification system (PALM-COEIN) for causes of abnormal uterine bleeding in nongravid women of reproductive age. *Int J Gynaecol Obstet*. 2011;113(1):3–13.
14. Reslova T, Tosner J, Resl M, Kugler R, Vavrova I. Endometrial polyps. A clinical study of 245 cases. *Arch Gynecol Obstet*. 1999;262(3-4):133–139.

n R, Onalan G, Tonguc E, Ozdener T, Dogan M, Mollamahmutoglu L. Body mass is an independent risk factor for the development of endometrial polyps in ts undergoing in vitro fertilization. *Fertil Steril*. 2009;91(4):1056–1060.

16. Gregoriou O, Konidaris S, Vrachnis N, et al. Clinical parameters linked with malignancy in endometrial polyps. *Climacteric*. 2009;12(5):454–458.
17. Linkov F, Edwards R, Balk J, et al. Endometrial hyperplasia, endometrial cancer and prevention: gaps in existing research of modifiable risk factors. *Eur J Cancer*. 2008; 44(12):1632–1644.
18. Epplein M, Reed SD, Voigt LF, Newton KM, Holt VL, Weiss NS. Risk of complex and atypical endometrial hyperplasia in relation to anthropometric measures and reproductive history. *Am J Epidemiol*. 2008;168(6):563–570 [discussion 71-6].
19. Horn LC, Schnurrbusch U, Bilek K, Hentschel B, Einenkel J. Risk of progression in complex and atypical endometrial hyperplasia: clinicopathologic analysis in cases with and without progestogen treatment. *Int J Gynecol Cancer*. 2004;14(2):348–353.
20. McCormick BA, Wilburn RD, Thomas MA, Williams DB, Maxwell R, Aubuchon M. Endometrial thickness predicts endometrial hyperplasia in patients with polycystic ovary syndrome. *Fertil Steril*. 2011;95(8):2625–2627.
21. Heller DS, Mosquera C, Goldsmith LT, Cracchiolo B. Body mass index of patients with endometrial hyperplasia: comparison to patients with proliferative endometrium and abnormal bleeding. *J Reprod Med*. 2011;56(3-4):110–112.
22. Ricci E, Moroni S, Parazzini F, et al. Risk factors for endometrial hyperplasia: results from a case–control study. *Int J Gynecol Cancer*. 2002;12(3):257–260.
23. Hartz AJ, Barboriak PN, Wong A, Katayama KP, Rimm AA. The association of obesity with infertility and related menstrual abnormalities in women. *Int J Obes*. 1979;3(1): 57–73.
24. Rowland AS, Baird DD, Long S, et al. Influence of medical conditions and lifestyle factors on the menstrual cycle. *Epidemiology*. 2002;13(6):668–674.
25. Wei S, Schmidt MD, Dwyer T, Norman RJ, Venn AJ. Obesity and menstrual irregularity: associations with SHBG, testosterone, and insulin. Obesity (Silver Spring). 2009; 17(5):1070–1076.
26. Castillo-Martinez L, Lopez-Alvarenga JC, Villa AR, Gonzalez-Barranco J. Menstrual cycle length disorders in 18- to 40-y-old obese women. *Nutrition*. 2003;19(4): 317–320.
27. Lambert-Messerlian G, Roberts MB, Urlacher SS, et al. First assessment of menstrual cycle function and reproductive endocrine status in Samoan women. *Hum Reprod*. 2011;26(9):2518–2524.
28. De Pergola G, Tartagni M, d'Angelo F, Centoducati C, Guida P, Giorgino R. Abdominal fat accumulation, and not insulin resistance, is associated to oligomenorrhea in non-hyperandrogenic overweight/obese women. *J Endocrinol Invest*. 2009;32(2): 98–101.
29. Douchi T, Kuwahata R, Yamamoto S, Oki T, Yamasaki H, Nagata Y. Relationship of upper body obesity to menstrual disorders. *Acta Obstet Gynecol Scand*. 2002;81(2): 147–150.
30. Clark AM, Thornley B, Tomlinson L, Galletley C, Norman RJ. Weight loss in obese infertile women results in improvement in reproductive outcome for all forms of fertility treatment. *Hum Reprod*. 1998;13(6):1502–1505.
31. Hollmann M, Runnebaum B, Gerhard I. Effects of weight loss on the hormonal profile in obese, infertile women. *Hum Reprod*. 1996;11(9):1884–1891.
32. Lake JK, Power C, Cole TJ. Women's reproductive health: the role of body mass index in early and adult life. *Int J Obes Relat Metab Disord*. 1997;21(6):432–438.
33. *NICE*. Clinical guideline 44: heavy menstrual bleeding. <http://www.nice.org.uk/nicemedia/pdf/CG44FullGuideline.pdf/>; 2007.

34. Warner PE, Critchley HO, Lumsden MA, Campbell-Brown M, Douglas A, Murray GD. Menorrhagia II: is the 80-mL blood loss criterion useful in management of complaint of menorrhagia? *Am J Obstet Gynecol.* 2004;190(5):1224–1229.
35. Palep-Singh M, Prentice A. Epidemiology of abnormal uterine bleeding. *Best Pract Res Clin Obstet Gynaecol.* 2007;21(6):887–890.
36. Critchley HO, Maybin JA. Molecular and cellular causes of abnormal uterine bleeding of endometrial origin. *Semin Reprod Med.* 2011;29(5):400–409.
37. Dossus L, Rinaldi S, Becker S, et al. Obesity, inflammatory markers, and endometrial cancer risk: a prospective case–control study. *Endocr Relat Cancer.* 2010;17(4): 1007–1019.
38. Lash MM, Armstrong A. Impact of obesity on women's health. *Fertil Steril.* 2009; 91(5):1712–1716.
39. Kulie T, Slattengren A, Redmer J, Counts H, Eglash A, Schrager S. Obesity and women's health: an evidence-based review. *J Am Board Fam Med.* 2011;24(1): 75–85.
40. Fakih M, Cherfan V, Abdallah E. Success rate, quality of life, and descriptive analysis after generalized endometrial ablation in an obese population. *Int J Gynaecol Obstet.* 2011;113(2):120–123.
41. Cheung AP. Ultrasound and menstrual history in predicting endometrial hyperplasia in polycystic ovary syndrome. *Obstet Gynecol.* 2001;98(2):325–331.
42. Azziz R, Marin C, Hoq L, Badamgarav E, Song P. Health care-related economic burden of the polycystic ovary syndrome during the reproductive life span. *J Clin Endocrinol Metab.* 2005;90(8):4650–4658.
43. Gambineri A, Pelusi C, Vicennati V, Pagotto U, Pasquali R. Obesity and the polycystic ovary syndrome. *Int J Obes Relat Metab Disord.* 2002;26(7):883–896.
44. Polson DW, Adams J, Wadsworth J, Franks S. Polycystic ovaries – a common finding in normal women. *Lancet.* 1988;1(8590):870–872.
45. Farquhar CM, Birdsall M, Manning P, Mitchell JM, France JT. The prevalence of polycystic ovaries on ultrasound scanning in a population of randomly selected women. *Aust N Z J Obstet Gynaecol.* 1994;34(1):67–72.
46. Fauser BC, Tarlatzis BC, Rebar RW, et al. Consensus on women's health aspects of polycystic ovary syndrome (PCOS): the Amsterdam ESHRE/ASRM-sponsored third PCOS consensus workshop group. *Fertil Steril.* 2012;97(1):28–38:e25.
47. Salehi M, Bravo-Vera R, Sheikh A, Gouller A, Poretsky L. Pathogenesis of polycystic ovary syndrome: what is the role of obesity? *Metabolism.* 2004;53(3):358–376.
48. Pettigrew R, Hamilton-Fairley D. Obesity and female reproductive function. *Br Med Bull.* 1997;53(2):341–358.
49. Laitinen J, Taponen S, Martikainen H, et al. Body size from birth to adulthood as a predictor of self-reported polycystic ovary syndrome symptoms. *Int J Obes Relat Metab Disord.* 2003;27(6):710–715.
50. Teede H, Deeks A, Moran L. Polycystic ovary syndrome: a complex condition with psychological, reproductive and metabolic manifestations that impacts on health across the lifespan. *BMC Med.* 2010;8:41.
51. Welt CK, Gudmundsson JA, Arason G, et al. Characterizing discrete subsets of polycystic ovary syndrome as defined by the Rotterdam criteria: the impact of weight on pheno- and metabolic features. *J Clin Endocrinol Metab.* 2006;91(12):4842–4848.
ein RM, Copperman NM, Jacobson MS. Effect of weight loss on menstrual func- in adolescents with polycystic ovary syndrome. *J Pediatr Adolesc Gynecol.* 24(3):161–165.

53. Clark AM, Ledger W, Galletly C, et al. Weight loss results in significant improvement in pregnancy and ovulation rates in anovulatory obese women. *Hum Reprod.* 1995;10(10): 2705–2712.
54. Lass N, Kleber M, Winkel K, Wunsch R, Reinehr T. Effect of lifestyle intervention on features of polycystic ovarian syndrome, metabolic syndrome, and intima-media thickness in obese adolescent girls. *J Clin Endocrinol Metab.* 2011;96(11):3533–3540.
55. Tang T, Glanville J, Hayden CJ, White D, Barth JH, Balen AH. Combined lifestyle modification and metformin in obese patients with polycystic ovary syndrome. A randomized, placebo-controlled, double-blind multicentre study. *Hum Reprod.* 2006;21(1): 80–89.
56. Rachon D, Teede H. Ovarian function and obesity – interrelationship, impact on women's reproductive lifespan and treatment options. *Mol Cell Endocrinol.* 2010;316(2): 172–179.
57. Iatrakis G, Tsionis C, Adonakis G, et al. Polycystic ovarian syndrome, insulin resistance and thickness of the endometrium. *Eur J Obstet Gynecol Reprod Biol.* 2006;127(2): 218–221.

39 Developing HRT Prescribing in Obese Women

Kate Maclaran[1] and Nick Panay[2]

[1]West London Menopause and PMS Centre, Queen Charlotte's & Chelsea and Chelsea & Westminster Hospitals, London, UK, [2]Imperial College, London, UK

Introduction

Menopause, defined as the last menstrual period, is a pivotal time in a woman's life, which can be associated with numerous physical, metabolic and psychological consequences. Obesity can have a profound influence on the post-menopausal hormonal milieu and how we manage symptoms of oestrogen deficiency. The average age of menopause in Western countries is currently 51–52 years, although there is evidence that obesity is associated with a later menopause [1].

After menopause, production of oestrogen from the ovaries declines, and peripheral aromatisation of androgens, such as androstenedione and dihydroepiandrosterone, to oestrone becomes the primary source of oestrogen. The effects of overweight and obesity on post-menopausal sex steroids are well documented (Table 39.1). In obese post-menopausal women, increased peripheral conversion of androgens in adipose tissue and a reduction in the production of sex-hormone-binding globulin (SHBG) from the liver result in an increased bioavailability of oestrogen and testosterone. Elevated body mass index (BMI) has been positively correlated with circulating concentrations of oestrogens and androgens and negatively correlated with SHBG [3].

As a result of higher endogenous oestrogen, it could therefore be expected that obese women would experience fewer vasomotor symptoms. However, elevated BMI [4,5] and particularly abdominal adiposity [6] have been consistently associated with more severe vasomotor symptoms. The pathophysiology of hot flushes remains poorly understood, but current thinking points to changes in core temperature acting within a narrowed thermoregulatory zone in symptomatic women [7] (Figure 39.1). It has been suggested that the thermoregulatory effects of obesity, with adipose tissue acting as a potent insulator, may explain the increase in vasomotor symptoms in obese women. Additionally, the metabolic effects of visceral fat deposition, such as hyperinsulinaemia, or alterations in leptin and other cytokines may also contribute to severity of vasomotor symptoms [8].

Obesity. DOI: http://dx.doi.org/10.1016/B978-0-12-416045-3.00039-X

Table 39.1 Mean Concentration of Post-Menopausal Hormones Categorised by BMI from Collaborative Re-Analysis of 13 Prospective Studies [2]

Hormone and Category	*n*	Mean[a] (95% CI)	*P*-het (trend)	Relative Mean[b] and 95% CI
Oestradiol (pmol/l)				
Below 55 years	757	57.4 (55.2–59.6)	<0.001	
55–59 years	1446	50.0 (48.6–51.5)	(0.001)	
60–64 years	1546	50.0 (48.7–51.4)		
65–69 years	1233	51.2 (49.7–52.8)		
70 years and above	632	50.9 (48.4–53.5)		
Calculated Free Oestradiol (pmol/l)				
Below 55 years	669	0.83 (0.79–0.87)	<0.001	
55–59 years	1346	0.74 (0.71–0.76)	(<0.001)	
60–64 years	1422	0.73 (0.70–0.75)		
65–69 years	1089	0.73 (0.71–0.76)		
70 years and above	477	0.71 (0.66–0.76)		
Oestrone (pmol/l)				
Below 55 years	467	115 (110–120)	0.001[c]	
55–59 years	1030	108 (105–112)	(0.001)	
60–64 years	1111	105 (102–108)		
65–69 years	798	103 (99–106)		
70 years and above	442	107 (102–113)		
Androstenedione (nmol/l)				
Below 55 years	673	2.62 (2.50–2.74)	<0.001	
55–59 years	1350	2.36 (2.28–2.43)	(<0.001)	
60–64 years	1515	2.30 (2.23–2.37)		
65–69 years	1147	2.11 (2.03–2.18)		
70 years and above	496	2.03 (1.92–2.16)		
Dihydroepiandrosterone (nmol/l)				
Below 55 years	723	2478 (2351–2612)	<0.001	
55–59 years	1456	2046 (1970–2125)	(<0.001)	
60–64 years	1575	1678 (1619–1739)		
65–69 years	1156	1515 (1453–1580)		
70 years and above	479	1399 (1297–1509)		

Forest plot axes (each hormone): 0.5 0.75 1.0 1.5 2.0

(*Continued*)

Table 39.1 (Continued)

Hormone and Category	*n*	Mean[a] (95% CI)	*P*-het (trend)	Relative Mean[b] and 95% CI
Testosterone (nmol/l)				
Below 55 years	756	0.89 (0.85–0.93)	0.0001[c]	
55–59 years	1517	0.83 (0.80 – 0.86)	(0.003)	
60–64 years	1604	0.80 (0.77–0.83)		0.5 0.75 1.0 1.5 2.0
65–69 years	1228	0.79 (0.76–0.82)		
70 years and above	561	0.84 (0.79–0.90)		
Calculated Free testosterone (pmol/l)				
Below 55 years	724	13.9 (13.1–14.7)	<0.001[c]	
55–59 years	1478	12.7 (12.2–13.2)	(<0.001)	
60–64 years	1569	12.1 (11.7–12.6)		0.5 0.75 1.0 1.5 2.0
65–69 years	1188	11.7 (11.2–12.2)		
70 years and above	535	11.5 (10.7–12.5)		
SHBG (nmoll)				
Below 55 years	754	39.4 (37.8–41.0)	<0.001[c]	
55–59 years	1530	40.6 (39.4–41.8)	(<0.001)	
60–64 years	1612	40.8 (39.7–42.0)		0.5 0.75 1.0 1.5 2.0
65–69 years	1226	42.4 (41.0–43.7)		
70 years and above	556	46.7 (44.2–49.4)		

[a]Mean values are scaled to the overall geometric mean concentration.
[b]Values are depicted as a proportion of the mean in the first subset (dotted lines).
[c]Significant interaction with study ($P < 0.01$).
Source: Reprinted with permission from Macmillan Publishers Ltd on behalf of Cancer Research UK.

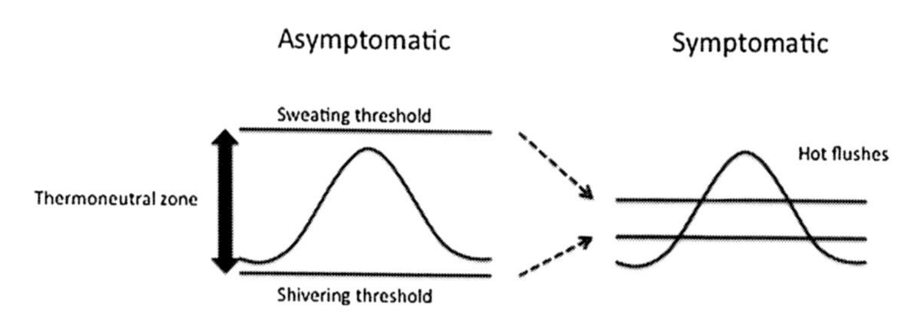

Figure 39.1 Mechanism of hot flushes. Small changes in core body temperature occurring within a narrowed thermoneutral zone trigger hot flushes in symptomatic post-menopausal women.
Source: Adapted from Ref. [7].

Effect of Menopause on Metabolic Factors

Overweight and obesity are associated with many long-term health consequences. The menopause transition is accompanied by a number of adverse physical and metabolic effects which can exacerbate these risks.

Menopause is associated with detrimental changes in body composition, particularly central body fat deposition, with an accumulation of abdominal android fat and a relative reduction in gynoid fat deposition around the thighs and hips [9]. Menopause is also associated with a deterioration of insulin sensitivity [10] and adverse changes to the lipid profile, with increases in total and low density lipoprotein (LDL) cholesterol and reductions in low density lipoprotein (HDL) cholesterol [11]. Furthermore, the sex steroid deficiency accompanying menopause can result in direct vascular effects, such as impaired endothelial function, which may further increase cardiovascular risk.

The importance of the hormonal changes associated with menopause have long been recognised in women experiencing premature loss of ovarian function, who have an increased risk of ischaemic heart disease [12,13] and possibly stroke [14]. Although recent studies have shown that the risk of cardiovascular disease does not increase precipitously at menopause [15], the prevalence of obesity, type-2 diabetes mellitus and metabolic syndrome all increase through the menopause transition, independently of age and other cardiovascular risk factors [16].

Managing Menopausal Symptoms in Obese Women

Many women will seek advice for managing menopausal symptoms and this should be seen as an opportunity to assess the risk factors for various health conditions. Lifestyle modifications should be recommended to all women at the time of menopause, both to reduce the severity of menopausal symptoms and to reduce risk of future morbidity such as cardiovascular disease, osteoporosis and breast cancer. Many of these measures will be particularly beneficial in overweight women. Reducing caffeine and alcohol, regular aerobic exercise and smoking cessation will have widespread benefits. Additionally, there is evidence that in overweight and obese women, weight loss can improve vasomotor symptoms [17].

Hormone Replacement Therapy

Hormone replacement therapy (HRT) is the most effective treatment for the symptoms of oestrogen deficiency and is particularly important in those with early menopause to prevent the long-term effects of oestrogen deficiency. The risks and
f HRT will differ depending on individual risk factors such as age, medi-
y and family history. The current evidence shows that HRT can improve
toms associated with oestrogen deficiency and reduce the risk of
sis and colon cancer. HRT is associated with increased risk of venous

thromboembolism (VTE), stroke, and with prolonged use, breast cancer. There is increasing evidence to support a window of opportunity for the primary prevention of cardiovascular disease, where greatest benefit in preventing the progression of atheroma is seen when HRT is initiated early in the post-menopause. In older women, HRT may cause adverse cardiovascular effects through coagulation activation and abnormal vascular remodelling, although the use of age-appropriate doses and transdermal routes can help minimise those risks.

Although oestrogen may have a benefit on many of the metabolic consequences of menopause, the shared risks of HRT and obesity have led to concerns about the use of HRT in overweight women.

Shared Effects of HRT and Obesity

Breast Cancer

In contrast to obese pre-menopausal women, it is well established that obese post-menopausal women are at increased risk of breast cancer. Observational studies have reported increased breast cancer rates of 30–50% in obese women [18–20]. A recent meta-analysis of 2.5 million women showed that a 5 kg/m^2 increase in BMI is associated with 12% increase in the incidence of breast cancer (95% CI 1.08–1.16) [21]. Furthermore, obesity is associated with poorer breast cancer survival [22], with a threefold increased mortality observed in women with a BMI greater than 40 kg/m^2 [23]. A recent cohort study demonstrated that patients with a BMI of 30 kg/m^2 or more have a 46% increased risk of developing distant metastases after 10 years [24]. The increased risk of breast cancer associated with obesity is thought to be due to both increased exposure to oestrogen as a result of increased peripheral aromatisation of androstenedione and additional metabolic risk factors such as hyperinsulinaemia.

The exact risks of breast cancer associated with HRT have been much debated. Large observational studies [25,26], which suggested an increased risk of breast cancer with HRT use, were followed by the Women's Health Initiative (WHI) studies. They were a series of randomised clinical trials and an observational study, designed to assess the primary prevention of cardiovascular disease. The trial had several different arms including an oestrogen plus progestin arm and an oestrogen-only arm for hysterectomised women. The oestrogen/progestin arm examined the effects of oral conjugated equine oestrogens (CEE) 0.625 mg and medroxyprogesterone acetate (MPA) 2.5 mg versus placebo in 16,608 post-menopausal women. This study halted early after 5.2 years due to a 26% increased risk of breast cancer in the treatment arm [27].

The oestrogen-only arm, involving 10,739 hysterectomised women receiving 0.625 mg CEE or placebo, was also terminated early, with average follow-up of 6.8 years. This study showed a reduction in breast cancer risk in the CEE-only arm (HR 0.77), although it did not reach statistical significance [28].

The WHI studies have been the subject of much critique as the study subjects were started on HRT at an advanced age (average 63 years), often with a significant delay following menopause. Furthermore, subjects were not using HRT for symptom relief (only 12–17% had moderate to severe vasomotor symptoms [29] and, of particular relevance, a high proportion of the women in the WHI studies were obese (35–45% had a BMI $\geq$ 30 kg/m^2).

The most recent WHI follow-up [30], after a mean of 11 years and an average of 5.6 years treatment, showed a hazard ratio of 1.25 (95% CI 1.07–1.46; $P = 0.004$) for the risk of breast cancer in oestrogen/progestin users compared to non-users. The authors equated this to an increase in breast cancer *mortality* for women randomised to oestrogen plus progestin compared with placebo of 1.3 per 10 000 women per year.

Unfortunately, most randomised controlled data for HRT breast cancer risk pertain to the use of CEE/MPA, a regimen now rarely used. We are lacking data regarding breast cancer risk associated with oestradiol and other progestogens, but there is evidence that progesterone and dydrogesterone may be associated with a lower breast cancer risk [31]. Therefore, although combined HRT is associated with an increase risk of breast cancer, the effect of obesity on breast cancer risk is much greater.

The breast cancer risk associated with HRT use appears to be more marked in slimmer women compared to overweight and obese women. Observational data have shown that HRT use does not increase breast cancer risk in post-menopausal women with a BMI above 24.4 kg/m^2 [32]. The Collaborative Group on Hormonal Influences in breast cancer examined 51 observational studies with 52,705 incident breast cancer cases and found that the relative risk associated with current or recent use of HRT decreased progressively with increasing BMI (Table 39.2) [26]. Therefore, although obese post-menopausal women have an increased risk of breast cancer, the use of HRT does not appear to further increase this risk [33].

It has been suggested that obese women on HRT may have worse tumour characteristics and poorer prognosis than women with normal BMI on HRT, although further data are needed [34].

Endometrial Cancer

Obesity and unopposed oestrogen therapy are the most important risk factors for endometrial cancer. Both the peripheral synthesis of oestrogen in adipose tissue and an independent effect of reduced SHBG (leading to increased bioavailability of oestrogen and increased androgens available for aromatisation) observed in obese women contribute to the pathogenesis of endometrial cancer.

Table 39.2 Relative Risk of Breast Cancer in HRT Users Compared to Similar Weight Non-Users

BMI (kg/m^2)	**<22.5**	**22.5–24.9**	**$\geq$ 25.0**
	1·73 [0·12]	1·29 [0·14]	1·02 [0·11]

k; SE, standard error.
om Ref. [26]

Obesity has long been recognised as a risk factor for endometrial cancer, and it is estimated that in Europe, 39% of endometrial cancer incidence can be explained by excess body weight [35]. Epidemiological studies have observed an estimated threefold elevated risk in women with BMI over 29 kg/m^2 compared to those with normal BMI [36] and relative risk of 1.59 (1.50–1.68) per 5 kg/m^2 increase in BMI [21].

The use of oestrogen-only HRT is associated with a two- to threefold increased risk of developing endometrioid-type endometrial adenocarcinoma. In contrast, the addition of progestogen reduces this risk. Data from the Million Women Study, an observational study of 716,738 post-menopausal women in the United Kingdom, showed that unopposed oestrogen was associated with a significant increased risk of endometrial cancer (RR 1.45, 95% CI 1.02–2.06) [37]. Women using cyclical oestrogen plus progestogen therapy had similar risk as non-users (RR 1.05, 95% CI 0.91–1.22), and continuous combined treatment was associated with a significantly reduced risk (RR 0.71, 95% CI 0.56–0.90).

Several cohort studies have investigated the combined effects of HRT and obesity. An Analysis of the California Teachers Study participants examined the risk of type-1 endometrial cancer in 19,482 women in a prospective cohort study [38]. They found that increased BMI above 30 kg/m^2 was associated with significantly increased risk of endometrial cancer (RR 1.9, 95% CI 1.5–2.5) compared to a BMI of below 25 kg/m^2. This risk was much more pronounced in those who never used HRT (RR 3.5, 95% CI 2.2–5.5).

These findings were consistent with those from other large cohort studies [39–41] and a meta-analysis [42], all of which suggest that combined oestrogen and progestogen therapy appears to mitigate the effects of obesity on endometrial cancer risk. A meta-analysis of 17,710 cases found that the association between BMI and endometrial cancer risk was significantly attenuated in HRT users compared to non-users [42]. It has been suggested that these findings signify a threshold effect of oestrogen on endometrial cancer risk rather than a multiplicative effect [38].

Cardiovascular Disease and Lipids

There is increasing evidence that HRT used around the time of the menopause confers cardiovascular benefit. However, the use of oestrogen in older women and in the presence of coronary artery disease can cause cardiovascular harm, as a result of plaque instability and rupture. Obesity is associated with many risk factors for cardiovascular disease including adverse changes in lipid profile, hypertension and insulin resistance. Furthermore, the presence of hot flushes, which are more common in obese women, has been associated with increased cardiovascular risk [43].

In obese women, there are a variety of potential mechanisms by which exogenous oestrogen therapy may be of cardiovascular benefit. Oestrogen therapy has a beneficial effect on body fat distribution, helping prevent the post-menopausal weight gain and the accumulation of central body fat [44]. Oral oestrogen therapy results in decreased LDL cholesterol and increased HDL cholesterol; however, this is at the expense of an increase in triglycerides [45]. Other beneficial effects include reductions in lipoprotein (a) [46], changes in LDL particle size and clearance, and inhibition of LDL oxidation [47].

Few studies have reported specifically on cardiovascular events in obese HRT users, but available data are not suggestive of an increased risk of coronary heart disease events or mortality. In the Californian Teachers Study, a large cohort study of 115,433 women, the association of increased cardiovascular mortality in obese women (RR 1.38 1.22–1.55) was not increased by HRT use [48]. In the oestrogen plus progestin arm of the WHI randomised controlled trial (RCT), there was no significant interaction between BMI and risk of cardiovascular events in HRT users ($P = 0.6$) [49].

Venous Thromboembolism

Both obesity and exogenous hormone therapy are important risk factors of VTE. There are both observational and RCT evidence that a synergistic effect exists between HRT and obesity in VTE risk, such as that has been observed with combined oral contraceptive usage [50].

Data from the WHI RCT reported a three- to fivefold increased risk with oestrogen/progestin therapy in overweight and obese women [51], although this risk was slightly lower in oestrogen-only arm [52].

In the ESTHER case–control study, oral therapy was associated with a 10-fold increase in VTE risk in overweight women [53]. In contrast, transdermal therapy has not been associated with increased VTE risk, even in obese women, who had a similar risk to obese non-HRT users (Table 39.3). Despite these reassuring data, further RCT data regarding the thrombotic risk with transdermal therapy is required.

Table 39.3 Summary of Risk of VTE with HRT Use Stratified by BMI in WHI and ESTHER Studies

	RRs (compared to non-obese non-HRT users)		
	Normal BMI ($BMI \leq 25\ kg/m^2$)	**Overweight ($25 < BMI \leq 30\ kg/m^2$)**	**Obese ($BMI > 30\ kg/m^2$)**
WHI			
Oral E and P	1.78	3.8	5.6
Oral E	1.92	2.32	4.4
ESTHER			
Oral	5.9	10.2	20.6
T[illegible]mal	1.2	2.9	5.4
[illegible]s	1.0	2.7	4.0

[illegible], progestogen.
[illegible]om Refs. [51–53]

It is thought that obesity and oral oestrogen affect coagulation through similar adverse effects on coagulation and fibrinolysis. Oral HRT is associated with activated protein C resistance and increases in plasma levels of prothrombin and other coagulation factors. In contrast, transdermal therapy avoids first-pass hepatic metabolism and therefore minimises any potentially deleterious effects on coagulation or fibrinolysis [54].

Stroke

Concerns exist regarding the risk of stroke with HRT. The WHI oestrogen-alone arm terminated 8 months early after almost 7 years due to an apparent increased risk of stroke with 0.625 mg CEE (HR 1.39, estimated at an excess of 12 cases per 10,000 patient years) [28]. However, data remain conflicting regarding the exact risk of stroke with HRT. In the Nurses Health Study, use of oestrogen plus progestin (RR 1.45, 95% CI 1.10–1.92) and higher doses of CEE alone was associated with increased risk of ischaemic stroke, whereas lower dose oestrogen conferred no increased risk of stroke [55]. A recent meta-analysis found that oral HRT was associated with a 32% increased risk of stroke (OR 1.32, 95% CI 1.14–1.53), and although no increased risk was seen with transdermal therapy, there were too few studies to draw firm conclusions [56]. Therefore, although the literature is suggestive of an increased risk of stroke with oral HRT, there is a dose-dependent relationship and the event rate is extremely low, especially in younger women.

The data regarding the impact of obesity on stoke risk in HRT users are conflicting. In the Nurses Health Study, HRT users with a BMI > 32 kg/m^2 had more than double the risk of stroke compared to HRT users of normal weight (RR 2.37) [55]. A further observational study in 16,906 Swedish women found no overall association between HRT use and stoke risk. However, stroke risk was increased among HRT users with elevated BMI (RR 1.37 per SD increase in BMI, 95% CI 1.08–1.75) even after adjustment for other risk factors including smoking, hypertension and diabetes [57]. The evidence from this study indicates that BMI appears to be a less important risk factor for stroke among HRT users than smoking (RR 2.5) and hypertension ≥160/100 (RR 5.78) (Table 39.4).

The risk of stroke associated with HRT is most likely due to thrombotic rather than atherosclerotic mechanisms [58], and, therefore, it is biologically plausible that obesity may further increase this risk. Despite this, stroke remains an uncommon event, especially in younger women, and the risk of stroke should theoretically be reduced by transdermal therapy, although further studies are needed. In obese women wishing to use HRT, it is particularly important to consider the presence or absence or other risk factors for stroke such as hypertension and smoking.

Insulin Resistance

Overweight and obesity frequently occur alongside insulin resistance. Although there is evidence that oestrogen can be beneficial to glucose and insulin

Table 39.4 Risk of Stroke in HRT Users

Baseline Characteristics[a]	Number of Subjects	Relative Risk for Stroke (95% CI)	
		Age-adjusted	Covariate- adjusted[b]
Age: per SD (6 years of increase)		1.89 (1.47–2.42)***	1.71 (1.29–2.28)***
Waist circumference: per SD (9 cm of increase)		1.48 (1.17–1.88)***	1.40 (1.09–1.80)**
BMI: per SD (3.58 of increase)		1.46 (1.16–1.84)***	1.37 (1.08–1.75)**
Smoking Status			
Non-smoker	1588/73.9	1.0	1.0
Current smoker	560/26.1	2.11 (1.16–3.85)**	2.50 (1.35–4.61)**
Blood Pressure (mmHg)			
SBP/SD (19 mmHg of increase)		1.72 (1.32–2.24)***	1.73 (1.29–2.30)***
DBP/SD (9 mmHg of increase)		1.73 (1.33–2.26)***	1.80 (1.34-2.41)***
BP < 140/90	1173/54.8	1.0	1.0
BP > 160/100	294/13.7	5.34 (2.31–12.31)***	5.78 (2.41–13.89)***

**$P < 0.05$.
***$P < 0.01$ compared to reference.
[a]Variables with statistical significance.
[b]Relative risks were adjusted for age, smoking, alcohol consumption, BMI, BP, diabetes, BP-lowering drugs, lipid-lowering drugs and/or aspirin (except the variable itself run in model).
Source: Data from Ref. [11]. Reproduced with kind permission from Elsevier; copyright 2006.

metabolism, diabetic women are prescribed HRT less frequently than non-diabetic women due to the perceived risks of HRT in this population.

Oestrogen can have varying effects on glucose and insulin metabolism depending on the route and type of oestrogen used. Oral administration of 17B-estradiol to post-menopausal women has been associated with an improvement in insulin resistance, whereas alkylated oestrogens, such as ethinyl estradiol, and high-dose CEE may result in impaired glucose tolerance. Current evidence suggests that transdermal oestrogen has no significant impact on glucose metabolism. Furthermore, HRT may have divergent effects on lipids, particularly HDL levels, depending on the presence or absence of insulin resistance [59].

There is observational and RCT evidence that oral HRT may help to prevent the onset of diabetes. In the Nurses Health Study ($n = 21{,}028$), after adjustment for age and BMI, HRT users had a significant reduction in the incidence of new onset
RR 0.80, 95% CI 0.67–0.96) compared to past and non-users [60]. The
progestin arm of WHI reported that the incidence of treated diabetes
tment arm (3.5%) was significantly less than the placebo arm (4.2%,
, even after adjustment for changes in BMI and waist circumference

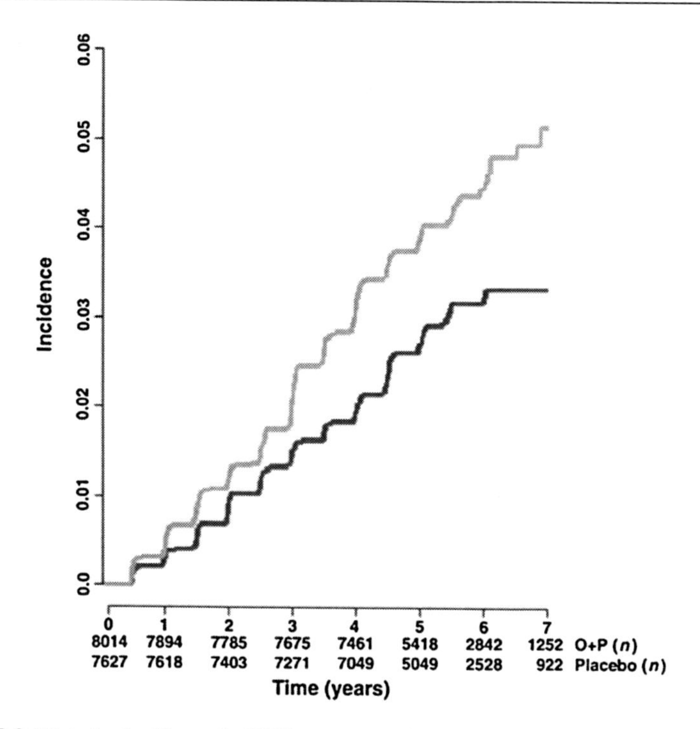

Figure 39.2 Diabetes incidence in WHI oestrogen plus progestogen arm (blue line) versus placebo (red line). Hazard ratio 0.79 (95% CI 0.67–0.93). (For interpretation of the references to colour in this figure legend, the reader is referred to the web version of this book.)

Source: Data from Ref. [61], figure 1. Reproduced with kind permission from Springer Science + Business Media; copyright 2004.

(Figure 39.2) [61]. Additionally, there was a trend towards more marked reduction in overweight women and those with waist circumference greater than 88 cm, although this did not reach statistical significance.

In post-menopausal women with established type-2 diabetes mellitus, HRT has been shown to significantly reduce serum cholesterol, fasting glucose and fasting insulin [59]. This improvement in diabetic control in women on HRT has also been observed in several other studies [62,63]. Thus, the presence of insulin resistance is not an absolute contraindication to HRT, and treatment must be individualised taking into account a global assessment of HRT risk factors.

Osteoporosis

Post-menopausal osteoporosis and subsequent fracture are increasingly important health burdens. Around 50% of women and 20% of men will sustain an osteoporotic fracture during their lifetime [64].

Women with low BMI tend to have lower bone density and increased risk of fracture. Obesity was previously thought to be protective for fracture due to increased soft-tissue protection; however, recent studies have contradicted these beliefs. Although obesity may confer some relative protection against hip and pelvic fractures, obese women do not have lower fracture rates than non-obese women, and indeed may be at increased risk for particular fractures such as those of the ankle and upper leg [65]. Bone density is not the sole determinant of fracture risk, and although obese women have relatively high bone density, the effect of obesity on bone strength is unknown. Furthermore, obesity is associated with poor mobility and falls.

HRT is one of the few bone-sparing agents to have been shown to prevent fracture in both osteoporotic and non-osteoporotic women. Additionally, there is evidence that even 'ultra-low' dose of transdermal oestrogen is effective to maintain bone density [66].

Few studies have examined whether HRT has varying effects on bone in the presence of overweight. Stratification of the WHI oestrogen/progestin arm fracture outcomes by BMI showed that HRT reduced the risk of hip fracture in women with a BMI of less than 25 kg/m^2 (HR 0.50, 95% CI 0.28–0.90) and with a BMI of 25–30 kg/m^2 (HR 0.67, 95% CI 0.37–1.20), but not in women with a BMI of 30 kg/m^2 or more (HR 1.11, 95% CI 0.52–2.39), although the interaction of hormone therapy with BMI was not statistically significant ($P = 0.41$) [67]. These data again point to the fact that in obese women, Bone Mineral Density (BMD) is not the sole predictor of fracture.

Prescribing HRT in Obese Women

The decision to use HRT must take into account the above-mentioned risks and benefits and will be individual for each woman. In those who wish to use HRT, certain measures can be taken to help minimise any risks.

Route of Administration

The route of administration can have a significant effect on the metabolic effects of oestrogen. Transdermal therapy will result in less impact on haemostasis and is therefore associated with lower risk of VTE. It may also be associated with a lower risk of stroke and coronary heart disease, although further data are needed. There is currently no evidence that transdermal therapy influences the risk of breast cancer. Current guidelines [68] recommend that transdermal therapy should be considered as first line in all obese women requesting HRT. Transdermal HRT has the additional benefits of fewer side effects and more stable serum hormone levels.

The route of administration will influence the effects of oestrogen on lipids and
stance. Oral therapy has been shown to reduce LDL and increase HDL
transdermal therapy; oral therapy tends to raise triglycerides whereas
therapy reduces them [69]. Although both oral and transdermal oestra-
fasting glucose and insulin, only oral therapy appears to benefit insulin

sensitivity [70]. It has, therefore, been suggested than in women with impaired insulin sensitivity, who are otherwise at low risk for VTE, oral therapy may have certain benefits above transdermal therapy.

Dose

Obese women using HRT have higher oestradiol levels than non-obese users [71], which may contribute to the increased risks associated with HRT and obesity. If this is the case, dose reduction should help minimise risk and, therefore, as with all women using HRT, the lowest effective dose should be sought.

Many of the effects of oestrogen are dose dependent, particularly the risk of stroke [55] and VTE. Furthermore, there is evidence that even low-dose oestrogen may control symptoms and provide bone protection while minimising the impact on mammographic density [72]. In order to reduce the risks of HRT, particularly in obese women, age-appropriate doses should be used. Younger post-menopausal women may need 1–2 mg of oral oestradiol or 50–100 μg transdermal oestrogen. However, in higher risk or older women, 0.5 mg oestradiol orally or 25 μg transdermal patches could be considered. Periodic trials of reduced doses, particularly in women over 60 years, should be considered.

Type of Oestrogen

There have been few studies that have directly compared the effects of conjugated oestrogens with oestradiol, and no RCTS have directly assessed the risk of cardiovascular disease or breast cancer using hard clinical endpoints. Trials to date have primarily focused on surrogate cardiovascular outcomes and so whether the type of oestrogen affects the risk of breast cancer is currently unknown.

CEE causes a greater rise in triglycerides than orally administered oestradiol [69] and may have differing effects on insulin and glucose metabolism [73]. Several studies have compared oral CEE with transdermal oestradiol on surrogate markers of cardiovascular disease [74,75]; however they are difficult to interpret as the dose and route of administration may influence the results more than the type of oestrogen. A recent meta-analysis did not find that the type of oestrogen had any influence on the risk of coronary event, stroke or VTE [56].

Type of Progestogen

In non-hysterectomised women, progestogen is required for endometrial protection, given in a sequential or continuous combined regimen. Progestogens have distinctive properties depending on their chemical derivatives and can, therefore, exert varying biological effects. The progestogenic component can significantly influence the consequences of HRT on the breast and cardiovascular system.

Comparison of the two arms of the WHI studies highlights clearly the marked difference in breast cancer incidence with the addition of MPA. There are currently no RCT data investigating the risk of breast cancer associated with the different progestogen preparations, but observational data from the French E3N cohort suggests that dydrogesterone and micronised progesterone may be associated with less breast cancer risk than other progestogens [31].

Several studies have shown that the cardioprotective effects of oestrogen are attenuated following the addition of a progestogen. In a RCT, the beneficial effects of CEE on endothelial function were reversed by the addition of MPA [76]. Data using hard clinical endpoints are limited, although meta-analysis has shown that addition of a progestogen doubled the risk of VTE but had no impact on the risk of cerebrovascular disease or coronary events [56].

In addition to oral or transdermal progestogens, the Lenonorgestrel - intrauterine system (LNG-IUS) is a particularly helpful device for peri-menopausal contraception, which controls peri-menopausal bleeding problems and acts as an effective method of endometrial protection for post-menopausal women on oestrogen therapy [77]. There is currently only limited data investigating its use specifically in obese women [78].

Alternatives to HRT

Various non-hormonal options exist for the management of menopausal symptoms. These include complementary preparations such as red clover isoflavones and pharmacological treatments such as SSRIs, SNRIs and gabapentin. The effect of obesity on the efficacy of these treatments on vasomotor symptoms has not been examined.

Conclusion

The menopause transition is associated with many adverse metabolic changes, which can dramatically influence a woman's future health. Obesity is not in itself a contraindication to HRT and indeed obese women may derive significant benefits from HRT. Obesity and HRT share some common risks, particularly oestrogen-dependent cancer, although there is evidence that HRT may attenuate the obesity associated risk of breast and endometrial cancers. As with all women, an individual risk assessment should be undertaken prior to HRT prescription, and consideration should be given to low-dose transdermal therapy to minimise the risk of VTE and stroke.

References

1. [illegible] M, Soda M, Nakashima E, et al. The effects of body mass index on age at men- [illegible] Int J Obes Relat Metab Disord. 2002;26:961–968.
[illegible]ous Hormones and Breast Cancer Collaborative Group, Key TJ, Reeves PN, [illegible] GK, et al. Circulating sex hormones and breast cancer risk factors in postmeno- [illegible]men: reanalysis of 13 studies. *Br J Cancer*. 2011;105:709–722.

3. Baglietto L, Severi G, English DR, et al. Circulating steroid hormone levels and risk of breast cancer for postmenopausal women. *Cancer Epidemiol Biomarkers Prev.* 2010;19:492–502.
4. Gold EB, Colvin A, Avis N, et al. Longitudinal analysis of the association between vasomotor symptoms and race/ethnicity across the menopausal transition: study of women's health across the nation. *Am J Public Health.* 2006;96:1226–1235.
5. Whiteman MK, Staropoli CA, Langenberg PW, McCarter RJ, Kjerulff KH, Flaws JA. Smoking, body mass, and hot flashes in midlife women. *Obstet Gynecol.* 2003;101:264–272.
6. Thurston RC, Sowers MR, Sutton-Tyrrell K, et al. Abdominal adiposity and hot flashes among midlife women. *Menopause.* 2008;15:429–434.
7. Freedman RR. Pathophysiology and treatment of menopausal hot flashes. *Semin Reprod Med.* 2005;23:117–125.
8. Alexander C, Cochran CJ, Gallicchio L, Miller SR, Flaws JA, Zacur H. Serum leptin levels, hormone levels, and hot flashes in midlife women. *Fertil Steril.* 2010;94:1037–1043.
9. Ley CJ, Lees B, Stevenson JC. Sex- and menopause-associated changes in body-fat distribution. *Am J Clin Nutr.* 1992;55:950–954.
10. Spencer CP, Godsland IF, Stevenson JC. Is there a menopausal metabolic syndrome? *Gynecol Endocrinol.* 1997;11:341–355.
11. Stevenson JC, Crook D, Godsland IF. Influence of age and menopause on serum lipids and lipoproteins in healthy women. *Atherosclerosis.* 1993;98:83–90.
12. Jacobsen BK, Knutsen SF, Fraser GE. Age at natural menopause and total mortality and mortality from ischemic heart disease: the Adventist Health Study. *J Clin Epidemiol.* 1999;52:303–307.
13. Rivera CM, Grossardt BR, Rhodes DJ, et al. Increased cardiovascular mortality after early bilateral oophorectomy. *Menopause.* 2009;16:15–23.
14. Rocca WA, Grossardt BR, Miller VM, Shuster LT, Brown Jr RD. Premature menopause or early menopause and risk of ischemic stroke. *Menopause.* 2012;19:272–7.
15. Vaidya D, Becker DM, Bittner V, Mathias RA, Ouyang P. Ageing, menopause, and ischaemic heart disease mortality in England, Wales, and the United States: modelling study of national mortality data. *BMJ.* 2011;343:d5170.
16. Janssen I, Powell LH, Crawford S, Lasley B, Sutton-Tyrrell K. Menopause and the metabolic syndrome: the study of women's health across the nation. *Arch Intern Med.* 2008;168:1568–1575.
17. Huang AJ, Subak LL, Wing R, et al. An intensive behavioral weight loss intervention and hot flushes in women. *Arch Intern Med.* 2010;170:1161–1167.
18. Ballard-Barbash R, Swanson CA. Body weight: estimation of risk for breast and endometrial cancers. *Am J Clin Nutr.* 1996;63(suppl 3):437S–441S.
19. Huang Z, Hankinson SE, Colditz GA, et al. Dual effects of weight and weight gain on breast cancer risk. *JAMA.* 1997;278:1407–1411.
20. Harvie M, Hooper L, Howell AH. Central obesity and breast cancer risk: a systematic review. *Obes Rev.* 2003;4:157–173.
21. Renehan AG, Tyson M, Egger M, Heller RF, Zwahlen M. Body-mass index and incidence of cancer: a systematic review and meta-analysis of prospective observational studies. *Lancet.* 2008;371:569–578.
22. Protani M, Coory M, Martin JH. Effect of obesity on survival of women with breast cancer: systematic review and meta-analysis. *Breast Cancer Res Treat.* 2010;123:627–635.

23. Petrelli JM, Calle EE, Rodriguez C, Thun MJ. Body mass index, height, and postmenopausal breast cancer mortality in a prospective cohort of US women. *Cancer Causes Control*. 2002;13:325–332.
24. Ewertz M, Jensen MB, Gunnarsdóttir KÁ, et al. Effect of obesity on prognosis after early-stage breast cancer. *J Clin Oncol*. 2011;29:25–31.
25. Beral V, Million Women Study Collaborators. Breast cancer and hormone-replacement therapy in the Million Women Study. *Lancet*. 2003;362:419–427.
26. Collaborative Group on Hormonal Factors in Breast Cancer. Breast cancer and hormone replacement therapy: collaborative reanalysis of data from 51 epidemiological studies of 52,705 women with breast cancer and 108,411 women without breast cancer. *Lancet*. 1997;350:1047–1059.
27. Rossouw JE, Anderson GL, Prentice RL, et al. Risks and benefits of estrogen plus progestin in healthy postmenopausal women: principal results from the women's health initiative randomized controlled trial. *JAMA*. 2002;288:321–333.
28. Anderson GL, Limacher M, Assaf AR, et al. Effects of conjugated equine estrogen in postmenopausal women with hysterectomy: the Women's Health Initiative randomized controlled trial. *JAMA*. 2004;291:1701–1712.
29. Rossouw JE, Prentice RL, Manson JE, et al. Postmenopausal hormone therapy and risk of cardiovascular disease by age and years since menopause. *JAMA*. 2007;297:1465–1477.
30. Chlebowski RT, Anderson GL, Gass M, et al. Estrogen plus progestin and breast cancer incidence and mortality in postmenopausal women. *JAMA*. 2010;304:1684–1692.
31. Fournier A, Berrino F, Clavel-Chapelon F. Unequal risks for breast cancer associated with different hormone replacement therapies: results from the E3N cohort study. *Breast Cancer Res Treat*. 2008;107:103–111.
32. Schairer C, Lubin J, Troisi R, Sturgeon S, Brinton L, Hoover R. Menopausal estrogen and estrogen-progestin replacement therapy and breast cancer risk. *JAMA*. 2000;283:485–491.
33. Kuhl H. Breast cancer risk in the WHI study: the problem of obesity. *Maturitas*. 2005;51:83–97.
34. Rosenberg L, Czene K, Hall P. Obesity and poor breast cancer prognosis: an illusion because of hormone replacement therapy? *Br J Cancer*. 2009;100:1486–1491.
35. Bergström A, Pisani P, Tenet V, Wolk A, Adami HO. Overweight as an avoidable cause of cancer in Europe. *Int J Cancer*. 2001;91:421–430.
36. Trentham-Dietz A, Nichols HB, Hampton JM, Newcomb PA. Weight change and risk of endometrial cancer. *Int J Epidemiol*. 2006;35:151–158.
37. Beral V, Bull D, Reeves G, Million Women Study Collaborators. Endometrial cancer and hormone-replacement therapy in the Million Women Study. *Lancet*. 2005;365:1543–1551.
38. Canchola AJ, Chang ET, Bernstein L, et al. Body size and the risk of endometrial cancer by hormone therapy use in postmenopausal women in the California Teachers Study cohort. *Cancer Causes Control*. 2010;21:1407–1416.
39. Chang SC, Lacey Jr JV, Brinton LA, et al. Lifetime weight history and endometrial cancer risk by type of menopausal hormone use in the NIH-AARP diet and health study. *Cancer Epidemiol Biomarkers Prev*. 2007;16:723–730.
40. Friedenreich C, Cust A, Lahmann PH, et al. Anthropometric factors and risk of endome- ancer: the European prospective investigation into cancer and nutrition. *Cancer Control*. 2007;18:399–413.
lough ML, Patel AV, Patel R, et al. Body mass and endometrial cancer risk by e replacement therapy and cancer subtype. *Cancer Epidemiol Biomarkers Prev*. :73–79.

42. Crosbie EJ, Zwahlen M, Kitchener HC, Egger M, Renehan AG. Body mass index, hormone replacement therapy, and endometrial cancer risk: a meta-analysis. *Cancer Epidemiol Biomarkers Prev*. 2010;19:3119–3130.
43. Pines A. Vasomotor symptoms and cardiovascular disease risk. *Climacteric*. 2011;14:535–536.
44. Gambacciani M, Ciaponi M, Piaggesi L, et al. Body weight, body fat distribution, and hormonal replacement therapy in early postmenopausal women. *J Clin Endocrinol Metab*. 1997;82:414–417.
45. The Writing Group for the PEPI Trial. Effects of estrogen or estrogen/progestin regimens on heart disease risk factors in postmenopausal women. The Postmenopausal Estrogen/Progestin Interventions (PEPI) Trial. *JAMA*. 1995;273:199–208.
46. Darling GM, Johns JA, McCloud PI, Davis SR. Estrogen and progestin compared with simvastatin for hypercholesterolemia in postmenopausal women. *N Engl J Med*. 1997;337:595–601.
47. Shwaery GT, Vita JA, Keaney Jr JF. Antioxidant protection of LDL by physiological concentrations of 17 beta-estradiol. Requirement for estradiol modification. *Circulation*. 1997;95:1378–1385.
48. Bessonova L, Marshall SF, Ziogas A, et al. The association of body mass index with mortality in the California Teachers Study. *Int J Cancer*. 2011;129:2492–2501.
49. Manson JE, Hsia J, Johnson KC, et al. Estrogen plus progestin and the risk of coronary heart disease. *N Engl J Med*. 2003;349:523–534.
50. Abdollahi M, Cushman M, Rosendaal FR. Obesity: risk of venous thrombosis and the interaction with coagulation factor levels and oral contraceptive use. *Thromb Haemost*. 2003;89:493–498.
51. Cushman M, Kuller LH, Prentice R, et al. Estrogen plus progestin and risk of venous thrombosis. *JAMA*. 2004;292:1573–1580.
52. Curb JD, Prentice RL, Bray PF, et al. Venous thrombosis and conjugated equine estrogen in women without a uterus. *Arch Intern Med*. 2006;166:772–780.
53. Canonico M, Oger E, Conard J, et al. Obesity and risk of venous thromboembolism among postmenopausal women: differential impact of hormone therapy by route of estrogen administration. The ESTHER Study. *J Thromb Haemost*. 2006;4:1259–1265.
54. Scarabin PY, Alhenc-Gelas M, Plu-Bureau G, Taisne P, Agher R, Aiach M. Effects of oral and transdermal estrogen/progesterone regimens on blood coagulation and fibrinolysis in postmenopausal women. A randomized controlled trial. *Arterioscler Thromb Vasc Biol*. 1997;17:3071–3078.
55. Grodstein F, Manson JE, Colditz GA, Willett WC, Speizer FE, Stampfer MJ. A prospective, observational study of postmenopausal hormone therapy and primary prevention of cardiovascular disease. *Ann Intern Med*. 2000;133:933–941.
56. Sare GM, Gray LJ, Bath PM. Association between hormone replacement therapy and subsequent arterial and venous vascular events: a meta-analysis. *Eur Heart J*. 2008;29:2031–2041.
57. Li C, Engström G, Hedblad B, Berglund G, Janzon L. Risk of stroke and hormone replacement therapy. A prospective cohort study. *Maturitas*. 2006;54:11–18.
58. Lobo RA, Clarkson TB. Different mechanisms for benefit and risk of coronary heart disease and stroke in early postmenopausal women: a hypothetical explanation. *Menopause*. 2011;18:237–240.
59. Crespo CJ, Smit E, Snelling A, Sempos CT, Andersen RE. Hormone replacement therapy and its relationship to lipid and glucose metabolism in diabetic and nondiabetic

postmenopausal women: results from the Third National Health and Nutrition Examination Survey (NHANES III):NHANES III. *Diabetes Care*. 2002;25:1675–1680.
60. Manson JE, Rimm EB, Colditz GA, et al. A prospective study of postmenopausal estrogen therapy and subsequent incidence of non-insulin-dependent diabetes mellitus. *Ann Epidemiol*. 1992;2:665–673.
61. Margolis KL, Bonds DE, Rodabough RJ, et al. Effect of oestrogen plus progestin on the incidence of diabetes in postmenopausal women: results from the Women's Health Initiative Hormone Trial. *Diabetologia*. 2004;47:1175–1187.
62. Ferrara A, Karter AJ, Ackerson LM, Liu JY, Selby JV, Northern California Kaiser Permanente Diabetes Registry.. Hormone replacement therapy is associated with better glycemic control in women with type 2 diabetes: The Northern California Kaiser Permanente Diabetes Registry. *Diabetes Care*. 2001;24:1144–1150.
63. Friday KE, Dong C, Fontenot RU. Conjugated equine estrogen improves glycemic control and blood lipoproteins in postmenopausal women with type 2 diabetes. *J Clin Endocrinol Metab*. 2001;86:48–52.
64. van Staa TP, Dennison EM, Leufkens HG, Cooper C. Epidemiology of fractures in England and Wales. *Bone*. 2001;29:517–522.
65. Compston JE, Watts NB, Chapurlat R, et al. Obesity is not protective against fracture in postmenopausal women: GLOW. *Am J Med*. 2011;124:1043–1050.
66. Ettinger B, Ensrud KE, Wallace R, et al. Effects of ultralow-dose transdermal estradiol on bone mineral density: a randomized clinical trial. *Obstet Gynecol*. 2004;104:443–451.
67. Cauley JA, Robbins J, Chen Z, et al. Effects of estrogen plus progestin on risk of fracture and bone mineral density: the Women's Health Initiative randomized trial. *JAMA*. 2003;290:1729–1738.
68. Lambrinoudaki I, Brincat M, Erel CT, et al. EMAS position statement: managing obese postmenopausal women. *Maturitas*. 2010;66:323–326.
69. Godsland IF. Effects of postmenopausal hormone replacement therapy on lipid, lipoprotein, and apolipoprotein (a) concentrations: analysis of studies published from 1974–2000. *Fertil Steril*. 2001;75:898–915.
70. Spencer CP, Godsland IF, Cooper AJ, Ross D, Whitehead MI, Stevenson JC. Effects of oral and transdermal 17beta-estradiol with cyclical oral norethindrone acetate on insulin sensitivity, secretion, and elimination in postmenopausal women. *Metabolism*. 2002;49:742–747.
71. Karim R, Mack WJ, Hodis HN, Roy S, Stanczyk FZ. Influence of age and obesity on serum estradiol, estrone, and sex hormone binding globulin concentrations following oral estrogen administration in postmenopausal women. *J Clin Endocrinol Metab*. 2009;94:4136–4143.
72. Grady D, Vittinghoff E, Lin F, et al. Effect of ultra-low-dose transdermal estradiol on breast density in postmenopausal women. *Menopause*. 2007;14(3 Pt 1):391–396.
73. Stevenson JC. HRT and cardiovascular disease. *Best Pract Res Clin Obstet Gynaecol*. 2009;23:109–120.
74. Sumino H, Ichikawa S, Kasama S, et al. Different effects of oral conjugated estrogen and transdermal estradiol on arterial stiffness and vascular inflammatory markers in [illegible]nopausal women. *Atherosclerosis*. 2006;189:436–442.
[illegible] Chen MJ, Sheu WH, et al. Differential effects of oral conjugated equine [illegible]ı and transdermal estrogen on atherosclerotic vascular disease risk markers and [illegible]lial function in healthy postmenopausal women. *Hum Reprod*. [illegible]2715–2720.

76. Wakatsuki A, Okatani Y, Ikenoue N, Fukaya T. Effect of medroxyprogesterone acetate on endothelium-dependent vasodilation in postmenopausal women receiving estrogen. *Circulation*. 2001;104:1773–1778.
77. Somboonporn W, Panna S, Temtanakitpaisan T, Kaewrudee S, Soontrapa S. Effects of the levonorgestrel-releasing intrauterine system plus estrogen therapy in perimenopausal and postmenopausal women: systematic review and meta-analysis. *Menopause*. 2011;18:1060–1066.
78. Wan YL, Holland C. The efficacy of levonorgestrel intrauterine systems for endometrial protection: a systematic review. *Climacteric*. 2011;14:622–632.

40 Incontinence and Prolapse in the Obese Woman

Douglas G. Tincello

Reproductive Science Section, CSMM University of Leicester, Leicester, UK

Introduction

This chapter will discuss the available literature relating to the impact of obesity on the prevalence and severity of pelvic floor dysfunction in women (urinary and anal incontinence and uterovaginal prolapse). The evidence for symptom resolution after weight loss will be presented, and recent data on the efficacy and safety of urogynaecology surgery will be examined. Throughout this chapter, unless specifically indicated, the definitions of overweight and obesity will be the standard WHO definitions, based on body mass index (BMI): overweight BMI 25–29 kg/m^2, obese BMI 30–35 kg/m^2, morbid obesity BMI $>$ 35 kg/m^2.

The International Continence Society publishes standardisation documents from time to time, with terminology and definitions of common symptoms, signs and diagnoses relating to incontinence [1]. Those definitions will be used throughout this chapter. Urinary symptoms related to continence in women can be divided into storage symptoms, voiding symptoms or post-voiding symptoms (Table 40.1). It is now recognised that these symptoms do not correlate well with the underlying cause of incontinence or findings during urodynamic testing. While women with the isolated symptom of stress urinary incontinence (SUI) are very likely to have urethral sphincter weakness, the relationship between other storage symptoms and diagnosis is less clear. Thus, the symptom syndrome 'overactive bladder' (OAB) is now defined as 'urgency, with or without urge incontinence, usually with frequency and nocturia' [1].

Incidence and Prevalence

Urinary incontinence is a cause of significant morbidity and cost, estimated at over £500 million representing 1% of the health care budget in the United

Obesity. DOI: http://dx.doi.org/10.1016/B978-0-12-416045-3.00040-6

Table 40.1 Selected Symptoms and Signs of Urinary Incontinence

	Definition
Storage symptoms	
Increased daytime frequency	Patient who considers he/she voids too often by day
Nocturia	Waking at night one or more times to void
Urgency	A sudden compelling desire to pass urine, which is difficult to defer
Urinary incontinence	Any involuntary leakage of urine
Stress urinary incontinence	Involuntary leakage on effort or exertion or on sneezing and coughing
Urge urinary incontinence	Involuntary leakage accompanied by or immediately preceded by urgency
Mixed urinary incontinence	Involuntary leakage associated with urgency and also with exertion, effort, sneezing or coughing
Nocturnal enuresis	The loss of urine occurring during sleep
Voiding symptoms	
Slow stream	Reported by the individual as her perception of reduced urine flow, usually compared to previous performance or in comparison to others
Splitting or spraying	. . .of the stream may be reported
Intermittent stream	When the individual describes urine flow which stops and starts, on more than one occasion, during micturition
Hesitancy	Difficulty in initiating micturition resulting in a delay in the onset of voiding after the individual is ready to pass urine
Terminal dribble	A prolonged final part of micturition, when the flow has slowed to a trickle or dribble
Post-micturition symptoms	
Incomplete emptying	A self-explanatory term for a feeling experienced by the individual after passing urine
Post-micturition dribble	The involuntary loss of urine immediately after he or she has finished passing urine, usually after rising from the toilet in case of a women

Source: Definitions from the International Continence Society Standardisation Document [1].

Kingdom [2], €400 billion in Europe [3] and between $25 and $50 billion in the United States [4,5]. Incontinence becomes more prevalent with increasing age [6] (Figure 40.1). It has recently been estimated that 11% of the global population suffer from OAB and 8% from urinary incontinence [7]. Known risk factors for urinary incontinence include vaginal childbirth [8], large babies [9], perineal ...perative delivery [10] and increasing age [11]. Detailed analysis of data ...ting the relationship between urinary symptoms and obesity will be dis... he below sections.

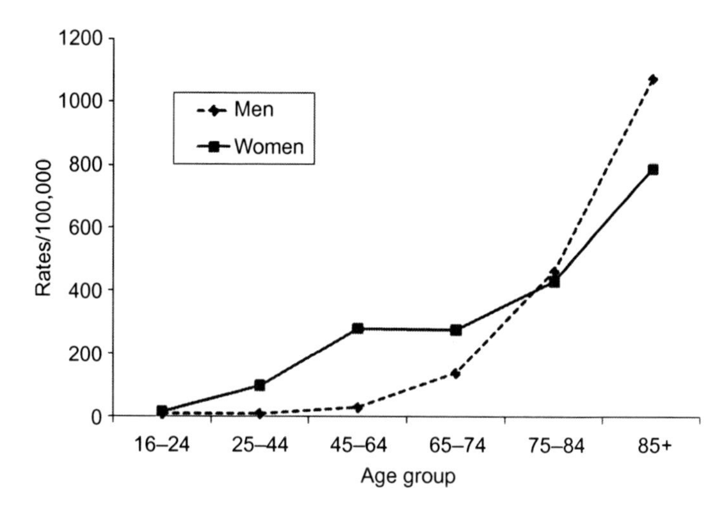

Figure 40.1 Prevalence of urinary incontinence by age in men and women
Source: Data from Hospital Episode Statistics, UK Department of Health.

Normal Bladder Function and Causes of Incontinence in Women

The bladder has two functions: to store urine and to then void the stored urine. During the storage phase of micturition, the detrusor muscle of the bladder is quiescent and accommodates increasing urine volumes with no increase in pressure. As bladder capacity is reached, sensory afferents (stretch receptors) in the bladder wall are triggered to give rise to increasing sensation of bladder filling. The sphincter mechanism (discussed later) is closed. Cortical inhibition of the spinal voiding reflex arc, learned during toilet training in infancy, allows delay of micturition until socially convenient. During voiding, cortical inhibition is removed and a co-ordinated relaxation of the pelvic floor and urethral sphincters occurs synchronously with detrusor contraction. Detrusor contraction is mediated by muscarinic cholinergic nerves of the parasympathetic nervous system, and urethral sphincter tone is maintained by noradrenergic neurons of the sympathetic nervous system and somatic fibres from the pudendal nerves.

In women, continence is achieved by a combination of the ligaments supporting the activity of bladder neck and pelvic floor muscle. The urethral sphincter mechanism in women is a functional system including the internal (smooth muscle) and external (striated muscle) sphincters, together with the muscles of the pelvic floor and the ligaments supporting the urethra (pubourethral ligament). In the normal situation, the ligaments and pelvic floor maintain the urethra above the urogenital hiatus, and thus within the abdomen. Increases in abdominal pressure are transmitted equally to the bladder and bladder neck (Figure 40.2A).

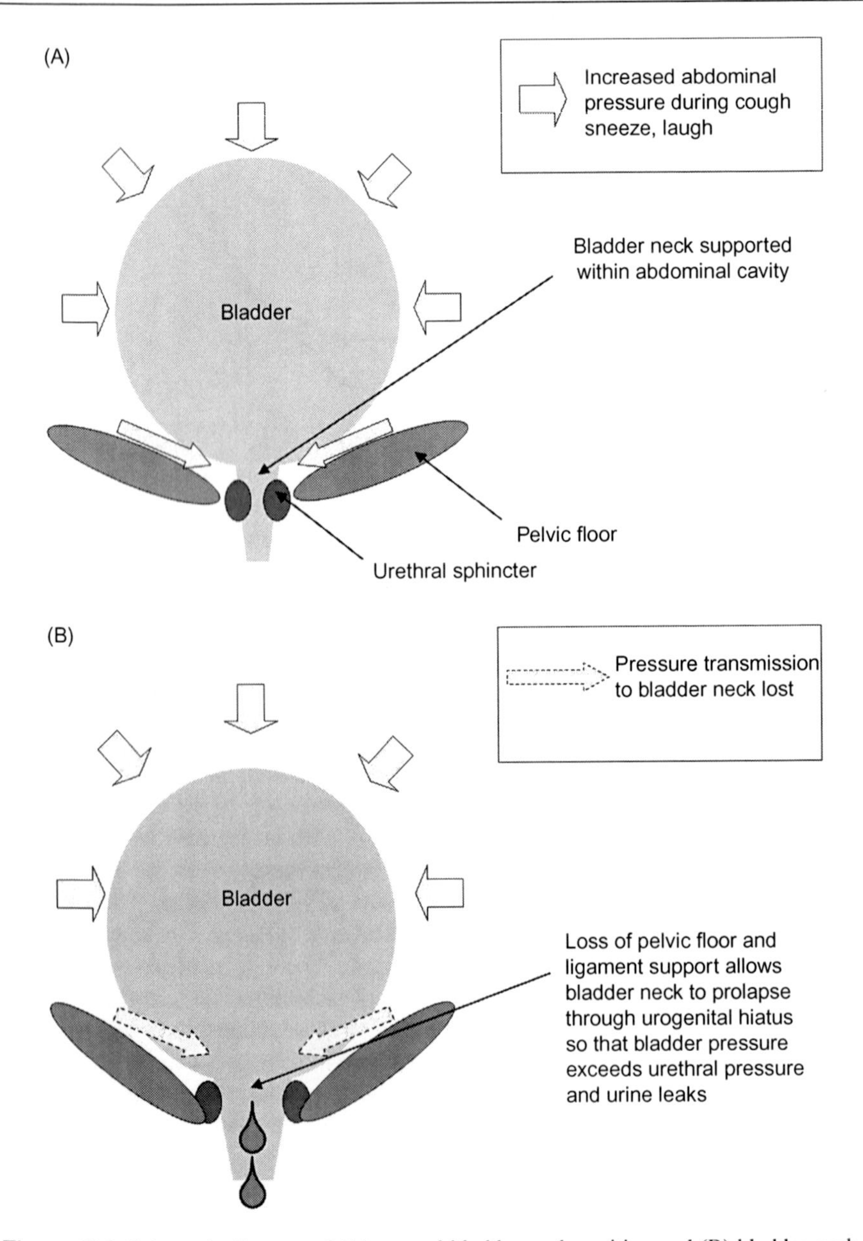

Figure 40.2 Schematic diagram of (A) normal bladder neck position and (B) bladder neck displacement through the urogenital hiatus causing stress incontinence.

ᵢic Stress Incontinence

ed above, SUI is both a symptom and a sign [1]. In isolation, SUI is a ;ood predictor of the presence of an incompetent urethral sphincter, leading to incontinence. Urethral sphincter weakness in most cases is due to hypermobility, where the pelvic floor and ligaments cannot retain the urethra in position and it falls through the urogenital hiatus during increases in abdominal pressure, leading to loss of pressure transmission to the urethra and hence leakage of urine (Figure 40.2). Intrinsic sphincter deficiency is less common and occurs where urethral closure pressure is low without any urethral mobility. Urethral incompetence can be demonstrated during urodynamic testing by involuntary leakage associated with increased abdominal pressure and no detrusor muscle contraction, and it is termed urodynamic stress incontinence (USI) [1].

Detrusor Overactivity

Detrusor overactivity (DO) is a condition characterised by the urodynamic observation of involuntary detrusor contractions during the filling phase of micturition [1]. DO is a poorly understood condition of uncertain aetiology. Women with DO will often complain of symptoms of OAB, but may not be incontinent unless the urethral sphincter function is compromised, or the detrusor contractions are of very high pressure amplitude and overcome urethral resistance. It should be remembered that the relationship between demonstrable DO and OAB syndrome is not absolute, and patients with OAB may demonstrate normal urodynamic investigations.

Obesity and Urinary Incontinence

Bearing in mind the changes of bladder neck movement and of the abdominal pressure transmission as mentioned above, it seems most likely that obesity will affect the risk of developing USI rather than OAB/DO. There exist some urodynamic data in obese women before and after surgery, which have been recently reviewed in a systematic review [12]to confirm this theory. Abdominal and intra-vesical pressures are higher in obese women [13], and successful weight loss is associated with reduced urethral mobility and intra-vesical pressure. Bariatric surgery (techniques to reduce stomach capacity) has been shown to reduce not only the weight dramatically but also the bladder pressure by almost 50% (17–10 cm H_2O) [14]. A recent study found that obese women had more severe incontinence as defined by a lower valsalva leak point pressure, in keeping with the theory that obesity increases the pressure load upon the bladder and bladder neck [15].

In terms of urinary symptoms, there is now a large body of evidence confirming that urinary symptoms are more prevalent in overweight and obese women. The Leicestershire Incontinence Study, funded by the UK Medical Research Council, surveyed over 12,000 community dwelling women aged 40 years or over with validated urinary symptom questionnaires [16]. Over 7000 respondents were also sent a food questionnaire. The multivariate analysis showed an increased risk of new onset SUI both for overweight (odds ratio (OR) 1.25, 95% confidence intervals (CI) 0.94, 1.67) and obese (OR 1.74; CI 1.22, 2.48) women. A questionnaire study of 1336 Swedish women revealed a 12% prevalence of weekly or more urinary incontinence [17]. Among those women, obesity was a major risk factor (OR 4.1; CI 2.6, 6.6), but overweight women also had increased risk (OR 1.8; CI 1.3, 2.5). More recently, obesity has been confirmed as an independent risk factor for SUI by other groups from the United States [18–20], central America [21], Korea [22], Taiwan [23] and the large EPINCONT study from Scandinavia [24] (OR 1.4; CI 1.2, 1.5 for overweight; OR 1.7; CI 1.6, 2.0 for obese) indicating that this effect is independent of ethnic origin. Interestingly, Lawrence et al. [18] found that obesity and diabetes were independent but additive risk factors for stress incontinence. Although there is a clear association between obesity and USI, it should be remembered that large epidemiological research studies often identify women with relatively mild and not bothersome symptoms. Gerten et al. [25] compared a cohort of women seeking bariatric surgery for weight loss and found the impact of their incontinence to be less than the women who attended a urogynaecology clinic, as assessed by the Incontinence Impact Questionnaire and Urogenital Distress Inventory. However, among women who underwent surgery for incontinence, obese women reported more distress and bother from their symptoms than women of normal BMI [13].

The association between obesity and OAB/DO appears as strong. The Leicestershire MRC study reported similar odds ratios for OAB as for SUI in overweight (OR 1.24; CI 0.98, 1.63) and obese (OR 1.46; CI 1.02, 2.09) women [16]. The EPINCONT study found similar risks for overweight (OR 1.1; CI 0.9, 1.3) and obese (OR 1.5; CI 1.2, 1.9) women [24]. A study of over 3000 women in the United States found not only a higher risk of OAB in the obese women (OR 2.67; CI 2.20, 3.22) but also that this risk was further increased in obese women who also had type-II diabetes [18]. This finding agrees with other work suggesting diabetes to be an independent risk factor for OAB [26,27]. A comprehensive systematic review presents all the evidence confirming the association, and readers are directed to this for a full discussion of the evidence [28].

Obesity and Faecal Incontinence

Faecal incontinence is also common in obese women. A survey of 256 women attending a bariatric surgery seminar in the United States reported a prevalence of flatus incontinence of 87%, liquid incontinence of 80% and incontinence of solid stool of 19% [29]. Among 551 women who attended a urogynaecology clinic, every 5 point increase in BMI carried increased risk of faecal incontinence (OR 1.21; CI 1.05, 1.40) and of constipation (OR 1.13; CI 0.98, 1.31) [30]. Sileri et al. [31] found severe faecal incontinence in 25% and constipation in 20% of 139 patients (93 women) who underwent bariatric surgery. A 2011 systematic review of rgery research (men and women) identified 13 studies that reported the cal incontinence between 16% and 68%, in each case higher than the ed for non-obese individuals [32]. This review also reported constipa- tween 17% and 29%.

Obesity and Prolapse

The increased intra-abdominal pressure associated with obesity will theoretically lead to increased pressure upon the pelvic floor, uterine and vaginal ligaments and connective tissue supports, and, therefore, increases the risk of developing symptomatic prolapse. Although many authors have examined the obstetric and gynaecology factors associated with prolapse, there are few studies which examine obesity. A large, community-based epidemiological survey of 17,000 women found a trend between increasing obesity and prevalence of prolapse [33]. A survey of over 21,000 women attending menopause clinics in Italy found a 5% prevalence of prolapse, and the risk of prolapse was higher in overweight (OR 1.4; CI 1.2, 1.7) and obese (OR 1.6; CI 1.3, 1.9) women [34]. Similar risks are reported for overweight women (OR 1.9; CI 1.2, 3.1) [35] in a study of 5000 women from Sweden.

A large US study analysed data from 16,608 women in a hormone replacement trial followed for 5 years [36]. They found that overweight and obese women had increased risks of progression of all prolapse types (assessed by pelvic examination). The excess risks for anterior vaginal prolapse were 32% and 48% in overweight and obese women, respectively, for posterior prolapse 37% and 58% and for uterine prolapse 43% and 69%. In this study, loss of weight was associated with no regression of prolapse and a suggestion that uterine prolapse may worsen.

Weight Loss and the Effects Upon Continence and Prolapse

Given that obesity is associated with incontinence and prolapse, in theory due to increased abdominal pressure, it makes sense that weight loss should reduce the severity and impact of these symptoms. Weight loss is a simple intervention with minimal side effects, so are there data which confirm that relief of pelvic floor dysfunction symptom occurs?

Two systematic reviews, of bariatric surgery studies and non-randomised weight reduction intervention studies, were published in 2008 and 2009 after analysing the literature [28,37]. Six studies from bariatric surgery were reviewed; in patients (predominantly women) who lost weight (typically 50% of excess weight or a fall in BMI of 15–20 points), urinary continence cure rates were between 30% and 64%. Two studies reported urinary continence outcomes after weight reduction programmes: loss of 5% of body weight (a typical target for such programmes) was associated with 50% or more improvement of symptom and reduction in leakage severity assessed by pad testing (median reduction of 19 g).

A pilot trial of 48 women randomised to a weight control programme, or delayed enrolment, followed for 6 months demonstrated a median weight loss of 16 kg in the intervention group (BMI from 35–28 kg/m^2) which was associated with significant reductions in mean incontinence episodes (60% vs 15%) [38]. The trial team subsequently completed a larger study (the PRIDE study) randomising

338 women to a 6-month weight control programme or an education programme [39]. Mean weight loss was 8% of body weight at baseline, and there were improvements in weekly incontinence episodes of 47.4% (CI 39.9, 54.0) in the intervention group compared to 28.1% (CI 12.6, 40.9) in the control group. This difference was dominated by changes in stress incontinence symptoms. Secondary analysis of the data showed that larger weight loss (up to 10% of baseline) was associated with greater improvements in symptoms. Women losing up to 10% weight were more likely to achieve at least 70% of symptom relief than those losing less weight (compared to controls): for 5–10% loss of weight, OR 2.4 (CI 1.1, 5.1) and for more than 10% loss, OR 3.3 (CI 1.7, 6.4) [40]. These beneficial effects of weight loss persisted up to 18 months after the intervention [41].

Data on improvements in anal incontinence after weight loss are sparser. Burgio et al. [42] found that the prevalence of incontinence to solids or liquids fell from 19.4% to 9.1% at 6 months and to 8.6% at 12 months after bariatric surgery [42]. A secondary analysis of the patients in the PRIDE study showed that 16% of women had at least monthly solid or liquid incontinence. The majority of these women (33 of 55) had improvement in their symptoms during the study follow-up [43].

Data on whether weight loss alters prolapse severity are also scarce. As mentioned above, a large prospective study found that weight loss had only minimal effects upon anatomical prolapse [36]. A study from Egypt assessed 400 women and found vaginal prolapse in 65% of the sample, although symptoms were much less common [44]. Among the identified risk factors, a history of significant weight loss actually was associated with an increased risk of prolapse. Cuicchi et al. [45] reported pelvic floor symptoms from 100 obese women before and after bariatric surgery [45]. Pelvic organ prolapse symptoms were reported by 56 women, and 15 women had documented anatomical prolapse. After surgery, 74% of the affected women had resolution of their prolapse symptoms.

Weight-loss programmes or bariatric surgery appear to be an effective way of reducing or even curing symptoms of urinary incontinence. A target weight loss between 5% and 10% seems the optimal target to balance achievability with significant and meaningful reductions in urinary symptoms [40]. More data are required on whether significant weight loss improves prolapse symptoms or may actually increase the prevalence and severity of these symptoms.

Continence and Prolapse Surgery in the Obese Woman

Surgery is a commonly used treatment for women with stress incontinence and also for pro[illegible]se if vaginal pessaries and pelvic floor exercises have failed. Surgery in [illegible] women may not only be technically more difficult, with higher rates of [illegible]ons, but may also carry a higher failure rate for both continence and pro- [illegible]ral studies address this issue, all of which are retrospective reviews with [illegible]gth of follow-up. A study of 242 women having retropubic tension-free vaginal tape (TVT) showed the subjective cure rate at 6 months to be 85%, 95% and 89% in women of normal BMI, the overweight and the obese women, respectively [46], with no difference in complication rate. Similar data were reported from a study of 285 Korean women, 45 of whom were obese and 159 overweight [47]. Rafii et al. [48] compared the outcomes after TVT in 187 women: 39 with BMI > 30 kg/m^2 and 62 with BMI between 25 and 30 kg/m^2. After follow-up of at least 6 months, the objective cure and subjective cure rates were statistically not different at between 82% and 93%. Persistent urge urinary incontinence (a complication of continence surgery) was more common in women with BMI > 30 kg/m^2 (17.9% compared to 6.4% or 3.4%, $P = 0.02$).

More recent retrospective studies of other continence procedures have demonstrated similar cure rates, irrespective of BMI for the transobturator tapes (TOT) [49–51] and fascial slings [52]. One study reported 10 year outcomes after TVT insertion and reported cure rates of 61% in women with BMI < 30 kg/m^2, compared to 50% in women with BMI > 30 kg/m^2, although no statistical comparisons were done [53]. In this cohort, the re-operation rate was 2% in the women with low BMI and 4.5% in those with BMI of >30 kg/m^2.

Only one study reported any difference in either short- or long-term complications. An analysis of 31 morbidly obese women (BMI > 40 kg/m^2) compared to 52 women with BMI < 30 kg/m^2 having TVT with a mean follow-up of 18.5 months reported similar cure rates but a statistically significant excess of short-term complications (overall 48.4% vs 38.5%), although actual numbers of clinically significant events (chronic retention, wound haematoma, deep vein thrombosis, pneumonia or cardiac events) were rare in either group [54].

The systematic review by Greer et al. [12] reviewed all the data available from surgical reports at the time (2007–2008). Pooled data from seven studies of TVT confirmed the low rate of bladder perforation in both obese and non-obese women which was actually less in the obese group (1.2% vs 6.6%; OR 0.277; CI 0.098, 0.782). Other complications, including new urgency, appeared similar in obese or non-obese women. The authors were able to conduct a meta-analysis of cure rate using data from 453 obese and 1186 non-obese women [12]. Pooled cure was 81% in the obese and 85% in the non-obese women (OR 0.576; CI 0.426–0.779).

Thus overall, the data suggest that mid-urethral sling surgery for stress incontinence (TVT and TOT) is slightly less effective in obese women than in those of normal body weight, at least in the medium term (2 years). There is limited evidence about long-term outcome, so this needs to be confirmed by more long-term studies. The complication rate and adverse event profile appear similar regardless of BMI.

For prolapse surgery, data on the relationship between obesity and the outcomes and complications are extremely limited. Five year follow-up of women having prolapse surgery revealed no association between obesity and the risk of prolapse recurrence [55]. Similarly, obesity was found not to be a risk factor for the development of prolapse after colposuspension [56]. A recent secondary analysis of data from a trial of prophylactic colposuspension with sacrocolpopexy found that although operating time was longer in the obese women, there were no difference in outcomes for incontinence, prolapse symptoms or satisfaction [57]. A small Spanish study reported higher failure rates of prolapse surgery at 1 year in women with higher BMI, with 17 of 69 women having surgical failure overall. BMI was greater in the women whose surgery failed (29.6 ± 2.03 vs 27.1 ± 3.32)[58].

We were able to identify only one study comparing surgical complications after prolapse surgery by different BMI [59]. Adverse events were rare; the need for blood transfusion (OR 2.46; CI 1.38, 4.39) and the incidence of long-term urinary retention (OR 2.20; CI 1.21, 4.03) were more common in the women of normal BMI.

So, overall, there are few data reporting on the relationship between the success of prolapse surgery, or complications, and obesity. It appears that obesity carries neither excess risk of surgical failure, nor complications, although data from larger studies with longer follow-up are needed to be confident of this conclusion.

Conclusion

Obesity and being overweight are associated with a clearly increased risk of urinary incontinence, both for stress incontinence and for OAB. Additionally, obesity is associated with a higher prevalence of symptoms of anal incontinence and of pelvic organ prolapse. Weight loss (either by diet and exercise or by bariatric surgery) is associated with large improvements in urinary, anal and prolapse symptoms. Achieving a target weight loss between 5% and 10% of baseline weight will bring about complete resolution of urinary incontinence, anal incontinence and prolapse symptoms in up to 70% of women. On this basis, weight loss should be considered the first-line management in the obese or overweight woman with pelvic floor symptoms, and the prospect of cure of these symptoms is likely to be a major source of motivation to comply with weight-loss strategies.

Where surgery is deemed necessary, women should be advised that both continence and prolapse surgery appear equally safe in the obese patient but that the cure rate of mid-urethral tapes for urinary incontinence is slightly compromised. Further information is required to confirm whether the long-term failure rate after prolapse surgery is greater in the obese woman.

References

1. Abrams P, Cardozo L, Fall M, et al. The standardisation of terminology of lower urinary tract function: report from the standardisation sub-committee of the international continence Society. *Neurourol Urodyn*. 2002;21:167–178.
2. [illegible]A, Shaw C, McGrother CW, Dallosso HM, Cooper NJ, The Leicestershire MRC [illegible]nce Study Team. The cost of clinically significant urinary storage symptoms for [illegible]y dwelling adults in the UK. *BJU Int*. 2004;93:1246–1252.
3. [illegible] Mungapen L, Milsom I, Kopp Z, Reeves P, Kelleher C. The economic impact [illegible]ve bladder syndrome in six Western countries. *BJU Int*. 2009;103:202–209.

4. Onukwugha E, Zuckerman IH, McNally D, Coyne KS, Vats V, Mullins CD. The total economic burden of overactive bladder in the United States: a disease-specific approach. *Am J Manag Care*. 2009;15:S90–S97.
5. Kannan H, Radican L, Turpin RS, Bolge SC. Burden of illness associated with lower urinary tract symptoms including overactive bladder/urinary incontinence. *Urology*. 2009;74:34–38.
6. McGrother CW, Donaldson MM, Shaw C, et al. Storage disorder of the bladder: prevalence, incidence and need for services in the UK. *BJU Int*. 2004;93:763–769.
7. Irwin DE, Kopp ZS, Agatep B, Milsom I, Abrams P. Worldwide prevalence estimates of lower urinary tract symptoms, overactive bladder, urinary incontinence and bladder outlet obstruction. *BJU Int*. 2011;108:1132–1138.
8. MacLennan AH, Taylor AW, Wilson DH, Wilson D. The prevalence of pelvic floor disorders and their relationship to gender, age, parity and mode of delivery. *BJOG*. 2000;107:1460–1470.
9. Glazener CM, Herbison GP, MacArthur C, et al. New postnatal urinary incontinence: obstetric and other risk factors in primiparae. *BJOG*. 2006;113:208–217.
10. Boyles SH, Li H, Mori T, Osterweil P, Guise JM. Effect of mode of delivery on the incidence of urinary incontinence in primiparous women. *Obstet Gynecol*. 2009;113:134–141.
11. Swithinbank LV, Donovan JL, du Heaume JC, et al. Urinary symptoms and incontinence in women: relationship between occurrence, age, and perceived impact. *Br J Gen Pract*. 2002;49:897–900.
12. Greer WJ, Richter HE, Bartolucci AA, Burgio KL. Obesity and pelvic floor disorders: a systematic review. *Obstet Gynecol*. 2008;112:341–349.
13. Richter HE, Kenton K, Huang L, et al. The impact of obesity on urinary incontinence symptoms, severity, urodynamic characteristics and quality of life. *J Urol*. 2010;183:622–628.
14. Sugarman H, Windsor A, Bessos M, Kellum J, Reines H, DeMaria E. Effects of surgically induced weight loss on urinary bladder pressure, sagittal abdominal diameter and obesity co-morbidity. *Int J Obest Relat Metab Disord*. 2008;22:230–235.
15. El-Hefnawy AS, Wadie BS. Severe stress urinary incontinence: objective analysis of risk factors. *Maturitas*. 2011;68:374–377.
16. Dallosso HM, McGrother CW, Matthews RJ, Donaldson MM, The Leicestershire MRC Incontinence Study Team. The association of diet and other lifestyle factors with overactive bladder and stress incontinence: a longitudinal study in women. *BJU Int*. 2003;92:69–77.
17. Uustal Fornell E, Wingren G, Kjolhede P. Factors associated with pelvic floor dysfunction with emphasis on urinary and fecal incontinence and genital prolapse: an epidemiological study. *Acta Obstet Gynecol Scand*. 2004;83:383–389.
18. Lawrence JM, Lukacz ES, Liu IL, Nager CW, Luber KM. Pelvic floor disorders, diabetes, and obesity in women: findings from the Kaiser Permanente Continence Associated Risk Epidemiology Study. *Diabetes Care*. 2007;30:2536–2541.
19. Markland AD, Richter HE, Fwu CW, Eggers P, Kusek JW. Prevalence and trends of urinary incontinence in adults in the United States, 2001 to 2008. *J Urol*. 2011;186:589–593.
20. Hawkins K, Pernarelli J, Ozminkowski RJ, et al. The prevalence of urinary incontinence and its burden on the quality of life among older adults with medicare supplement insurance. *Qual Life Res*. 2011;20:723–732.

21. Lopez M, Ortiz AP, Vargas R. Prevalence of urinary incontinence and its association with body mass index among women in Puerto Rico. *J Womens Health*. 2009;18:1607–1614.
22. Ham E, Choi H, Seo JT, Kim HG, Palmer MH, Kim I. Risk factors for female urinary incontinence among middle-aged Korean women. *J Womens Health*. 2009;18:1801–1806.
23. Hsieh CH, Hsu CS, Su TH, Chang ST, Lee MC. Risk factors for urinary incontinence in Taiwanese women aged 60 or over. *Int Urogynecol J Pelvic Floor Dysfunct*. 2007;18:1325–1329.
24. Hannestad YS, Rortveit G, Daltveit AK, Hunskaar S. Are smoking and other lifestyle factors associated with female urinary incontinence? The Norwegian EPINCONT Study. *Br J Obstet Gynaecol*. 2005;110:247–254.
25. Gerten KA, Richter HE, Burgio KL, Wheeler TL, Goode PS, Redden DT. Impact of urinary incontinence in morbidly obese women versus women seeking urogynecologic care. *Urology*. 2007;70:1082–1085.
26. Liu RT, Chung MS, Lee WC, et al. Prevalence of overactive bladder and associated risk factors in 1359 patients with type 2 diabetes. *Urology*. 2011;78:1040–1045.
27. Uzun H, Zorba OU. Metabolic syndrome in female patients with overactive bladder. *Urology*. 2012;79:72–75.
28. Hunskaar S. A systematic review of overweight and obesity as risk factors and targets for clinical intervention for urinary incontinence in women. *Neurourol Urodyn*. 2008;27:749–757.
29. Wasserberg N, Haney M, Petrone P, et al. Fecal incontinence among morbid obese women seeking for weight loss surgery: an underappreciated association with adverse impact on quality of life. *Int J Colorectal Dis*. 2008;23:493–497.
30. Erekson EA, Sung VW, Myers DL. Effect of body mass index on the risk of anal incontinence and defecatory dysfunction in women. *Am J Obstet Gynecol*. 2008;198:596–604.
31. Sileri P, Franceschilli L, Cadeddu F, et al. Prevalence of defaecatory disorders in morbidly obese patients before and after bariatric surgery. *J Gastrointest Surg*. 2012;16:62–66.
32. Poylin V, Serrot FJ, Madoff RD, et al. Obesity and bariatric surgery: a systematic review of associations with defecatory dysfunction. *Colorectal Dis*. 2011;13:e92–e103.
33. Mant J, Painter R, Vessey M. Epidemiology of genital prolapse: observations from the oxford family planning association study. *Br J Obstet Gynaecol*. 1997;104:579–585.
34. Progetto Menopausa Italia Study Group. Risk factors for genital prolapse in non-hysterectomized women around menopause. Results from a large cross-sectional study in menopausal clinics in Italy. *Eur J Obstet Gynecol Reprod Biol*. 2000;93:135–140.
35. Miedel A, Tegerstedt G, Maehle-Schmidt M, Nyren O, Hammarstrom M. Nonobstetric risk factors for symptomatic pelvic organ prolapse. *Obstet Gynecol*. 2009;113:1089–1097.
36. Kudish BI, Iglesia CB, Sokol RJ, et al. Effect of weight change on natural history of pelvic organ prolapse. *Obstet Gynecol*. 2009;113:81–88.
37. Subak LL, Richter HE, Hunskaar S. Obesity and urinary incontinence: epidemiology [illegible]ical research update. *J Urol*. 2009;182:S2–S7.

[illegible]L, Whitcomb E, Shen H, Saxton J, Vittinghoff E, Brown JS. Weight loss: a [illegible]d effective treatment for urinary incontinence. *J Urol*. 2005;174:190–195.

[illegible]L, Wing R, West DS, et al. Weight loss to treat urinary incontinence in over[illegible]d obese women. *N Engl J Med*. 2009;360:481–490.

40. Wing RR, Creasman JM, West DS, et al. Improving urinary incontinence in overweight and obese women through modest weight loss. *Obstet Gynecol.* 2010;116:284–292.
41. Wing RR, West DS, Grady D, et al. Effect of weight loss on urinary incontinence in overweight and obese women: results at 12 and 18 months. *J Urol.* 2010;184:1005–1010.
42. Burgio KL, Richter HE, Clements RH, Redden DT, Goode PS. Changes in urinary and fecal incontinence symptoms with weight loss surgery in morbidly obese women. *Obstet Gynecol.* 2007;110:1034–1040.
43. Markland AD, Richter HE, Burgio KL, Myers DL, Hernandez AL, Subak LL. Weight loss improves fecal incontinence severity in overweight and obese women with urinary incontinence. *Int Urogynecol J.* 2011;22:1151–1157.
44. Gomman HM, Nossier SA, Fotohi EM, Kholeif AE. Prevalence and factors associated with genital prolapse: a hospital-based study in Alexandria (Part I). *J Egypt Public Health Assoc.* 2001;76:313–335.
45. Cuicchi D, Lombardi R, Cariani S, Leuratti L, Lecce F, Cola B. Clinical and instrumental evaluation of pelvic floor disorders before and after bariatric surgery in obese women. *Surg Obes Relat Dis.* 2011. Aug 38 (epub ahead of print). http://dx.doi.org/10.1016/j.soard.2011.08.013.
46. Mukherjee K, Constantine G. Urinary stress incontinence in obese women: tension-free vaginal tape is the answer. *BJU Int.* 2001;88:881–883.
47. Ku JH, Oh JG, Shin JW, Kim SW, Paick JS. Outcome of mid-urethral sling procedures in Korean women with stress urinary incontinence according to body mass index. *Int J Urol.* 2006;13:379–384.
48. Rafii A, Darai E, Haab F, Samain E, Levardon M, Deval B. Body mass index and outcome of tension-free vaginal tape. *Eur Urol.* 2003;43:288–292.
49. Tchey DU, Kim WT, Kim YJ, Yun SJ, Lee SC, Kim WJ. Influence of obesity on short-term surgical outcome of the transobturator tape procedure in patients with stress urinary incontinence. *Int Neurourol J.* 2010;14:13–19.
50. Rechberger T, Futyma K, Jankiewicz K, Adamiak A, Bogusiewicz M, Skorupski P. Body mass index does not influence the outcome of anti-incontinence surgery among women whereas menopausal status and ageing do: a randomised trial. *Int Urogynecol J.* 2010;21:801–806.
51. Liu PE, Su CH, Lau HH, Chang RJ, Huang WC, Su TH. Outcome of tension-free obturator tape procedures in obese and overweight women. *Int Urogynecol J.* 2011;22:259–263.
52. Haverkorn RM, Williams BJ, Kubricht III WS, Gomelsky A. Is obesity a risk factor for failure and complications after surgery for incontinence and prolapse in women? *J Urol.* 2011;185:987–992.
53. Aigmueller T, Trutnovsky G, Tamussino K, et al. Ten-year follow-up after the tension-free vaginal tape procedure. *Am J Obstet Gynecol.* 2011;205:496.e1–5.
54. Skriapas K, Poulakis V, Dillenburg W, et al. Tension-free vaginal tape (TVT) in morbidly obese patients with severe urodynamic stress incontinence as last option treatment. *Eur Urol.* 2006;49:544–550.
55. Clark AL, Gregory T, Smith VJ, Edwards R. Epidemiologic evaluation of reoperation for surgically treated pelvic organ prolapse and urinary incontinence. *Am J Obstet Gynecol.* 2003;189:1261–1267.
56. Auwad W, Bombieri L, Adekanmi O, Waterfield M, Freeman R. The development of pelvic organ prolapse after colposuspension: a prospective, long-term follow-up study on the prevalence and predisposing factors. *Int Urogynecol J Pelvic Floor Dysfunct.* 2006;17:389–394.

57. Bradley CS, Kenton KS, Richter HE, et al. Obesity and outcomes after sacrocolpopexy. *Am J Obstet Gynecol*. 2008;199:690–698.
58. Diez-Calzadilla NA, March-Villalba JA, Ferrandis C, et al. Risk factors in the failure of surgical repair of pelvic organ prolapse. *Actas Urol Esp*. 2011;35:448–453.
59. Nam KH, Jeon MJ, Hur HW, Kim SK, Bai SW. Perioperative and long-term complications among obese women undergoing vaginal surgery. *Int J Gynaecol Obstet*. 2010;108:244–246.

41 Laparoscopic Surgery in Obese Women

Christy Burden[1,2] and Sanjay Vyas[3]

[1]Obstetrics and Gynaecology, Gloucester Royal Hospital, Gloucestershire, UK, [2]Research into Safety and Quality (RiSQ), Southmead Hospital, Bristol, UK, [3]Southmead Hospital, Bristol, UK

Introduction

Obesity

Obesity is irrefutably one of the biggest challenges for healthcare in the next millennium. As clinicians, we are increasingly required to operate on obese patients requiring surgery, for a variety of benign and malignant conditions. Obesity has reached 'epidemic portions' in the Western World, with at least 300 million clinically obese (Body Mass Index (BMI) >30 kg/m^2), and with more than one billion adults overweight (BMI >25 kg/m^2) [1].

In England, the most recent Health Survey for England (HSE) data demonstrated that, in 2009, 61.3% of adults (aged 16 or over) and 28.3% of children (aged 2–10) were overweight or obese, of these, 23.0% of adults and 14.4% of children were obese [1]. In 2007, the UK Government commissioned foresight report envisaged that if no strategy were undertaken to reduce this epidemic, 60% of men, 50% of women and 25% of children would be obese by 2050. There is already a major financial impact on the NHS – direct costs caused by obesity are predicted to be £4.2 billion per year and envisaged to more than double by 2050 if no action is taken [1].

Obesity is associated with various conditions including diabetes mellitus, hypertension, hypercholesterolaemia, heart disease, asthma and arthritis, all which contribute to increased morbidity and mortality. It also predisposes women to a higher risk of cancer, and most importantly in gynaecology, it is one of the biggest risk factors for endometrial cancer. Women who are 9–22 kg above their healthy body weight have a threefold increase in having endometrial cancer, rising to ninefold if they are 22 kg over their ideal health weight.

Women with mild obesity (BMI 30–35 kg/m^2) may create few additional challenges for perioperative management; however, it is important to acknowledge the difference between overweight patients (BMI > 25 kg/m^2), and those who are

Obesity. DOI: http://dx.doi.org/10.1016/B978-0-12-416045-3.00041-8

significantly obese (BMI > 35 kg/m^2). Various issues such as theatre resources and perioperative management will be comparable for these two groups of patients, but there are several other vital considerations especially for those who are morbidly obese (BMI > 40 or > 35 kg/m^2 with obesity associated co-morbidities).

In this chapter, we aim to review the current literature regarding evidence of the role of laparoscopic surgery in obese women in gynaecology and to discuss techniques used to overcome surgical challenges pre-, post- and intra-operatively.

Physiological Changes in the Obese Patient

Obesity leads to several cardiovascular and haemodynamic changes in the body with associated physiological abnormalities. Increased body mass increases metabolic demand, therefore, causing cardiovascular adaptation, resulting in cardiac disease [2]. Raised oxygen demand in turn leads to increased cardiac output, larger stroke volume, decreased vascular resistance and increased cardiac work, therefore, causing hypertension, cardiomegaly and arrhythmias.

For the respiratory system, oxygen consumption and carbon dioxide are also increased in obese patients. Increased weight around the ribs and intra-abdominally exerts pressure on the diaphragm which in turn reduces chest wall compliance. Both increased oxygen consumption, and reduced chest wall compliance leads to lower expiratory reserve volume and decreased functional residual capacity (FRC). In the obese patient, lying flat (supine position), as required for laparoscopic surgery, and anaesthesia can decrease FRC to levels lower than closing capacity resulting in airway closure and hypoxaemia.

Gastric function is also altered in obese patients. They have delayed gastric emptying, lower gastric pH and larger gastric volumes, all which predispose them to an increased risk for gastric aspiration during laparoscopic surgery.

A further consideration in obese patients is the pharmacodynamics. Patients will have greater body adipose content, less water content, increased blood volume and distribution and renal blood flow, thus drugs may be metabolised differently intra- and post-operatively.

Benefits of Laparoscopic Surgery

The clinical benefits of laparoscopic surgery for patients are well documented; smaller incisions, reduced post-operative pain, quicker mobilisation, and lower infection rates [3]. For healthcare providers and society, there are the advantages of shorter hospital stay, faster recovery rates and quicker return to work, all of which reduce inpatient and social costs [3]. There are also benefits for the surgeon with bette[illegible]lisation of organs and improved tissue approximation [3].

[illegible]opic surgery provides further advantages over open surgery in obese [illegible]uced wound infection and hernia rates, as well as reduced post-opera[illegible]d fever [4,5]. The most significant benefits in the obese population are [illegible] quicker mobilisation and therefore, reduced risk of susceptibility to thromboembolic events, a major cause of morbidity and mortality in obese post-operative patients [6].

Eltabbakh et al. [7] undertook a prospective study on overweight women with BMI between 28 and 60 kg/m^2 comparing 42 women who underwent laparoscopic hysterectomy, bilateral salpingo-oophorectomy and lymph node dissection for stage 1 endometrial carcinoma with 40 women who underwent laparotomy. They confirmed the laparoscopic group had shorter hospital stay (2.5 vs 5.6 days, $P < 0.01$), less pain (32.3 vs 124.1 mg pain medication, $P < 0.01$) and earlier return to normal activity.

Laparoscopic Surgery and Obesity

Minimally, invasive surgery can now be used for the diagnosis and treatment of both acute and chronic conditions in gynaecology [8]; laparoscopic hysterectomy (vaginal assisted or total), oophorectomy, ovarian cystectomy, salpingectomy, abscess drainage, excision of endometriosis, sacro-colpoplexy and myomectomy, all have a proven track record.

Obesity was once considered a contraindication to laparoscopic surgery. Difficulties on entering the peritoneal cavity (principally with the Veress needle), impaired tolerance to the Trendelenburg position, impaired manipulation of laparoscopic instruments, difficulty in the maintenance of a pnuemoperitoneum, increased accumulation of fat in the omentum and therefore, lack of clear vision of the operative field, all contribute to the challenges of laparoscopic surgery in obese women. The disadvantages of laparoscopic surgery include, increased complication rates, particularly ureteric damage, and a longer operating time [8]. These difficulties are potentially increased further in obese patients, due to the challenges discussed earlier. The use of the laparoscopic approach in haemdynamically unstable obese women with a haemoperitoneum for an emergency procedure, is controversial, but not contraindicated [9].

A small study reviewing women undergoing laparoscopic surgery for benign conditions in gynaecology confirmed an association between increasing BMI and reduced completion of planned laparoscopy, an increased number of attempts of entry and increased difficulty in surgical landmark identification [10]. Thomas et al. [11] in a study of 170 women with benign adnexal masses reported a significantly increased risk of conversion to laparotomy for obese patients compared to those with a normal weight. A large randomised controlled trial also reinforced the increased risk of conversion to laparotomy in obese compared to normal weight patients during laparoscopic endometrial cancer staging (23% vs 36%) [4].

However, in contrast to these studies, there is a growing body of evidence advocating the use of laparoscopic rather than open surgery in obese individuals in gynaecology, demonstrating, in fact excellent outcomes with low conversion rates to laparotomy [5,9,12–14]. A large retrospective case–controlled study of laparoscopic surgery in benign and malignant conditions, showed no difference in surgical outcomes, and intra- and post-operative complications [9]. A case series in Europe suggests that surgical and anaesthetic complications are reduced whe laparoscopy is used preferentially to the abdominal approach in obese women in gynaecological surgery [13].

A large majority of the data on outcomes in obese women comes from retrospective reviews in gynaecological oncology patients. The traditional approach to surgery in the management of early endometrial cancer is total hysterectomy, bilateral salpingo-oophorectomy and pelvic node dissection by open technique (laparotomy). However, the laparoscopic approach is now also used increasingly in women with endometrial cancer, with excellent outcomes [15–18]. This wealth of evidence confirms that a laparoscopic procedure in obese gynaecological oncology patients is both safe and feasible when performed by skilled laparoscopic surgeons in specialised centres [19–23].

As discussed, one main drawback of laparoscopic surgery over open surgery is a longer operating time. Once again the literature is inconclusive to whether this time is further increased in obese women. One retrospective case-note study of 533 women showed no difference in laparoscopic operating time, in obese patients [5], whereas Ghezzi et al. [24] found there was the duration of surgery were comparable irrespective of a BMI < 30 or > 30 kg/m^2.

Although the data is conflicting, likely to be due to the heterogeneity of the studies, most recent research concludes that obese patients are likely to benefit from laparoscopic surgery, assuming careful pre-operative planning is undertaken, and that technical challenges can be overcome by trained, skilled theatre staff, surgeons and anaesthetists.

Morbid Obesity

Morbid obesity is likely to pose the greatest challenge to surgeons and anaesthetists if surgery is required. A small study has reported the outcome of laparoscopy versus open surgery for endometrial cancer in women with morbid obesity. They have shown mean operating time was similar (142.5 min laparoscopic vs 153.8 min open), with no increased morbidity [25]. The hospital stay was longer in the open surgery group (4 vs 11 days) [25]. Although the numbers in the study were small, this demonstrates that laparoscopic surgery is feasible, safe and beneficial even in the morbidly obese.

Alternatives

There are various alternatives to surgery in many benign chronic conditions in gynaecology, and these should be explored thoroughly in obese patients, prior to surgery. The progesterone intra-uterine releasing system (Mirena, Shering H.C. Ltd., West Sussex, UK) is excellent for treating menorrhagia, and commonly, there are [illegible] effective non-surgical medications available. Similarly, drugs such as [illegible] hormone-releasing hormone agonists can be used effectively in endo[illegible]d fibroids. Ring pessaries are frequently very effective for prolapse and [illegible]ly low morbidity.

Pre-Operative Preparation

The Patient

The key to safe and successful laparoscopic surgery on obese patients is extensive pre-operative evaluation, counselling and theatre preparation.

As comprehensive pre-operative assessment is vital, individual units should consider the development of guidelines for the management of surgery on obese patients and the importance of team working should be emphasised. Adherence to guidelines and review of cases highlighting challenges or deficiencies found in the surgical care of obese women will inform and improve future practice.

Surgeons should judge whether or not the operation is essential and likely to be of benefit. Women should be informed of the risks and should always be involved in the decision-making. To neglect to discuss their obesity is paternalistic and should be avoided, as women are usually aware of the issue. Staff involved in their care should strive to be sensitive and responsive to their needs and treat them with dignity and respect. The ideal approach for obese patients is pre-operative weight loss, though this is often difficult to achieve [4].

Extensive patient counselling is required in morbidly obese patients, and their care and treatment should be consultant led. They must understand the challenging technical and practical difficulties that may be encountered by the staff and surgeons during the laparoscopic procedure. This may include difficulty with intravenous access and therefore, the need for central lines and the possibility of conversion to laparotomy.

All patients should have their height, weight and the BMI calculated and recorded. The day-surgery operational guide, developed by the department of health suggests that patients up to BMI 40 kg/m^2 could be suitable for day case procedures. This is likely, however, to depend on the procedure to be undertaken, associated co-morbidities and guidelines in each individual unit or hospital should be checked.

As obese patients often have several associated co-morbidities, a detailed pre-operative evaluation is imperative, including a cardiovascular and respiratory history. Patients should be assessed for hypertension, smoking status, sleep apnoea, obstructive pulmonary disease and peripheral vascular disease as these may substantially affect the operative anaesthetic risk of the patient. Ideally, a consultant anaesthetist, with expertise in bariatric anaesthesia should review patients in a pre-operative assessment clinic, especially if the patient is morbidly obese. The airway should be examined as intubation difficulties can be caused by excessive fat around the face and a short neck, and a pre-operative intubation plan formulated. Continuous positive airway pressure may be required for the obese patient with severe sleep apnoea and is ideally started pre-operatively.

Baseline blood tests should be undertaken including full blood count, serum electrolytes, glucose and renal and liver function tests. In all obese patients, a pre-operative electrocardiogram and chest X-ray are indicated with an additional echocardiogram and/or arterial blood gas if there is significant cardiopulmonary disease.

Ischaemia, arrhythmias and ventricular hypertrophy can be detected on the electrocardiogram. Cardiomegaly and pulmonary abnormalities can be revealed on the chest X-ray.

The evidence to suggest the routine use of pulmonary function tests on all healthy obese patients is controversial and, therefore, may be used depending on associated co-morbidities (smokers, asthma, lung disease, sleep apnoea).

A thorough examination of the patient's body type and panniculus is essential for determining ease of intravenous access, trochar placement and positioning during the operative procedure. Patients with increased waist versus hip circumference are likely to be a greater challenge surgically especially for entering the peritoneal cavity and port placement.

Staff and Operating Room

It is essential in a pre-operative assessment to liase with the theatre team. In cases of morbid obesity, the theatre staff should be informed if the need for specialised equipment is anticipated. It is crucial that the surgical team and theatre staff are familiar with the additional operative requirements and instruments required for laparoscopic surgery on obese patients, so that slow set-up and lack of equipment do not delay a surgical procedure. The surgical team and theatre staff must also be able to swiftly troubleshoot any equipment failures to avoid further prolonging the operating and anaesthetic time. Appropriate staff training can reduce the chance of damage to specialised equipment or more importantly injury to patients and staff.

A body weight of 130–160 kg is taken by the average operating table (Maquet Gmbh & Co., Rastatt, Germany). To support extra width in some circumstances, side extensions are available. Tables are available supporting weights of between 225 and 360 kg, which will also include lithotomy stirrups (Allen Medical Systems, Acton, MA). It is imperative to check the weight limit of the table in each individual hospital or unit. Many day-surgery facilities utilise trolleys for surgery, therefore the upper weight-bearing limit must be available, so alternative arrangements can be made.

Extra pressure reducing mattresses, blankets and sheets, padding and lifting devices may also be required for the positioning of obese patients. Operating rooms will need to be equipped with large cuff blood pressure and monitoring devices, with large compression lower extremity stockings and pneumatic boots.

Central venous access equipment should be available if peripheral access is not feasible, alongside equipment necessary for managing difficult intubations.

Anaesthesia in Obesity

Obesity is often also seen as a relative anaesthetic contraindication to laparoscopic surgery, particularly for gynaecological pelvic surgery, and extended time for administration of anaesthesia must be anticipated.

CO_2 insufflation of the peritoneal cavity exerts pressure on the diaphragm, resulting in decreased lung volumes (reduced FRC), reduced pulmonary compliance, raised resistance and ventilation–perfusion mismatch and increased peak inspiratory pressure. Consequently, patients undergoing laparoscopy are at risk of decreased oxygen saturations and lung atelectasis. Induction of anaesthesia, mechanical ventilation and loss of muscle tone due to paralysis all contribute to further reduced lung function even in normal weight patients. Moreover, the Trendelenburg position, essential for a good surgical view of the pelvis, exerts further pressure on the diaphragm and reduces cardiac preload. All these effects are heightened in obese patients and are the main reasons thought to restrict the use of laparoscopic surgery in these patients.

These physiological changes, however, are usually well tolerated in healthy women of normal weight and a recent study has found that obese patients are as haemodynamically stable during their laparoscopic procedure as the normal weight controls [26]. Sprung et al. [38] also studied respiratory mechanics in morbidly obese patients (mean BMI 46.6 kg/m^2) and normal control subjects (mean BMI 22.6 kg/m^2) under anaesthesia, in the supine, Trendelenburg and reverse Trendelenburg body positions before and after the insufflation of the abdomen with 20 mmHg of carbon dioxide. Inspiratory resistance was 68% higher, and lung compliance was 30% lower in morbidly obese patients in the supine position. The insufflation of a pnuemoperitoneum further reduced lung compliance in the obese group. Interestingly, however, pulmonary artery saturation levels were affected by the pre-operative weight difference between the patient groups but did not change significantly in either group after insufflation or repositioning of the patient into the Trendelenburg position. During the anaesthetic, the obese patients did need 15% higher minute ventilation in the supine position before insufflation. It was concluded therefore that the most strain on the respiratory system was due to pre-operative weight and supine positioning. The pneumoperitoneum adds to reduced lung compliance, but this, and the Trendelenburg position does not affect overall oxygenation.

Although still an indisputable challenge, the increased anaesthetic risks of laparoscopic surgery itself in obese patients may be overestimated. Those patients, who are able to tolerate the induction of anaesthesia and supine positioning, are likely to tolerate the Trendelenburg position and pneumoperitoneum.

Patient Positioning

Little data is available on the optimum position for the obese patient during a laparoscopic procedure. In our experience, the woman should be placed in the Lloyd–Davies position, as used in normal weight patients for laparoscopic gynaecology procedures, with her buttocks well down the operating table. A foley catheter should be placed in the bladder and a uterine manipulator in the uterus, unless there is the possibility of an intra-uterine pregnancy.

After pneumatic boots and compression stockings have been placed, the use of padded stirrups with extra padding around the pressure points (ankles and knees) should be used. A gel pad can be positioned under the patient to reduce pressure on the lower back, and stop the patient from slipping. Shoulder blocks placed at the acromio-clavicular joints can further reduce the chance of the patient slipping whilst in the Trendelenburg position. If shoulder blocks are used, then the arms should be placed straight down by the patients' side to reduce the risk of brachial plexus injury.

The panniculus in the obese patient can be manipulated and repositioned to aid body landmark identification (ischial spines, the xiphoid process and the costovertebral edge), port placement and the manipulation of laparoscopic instruments. In obesity, the umbilicus can be located 3–6 cm caudal to the aortic bifurcation; therefore, the umbilicus cannot be used accurately for the placement of ports, thus relying on these other landmarks. The panniculus can be weighted or taped away from the operating field, with gentle traction. Care should be taken however as excessive traction can cause skin trauma and tissue necrosis.

Medication

Obese patients should all have intravenous intra-operative antibiotics. Depending on the complexity of the laparoscopic surgery and associated co-morbidities, the antibiotics may be continued for at least 24 h post-operatively. Research has shown in obese patients, antibiotics are effective in reducing the incidence of wound infection from 21% to 4% [27,28]. To reduce gastric pH, H_2 receptor antibodies should also be administered 60–90 min prior to surgery [29,30].

Laparoscopic Surgical Techniques

The team required for any laparoscopic procedure must be defined and all informed of the specific surgical or anaesthetic considerations. The team should include the surgeon, anaesthetist, surgical assistant, anaesthetic assistant, scrub nurse, operating department practitioner and runner. Should the laparoscopy need to be converted to open surgery, the same team would be involved in the laparotomy.

In obese women, access to the abdominal cavity is the main challenge in laparoscopic surgery. The open (Hassan) technique is recommended by the Royal College of Obstetricians and Gynaecologists green top guidelines as the primary entry technique in morbidly obese patients [31]. However, this entry may not be ideal or necessary in all obese women, especially as not all current gynaecologists have the expertise to perform this technique. A closed entry (Veress needle), more commonly used in gynaecology, is an alternative in obese women. At the base of the skin in the umbilicus, the deep fascia and parietal peritoneum of the anterior abdominal wall are fused, therefore, it is essential to make the incision as close to the base as possible. In this area, there is less chance of the parietal peritoneum 'moving' away from the Veress needle causing extra-peritoneal insufflation and surgical emphysema. If the needle is placed vertically, the average distance from the base of the umbilicus to the peritoneum is 6 cm (+/−3 cm). Therefore, a standard length needle can be used even in extremely obese women [32]. Insertion at 45° is not recommended in obese patients, as even from within the umbilicus the needle has to traverse distances of 11–16 cm; too long for a standard Veress needle [33]. An extra long Veress needle is available if required.

Grasping the abdominal wall can be difficult in obese patients and it can also increase the distance from the skin to the fascia. A single randomised trial has studied the effect of elevating or not elevating the abdominal wall prior to entry with the Veress needle and found increased failure to enter the peritoneal cavity when the abdominal wall is elevated [34,35]. It is therefore not routinely recommended [31]. Alternatively, the skin incision could be extended to the fascia. This is then grasped and lifted, stopping the needle passing through the thick sub-cutaneous tissue. Other useful entry techniques in the obese patients include Palmers point (left upper quadrant) entry, and the use of newer technologies including visual access systems, radially expanding trochars and second generation Endotip® systems (Karl Storz, Tunlingen, Germany) or a direct vision optical Veress needle. The efficacy of these newer devices currently is still under review. Direct trochar entry is not recommended at present [31]. Ultimately, the safest entry method will depend on the equipment available at each unit and the individual surgeons' skill and preference.

During the operative procedure, trochars should be placed towards the operation site to avoid rotation or slippage, or they may be stitched in place. Longer cannulaes and instruments are available and may be used. To further facilitate ease of operating, additional ports may be necessary in the overweight patient.

Prior to lateral ports placement, it is crucial that the inferior epigastric vessels are visualised, and the secondary ports are inserted away from these vessels. Found alongside them are the deep epigastric arteries and the venae comitantes which can be seen just lateral to the umbilical ligaments (the obliterated hypogastric arteries) [31]. These can be difficult to visualise in the morbidly obese patient. The incision should therefore be made extremely lateral to the edge of the rectus sheath, ensuring avoidance of injury to the pelvic sidewall vessels.

It is essential to ensure adequate haemostasis intra-operatively. It is challenging to assess obese women post-operatively for intra-abdominal bleeding, and additionally, a post-operative haematoma increases the risk of pelvic infection. Consideration should be given to leaving a drain in situ if haemostasis is a concern. Subsequent laparotomy carries with it substantial risks including those of further anaesthesia.

Post-operative hernias are more likely in obese patients; therefore, it is imperative that any 'non-midline port over 7 mm and any midline port greater than 10 mm requires formal deep sheath closure' [31]. One trial has studied the use of 5 mm ports only for management of endometrial cancer with laparoscopic hysterectomy, salpingo-ophorectomy and pelvic lymphadenectomy. They showed it was safe and feasible and did not compromise surgical efficacy even in the obese women [36]. It could therefore be recommended to use the smallest ports feasible when operating on obese women and aiming to remove the specimen from the umbilical port if required, thus reducing the risk of a post-operative hernia.

Post-Operative

Post-operative care should involve close observation of vital signs. With the aim of early identification of wound infection and other post-operative problems, these women require daily review.

Adequate thromboprophylaxis is vital, venous thromboembolisms occur in 5–12% of obese patients who undergo surgery [37]. All women should be fitted with thromboembolic deterrent stockings, and advised on rehydration and early mobilisation. Standard prophylaxis of sub-cutaneous low-molecular-weight heparin will not be adequate, and larger doses are often necessary. This dose will depend on the individual's BMI and clinical condition.

Obese patients are also at risk of hypoxaemia due to reduced FRC and atelectasis. Supplemental oxygen, semi-recumbent positioning and chest physiotherapy may be required. It is important to ensure adequate analgesia as abdominal pain may further restrict ventilation and prevent mobilisation. Regional anaesthesia may be an excellent option for post-operatively pain control in obese patients.

If patients have significant medical co-morbidities or have had complex surgery, it may be necessary to involve intensivists and anaesthetists, in their post-operative care. For the extremely high risk patient admission to the intensive care unit for further invasive monitoring may be required.

Conclusion

It is essential doctors caring for overweight patients understand the physiological abnormalities that are associated with obesity. Technical difficulties can be overcome by technique modification, alongside skilled surgeons and anaesthetists. The minimally invasive approach of laparoscopic surgery allows shorter hospital stay, faster recovery, improved quality of life and reduced complication rate in obese women. The laparoscopic approach may also give better visualisation of the pelvis. Patients with a high BMI are most importantly likely to benefit from laparoscopic surgery due to a lower risk of surgical site infection, wound dehiscence, pelvic abscess and venous thromboembolism. With comprehensive preparation and thorough pre-operative assessment, laparoscopy can be undertaken safely and is now the favoured surgical approach in obese women in gynaecology.

References

1. *WHO*. Obesity and overweight. <www.who.int/dietphysicalactivity/publications/facts/obesity/en/print.html/>; 2010 Accessed 20.10.11.
2. Shenkman Z, Shir Y, Brodsky JB. Perioperative management of the obese patient. *Br J Anaesth*. 1993;70:349–359.
3. A consensus document concerning laparoscopic entry techniques: Middlesborough. *Gynaecol Endosc*. 1999;8:403–406.

4. Walker J, Mannel R, Piedmonte. MR, et al. Phase III trial of laparoscopy (scope) vs laparotomy (open) for surgical resection and comprehensive surgical staging of uterine cancer: a gynaecologic oncology group study (abstract). *Gynaecol Oncol.* 2006;101: S11–S12.
5. Camanni M, Bonino L, Delpiano EM, Migliaretti G, Berchialla P, Deltetto F. Laparoscopy and body mass index: feasibility and outcome in obese patients treated for gynecologic diseases. *J Minim Invasive Gynecol.* 2010;17/5:576–582.
6. Lamvu G, Zoloun D, Boggess J. Obesity: physiological changes and challenges during laparoscopy. *Am J Obstet Gynaecol.* 2004;191(2):669–674.
7. Eltabbakh GH, Shamonki MI, Moody JM, Garafano L. Hysterectomy for obese women with endometrial cancer: laparoscopy or laparotomy? *Gynaecol Oncol.* 2000;78: 329–335.
8. Magos AL, Baumann R, Turnbull AC. Managing gynaecological emergencies with laparoscopy. *BMJ.* 1989;299:371–374.
9. Hsu S, Mitwally MF, Aly A. Laparoscopic management of tubal ectopic pregnancy in obese women. *Fertil Steril.* 2004;81:198–202.
10. McIlwaine K, Cameron M, Readman E, Manwaring J, Maher P. The effect of patient body mass on surgical difficulty in gynaecological laparoscopy. *Gynaecol Surg.* 2011;8(2): 145–149.
11. Thomas D, Ikeda M, Deepkia K, Medina C, Takacs P. Laparoscopic management of benign adnexal mass in obese women. *J Minim Invasive Gynecol.* 2006;13:311–314.
12. Siddiqui A, Livingston E, Huerta S. A comparison of open and laparoscopic Roux-en-Y gastric bypass surgery for morbid and super obesity: a decision-analysis model. *Am J Surg.* 2006;e-1:192–195.
13. Raiga J, Barakat P, Diemunch P, Calmelet P, Brettes JP. Laparoscopic surgery and massive obesity. *J* Gynecol Obstet Biol Reprod *(Paris).* 2000;29:154–160.
14. Chopin N, Malaret JM, Lafay-Pillet MC, Fotso A, Foulot H, Chapron C. Total laparoscopic hysterectomy for benign uterine pathologies: obesity does not increase the risk of complications. *Hum Reprod.* 2009;24(12):3057–3062.
15. O'Hanlan KA, Lopez L, Dibble S, Garnier AC, Huang GS, Leuchtenberger M. Total laparoscopic hysterectomy: body mass index and outcomes. *Obstet Gynaecol.* 2003;102(**6**): 1384–1392.
16. Obermair A, Manolitas TP, Laung Y, Hammand IG, McCartney AJ. Total laparoscopic hysterectomy for obese women with endometrial cancer. *Int J Gynaecol Cancer.* 2005;15:319–324.
17. Matory Jr WE, O'Sullivan J, Fudem G, Dunn R. Abdominal surgery in patients with severe morbid obesity. *Plast Reconstr Surg.* 1994;94(7):976–987.
18. Santi A, Kuhn A, Gyr T, et al. Laparoscopy or laparotomy? A comparison of 240 patients with early-stage endometrial cancer. *Surg Endosc.* 2010;24(4):939–943.
19. Gemignani ML, Curtin JP, Zelmanovich J, Patel DA, Venkatraman E, Barakat RR. Laparoscopic-assisted vaginal hysterectomy for endometrial cancer: clinical outcomes and hospital charges. *Gynecol Oncol.* 1999;73(1):5–11.
20. Boike G, Luraine J, Burke J. A comparison of management of endometrial cancer with traditional laparotomy. *Gynaecol Oncol.* 1994;52:105.
21. Holub Z, Voracek J, Shomani A. A comparison of laparoscopic surgery with open procedure in endometrial cancer. *Eur J Gynaecol Oncol.* 1998;19:294–296.
22. Occeli B, Samouelien V, Narduzzi F, LeBlanc E, Querleu D. The choice of approach in the surgical management of endometrial cancer: a retrospective series of 155 cases. *Bull Cancer.* 2003;90:347–355.

23. Pellegrino A, Signorelli M, Fruscio R, et al. Feasibility and morbidity of total laparoscopic radical hysterectomy with or without pelvic limphadenectomy in obese women with stage I endometrial cancer. *Arch Gynecol Obstet.* 2009;279(5):655–660.
24. Ghezzi F, Cromi A, Bergamini V, et al. Laparoscopic assisted vaginal hysterectomy versus total laparoscopic hysterectomy for the management of endometrial cancer. A randomised clinical trial. *J Minim Invasive Gynaecol.* 2006;13:114–120.
25. Yu CKH, Cutner A, Mould T, Olaitan A. Total laparoscopic hysterectomy as a primary surgical treatment for endometrial cancer in morbidly obese women. *BJOG.* 2005;112: 115–117.
26. Aloni T, Everon S, Ezri. T, et al. Morbidly obese patients are haemodynamically stable during laparoscopic surgery: a thoracic bioimpedance study. *J Clin Monit Comp.* 2006;20:261–266.
27. Gallup DC, Gallup DG, Nolan TE, Smith RP, Messing MF, Kline KL. Use of subcutaneous closed drainage system and antibiotics in obese gynaecologic patients. *Am J Obstet Gynecol.* 1996;75:358–361.
28. Pories WJ, VanRij AM, Burlingham BT, Fulghum RS, Meelheim D. Prophylactic cefazolin in gastric bypass surgery. *Surgery.* 1981;90:426–432.
29. Vaughan RW, Bauer S, Wise L. Volume and pH of gastric juice in obese patients. *Anesthesiology* 1975;43:686–689.
30. Lam AM, Grace DM, Peny F, Vezina WC. Prophylactic intravenous cimetidine reduces the risk of acid aspiration in morbidly obese patients (abstract). *Anesthsiology* 1983;59: A242.
31. RCOG. *Preventing Entry-Related Gynaecological Laparoscopic Injuries – RCOG Green Top Guideline No 49.* London: RCOG; 2008.
32. Holtz G. Insufflation of the obese patient. In: Diamond MP, Corfman RS, DeCherney AH, eds. *Complication of Laparoscopy and Hysteroscopy.* 2nd ed. Oxford: Blackwell Science; 1997:22–25.
33. Hurd WH, Bude RO, DeLancey JO, Gauvin JM, Aisen AM. Abdominal wall characteristics with magnetic resonance imaging and computed tomography. The effect of obesity on the laparoscopic approach. *J Reprod Med.* 1991;36:473–476.
34. Ahmad G, Duffy JMN, Phillips K, Watson A. Laparoscopic entry techniques (protocol). *Cochrane Database Syst Rev.* 2008;3:CD006583. doi:10.1002/14651858.CD006583.
35. Briel JW, Plaisier PW, Meijer WS, Lange JF. Is it necessary to lift the abdominal wall when preparing a pneumoperitoneum? A randomised study. *Surg Endosc.* 2000;14: 862–864.
36. Ghezzi F, Cromin A, Bergamini V, et al. Laparoscopic management of endometrial cancer in nonobese and obese women: a consecutive series. *J Minim Invasive Gynaecol.* 2006;13:269–275.
37. Wilson AT, Reilly CS. Anaesthesia and the obese patient. *Int J Obes.* 1993;17: 427–435.
38. Sprung J, Whalley DG, Falcone T, Warner DO, Hubmayr RD, Hammel J. The impact of morbid obesity, pneumoperitoneum, and posture on respiratory system mechanics and oxygenation during laparoscopy. *Anesth Analg.* 2002;94(5):1345–1350.

Section 9

The Future Research and Health Service Planning

42 Obese Women and Quality of Life

Rhona J. McInnes[1] *and Cindy M. Gray*[2]

[1]Senior Lecturer in Midwifery, School of Nursing, Midwifery & Health, University of Stirling, Scotland, [2]Research Fellow, Institute of Health and Wellbeing, University of Glasgow, Scotland

Quality of Life

Health should be 'a state of physical, mental, social and spiritual well-being' rather than merely the absence of disease [1]. Thus, quality of life refers to the impact of a health condition, such as obesity, on a person's functional status and well-being. The term encompasses not only health status, but also environmental and economic factors, such as cultural and socioeconomic status [2]. Precise characterisation of quality of life varies according to the conceptual framework within which it is being considered, but when patient experience is key, health-related quality of life is generally felt to be appropriate. Health-related quality of life is a multidimensional construct that uses a person's assessment of their emotional, physical, social and subjective well-being to reflect how health or ill-health affects their daily functioning [3].

Obesity and Quality of Life

Obesity impacts on many aspects of a person's sense of well-being. A direct relationship has been found between quality of life and severity of obesity, with people who are most obese reporting poorest quality of life [2], and those experiencing pain showing the greatest deficit [4]. A number of studies report gender differences, with obese women generally showing greater quality of life impairments than obese men [3] in both community and clinical settings [5,6]. As gender may have an important mediating role on the impact of obesity on quality of life, this chapter takes a gendered approach and focuses on obese women's experiences of the physical, psychological and social domains of health-related quality of life.

Physical Functioning

Obesity is associated with a number of long-term and debilitating health conditions [7]. Increased body mass index (BMI) leads to greater risk of co-morbidity

Obesity. DOI: http://dx.doi.org/10.1016/B978-0-12-416045-3.00042-X

and impairment in both men and women [8,9], but women are seven times more likely than men to experience excess quality-adjusted life years lost to being overweight [10]. Self-rated health has been shown to be poorer among obese, compared to non-obese, women across the adult lifespan [11,12], particularly among women with a low educational level [13]. Motility and functional disability may be problematic for obese women, with reports of strain and pain in undertaking sporting activities, strenuous tasks at work, moderate housework and in walking outdoors or climbing stairs [14]. The impact may be greater in women than men; research showed that physical quality of life impairment was reduced among obese men with strong social support, but remained constant in women regardless of social support [15].

Obesity is associated with pelvic floor dysfunction and urinary incontinence (UI) in women, with severity of obesity showing a positive relationship with frequency of incontinent episodes and symptom distress, and a negative relationship with quality of life [16,17]. UI is also associated with impaired sexual functioning in obese women [18], who report specific problems with sexual desire, arousal, orgasm and satisfaction [19], and show greater impairment than obese men [20]. UI may be associated with faecal incontinence in women who are severely obese. Over 60% of respondents to a survey of women seeking information about weight loss surgery reported symptoms of faecal incontinence (compared with a prevalence of 4–19% in the general population), causing embarrassment and poorer lifestyle functioning [21].

Psychological Functioning

While obesity-related impairment in physical functioning is consistently reported by women and men [2,3], the relationship between obesity and psychological well-being appears more complicated. Early research found little or no association between negative psychological outcomes and obesity [22], but later studies suggest that a relationship does exist and may be stronger in women than in men. For example, US population survey data has demonstrated a positive relationship between depression and obesity in women but not in men [23]. Indeed, some studies have found that obese men demonstrate lower levels of depression than their normal weight counterparts [24,25].

Obese women report lower levels of self-esteem than obese men, although this gender disparity reduces as BMI increases [26]. They may also experience problems with increased anxiety compared with women whose weight is in the normal or overweight range. There is mixed evidence on whether gender plays a mediating role between obesity and anxiety: a study of patients referred to specialist obesity treatment clinics in the United Kingdom reported that women were significantly more likely to be suffering from anxiety than men [27], but an epidemiological study in the United States found no evidence of differences in lifetime prevalence of anxiety disorders between obese women and obese men [28].

Socio-demographic factors, such as age, socioeconomic status, education level and ethnicity, may also impact on the psychological well-being of obese women [29]. Caucasian women seeking bariatric surgery reported lower self-esteem and higher levels of depression than African–American women, even though BMIs were higher among the African–Americans [30]. Other research suggests that depression may be more prevalent among overweight and obese women who are more highly educated [31]. Finally, problems with physical health may impact on psychological well-being: for example, UI is associated with increased depression in obese women [17].

Social Functioning

Western cultures have a long history of stigmatisation of excess body weight [32], which is often perceived to be associated with some character flaw (e.g. laziness, stupidity, sloppiness, lacking self-discipline) or moral failing on the part of the individual [33]. Discriminatory behaviour has been shown to develop in early childhood [34]; it has persisted as the societal prevalence of obesity has increased, and may even be getting worse [35]. Recent evidence suggests that weight-related stigmatisation may be more socially acceptable than other forms of discrimination [36].

Weight-related stigmatisation has a negative effect on psychological health [37,38]. It is associated with feelings of shame, worthlessness, disengagement and social isolation [39], and the frequency of stigmatising experiences has been shown to be positively associated with depression and negatively associated with self-esteem [40]. Discrimination may impact more severely on Caucasian than on African–American women [41], and more severely on women than on men [42,43]. It has also been argued that the stress associated with weight-related stigmatisation may lead to increased mortality among women [44].

Gender inequality is particularly evident in personal relationships, which are often the source of the worst stigmatising experiences [45]. Puhl and Brownell [46] found that 47% of obese women reported weight-related discrimination from their spouses, and an inverse relationship has been demonstrated between BMI in women and relationship satisfaction [47]. In a study of heterosexual couples in New Zealand, overweight and obese women reported having partners who were less desirable and lower quality relationships that were more likely to end. In contrast, BMI in men was not associated with relationship functioning; instead, those with heavier female partners felt them to be a poorer match to their attractiveness ideal than men whose partners had a lower BMI [48]. Poor relationship functioning is often associated with low self-esteem, sexual problems and impaired psychological functioning in obese women [49,50].

Weight-related stigmatisation also occurs in many different social situations [3]. Obese women have reported being singled out at work. A review of job-based weight discrimination found that women tended to be judged more harshly than men for being overweight and that this had an influence on employment-related decisions [51]. Women have also reported being fired because of their weight [52] and experiencing discrimination when applying for jobs [46], and being overweight is associated with lower income in women but not in men [51].

In health care settings, women are more likely to experience weight-related discrimination than men [53], with doctors being reported as being the most frequent source of stigma [46]. Although health professionals believe they can put aside their negative attitudes and beliefs and deliver appropriate screening and care [54,55], obese patients' experiences of health service interactions indicate that health care providers are often unaware of the negative impact of their weight-related beliefs and attitudes on their interpersonal skills. Health professionals suggest that obese patients are more likely to avoid appointments, give incorrect accounts of themselves and are less likely to follow advice [55,56]. Such judgements are reflected in the accounts of obese patients who describe being misunderstood, ignored and not trusted by health professionals [57]. The tendency of clinicians to hold obese patients more responsible for their condition or any presenting complaints that could potentially be weight related [55] may be viewed by patients as 'blaming', especially if the link between their health problem and their weight is not apparent. So although care practices might not be affected by a patient's weight, the quality of the clinician–patient relationship may suffer.

Some obese women's negative accounts of accessing health services reflect unease about trying to fit in to a relatively small space [58]. The furniture and medical equipment used routinely for clinical assessment may be inappropriately sized and therefore of limited utility. In one study, women described apprehension around being unable to find a chair without arms in waiting rooms [58]. Physical size may lead to embarrassment about undressing for examinations [59], which can then be compounded by concerns that clinical gowns may not fit [57–59].Of concern is evidence that obese women may not engage with some important health screening services. Compared to women of normal weight, women who are obese or overweight are less likely to access cervical and breast screening services; however, uptake for other screening services (e.g., colorectal cancer) do not seem to be affected by weight status [60,61]. Obese women's discomfort with health care interactions also relates to assessment of their weight and discussions around weight or weight loss [62]. Being weighed in itself can be highly embarrassing [59], but can be made worse by insensitivity about the way in which weight-related discussions are conducted; for example, in front of relatives [57]. Issues around weight are not always raised in a helpful manner [63] and can dominate a clinical consultation, with patients feeling that their health complaints are being dismissed in favour of advice on weight loss, when this is not their reason for attending [58,59].

Body Image

Thinness is closely related to perceived physical attractiveness in women especially in Westernised cultures leading to societal and media pressure to be thin.

Women are likely to internalise these cultural norms and experience some degree of body image dissatisfaction, which research has confirmed is more prevalent in women than men [27,64], persists across the lifespan [65,66] and increases with BMI [67]. Poor body image doubtless contributes to difficulties surrounding health care encounters, and it may also be a key factor in impairment of obese women's overall quality of life [68,69]. Body image dissatisfaction is linked to poorer psychosocial functioning [70], lower self-esteem and higher dysphoria [71] and has been shown to mediate the relationship between obesity and depression in some women [72].

There is a clear relationship between body image dissatisfaction and binge eating disorder (BED) [73,74]. BED is associated with severe obesity, a history of marked weight fluctuations and is more common among women than men [75]. An inverse relationship exists between severity of BED and psychological well-being; obese women with BED show higher levels of depression, lower self-esteem and greater impairment in social functioning than those without BED [73,74].

General well-being can also be improved through physical activity. Obese women who exercise regularly report greater life satisfaction, lower levels of depression and better physical health than obese women who are inactive [76]; and becoming active can significantly improve body image satisfaction, depression, tension and fatigue [77,78]. However, having a negative body image can prevent obese women from taking part in exercise [79,80]. Compared to overweight men, overweight women are much more reluctant to exercise and more likely to cite embarrassment, feeling intimidated by exercise and not liking exercising around fit people or the opposite sex as barriers to joining a health club [81]. Other barriers for obese women include physical difficulty in exercising [39], lack of motivation [82] and lack of enjoyment [83].

Obesity and Quality of the Childbearing Experience

Obese women who are pregnant do not fit the 'normal pregnant mould' [84]; they are more likely to experience complications, medical intervention and poorer pregnancy outcomes [85]. Obesity is often associated with poorer mental health and feelings of isolation during pregnancy [86,87]. Unwanted negative comments from third parties who assume that an obese woman is simply getting larger rather than being pregnant may mean that she misses out on some positive aspects of her pregnancy [57,88]. However, for some obese women, pregnancy is a time of liberation from weight-related stigma and is associated with lower levels of body image dissatisfaction and self-consciousness and increased self-confidence [57,86]. They may be comfortable with the idea of gaining weight during pregnancy, and while they recognise that being physically active during pregnancy can help general well-being, they view being active during pregnancy as a personal choice [89].

Childbearing is a period of intense contact with health care providers. Examinations can be very intimate; exposure of the body is common during pregnancy and childbirth and to some degree necessary. Perceptions of blaming by health professionals and poor body image can combine to intensify feelings of humiliation and embarrassment about body exposure during routine clinical examinations [88], unless health professionals are trained to support and manage obese patients on an individual level with kindness and respect [57]. Feelings of guilt and concern for their infant may push pregnant women who are obese to accept otherwise unwelcome treatment in order to ensure 'adequate care' for their unborn child [57]. Lack of sensitivity from health professionals can therefore easily evoke negative emotions, such as anger, sadness, lack of trust and guilt, and lead obese women to feel a lack of recognition or respect and less worthy of her pregnancy than women of normal weight [57,86]. Unease surrounding interactions with health professionals during pregnancy can be compounded by inadequate equipment and facilities. Ultrasound scanning equipment is recognised as a particular problem by women and clinicians [57,86,88,90,91]. Body weight, which can become the focus of maternity care, can be a particular source of embarrassment for some women, especially when they are being weighed and their weight is spoken out loud or recorded in hand-held records [86].

Current health policy emphasises the importance of women being involved in their maternity care, through shared decision-making, active birthing, intervention reduction and maternal choice. However, obese pregnant women experience less holistic care, reduced choice (e.g. access to water birth, home birth, types of pain relief) and less involvement in decision-making [57,88,91]. Obesity may prevent women adopting active birthing positions, and lack of support from health professionals in trying to achieve a normal birth can lead to further feelings of disappointment and a sense of not being understood [57]. If obese women are not involved in discussions about managing both their pregnancy and weight, then they can feel invisible, guilty and not trusted. They may sense that decisions about them are driven by fear and some indicated that attending high-risk doctors' clinics meant they missed out on midwifery-led care. Fewer opportunities to see a midwife resulted in reduced options for normal birth [88,92].

While it must be recognised that there are increased risks associated with being obese during pregnancy, it is important that these are raised sensitively and at a time that is appropriate for each individual patient. If weight is ignored early on, some women may experience distress when their plans for a normal birth are overruled and their high-risk status disclosed. However, early designation of high-risk status means that the focus of maternity care becomes foetal (rather than maternal) well-being, with additional, and often more invasive, screening [91]. Being classified as high risk can also be difficult for the woman to deal with emotionally; most obese women just want to be treated like everyone else during their pregnancy [92].

Post-partum being obese is associated with a lower prevalence of breastfeeding. There is evidence that women who are obese are less likely to initiate breastfeeding, experience delayed lactogenesis and more breastfeeding challenges, and tend to breastfeed exclusively for shorter durations than non-obese women [93]. Lack of success with breastfeeding can affect a woman's image of herself as a mother [94], and can perpetuate a cycle of poor nutrition.

Conclusion

This chapter has described quality of life impairments that are commonly reported by women who are obese, and who may experience greater reductions in physical, psychological and social functioning than their male counterparts. This gender inequality may in part result from increased cultural pressure on women to be thin, leading to them experiencing more stigmatisation and greater self-deprecation. Obese women may find health care encounters particularly difficult, as health professionals' beliefs and attitudes about overweight and obesity may undermine the quality of the clinician–patient relationship. Embarrassment about physical examinations and discussion of weight status can be extremely distressing, particularly in pregnancy, when both may be unavoidable. Care should be taken to avoid unnecessary medicalisation of excess body weight during pregnancy and to ensure that obese women receive the levels of support they need to make appropriate choices to enhance the quality of their childbearing experience.

References

1. World Health Organization. *Constitution of the World Health Organization*. Geneva: World Health Organization;1946.
2. Fontaine KR, Barofsky I. Obesity and health-related quality of life. *Obes Rev*. 2001;2:173–182.
3. Kolotkin RL, Meter K, Williams GR. Quality of life and obesity. *Obes Rev*. 2001;2:219–229.
4. Barofsky I, Fontaine KR, Cheskin LJ. Pain in the obese: impact on health-related quality-of-life. *Ann Behav Med*. 1997;19:408–410.
5. Kolotkin RL, Crosby RD. Psychometric evaluation of the impact of weight on quality of life-lite questionnaire (IWQOL-Lite) in a community sample. *Qual Life Res*. 2002;11:157–171.
6. Kolotkin RL, Crosby RD, Pendleton R, Strong M, Gress RE, Adams T. Health-related quality of life in patients seeking gastric bypass surgery vs non-treatment-seeking controls. *Obes Surg*. 2003;13:371–377.
7. Government Office for Science. *Foresight Tackling Obesities: Future Choices – Project Report*. 2nd ed. London: Department of Innovation Universities and Skills; 2007.
8. Seo DC, Torabi MR. Racial/ethnic differences in body mass index, morbidity and attitudes toward obesity among U.S. adults. *J Natl Med Assoc*. 2006;98:1300–1308.
9. Anandacoomarasamy A, Caterson I, Sambrook P, Fransen M, March L. The impact of obesity on the musculoskeletal system. *Int J Obes*. 2008;32:211–222.
10. Muennig P, Lubetkin E, Jia H, Franks P. Gender and the burden of disease attributable to obesity. *Am J Public Health*. 2006;96:1662–1668.
11. Rohrer JE, Young R. Self-esteem, stress and self-rated health in family planning clinic patients. *BMC Fam Pract*. 2004;5:11.
12. Jones GL, Sutton A. Quality of life in obese postmenopausal women. *Menopause Int*. 2008;14:26–32.

13. Garcia-Mendizabal MJ, Carrasco JM, Perez-Gomez B, et al. Role of educational level in the relationship between body mass index (BMI) and health-related quality of life (HRQL) among rural Spanish women. *BMC Public Health*. 2009;9:120.
14. Larsson UE, Mattsson E. Perceived disability and observed functional limitations in obese women. *Int J Obes Relat Metab Disord*. 2001;25:1705–1712.
15. Wiczinski E, Doring A, John J, von Lengerke T. Obesity and health-related quality of life: does social support moderate existing associations? *Br J Health Psychol*. 2009;14:717–734.
16. Richter HE, Kenton K, Huang L, et al. The impact of obesity on urinary incontinence symptoms, severity, urodynamic characteristics and quality of life. *J Urol*. 2010;183:622–628.
17. Sung VW, West DS, Hernandez AL, Wheeler TL, Myers DL, Subak LL. Association between urinary incontinence and depressive symptoms in overweight and obese women. *Am J Obstet Gynecol*. 2009;200:557e1–557e5.
18. Melin I, Falconer C, Rossner S, Altman D. Sexual function in obese women: impact of lower urinary tract dysfunction. *Int J Obes*. 2008;32:1312–1318.
19. Assimakopoulos K, Panayiotopoulos S, Iconomou G, et al. Assessing sexual function in obese women preparing for bariatric surgery. *Obes Surg*. 2006;16:1087–1091.
20. Ostbye T, Kolotkin RL, He H, et al. Sexual functioning in obese adults enrolling in a weight loss study. *J Sex Marital Ther*. 2011;37:224–235.
21. Wasserberg N, Haney M, Petrone P, et al. Fecal incontinence among morbid obese women seeking for weight loss surgery: an underappreciated association with adverse impact on quality of life. *Int J Colorectal Dis*. 2008;23:493–497.
22. Friedman MA, Brownell KD. Psychological correlates of obesity: moving to the next research generation. *Psychol Bull*. 1995;117:3–20.
23. Onyike CU, Crum RM, Lee HB, Lyketsos CG, Eaton WW. Is obesity associated with major depression? Results from the Third National Health and Nutrition Examination Survey. *Am J Epidemiol*. 2003;158:1139–1147.
24. Palinkas LA, Wingard DL, Barrett Connor E. Depressive symptoms in overweight and obese older adults: a test of the "jolly fat" hypothesis. *J Psychosom Res*. 1996;40:59–66.
25. Carpenter KM, Hasin DS, Allison DB, Faith MS. Relationships between obesity and DSM-IV major depressive disorder, suicide ideation, and suicide attempts: results from a general population study. *Am J Public Health*. 2000;90:251–257.
26. Kolotkin RL, Crosby RD, Kosloski KD, Williams GR. Development of a brief measure to assess quality of life in obesity. *Obes Res*. 2001;9:102–111.
27. Tuthill A, Slawik H, O'Rahilly S, Finer N. Psychiatric co-morbidities in patients attending specialist obesity services in the UK. *Int J Med*. 2006;99:317–325.
28. Simon GE, von Korff M, Saunders K, et al. Association between obesity and psychiatric disorders in the US adult population. *Arch Gen Psychiatry*. 2006;63:824–830.
29. van der Merwe M-T. Psychological correlates of obesity in women. *Int J Obes*. 2007;31:S14–S18.
30. Mazzeo SE, Saunders R, Mitchell KS. Binge eating among African American and Caucasian bariatric surgery candidates. *Eat Behav*. 2005;6:189–196.
31. Siegel JM, Yancey AK, McCarthy WJ. Overweight and depressive symptoms among African–American women. *Prev Med*. 2000;31:232–240.
32. Stunkard A, Mendelson M. Obesity and body image. 1. Characteristics of disturbances in body image of some obese persons. *Am J Psychiatry*. 1967;123:1296–1300.

33. Crandall CS, Schiffhauer KL. Anti-fat prejudice: beliefs, values, and American culture. *Obes Res*. 1998;6:458–460.
34. Cramer P, Steinwert T. Thin is good, fat is bad: how early does it begin? *J Appl Dev Psychol*. 1998;19:429–451.
35. Latner JD, Stunkard AJ. Getting worse: the stigmatization of obese children. *Obes Res*. 2003;11:452–456.
36. Latner JD, O'Brien KS, Durso LE, Brinkman LA, MacDonald T. Weighing obesity stigma: the relative strength of different forms of bias. *Int J Obes*. 2008;32:1145–1152.
37. Myers A, Rosen JC. Obesity stigmatization and coping: relation to mental health symptoms, body image, and self-esteem. *Int J Obes*. 1999;23:221–230.
38. Puhl RM, Moss-Racusin CA, Schwartz MB. Internalization of weight bias: implications for binge eating and emotional well-being. *Obesity*. 2007;15:19–23.
39. Conradt M, Dierk JM, Schlumberger P, Rauh E, Hebebrand J, Rief W. Who copes well? Obesity-related coping and its associations with shame, guilt, and weight loss. *J Clin Psychol*. 2008;64:1129–1144.
40. Friedman KE, Reichmann SK, Costanzo PR, Zelli A, Ashmore JA, Musante GJ. Weight stigmatization and ideological beliefs: relation to psychological functioning in obese adults. *Obes Res*. 2005;13:907–916.
41. Quinn DM, Crocker J. Vulnerability to the affective consequences of the stigma of overweight. In: Swim JK, Stangor C, eds. *Prejudice: The Target Perspective*. San Diego, CA: Academic Press; 1998:125–143.
42. Cossrow NHF, Jeffery RW, McGuire MT. Understanding weight stigmatization: a focus group study. *J Nutr Educ*. 2001;33:208–214.
43. Ferguson C, Kornblet S, Muldoon A. Not all are created equal: differences in obesity attitudes between men and women. *Womens Health Issues*. 2009;19:289–291.
44. Muennig P, Jia HM, Lee RF, Lubetkin E. I think therefore I am: perceived ideal weight as a determinant of health. *Am J Public Health*. 2008;98:501–506.
45. Puhl RM, Moss-Racusin CA, Schwartz MB, Brownell KD. Weight stigmatization and bias reduction: perspectives of overweight and obese adults. *Health Educ Res*. 2008;23:347–358.
46. Puhl RM, Brownell KD. Confronting and coping with weight stigma: an investigation of overweight and obese adults. *Obesity*. 2006;14:1802–1815.
47. Sheets V, Ajmere K. Are romantic partners a source of college students' weight concern? *Eat Behav*. 2005;6:1–9.
48. Boyes AD, Latner JD. Weight stigma in existing romantic relationships. *J Sex Marital Ther*. 2009;35:282–293.
49. Kinzl JF, Trefalt E, Fiala M, Hotter A, Biebl W, Aigner F. Partnership, sexuality, and sexual disorders in morbidly obese women: consequences of weight loss after gastric banding. *Obes Surg*. 2001;11:455–458.
50. Hill AJ, Williams J. Psychological health in a non-clinical sample of obese women. *Int J Obes*. 1998;22:578–583.
51. Roehling MV. Weight-based discrimination in employment: psychological and legal aspects. *Pers Psychol*. 1999;52:969–1016.
52. Rothblum ED. Does overweight hold you back? *WeightWatchers*. 1996;28:46–48.
53. Hansson LM, Naslund E, Rasmussen F. Perceived discrimination among men and women with normal weight and obesity. A population-based study from Sweden. *Scand J Public Health*. 2010;38:587–596.
54. Budd GM, Mariotti M, Graff D, Falkenstein K. Health care professionals' attitudes about obesity: an integrative review. *Appl Nurs Res*. 2011;24:127–137.

55. Persky S, Eccleston C. Medical student bias and care recommendations for an obese versus non-obese virtual patient. *Int J Obes*. 2011;35:728–735.
56. Hansson LM, Rasmussen F, Ahlstrom GI. General practitioners' and district nurses' conceptions of the encounter with obese patients in primary health care. *BMC Fam Pract*. 2011;12:7.
57. Nyman VM, Prebensen AK, Flensner GEM. Obese women's experiences of encounters with midwives and physicians during pregnancy and childbirth. *Midwifery*. 2010;26:424–429.
58. Merrill E, Grassley J. Women's stories of their experiences as overweight patients. *J Adv Nurs*. 2008;64:139–146.
59. Drury CA, Louis M. Exploring the association between body weight, stigma of obesity, and health care avoidance. *J Am Acad Nurse Pract*. 2002;14:554–561.
60. Cohen SS, Palmieri RT, Nyante SJ, et al. Obesity and screening for breast, cervical, and colorectal cancer in women: a review. *Cancer*. 2008;112:1892–1904.
61. Littman AJ, Koepsell TD, Forsberg CW, Boyko EJ, Yancy WS. Preventive care in relation to obesity: an analysis of a large, national survey. *Am J Prev Med*. 2011;41:465–472.
62. Gray CM, Hunt K, Lorimer K, Anderson AS, Benzeval M, Wyke S. Words matter: a qualitative investigation of which weight status terms are acceptable and motivate weight loss when used by health professionals. *BMC Public Health*. 2011;11:513.
63. Aston M, Price S, Kirk SFL, Penney T. More than meets the eye. Feminist poststructuralism as a lens towards understanding obesity. *J Adv Nurs*. 2012;68:1187–1194.
64. Grilo CM, Masheb RM, Brody M, Burke-Martindale CH, Rothschild BS. Binge eating and self-esteem predict body image dissatisfaction among obese men and women seeking bariatric surgery. *Int J Eat Disord*. 2005;37:347–351.
65. Bentovim DI, Walker MK. The influence of age and weight on women's body attitudes as measured by the body bttitudes questionnaire (BAQ). *J Psychosom Res*. 1994;38:477–481.
66. Ferraro FR, Muehlenkamp JJ, Paintner A, Wasson K, Hager T, Hoverson F. Aging, body image and body shape. *J Gen Psychol*. 2008;135:379–392.
67. Hrabosky JI, Grilo CM. Body image and eating disordered behavior in a community sample of Black and Hispanic women. *Eat Behav*. 2007;8:104–106.
68. Cox TL, Ard JD, Beasley TM, Fernandez JR, Howard VJ, Affuso O. Body image as a mediator of the relationship between body mass index and weight-related quality of life in black women. *J Womens Health*. 2011;20:1573–1578.
69. Minniti A, Bissoli L, di Franesco V, et al. The relationship between body image and quality of life in treatment-seeking overweight women. *Eat Weight Disord*. 2004;9:206–210.
70. Mond JM, Rodgers B, Hay PJ, et al. Obesity and impairment in psychosocial functioning in women: the mediating role of eating disorder features. *Obesity*. 2007;15:2769–2779.
71. Foster GD, Wadden TA, Vogt RA. Body image in obese women before, during, and after weight loss treatment. *Health Psychol*. 1997;16:226–229.
72. Gavin AR, Simon GE, Ludman EJ. The association between obesity, depression, and educational attainment in women: the mediating role of body image dissatisfaction. *J Psychosom Res*. 2010;69:573–581.
73. Clark MM, Forsyth LH, Lloyd-Richardson EE, King TK. Eating self-efficacy and binge eating disorder in obese women. *J Appl Biobehav Res*. 2000;5:154–161.
74. Fassino S, Leombruni P, Piero A, Abbate-Daga G, Rovera GG. Mood, eating attitudes and anger in obese women with and without binge eating disorder. *J Psychosom Res*. 2003;54:556–559.
75. Spitzer RL, Devlin M, Walsh BT, et al. Binge eating disorder: a multisite field trial of the diagnostic criteria. *Int J Eat Disord*. 1992;11:191–203.

76. Menzyk K, Cajdler A, Pokorski M. Influence of physical activity on psychosomatic health in obese women. *J Physiol Pharmacol*. 2008;59:441–448.
77. Annesi JJ, Unruh JL. Correlates of mood changes in obese women initiating a moderate exercise and nutrition information program. *Psychol Rep*. 2006;99:225–229.
78. Annesi JJ, Unruh JL. Relations of exercise, self-appraisal, mood changes and weight loss in obese women: testing propositions based on Baker and Brownell's (2000) model. *Am J Med Sci*. 2008;335:198–204.
79. Ball K, Crawford D, Owen N. Too fat to exercise? Obesity as a barrier to physical activity. *Aust N Z J Public Health*. 2000;24:331–333.
80. Thomas AM, Moseley G, Stallings R, Nichols-English G, Wagner PJ. Perceptions of obesity: Black and White differences. *J Cult Divers*. 2008;15:174–180.
81. Miller WC, Miller TA. Attitudes of overweight and normal weight adults regarding exercise at a health club. *J Nutr Educ Behav*. 2010;42:2–9.
82. Genkinger JM, Jehn ML, Sapun M, Mabry I, Young DR. Does weight status influence perceptions of physical activity barriers among African–American women? *Ethn Dis*. 2006;16:78–84.
83. Ekkekakis P, Lind E, Vazou S. Affective responses to increasing levels of exercise intensity in normal-weight, overweight, and obese middle-aged women. *Obesity*. 2010;18:79–85.
84. Keenan J, Stapleton H. Bonny babies? Motherhood and nurturing in the age of obesity. *Health Risk Soc*. 2010;12:369–383.
85. Jungheim ES, Moley KH. Current knowledge of obesity's effects in the pre- and periconceptional periods and avenues for future research. *Am J Obstet Gynecol*. 2010;203:525–530.
86. Richens Y, Lavender T, eds. *Care for Pregnant Women Who Are Obese*. London: Quay Books; 2010.
87. Lacoursiere DY, Baksh L, Bloebaum L, Varner MW. Maternal body mass index and self-reported postpartum depressive symptoms. *Matern Child Health J*. 2006;10:385–390.
88. Furber CM, McGowan L. A qualitative study of the experiences of women who are obese and pregnant in the UK. *Midwifery*. 2011;27:437–444.
89. Weir Z, Bush J, Robson SC, McParlin C, Rankin J, Bell R. Physical activity in pregnancy: a qualitative study of the beliefs of overweight and obese pregnant women. *BMC Pregnancy Childbirth*. 2010;10:18.
90. Furness PJ, McSeveny K, Arden MA, Garland C, Dearden AM, Soltani H. Maternal obesity support services: a qualitative study of the perspectives of women and midwives. *BMC Pregnancy Childbirth*. 2011;11.
91. Heslehurst N, Lang R, Rankin J, Wilkinson JR, Summerbell CD. Obesity in pregnancy: a study of the impact of maternal obesity on NHS maternity services. *BJOG*. 2007;114:334–342.
92. Mills A, Schmied VA, Dahlen HG. 'Get alongside us', women's experiences of being overwight and pregnant in Sydney, Australia. *Matern Child Nutr*. 2011;10:1111.
93. Amir LH, Donath S. A systematic review of maternal obesity and breastfeeding intention, initiation and duration. *BMC Pregnancy Childbirth*. 2007;7:9.
94. Murphy E. 'Breast is best': infant feeding decisions and maternal deviance. *Sociol Health Illn*. 1999;21:187–208.

43 Understanding Eating Behaviour and Lifestyle Issues in Women – Implications for Obesity Development and Prevention

Annie S. Anderson and Angela M. Craigie

Centre for Public Health Nutrition Research, Population Health Sciences, Medical Research Institute, Ninewells Hospital and Medical School, The University of Dundee, Dundee, UK

Introduction

The lifelong nutritional status of a woman, from her own conception and throughout every life stage, will impact on her ability to conceive, achieve successful pregnancy outcomes and nurture healthy children as well as affecting her own health and well-being. Dietary intake is the main environmental influence on nutritional status, but lifestyle behaviours including physical activity (particularly with respect to body composition), alcohol (as an additional energy contributor and as displacement for nutrient-dense foods) and smoking (impacting on metabolic handling) will also have an impact on nutrition and wellbeing.

The global obesity epidemic has shifted the focus of concern about malnutrition in the form of undernutrition to the bigger problem of overnutrition and indeed the co-existence of both nutritional challenges across the developed and developing world. The biological basis of obesity has been detailed in Chapter 3, and the implications of obesity in women and reproductive health have been well described in Chapter 8. The challenge of finding solutions to the obesity crisis necessitates a greater understanding of the development of a woman's dietary and health behaviours across the lifespan, examining both population (socio-cultural) influences as well as individual (psycho-social) factors.

Current Diet and Lifestyle Issues

All food and (most) drinks provide kilocalories and hence potentially contribute to the development of weight gain and obesity. However, in a systematic review of

Obesity. DOI: http://dx.doi.org/10.1016/B978-0-12-416045-3.00043-1

determinants of weight gain, overweight and obesity [1] undertaken by the World Cancer Research Fund, the main food and drink habits that promoted weight gain were identified as high intakes of energy-dense foods (e.g. 225–275 kcals/100 g), sugary drinks (including colas, squashes and fruit juices) and regular consumption of 'fast foods' (readily available convenience foods that tend to be high in energy). In addition, longitudinal data from the US Nurses Health studies [2] have demonstrated that consumption of potato crisps, chips, processed meat, red meat, butter, sweets and desserts, refined grains, sugar-sweetened beverages and fruits juices were significantly associated with weight gain over each 4-year study period.

The UK Scientific Advisory Committee on Nutrition (SACN) [3] highlight that, in the context of reproduction, the impact of energy-dense, nutrient-poor diets on the health of women and children is of particular concern. It is notable that energy-dense diets which are associated with the development of obesity contain high intakes of refined carbohydrate and saturated fat as well as low intakes of foods rich in micronutrients such as vegetables, fruit and oil-rich fish. Thus, the metabolic impact of obesity will also be influenced by overall nutritional status, particularly during periods of growth when nutrient demands are highest.

In the UK National Diet Nutrition Survey, women aged 19–24 years were identified as being at particular risk of poor dietary variety, low nutrient intake and biochemical status [4]. For example, 98% of young adults failed to consume five portions of fruits and vegetables per day, while cans of soft drinks averaged 8–9 per week. That age group were the least likely of all adults to have salt intakes within the desirable level (83% of young women had intakes greater than 6 g/day). In case of both men and women, more than 20% of young adults failed to have intakes of potassium, iron and magnesium above the Lower Reference Nutrient Intake. In addition, 28% of young women had low biochemical status for vitamin D. The latter is of particular interest given that obese women are more likely to have lower vitamin D status [5]. Many of these findings on dietary intake and nutrient status were more marked in low-income groups (which also have a higher incidence of obesity) as demonstrated by the Low Income Diet and Nutrition Survey [6]. Little is known about the current nutritional status in women from ethnic minority groups.

Across the social spectrum, a recent prospective study [7] of 12,445 non-pregnant women aged 20–34 years reported that only a small proportion of the sample was following current nutrition and lifestyle recommendations for pre-pregnancy. For example, among the 238 women who became pregnant within 3 months of the interview, only 2.9% were taking 400 μg or more of folic acid supplements a day and drinking four or fewer units of alcohol per week. Of those 238 women, a quarter (26%) smoked, 47% reported not eating 5 portions of fruit and vegetables per day and 43% reported not taking any strenuous exercise in the past 3 months.

In 2011, The Office for National Surveys in England [8] reported that 20% of women smoke which rises to 30% among lowest income households. A quarter (26%) of mothers in the United Kingdom smoked at some point in the 12 months immediately before or during their pregnancy. Of those mothers, just over half (54%) gave up at some point before the birth [9].

Alcohol intake in women has increased in recent decades and is of particular concern with respect to pregnancy outcome and as a risk factor for breast cancer. The average weekly alcohol consumption is estimated at 8 units for women. With 55% of women taking at least one drink per week and 13% reporting drinking over 6 units on at least 1 day in the previous week [10]. Of women who drank before pregnancy, 34% gave up while they were pregnant and 61% said they drank less during their pregnancy. There is notable concern about current UK guidance on upper levels for alcohol intake (that women should not regularly drink more than 2–3 units a day which may equate to 14–21 units per week) given that risk for breast cancer starts to increase with intakes greater than 10 g/day (approximately 8 units per week).

Socio-Cultural Influences on Eating and Lifestyle Behaviours

Throughout a woman's life, one of the greatest influences on eating habits, nutrient intake, development of obesity and overall morbidity is socio-economic position [11]. Living in conditions of relative poverty can exert a lifelong (and often intergenerational) effect on economic opportunities and life circumstances. Individuals from relative poverty have lower educational attainment and are more likely to be unemployed. In addition, poorer women have worse levels of physical and mental health including coping capacity to deal with life circumstances.

However, while obesity and related co-morbidities are more prevalent in women from more disadvantaged backgrounds, these condition are widespread across all social groups. The Foresight Obesity publication [12] highlights the overwhelming influence of society on eating behaviours noting that 'People in the UK today don't have less willpower and are not more gluttonous than previous generations. Nor is their biology significantly different to that of their forefathers. Society, however, has radically altered over the past five decades, with major changes in work patterns, transport, food production and food sales. These changes have exposed an underlying biological tendency, possessed by many people, to both put on weight and retain it'.

Early cultural influences on eating behaviours will include parental and wider family members, but habits evolve rapidly as children are exposed to the school environment (including peer influences) which in turn will impact on food knowledge, skills, exposure to widening cuisines and rejection of traditional eating patterns.

Our food cultures, both at micro-level (family) and macro-level (social group or region) are made up of a range of cultural components which include taste preference, meal timing, eating situations, gender associations with food and values (lay and scientific) attached to specific foods [13]. Many of these factors will be influenced by family practices.

In a recent meta-analysis of family-eating habits [14], children and adolescents who shared family meals three or more times per week were more likely to be in the normal weight range, have healthier dietary and eating patterns and were less likely to engage in disordered eating. In turn, family-eating practices are associated with home meal preparation. However, family food preparation appears to be diminishing with an increase in the growth of take-away meals (e.g. the fast foods associated with weight gain). Both income and social class are highlighted as major influences on home cooking and food choices. For example, in the 'Growing up in Scotland' [15] report, 41% of children from low income families ate take-away food during the week of study compared to 23% of children in more affluent areas.

There are significant gender differences in the frequency and confidence of cooking meals. In one report, 68% of women said they cooked every day compared with only 18% of men. On average, women cooked on 5.8 days a week and men on only 2.5 days [16]. For working women (65.4% of the female population), this gender imbalance (which may also be evident in other domestic chores) may contribute to additional tiredness and lack of enthusiasm about food preparation. These findings are a reminder of the challenge faced by women in balancing the demands of external employment and carrying the main responsibility for domestic food procurement and preparation.

Home-food preparation does not always equate with optimal nutritional intake but, theoretically, the practice allows consumers to alter the ingredients to healthier options/proportions, omit energy-dense ingredients and control portion sizes. While cooking-skills intervention programmes have been tested (e.g. on pregnant teenage women), these have had low uptake [17] and have shown very modest impact on changing dietary intake [18].

Parental eating and activity habits are clearly associated with those of their children. Both maternal and paternal body mass index (BMI) have been shown to be positively associated with offspring BMI in childhood and mid-adulthood [19]. Longitudinal studies such as the 1958 British birth cohort [20] have reported increased odds for offspring obesity (at age 33) of 8.4 and 6.8 for male and female offspring, respectively, when both parents are obese, compared to offspring with two parents of normal BMI. In turn, there are strong relationships between excess body weight in childhood and obesity in adulthood. It has been estimated that about one-third (26–41%) of obese pre-school children go on to become obese adults and about half (42–63%) of obese school-age children go on to become obese adults [21]. A review on this topic by Singh et al. [22] reported that persistence of overweight was greater with increasing level of overweight. Others [23] have noted that while the majority of overweight youth remained overweight as adults, the majority of overweight adults were not overweight youth highlighting the importance of lifestyle factors beyond childhood as well as within childhood. A recent review has also reported evidence of tracking of both physical activity and diet between childhood and adulthood with similar magnitude [24].

The potential influence of school food on the development of eating habits in young adults should not be ignored. In the United Kingdom, the nutritionally balanced school meal virtually disappeared in 1980 when school meal standards were abandoned. The 1980 regulation change was accompanied by a significant globalisation of UK food culture with cheap, energy-dense school menus (epitomised as the burger culture) which has undoubtedly contributed to the poor eating habits and rising obesity levels observed in young women throughout the United Kingdom in the following decades.

New regulations on school food [25] introduced in 2007 may need to be in place for many years before significant change in eating habits among young adults are seen. In adolescents, inappropriate dieting, skipping meals, snacking, eating away from home, consuming fast foods and trying unconventional diets make the achievement of appropriate dietary intake challenging. These issues become increasingly relevant in teenage pregnant women, where attempts to change dietary habits need to move beyond standard counselling practices to approaches which take account of increased independence, search for self-identity, peer group conformity and body image dissatisfaction [26].

Beyond school and family, women's eating habits are also influenced by life transitions including marriage and cohabitation. Anderson et al. [27] reported that changes after cohabitation include an increase in shared meal occasions, increased likelihood of including alcohol at meal times and social support for dietary temptations (and restrictions). Over a 3-month study period in 22 couples, they reported that body weight increased significantly in women ($58.3 \pm 7.1-59.8 \pm 7.9$ kg) and in men ($76.7 \pm 12.0-78.4 \pm 12.5$ kg).

Surveys of alcohol consumption [10] have also demonstrated that those who were cohabiting/ married were more likely to have drunk alcohol in the previous week (67%) compared to the single, divorced/separated or widowed. Married/cohabiting men and women were also the most likely to report drinking on five or more days in the previous week (18%) while single adults were the least likely (8%).

Concerns about body weight are often cited by women as a reason for smoking, and it has been noted that fashion-conscious women are significantly more likely to smoke [28]. Around half of the female smokers say that concerns about weight gain discourage them from trying to stop [29].

Food access and affordability are thought to exert significant influences on day to day food purchases and are particularly relevant for socially disadvantaged consumers. A national mapping study of food retailing from Scotland reported that the overall number of healthy items available per shop was weakly negatively correlated with area deprivation (i.e. less healthy items in more deprived areas), and there was a tendency for prices to be lower in larger shops and in areas with a low level of social and economic deprivation [30].

As the Foresight Obesity reports [12] highlights, the role of the food industry and its powerful retail and marketing sectors are the major influences on eating behaviours and on our obesogenic environment. Attempts to alter lifestyle habits in women by health practitioners will always compete with well-funded industry initiatives aimed at increasing purchase and consumption of inappropriate food and drink. There is growing evidence that food advertising and marketing, generally for processed, fast foods and sugary drinks, influence the choice of food and drinks. The evidence is strongest for causal links between television advertising and the choice of processed foods in children [31]. A review undertaken by Hastings et al. [32] reported that children in both the developed and developing world have extensive recall of food advertising. Parents – especially those from disadvantaged backgrounds – frequently respond to such requests by purchase. Adults have also been shown to respond to television food advertisements by consuming more snack foods (both healthy and less healthy), irrespective of whether they were the same foods being promoted and irrespective of hunger suggesting that advertising has an effect beyond brand influence [33].

The International Association of Consumer Food Organizations [34] has estimated that for every dollar spent by the World Health Organization trying to reduce the incidence of diseases associated with a Western diet, more than $500 is spent by the food industry promoting this same diet.

Advertising targeted at women to promote smoking [35,36] and alcohol intake has also been widely identified as strong influences on these health behaviours (notably in young women).

Individual-level Influences

Individual-level influences on eating habits span a wide range of external food-related factors (e.g. taste, preference, flavour and colour), early exposure and conditioning (e.g. importance of repeated exposure in food acceptability) and internal factors (e.g. mood, stress, dietary restraint characteristics) [37]. In addition, many theoretical models of health behaviours have been described and applied to food choices to demonstrate the impact of psycho-social variables [38].

These models are particularly relevant with respect to the development of intervention programmes and the identification of salient modifiable factors which have the potential to initiate and maintain changes in dietary choices. Psychological models draw upon perceived susceptibility to ill health, perceived severity of conditions, benefits, barriers, cues to action and self-efficacy. This final component is thought to be particularly important with respect to changing behaviours. The sense of control or belief in one's ability to change and succeed may be an important underlying determinant of health behavioural change. For example, recent research from a weight-loss clinic [39] has shown unrealistic expectations for weight-loss goals with women describing their 'dream' goal as losing about 30% body weight, 'happy' goal to lose 25%, 'acceptable' goal as losing 20% and 'disappointed' if losing 10%. This is in contrast with current recommendations which aim to achieve modest amounts of weight loss, e.g. 5–10% [40]. These types of behavioural aspects of intervention work are highlighted by NICE in their guideline on 'Dietary interventions and physical activity interventions for weight management before, during and after pregnancy' [41]. This guideline highlights the need to move beyond giving women simple knowledge on diet and physical activity, and should include:

- Helping women to feel positive about the benefits of health-enhancing behaviours and changing their behaviours;
- Helping plan women's changes in terms of easy steps over time;
- Identifying and planning situations that might undermine the changes women are trying to make and plan explicit 'if–then' coping strategies to prevent relapse.

However, many programmes aimed at changing dietary behaviour have not embraced theoretical frameworks and have focussed more on ways to increase nutrition knowledge as a route to change motivation and behaviour. There is little evidence to support such education programmes per se as a major influence on food and drink consumption. For example, in the 2010 Scottish Health Survey, only 22% of adults reported eating the recommended number of portions of fruit and vegetables, despite, 87% reporting an awareness of the '5 a day' message [42].

Pregnancy has often been considered an ideal time to promote dietary education [43], but dietary advice alone has not been shown to have a significant impact on nutrient intake during pregnancy [44]. These findings lend support to the conclusion that giving written advice (which is the most widespread method of giving information) can influence knowledge about healthier eating when provided in a systematic manner, but does not seem to alter attitudes, or indeed behaviour.

The use of folic acid supplements by women wishing to become pregnant has been widely described in terms of socio-demographics, but recent work has highlighted reasons why knowledge of recommended behaviours per se did not result in appropriate use of the supplement [45]. Focus group discussions indicated that women were aware of current guidance but often linked folic acid use with morning sickness, and indicated busy lives, competing priorities for concern, and forgetfulness in accounting for low usage. Building a 'lay evidence base' from their own experiences, many cited healthy pregnancy outcomes without supplement use and also expressed scepticism about its importance [45]. These results indicate the importance of discussing dietary actions with women to ascertain perceived barriers.

The limited impact of education per se does not mean that we should ignore educational efforts as one part of a portfolio of efforts to impact on diet and other health behaviours. For example, in many individuals, improving knowledge and skills such as in nutrition label reading or identifying low fat products can be a useful implementation tool for changing choices. Widespread awareness about nutritional harms (or benefits) can also help to frame public understanding, opinions and help to generate support for healthy public policy to create healthy environments.

Michie et al. [46] bring together these socio-cultural and individual-level issues for understanding health behaviours (and hence potential intervention approaches) within the *Behaviour Change Wheel*. Within this model, they describe three broad components that combine to influence behaviour notably capability, opportunity and motivation. Capability is further divided into physical (e.g. skills to prepare appropriate foods) and psychological capability (e.g. the capacity to reason the healthiest choices). Likewise, the opportunity component includes the physical opportunity present in a given environment (e.g. access to healthy foods) and social opportunity afforded (e.g. cultural norms that dictates the way that we think about things). The authors describe two motivational components that being reflective mechanisms (involving evaluations and considerations) and automatic processes (involving impulses that might arise from specific responses and cues). Such models offer scope to develop targeted intervention programmes throughout the life course that take account of both individual level and societal influences on food choice involving both policy makers and practitioners.

Conclusion

In conclusion, the challenge of optimising diet and other lifestyle habits is lifelong. Many stakeholders have the power to influence women's health behaviours. Combined efforts from parenting and schools in early life as well as our wider socio-cultural background in adult years provide the backdrop for educational efforts from health professionals and public policy. Combined approaches and strategies are required to significantly alter health and obesity status [31].

Acknowledgements

Thanks to Dr Maureen Macleod and Ms Lyndsay Watkins for their help with manuscript preparation.

References

1. World Cancer Research Fund/American Institute for Cancer Research. Chapter 8: Determinants of weight gain, overweight and obesity. In: *Food, Nutrition, Physical Activity and the Prevention of Cancer: A Global Perspective*. Washington DC: AICR; 2007.
2. Mozaffaraian D, Tao H, Eric B, et al. Changes in diet and lifestyle and long-term weight gain in women and men. *N Engl J Med*. 2011;364(25):2392–2404.
3. Scientific Advisory Committee on Nutrition. *The Influence of Maternal, Fetal and Child Nutrition on the Development of Chronic Disease in Later Life*. London: TSO; 2011: <http://www.sacn.gov.uk/reports_position_statements/reports/the_influence_of_maternal_fetal_and_child_nutrition_on_the_development_of_chronic_disease_in_later_life.html/>.
4. Scientific Advisory Committee on Nutrition. *The Nutritional Wellbeing of the British Population*. London: TSO; 2008:<http://www.sacn.gov.uk/pdfs/nutritional_health_of_the_population_final_oct_08.pdf/>
5. Bodnar LM, Catov JM, Roberts JM, Simhan HN. Prepregnancy obesity predicts poor vitamin D status in mothers and their neonates. *J Nutr*. 2007;137(11):2437–2442.
6. Nelson N, Erens B, Bates B, Church S, Boshier T. *Low Income Diet and Nutrition Survey: Summary of Key Findings*. London: TSO; 2007:http://www.food.gov.uk/multimedia/pdfs/lidnssummary.pdf/>.
7. Inskip HM, Crozier SR, Godfrey KM, Borland SE, Cooper C, Robinson SM. Women's compliance with nutrition and lifestyle recommendations before pregnancy: general population cohort study. *BMJ*. 2009;338:b481.
8. The NHS Information Centre, Lifestyle Statistics. *Statistics on Smoking: England*. ONS: Leeds; 2011:<http://www.ic.nhs.uk/webfiles/publications/003_Health_Lifestyles/Statistics%20on%20Smoking%202011/Statistics_on_Smoking_2011.pdf>.
9. The NHS Information Centre. *Infant Feeding Survey. 2010: Early Results*. ONS: Leeds; 2011: <http://www.ic.nhs.uk/webfiles/publications/003_Health_Lifestyles/IFS_2010_early_results/IFS_2010_headline_report_tables2.pdf>.

10. The NHS Information Centre. Statistics on Alcohol: England, 2011. ONS: Leeds. 2011 <http://www.ic.nhs.uk/webfiles/publications/003_Health_Lifestyles/Alcohol_2011/NHSIC_Statistics_on_Alcohol_England_2011.pdf>.
11. Anderson AS. Nutrition interventions in women in low-income groups in the UK. *Proc Nutr Soc*. 2007;66(1):25–32.
12. *Government Office for Science*. Foresight tackling obesities: future choices – project support. <http://www.bis.gov.uk/assets/bispartners/foresight/docs/obesity/17.pdf/>; 2007. Accessed 25th July 2012.
13. Fitzpatrick I, MacMillan T, Hawkes C, Anderson AS, Dowler E. *Understanding Food Culture in Scotland and its Comparison in an International Context: Implications for Policy Development*. Edinburgh: NHS Health Scotland; 2010. Available from: <http://www.communityfoodandhealth.org.uk/fileuploads/nhshsfoodculturescotlandreport-2443.pdf/>.
14. Hammons AJ, Fiesse BH. Is frequency of shared family meals related to the nutritional health of children and adolescents? *Pediatrics*. 2011;127(6):e1565–e1574.
15. Scottish Government. *Growing up in Scotland: Sweep 3 Food and Activity Report*. Edinburgh: Scottish Executive; 2009.
16. Caraher M, Dixon P, Lang T, Carr-Hill R. The state of cooking in England: the relationship of cooking skills to food choice. *Br Food J*. 1999;101(8):590–609.
17. Wrieden WL, Symon A. The development and pilot evaluation of a nutrition education intervention programme for pregnant teenage women (food for life). *J Hum Nutr Diet*. 2003;16(2):67–71.
18. Wrieden WL, Anderson AS, Longbottom PJ, et al. The impact of a community-based food skills intervention on cooking confidence, food preparation methods and dietary choices – an exploratory trial. *Public Health Nutr*. 2007;10(2):203–211.
19. Cooper R, Hypponen E, Berry D, Power C. Associations between parental and offspring adiposity up to midlife: the contribution of adult lifestyle factors in the 1958 British birth cohort study. *Am J Clin Nutr*. 2010;92(4):946–953.
20. Lake JK, Power C, Cole JT. Child to adult body mass index in the 1958 British birth cohort: associations with parental obesity. *Arch Dis Child*. 1997;77:376–380.
21. Serdula MK, Ivery D, Coates RJ, Freedman DS, Williamson DF, Byres T. Do obese children become obese adults? A review of the literature. *Prev Med*. 1993;22(2):167–177.
22. Singh AS, Mulder C, Twisk JWR, van Mechelan W, Chinpaw MJM. Tracking of childhood overweight into adulthood: a systematic review of the literature. *Obes Rev*. 2008;9(5):474–488.
23. Herman KM, Craig CL, Gauvin L, Katzmarzyk PT. Tracking of obesity and physical activity from childhood to adulthood: the physical activity longitudinal study. *Int J Paediatr Obes*. 2009;4(4):281–288.
24. Craigie AM, Lake AA, Kelly SA, Adamson AJ, Mathers JC. Tracking of obesity-related behaviours from childhood to adulthood: a systematic review. *Maturitas*. 2011;70(3):266–284.
25. Scottish Government. *The Schools (Health Promotion and Nutrition) (Scotland) Act 2007* <http://www.scotland.gov.uk/Topics/Education/Schools/HLivi/foodnutrition/>; Scottish Government, 2008; Accessed 25th July 2012.
26. Anderson AS, Wrieden WL. Teenage pregnancies. In: Symonds ME, Ramsey MM, eds. *Maternal–Fetal Nutrition During Pregnancy and Lactation*. Cambridge: Cambridge University Press; 2010.
27. Anderson AS, Marshal DW, Lea EJ. Shared lives – an opportunity for obesity prevention? *Appetite*. 2004;43:327–329.

28. O'Connor EA, Friel S, Kelleher CC. Fashion consciousness as a social influence on lifestyle behaviour in young Irish adults. *Health Promot Int*. 1997;12(2):135–139.
29. Spring B, Howe D, Berendsen M, et al. Behavioural intervention to promote smoking cessation and prevent weight gain: a systematic review and meta-analysis. *Addiction*. 2009;104:1472–1486.
30. Dawson J, Marshall D, Taylor M, Cummins S, Sparks L, Anderson A. *Accessing Healthy Food: A Sentinel Mapping Study of Healthy Food Retailing in Scotland*. Scotland: Food Standards Agency; 2008. <http://www.food.gov.uk/multimedia/pdfs/accessfoodscotexec.pdf/>.
31. World Cancer Research Fund. *Policy and Action for Cancer Prevention*. London: WCRF; 2009.
32. Hastings, G, McDermott L, Angus K, Stead M, Thomson S. The extent, nature and effects of food promotion to children: a review of the evidence: a technical paper prepared for the World Health Organisation: Geneva; 2006. <http://whqlibdoc.who.int/publications/2007/9789241595247_eng.pdf>.
33. Harris J, Bargh JA, Brownell KD. Priming effects of television food advertising on eating behaviour. *Health Psychol*. 2009;28(4):404–413.
34. IACFO. In: Dalmeny K, Hanna E, Lobstein T, eds. *Broadcasting Bad Health: Why Food Marketing to Children Needs to be Controlled*. London: International Association of Consumer Food Organizations; 2003.
35. Amos A, Haglund M. From social taboo to torch of freedom: the marketing of cigarettes to women. *Tob Control*. 2000;9:3–8.
36. Hastings G, Anderson S, Cooke E, Gordon R. Alcohol marketing and young people's drinking: a review of the research. *J Public Health Policy*. 2005;26(3):296–311.
37. Cox DN, Anderson AS. Food choice. In: Gibney MJ, Margetts BM, Kearney JM, Arab L, eds. *Public Health Nutrition*. Oxford: Blackwell Publishing; 2004:144–166.
38. National Cancer Institute. *Theory at a Glance: a Guide for Health Promotion Practice*. Washington: National Institute of Health; 2005:<http://www.cancer.gov/cancertopics/cancerlibrary/theory.pdf>.
39. Dutton GR, Perri MG, Dancer-Brown M, Goble M, Van Vessem N. Weight loss goals of patients in a health maintenance organization. *Eat Behav*. 2010;11(2):74–78.
40. Scottish Intercollegiate Guidelines Network. *Management of Obesity: a National Clinical Guideline*. Edinburgh: SIGN; 2010:<http://www.sign.ac.uk/guidelines/fulltext/115/index.html/>.
41. *National Institute for Health and Clinical Excellence*. Dietary interventions and physical activity interventions for weight management before, during and after pregnancy; 2010: <http://www.nice.org.uk/nicemedia/live/13056/49926/49926.pdf/>.
42. Bromley C, Graham H, Sharp C. *Knowledge, Attitudes and Motivations to Health: A Module of the Scottish Health Survey*. Edinburgh: NHS Health Scotland; 2011:<http://www.scotpho.org.uk/nmsruntime/saveasdialog.asp?lID=8398&sID=6084/>
43. Anderson AS. Symposium on 'nutritional adaptation to pregnancy and lactation'. Pregnancy as a time for dietary change? *Proc Nutr Soc*. 2001;60(4):497–504.
44. Anderson AS, Campbell DM, Shepherd R. The influence of dietary advice on nutrient intake during pregnancy. *Br J Nutr*. 1995;73(2):163–177.
45. Barbour RS, Macleod M, Mires G, Anderson AS. Uptake of folic acid supplements before and during pregnancy: focus group analysis of women's views and experiences. *J Hum Nutr Diet*. 2012;25(2):140–147.
46. Michie S, van Stralen MM, West R. The behaviour change wheel: a new method for characterising and designing behaviour change interventions. *Implement Sci*. 2011;6:42.

44 Planning for the Future: Maternity Services in 2035

Amanda Jefferys[1], Tim Draycott[1] and Tahir Mahmood[2]

[1]North Bristol NHS Trust, Bristol, UK, [2]Victoria Hospital, NHS Fife, Kirkcaldy, UK

Planning for the Future: Maternity Services in 2035

In order to plan the provision of maternity services in the future, we first need to have an understanding of the likely magnitude of the obesity epidemic. It is essential that we identify the key health burdens faced by the obstetric obese population, so that an evidence-based effective package of care can be provided for women within the funding framework available.

Projected Prevalence of Obesity in 2035

Obesity is an epidemic and will pose significant problems for obstetric services in the future. Our ability to provide services for obese pregnant women will depend on a number of factors including the projected size of the population, availability of specialist services, effective interventions and also specialist staff – all within the health economics of the country.

Traditionally, obesity has been deemed a developed world problem, but it is a global problem. In the United Kingdom and the United States, 25% and 35% of women are obese, respectively (International Obesity Taskforce (IOTF); (www.iaso.org/)). Obesity rates are also increasing in the developing world – currently running at 5–10% (IOTF) of the adult female population for many of the sub-Saharan African countries and expected to rise. The rise in the levels of obesity is proving a particular problem in the Middle East, the Pacific Islands, Southeast Asia and China [1], where rates of obesity match those in the West, ranging from 5% to 40% (IOTF). This rise can be attributed perhaps to the close association between obesity, low socio-economic status and shifts in lifestyle with an increasing adoption of western lifestyles. Some of these countries will have the most limited resources to deal with their obesity epidemic.

Obesity. DOI: http://dx.doi.org/10.1016/B978-0-12-416045-3.00044-3

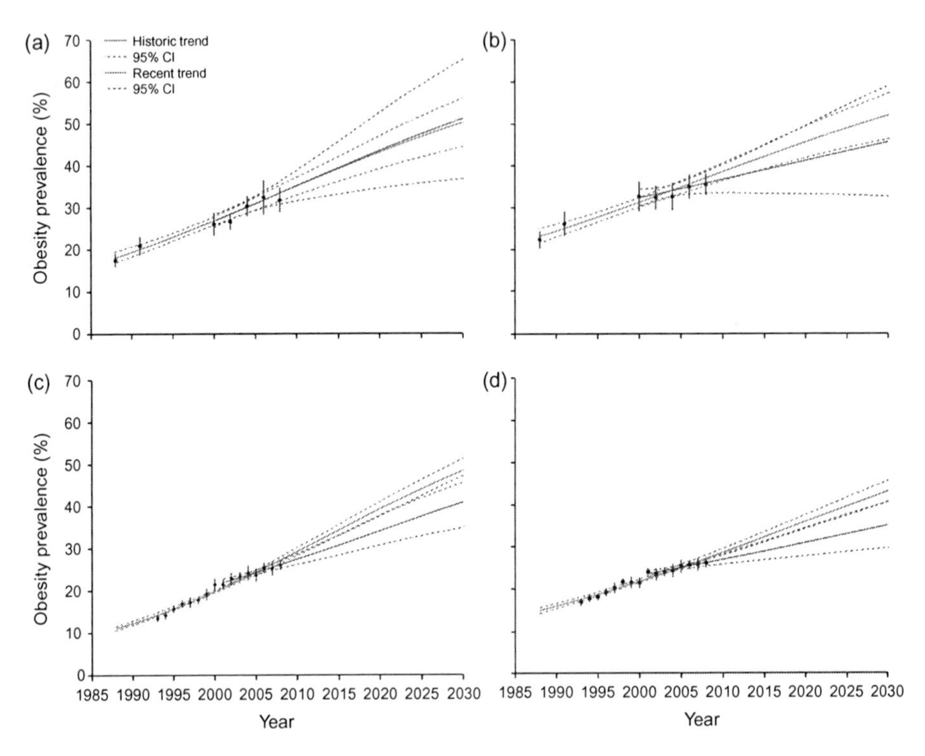

Figure 44.1 Historic and recent trends in adult obesity prevalence in men and women in the United States and the United Kingdom. (a) = US men, (b) = US women, (c) = UK men, (d) = UK women. Black dots (bars = 98% CI) show recorded prevalence from national surveys; each dot = one data point. Historic trend used all data points; recent trend used data points after 2000.
Source: Data from Ref. [2].

But what of future levels of obesity? Rates of obesity are expected to rise even further in both the developed and the developing world. It is predicted that in the United Kingdom and the United States, approximately 40% of women will be obese by 2030 [2] (Figure 44.1). In the developing world, the prevalence of obesity has tripled over the past 20 years, and with widespread adoption of an increasingly western lifestyle, this is a trend that can only be expected to continue.

Rises in the prevalence of obesity generally in women will be reflected in the obstetric population worldwide. Currently in the United Kingdom, 20% women of childbearing age are known to be obese [3] (Figure 44.2) with rates of extreme obesity (BMI $\geq 50\ kg/m^2$) reported at approximately 1 per 1000 [4]. In the United States, obesity rates are estimated to be 20% [5], and in Australia 35% of the obstetric population are currently recognised as overweight or obese [6]. Figures for the developing world are harder to come by because of less-robust

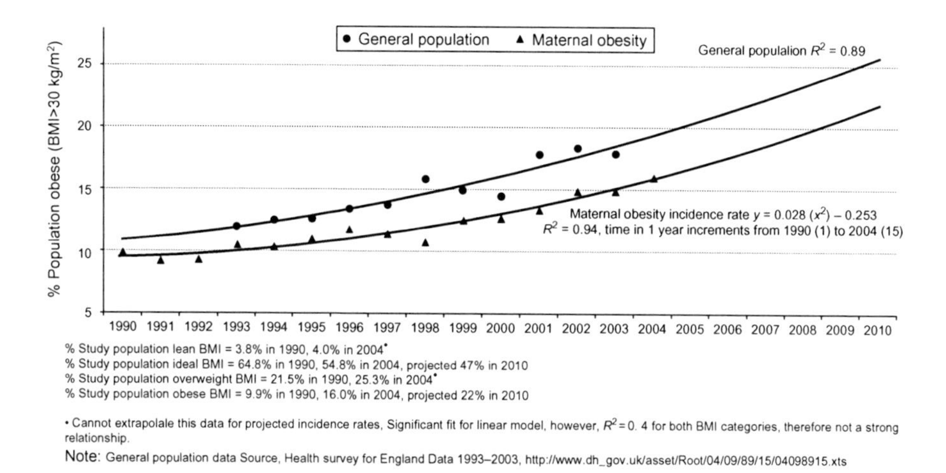

Figure 44.2 Incidence of maternal obesity in 36,821 women over a 15-year period, the projected incidence of maternal obesity by 2010, and the prevalence of obesity in women of childbearing age (16–44 years) in Englands general population.
Source: Data from Ref. [3].

systems of recording; however, obesity among the women of childbearing age has been found to be a particular problem in Latin America, the Caribbean, the Middle East, North Africa and central eastern Europe, where obesity rates range from 10% to 20% [7].

Key Health Burdens of Obesity in Pregnancy

Currently obesity accounts for 16% of the global burden of disease [1]. The significant health burden of obesity is also seen in obstetric practice. When considering the burden obesity in pregnancy places on a health service, it is perhaps easiest to divide these into direct and indirect health burdens.

Direct Health Burdens

Maternal obesity creates significant problems in delivering the desired standard of care. Obese women often require additional equipment, additional staff and additional time for their care across all stages of pregnancy.

Structural Issues

In the antenatal setting, additional equipment is often necessary to deliver what would usually be considered 'routine' care, including large blood pressure cuffs, scales able to take extremes of weight and large weight-bearing examination couches (>200 kg). In future, these equipments will need to be provided in all care settings, including community and hospital locations at significant cost. This additional cost burden can also be seen in routine scanning, where more time is required to perform a technically ‘adequate’ scan. Obese women are also at increased risk of missed diagnoses on scan because of the technical difficulties in achieving an optimal view. This would require special expertise in scanning, probably a service delivered by the experienced consultants and sonographers rather than senior trainees.

Process Issues/Intrapartum Care

These difficulties in delivering care extend into the labour ward and intrapartum care. Difficulties in manual handling of obese women is a particular risk to staff in terms of potential injury and the associated cost of lost days at work. Again, delivery of routine obstetric and anaesthetic care such as obtaining IV access, foetal monitoring and insertion of regional anaesthetics is often difficult and more time consuming adding to the costs. These difficulties may require delivery of such service to be continuously consultant delivered. Equipment issues also extend into the labour ward with the additional cost of operating theatre weight-bearing tables for Caesarean section (CS).

Outcome

Operative morbidity is increased in this group of women who are known to be at higher risk of operative intervention. Complications including post-operative infections [8] and post-partum haemorrhage [9] are strongly associated with maternal obesity adding significantly to the health and cost (longer hospital stays, Intensive Care Unit (ITU) care, antibiotics, return to theatre) burden. Additionally, procedures tend to be longer and more technically demanding, again requiring experienced obstetricians. These operative difficulties and increased potential for complications in the second stage mean that almost the continuous presence of the consultant in the labour ward may be required. It is well documented that regional anaesthetic techniques are more likely to fail in such patients requiring increased time, expertise and the possibility of the need for general anaesthesia, with its incumbent risks including airway problems and intra-operative cardiopulmonary complications [10]. This increase in the risk of obstetric morbidity [4] (Table 44.1) will result in increased hospital stays, and, therefore, would require an increased number of inpatient beds.

These health burdens extend to the provision of neonatal services. Many of the complications associated with pregnancy in obese women such as gestational diabetes and hypertension result in iatrogenic prematurity; this problem is compounded by the difficulties in foetal surveillance both antenatally and during labour. The increased burden on neonatal units as a result of these additional complications also needs to be considered.

Table 44.1 Pregnancy Complications in Extremely Obese and Comparison Women

Pregnancy Outcome or Complication	No. (%[a]) of Women in the Case Group ($n = 659$)	No. (%[a]) of Women in the Comparison Group ($n = 634$)	Unadjusted OR (95% CI)	Adjusted OR[b] (95% CI)
Any gestational hypertensive disorder				
No	507 (77)	599 (95)	1[c]	1[c]
Yes	151 (23)	34(5)	5.25 (3.55–7.75)	5.49 (3.65–8.25)
Pregnancy-induced hypertension				
Multiparous				
No	387 (59)	343 (54)	1[c]	1[c]
Yes	41 (6)	11 (2)	3.30 (1.67–6.53)	3.19 (1.60–6.37)
Primiparous				
No	182 (28)	271 (43)	0.60 (0.47–0.75)	0.76 (0.59–0.98)
Yes	46 (7)	7 (1)	5.82 (2.60–13.10)	7.35 (3.07–17.60)
Pre-eclampsia				
No	595 (91)	618 (98)	1[c]	1[c]
Yes	61 (9)	14 (2)	4.55 (2.50–8.18)	4.46 (2.43–8.16)
Eclampsia				
No	656 (100)	630 (100)	1[c]	1[c]
Yes	0 (0)	2 (0.3)	0 (0–1.85)	NI
Thrombotic event				
No	655 (100)	633 (100)	1[c]	1[c]
Yes	3 (0.5)	0 (0)	∞ (0.75–∞)	NI
Gestational diabetes				
No	555 (89)	623 (98)	1[c]	1[c]
Yes	68 (11)	10 (2)	7.63 (3.89–15.00)	7.89 (3.94–15.80)
Gestational diabetes requiring insulin				
White ethnicity				
No	503 (81)	529 (84)	1[c]	1[c]
Yes	45 (7)	3 (0.5)	15.80 (4.87–51.10)	15.7 (4.75–51.80)

(Continued)

Table 44.1 (Continued)

Pregnancy Outcome or Complication	No. (%[a]) of Women in the Case Group ($n = 659$)	No. (%[a]) of Women in the Comparison Group ($n = 634$)	Unadjusted OR (95% CI)	Adjusted OR[b] (95% CI)
BME groups				
No	71 (11)	98 (15)	0.76 (0.55–1.06)	0.72 (0.52–1.02)
Yes	3 (0.5)	3 (0.5)	1.01 (0.20–5.23)	0.93 (0.18–4.70)
Induction of labour				
No	414 (63)	487 (77)	1[c]	1[c]
Yes	241 (37)	147 (23)	1.93 (1.51–2.46)	1.97 (1.53–2.54)
Labour				
No	217 (33)	93 (15)	1[c]	1[c]
Yes	437 (67)	541 (85)	0.35 (0.26–0.46)	0.38 (0.28–0.50)
Preterm delivery at less than 37 weeks				
No	585 (90)	587 (93)	1[c]	1[c]
Yes	65 (10)	43 (7)	1.52 (1.01–2.27)	1.58 (1.04–2.40)
Preterm delivery at less than 32 weeks				
No	638 (98)	625 (99)	1[c]	1[c]
Yes	12 (2)	5 (1)	2.35(0.82–6.71)	2.82 (0.89–8.95)
Post-term delivery 42 or more weeks				
No	618 (95)	606 (96)	1[c]	1[c]
Yes	32 (5)	24 (4)	1.31 (0.76–2.25)	1.35 (0.77–2.37)
Shoulder dystocia				
No	315 (96)[d]	481 (97)	1[c]	1[c]
Yes	14 (4)	14 (3)	1.64 (0.76–3.55)	1.89 (0.82–4.34)
Caesarean delivery				
No	327 (50)	494 (78)	1[c]	1[c]
Yes	328 (50)	140 (22)	3.54 (2.78–4.51)	3.50(2.72–4.51)

Problems or failure with epidural anaesthesia				
No	152 (83)	123 (95)	1[c]	1[c]
Yes	32 (17)	7 (5)	3.70 (1.53–10.20)	3.54 (1.49–8.42)
Problems or failure with spinal anaesthesia				
No	176 (88)	110 (88)	1[c]	1[c]
Yes	25 (12)	2 (2)	7.81 (1.88–69.00)	9.10 (2.02–41.00)
Problems or failure with combined spinal–epidural anaesthesia				
No	41 (89)	12 (100)	1[c]	1[c]
Yes	5 (11)	0(0)	∞ (0.35–∞)	NI
General anaesthesia for delivery				
No	618 (94)	628 (99)	1[c]	1[c]
Yes	37 (6)	6 (1)	6.27 (2.63–15.00)	6.35 (2.63–15.30)
Intensive care unit admission				
No	639 (97)	629 (99)	1[c]	1[c]
Yes	18 (3)	5 (1)	3.54 (1.31–9.60)	3.86 (1.41–10.60)
Post-partum haemorrhage requiring transfusion				
No	645 (98)	630 (99)	1[c]	1[c]
Yes	12 (2)	4 (1)	2.93 (0.94–9.13)	3.04 (0.96–9.67)

OR, odds ratio; CI, confidence interval; NI, variable was not included in the multivariate model due to low event numbers; BME, black or other minority ethnic.
Complications are defined according to standard hospital definitions; shoulder dystocia is defined as difficulty in delivery of the shoulders requiring any additional maneuver to assist delivery.
[a]Percentage of women with complete data.
[b]Adjusted for age, socio-economic group, parity, ethnicity, and smoking.
[c]Reference group.
[d]Includes only women delivering vaginally.
Source: Data from Ref. [4].

Indirect Health Burdens

There is a strong association between maternal obesity and gestational diabetes, hypertensive disorders, venous thromboembolism (VTE), delivery by CS and stillbirth. A recent CMACE report has confirmed that obese women are at twice the risk of stillbirth, four times the risk of developing gestational diabetes and hypertensive disorders in pregnancy, and one and a half times more likely to require Caesarean delivery [11]. The increased rates of Caesarean delivery are likely to stem from ineffective myometrial contraction resulting in a higher incidence of failure to progress in labour [12], higher rates of induction of labour due to prolonged pregnancy, obstetric intervention [13] and higher rates of post-partum haemorrhage [14].

Maternal obesity and its incumbent additional obstetric demands makes the current largely trainee-delivered delivery model of service, ethically and legally unsustainable. Services instead should be largely delivered by the onsite consultant teams of obstetricians, anaesthetists and neonatologists. Moreover, additional midwifery staffing will be required.

We need to consider how obstetric services in the future are going to be able to accommodate the increase in these health burdens and the necessary shift in staffing structure.

Current Model of Care

A recent report [15] confirmed that there is no consistent approach in the United Kingdom for the management of obesity in pregnancy; maternity services for obese women are currently provided by a combination of consultant-led units, midwifery units attached to consultant-led units and standalone midwifery units. A significant proportion (44%) of these units did not have local guidelines for the management of obese women. A significant proportion of obstetric units did not have appropriate basic equipment to accommodate the needs of obese women such as chairs, beds and theatre tables. Furthermore, provision of pre-conceptual care and antenatal anaesthetic assessment provided for these women was patchy. The results of this survey have prompted publication of a joint CMACE/RCOG guideline [16]. However, with the predicted rise in maternal obesity, a national model of care urgently needs to be put in place based on this national guidance that can be implemented locally using resources available.

Future Model of Care

When considering a model of care for obese women in pregnancy, we need to consider not only what care needs to be provided but also who this care should be provided for, by whom and where (Figure 44.3).

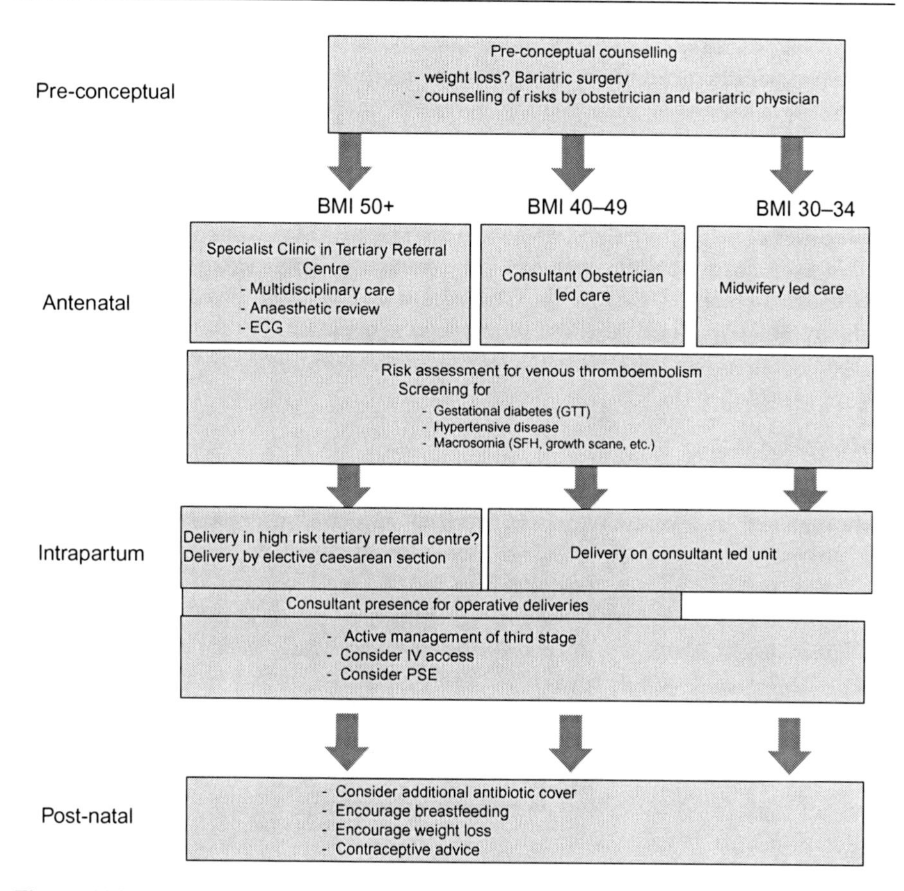

Figure 44.3 Flow chart of future model of care.

Pre-Conceptual Care

Care of obese women needs to start pre-conceptually with advice about the increased risks associated with obesity in pregnancy including miscarriage, congenital abnormality, gestational diabetes, hypertensive disorders, VTE, foetal macrosomia, iatrogenic prematurity, delivery by CS, post-partum haemorrhage and post-operative wound infections.

Weight-Loss Strategies

Women should be encouraged to delay pregnancy and lose weight prior to embarking on a pregnancy. These clinics should therefore be able to provide contraceptive advice. Guidance on weight-loss strategies exists [17,18] and weight-loss strategies have been extensively researched in parallel with the sharp rise on obesity rates. A recent randomised control trial [19] assessed the success of a variety of weight-loss programmes and found that commercial group weight-loss programmes resulted in a significant increase in weight loss after a year compared to one-to-one counselling by health care professionals or provision of gym vouchers.

For extremely obese women where weight-loss strategies have failed, it may be that referral for consideration of weight-loss surgery may be appropriate pre-conceptually.

It would seem sensible that this pre-conceptual care package is delivered in specialist-combined clinics by an obstetrician and bariatric physician, but clearly there are few physicians qualified in this field at present.

Antenatal Care

Multidisciplinary teams should provide the antenatal care in these women, probably consultant led in a continuum from the pre-conceptual care package. Because of the predicted scale of the problem, although it would be desirable to provide specialist multidisciplinary antenatal care for all obese women, in reality, this may be impractical. Consideration needs to, therefore, be given to the inclusion criteria for these clinics. Currently, specialist obstetric care tends to be reserved for those with a BMI greater than or equal to 40 kg/m^2; however, it may be that in the future, this cut-off needs to be raised further – possibly to a BMI of 50 kg/m^2 (known as the morbidly obese) and above, as this is the level at which there may be significant implications and changes in management of the pregnancy particularly in terms of mode of delivery and anaesthetic management. In order to maximise the health economic benefits, obese women of a lower BMI ($<40\ kg/m^2$) could be managed in other settings (e.g. a non-obesity specialist obstetric setting or by midwives in the community) provided robust guidelines and onward referral criteria are agreed and continually audited.

Centralisation of specialist obesity services for morbidly obese ($BMI > 50\ kg/m^2$) women may be most cost effective, allowing specialist equipment such as large couches, large blood pressure cuffs and specialist operating tables to be concentrated in one place and reducing the need for such equipment in peripheral care settings. Centralisation of services should include a specialist multidisciplinary team that is also available close by: a specialist midwife, an obstetrician (with an interest in maternal medicine), a bariatric physician and a dietician with other facilities such as an obstetric anaesthetist (preferably one with an interest in bariatrics). Concentration of these services in one place increases expertise and reduces costs. Specialist clinics could allow appropriate structure of services, e.g. longer allocation of time for scans. It may also be an opportunity to introduce novel approaches to weight loss, e.g. the provision of exercise bikes in the waiting room or arranging walking groups with the specialist midwife. However, care must be taken when designing such a service for morbidly obese pregnant women as attendance at such a clinic could easily become stigmatised which may affect attendance and compliance. A care pathway should be developed based on national guidance [16,20].

Weight Management During Pregnancy

Obese women should be advised that they can safely lose up to 5 kg of weight during pregnancy and that such weight loss will result in improved obstetric and neonatal outcomes [21]. Weight loss should be encouraged via on-going dietary advice provided by a specialist dietician and by an encouragement of regular aerobic exercise. As reported previously, there may also be a place for funding specialist commercial weight-loss programmes as part of the care pathway.

Risk Management

The clinic should be used as an opportunity to screening for potential complications of obesity during pregnancy, including gestational diabetes, hypertensive disorders and foetal macrosomia.

Particular attention should also be paid in this group of women to the need for prophylaxis against VTE – as already stated these women represent a high risk group – frequent risk assessments should, therefore, be performed and thromboprophylaxis and graduated elasticated compression stockings should be provided where appropriate. All women attending the clinic should receive a specialist anaesthetic review during pregnancy including electrocardiogram so that potential anaesthetic problems can be pre-empted, and risks and difficulties are discussed with the patient.

Previous Gastric Banding

Pregnancies in women who have undergone gastric banding are increasing, and this is likely to increase further with the rising trend in morbid obesity and weight-loss surgery. Currently, no advice exists on the best way to manage gastric bands in pregnancy (i.e. leave the band inflated to reduce gestational weight gain and obstetric and perinatal complications or leave deflated for the duration of pregnancy due to theoretical concerns over band complications, nausea and vomiting and intrauterine growth restriction). It is vital with the sharp rise in the obese obstetric population and the incidence of pregnancies in those who have undergone gastric banding, therefore, that further research is performed in this area to provide evidence-based guidance.

Intrapartum Care

The risks of obesity during the intrapartum period are well documented. Therefore, obese women, and, in particular, morbidly obese women are considered to be at high risk. Again centralisation of intrapartum care of these women to high-risk tertiary referral centres with specialist obesity services may be sensible in terms of provision of equipment (e.g. theatre tables) and specialist staff (obstetricians, anaesthetists, neonatologists). Intrapartum care should be consultant led (both obstetrically and anaesthetically) to minimise complications and deal effectively with complications should they arise. This includes the presence of a consultant obstetrician for CS for women with a BMI of greater than 40 kg/m^2[22]. This may also mean that that the tertiary units managing women with BMI > 50 kg/m^2 should have consultant obstetricians available continuously.

Obese women are known to be at greater risk of failure to progress in labour and subsequent CS – it has been suggested that this may be due to fatty deposition within the myometrium causing ineffective myometrial contractility [12]. Consequently, they are also at increased risk of post-partum haemorrhage secondary to uterine atony. Obese women in labour should be managed actively with close attention to the partogram and early intravenous access. Difficulties with monitoring foetal wellbeing due to body habitus are common; therefore, there should be a low threshold for application of a foetal scalp electrode.

As already discussed, obesity is known to be associated with higher rates of both induction of labour and failed induction of labour. Knowing the increased risks of emergency CS for failure to progress or failed induction of labour and the associated risks this carries (failed regional anaesthesia, failed intubation, more junior staff performing the procedure, and operative difficulties resulting in longer decision to delivery intervals and increased perinatal and maternal morbidity and mortality), it may be reasonable to propose a policy of elective CS for morbidly obese women (BMI > 50 kg/m^2). This would permit delivery to be performed by senior staff (obstetricians and anaesthetists) in a controlled and non-urgent manner within daytime hours. Interestingly, in a recent prospective cohort study [23] in which a total of 591 women planned either a vaginal birth (417) or an elective CS (174), there was no statistically significant reduction in anaesthetic (failed regional anaesthesia, general anaesthesia), post-natal (wound infection or wound complication, intensive care unit admission, major maternal morbidity) or neonatal (birthweight of 4.5 kg or greater, shoulder dystocia, neonatal intensive care unit admission, neonatal death) outcomes in extremely obese women (BMI > 50 kg/m^2) who planned vaginal birth. Therefore, it may be reasonable to offer these women delivery by elective CS until 24 h consultant presence is available.

Post-Natal Care

A package of post-natal care provision is as important as the antenatal and intrapartum care for these women to reduce the risk of relatively common complications including infection and VTE. Obesity is a risk factor for surgical site infection [24], and therefore, serious consideration should be given to the provision of additional antibiotic cover to these obese women post-surgery, particularly those who are morbidly obese and who face a particularly high risk of wound infection. Breastfeeding should also be encouraged in these women not only for the neonatal benefits but also for the proven improvements in post-partum weight loss [25].

As these women are often inpatient for some days after delivery and subsequently in contact with community midwives and General Practitioner (GPs) up to the point of their post-natal check, the post-natal period should be seen as a key opportunity for health promotion. Women should be advised on reliable forms of long-acting contraception – the intrauterine system, the depot injection and the sub-dermal contraceptive implants would all be appropriate, reliable, long acting but reversible options for these women. This period should also be seen as an opportunity to encourage a programme of weight loss. It may be during this period that subsidised weight-loss groups and exercise programmes may be beneficial. If these interventions are put in place, significant weight reduction may be achieved prior to the next pregnancy resulting in a significant reduction in obstetric and perinatal morbidity and mortality.

How Can We Deliver this Care?

Predicted New Service Delivery Model

Delivering this level of care to such numbers of women is going to take commitment at all levels if it is going to be successfully implemented. There is no doubt that provision of this level of care to the expanding number of obese women is going to take increased investment in the area by the Department of Health (DH). It will be the responsibility of the RCOG to produce guidance for health professionals and plan services and workforce to guide budget allocation by the DH. More locally provision of the service will have to be commissioned, guidelines implemented and audited and training should be provided for health care professionals. If much of the care for morbidly obese women is to be centralised, then funding provision must be increased accordingly to provide the additional staff and expertise required.

It may be that we move to a system similar to that adopted by neonatal services where units are graded and consequently funded and staffed according to the level of care they provide (e.g. level I, II and III units), with level II units able to care for those with a BMI of 40 kg/m^2 or more, but care of those who are morbidly obese (BMI > 50 kg/m^2) being restricted to level III units.

Health Service Planning Assumptions and Inpatient Beds Requirements

As it is impossible to know exactly what obesity levels will be among the obstetric population in 2035, we cannot make an accurate prediction of increased inpatient bed requirements and costs. However, based on certain assumptions, we can determine the likely magnitude of the problem: if we assume based on current figures [3] that 20% of the obstetric population have a BMI of >35 kg/m^2, it can be predicted that the obstetric population may rise to approximately 40% by 2035. Knowing that the CS rate is higher at 37% for this population of women [15] than the general obstetric population of 25% [26], and assuming that the average length of stay following CS is 2.5 days compared to 0.5 day for spontaneous vaginal delivery, we can calculate that in an obstetric population of 700,000, this will result in the need for 30,000 additional hospital bed days annually in the United

Kingdom alone. The additional admission costs for obese and morbidly obese women are thought to be £550 and £1035 per patient, respectively [27].

Research Needed to Inform Future Planning

There can be no doubt that extensive research is necessary to direct service provision for the obese obstetric population, as many questions about best practice in terms of both clinical outcomes and costs for this population remain unanswered.

Research is currently ongoing to prospectively identify the effects of obesity on pregnancy outcome. A multicentre randomised trial is also underway to investigate dietary and physical activity interventions to reduce glycaemic index and weight in pregnancy to improve obstetric and perinatal outcomes (UK Pregnancies Better Eating and Activity Trial (UPBEAT); (www.ukctg.nihr.ac.uk/trialdetails/ISRCTN89971375) and www.kcl.ac.uk/medicine/research/divisions/wh/.../upbeat.aspx).

Moreover, the ideal management of women after weight-loss surgery is unclear, as is the ideal mode of delivery of morbidly obese women.

Summary

There is no doubt that obesity and particularly morbid obesity in the obstetric population will be a key health concern for the future. The significant health and cost burdens that these pregnancies incur are going to be problematic, particularly if not adequately managed and planned for. We suggest that guidance needs to be clear on the best clinical management of these women and that there also needs to be a clear structure as to who provides this care with a centralisation of services for the morbidly obese with designated specialist obesity services to maximise and concentrate expertise and reduce costs, possibly moving to three tiers of obstetric service to match neonatal services. However, none of this will be possible without extensive and ongoing research and investment in the area, nationally in the United Kingdom and internationally.

References

1. Hossain P, Kawar B, El Nahas M. Obesity and diabetes in the developing world—a growing challenge. *N Engl J Med*. 2007;356(3):213–215.
2. Wang CY, McPherson K, Marsh T, Gortmaker SL, Brown M. Health and Economic burden of the projected obesity trends in the UK and USA. *Lancet*. 2011;378:815–825.
3. Simmons D. Diabetes and obesity in pregnancy. *Best Practice and Research Clinical Obstet Gynaecol*. 2011;25(1):25–36.
4. Knight M, Kurinczuk JJ, Spark P, Brocklehurst P. Extreme obesity in pregnancy in the United Kingdom. *Obstet Gynaecol*. 2010;115(5):989–997.
5. Kim SY, Dietz PM, England M, Morrow B, Callaghan WM. Trends in pre-pregnancy obesity in nine states, 1993–2003. *Obesity*. 2007;15(4):986–993.
6. Callaway LK, Prins JB, Chang AM, McIntyre HD. The prevalence and impact of overweight and obesity in an Australian obstetric population. *Med J Aust*. 2006;104(2):56–59.

7. Martorell R, Kettel Kahn R, Hughes ML, Grummer-Strawn LM. Obesity in women from developing countries. *Eur J Clin Nutr*. 2000;54:247–252.
8. Myles T, Gooch J, Santolaya J. Obesity as an independent risk factor for infectious morbidity in patients who undergo cesarean delivery. *Obstet Gynaecol*. 2002;100(5):959–964.
9. Usha Kiran TS, Hemmadi S, Bethel J, Evans J. Outcome of pregnancy in a woman with an increased body mass index. *Br J Obstet Gynaecol*. 2005;112(6):768–772.
10. Saravanakumar K, Rao SG, Cooper GM. The challenges of obesity and obstetric anaesthesia. *Curr Opin Obstet Gynaecol*. 2006;18(6):631–635.
11. CMACE. *Maternal Obesity in the UK: Findings from a National Project*. London: CMACE/NPSA; 2010.
12. Zhang J, Bricker L, Wray S, Quenby S. Poor uterine contractility in obese women. *Br J Obstet Gynaecol*. 2007;114(3):343–348.
13. Wolfe KB, Rossi RA, Warshak CR. The effect of maternal obesity on the rate of failed induction of labor. *Am J Obstet Gynaecol*. 2011;205(2):128.e1–128.e7.
14. Arrowsmith S, Wray S, Quenby S. Maternal obesity and labour complications following induction of labour in prolonged pregnancy. *Br J Obstet Gynaecol*. 2011;118 (5):578–588.
15. CMACE. *Survey on NHS Maternity Provision for Obese Women*. London: CMACE/ NPSA; 2008.
16. CMACE/RCOG. *Joint Guideline: Management of Women with Obesity in Pregnancy*. London: CMACE/NPSA; 2010.
17. NICE. *Guideline CG43: Obesity*. London: The National Institute of Clinical Excellence; 2006.
18. SIGN. *Management of Obesity: A National Clinical Guideline*. Edinburgh: The Scottish Intercollegiate Guidelines Network; 2010.
19. Jolly K, Lewis A, Beach J, et al. Comparison of range of commercial or primary care led weight reduction programmes with minimal intervention control for weight loss in obesity: lighten up randomised controlled trial. *BMJ*. 2011;343:d6500.
20. NICE. *Guideline PH27: Weight Management before, During and After Pregnancy*. London: National Institute of Clinical Excellence; 2010.
21. Kiel DW, Dodson EA, Artal R, Boehmer TK, Leet TL. Gestational weight gain and pregnancy outcomes in obese women-how much is enough? *Obstet Gynaecol*. 2007;110 (4):752–758.
22. RCOG. *Good Practice No. 8: Responsibility of Consultant on Call*. London: The Royal College of Obstetricians and Gynaecologists; 2009.
23. Homer C, Kurinczuk J, Spark P, Brocklehurst P, Knight M. Planned vaginal delivery or planned caesarean delivery in women with extreme obesity. *BJOG*. 2011;118: 480–487.
24. SIGN. *Antibiotic Prophylaxis in Surgery: a National Clinical Guideline*. Edinburgh: The Scottish Intercollegiate Guidelines Network; 2008.
25. Baker JL, Gamborg M, Heitmann BL, Lissner L, Sorensen TIA, Rasmussen KM. Breastfeeding reduces postpartum weight retention. *Am J Clin Nutr*. 2008;88 (6):1543–1551.
26. Bragg F, Cromwell DA, Edozien LC, et al. Variation in rates of Caesarean section among English NHS trusts after accounting for maternal and clinical risk: cross sectional study. *BMJ*. 2010;341:c5065.
27. Dennison F, Battacharya S, Norman J, et al. *The Clinical and Short-Term NHS Costs of Maternal Obesity for Maternity Services in Scotland*. Scottish Governments Chief Scientist Office–Focus on Research.

45 Providing Infertility Services for Obese Women

Sarah McRobbie[1] *and Abha Maheshwari*[2]

[1]Specialist Trainee, Gynaecology and Obstetrics, NHS Grampian
[2]Senior Lecturer, Reproductive Medicine University of Aberdeen

Introduction

Approximately half of the United Kingdom's female population is either overweight or obese [1]. Hence, the burden on Health Services is stretched not only with increased cardiovascular risk and other morbidities but with reproductive consequences as well.

This chapter will review the provision of services based on body mass index (BMI), obstacles that may be encountered in providing services to obese women and propose an algorithm for offering these services.

What Is the Current Situation?

There is a wide variation in the BMI cut-off used in various clinics as criteria to access treatment; this further varies depending on whether the treatment is funded by the state or by the couple themselves. For instance, most assisted reproduction units have set BMI limits between 30 and 35 kg/m^2 for access to fertility services, but thresholds vary between 25 and 40 kg/m^2[2]. In fact, a national survey in England in 2008 by the UK Department of Health revealed a postcode lottery in relation to BMI and access to fertility services [3]. The National Institute for Health and Clinical Excellence (NICE) guidelines suggest achieving a BMI $<$ 29 kg/m^2 is desirable [4], and the British Fertility Society guideline suggests that treatment should only be started once a BMI $<$ 35 kg/m^2 has been achieved. However, this guideline also states that a BMI $<$ 30 kg/m^2 in a female under the age of 37 with a normal follicle stimulating hormone level would be preferable [1]. Furthermore, The European Society of Human Reproduction and Embryology Task Force on Ethics and Law, in an article in 2010, stated that 'making assisted reproduction conditional upon prior lifestyle modification (efforts) is really to discriminate against people with certain values and beliefs' [5].

Hence, there is no consensus on what should be the upper limit to access fertility services.

Obesity. DOI: http://dx.doi.org/10.1016/B978-0-12-416045-3.00045-5

Evidence for BMI Restriction to Fertility Services and Not Providing Service

The arguments commonly presented for placing restrictions on BMI are concerns about poor outcomes from treatment, cost-effectiveness and increased risks to health; and if a pregnancy is achieved, the increased morbidity that may ensue [1,6,7].

The Centre for Maternal and Child Enquiries has placed obesity as its primary focus area for 2008–2011 [8]. One of the main reasons for this can be recognised by the fact that when one reviews the findings of The Confidential Enquiry into Maternal Death; more than half of all maternal morbidity from direct or indirect causes was in women who were either overweight or obese. Furthermore, over 15% of all women who died were at least morbidly obese [9]. Thus, these are significant findings and perhaps reiterate that pregnancy is not without risks and we as clinicians should recall the phrase '*Primum non nocere*' (First, do no harm) and the fact that we have a duty of care for non-maleficence to our patients.

What Is the Need?

Obesity impacts on menstrual disorders, infertility and complications in pregnancy [10]. Some observational studies suggest that the highest proportion of women seeking help for fertility problems are those with a BMI > 30 kg/m^2; however, this is not necessarily the group that receive fertility treatment [11]. There is an ongoing debate whether there should be restrictions on access to fertility treatment based on BMI [12]. However, denying fertility advice and treatments when a high proportion of overweight and obese women conceive spontaneously does not seem to be appropriate. Hence, there is a need for providing services which are specifically tailored to the needs of these women.

What Are the Potential Barriers?

There are several potential barriers to provide infertility services to obese women.

Awareness of the Problem Both by Patients and Clinics

A questionnaire-based study of 412 women in early pregnancy in Brisbane showed that 305 were overweight or obese pre-pregnancy. Approximately a third of the women (23 out of 65) in the overweight range recognised they were overweight and only 16% (8 out of 50) recognised they were obese [13]. In addition, this study also highlighted that it is not just the patient who can be a barrier to weight loss, it can be the assessor and advisor themselves, in the sense that they may be worried about causing offence and damaging rapport, and therefore not appropriately advise on weight loss [13].

Focusing on Short-Term Gain

Most infertility clinics only focus on the chances of pregnancy and at most the chance of achieving a live birth. If a patient is not undergoing treatment due to obesity, they do not tend to be followed up.

Limited Advice

The advice most women get from fertility clinics is to lose weight. However, this is predominantly poorly managed as most of the time no detailed information is given on how much, over what duration and what is the best method to achieve this; usually no regular follow-up is put in place to review this. In addition, these women are penalised by going back on the waiting list for treatment. Hence, these patients who are already disheartened due to infertility feel further dejected and are less likely to comply with advice and lose weight.

Limitation of Resources

Even if clinics wish to provide support and follow-up to these women, most of the time there are no resources to do so because of an already constrained economic environment.

Generalised Rather than Individualised Approach

As is evident from the literature, BMI is often evaluated in isolation to determine access criteria to fertility treatment. However, it is also apparent from the literature that several other factors including age, smoking and duration of infertility play a role – age potentially being the most important. A retrospective study by Sneed et al. [14] suggests that BMI alone does not appear to have a significant impact on the outcomes of in vitro fertilisation (IVF) treatment; however, when you look at the effects of BMI and age interaction, there is a notable effect. For instance, the study showed that with younger patients undergoing IVF, a high BMI had a negative influence and markedly decreased pregnancy rates, but the effect was attenuated with increasing age, and after age 36, minimal impact was demonstrated secondary to BMI; age appeared to be the predominant factor affecting fertility (Figure 45.1) [14].

What Can We Offer?

To provide services, the approach needs to address and consider the medical, social, psychological and ethical aspects of the individual.

We know that there are a number of recognised ways to lose weight which may need to be adopted including diet (and folic acid), exercise, psychological measures including cognitive behavioural therapy, pharmacological treatment and at times bariatric surgery.

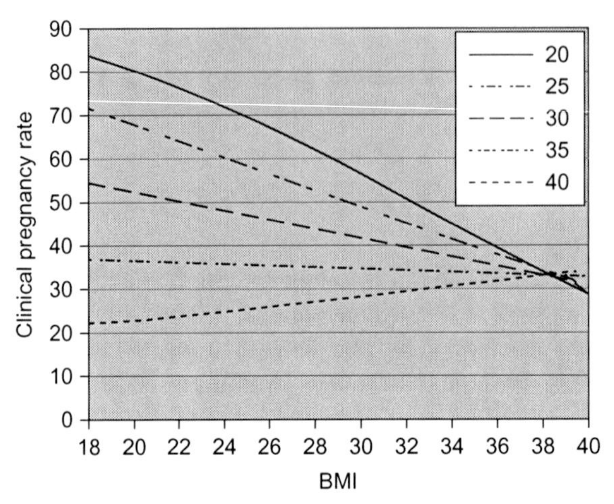

Figure 45.1 The effect of age and BMI on clinical pregnancy rates. Lines illustrate the best-fit regression lines derived by multiple logistic regression for clinical pregnancy rates as a function of BMI and age at 5-year intervals: 20, 25, 30, 35 and 40.
Source: From Sneed et al. [14].

Individualised Care

As opposed to a solely BMI-based approach, one should take into consideration age, tests of ovarian reserve, smoking and any relevant co-morbidities. Based on these factors, it should then be determined (according to the algorithm in Figure 45.2), if a woman is suitable to be treated immediately or needs to wait.

If one decides to defer the treatment, it is important to outline the therapeutic measures that will be employed to help lose weight. Table 45.1 from the American Congress of Obstetricians and Gynecologists (ACOG) paper in 2005 highlights the approach the ACOG would advise adopting on the management of weight loss in patients according to BMI and other factors [15].

It is important to take an individualised approach to the patient and not have a rigid protocol which is unlikely to be helpful.

Education and Awareness

There is a need for educating and ensuring women understand and are aware of the concerns, issues and implications surrounding their weight. Moreover, it may also be useful to have charts of success rates according to BMI available in waiting rooms to encourage and motivate women. In addition, weight-loss strategies should be well publicised and information must be readily available in various formats (written and internet sites) to decrease apathy towards this issue. Furthermore, implications of increased BMI on the whole life course should be emphasised as well.

Preconception Counselling

Instead of refusing to see women with a high BMI in the infertility clinic, we can use this opportunity to undertake preconception counselling to emphasise the impact of obesity on reproductive outcomes and the need to lose weight for both potential

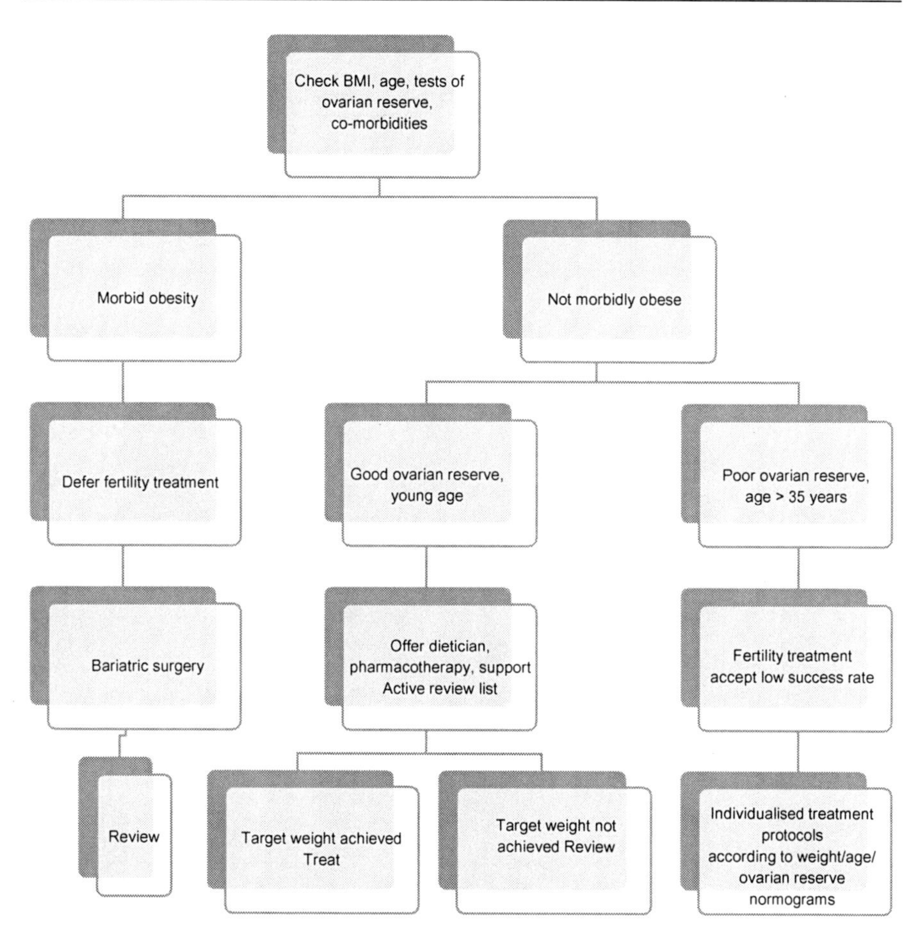

Figure 45.2 Recommended algorithm for the management of the obese women presenting with infertility.

short- and long-term benefits. In terms of weight loss in the obese woman planning pregnancy, it is strongly advised that this is tackled at the preconception stage prior to receiving infertility treatment. Perhaps, a triad of weight loss, contraception and high-dose folic acid (Figure 45.3) should be the initial basis of preconceptional care for these women [16]. Contraception should be carefully reviewed; ovulation can improve with minimal weight loss, and contraception is of importance to prevent undue risks and increased morbidity from pregnancy until a target weight is achieved.

NICE public health guidance states that 'Health professionals should use any opportunity, as appropriate, to provide women with a BMI of 30 or more with information about the health benefits of losing weight before becoming pregnant (for themselves and the baby they may conceive). This should include information on the increased health risks their weight poses to themselves and would pose to their unborn child' [4].

Table 45.1 A Guide to Selecting Treatment for Weight Reduction Based on BMI

Treatment	**BMI Category (kg/m^2)**				
	25–26.9	**27–29.9**	**30–34.9**	**35–39.9**	**≥40**
Diet – physical activity and behaviour therapy	If associated with co-morbidities	If associated with co-morbidities	+	+	+
Pharmacotherapy		If associated with co-morbidities	+	+	+
Surgery			If associated with co-morbidities	If associated with co-morbidities	If associated with co-morbidities

Source: From ACOG [15].

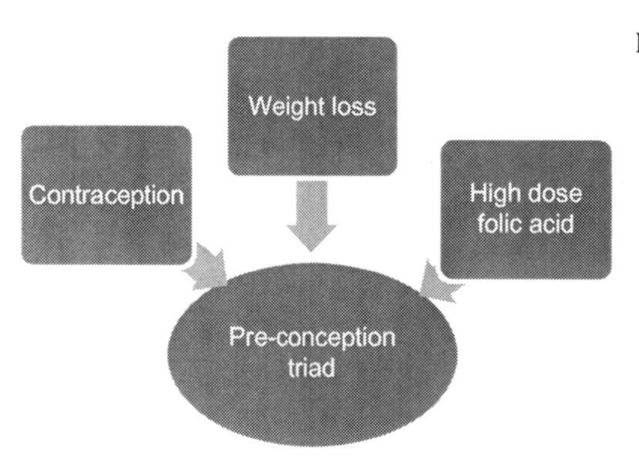

Figure 45.3 The preconception triad.

Multidisciplinary Approach

There must be consideration of a woman's needs, ideas, concerns and expectations in order to try to maximise the overall outcome. We need to regard this as a multidisciplinary approach with GP involvement and good nursing support initially at the primary care level, as well as close liaison with clinic staff, dieticians, surgeons and other health professionals as deemed appropriate.

In addition to weight loss, it will be important to tackle other lifestyle factors including smoking cessation and reduction in alcohol intake [4].

Group Therapy

NICE guidelines in 2004 advised that participation in a group programme involving exercise and dietary advice had better conception rates than just solely weight-loss advice [17]. Furthermore, if there are a group of women with similar need and goals, it may be possible that they could be put in touch with each other to setup and establish a group programme. However, a coordinator from the clinic will be essential to facilitate this.

Pharmacotherapy

Orlistat is the only oral medication that is licensed for weight loss in obese women. Its primary action is to reduce the absorption of fat by acting as a gastrointestinal lipase inhibitor [15,18]. In April 2009, this medication was made available as a 'pharmacy only medication' and as a result can be purchased over-the-counter by patients with a BMI ≥ 28 kg/m^2 and over 18 years of age [18]. Furthermore, in 2006 the meta-analysis by NICE found that orlistat in combination with a weight-reducing diet is more effective for weight-loss maintenance than placebo and diet at 12 months. Median weight loss across the 15 studies was 5.4 kg with orlistat and 2.7 kg with placebo. Moreover, there is some evidence to suggest that orlistat in combination with lifestyle measures had a significant positive effect on the occurrence of type 2 diabetes compared to placebo alone [19]. However, it is important to bear in mind that the dose available to purchase is 60 mg three times a day as opposed to 120 mg three times a day that was used in the NICE meta-analysis studies [18]. Furthermore, side effects have been reported with orlistat and these are mainly mild and short-lived gastrointestinal upset such as faecal urgency and oily rectal discharge; these are potentially related to and triggered by a caloric intake that is high in fat. Thus, minimising and distributing fat intake throughout any one day is advised to reduce these side effects [15,18]. Moreover, there is some evidence to suggest that stopping medication may result in prompt regain of any weight loss that may have already been achieved [18]. Therefore, in some women surgical management may be a more viable option for longer term sustained weight loss as we discuss later [15].

Bariatric Surgery

A recent review has suggested that bariatric surgery has 'positive effects on fertility and reduced the risk of gestational diabetes and pre-eclampsia' [20]. There is some evidence to suggest that there may even be reduced foetal macrosomia following the procedure [21,22]. However, it should still be managed as a high-risk pregnancy due to findings in other studies such as that by Sheiner et al. [23] which showed a statistically significant ($P < 0.001$) risk of intrauterine growth restriction if a woman had a history of bariatric surgery [23]. Yet, this is not consistent in all studies and not all studies show reduction in macrosomia either [24].

NICE guidelines in 2006 suggested that surgical management of obesity is more effective with a BMI > 38 kg/m^2 over the longer term than a non-surgical approach [19]. A retrospective study by Teitelman et al. [25] showed that in approximately 50% of women who had undergone bariatric surgery, there was return to a regular menstrual cycle in 71.4% and this correlated with weight loss [25]. However, large studies are required to review fertility pre- and post-procedure, and small retrospective studies have indicated an improvement [26]. There is no conclusive evidence either way on whether there is improvement or adverse effects in perinatal outcomes following bariatric surgery [20].

Moreover, further studies are needed to review whether there is an increase in congenital malformations associated with gastric banding as observational studies have suggested that there is a potential increase but this remains to be confirmed as to whether there is cause and effect [27]. Perhaps, one needs to consider whether any foetal anomalies and complications may be linked to the potential for nutritional deficiencies with this surgery. Therefore, these women need very close observation throughout pregnancy for complications and also particular attention paid to their nutritional needs; hence, they should be regarded as a high-risk pregnancy.

There is some evidence to suggest that pregnancy following bariatric surgery does not have an optimum time frame. One study has shown no difference in preterm labour and birthweight regardless of whether the timing of pregnancy was less than or greater than 1 year following the procedure [28].

Evidence on maternal and foetal outcomes after bariatric surgery are limited to observational case–control and cohort studies. No randomised controlled trials (RCTs) exist, nor are they likely to be feasible [24]. However, further information is needed to evaluate better the risks and benefits of bariatric surgery and the subsequent management of women throughout pregnancy and conception following such treatment.

Follow-Up/Support from the Clinic

These women should not be forgotten but should be under regular review as has occurred in New Zealand since the year 2000. The arrangements in New Zealand are based on a system to 'provide a rationing basis for public access to treatment for couples who were most in need but balanced by those who would benefit most from treatment' [6]. Multiple factors are included such as age, smoking status and duration of infertility to determine a 'prognostic score' and this then ultimately determines whether access to treatment is granted. If a woman has a qualifying score but her BMI is out with the eligibility criteria (only women with a BMI between 18 and 32 kg/m^2 are eligible for fertility treatment), she is placed on 'active review' and advised regarding weight loss and will only receive treatment when she meets the set BMI targets [6].

This gives women a goal and an incentive to lose weight.

There are resources available as outlined in the NICE guidance: these include information from the Department of Health in 2009 and also websites, including the 'Eat well' website (www.eatwell.gov.uk), that allow women to access information on diet and exercise from preconception to post-natal and beyond [4]. To achieve this, we should be offering a 'weight-loss support programme' and follow the principles set out as per NICE (Figure 45.4) [4].

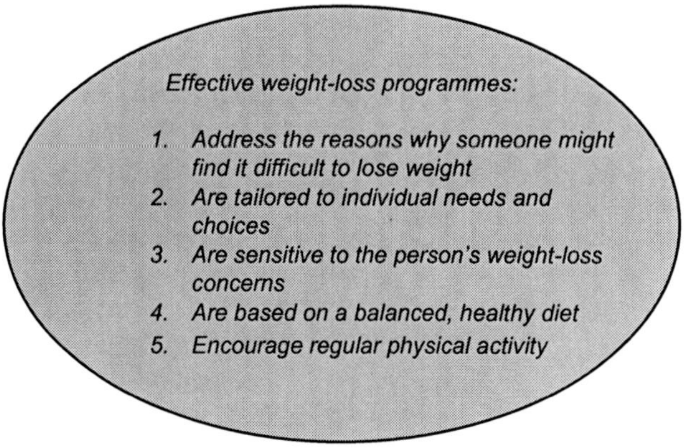

Effective weight-loss programmes:

1. *Address the reasons why someone might find it difficult to lose weight*
2. *Are tailored to individual needs and choices*
3. *Are sensitive to the person's weight-loss concerns*
4. *Are based on a balanced, healthy diet*
5. *Encourage regular physical activity*

Figure 45.4 How to organise an effective weight-loss programme.
Source: NICE [4].

How Much Weight Do You Need to Lose?

Weight loss with or without exercise has been shown to improve insulin sensitivity and ovulation frequency as previously discussed in other chapters. Kiddy et al. [29] described 11 anovulatory obese women losing more than 5% of their pre-treatment weight, and 9 out of the 11 showing improvement in reproductive function in terms of either becoming pregnant or developing a more regular menstrual pattern. However, the group who lost less than 5% of their pre-treatment weight had little improvement in reproductive function [29]. These findings were reiterated by Hollman et al. [30], where it was shown that in a group of 35 women undergoing a weight-reduction programme over 32 weeks with a weight loss of less than 10%, there was improvement in menstrual cycle in 80% and a clinical pregnancy occurred in 29% [30].

Moreover, NICE public health guidance advises that women should not be expected to lose more than 0.5–1 kg weekly [4].

Furthermore, NICE guidelines advise that health professionals should encourage women with a BMI of 30 kg/m^2 or above to lose weight prior to becoming pregnant, advise that 5–10% loss of body weight would improve both general health and also improve their chances of pregnancy as we have already discussed. Further weight loss should be encouraged to attain a BMI between 18.5 and 24.9 kg/m^2; however, 'losing weight to within this range may be difficult and women will need to be motivated and supported' [4].

However, it is important to ensure that timely access to diagnostic tests and effective treatments are set out in units. Zachariah et al. [2] showed that out of 86 Human Embryology and Fertilization Authority (HEFA)-licensed fertility units surveyed in the United Kingdom, 37% applied a time limit for weight loss.

Maintenance of Weight Loss

Although maintenance of weight loss initiated due to fertility reasons strictly does not come in the remit of services from fertility clinics, assessment and management should also focus on determining whether the woman has continued motivation to lose weight. ACOG [15] shows the stages of change model to assess this through the stages of pre-contemplation to maintenance.

The Public Health Interventions Advisory Committee suggests that weight-loss management takes a 'life-course approach' and recognises that the period before, during and after pregnancy provides an opportunity to give advice on diet, improving physical activity and effective weight management; health professionals need to ensure that opportunities are not missed. We should also take into account the individual woman as a whole and her social and economic circumstances and try to involve the whole family to improve health benefits globally [4].

Prepregnancy BMI is a greater determinant of healthy outcomes for mothers and babies than any weight gain during pregnancy[4].

Cost-Effectiveness of Weight-Loss Strategies

Moreover, the current LIFESTYLE study that is being undertaken in the Netherlands is a multicentre RCT looking at the cost-effectiveness of a 6 month structured lifestyle programme (diet adaptation, exercise and behavioural therapy) aiming at weight loss of at least 5–10% in sub-fertile women with a BMI between 29 and 40 kg/m^2, followed by conventional fertility care and the control group with conventional fertility care without recruitment to a weight-reduction programme. The aims are to review the costs and effects of intervention prior to conception; the results are expected in 2014 [31].

How Can We Improve the Services and Promote Weight Loss?

Taking into consideration all the above principles, we propose steps for providing infertility services for obese women as shown in Figure 45.5.

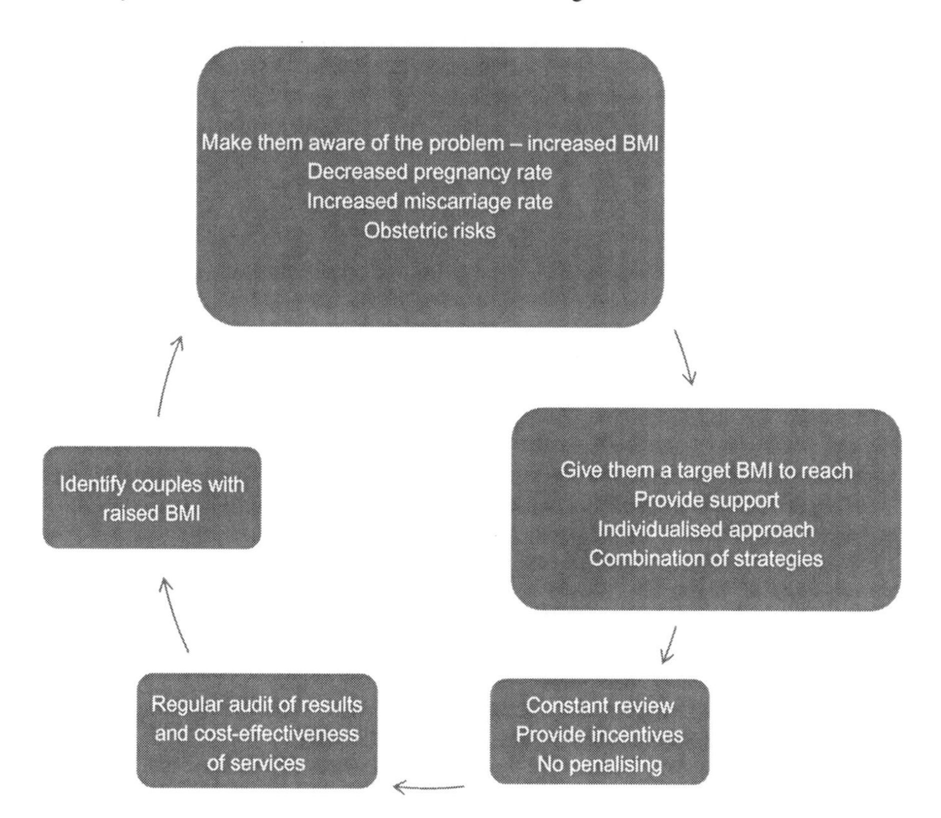

Figure 45.5 Management of the infertile couple presenting with increased BMI.

Conclusions

The increasing number of obese individuals in the United Kingdom and worldwide is impacting significantly on the public health burden. There is no consensus at present on upper limit of BMI cut-off to provide fertility treatment. However, individual UK fertility units have their own guidelines. An individualised multi-disciplinary approach rather than a one size fit for all is needed to provide fertility services to obese women and motivate them to lose weight. Regular audit of services for results and cost-effectiveness is needed.

References

1. Balen AH, Anderson RA. Impact of obesity on female reproductive health: British fertility society, policy and practice guidelines. *Hum Fertil*. 2007;10:195–206.
2. Zachariah M, Fleming R, Acharya U. Management of obese women in assisted conception units: a UK survey. *Hum Fertil*. 2006;9:101–105.
3. The United Kingdom Department of Health. *Primary Care Trust Survey – Provision of IVF in England 2007*. London: Department of Health; 2008:<http://www.dh.gov.uk/en/Publicationsandstatistics/Publications/PublicationsPolicyAndGuidance/DH_085665>
4. NICE. *Dietary Interventions and Physical Activity Interventions for Weight Management Before, During and After Pregnancy – NICE Public Health Guidance No: 27*. London: National Institute for Health and Clinical Excellence; 2010.
5. ESHRE Task Force on Ethics Law, Dondrop W, de Wert G, Pennings G, et al. Lifestyle-related factors and access to medically assisted reproduction. *Hum Reprod*. 2010;25:578–583.
6. Gillett WR, Putt T, Farquhar CM. Prioritising for fertility treatments – the effect of excluding women with a high body mass index. *BJOG*. 2006;113:1218–1221.
7. Nelson SM, Fleming R. Obesity and reproduction: impact and interventions. *Curr Opin Obstet Gynecol*. 2007;19:384–389.
8. Centre for Maternal and Child Enquiries (CMACE) and the Royal College of Obstetricians and Gynaecologists (RCOG). *Joint Guideline – Management of Women with Obesity in Pregnancy*. London: CMACE/RCOG; 2010.
9. Lewis G, ed. *Saving Mothers' Lives – Reviewing Maternal Deaths to Make Motherhood Safer 2003–2005*. London: CEMACH; 2007.
10. Sharpe RM, Franks S. Environment, lifestyle and infertility – an intergenerational issue. *Nat Cell Biol*. 2002;4(suppl):33–40.
11. Vahratian A, Smith YR. Should access to fertility-related services be conditional on body mass index? *Hum Reprod*. 2009;24:1532–1537.
12. Pandey S, Maheshwari A, Bhattacharya S. Should access to fertility treatment be determined by female body mass index? *Hum Reprod*. 2010;25:815–820.
13. Callaway LK, O'Callaghan MJ, McIntyre HD. Barriers to addressing overweight and obesity before conception. *Med J Aust*. 2009;191:425–428.
14. Sneed ML, Uhler ML, Grotjan HE, Rapisarda JJ, Lederer KJ, Beltsos AN. Body mass index: impact on IVF success appears age-related. *Hum Reprod*. 2008;23:1835–1839.

15. ACOG Commitee on Gynecologic Practice: Committee Opinion. *The Role of the Obstetrician–Gynecologist in the Assessment and Management of Obesity*. London: ACOG; 2005:Number 319.
16. Nelson S, Fleming R. The preconceptual contraception paradigm: obesity and infertility. *Hum Reprod*. 2007;22:912–915.
17. NICE. *Fertility: Assessment and Treatment for People with Fertility Problems – NICE Clinical Guideline No: 11*. London: National Institute for Health and Clinical Excellence; 2004.
18. SIGN. *Management of Obesity – SIGN Guideline No. 115*. Edinburgh: Scottish Intercollegiate Guidelines Network; 2010.
19. NICE. *Obesity: Guidance on the Prevention, Identification, Assessment and Management of Overweight and Obesity in Adults and Children – NICE Clinical Guideline No: 43*. London: NICE; 2006.
20. Hezelgrave NL, Oteng-Ntim E. Pregnancy after bariatric surgery: a review. *J Obes*. 2011;1–5.
21. Wittgrove AC, Jester L, Wittgrove P, Clark GW. Pregnancy following gastric bypass for morbid obesity. *Obes Surg*. 1998;8:461–464.
22. Patel JA, Patel NA, Thomas RL, Nelms JK, Colella JJ. Pregnancy outcomes after laparoscopic Roux-en-Y gastric bypass. *Surg Obes and Relat Dis*. 2008;4(39–45):2005.
23. Sheiner E, Levy A, Silverberg D, et al. Pregnancy after bariatric surgery is not associated with adverse perinatal outcome. *Am J Obstet Gynecol*. 2004;190:1335–1340.
24. Dixon JB, Dixon ME, O'Brien PE. Birth outcomes in obese women after laparoscopic adjustable gastric banding. *Obstet Gynecol*. 2005;106:965–972.
25. Teitelman M, Grotegut CA, Williams NN, Lewis JD. The impact of bariatric surgery on menstrual patterns. *Obes Surg*. 2006;16:1457–1463.
26. Bilenka B, Ben-Shlomo I, Cozacov C, Gold CH, Zohar S. Fertility, miscarriage and pregnancy after vertical banded gastroplasty operation for morbid obesity. *Acta Obstet Gynecol Scand*. 1995;74:42–44.
27. Friedman D, Cuneo S, Valenzano M, et al. Pregnancies in an 18-year follow-up after biliopancreatic diversion. *Obes Surg*. 1995;5:308–313.
28. Dao T, Kuhn J, Ehmer D, Fisher T, McCarty T. Pregnancy outcomes after gastric-bypass surgery. *Am J Surg*. 2006;192:762–766.
29. Kiddy DS, Hamilton-Fairley D, Bush A, et al. Improvement in endocrine and ovarian function during dietary treatment of obese women with polycystic ovary syndrome. *Clin Endocrinol*. 1992;36:105–111.
30. Hollman M, Runnebaum B, Gerhard I. Effects of weight loss on the hormonal profile in obese, infertile women. *Hum Reprod*. 1996;11:1884–1891.
31. Mutsaerts MA, Groen H, ter Bogt NC, et al. The LIFESTYLE study: costs and effects of a structured lifestyle program in overweight and obese subfertile women to reduce the need for fertility treatment and improve reproductive outcome. A randomised controlled trial. *BMC Womens Health*. 2010;10(22):1–9.

46 Summary and Research Recommendations

Tahir Mahmood[1] *and Sabaratnam Arulkumaran*[2]

[1]Victoria Hospital, NHS Fife, Kirkcaldy, UK, [2]St George's University of London, London, UK

Introduction

Obesity is derived from the Latin term 'Obesus', which means 'one who has become plump through eating'. The negative effect of obesity on an individual's health has been known for a longer time and can be found in the writings of Hippocrates, Galen and Avicenna [1]. The rising prevalence of obesity is a pandemic of public concern affecting all age groups and not only confined to developed countries. **Haththotuwa et al. (Chapter 1)** point out that low- and middle-income countries are facing a 'double burden of disease' since undernutrition also poses a significant problem in these countries. Obesity increases the risk of developing type 2 diabetes, hypertension (HTN), ischaemic heart disease, cancer (particularly colon and breast), cerebrovascular disease and osteoarthritis. Furthermore, a recent meta-analysis including data from 57 prospective studies with 894,576 participants reported that for every increase in body mass index (BMI) of 5 kg/m^2 there was a 30% overall higher mortality with a 40% increase in vascular mortality, a greater than 50% increase in diabetic, renal and hepatic mortality, a 10% increase in neoplastic mortality and 20% increase in respiratory and other mortality [2].

In addition, overweight and obesity increases the risk of many other conditions including cerebrovascular disease, gallstones, osteoarthritis, sleep apnoea, impaired physical functioning, psychological illness and social discrimination. The impact of obesity on fertility and adverse outcomes of pregnancy are well recognised.

Environmental, Genetic and Psychological Factors

In order to develop focused strategies, we need to have a clear understanding of the interplay of genetic, environmental and psychological factors which lends to increased propensity to obesity. The age of onset and the progression of weight gain differs between individuals: with sex, ethnicity and baseline BMI being

Obesity. DOI: http://dx.doi.org/10.1016/B978-0-12-416045-3.00046-7

predictors of progressive weight gain. Physical activity levels are major determinants for obesity. There is marked variation in level of physical activity among ethnic minority populations in United Kingdom. **Lim and Thanoon (Chapter 2)** have therefore questioned whether it is possible that a combination of personal, socio-economic, cultural and environmental barriers may discourage people from ethnic minority groups from engaging in physical activity?

Hasan and Mahmood (Chapter 3) point out that fat mass is genetically influenced among individuals who are more genetically similar such as monozygotic twins than dizygotic twins. Even for twins reared apart, the estimated heritability was at 65–75% for BMI. Furthermore, results of longitudinal behaviour genetic studies suggest that different obesity-promoting genes may become active at different ages across the lifespan. The 'Human Obesity Gene Map' has shown that the number of genes associated with obesity-related traits have increased dramatically over the past decade, thereby suggesting that the platform of specific genes that might contribute to obesity is large and involves loci throughout the genome. The effect of new environments on propensity to develop obesity has been examined in two studies: one on 247 Punjabi immigrants in London against their siblings still living in Punjab, India, and the other one on Pima Indians living in remote rural regions of Mexico compared with those living in Arizona have clearly demonstrated how new environmental conditions could modulate genetic propensity for the development of obesity [3,4].

However, **Kökönyei et al. (Chapter 4)** provide a succinct account of the role played by the psychological factors from both cross-sectional and prospective studies describing a bi-directional relationship between depression, anxiety and obesity, though certain moderators such as gender and age may modify the relationship between psychopathology and obesity. Both early relational traumas such as abuse and neglect and personality traits such as high neuroticism and low conscientiousness relate to the increased risk of obesity. However, a clear differentiation should be made in between the childhood and adulthood obesity as underlying factors are quite diverse. In addition, economical and psychological factors related to poverty also influence food intake as 'The nutrition paradox' leads to the observation that among people with lower socio-economic status, food insecurity might also be associated with obesity, especially in women. Besides mindless eating and external eating, emotional eating can also play a part in obesity. It is also a well-documented observation that childhood traumas, e.g. sexual abuse, neglect and interpersonal aggression represent a risk factor to obesity in both childhood and adult obesity [5–7].

Busby and Seif (Chapter 5) report that based on World Health Organization criteria [8] for weight (>2 SD above median) the worldwide estimate for overweight and obese preschool children aged from birth to 5 years old was 43 million. The prevalence of overweight and obesity in developed countries is about double than in developing countries (11.7% and 6.1%, respectively); the majority of affected children however live in developing countries (35 million). The adolescent obesity has major impact on the long-term health of individuals as a number of them would develop type 2 diabetes and HTN in their 30s.

Sexual Health

Damodaran and Swaminathan (Chapter 6) recommends that where possible the most effective methods of contraception should be provided for obese women and advised on prevention of sexually transmitted infections (STI) reinforced for both sexes at all opportunities. Health professionals working in sexual health services should also be encouraged to have sensitive discussions around weight loss and to signpost individuals to appropriate services for weight reduction [9,10].

The challenge for health care planners is to develop focused education strategies for this group of women not only to reduce the incidence of untimed pregnancies but also to address sexual risk-taking behaviour.

Cameron (Chapter 7) points out that sexual health is impaired among obese men and women and this was clearly demonstrated by a US study where obesity was considered to be associated with a high prevalence of lack of sexual enjoyment, lack of sexual desire, difficulty with sexual performance and a avoidance of sexual encounters. She points out that there are reports of higher incidences of STI which appeared to be 10 times greater for obese men than for men of normal weight. In addition, the odds of contracting STI among women were higher as well. It can only be concluded that obesity affects sexual risk-taking behaviour. Obese women, as previously noted, are less likely to be using effective contraception than normal weight counterparts, and that also increases the risk of STI and risks related to untimed pregnancy.

Reproduction

Messinis et al. (Chapter 8) explore the role of hyperandrogenism, hyperinsulinaemia and adipokines. Leptin levels were found to be positively correlated with insulin resistance in women, as confirmed by the in vivo and in vitro experiments in animals where high levels of leptin may inhibit folliculogenesis.

They also remind us that although obesity is not included in the diagnostic criteria for polycystic ovary (PCO), but obesity may intensify the severity of the phenotypic characteristics of the PCO, including disturbed menstrual cycle. The success of the infertility treatment in women with PCO is highly influenced by the increased body weight. Diet and lifestyle changes result in the improvement of various parameters and are recommended as first-line approach in obese women with PCO.

Bhandari and Quenby (Chapter 9) reiterate the relationship between BMI >25 and recurrent miscarriage before the first live-born child and states that this risk persists in subsequent pregnancies. Although maternal age is a strong predictor, it is now apparent that an increase in BMI acts independent of age. This observation is important for counselling the ever-rising number of women requesting detailed investigations into their history of recurrent miscarriages with a little insight that they could help themselves by weight management [11].

Hamilton and Maheshwari (Chapter 10) have observed that the proportion of patients accessing infertility services who are obese is increasing. They point out that natural fecundity; responses to treatment and pregnancy outcome are sub-optimal in this group of patients. Furthermore, increased doses of gonadotrophin are required to elicit an ovarian response, fertilisation rates of oocytes may be impaired, the ultimate yield of cumulus–oocyte complexes may be less than in normal weight controls and there may be less available surplus embryos for cryostorage thereby potentially having an impact on cumulative pregnancy rates per episode of ovarian stimulation. The increased risks of miscarriage in the obese after in vitro fertilisation may be attributed to qualitative effects on oocytes leading to aberrant embryo development [12,13].

All these observations lead to an ethical question: Should there be a BMI cut-off, when women are not offered assisted-conception treatment?

Male Reproduction

Busby and Seif (Chapter 5) point out that unlike their female counterparts, pubertal development in obese boys may be delayed. This may be related to increased aromatisation of androgens to oestrogens in adipose tissue and feedback inhibition of gonadotrophin secretion may be involved. These factors are aggravated by low esteem and lack of physical activity. Regrettably, a significant proportion of these adolescents become grossly obese in their 30s. Obese adolescents are more likely to develop diabetes mellitus (DM) which may be an insidious presentation, and have a threefold-increased risk of developing HTN due to sodium retention, increased sympathetic tone or increased angiotensin system activity and lipid metabolic abnormalities [14].

Male sexual dysfunction has been extensively explored by **Paduch et al. (Chapter 11)** clearly demonstrating how obesity is known to affect hormonal levels specifically increase in circulating oestradiol (E_2) level and decrease in total testosterone level. In addition, obesity-related medical co-morbidities like HTN, dyslipidaemia and DM are linked to erectile dysfunction.

Erectile and endothelial dysfunction may have similar pathways through impaired nitric oxide activity; in obese men, the endothelial dysfunction may be further impaired through increase in interleukins (IL-6, IL-8, IL-18) and CRP (C-reactive protein). Obese men with erectile dysfunction had significant increase in CRP and inflammatory procytokines. Their scholarly review also contributes towards our understanding of male sexual dysfunction and male factor infertility.

Kay and Martins da Silva (Chapter 12) report that sperm parameters that demonstrate the most convincing correlation with obesity are a low total sperm concentration, a decrease in sperm motility, an increase in DNA fragmentation and sperm dysfunction. Endocrine abnormalities (including increased plasma levels of oestrogen, leptin, insulin resistance and reduced androgens and inhibin B levels) are likely to be central in the aetiology of sperm dysfunction in obese men. However, other factors may also contribute, including genetic abnormalities, sexual dysfunction, testicular hyperthermia and oxidative stress [15,16].

George and Anderson (Chapter 13) examine the role of bariatric surgery for the management of obese male and its impact on endocrine profile. What is clear that although following bariatric surgery, there is evidence of improvement in endocrine profile and sexual function but no such dramatic change in sperm function has been reported [17].

Impact of Pre-Pregnancy Obesity

Gluckman and co-workers (Chapter 14) explain that maternal obesity has profound effects on the intrauterine environments of the baby, and the offspring's BMI is directly correlated with maternal BMI rather than the paternal BMI. Clearly, this effect is long lasting as maternal BMI was noted to be associated with offspring adiposity at 12 and 24 months while maternal glycaemia was correlated with offspring adiposity at birth, although not at 12 and 24 months, suggesting a link between maternal obesity and offspring adiposity. Foetuses of obese mothers not only have greater percentage body fat but have elevated cord blood leptin, IL-6 and more insulin resistance at birth. A strong correlation is present between foetal adiposity, maternal insulin resistance and maternal BMI, which persists despite correcting for confounding factors. Intrauterine environments can be improved by weight management as demonstrated by a study where the cardiometabolic profile of offspring born before or after the mother underwent bariatric surgery demonstrated a significantly lower risk of obesity in offspring of those obese mothers who achieved substantial weight reduction through surgery.

The Southampton Women's Survey clearly provides an opportunity to further explore strategies for managing excessive maternal weight gain during pregnancy in order to curtail the long-term effects of overnutrition as opposed to maternal obesity per se [18].

Issues Related to Pregnancy

Walsh and McAuliffe (Chapter 15) have reviewed literature on screening for foetal abnormalities during pregnancy and reported an elevated risk of congenital anomalies, in particular neural tube defects and congenital heart defects. Furthermore, other congenital defects, including omphalocoele, hypospadias, renal anomalies and hydrocephalus, are over represented in the obese population. The sensitivity of diagnosing foetal abnormality decreases with incremental BMI and this is especially important when the BMI is 35 kg/m^2 or greater. Similar concerns have been expressed as regards foetal growth assessment during the third trimester [19].

Of late, it has been reported that there is increased risk for obese women being post-dates. There are serious implications for clinical practice [20].

Malarselvi and Quenby (Chapter 16) explain that it seems likely that cortisol is important in the onset of parturition and the disregulation of this in obese women may contribute to the increased incidence of post-dates pregnancy. Alternatively, the alteration in the oestrogen/progesterone ratio which is known to occur prior to the onset of labour may be upset by a difference in circulating oestrogens in obese women given the concentration of oestrogen in the excess adipose tissue. It has been postulated that perhaps some of the so-called prolonged pregnancies might be due to wrong dating in early pregnancy. Further research is warranted in this area.

Ellison and Thomson (Chapter 17) point out that the exponential rise in obesity seen in the general population over recent years has been mirrored in the obstetric population which has seen a >50% increase in the prevalence of obesity between 1990 and 2004. Regrettably, in the United Kingdom approximately 20–30% of deliveries are affected by caesarean section (CS). Women delivered by emergency CS have double the risk of post-partum venous thromboembolism (VTE) compared with elective CS. Women who deliver by emergency CS have a roughly fourfold increased risk of post-partum VTE compared with those who deliver vaginally. These risks are higher again for obese women. They justifiably make a case for screening women throughout pregnancy for the increased risk of venous thromboembolic disease and considering prophylactic treatment with suitable agents [21].

The relationship between super obesity and the development of pre-eclampsia has been explored by **Broughton-Pipkin (Chapter 18)** demonstrating that a raised BMI both pre-pregnancy and at booking was a consistently significant risk factor for pre-eclampsia. Biochemical changes in obesity are suggestive of a type of chronic inflammatory condition. While the aetiology of pre-eclampsia is still being explored, is it possible that an obese woman is having an exaggerated inflammatory response to pregnancy? It is equally possible that pregnancy is simply unmasking a tendency to underlying cardiovascular disease (just like gestational diabetes), or are gestational HTN and pre-eclampsia truly pregnancy specific or is there a spectrum of gestational cardiovascular disease, between a pure form and that driven by classic predisposal to non-pregnant cardiovascular disease? Is it possible that this lipidaemia in obese women triggers the development of placental blood atherosis and pre-eclampsia? Perhaps this lipidaemia together with increased inflammation explains the strong epidemiological association of pre-eclampsia with pre-pregnancy BMI [22].

It is so intriguing to read **Doshan and Konje (Chapter 19)** as they provide an insight into placental function in women with obesity. The human placenta virtually expresses all known cytokines, including tumour necrosis factor (TNF), resistin and leptin, which are also produced by adipose tissue. Similarly, a two- to threefold increase in placental macrophages population, characterised by increased expression of the pro-inflammatory cytokines in obese women compared with non-obese women, has been reported. However, there are no published studies examining the process of trophoblastic invasion in relation to foetal growth in obese pregnant women. Adipokines also modulate placental function, e.g. leptin has been shown to regulate placental angiogenesis protein synthesis and growth and cause immunomodulation. Thus, it is possible that the increase in local and/or circulating levels of leptin in maternal obesity may modulate placental inflammation and function.

Obesity is also associated with pro-thrombotic state. Excess adipose tissue contributes directly to the pro-thrombotic state by impaired platelet function via low-grade inflammation and increase in circulating leptin, impaired fibrinolysis by production of plasminogen activator inhibitor-1 (PAI-1) and possibly thrombin activatable fibrinolysis inhibitor, impaired coagulation by release of tissue factor and affecting hepatic synthesis of coagulation factors.

They have postulated that foetal overgrowth may be associated with highly successful trophoblastic invasion, increasing subsidiary transfer by overcoming flow limitations and providing subsidiaries for placenta transport processes. Conversely, placental trophoblastic invasion may be disturbed in that subset of obese women who are destined to deliver growth restricted babies [23].

Gestational Diabetes and Metabolic Syndrome

Hornnes and Lauenborg (Chapter 20) remind us that several studies have demonstrated that there is no clear threshold for having or not having gestational diabetes mellitus (GDM). The association between blood glucose levels and different adverse outcomes is linear and gradual. The higher the glucose values, the higher the risk for an adverse outcome. Not all obese women develop GDM but obese women are four times more likely and severely obese women almost nine times more likely to develop gestational diabetes than lean women. Women who develop GDM are predisposed to developing type 2 DM in later life with the risk greatest in those who are obese. However, there remains considerable debate when to perform an oral glucose tolerance test and what is an ideal test [24].

The underlying mechanism of metabolic syndrome of pregnancy has been extensively debated by **Visser and Yogev (Chapter 21)**, **Balani et al. (Chapter 22)** and **Huda and Nelson (Chapter 23)**. The consensus of opinion is that adipose tissue should no longer be considered a storage organ of excess triglyceride. In particular, fat cells secrete factors involved in inflammation, haemostasis (TNF-α, IL-16, PAI-1), insulin sensitivity (adiponectin), in energy balance and control of appetite (leptin). A state of chronic positive energy balance such as obesity leads to increased storage of triglycerides which results firstly in adipocytes hypertrophy and subsequently hyperplasia through adipogenesis. This in turn leads to cellular dysfunction resulting in disregulatory release of adipokines, increased release of free fatty acids and inflammation. It has been reported that levels of IL-6 was 50% higher and CRP was almost double, in the obese pregnant women, relative to the lean group. High CRP is well known to correlate with endothelial dysfunction and impaired insulin sensitivity in non-pregnant population and predicts the risk for type 2 diabetes in women.

Huda and Nelson (Chapter 23) also expand on the physiological hyperlipidaemia of pregnancy which appears to be exaggerated in obese women especially those with higher visceral fat content, with higher serum triglycerides, very low-density lipoprotein cholesterol and free fatty acid concentrations than those observed in lean women. This is seen together with lower levels of the endothelial protective high-density lipoprotein cholesterol, although low-density lipoprotein (LDL) cholesterol and total cholesterol concentrations appear similar. Despite the similar concentration of LDL, its susceptibility to oxidation, classic associate of endothelial dysfunction, arthrosclerosis and cell toxicity appears to be exaggerated by maternal obesity. Collectively, this pattern of dyslipidaemia is similar to that of the metabolic syndrome in the non-pregnant population. The common pathway in which inflammation leads to insulin resistance is considered to be through modulation of insulin signalling and in particular induces serene phosphorylation relation of the insulin receptor IRS1.

Sattar and Greer [25] have proposed that the development of metabolic syndrome among obese women during pregnancy may be a stress response to maternal metabolic changes. Women who develop adverse pregnancy outcomes such as pre-eclampsia and GDM make greater excursions into metabolic disturbances during pregnancy. Obesity brings women closer to the threshold over which metabolic disease in pregnancy manifests itself.

Visser and Yogev (Chapter 21) ponder about the occurrence of intrauterine growth restriction among the obese pregnant despite the fact that there are higher circulating concentrations of nutrients leading to enhanced placental transfers. It is plausible that obese women may lack one or more major or micronutrients essential for foetal growth despite overnutrition. Obese women may also limit their calorie intake, thus the consumed nutrients may be directed to supporting maternal metabolism rather than growth of foetus. It is also possible that a post-receptor insulin signalling defect may reduce the insulin-induced stimulation of placental transport systems such as system A, resulting in reduced nutrient transfer and foetal growth restriction. It should be noted that a combination between maternal obesity and gestational diabetes have a symbiotic and greater effect than either condition alone.

Metformin is a remarkable drug and is widely used in GDM. Its use may be associated with reduced risk of cancer, most likely by killing tumour initiating stem cells with anti-angiogenesis' effects, anti-inflammatory effects, growth-inhibitory effects and anti-oxidative effects. *Visser and Yogev (Chapter 21)* cautiously say that although this drug appears to be good for the prevention and/or treatment of cancer but what about a 9-month exposition to the foetus? In other words, we still do not know enough about possible adverse effects of this drug in the embryo and foetus, therefore animal and basic human studies seem mandatory before it might be declared safe.

Balani et al. (Chapter 22) have proposed that by controlling/restricting gestational weight gain to less than 15 lb among overweight or obese pregnant women, could lead to a significantly lower risk of Pre eclampsia, Caesarean section rates and babies who are large for dates. The authors concluded that limited or no weight gain during pregnancy in obese pregnant women results in a more favourable pregnancy outcome. There is conflicting advice as regards the type of food which should be consumed by the obese pregnant women, but a Cochrane review on the effects of physical activity on pregnancy concluded that the available data were 'insufficient to infer important risks or benefits to the mother or infant'. Indeed, Guidelines from the American College of Obstetricians and Gynaecologists (2002) [26] recommended that pregnant women should exercise for 30 min or more on most days of the week and participate in moderate-intensity exercise unless there were medical or obstetric complications. Whilst these recommendations have been widely adopted, they are consensus rather than evidence based.

Intrapartum Care and Pregnancy Outcome

Chiswick and Denison (Chapter 24) observe that obese women are more likely to require induction of labour for a variety of reasons. National Institute for Health and Clinical Excellence (NICE) recommends the use of membrane sweeping and vaginal prostaglandin E_2 followed by artificial rupture of membranes and syntocinon augmentation where necessary. However, no studies to date have specifically examined the suitability of these various methods in relation to maternal BMI. Dose and method of delivery of prostaglandin used for cervical ripening varies between countries, but again no studies have specifically looked at whether a variable dose is required depending on weight.

It has been reported that obese women with BMI $>30\ \text{kg/m}^2$ and extremely obese BMI $>40\ \text{kg/m}^2$ women were more likely to require pre-delivery oxytocin for labour augmentation following prostaglandin. The total dose of oxytocin required was also proportional to BMI. Therefore, it was hypothesised that increased requirement of oxytocin may be explained by impaired myometrial contractility in obese women and the relative increase in volume of distribution in obese women which may have a dilutional effect on both the prostaglandin ripening agent and the oxytocin. This is indeed biologically plausible [27].

Issues around intrapartum care and place of birth have been fully explored by **Edwards and Lim (Chapter 25)**, suggesting that in line with the established guidelines, a detailed risk assessment of each women with BMI $>30\ \text{kg/m}^2$ should be carried out and a multidisciplinary team should be involved in the care of these women. A policy of planning delivery in a regional or tertiary centre, rather than in small, rural maternity units, should be considered as not all birthing facilities have the equipment to enable health professionals to care appropriately and safely for obese women in labour. The implementation of a 'bariatric protocol' to identify and mobilise the necessary equipment and resources for women with BMI $>50\ \text{kg/m}^2$ is suggested for maternity units that care for the obese parturient [28,29].

Richens et al. (Chapter 26) provide a detailed account on the role of midwives in supporting these high-risk group women providing continuity of care during antenatal, intrapartum period and post-natal phase. Central to effective maternity care is an understanding of the women receiving care and their needs. Their views and beliefs are essential, and measures should be in place to ensure that each key point is communicated to them. Each antenatal clinic visit should have a clearly defined objective. Midwives are in unique position to engage with women and gain their confidence as regards weight management, smoking cessation and healthy eating. The role of midwife should continue during labour and post-natal care. During that phase, they may be in a position to help the women lose weight by encouraging breastfeeding.

Milne and Lee (Chapter 27) point out that obesity is associated with increased maternal mortality and is a specific risk factor for anaesthesia-associated deaths. Obesity is associated with increased risk of cardiovascular system malfunction because of associated co-morbidities (pre-eclampsia, diabetes, etc.), airway difficulties, aspiration risk and technical challenges to regional anaesthetic techniques.

There should be clearly agreed guidelines between obstetric and anaesthetic teams to ensure that all high-risk women are identified antenatally and they had been appropriately assessed by the anaesthetic team as early as possible during pregnancy and a clear plan laid out for intrapartum care. Early anaesthetic involvement in labour may permit time for epidural or combined spinal/epidural techniques which may be time consuming in the emergency situation. General anaesthesia is potentially hazardous and should only be undertaken by senior, experienced personnel. Specialised equipment must be available. Management in the immediate post-operative period following CS should be considered in advance. There is a risk of inadequate ventilation or airway compromise post-operatively leading to hypoxaemia. There should be a low threshold for admitting these women to a high-dependency area capable of managing obese patients. Maintenance of epidural analgesia into the post-operative period may limit respiratory depression from opioid use and permit rapid establishment of anaesthesia [30].

Almost 3 million stillbirths happen worldwide every year [31]. After excluding chromosomal abnormalities, maternal associations include obesity, smoking and diseases of the endocrine, renal, cardiac and haematological systems. Infections account for some foetal losses as well. **Black and Bhattacharya (Chapter 28)** elucidate the evidence around stillbirths and obesity. In the economically rich countries, obesity makes the largest contribution as it has been confirmed by the recent meta-analysis of the publications from five high-income countries with the highest stillbirth rates which reveal overweight and obesity as the highest ranking modifiable risk factors with a population-attributable risk of 8–18%. Most recently published evidence suggests that obese women are at least twice more likely to experience a stillbirth than those of normal weight. It is plausible that obesity is a potential explanation for the recent rise in stillbirth rates as the soaring levels of obesity coincide with this trend. Furthermore, given the variation in strategies adopted to identify gestational diabetes and the cause of stillbirths, it is possible that undiagnosed gestational diabetes may form a significant proportion of those stillbirths which are classified as unexplained. Recent research is indicative of the role of placental dysfunction in a sizeable proportion of stillbirths in high-income countries with a population-attributable risk of 23% in offspring, small for gestational age and 15% in placental abruption. As the average maternal age on first childbearing is increasing, this risk factor along with obesity becomes increasingly important over time. High rates of obesity may partially explain the higher stillbirth rates associated with deprivation, which is further complicated by poor dietary habits with sub-optimal physical activity which are common both in obese women and those of low socio-economic class. This chapter clearly lays out various preventive strategies to address this challenge.

Examination of placenta is highly relevant in order to elucidate the impact of biochemical changes during pregnancy and how it affects placental structure and the foetal well-being. **Evans (Chapter 29)** provide a pathologist's perspective in this debate. She describes specific features of the placenta and the foetal anomalies seen in association with diabetes, in recognition that this is the single most common presentation of the obese and overweight woman during pregnancy.

There are obvious markers of placental impaired function seen in the placentae of obese women with GDM. More work is necessary to improve our understanding the role of placenta in cases of unexplained stillbirths.

Bick and Beake (Chapter 30) reiterate that it is important to consider how being overweight or obese during pregnancy can impact on the post-natal health of a woman and her baby. The most recent triennial report of maternal deaths in the United Kingdom highlighted in stark terms why obesity is such a crucial public health issue: obesity remained a common factor among women who died during 2006–2008, and was associated with deaths from thromboembolism, sepsis and cardiac disease; 78% of the women who died following a thromboembolic event were overweight or obese [32]. It is important not to overlook these risk factors during post-natal period. Furthermore, obese women are at more risk of experiencing commonly experienced health problems after giving birth, such as backache, fatigue, urinary or faecal incontinence and depression, and perineal pain is equivocal. Excessive weight gain and persistent weight retention during the first post-natal year are strong predictors of being overweight a decade later. Focused strategies are urgently required to advice women how to achieve gradual weight loss, and graded exercise programmes could impact on subsequent pregnancy outcomes.

Interventions to Improve Care of Women During Pregnancy

Lifestyle modifications including dietary control and exercise remain as the first-line treatment for weight reduction. In individuals who experience difficulty in reducing significant weight with lifestyle intervention alone, the use of anti-obesity drugs is a reasonable adjunctive measure. **Li et al. (Chapter 31)** have reviewed current evidence as regards the use of orlistat and metformin as they are the only preparations available currently, and their use is probably safe for women planning for pregnancy. These weight-loss interventions should precede any planned conception or fertility treatment. It has been suggested in randomised controlled trials (RCTs) that a combination of medication and lifestyle-behavioural modifications is more effective than either invention alone in weight reduction. Orlistat inhibits intestinal fat absorption. A more recent Cochrane review including 16 trials on long-term orlistat therapy showed a placebo-subtracted significant weight reduction, a significantly higher likelihood in achieving 5–10% weight loss in patients receiving treatment with orlistat and a significant improvement in the lipid profile. There have been no adequate well-controlled studies for the use of orlistat in human pregnancy, and hence its use in pregnant or breastfeeding women is not recommended.

A recent meta-analysis showed that metformin treatment resulted in a statistically significant but modest decrease in BMI compared to placebo, only with high dose (>1500 mg/day), and in those treated for more than 8 weeks. The use of metformin during pregnancy is safe and does not impose any adverse foetal effects. However, treatment with these drugs does not lower the rate of miscarriage or obstetric complications [33].

Deonarine et al. (Chapter 32) argue that among the therapeutic interventions for weight loss, bariatric surgery has the most profound effect on weight loss and is being increasingly used to treat severe obesity. Although patients in the past sought bariatric surgery for medical and physical reasons, it is now increasingly used to improve fertility among severely obese women. Data from the United States suggest that there are long-term, lifelong health economic benefits from bariatric surgery for the individual and the society. NICE has issued its guidance on this important intervention [34].

Although these procedures are becoming safer now the impact of the procedures on absorption of essential nutrients during pregnancy and the foetal growth needs to be carefully monitored.

Thangaratinam and Khan (Chapter 33) provide evidence-based advice about the effectiveness of various strategies. They observe that perhaps the greatest challenge for obese women however is to achieve any behavioural changes in diet or physical activity.

Physical activity—based interventions in obese and overweight women do not show significant benefit in minimising weight gain in pregnancy, birth weight or obstetric complications.

Pregnant obese women should be advised to increase their folate intake to 5 mg daily in the pre-conception period and first trimester to reduce the risk of neural tube defect. Vitamin D supplementation of 10 mg/day is recommended to prevent adverse outcomes associated with vitamin D deficiency. Obese women are at risk of thromboembolism; therefore, there should be a clearly outlined policy for antenatal and post-natal thromboprophylaxis. NICE guidance has recommended a life-course approach by focusing on pregnancy and 1 year childbirth as the crucial period to target weight-management interventions based on behavioural change, dietary and physical activity [35].

Edozien (Chapter 34) has reviewed the association of severe obesity with increased maternal morbidity and mortality. Obese women have a higher than average risk of dying during childbirth. They often begin pregnancy with co-morbidities or develop these, particularly gestational diabetes and HTN, in the course of pregnancy. They are at significant risk of psychological morbidity as well. Unfortunately, the development of strategies for addressing the burden of maternal obesity has lagged well behind the rising prevalence of the condition. National guidelines are now available, but these are mostly focused on clinical care during pregnancy or weight reduction interventions. A more comprehensive strategy is required for the achievement of long-term reduction in obesity-related maternal mortality and morbidity. A multimodal framework for meeting this challenge encompasses primary, secondary and tertiary prevention programmes and calls for coordinated involvement of the three health sector tiers (primary, secondary and tertiary), and sectors such as education, social services and employment. A life-course approach underpinned by a family-centred philosophy is the way forward [32].

Churchill et al. (Chapter 35) lay out a set of standards for the provision of high-quality obstetric services. They argue that the quality of care can only improve when the health care systems and professionals working in them provide equitable, effective, evidence-based and safe, consistent and humane care driven by high-quality standards. Setting standards of care is a new concept which has been led by the Royal College of Obstetricians and Gynaecologists (RCOG) [36,37]. The standards are defined as 'A standard or more commonly set of standards should underpin an overarching statement of purpose'. A standard can be a measure of outcome, e.g. a predetermined disease end point such as a pulmonary embolism or a process to which professionals and through them the health care systems must conform. In meeting the standard whether that is an outcome or a process, it informs the patient or clinician of the level of quality of health care being provided. Some standards are fixed and do not change, they can be likened to the Ten Commandments but most will change over time, responding to events, developments and social challenges. In this way, standards evolve to meet the prevailing scientific, social or political conditions in which health care systems are operating. It is likely that by setting and applying high standards of care for obese people we will have an impact on the secondary problems caused by obesity in pregnancy.

Long-Term Impact of Obesity

What is becoming more apparent is the important role played by the maternal condition before and during gestation. **Ioannis and Pinelopi (Chapter 36)** have explored how maternal obesity affects the growth potential of the baby and casts its shadow on cardiometabolic health of the offspring during adolescence and the adult life. Maternal environment especially during early foetal life seems to have a determinant role in the tendency to develop obesity, diabetes and cardiovascular risk of the progeny. In recent years, strong evidence in support of prenatal programming of offspring obesity, diabetes and cardiovascular risk has accrued, whereby factors mainly related to maternal health are believed to alter foetal development in utero in a manner that promotes excess adipose tissue deposition during pre- or post-natal life.

The 'developmental overnutrition hypothesis' has been proposed as a potential pathophysiological link between maternal obesity and offspring obesity, as well as other metabolic abnormalities (e.g. diabetes and cardiometabolic diseases). According to this *hypothesis*, high maternal glucose and high free fatty acid and amino acid plasma concentrations result in permanent changes in appetite control, neuroendocrine function and energy metabolism in the developing foetus, leading to risk of adiposity in later life. This hypothesis is based on the concept of the association of maternal adiposity with a greater risk of insulin resistance and glucose intolerance, which result in higher concentrations of glucose and free fatty acids plasma concentrations. The offspring of these mothers would be expected to be programmed to become more obese themselves. Consequently, the obesity epidemic would be accelerated through successive generations independent of further genetic or environmental factors [38,39].

Obesity and Gynaecology

Obesity also casts shadows on the long-term health of the women. **Mehasseb and Shafi (Chapter 37)** explain that obesity affects the production of peptides (e.g. insulin, Insulin-like-growth factor 1 (IGF-1) and sex hormone-binding globulin (SHBG)) and steroid hormones (i.e. oestrogen, progesterone and androgens). It is likely that prolonged exposure to high levels of oestrogen and insulin associated with obesity may contribute to the development of female malignancies.

Obesity in the menopause produces a state of excess oestrogen production. Adipose tissue cells express various steroid hormone—metabolising enzymes and are an important source of circulating oestrogens, especially in post-menopausal women. This is due to the peripheral conversion of androgens secreted from the adrenal glands and ovaries into oestrone by the enzyme aromatase in the fat cells. The situation is aggravated by the fact that increased body fat is associated with decreased circulating levels of both progesterone and SHBG. With lower SHBG, there is a higher circulating level of free active oestrogens.

Excess weight, increased plasma triglyceride levels and low levels of physical activity can all raise circulating insulin levels leading to chronic hyperinsulinaemia which has been associated with cancers of the breast and the endometrium. Proteins secreted by adipose tissue (adipokines) also contribute to the regulation of immune response (leptin), inflammatory response (TNF-α, IL-6 and serum amyloid A), vasculature and stromal interactions and angiogenesis (vascular endothelial growth factor 1) and extracellular matrix components (type VI collagen). Physical activity among post-menopausal women at a level of walking about 30 min/day was associated with a 20% reduction in breast cancer risk, mainly among women who were of normal weight. The protective effect of physical activity was not found among overweight or obese women [40].

Critchley et al. (Chapter 38) explore relationship between obesity and menstrual bleeding abnormalities. There is a clear association with obesity and the development of endometrial polyps, endometrial hyperplasia and ovulatory dysfunction. PCO is linked to abnormal uterine bleeding secondary to these same factors. As PCO is associated with obesity and obesity augments its development, many of the effects of obesity on menstrual disorders may be manifest through anovulation and PCO. Although there are PCOs-independent effects of obesity on menstruation, in a fertile female obese population, they found no difference in the length of menstrual cycle and no correlation between BMI and menstrual loss. As the rate of obesity increases one clinical impact is more menstrual disorders and abnormal uterine bleeding. This is likely to be particularly linked to the effect of obesity on the expression of the PCO phenotype [41].

Maclaran and Panay (Chapter 39) give us advice as regards the use of hormone replacement treatment (HRT). Obesity is associated with increased reporting of vasomotor symptoms and can have a profound effect on how we manage the menopause. Although oestrogen may have a benefit on many of the metabolic consequences of menopause, the shared risks of HRT in obesity have led to concerns about the use of HRT in this population. In obese women, HRT has been associated with improvements in cardiovascular risk factors and incidence of type 2 diabetes. The risk of endometrial and breast cancers is increased by obesity but HRT does not appear to increase this risk further. HRT is associated with increased risk of VTE and stroke and in obese women using HRT, certain strategies should be adopted to minimise these risks such as low-dose transdermal therapy. Obesity is not in itself a contraindication to HRT and indeed obese women may derive significant benefits from HRT. Obesity and HRT share some common risks, particularly oestrogen-dependent cancer, although there is evidence that HRT may attenuate the obesity-associated risk of breast and endometrial cancers [42].

Tincello (Chapter 40) explains that obesity is associated with higher abdominal pressure and this leads to an approximately twofold increased risk of urinary incontinence, both stress incontinence and overactive bladder symptoms (urgency, frequency and urgency incontinence). Faecal incontinence and prolapse symptoms are subject to similar increased risks in the obese woman. Weight loss has a profound effect upon urinary, anal and prolapse symptoms; achieving a target weight loss of 5–10% of baseline body weight is associated with cure of urinary leakage of 30–70%, improvement of anal symptoms of around 65% and resolution of prolapse symptoms in 75%. Therefore, weight loss either by diet and exercise or by bariatric surgery should be recommended as first-line intervention for obese-incontinent women. Surgery for incontinence carries a slightly lower cure rate in obese women but no higher risk of complications. Data on prolapse cure are not available [43].

On this basis, weight loss could be considered the first-line management in obese and overweight women with pelvic floor symptoms, and the prospect of cure of these symptoms is likely to be a major source of motivation to comply with weight-loss strategies.

Where surgery is deemed necessary, women should be advised that both continence and prolapse surgery appear equally safe in obese patients but the cure rate of mid-urethral tape for urinary incontinence is slightly compromised. Further information is required to confirm whether the long-term failure rate after prolapse surgery is greater in obese women.

The traditional surgical techniques are now being replaced with advanced laparoscopic surgical approach, and its role has been explored by **Burden and Vyas (Chapter 41)**. The introduction of advanced laparoscopic surgery has made it possible to perform most operations on obese patients with minimal invasive techniques. The clinical benefits of laparoscopic surgery for patients include smaller incisions, reduced post-operative pain, quicker mobilisation and low infection rates. For health care providers and society there are the advantages of shorter hospital stay, faster recovery rates and quicker return to work, all of which reduce hospital inpatient and social costs [44,45].

Health Service Planning

McInnes and Gray (Chapter 42) have investigated the impact of obesity on the quality of life in obese women. Our understanding of this crucial effect is important as it would help health planners to develop strategies to take a broader view in addressing health service provision and accessibility by this group. Obesity is associated with a reduction in quality of life, with women often experiencing greater impairment than men. General health problems, mobility, functional disability, depression and low self-esteem are commonly reported. Weight-related stigmatisation and society pressure on women to be thin causes gender disparity and body image dissatisfaction. This can contribute to discomfort about health care encounters both generally and during childbearing, where the perceived attitudes of health professionals, physical examinations and discussions of weight status all have potential to cause distress. The tendency of clinicians to hold obese patients more responsible for their condition or any presenting complaints that could potentially be weight related may be viewed by patients as 'blaming', especially if the link between their health problem and their weight is not apparent. Although care practices might not be affected by a patient's weight, the quality of the clinician–patient relationship may suffer. Obese women may find health care encounters particularly difficult, as health professionals' beliefs and attitudes about overweight and obesity may undermine the quality of the clinician–patient relationship [46].

Anderson and Craigie (Chapter 43) explain that we must try to understand factors modulating eating behaviour and lifestyle issues among obese women if our strategies are going to make any inroads in the management of this group of women. Eating behaviour and other lifestyle habits have a major role to play in optimising the health (and obesity control) of women. Wide socio-cultural influence of income, gender, employment, ethnicity, marital status as well as parenting have lifelong influences on eating behaviours alongside the efforts of the food industry to promote consumption. Indigenous-level influences including many psychosocial factors do impact on immediate health behaviours, and intervention programmes needed to embrace factors such as excess, perceived control and social support as well as educational initiatives to help women achieve healthy lifestyles. Many stakeholders have the power to influence women's health behaviours. Combined efforts from parenting and schools in earliest life as well as a wide social culture background in adult years provide the backdrop for educational efforts from health professionals and public policy. Combined approaches and strategies are required to significantly alter health and obesity status [47].

We would be facing a huge challenge to provide safe and quality-assured maternity services to an obstetric population where obese women would be presenting with challenging co-morbidities during pregnancy and labour. **Jefferys et al. (Chapter 44)** have made a bold attempt to look into this uncertain future. In this chapter, they have clearly outlined all the risk factors affecting an obese women's pregnancy and have proposed practical solutions in order to prepare the profession and the health care planners to deal with the future challenges. There would be additional financial burden for the National Health Service in United Kingdom (NHS) in order to minimise risks related to diagnostic services and antenatal surveillance and huge issues for the provision of intrapartum care. It is important to consider how obstetric services in the future are going to be able to accommodate the increase in these health burdens and the necessary shift in staffing structure.

The authors recommend that it would be desirable to provide specialist multidisciplinary antenatal care for all obese women. Currently, specialist obstetric care tends to be reserved for those with a BMI greater than or equal to 40 kg/m^2; however, it may be that in the future this cut-off needs to be raised further – possibly to a BMI of 50 kg/m^2 (known as the morbidly obese). Centralisation of specialist obesity services for morbidly obese (BMI >50 kg/m^2) women may be most cost-effective, allowing specialist equipment to be concentrated in one place and reducing the need for such equipment in peripheral care settings. There is no doubt that provision of this level of care to the expanding number of obese women is going to require increased investment by the department of health. It will be the responsibility of the RCOG to produce guidance for health professionals and plan services and workforce [48,49].

Finally, **McRobbie and Maheshwari (Chapter 45)** remind us that an association between obesity and reproductive function has long been recognised. Hippocrates wrote: *People of such constitution cannot be prolific ... fatness and flabbiness are to blame. The womb is unable to receive the semen and they menstruate infrequently and little.*

Currently, there is no consensus on the optimal BMI (<35 or <30 kg/m^2) to provide service but obesity is associated with higher risk in pregnancy. Currently, women referred with raised BMI that is out with an Individual Fertility Units' treatment guidelines are generally advised to lose weight with no plan for how and over what time frame to do this. Moreover, no regular follow-up is put in place.

The authors have proposed that for sub-fertile women, with increased BMI, should be fully assessed and educated on the implications of increased BMI, and any other co-morbidities and concerns identified. An individualised plan, incorporating appropriate treatment (diet + exercise ± psychotherapy ± pharmacotherapy ± bariatric surgery) should be made. The plan should take into consideration factors such as age, ovarian reserve and so on [50,51].

Research Agenda

- In order to develop research-based focused strategies to tackle health risks associated with obesity in different ethnic groups, the researchers should work towards developing ethnically adjusted BMI charts taking account of waist circumference measurements.
- Any future research programmes should investigate the mitigating factors leading to this increased risk and how behaviour modifying strategies can be encouraged.
- Understanding the biochemical and molecular process involved in energy regulation would facilitate development of innovative preventive/treatment measures in susceptible individuals.
- Prospective studies focusing on the critical growth periods for obesity such as growth in the intrauterine environment, adiposity in early childhood and adolescence and life experiences specific to those periods would be of immense interest.
- Studies should focus on the role played by the psychological factors which promotes 'mindless eating' behaviour.

Sexual Health and Fertility

- Studies should identify strategies to increase awareness among obese women about the risks associated with untimed pregnancy as they do not readily access contraceptive services.
- Well-designed studies are urgently required to assess the effect of obesity on drug pharmacokinetics and phamacogenetics, especially steroidal contraceptives.
- Long-term follow-up studies are required to assess the safety and efficacy of different hormonal contraceptive methods among women who had undergone bariatric surgery.
- How should we manage infertility treatment of obese women with polycystic ovaries? A steady decrease of inter-abdominal fat is associated with restoration of ovulation. Therefore, more research is required to ascertain what type of food and what specific types of exercise are more effective in this group of women.

Infertility

- Is there a dose-dependent effect of obesity on the qualitative development of oocytes, the fertilisation rates and aberrant development in assisted-conception treatment cycles?
- Why heavier women may need more hormones to induce ovulation or for controlled hyperstimulation?
- Does weight-loss management by dieting and bariatric surgery improve sperm parameters and fertility potential?
- Sexual dysfunctions in male and the role of bariatric surgery should be investigated through long-term follow-up studies.
- Should NHS offer bariatric surgery to a selective group of women who are obese and have failed to respond to dietary regime as long-term health gains outweigh initial costs both for the mother and the newborn?

Pregnancy Care

- An individual patient-based data analysis on the effects of interventions on maternal and foetal outcomes will ascertain the magnitude of incremental benefit associated with the reduction in gestational weight gain and identify groups that benefit the most from interventions.
- Comparison of cost-effectiveness of the measures to tackle obesity in non-pregnant women of the childbearing age with those who are already pregnant will be the first step in forming health policies and measures for this problem.
- The role of anti-obesity medication and exercise on weight loss needs to be investigated in large randomised studies to ensure that weight reduction by calorific restriction or pharmacological intervention during the periconceptual period has no negative impact on the conceptus.
- A large-scale RCT of weight management and metformin is desirable to provide robust evidence of management of obese women with recurrent miscarriage. It would also be important to consider ethnicity.

Intrapartum Care

- There is a need for further trials to establish whether a weight-related dose of prostaglandins would be beneficial without an increase in harm, particularly in women who are extremely obese.
- Although induction of labour is often required and there is good evidence to support its use, further studies are required to determine the optimal method used to induce labour in women who are obese.
- Adherence of NICE/RCOG guidelines on risk assessment for VTE at different stages of pregnancy and following birth should constitute an outcome measure.
- Is it possible that some diagnoses of pre-eclampsia in the obese relate to the unmasking of underlying HTN by pregnancy, in the same way that pregnancy can mask underlying diabetes?
- Of late there has been a considerable interest in the use of metformin as a drug of choice in maternal obesity. Long-term safety data are required about the effects of this drug in the embryo and foetus.
- National audits are established to ensure that a high-quality care is delivered for all obese women during pregnancy.
- Should we start routinely weighing all women who have a BMI $>30\ kg/m^2$ at their first antenatal care visit during third trimester as this would allow patient-focused care during labour and following birth?
- Is it possible to develop an algorithm taking into account of all the risk factors for increased stillbirths in a population sample controlled for ethnicity which can help to develop focused antenatal care strategies?
- Large randomised studies are required to assess the effectiveness of interventions during antenatal period such as the use of kick charts, uterine artery Doppler flow and assessment of liquor volume during pregnancy among the obese population.
- Strategies to minimise weight gain during pregnancy should be carefully evaluated.

Gynaecological Disorders

- Further research to define the causal role of obesity in gynaecological malignancies is needed.
- Future trials may involve studies on the effect of dietary changes on weight gain and cancer risk, the effect of patterns of physical activity in relation to weight gain and cancer risk, the combined effects of changes in diet and physical activity on obesity and female cancer risk.
- Further research is needed into the differential impact of obesity on endometrial and ovarian function. Studies should address the optimal lifestyle, medical and surgical management of abnormal bleeding in obese women.
- Pelvic floor dysfunction is quite common. Long-term follow-up studies are indicated to follow-up women treated with newer operations.
- The safety of HRT should remain a key priority as an audit tool when prescribed to obese women.
- Laparoscopic surgery has definite place for the care of these women. National audits should be set up to ensure that outcomes are recorded for each surgeon and adverse events are notified.

The Future Service Planning

- Future research should focus on the prevention of maternal obesity. This may involve behavioural, psychological, medical or surgical therapy in obese women planning a first or subsequent pregnancy.
- The future organisation of maternity services on a regional network basis be evidence based and supported by national audits.
- More research is urgently required to ensure that women with excessive BMI and seeking infertility treatment are appropriately managed through nationally agreed protocols and the individual unit's practice is audited.

References

1. Mahmood TA. Obesity: a reproductive hurdle. *Br J Diab Vasc Disease*. 2009;9(1):3–4.
2. Whitlock G, Lewington S, Sherliker P, et al. Body mass index and cause-specific mortality in 900,000 adults: collaborative analyses of 57 prospective studies. *Lancet*. 2009;373(9669):1083–1096.
3. Bhatnagar D, Anand IS, Durrington PN, et al. Coronary risk factors in people from Indian subcontinent living in West London and their siblings in India. *Lancet*. 1995;345 (8947):405–409.
4. Ravussin E, et al. Effects of a traditional lifestyle on obesity in Pima Indians. *Diabetes Care*. 1994;17(9):1067–1074.
5. Wilding JPH. Pathophysiology and aetiology of obesity. *Medicine*. 2011;39(1):6–10.
6. McTigue KM, Garrett JM, Popkin BM. The natural history of the development of obesity in a cohort of young U.S. adults between 1981 and 1998. *Ann Intern Med*. 2002;136(12):857–864.
7. Kahn HS, Cheng YJ. Longitudinal changes in BMI and in an index estimating excess lipids among white and black adults in the United States. *Int J Obes*. 2007; 32(1):136–143.
8. World Health Organization. Global database on body mass index. <http://apps.who.int/bmi/index.jsp?introPage=intro_3.html/>; 2012, Accessed 26th May 2012.
9. Damodaran S, Swaminathan K, Mahmood TA. What is the best contraceptive method for obese women? *Br J Sex Med*. 2008;31(4):12–14.
10. Thanoon O, Mahmood T. Contraception in patients with medical conditions. *Obstet Gynecol Reprod Med*. 2011;9:263–267.
11. Royal College of Obstetricians and Gynaecologists. *The Investigation and Treatment of Couples with Recurrent First-Trimester and Second-Trimester Miscarriage – Green-Top Guideline No. 17*. London: RCOG Press; 2011:1–18.
12. ESHRE Task Force on Ethics and Law, Dondorp W, de Wert G, et al. Lifestyle-related factors and access to medically assisted reproduction. *Hum Reprod*. 2010;25:578–583.
13. Maheshwari A, Stofberg L, Bhattacharya S. Effect of overweight and obesity on assisted reproductive technology a systematic review. *Hum Reprod*. 2007;1:433–444.
14. Ben-Sefer E, Ben-Natan M, Ehrenfeld M. Childhood obesity: current literature, policy and implications for practice. *Int Nurs Rev*. 2009;56(2):166–173.
15. Kay VJ, Barratt CLR. Male obesity: impact on fertility. *Br J Diab Vasc Dis*. 2009; 9(5):237–241.

16. Mahmood TA. Influence of excess adiposity on reproductive function. *Br J Diab Vasc Dis*. 2009;9(5):197–199.
17. Corona G, et al. Is obesity a further cardiovascular risk factor in patients with erectile dysfunction? *J Sex Med*. 2010;7(7):2538–2546.
18. Gluckman PD, Hanson MA, Cooper C, Thornburg KL. Effect of in utero and early-life conditions on adult health and disease. *N Engl J Med*. 2008;359:61–73.
19. Blomberg MI, Källén B. Maternal obesity and morbid obesity: the risk for birth defects in the offspring. *Birth Defects Res A Clin Mol Teratol*. 2010;88(1):35–40.
20. Denison FC, Price J, Graham C, Wild S, Liston WA. Maternal obesity, length of gestation, risk of postdates pregnancy and spontaneous onset of labour at term. *BJOG*. 2008;115:720–725.
21. Royal College of Obstetricians and Gynaecologists. *Reducing the Risk of Thrombosis and Embolism During Pregnancy and the Puerperium – Green-Top Guideline No. 37a*. London: RCOG Press; 2009.
22. Ovesen P, Rasmussen S, Kesmodel U. Effect of prepregnancy maternal overweight and obesity on pregnancy outcome. *Obstet Gynecol*. 2011;118(2):305–312.
23. Higgins, Greenwood SL, Wareing M, Sibley CP, Mills TA. Obesity and the placenta: a consideration of nutrient exchange mechanisms in relation to aberrant fetal growth. *Placenta*. 2011;32(1):1–7.
24. Tieu J, Middleton P, McPhee AJ, Crowther CA. Screening and subsequent management for gestational diabetes for improving maternal and infant health. *Cochrane Database Syst Rev*. 2010;7:CD007222.
25. Sattar N, Greer IA. Pregnancy complications and maternal cardiovascular risk: opportunities for intervention and screening? *BMJ*. 2002;325(7356):157–160.
26. American College of Obstetricians & Gynecologists. ACOG committee opinion. Exercise during pregnancy and the postpartum period. Number 267, January 2002. American College of Obstetricians and Gynecologists. Committee on Obstetric Practice. *Int J Gynaecol Obstet*. 2002;77:79–81.
27. Bhattacharya S, Campbell DM, Liston WA, Bhattacharya S. Effect of body mass index on pregnancy outcomes in nulliparous women delivering singleton babies. *BMC Public Health*. 2009;7:168.
28. CMACE/RCOG Joint Guideline. *Management of Women with Obesity in Pregnancy*. London: Centre for Maternal and Child Enquiries and Royal College of Obstetricians and Gynaecologists; 2010.
29. National Institute for Health and Clinical Excellence. *Intrapartum Care: Care of Healthy Women and Their Babies During Childbirth*. London: National Institute for Health and Clinical Excellence; 2007.
30. Saravanakumar K, Rao SG, Cooper GM. The challenges of obesity and obstetric anaesthesia. *Curr Opin Obstet Gynecol*. 2006;18:631–635.
31. Vais A, Kean L. Stillbirth – Is it a preventable public health problem in the 21st century? *Obstet Gynecol Reprod Med*. 2012;22(5):129–134.
32. Centre for Maternal and Child Enquiries. Saving mothers' lives: reviewing maternal deaths to make motherhood safer: 2006–2008. The eighth report on confidential enquires into maternal deaths in the United Kingdom. *BJOG*. 2011;118(suppl 1):1–203.
33. Metwally M, Amer S, Li TC, Ledger WL. An RCT of metformin versus orlistat for the management of obese anovulatory women. *Hum Reprod*. 2009;24(4):966–975.
34. NICE. *Obesity: Guidance on the Prevention, Assessment and Management of Overweight and Obesity in Adults and Children*. London: National Institute for Health and Clinical Excellence; 2006. <http://www.nice.org.uk/CG43/>.

35. NICE. *Dietary Interventions and Physical Activity Interventions for Weight Management before, During and After Pregnancy – NICE Public Health Guidance 27.* London: National Institute for Health and Clinical Excellence; 2010.
36. RCOG. *Standards for Maternity Care – Report of a Working Party.* London: RCOG Press; 2008 (ISBN 978-1-904752-63-9).
37. Mahmood TA, Owen P, Arulkumaran S, Dhillon C, eds. *Models of Care in Maternity Services.* London: RCOG Press; 2010 (ISBN 978-1-906985-38-7).
38. Taylor PD, Poston L. Developmental programming of obesity in mammals. *Exp Physiol.* 2007;92(2):287–298.
39. Davey Smith G, et al. Is there an intrauterine influence on obesity? Evidence from parent child associations in the Avon Longitudinal Study of Parents and Children (ALSPAC). *Arch Dis Child.* 2007;92(10):876–880.
40. Zaninotto P, Head J, Stamatakis E, Wardle H, Mindell J. Trends in obesity among adults in England from 1993 to 2004 by age and social class and projections of prevalence to 2012. *J Epidemiol Community Health.* 2009;63:140–146.
41. Munro MG, Critchley HO, Broder MS, Fraser IS. FIGO classification system (PALM-COEIN) for causes of abnormal uterine bleeding in nongravid women of reproductive age. *Int J Gynaecol Obstet.* 2011;113(1):3–13.
42. Thurston RC, Sowers MR, Sutton-Tyrrell K, et al. Abdominal adiposity and hot flashes among midlife women. *Menopause.* 2008;15:429–434.
43. Irwin DE, Kopp ZS, Agatep B, Milsom I, Abrams P. Worldwide prevalence estimates of lower urinary tract symptoms, overactive bladder, urinary incontinence and bladder outlet obstruction. *BJU Int.* 2011;108:1132–1138.
44. Liga J, Barakat P, Diemunch P, Calmelet P, Brettes JP. Laparoscopic surgery and 'massive' obesity. *J Gynecol Obstet Biol Reprod (Paris).* 2000;29:154–160.
45. Santi A, Kuhn A, Gyr T, et al. Laparoscopy or laparotomy? A comparison of 240 patients with early-stage endometrial cancer. *Surg Endosc.* 2010;24(**4**):939–943.
46. Hansson LM, Naslund E, Rasmussen F. Perceived discrimination among men and women with normal weight and obesity. A population-based study from Sweden. *Scand J Public Health.* 2010;38:587–596.
47. Herman KM, Craig CL, Gauvin L, Katzmarzyk PT. Tracking of obesity and physical activity from childhood to adulthood: the Physical Activity Longitudinal Study. *Int J Paediatr Obes.* 2009;4(4):281–282.
48. Dennison F, Battacharya S, Norman J, et al. *The Clinical and Short-Term NHS Costs of Maternal Obesity for Maternity Services in Scotland.* Edinburgh: Scottish Governments Chief Scientist Office – Focus on Research; 2012.
49. Knight M, Kurinczuk JJ, Spark P, Brocklehurst P. Extreme obesity in pregnancy in the United Kingdom. *Obstet Gynecol.* 2010;115(5):989–997.
50. Balen AH, Anderson RA. Impact of obesity on female reproductive health: British fertility society – policy and practice guidelines. *Hum Fertil.* 2007;10:195–206.
51. Pandey S, Maheshwari A, Bhattacharya S. Should access to fertility treatment be determined by female body mass index? *Hum Reprod.* 2010;25:815–820.

Lightning Source UK Ltd.
Milton Keynes UK
UKOW040646131112

202077UK00001B/89/P